THE ROUTLEDGE HANDBOOK OF POLITICAL ECOLOGY

The Routledge Handbook of Political Ecology presents a comprehensive and authoritative examination of the rapidly growing field of political ecology. Located at the intersection of geography, anthropology, sociology, and environmental history, political ecology is one of the most vibrant and conceptually diverse fields of inquiry into nature–society relations within the social sciences. With contributions from over 50 leading scholars, the Handbook presents a systematic overview of political ecology's origins, practices, and core concerns, and aims to advance both ongoing and emerging debates. While there are numerous edited volumes, textbooks, and monographs under the heading "political ecology" these have tended to be either collections of empirically based (mostly case study) research on a given theme, or broad overviews of the field aimed at undergraduate audiences. *The Routledge Handbook of Political Ecology* is the first systematic, comprehensive overview of the field. With authors from North and South America, Europe, Australia, and elsewhere, the Handbook provides a state-of-the-art examination of political ecology; addresses ongoing and emerging debates in this rapidly evolving field; and charts new agendas for research, policy, and activism.

The Handbook opens with several chapters that critically reflect on political ecology and situate it within the broader scope of nature–society scholarship. These are followed by a section on the practice of political ecology: ethics, methods, activism, and policy. The remainder of the book comprises five sub-sections that examine fundamental concepts at the heart of political ecology: environmental knowledge, environmental change, environmental governance, environmental identities, and environmental politics.

The Routledge Handbook of Political Ecology introduces political ecology as an interdisciplinary academic field. It will serve as an excellent resource for graduate and advanced undergraduate teaching, and as a key reference text for geographers, anthropologists, sociologists, environmental historians, and others working in and around the fields of political ecology, environmental politics, and the political economy of environmental change.

Tom Perreault is Professor of Geography at Syracuse University, USA. His research focuses on resource governance (particularly water and mining), agrarian change, indigenous social movement politics, and rural development in the central Andean region.

Gavin Bridge is Professor of Economic Geography at Durham University, UK. His research focuses on the political economy of extractive industries and how firms, states, and raw materials shape the political ecologies of oil, gas, and mining.

James McCarthy is Professor in the Graduate School of Geography at Clark University, USA. His research centers on the intersections of political economy and environmental politics, with particular emphases on the relationships between neoliberalism and environmental governance, and on rural areas and industries.

"*The Routledge Handbook of Political Ecology* truly is worth keeping on hand. Not only does it show how much political ecology has contributed so far to our understanding of nature–society relations; it also pushes the field in exciting directions. A compendium of top-notch scholarship, the Handbook promises to become both an essential reference and an inspiration for important new work."
Professor Susanne Freidberg, Department of Geography, Dartmouth College, USA

"In this compelling volume, three outstanding political ecology scholars have produced a timely and discerning resource that contributes method, conceptual insight, and empirical richness to one of today's most dynamic fields of enquiry. This is the first systematic overview of political ecology, comprising contributions by around 50 eminent scholars that bring the reader up to date with key intellectual and policy debates. The book is set to become a point of reference not only for academics and students in geography, anthropology, sociology, and environmental history, but also for activists and policy makers who want to add a critical dimension to their way of thinking and acting upon environmental and social issues."
Professor Maria Kaika, University of Manchester, UK

"This is a comprehensive volume that every political ecologist must have. Not only does it take stock of the past accomplishments of the field, but it also opens up new debates and new research frontiers."
Professor Giorgos Kallis, Autonomous University, Barcelona, Spain

"A rich feast of original essays; this already vibrant field just took a huge leap forward. In a crowded arena of apparently expanding green sensitivity, political ecology has just renewed its critical edge. Bravo."
Professor Tania Li, Department of Anthropology, University of Toronto, Canada

THE ROUTLEDGE HANDBOOK OF POLITICAL ECOLOGY

Edited by
Tom Perreault, Gavin Bridge and James McCarthy

Routledge
Taylor & Francis Group
LONDON AND NEW YORK

First published in paperback 2020
First published 2015
by Routledge
2 Park Square, Milton Park, Abingdon, Oxon OX14 4RN

and by Routledge
52 Vanderbilt Avenue, New York, NY 10017

Routledge is an imprint of the Taylor & Francis Group, an informa business

British Library Cataloguing-in-Publication Data
A catalogue record for this book is available from the British Library

Library of Congress Cataloging-in-Publication Data
Routledge handbook of political ecology / edited by Thomas Perreault, Gavin Bridge and James McCarthy.
pages cm – (Routledge international handbooks)
Includes bibliographical references and index.
ISBN 978-1-138-79433-7 (hardback) – ISBN 978-1-315-75928-9 (ebook) 1. Political ecology.
I. Perreault, Thomas Albert, editor of compilation. II. Bridge, Gavin, editor of compilation.
III. McCarthy, James (James P.) editor of compilation. IV. Title: Handbook of political ecology.
JA75.8.R69 2015
304.2–dc23
2015005140

ISBN: 978-1-138-79433-7 (hbk)
ISBN: 978-0-367-40760-5 (pbk)
ISBN: 978-1-315-75928-9 (ebk)

Typeset in Bembo and Minion Pro
by Saxon Graphics Ltd, Derby

To our advisors, Tony Bebbington, Jody Emel, and Michael Watts, for their contributions to this field, and for what they have taught us.

CONTENTS

FIGURES

TABLES

CONTRIBUTORS

About the editors

Gavin Bridge is Professor of Economic Geography at Durham University, UK and has research expertise in the political economy of natural resources. His research centers on the spatial and temporal dynamics of extractive industries – oil, gas, and mining – and has been funded by the US National Science Foundation, the National Geographic Society, the European Commission, and British Academy. He is editor of *Transactions of the Institute of British Geographers* and co-author, with Philippe Le Billon, of *Oil* (Polity Press, 2013).

James McCarthy is Professor in the Graduate School of Geography at Clark University, USA. His research centers on the intersections of political economy and environmental politics, with particular emphases on the relationships between neoliberalism and environmental governance, and on rural areas and industries, property relations, and social movements in North America. His most recent research examines the geographic and political economic implications of potential renewable energy transitions.

Tom Perreault is Professor of Geography at Syracuse University, USA. His research focuses on resource governance (particularly water and mining), indigenous/campesino social movement politics, and rural development in the central Andean region. His work has been funded by Fulbright, the Inter-American Foundation, and the United Nations Research Institute for Social Development. He is editor of *Geoforum* and of *Minería, Agua y Justicia Social en los Andes: Experiencias Comparativas de Perú y Bolivia* (Centro Bartolomé de las Casas/PIEB, 2014).

About the authors

Karen Bakker is the Canada Research Chair in Political Ecology and Director of the Program on Water Governance at the University of British Columbia (Vancouver, Canada). Her water research is situated at the nexus of political economy, political ecology, and environmental science. Her most recent sole-authored monograph is *Privatizing Water: Governance Failure and the World's Urban Water Crisis* (Cornell University Press, 2010).

Stefania Barca is an environmental historian with research interests in industrialization/de-industrialization, the commons, and labor. Her book *Enclosing Water: Nature and Political Economy in a Mediterranean Valley* (Cambridge, 2010) received the Turku Book Prize in 2011.

Anthony Bebbington is Milton P. and Alice C. Higgins Professor of Environment and Society, and Director of the Graduate School of Geography at Clark University. He is also Professorial Research Fellow in the School of Environment, Education and Development at the University of Manchester and a Fellow of the National Academy of Sciences and of the American Academy of Arts and Sciences.

Tor A. Benjaminsen is a human geographer and professor of development studies at the Norwegian University of Life Sciences. He carries out research on environmental change, land use conflicts, and conservation policies and narratives, mostly in Africa, but also more recently in the Norwegian Arctic.

Bruce Braun is Professor of Geography, Environment and Society at the University of Minnesota, USA. He is the author of *The Intemperate Rainforest: Nature, Culture and Power on Canada's West Coast* (University of Minnesota Press), and co-editor of *Social Nature: Theory, Practice and Politics* (with Noel Castree, Blackwell Publishers) and *Political Matter: Technoscience, Democracy and Public Life* (with Sarah Whatmore, University of Minnesota Press). His recent work includes essays on the biopolitics of biosecurity, new materialisms, and the politics of urban resilience.

Joe Bryan is Assistant Professor of Geography at the University of Colorado, Boulder. He is the author, with Denis Wood, of *Weaponizing Maps: Indigenous Peoples and Counterinsurgency in the Americas* (Guilford, 2015). He has worked extensively on participatory mapping projects with indigenous peoples in the Americas, including projects in the US, Nicaragua, and Honduras.

Edward R. Carr is Associate Professor of Geography and Director of the Humanitarian Response and Development Lab (HURDL) at the University of South Carolina. His research interests are situated at the intersection of development and adaptation to global change, and focus on finding ways to meet the shifting needs of vulnerable populations in the context of a changing global environment.

Noel Castree is Professor of Geography in the Department of Geography and Sustainable Communities, University of Wollongong, New South Wales, and Professor of Geography, SEED, University of Manchester, England.

Rosemary-Claire Collard is Assistant Professor of Geography at Concordia University in Montreal. She is interested in the intersection between capitalism and animal life. Her most recent research project tracked the exotic pet trade across six countries and is forthcoming as a book, *Animal Traffic*, with Duke University Press.

Diana K. Davis, a geographer and veterinarian, is Associate Professor of History at the University of California, Davis, where she teaches environmental history and geography. Her first book, *Resurrecting the Granary of Rome: Environmental History and French Colonial Expansion in North Africa*, has been translated into French and received the AAG's Meridian Book Award and the Marsh Prize from the ASEH. She has conducted research with Moroccan and Afghan nomads and in the French and British archives. She is currently completing an historical political ecological study of deserts, drylands, and desertification.

David Demeritt is Professor of Geography at King's College. His research focuses on the role of science in environmental politics and policymaking.

Michael Ekers is Assistant Professor of Geography at the University of Toronto, Scarborough. His teaching and research examines the cultural politics of farm and forestry work. He is also interested in the significance of Antonio Gramsci's writings for debates on space, nature, and social difference.

Rebecca Elmhirst is a Human Geographer based at the University of Brighton in the United Kingdom. Her research interests lie in feminist political ecology, and the politics of natural resource governance, displacement and migration, and she has extensive field experience in Southeast Asia (especially Indonesia). She is editor of the journal *Gender, Technology and Development* (Sage, New Delhi).

Denis Gautier is a geographer with interests in the governance and management of natural and renewable resources in tropical dryland areas. Trained as an agronomist, specialized in tropical forestry, he holds a PhD (Université d'Avignon) and a "*habilitation à diriger des recherches*" (Université de Paris 1 Panthéon-La Sorbonne) in geography. His research interests are the geographical analysis of resource management and practices, with a focus on the issues of access and use rights and on the territorialization processes, and on the links between ecosystem resilience and the vulnerability of woodland product-dependent people.

Julie Guthman is Professor of Social Sciences at the University of California, Santa Cruz. She is the author of *Weighing In: Obesity Food, Justice and the Limits of Capitalism* and *Agrarian Dreams: The Paradox of Organic Farming in California*, as well as many journal articles on the race, class, and body politics of alternative food movements.

Derek Hall is Associate Professor in the Department of Political Science and the Balsillie School of International Affairs at Wilfrid Laurier University. He is the author of *Land* (Polity, 2013) and co-author of *Powers of Exclusion: Land Dilemmas in Southeast Asia* (NUS Press and University of Hawai'i Press, 2011, with Philip Hirsch and Tania Murray Li).

Nik Heynen is Professor of Geography and Director of the Integrative Conservation (ICON) PhD Program at the University of Georgia. His research interests include urban political ecology, social movements, and food studies.

Ryan Holifield is an associate professor of Geography at the University of Wisconsin-Milwaukee. His current research interests include environmental justice policy and practice, socio-political dimensions of urban environmental change, and stakeholder participation in environmental governance.

Alf Hornborg is an anthropologist and Professor of Human Ecology at Lund University, Sweden. He is the author of *The Power of the Machine* (Alta Mira Press, 2001) and *Global Ecology and Unequal Exchange* (Routledge, 2011) and editor of several volumes at the interface of political ecology, environmental history, and ecological economics.

Leah S. Horowitz is Associate Professor of Geography at the Hawai'i Pacific University. Her research, mainly conducted in New Caledonia and the US, examines ways that social relationships and networks shape grassroots engagements with environmental governance, with a focus on mining and both urban and rural biodiversity conservation.

Matt Huber is Assistant Professor of Geography at Syracuse University. He writes and teaches about the political ecology of energy. He is the author of *Lifeblood: Oil, Freedom and the Forces of Capital* (University of Minnesota Press, 2013).

Sara Keene is a PhD candidate in the department of Development Sociology at Cornell University. Her research explores questions of state formation, governance, and social struggle in northern California's medical marijuana industry, as well as the politics and practice of back to the land movements in rural California.

Brian King is Associate Professor in the Department of Geography and Fellow at the Social Science Research Institute at the Pennsylvania State University. His research interests concentrate upon the impacts of conservation and development within southern Africa, social and environmental justice, and health–environment interactions. He is co-editor, with Kelley A. Crews, of *Ecologies and Politics of Health* (Routledge, 2013).

Christian A. Kull is Professor at the Institute for Geography and Sustainability, University of Lausanne, Switzerland. His research interests include the political and social aspects of resource management issues like fire, invasive species, small-scale farming, deforestation, tree planting, and protected areas conservation, particularly in developing country contexts. Christian is author of *Isle of Fire: The Political Ecology of Landscape Burning in Madagascar* (University of Chicago Press, 2004).

Rebecca Lave is Associate Professor at Indiana University. Her critical physical geography research approach combines political ecology, STS, and fluvial geomorphology. She has published in journals ranging from *Science* to *Social Studies of Science*, and is the author of *Fields and Streams: Stream Restoration, Neoliberalism, and the Future of Environmental Science* (2012).

Philippe Le Billon is Professor in the Department of Geography and the Liu Institute for Global Issues, University of British Columbia. Working on resource governance, armed conflicts, and corruption, his two latest books are *Wars of Plunder: Conflicts, Profits and the Politics of Resources* (Oxford, 2013) and *Oil* (with Gavin Bridge, Polity, 2012).

Enrique Leff is a Mexican environmental sociologist. He received his *doctorat de troisième cycle* in development economics Paris I-Sorbonne in 1975. He is currently a full-time researcher at the Instituto de Investigaciones Sociales (Institute of Social Research) and professor of political ecology and environmental sociology at the National Autonomous University of Mexico. He works in the fields of environmental epistemology and philosophy, ecological economics, environmental sociology, political ecology, and environmental education.

Diana Liverman is Regents Professor of Geography and Development and co-director of the Institute of the Environment at the University of Arizona. Her research focuses on the human dimensions of global environmental change including climate vulnerability, adaptation, and governance, especially in the Americas.

Alex Loftus is Senior Lecturer in Environment and Development in the Department of Geography at King's College London. He is the author of *Everyday Environmentalism: Creating an Urban Political Ecology* (University of Minnesota Press, 2012) and co-editor of *Gramsci: Space, Nature, Politics* (Wiley-Blackwell, 2012).

Celia Lowe is Associate Professor of Anthropology and International Studies at the University of Washington. She works in Southeast Asia, especially Indonesia, and her main interest is in the travels of biological and other forms of scientific knowledge between EuroAmerica and Southeast Asia. She is author of *Wild Profusion: Biodiversity Conservation in an Indonesian Archipelago* (Princeton, 2006).

Becky Mansfield is Professor of Geography at the Ohio State University. She is the author of articles on biopolitics, neoliberalism, and nature in journals such as the *Annals of the Association of American Geographers, Environment and Planning D: Society and Space*, and *BioSocieties*, and is editor of *Privatization: Property and the Remaking of Nature-Society Relations* (Wiley-Blackwell, 2011).

Brent McCusker is Associate Professor of Geography at West Virginia University. His research interests include political economy, political ecology, and development studies. His current research involves comparing scenarios of climate change in Malawi to potential human adaptation strategies. He is also leading a team that is conducting national scale mapping of vulnerability to natural and financial shocks amongst Nigerien households.

Tad Mutersbaugh is Professor of Geography at the University of Kentucky. He has written extensively on questions of certification, organic coffee, and gender in development issues with a focus on Mexico.

Abigail H. Neely is Assistant Professor of Geography at Dartmouth College. She conducts research in South Africa on present and past political ecologies of health. She's working on a book that examines the changing relationships between health and environment to trouble the divide between human and non-human agency.

Roderick P. Neumann is Professor of Geography and Chair of the Department of Global and Sociocultural Studies at Florida International University. He is the author of numerous publications on the political ecology of nature conservation including *Imposing Wilderness* (University of California Press, 1998) and *Making Political Ecology* (Edward Arnold, 2004).

Thokozile Nguse grew up in a disadvantaged family in rural area in South Africa. She is a single mother of two children and currently a Resourcer for a leading recruitment company in South Africa.

Jonathan Otto is a PhD candidate at the University of Kentucky. He is based in Mexico and his research is centered on environmental conservation, rural livelihoods, and economic development.

Scott Prudham is a Professor in the Department of Geography and School of the Environment at the University of Toronto. He is a former editor of the journal *Geoforum*, and President of the University of Toronto Faculty Association. His research is situated at the intersection of environmental politics, environmental change, and political economy. He is author of *Knock on Wood: Nature as Commodity in Douglas-fir Country* (Routledge, 2005) and co-editor of *Neoliberal Environments: False Promises and Unnatural Consequences* (Routledge, 2007).

Paul Robbins is Professor and Director of the Nelson Institute for Environmental Studies at the University of Wisconsin-Madison. His research investigates human/environment interactions and the politics of natural resource management in India and the United States. He is author of *Lawn People: How Grasses, Weeds, and Chemicals Make Us Who We Are* (Temple, 2007), and co-editor of *Global Political Ecology* (with Richard Peet and Michael Watts, Routledge, 2011).

Morgan Robertson is Associate Professor of Geography at the University of Wisconsin-Madison. He researches market-based environmental policy, efforts to commodify nature in the form of ecosystem services and pollution credits, and wetland protection and restoration.

Nathan F. Sayre is Associate Professor and Chair of the Department of Geography at the University of California-Berkeley. He studies the political economy and ecology of rangelands, with a focus on the southwestern United States. His current project is a book on the history of efforts to understand rangelands scientifically.

Juanita Sundberg is Associate Professor of Geography and Latin American Studies at the University of British Columbia. Her on-going research, funded by the Canadian Social Science and Humanities Research Council, is a collaborative bi-national project analyzing how militarized approaches to border governance and territorial control impact everyday life for residents of the United States and Mexico borderlands.

Erik Swyngedouw is Professor of Geography at Manchester University. His research interests include political ecology, urban governance, democracy and political power, and the politics of globalization. His was previously Professor of Geography at Oxford University and held the Vincent Wright Visiting Professorship at Science Po, Paris, 2014. His forthcoming book with MIT Press focuses on water and social power in twentieth-century Spain.

Astrid Ulloa is Professor of Geography at the Universidad Nacional in Colombia. She holds a PhD in Anthropology (2003) from the University of California, Irvine. Her research focuses on indigenous peoples' movements, ethnicity, territoriality, and the anthropology of the environment, and she has authored several books on the cultural and environmental politics of indigenous peoples in Colombia and elsewhere in Latin America.

Gabriela Valdivia is Assistant Professor in the Geography Department at UNC-Chapel Hill. She examines resource governance, environmental mobilization, and frontiers of nature–society relations in Latin America, with attention to everyday formations of national and regional communities. Her most recent work focuses on oil and environmental injustices in Ecuador.

Levi Van Sant is a PhD candidate in Geography and Integrative Conservation at the University of Georgia. His dissertation, tentatively titled "Plantation Geographies: Agriculture and White Supremacy in the South Carolina Lowcountry," analyzes the ways that shifting forms of state power and dynamic racial ideologies reproduce uneven agricultural landscapes in the colonial present.

Michael J. Watts is the Class of '63 Professor of Geography and Development Studies at the University of California where has taught for 35 years and where he founded the Environmental Politics Workshop in 1994. He has served on the Board of Trustees of the SSRC for a decade and on the Board of the Pacific Institute in Oakland for over two decades. He is currently working on the 2010 Gulf of Mexico oil spill.

James L. Wescoat Jr. is the Aga Khan Professor in the Department of Architecture at the Massachusetts Institute of Technology. He received his BLA degree in landscape architecture from Louisiana State University and his MA and PhD degrees in geography from the University of Chicago. His research has concentrated on water systems in South Asia and the United States from the site to river basin scales.

Ben Wisner has spent nearly 50 years seeking order in the murky-messy-muddy heart of Venn, where planet, people, and power intersect. Beginning with two years in a Tanzanian village (1966–8), he continues to return there when not working elsewhere in Africa, Asia, and Latin America, or writing stories for his grandson about donkeys and mice.

Wendy Wolford is the Robert A. and Ruth E. Polson Professor of Development Sociology at Cornell University and Faculty Director of Economic Development for Cornell's Atkinson Center for a Sustainable Future (ACSF). Wendy also co-leads the Theme Project on Contested Global Landscapes, organized by Cornell's Institute for the Social Sciences (ISS), and is a founding member of the Land Deals Politics Initiative (LDPI). Her current research on land issues in Mozambique builds on her over 15 years of research experience in Brazil and Ecuador.

Emily T. Yeh is Associate Professor of Geography at the University of Colorado, Boulder. She conducts research on nature–society relations in Tibetan parts of the People's Republic of China, including on the relationship between ideologies of nature and nation, the political ecology of pastoralism, and the emergence of environmental subjectivities. She is author of *Taming Tibet: Landscape Transformation and the Gift of Chinese Development* (Cornell, 2013).

Karl S. Zimmerer is Professor of Geography at Pennsylvania State University, at the Earth-Environmental Systems Institute (EESI) and Institutes for Energy and the Environment (PSIEE). He has authored more than 100 articles and chapters, and has published five books, including *Changing Fortunes: Biodiversity and Peasant Livelihoods in the Peruvian Andes*. He is the Director of the new GeoSyntheSES Lab for Social-Ecological Sustainability.

ACKNOWLEDGEMENTS

The editors gratefully acknowledge the many authors who contributed to this volume. We would also like to thank Sarah Gilkes and Andrew Mould for their encouragement, patience, and support. Gavin gratefully acknowledges the support of the People Program (Marie Curie Actions) of the European Union's Seventh Framework Program, under REAS agreement No 289374 "ENTITLE" (www.politicalecology.eu) and the discussions with colleagues and ENTITLE fellows this support has enabled. He also thanks Sarah, Callum, Max, and Giles for their support and patience. James wishes to thank Jennifer, Katelyn, and Alana for their patience and perspective. Tom is eternally indebted to Meredith, as ever, for putting up with him.

PART I

Introduction

1
EDITORS' INTRODUCTION

Gavin Bridge, James McCarthy, and Tom Perreault

Developed through academic inquiry and engaged political practice, political ecology has experienced a meteoric rise. Its growth as an academic field is perhaps most evident in Anglophone geography in North America, where political ecology constitutes one of the largest and fastest growing specialty groups of the Association of American Geographers. Growth has been both rapid and uneven, and at times contested by longer-established fields. Matching political ecology's rise in popularity has been its diversification, such that the term is now applied to a very broad set of concerns that revolve around societies' relationships with the non-human environment. Even a cursory look at journal titles and conference presentations shows that the label "political ecology" is applied to research topics as seemingly disparate as water access in India, land grabs in the Amazon, Sahelian pastoralism, lawn care in the United States, fisheries management, wetland markets, indoor air quality, AIDS, and obesity. And of course the Anglophone tradition is but one stream of political ecological thought, which has barely engaged with the Francophone, Spanish, and other literatures, particularly as developed in Africa, Latin America, and elsewhere in the global South. Political ecology also extends beyond academic enquiry to the knowledge claims and political practices advanced by people, many of them poor, who are subject to rationalities of resource management, environmental projects and/or pollution to which they do not consent. When viewed from this broad perspective, then, political ecology is a riotously diverse field, with origins and trajectories resembling more closely a tangled evolutionary lineage than a neat family tree.

The Routledge Handbook of Political Ecology attempts to make sense of this growing body of research and practice. The chapters we have assembled provide a critical overview of political ecology's geneaologies, some of its most important research foci, and emerging research agendas. Taken as a whole the volume provides a critical assessment of political ecology as a field, although we make no claim to be either comprehensive or definitive. Its chapters should be read as focused assessments of specific strands and conversations in political ecology – more field reports than final results. This introductory chapter probes the intellectual trajectory, current status, and possible future directions of political ecology. In the next section, we revisit the intellectual and political origins of the field, in an attempt to broaden – both historically and conceptually – our understanding of the various influences that have shaped it. We argue that the stereotypical "origin myth" of political ecology (as least as it exists in Anglophone geography), as emerging in the early 1980s from cultural ecology, the hazards tradition, and agrarian political ecology, and given form by Piers Blaikie, Harold Brookfield, and Michael Watts in a handful of books, is too narrow in its scope, and excludes a diversity of influences both within and beyond the academy. To be sure, these works were paramount in the development of political ecology, and remain vital texts today. However, we argue that political ecology's roots are both deeper and broader than commonly acknowledged, and that the field is but one manifestation of a critical re-thinking of nature and nature–society relations that also took expression elsewhere (notably in the work of David Harvey and Neil Smith, but also in the environmental movements of North America and Western Europe), and which finds its roots in the intellectual and political Zeitgeist of the late 1960s and early 1970s.

This discussion is followed by an effort to situate political ecology relative to allied fields in nature–society studies. These include other fields within the discipline of geography, such as land use/land cover change, environmental history, cultural ecology, and hazards and vulnerability studies, as well as other fields such as ecological anthropology, environmental sociology, sustainability studies, and ecological economics. Crucially, this effort requires a consideration of the geographies of political ecology, that is, the various theoretical and empirical directions the field has taken in different places and institutional settings. Whereas political ecology is arguably the dominant form of nature–society geography in the North American Anglophone academy, it remains at the radical margins in the UK and especially in continental Europe. More than a mere observation about academic fashion, this reality demands that we consider the implications of this differential status for radical scholarship in different locations. Whereas political ecologists in, say, Germany, Spain, or the UK may reasonably consider themselves at the vanguard of nature–society scholarship, political ecologists in the United States and Canada occupy a very different position. Such scholars may find themselves asking what is at stake, intellectually and politically, if we are all political ecologists now. Following this discussion, we outline the structure of the volume and its principal themes, and briefly introduce the chapters.

Revisiting the origins of political ecology

What, then, is political ecology? Most academic treatments of the concept – at least those in the Anglophone tradition of North America, the UK, Ireland, Australia, and New Zealand – trace the concept primarily to the 1980s and the seminal works of Watts (1983a, 1983b), Blaikie (1985), and Blaikie and Brookfield (1987). These authors were at once influenced by, and reacting against, an array of intellectual traditions, including the environmentalism of the 1960s and 1970s and its obsession with "over-population" and the depletion of (supposedly) finite resources, an intellectual current that was exemplified by the influential work of biologists Garret Hardin (1968) and Paul Ehrlich (1968), and the publication in 1972 of *The Limits to Growth* (Meadows et al. 1972). It was precisely this supposedly *apolitical* ecology, whose class commitments are concealed beneath a veil of techno-scientific "objectivism," that Enzensberger (1974) critiqued in his early use of the term "political ecology." Somewhat confusingly, however, while Enzensberger's radical critique – rooted in historical materialist analysis of demography, social relations, and nature – aligns politically with contemporary political ecological thought, he uses the term "political ecology" to refer to that which he critiques: the political nature of ecological science and the ecology movement of the 1960s and early 1970s, what he labels the "eco-industrial complex" (Enzensberger 1974: 10).

In the Anglophone academy, in which Watts, Blaikie, Brookfield, and other early political ecologists worked and published (for instance, see work by Susanna Hecht [Hecht and Cockburn 1989], Tom Bassett [1988], and Ben Wisner [1978]), political ecology also emerged as a reaction to the apolitical nature of the fields of cultural ecology and hazards studies (Watts 1983b). Trained in these fields, with their commitment to intensive field-based research and rigorous empiricism, these authors were similarly influenced by the resurgent Marxism of 1960s agrarian political economy and peasant studies, as well as the dependency and world systems thinking of, *inter alia*, Andre Gunder Frank, Samir Amin, and Immanuel Wallerstein. Perhaps more than anything else, political ecology was (and is) an epistemological project, which set out to shatter comfortable and simplistic "truths" about the relationship between society and its natural environment. Thus, early academic work in political ecology sought to deconstruct the dominant explanations of famine in Nigeria, soil erosion in Nepal, and deforestation in Brazil as rooted in over-population, improper land management, and brute ignorance. In their place, these authors erected alternative explanations for these phenomena, rooted in political economy, marginalization, colonial capitalism, and the abuses of predatory states. Political ecology thus grew to be distinctly catholic in both theory and method. Theoretically, the field was arguably oriented more towards understanding particular sets of dynamics in specific places than towards generating and answering the "next" question in a discipline-oriented epistemological framework; it thus drew from multiple theoretical frameworks from multiple fields that seemed to speak to those dynamics. Methodologically, the commitment to understanding dynamics in particular locations, combined with deep roots in cultural ecology and hazards studies, meant that political ecologists drew from a wide range of primarily field-based research methods, particularly ethnographic ones, usually supplemented with in-depth archival analysis.

These intellectual currents developed in relation to, and against a backdrop of, widespread social tumult during the 1960s and early 1970s (Watts 2001). This was a period characterized by widespread anti-authoritarianism and restless activism (the ten-month period from October 1967 to July 1968 witnessed over 2,000 recorded student protests worldwide). Violence in the streets of Mexico City, Paris, and Los Angeles, political assassinations and the anti-war, civil rights, women's rights and environmental movements of the 1960s and 1970s together shaped the individual experiences of academics and their research programs (see, for example, Chapter 3, this volume), and formed the social and political contexts from which political ecology emerged. The Catholic Church's opening to the left, and especially the emergence of Liberation Theology from the 1968 Medellín conference, had particular influence on scholars and activists working in Latin America. War in Southeast Asia and the 1959 Cuban Revolution spurred a resurgence in peasant studies and interest in the so-called "agrarian question," which in turn inspired a wave of radical political and academic activity (Wolf 1969). Meanwhile, many of the places political ecologists studied were profoundly transformed by the wave of decolonization that transformed the map of the world between the 1950s and the 1970s. The combination of formal decolonization (whether peacefully or by revolution), and the Cold War emphasis on proxy wars and spheres of influence in the formerly colonized world, led to the rise of "area studies" and intense academic and policy interest in peasant studies and agrarian political economy. It was in this context that state-funded research initiatives, such as the US government's Fulbright programs, both encouraged and facilitated international research. It is worth noting that the US Department of Education's area studies programs (known as "National Resource Centers," or "Title VI" programs) were initially established by Title VI of the National Defense Education Act of 1958, and were thus closely aligned with US foreign policy objectives. Political ecology – both its major early topics of concern and its radical political orientation towards them – emerged directly from this milieu (e.g. Blaikie 1985; Scott 1976; Watts 1983b). The restive period of the late 1960s gave way to political and economic convulsion and conservative retrenchment in the early 1970s. This period witnessed US defeat in Viet Nam, the OPEC oil embargo and ensuing "oil shock," Pinochet's CIA-backed coup in Chile, and the implosion of the Nixon presidency. The 1970s also saw broad-based environmental mobilization and landmark environmental legislation, much of it in direct reaction to deteriorating environmental conditions of North American and European cities. The pioneering work of Rachel Carson figures prominently in this regard, as did the burning of the Cuyahoga River in Cleveland, the Santa Barbara oil spill, the declared "death" of Lake Erie, and the partial meltdown of the Three Mile Island nuclear plant in Pennsylvania in 1979.

It was from this intellectual and political ferment that political ecology, in its various guises, emerged. Although differing somewhat from its contemporary valence, the term "political ecology" was already in use by the early 1970s. Eric Wolf (1972) used it (in his title, although curiously not in the body of his paper) to refer to landed property relations and the politics of resource management. Enzensberger (1974) used the term (and the shorthand "ecology") to refer to the bourgeois European and North American environmental movements of the 1960s and early 1970s, which he saw as fundamentally rooted in capitalist techno-science and therefore incapable of addressing the structural causes of environmental crises. In a similar vein, Walker (1973, 1974) critiqued the role of science in environmental policy, arguing that wetlands management cannot be separated from the political economic context and power relations within which such management takes place. Similar and simultaneous arguments were made by Harvey (1974) in his critique of the dominant neo-Malthusianism of the liberal environmental movement. Harvey's analysis extended to a discussion of the capitalist production of natural resources, which, he points out, cannot be understood apart from the social relations of production through which they are given meaning and value. The core principles of contemporary political ecology were thus already in place nearly a decade before the publication of *Silent Violence*. Harvey further developed his Marxist view of nature as produced by capitalist relations of production in *The Limits to Capital* (Harvey 1982), a thesis also at the core of Neil Smith's treatise *Uneven Development*, a work of enormous (though often unacknowledged) influence in political ecology (Smith 1984). Political ecology represents, in many regards, precisely the sorts of efforts on the part of radical geography to come to grips with the "matter of nature," as called for by Margaret FitzSimmons (1989). Our argument here – and as we further develop in the Handbook's concluding chapter – is that the intellectual roots of political ecology are both older and more diverse than is commonly acknowledged in the literature, and stem from a general turn toward Marxist scholarship, post-positivist approaches to nature–society relations, and a broad and growing acceptance of the central elements of feminist and postcolonial scholarship and politics. This chronology also places the origins of political ecology upstream from *Uneven Development* and *The Limits to Capital*. The seeds of political ecology and those of Marxist geography took root in the same fertile soil and were watered by the same social and political currents.

The Anglophone academy holds no patent on political ecology, however, and the term has also been used to describe the European green parties (Lipietz 1999) and the liberal environmental movement in Europe and North America more generally (Enzensberger 1974). And whereas much political ecology emerging from the global North (albeit largely about the global South) has been concerned with agrarian political economy, indigenous livelihoods and resource governance, political ecology in the global South has developed in response to colonial histories, largely as a politics of difference rooted in ecological and cultural conditions (Leff 2014). In this sense, then, the political ecology of the South moves beyond the academic to comprise a political program rooted in decolonization, emancipation, cultural reinvention, and the re-appropriation of nature. There are also growing bodies of scholarship regarding human–environment relations in other linguistic and national academic traditions developing under the sign of "political ecology." Increasing direct conversation between these traditions and the Anglophone one is a central goal of this volume.

This capacious intellectual framework allows much room for diversity and change. If authors agree on anything it is that political ecology evades simple definition (Neumann 2005). The field has been defined, variously, as integrating "the concerns of ecology and a broadly defined political economy" (Blaikie and Brookfield 1987: 17), as "a development discourse" for the times (Peet and Watts 1996: 3); as "an explicit alternative to apolitical ecology" (Robbins 2004: 5); and as "the study of power relations and political conflict over ecological distribution and the social struggles for the appropriation of nature" (Leff n.d.: 5). Some authors have placed greatest emphasis on development and agrarian conflict (e.g. Bryant and Bailey 1997) while others have emphasized the importance of bio-physical processes and ecological change (Zimmerer and Bassett 2003). Drawing explicitly on Blaikie and Brookfield, Bassett (1988: 455) employs a political ecology that emphasizes the "interrelationships between agricultural ecology, peasants and the state, and the accumulation strategies of different groups." Bryant and Bailey (1997) were explicitly concerned with "Third World" political ecology, while McCarthy's (2002) application of political ecological principles to the "First World" helped establish a new direction for political ecological research. Whereas most of these studies were primarily or exclusively concerned with rural spaces and agrarian economies, Swyngedouw (2004), Kaika (2006), Heynen (2014), Huber (2013), and others have shifted the focus of political ecology to the urban and the industrial. This shift in analytical scale and focus continues, with Biehler and Simon (2011) considering the political ecologies of the "great indoors," and Guthman (2011) and Mansfield (2012) training their attention on the political ecologies of the body. Despite having a strong origin story, political ecology has been relatively unconstrained by its history and it continues to evolve by exploring new spaces, scales, and themes. The diversity to which this restlessness gives rise begs the question, however, of what, if anything, lends coherence to these diverse approaches?

We hold that political ecology is best characterized not by research topic (e.g. agrarian dynamics, resource conflict, deforestation, conservation, resource governance), or scalar or socio-spatial focus (landscape, community, household, rural versus urban, Third or First World). Such boundary-making would ultimately prove fruitless, as the restless nature of academics (and the demands of the academy) would inevitably push these boundaries outward into new theoretical, empirical, methodological, and spatial/scalar frontiers. Rather, as we discuss below, the field's coherence derives from a set of commitments that are held in common – in various ways and to varying degrees – in all political ecological work. These are, first, a *theoretical commitment* to critical social theory and a post-positivist understanding of nature and the production of knowledge about it, which views these as inseparable from social relations of power. Much political ecological work is theoretically catholic, with roots in Marxist political economy, but also often influenced by poststructuralism, postcolonialism, and feminist geography. Thus, political ecology is closely associated with radical scholarship and a rejection of positivist approaches to social relations and environmental science. As Robbins (Chapter 6, this volume) notes, political ecology plays the role of "trickster," relentlessly critiquing conventional science and policy in order to formulate alternative understandings of the world. Second, political ecology retains a *methodological commitment* to in-depth, direct observation involving qualitative research of some sort, often in combination with quantitative methods and/or document analysis. As Davis (Chapter 20, this volume) points out, most political ecology is at least partly historical in its analysis. Some understanding of the place-based, historically sedimented social relations of production and exchange, and environmental practices is crucial for political ecological analysis. Underpinning these methodological commitments is a conviction that there are vital elements of nature–society relations that cannot be read from a social or spatial distance (e.g. via remote sensing, extensive surveys, soil or water samples, etc.), but can only be ascertained through intensive, open-ended, qualitative methods, most often in combination. Political ecology is thus methodologically plural, and most studies employ some combination of qualitative, broadly ethnographic methods (interviewing, direct observation) with historical documentary analysis and frequently quantitative analysis using GIS, survey methods, and an array of methods common in ecological science. Third, political ecology is characterized by a normative *political commitment* to social justice and structural political change. Political ecology is an explicitly normative intellectual project, which has from its beginning highlighted the struggles, interests, and plight of marginalized populations: peasants, indigenous peoples, ethnic and religious minorities, women, the poor. In this sense, and in contrast to many other approaches social and environmental analysis (e.g. cultural ecology, land use/land cover change analyses, etc.), political ecology is explicitly normative in its approach. Political ecologists thus seek not just to explain social and environmental processes, but to construct an alternative understanding of them, with an orientation toward social justice and radical politics.

Thus, to paraphrase Marx, the point of political ecology is not merely to understand the world; the point is to change it (see Chapter 13, this volume). As demonstrated by the chapters in this Handbook, these theoretical, methodological, and political commitments run throughout work in political ecology, lending coherence to an immensely diverse body of work (see also Chapter 48, this volume).

Placing political ecology

Political ecology has no natural monopoly on the study of nature–society relations, and its defining commitments have developed not in a vacuum but in conversation with – and reaction to – other areas of inquiry. Political ecology occupies but one patch of a broad field that is populated, in the main, by other social science disciplines. Environmental anthropology, environmental sociology, ecological economics, and environmental economics among others are, like political ecology, dedicated to understanding societies' relations with the non-human world: and through the knowledge these disciplines create, they also seek to inform and give shape to environmental futures. In addition, the general intellectual terrain which political ecology occupies also includes explicitly interdisciplinary fields like sustainability science, as well as more applied work such as hazards research, rural development, climate change mitigation, and urban environmental planning. Political ecology has developed alongside these areas of inquiry and in generative tension with them, finding intellectual allies while also actively "positioning for difference" in order to highlight what political ecology puts at stake, as a distinctive mode of knowledge production. There is, for example, a long-standing and productive conversation with ecological economics centered on the dissipative character of agricultural and industrial systems in energetic terms and its implications for uneven development. In political ecology this emerges early on in Bunker's work (1985) on extractive regimes in the Brazilian Amazon and is developed, in conversation with economic sociology and ecological economics, through work on social metabolism (Heynen et al. 2006), the "metabolic rift" (Foster 1999; Moore 2011) and ecologically unequal exchange (see Chapter 29, this volume). It continues too in recent contributions of political ecology to an emerging body of work on de-growth (Martinez-Alier 2012; Demaria et al. 2013; Kallis 2011). Here the distinctiveness of political ecology is both less obvious and less significant because it shares a common heterodox position with ecological economics vis-à-vis dominant forms of economic and environmental knowledge.

In relation to a number of other perspectives, however, political ecology has sought to "occupy" the intellectual terrain in a strategic sense, as a deliberate move to stake out and realize the possibility of an alternative position from which to challenge and contest dominant approaches. Political ecology has explicitly positioned itself as "other" to the environmental and resource management agendas of state and corporate elites, for example, examining and challenging these avowedly apolitical modes of "knowing nature" for the political work they enable. In a similar way, political ecology's engagement with environmental economics is marked by an arm's-length and critical engagement with the models and assumptions underpinning market-modes of socio-natural ordering. The process of critical distancing serves to constitute political ecology as a field, defining the "other" to which political ecology emerges as an important – even necessary – alternative mode of inquiry (see Chapter 6, this volume).

The relations among political ecology and its epistemological others, then, are fluid and dynamic and, in a non-trivial way, what is understood as political ecology depends very much on one's own training. For example, to some whose scientific training in the study of nature–society relations cleaves strongly to the "modern constitution" (Latour 1991), which purifies and separates science from the political, political ecology is what happens when scientific questions become contaminated by politics. This perspective is an institutional reality in countries in the global South where, paradoxically, many political ecologists funded and trained in the global North have developed their craft. In such circumstances research often sails under flags of convenience, such as rural development or economic geography, although its content negotiates the methodological, theoretical, and normative commitments outlined above. To other researchers who share "nature–society relations" as a general designation but who are primarily schooled in methods of modeling or environmental data analysis, political ecology describes a loose assemblage of (largely interchangeable) people whose stock in trade is theory rather than data, and for whom its internal differentiation and debates are largely irrelevant.

To adopt the epistemology of political ecology then, is to choose from among a range of possibilities for understanding nature–society relations. The way in which political ecology is situated relative to other modes of inquiry can have significant implications for the conduct of research, from ease of interdisciplinary collaboration and issues of research access, to the opportunities for experimentation with non-academic partners in "co-producing" knowledge (Forsyth 2008; see also Chapter 40, this volume). The constitutive character of critical theory for political ecology, for example, means that it more readily makes common cause with organizations challenging institutional power rather than those who would seek to uphold it. Finally, and notwithstanding the common commitments outlined above, political ecology research is frequently characterized more by pragmatism than by pedigree, and a sense that what matters is what works, given the questions posed. In the context of research undertaken by and with groups seeking to speak truth to power, or demanding greater control over their environmental futures, for example, political ecology regularly borrows methods of evidence gathering and environmental analysis widely used in "apolitical" fields of resource and environmental management – for example, pairing remotely sensed data on land use change with the direct experience of land users (Nightingale 2003; Robbins 2001), measuring soil fertility (Benjaminsen et al. 2010), mapping the distribution of environmental contamination (*Environmental Justice Atlas* n.d.), or numerically calculating material and waste flows (D'Alisa and Armiero 2013). In this sense one can say that political ecology continues to be radically experimental in the field, mixing conceptual genres and methodological registers in an effort to understand and transform socio-ecological relations.

Structure of the volume

While the chapters in this volume speak to the breadth of political ecology as a field, the volume is by no means comprehensive. As editors, we decided early on that it would be futile to try to include a chapter on every conceivable topic of interest to political ecologists. As a result, some readers will no doubt be disappointed that the volume does not include a chapter on deforestation, for instance, nor on water governance, resource extraction, or herding and range management. Instead, we have organized the volume primarily around analytical concepts, bodies of theory, and problems of praxis that characterize political ecology. The book is organized into four parts, of which the fourth is by far the longest and most diverse. The first part of the book is introductory in nature, and contains the present chapter as well as a sweeping historical analysis of the field by Michael Watts ("Now and then: The origins of political ecology and the rebirth of adaptation as a form of thought"). Watts traces the varied intellectual and political currents that have contributed to contemporary political ecology, and what he terms second-generation adaptive theory. After his forensic investigation of the ways that adaptation has been conceptualized in political ecology, cultural ecology, human ecology, ecological anthropology, and other fields, Watts examines the emergent concept of "resilience" as the latest manifestation of adaptation thinking. The chapter provides a comprehensive and novel assessment of political ecology, its intellectual history, and the continued salience of its core analytical and political critiques.

Part II of the book ("Origins, trajectories, and futures"), contains five chapters that together critically assess political ecology as an academic and intellectual field. The section begins with a chapter by Ben Wisner ("Speaking truth to power: A personal account of activist political ecology") in which the author recounts his personal history as a student, researcher, and activist, and what led him to pursue an academic life in what would become the field of political ecology. The chapter speaks – both literally and figuratively – to the concerns of contemporary students and the very different social, political, and economic contexts in which they pursue their studies. It is followed by a chapter by Enrique Leff ("The power-full distribution of knowledge in political ecology: A view from the South"), which examines the diverse traditions of political ecology, and contrasts perspectives from the global South with those prevalent in the global North. Leff makes a convincing case for distinct regionalizations of the field, and calls for attention to intellectual currents that characterize political ecology in Latin America and elsewhere in the formerly colonized world. The following chapter, by Denis Gautier and Christian Kull ("French research traditions on peasant agricultural systems: A convergence with political ecology?"), considers the French academic study of *systèmes agraires* as an example of parallel and convergent evolution with political ecology: this tradition and Anglophone political ecology have important common origins in the 1970s, examine some of the same central problematics, and have intersected with and influenced each other at some key junctures. The French research tradition on *systèmes agraires* has been particularly influential in Africa and has informed both academic research and development policy. This chapter is followed by one by Paul Robbins ("The Trickster science") in which he provocatively argues that intellectually, political ecology plays the role of the trickster – relentlessly critiquing dominant forms of knowledge and power on the way to formulating an alternative truth. In playing this role, however, political ecology often tries to "have it both ways," for instance by using science to critique science. Political ecology, then, only exists in relation to apolitical ecology. The final chapter in this section, by Bruce Braun ("From critique to experiment? Rethinking political ecology for the Anthropocene"), focuses on the future of political ecology, arguing for a shift of emphasis away from critique and toward experimentation as a form of active, direct engagement in environmental politics. This shift, Braun argues, is well suited both to the social and environmental dilemmas of the Anthropocene, and to the need for democratizing radical scholarship.

Part III, "Doing political ecology," considers the problems inherent in political ecology as a form of practice. The section contains eight chapters, organized around four problematics:

ethics, methods, activism, and policy and practice. The first pair of chapters in this section, by Juanita Sundberg ("Ethics, entanglement and political ecology") and Rosemary-Claire Collard ("Ethics in research beyond the human") both draw on feminist and postcolonial theory to examine some of the ethical dilemmas that inhere in political ecological research. Both authors provide first-person accounts of the complicated relationships of which they have become part, and the ethical decisions they have had to make, in the course of conducting research. Sundberg discusses the ethical dissonance inherent in being a person of privilege conducting research in the global South. In a field closely identified with field-based research in the global South, such attention to the legacies of empire, and the ethical obligations they entail, is of paramount importance. Collard, for her part, considers the ethical dilemmas of carrying out research on and with animals. Her discussion probes the very nature of the relationship between the human and the non-human, as well as the inadequacies of institutionalized ethical review standards that have become a nearly ubiquitous part of university-sponsored research. The next pair of chapters examines research methods in political ecology. Abigail H. Neely and Thokozile Nguse ("Relationship and research methods: Entanglements, intra-actions, and diffraction") and Karl Zimmerer ("Methods and environmental science in political ecology") take very different approaches to their discussions of research methods. Drawing primarily on feminist theory, Neely and Nguse explore the role of personal relationships involved in the practice of research. They highlight the often complicated power relations involved in fieldwork, and the fact that these are occasionally at odds with the static portrayal of power in research as presented in the literature. For his part, Zimmerer examines the complexities of integrating social and environmental science in political ecology. He argues that a greater use of scientific methods by political ecologists would likely yield important and novel results, necessary to understanding contemporary socio-environmental problems.

The following two chapters address questions of activism in political ecology. The first of the two, by Nik Heynen and Levi Van Sant ("Political ecologies of activism and direct action politics") argues strongly that political ecology and direct action environmental activism would both benefit from more direct engagements between the two. Drawing on personal experience in environmental activism, Heynen and Van Sant argue that political ecologists should engage directly in the environments and societies they study, and that direct action offers a path beyond the frequently muddled politics of academia. In the following chapter ("Political ecology as praxis"), Alex Loftus presents a re-reading of Marx's *Theses on Feuerbach*, in order to make a case for political ecology as engaged praxis. Through careful exegesis, Loftus argues that political ecology can provide an intellectual space where the roles of practice and theory may be contested and ultimately reconciled. This section ends with two chapters on policy and practice, by Brent McCusker ("Political ecology and policy: A case study in engagement"), and Anthony Bebbington ("At the boundaries of *la política*: Political ecology, policy networks and moments of government"). Both chapters provide personal, "inside-the-sausage-factory" accounts of the institutions and practices of policy making. McCusker discusses his experience with the United State Agency for International Development (USAID), and the ethical dilemmas and professional difficulties it entailed. In spite of his evident frustrations with the state policy apparatus, McCusker suggests that political ecologists individually, and political ecology as a field, have much to gain from a closer engagement with policy and policy makers. Bebbington draws on his experience working in various capacities with policy makers in Latin America, most recently regarding mining policy in El Salvador. Using as his starting point the Spanish word *política* – which translates both as politics and policy – he explores the complicated relationship between these two realms, and argues that ultimately political ecologists cannot help but engage with policy writ large.

Part IV of the volume, "Core questions in political ecology," is by far the largest section of the book. It is organized around five fundamental analytical concepts, and questions that political ecologists ask about them: (1) Environmental knowledge (*How do we come to know nature, and what differences do forms of environmental knowledge make?*), (2) Environmental change (*In what ways are nature and society transformed through economic activity, and how does this metabolic relationship affect different social groups in different ways?*), (3) Environmental governance (*Through what sorts of social arrangements and forms of rule do people 'manage' nature, and to what effect?*), (4) Environmental identities (*How are social subjectivities shaped through, and reflected by, differential access to and control over nature?*), and (5) Environmental politics (*In what ways and for what reasons do people mobilize politically around nature and natural resources?*). The first of these sections (Environmental knowledge) addresses one of the core questions of political ecology: how we come to know nature, and the political consequences of various forms of environmental knowledge. The first chapter in this section, by Rebecca Lave ("Reassembling the structural: Political ecology and Actor-Network Theory") provides a trenchant critique of ANT, and argues that its denial of structural power is ultimately irreconcilable with political ecology's theoretical and political commitments. Lave reviews the various attempts to incorporate ANT into political ecological analysis, concluding that such efforts are doomed to failure. The following chapter, by David Demeritt ("The promise of participation in science and political ecology") examines practices of public participation in scientific research and the potential role of political ecology in publicly oriented science. Demeritt demonstrates that normative claims for participation in science sit uncomfortably with claims that public involvement increases the legitimacy of such research. This recognition points to tensions between public involvement in research and the desire of researchers for autonomy and control over scientific process and results. Questions of diverse forms of environmental knowledge are addressed in the next chapter, by Leah Horowitz ("Local environmental knowledge"). In it, Horowitz reviews the burgeoning political ecology literature on indigenous and local environmental knowledge, and points to new directions of investigation, arguing that local knowledge plays a critical role in addressing emerging environmental governance. In the next chapter ("Participatory mapping"), Joe Bryan examines historical and contemporary practices in popular and participatory cartography. The chapter traces the history of these methods to early ecological anthropology and cultural ecology, in which maps were viewed unproblematically as data, and through more critical approaches in political ecology. Bryan also reviews recent controversies in participatory mapping in Mexico and elsewhere. As he demonstrates, these methods entail thorny, open-ended ethical questions that researchers and practitioners must confront, and which demand careful attention if maps are to be tools of liberation rather than oppression. The following chapter, by Diana Davis ("Historical approaches to political ecology"), examines the relationship between political ecology and the allied fields of environmental history and historical geography. Through a detailed, comparative review of these literatures, Davis argues that a critical approach to history should lie at the heart of political ecological research, and exhorts political ecologists to be more rigorous in their treatment of historical processes.

The next section ("Environmental change") addresses another of political ecology's core themes: the processes and social implications of environmental transformation. The first of these chapters, by Noel Castree ("Capitalism and the Marxist critique of political ecology") examines the Marxist literature on capitalism and nature via the "production of nature" thesis of the late Neil Smith. In doing so, Castree enriches our understanding of how political ecology evolved in relation to broader Marxist efforts to grapple with environmental questions. This is followed by a chapter on the political ecologies of natural hazards ("Political ecology of risk, hazards, vulnerability, and capacities") by Jim Wescoat. This chapter reviews one of the foundational intellectual currents of political ecology, pointing out that, far from being a thing of the past, work on hazards, risk, and vulnerability is on the rise, and remains a vital direction of research for political ecologists, particularly in the era of climate change and its many social and environmental implications. These themes are continued in the following chapter, by Diana Liverman ("Reading climate change and climate governance as political ecologies"). Here, Liverman provides a wide-ranging review of the recent (and massive) literature on climate change. She makes a strong case that political ecologists have much to contribute to research on, and debates over, climate change, particularly with regard to the differential social effects of climate change. The next chapter, by Astrid Ulloa ("Environment and development: Reflections from Latin America"), examines the political ecology of development practice in the global South. Ulloa argues that Latin American political ecologists, social movements, and activists have re-conceptualized Euro- and US-centric notions of development, recasting it in light of the political and cultural struggles of indigenous, peasant, and other marginalized peoples. In the following chapter Ed Carr also considers development practice ("Political ecology and livelihoods"), focusing specifically on the concept of livelihoods which has been one of political ecology's foundational concepts. Carr argues that most livelihoods frameworks – that is, the conceptual framing of livelihoods, as employed by development theorists and practitioners – are overly narrow and economistic, and view poor people's livelihoods problematically as tied to place. Such framings, Carr argues, are largely apolitical and ultimately do little to aid our understandings of how poor people may improve their lives. These themes are carried into the subsequent chapter, by Brian King ("Political ecologies of disease and health"), which examines the growing political ecology literature on human health. King points out that diseases such as AIDS are fundamental political ecological problems, which affect environments and how people interact with them. Similarly, he argues that political ecologists have much to contribute to the understanding of disease and its social and environmental implications. The next chapter, by Tor Benjaminsen ("Political ecologies of environmental degradation and marginalization"), also elaborates some of political ecology's core themes. The chapter reviews literature in the neo-Malthusian and Marxist traditions, and considers how political ecologists have engaged with ecological processes in their studies of environmental degradation. Drawing on his own work in Africa, Benjaminsen highlights the ecological and social complexity of environmental degradation, and how narratives of degradation may sometimes be used as tools of dispossession. This is followed by a chapter by Stefania Barca and Gavin Bridge ("Industrialization and environmental change"), which explores how political ecology has sought to understand the process of industrialization and its wrenching socio-ecological transformations. Explicit and sustained attention to industry within political ecology has been quite limited, relative to the interest in agrarian systems for example. Barca and Bridge argue the case for political ecologies of industrialization, centering their account on the mechanization of production and its associated intensification of control over labor and increase in social metabolism. In doing so they draw on both existing work in political ecology and from research in the borderlands of political ecology, environmental history, and ecological economics. The section's final chapter, by Alf Hornborg ("Conceptualizing ecologically unequal exchange: Society and nature intertwined"), considers how, within the capitalist world system, the interconnections of ecology and economics result simultaneously in accumulation and impoverishment. Through a careful review of the literatures in ecological economics and Marxist political economy, Hornborg examines the complex interrelationship between nature, energy, and unequal exchange.

The following section of the book ("Environmental governance") examines the multi-scalar and multi-institutional arrangements through which people manage environments and resources. Covering a wide range of empirical contexts, from biodiversity and agriculture to biosecurity and energy, the chapters in this section reflect on the ways in which economic and political power can be constituted through environments and ecologies. The first chapter ("Nature conservation") is by Rod Neumann and focuses on a topic central to the development of political ecology's critical perspective on the matter of nature, the conservation of biodiversity. The author examines an archetype of biodiversity conservation – state-protected areas – and shows how they take form through territorial and institutional practices of state formation and are constitutive of economies and identities. It is followed by a chapter by Derek Hall ("The political ecology of international agri-food systems") focused on the spectacular expansion since the 1970s of high-value export crops from the global South. Hall argues that the biographies of commodities like fresh fruits, flowers, and vegetables exemplify many core themes of a global political ecology. He also suggests that the international political economy of agri-food production may be changing in significant ways, highlighting the emergence of South–South funding and trade flows associated with "land grabs" for agricultural production. In the chapter that follows, Jon Otto and Tad Mutersbaugh ("Certified political ecology"), consider the rapid expansion of environmental commodity certification schemes. They argue that, in the absence of a strong concern for social justice, certification can effectively transfer environmental risks away from consumers and toward producers in the global South, and call for much greater attention to the knowledge politics involved in environmental certification. In the next chapter ("Property and commodification"), Scott Prudham places property at the center of material and semiotic relations between humans and the non-human world, and examines connections between arrangements of property and processes of commodification. The chapter reconsiders the legacies in political ecology of Malthus and Marx as a way of foregrounding questions of property over those of resource availability, and linking the commodification of nature with the commodification of labour. Karen Bakker's chapter ("Neoliberalization of nature") addresses one of the most socially contested approaches to environmental and resource management, and examines the extensive contributions of political ecology to its analysis. She argues for a more comprehensive understanding of neoliberalisation, informed not only by a broader range of "natures" but also by the currents of posthumanism so as to better acknowledge the political capacity of non-humans. The next chapter is by Morgan Robertson ("Environmental governance: political ecology and the state") and it explores how political ecology has chosen to engage with theorizations of the state and state power. Robertson argues that political ecology has an ambivalent relationship to the state: while the state is frequently referred to in research accounts, only rarely are political ecologists explicit about how they are conceptualizing and theorizing this significant environmental management institution.

The following chapter is by Gabriela Valdivia ("Eco-governmentality"), who explores the ways in which political ecology has taken up Foucault's writings on governmentality and re-worked them in the context of diverse environmental rationalities such as preventing climate change or protecting nature. In work on environmental regimes and eco-governmentality, she argues, political ecology is living out Foucault's methodological challenge of continuous empirical experimentation in order to understand how power takes shape through its capillary forms. The next chapter in this section is by Matthew Huber ("Energy and social power: From political ecology to the ecology of politics"). Huber reflects on the oblique way in which energy has been approached and understood within political ecology, and contrasts this with cultural ecology's earlier interest in energy expenditures and cultural practice. He argues that political ecology should place energy more centrally in its analysis, and that doing so expands significantly what one considers the ecologically political. The penultimate chapter in this section ("From biodiversity to biosecurity") is by Celia Lowe, and it examines what is at stake in the emergence of biosecurity as a rationality of environmental and social governance. Lowe argues that the rise of biosecurity, part of a broader expansion in practices of securing life through the registers of risk and preparedness, opens up new terrains for political ecological research. The final chapter in this section, by Nathan Sayre ("Scales and polities"), considers the central role of scale in environmental governance. Sayre reflects on meanings of scale that have emerged from several decades of political ecological research. He argues that these diverse engagements with scale now provide political ecology with a series of epistemological and methodological research guidelines.

The book's penultimate section ("Environmental identities") considers how social subjectivities are formed in relation to nature and natural resources. The opening chapter by Rebecca Elmhirst ("Feminist political ecology") outlines the extensive contributions made by feminist political ecology and the salience of gender to struggles over resources and environments. She argues that feminist political ecology's growing internal differentiation is a significant strength, and highlights its commitment to collaborating with others outside of academia – in policy, practice, and activism – in the production of environmental knowledge. In the next chapter ("Indigeneity"), Emily Yeh and Joe Bryan examine the intellectual and political work done by the concept of "indigeneity." Questions of indigeneity have been the primary route by which postcolonialism has entered into political ecology, although Yeh and Bryan emphasize the further generative possibilities of indigeneity as a concept, and the opportunities it presents for political ecology to think with the ontologies and cosmologies mobilized by indigenous social movements. The chapter by Michael Ekers ("On the concreteness of labor and class in political ecology") focuses on class as a social identity produced through material and representational relationships with the non-human world. Ekers argues that like race, gender, and other forms of social difference, class is not a category that pre-exists a relation to nature: instead, class is constituted through diverse forms of labor, including both waged and non-waged work. The final chapter in this section ("Nature, difference and the body") is by Julie Guthman and Becky Mansfield, who use the scale of the human body to explore embodied identities of health and illness. The authors query prevailing accounts of changes in bodily morphology and functionality – such as an "obesity epidemic," and an upsurge in autism – and call for a "critical political ecology of the body" that acknowledges the possibility of "environmentally-induced difference" at the scale of the body. Advances in epigenetics, they argue, demonstrate how biologically and chemically-induced differences arise as a result of how bodies interact with diverse environments.

The book's final section ("Environmental politics") examines the ways that processes of environmental change, degradation, or dispossession become politicized (or, de-politicized), a major topic of research in political ecology. The chapters consider both how engagements with these questions have shaped political ecology, and how political ecology's understandings of them have differed from cognate and dominant explanations, in ways that have helped to constitute and define the field. The first chapter in this section, by Wendy Wolford and Sara Keene ("Social movements"), argues that while political ecology has a foundational commitment to understanding marginalization and contestation from the perspective of the marginalized, it has focused largely on informal and unorganized politics rather than on social movements. Wolford and Keene argue for more research on organized social movements, but also explain the field's research emphases via explication of what are often implicit theories of politics, and of the state and civil society in particular, in political ecology. The next chapter, by Ryan Holifield ("Environmental justice and political ecology"), delineates the overlaps and distinctions between political ecology and environmental justice: two fields that often seem substantively very similar and convergent – both focused on environmental marginalization and inequality – but that have evolved largely separately due to distinctions of region, theory, method, and purpose. After reviewing the reasons for this history, Holifield argues for contemporary and future convergence between the fields, emphasizing what each stands to gain. In the next chapter ("Environmental conflict"), Philippe Le Billon situates the study of environmental conflicts at the heart of political ecology, and contends that one of the field's major impetuses and contributions has been to at a minimum complicate, and sometimes directly reject and refute, overly simplistic and deterministic explanations of environmental conflict. The chapter explores neo-Malthusian renderings of the "resource curse" as a paradigmatic example of this dynamic, while also considering the broader implications of using a "conflict lens" to examine human–environment relations. Finally, Erik Swyngedouw ("Urbanization and environmental futures: Politicizing urban political ecologies") considers the politics of the planetary-scale urbanization of nature. He argues that while such expanded metabolism generates concerns with "sustainability" and the like, such framings can easily lead in reactionary directions that stifle true political engagement through a focus on expertise and consensus; true environmental politics in this sense must include agonistic struggles over content and direction of human–environment relations.

In the book's concluding chapter, we revisit and more fully elaborate our claims regarding political ecology's intellectual roots and core commitments. While arguing that political ecology is rooted squarely in Marxist scholarship, the theoretical and methodological perspectives of feminist and postcolonial scholarship are now also widely embraced, and nearly ubiquitous in the field. We also point to exciting new directions for research and praxis in political ecology. These include a deepening of engagements across linguistic and academic traditions, a furthering of research into post-humanist ecologies, and further engagement with policy and practice. While refusing to be prescriptive in our analysis of these trends, we argue that in all its present and future diversity, political ecology is held together by its fundamental commitment to radical politics and social justice.

References

Bassett, T.J. 1988. The political ecology of peasant-herder conflicts in the northern Ivory Coast. *Annals of the Association of American Geographers*, 78(3): 453–472.

Benjaminsen, T., J. Aune and D. Sidibé. 2010. A critical political ecology of cotton and soil fertility in Mali. *Geoforum*, 41: 647–656.

Biehler, D. and G. Simon. 2011. The great indoors: Research frontiers on indoor environments as active political ecological spaces. *Progress in Human Geography*, 35: 172–192.

Blaikie, P. 1985. *The Political Economy of Soil Erosion in Developing Countries*. London: Routledge.

Blaikie, P. and H. Brookfield. 1987. *Land Degradation and Society*. London: Methuen.

Bryant, R.L. and S. Bailey. 1997. *Third World Political Ecology*. London: Routledge.

Bunker, S. 1985. *Underdeveloping the Amazon: Extraction, Unequal Exchange, and the Failure of the Modern State*. Chicago: University of Chicago Press.

D'Alisa, G. and M. Armiero. 2013. What happened to the trash? Political miracles and real statistics in an emergency regime. *Capitalism Nature Socialism*, 24(4): 29–45.

Demaria, F., F. Schneider, F. Sekulova and J. Martínez-Alier. 2013. What is degrowth? From an activist slogan to a social movement. *Environmental Values*, 22(2): 191–215.

Ehrlich, P. 1968. *The Population Bomb*. New York: Ballantine Books.

Environmental Justice Atlas. n.d. Available online at: http://ejatlas.org. Produced by the Environmental Justice Organisations, Liabilities and Trade research project.

Enzensberger, H.M. 1974. A critique of political ecology. *New Left Review*, 84: 3–32.

FitzSimmons, M. 1989. The matter of nature. *Antipode*, 21(2): 106–120.

Forsyth, T. 2008. Political ecology and the epistemology of social justice. *Geoforum*, 35: 756–764.
Foster, J. 1999. Marx's theory of metabolic rift: Classical foundations for environmental sociology 1. *American Journal of Sociology*, 105: 366–405.
Guthman, J. 2011. *Weighing In: Obesity, Food Justice, and the Limits of Capitalism*. Berkeley: University of California Press.
Hardin, G. 1968. The tragedy of the commons. *Science*, 162: 1243–1248.
Harvey, D. 1974. Population, resources and the ideology of science. *Economic Geography*, 50(3): 256–277.
Harvey, D. 1982. *The Limits to Capital*. London: Verso.
Hecht, S. and A. Cockburn. 1989. *The Fate of the Forest: Developers, Destroyers, and Defenders of the Amazon*. London: Verso.
Heynen, N. 2014. Urban political ecology I: The urban century. *Progress in Human Geography*, 38(4): 598–604.
Heynen, N., M. Kaika and E. Swyngedouw, eds. 2006. *In the Nature of Cities: Urban Political Ecology and the Politics of Urban Metabolism*. New York: Taylor & Francis.
Huber, M. 2013. *Lifeblood: Oil, Freedom, and the Forces of Capital*. Minneapolis: University of Minnesota Press.
Kaika, M. 2006. Dams as symbols of modernization: The urbanization of nature between geographical imagination and materiality. *Annals of the Association of American Geographers*, 96(2): 276–301.
Kallis, G. 2011. In defence of degrowth. *Ecological Economics*, 70(5): 873–880.
Latour, B. 1991. *We Have Never Been Modern*. Cambridge, MA: Harvard University Press.
Leff, E. n.d. Political ecology: A Latin American perspective. Mimeo, 51pp.
Leff, E. 2014. *Apuesta por la Vida: Imaginación Sociológica e Imaginarios Sociales en los Territorios Ambientales del Sur*. Buenos Aires: Siglo Veintiuno Editores.
Lipietz, A. 1999. *Qu'esst-ce que l'écologie politique. La grande transformation du XX Siècle*. Paris: La Décourverte.
Mansfield, B. 2012. Race and the new epigenetic biopolitics of environmental health. *BioSocieties*, 7: 352–372.
Martinez-Alier, J. 2012. Environmental justice and economic degrowth: An alliance between two movements. *Capitalism Nature Socialism*, 23(1): 51–73.
McCarthy, J. 2002. First World political ecology: Lessons from the Wise Use movement. *Environment and Planning A*, 34: 1281–1302.
Meadows, D.H., J. Randers, D.L. Meadows and W.W. Behrens. 1972. *The Limits to Growth: A Report for the Club of Rome's Project on the Predicament of Mankind*. New York: Universe Books.
Moore, Jason W. 2011. Transcending the metabolic rift: A theory of crises in the capitalist world-ecology. *The Journal of Peasant Studies*, 38(1): 1–46.
Neumann, R.P. 2005. *Making Political Ecology*. New York: Hodder Arnold.
Nightingale, A. 2003. A feminist in the forest: Situated knowledges and mixing methods in natural resource management. *ACME: An International E-Journal for Critical Geographies*, 2(1): 78.
Peet, R. and Watts, M., eds. 1996. *Liberation Ecologies: Environment, Development, Social Movements*. London: Routledge.
Robbins, P. 2001. Tracking invasive land covers in India, or why our landscapes have never been modern. *Annals of the Association of American Geographers*, 91(4): 637–659.
Robbins, P. 2004. *Political Ecology: A Critical Introduction*. Oxford: Blackwell.
Scott, J.C. 1976. *The Moral Economy of the Peasant: Rebellion and Subsistence in Southeast Asia*. New Haven: Yale University Press.
Smith, N. 1984. *Uneven Development*. Oxford: Blackwell.
Swyngedouw, E. 2004. *Social Power and the Urbanization of Water: Flows of Power*. Oxford: Oxford University Press.
Walker, R.A. 1973. Wetland preservation and management on Chesapeake Bay: The role of science in natural resource policy. *Coastal Zone Management Journal*, 1(1): 75–101.
Walker, R.A. 1974. Wetland preservation and management: A rejoinder – economics, science and beyond. *Coastal Zone Management Journal*, 1(2): 227–233.
Watts, M.J. 1983a. On the poverty of theory: Natural hazards research in context. In Kenneth Hewitt (ed.), *Interpretations of Calamity: From the Viewpoint of Human Ecology*. Boston: Allen & Unwin, pp. 231–262.
Watts, M.J. 1983b. *Silent Violence: Food, Famine, and Peasantry in Northern Nigeria*. Berkeley: University of California Press.
Watts, M.J. 2001. 1968 and all that … *Progress in Human Geography*, 25(2): 157–188.

Wisner, B. 1978. Does radical geography lack an approach to environmental relations? *Antipode*, 10(1): 84–94.

Wolf, Eric. 1969. *Peasant Wars of the Twentieth Century*. New York: Harper & Row.

Wolf, Eric. 1972. Ownership and political ecology. *Anthropological Quarterly*, 45(3): 201–205.

Zimmerer, K.S. and Bassett, T.J., eds. 2003. *Political Ecology: An Integrative Approach to Geography and Environment-Development Studies*. New York: The Guilford Press.

2

NOW AND THEN

The origins of political ecology and the rebirth of adaptation as a form of thought

Michael J. Watts[1]

> Nature itself has been rediscovered to function as a market.
>
> *(Duffield 2011: 763)*

Introduction

Released formally in March 2014, the second component of the *Fifth Assessment Report of the Intergovernmental Panel on Climate Change* (IPCC) covering impacts, adaptation, and vulnerability (IPCC 2014) makes for grim reading.[2] "We're now in an era where climate change isn't some kind of future hypothetical," said Chris Field, co-chair of the IPCC Working Group II: "We live in an era where impacts from climate change are already widespread and consequential" (www.climatesciencewatch.org/2014/03/31/ipcc-impacts-assessment-poses-urgent-challenge-for-risk-management/). In the report summary the word "risk" is mentioned an average of about six times per page. Eight core risks – most are potentially catastrophic – are identified "with high confidence", each "spans sectors and regions". Two months later a study by researchers at NASA and the University of California, Irvine, announced that a rapidly melting section of the West Antarctic Ice Sheet appears to be in an irreversible state of decline (www.jpl.nasa.gov/news/news.php?release=2014-14). Collectively it all looks like an ecological Armageddon. Coincident with the release of IPCC's assessment report was the arrival of the World Bank's annual development report – *Risk and Opportunity: Managing Risk for Development* (World Bank 2014) – devoted entirely to constellations of old and new risks which threatened to "reverse hard-won gains" (2014: 4). Inhabiting a world of radical precarity, poor communities, households, and states in the Global South confront a veritable avalanche of life-threatening and often interconnected shocks constituted by financial, economic, ecological, and other sorts of systemic risks. Climate change, seen by the Bank as simultaneously a burden and an opportunity, figures centrally in the risk portfolio that the World Bank's customary constituency – the global poor – are now required to manage in the name of sustainable development.

Implicit in the current epistemology of global climate science is a worldview somewhat at odds with the Darwinian orthodoxy of evolutionary gradualism (Boal 2009). Climate has obviously changed historically but in regard to patterns of human occupation and livelihood such oscillations represented a deep historical time – the very *longue durée*. What is on offer now is something unimaginable until relatively recently: namely abrupt, radical life-threatening shifts framed in the language of uncertainty, unpredictability, and contingency. The reports by the IPCC and the World Bank – there are a raft of others of course – and an ascendant recognition that we all now inhabit a human-made geologic era called the Anthropocene,[3] all point to an emergent science of impending global (terrestrial and oceanic) disaster – what one might call catastrophism. It is a crisis demanding a public response – political, policy, civic, and business – of an equal magnitude and gravity.

In a discursive sense, then, climate change is represented as a symptom of a planetary emergency. Global warming encompasses, and has direct consequences for, three fundamental human provisioning systems – water, food, and energy. But, to simply take the latest encomium produced by the World Economic Forum in its annual *Global Risks* report (WEF 2014), such threats are configured into complex assemblages of risk including non-state violence, critical infrastructures and cyber attacks, economic inequality and unemployment, and systemic financial risk, all of which are now seen to be inseparably and organically linked in complex networks of teleconnected effects (OECD 2003; WEF 2014).[4] A global risk portfolio mobilizes and enrolls powerful actors around the threat of massive, catastrophic, and systemic risks and uncertainties. Central to this vision of global systemic risk is the very nature of life (Kauffman 2000) itself drawing upon the molecular and digital sciences – complexity, self-organization, and adaptive agents are its avatars (Miller and Page 2007; Mitchell 2009) – which shapes the nature of what is to be governed and how. If life is constituted through complex and continual adaptation and emergence, life rests upon, and is composed of, radical uncertainty in which permanent danger and security form an unstable, unpredictable present – what Dillon and Reid (2009: 85) call a life "continuously becoming dangerous". Ash Amin (2010: 138) sees this as the condition of calamity, or "catastrophism":

> The recurrence, spread, severity and mutability of the world's natural and social hazards are considered as symptomatic of this state (of permanent risk), and its latent conditions are understood to be too volatile or random and non-linear to permit accurate prediction and evasive action. In the apocalyptic imaginary, hazard and risk erupt as unanticipated emergencies, disarming in every manifestation and in every way.

What is so striking about the climate change talk as a form of catastrophism is the ubiquity of the language of adaptation and its cognates: adaptive capacity, adaptive strategies, adaptive governance. Since the first IPCC report in 1991, adaptation – defined by the IPCC (2014) as "the process of adjustment to actual or expected climate and its effects" – has emerged as the lodestar of public and development policy coincident with the realization that mitigation has receded into a distant future. "Adapt Now" is the rallying cry of the moment (or one might say, "adapt or die"). Bassett and Fogelman (2013) show how pervasive is the adaptation lexicon not just within IPCC but in the citational world of key research journals. It is a term that has "gone viral" (Ribot 2011).[5] Historically, the IPCC has worked with a conceptual understanding of adaptation as adjustment, an idea that actually harkens back to 1960s cultural ecology. The focus is on proximate rather than structural processes regarding adaptation in social systems, and on passive, reactive, or anticipatory adjustments. At the same time, a body of academic research associated with Barry Smit, Carl Folke, and Neil Adger has pushed strenuously, not least in climate policy circles and within IPCC itself, for a more structural rendering of adaptation, one that draws upon ideas of vulnerability, resilience, and the insights of political ecology. In the latest IPCC report, for example, there is talk of climate-resilient pathways, of the "limits to adaptation" and the need for "transformational adaptation" (Summary, IPCC 2014: 24–25). Nothing here necessarily challenges Bassett and Fogelman's overall assessment that IPCC operates with a pedestrian, and in many respects, old-fashioned notion of adaptation as "adjustment to climate stimuli" (2014: 49). What is incontestable and quite striking in all of this work, is the fact of adaptation's revival and rehabilitation after a period – especially the 1970s and 1980s – in which it fell from grace. The call for, and ubiquity of, "adaptation speak" suggests adaptation is more than a keyword: it resembles a hegemonic discourse, anchored currently in equally powerful discourses of security, risk management, and resilient social systems.

From the political ecology vantage point, the "adapt now" *mentalité* is something of a paradox. Climate change adaptation work is unequivocal in identifying the concept's origins in evolutionary biology (Smit and Wandel 2006: 286) but it was precisely the flaws of organic analogies that political ecology sought to address. Acknowledging that its definition is disputed and semantically slippery, climate adaptation refers to "processes, actions or outcomes in a system in order for the system to better cope with, manage or adjust to some changing condition" (Smit and Wandel 2006: 282). At the very least there are striking resemblances here to earlier geographical research on natural hazards, and to a class of behaviorist stimulus–response models. What appears to be on offer is a recycled version of adaptation thinking of 1960s associated most closely, as I shall show, with cultural ecology, ecological anthropology, and general systems theory. If the measure of adaptive fitness is now "success or survival of a culture" (Smit and Wandel 2006: 282), it has nevertheless been repurposed and rebooted with a rather new conceptual vocabulary: security, risk, vulnerability, exposure, resilience, adaptive management, and governance (see Adger 2006a; Adger et al. 2009; Pelling 2011; Smit and Wandel 2006).

Embedded in the new vocabulary – what I shall call second generation adaptive theory – are two differing lineages. The first is a theory of complex systems – characterized by signaling and information processing, complex collective behavior, non-linearity, and "thoughtful (but perhaps not brilliant) adaptive agent" (Miller and Page 2007: 3) – that assure "continual adaptation and the emergence of cross-level organization" (Folke 2006: 257). Resilience provides both a normative and conceptual frame in complexity theory's deployment in global climate change analysis: adaptive capacity builds enhanced resilience. A four-phase "adaptive renewal cycle" (panarchy so-called) undergirds a capacious model of "socio-ecological systems analysis" drawing within its circumference, according to its in-house theoreticians, all that has gone before (Smit and Wandel 2006; Gunderson and Holling 2002).[6] And yet it was precisely the *limits* of adaptation as a form of thought which constituted the very ground on which political ecology emerged during the 1970s and 1980s. Minimally one needs to ask: is this old wine in new bottles? How and in what ways does "adaptation 2.0" address the weaknesses of "adaptation 1.0"? It is to this question of how adaptation is now put to work – claiming to be part of a commodious approach subsuming critical political ecology and incorporating social vulnerability and political economic context (see Adger, Eakin and Winkela 2010: 150) – that I want to address my remarks. Situating adaptation in this way affords an opportunity to not only think about how the concept was deployed "back then" (at a time, incidentally, when neoliberalism had yet to be born and climate change was cast as "inadvertent climate modification") but also to reflect genealogically about political ecology, that is to say its conditions of possibility, its intellectual lineages and its sites of genesis. Thinking with adaptation permits us, I believe, to consider how political ecology and its conceptual apparatuses emerged as a distinctive approach and what that approach might have to offer towards an understanding of what counts today as adaptive capacity or adaptive governance with respect to the massive challenges of global climate change (and in a world, of course, in which neoliberalism has run amok).

Cultural ecology, ecological anthropology, and adaptation as a form of thought

The Berkeley School and cultural ecology

The Berkeley School of Cultural Geography is inextricably associated with the name of Carl Sauer, a child of the Ozarks who trained in Geology and Geography at the University of Chicago (his PhD was awarded in 1915) and who, after a seven-year sojourn at the University of Michigan, migrated west to California where he presided over the Department of Geography at the University of California, Berkeley for over three decades from 1923 to 1957. Much ink has been spilled over Sauer's work, his legacy, and his theoretical project (see Mitchell 2001; Mathewson 2009), and particularly over the concept of culture as he deployed it. Between 1925, when he delivered his famous essay entitled the *Morphology of Landscape*, and his 1940 address to the Association of American Geographers, *Foreword to Historical Geography*, Sauer laid out a research program of how to think about human agency in relation to the transformation of the earth. Sauer's research program rested on four pillars: culture, Anthropology, the *longue durée* of history (what he called genetic history), and biophysical systems. Geography's proper domain was "territorial localization" through the comparative study of "modes of living". The human landscape was seen as the product of human agency and "practical experience", of accumulated residues as he put it, quoting Vilfredo Pareto. Fossils, ruins, and palimpsests were the forms Sauer pursued. Unreservedly anti-deterministic, anti-evolutionary, and anti-positivist, Sauer was fully resistant to forms of universalist argument. One might say he was resolutely materialist and historical (but certainly not a historical materialist). Human geography had little to do in his account with individuals, "only with human institutions or cultures" (Sauer 1941: 2).

There is good reason to be critical of Sauer's early ruminations on geographic observation and fieldwork, on the manner in which he construed culture narrowly as material form, and of a view of history understood as sequence. Perhaps most troublesome was the degree to which human agency – the driving force in the transformation of the earth – was not understood in social terms. Despite his cosmopolitan intellectualism and immersion in Franco-German ideas (Eduard Hahn, Freidrich Ratzel, Vidal de la Blache, Alfred Hettner), there is little evidence that he read or seriously thought about the work of Karl Marx, or Emile Durkheim or Max Weber. When all is said and done, nevertheless, if one places Sauer's work alongside the two other giant intellects of 1950s Berkeley Geography – Clarence Glacken and Paul Wheatley[7] – both of whom Sauer hired, then one can identify a broad approach, a Berkeley School, that integrated history, environment, culture, space, and economy into a distinctive and compelling research program. Glacken, of course, charted a deep history of ideas and beliefs about nature and culture in the Western tradition, while Wheatley, in his pursuit of pre-industrial urbanism, linked city and symbol to the production of social surpluses and forms of market behavior derived from the work of Karl Polanyi. Both exhibited close affinities with Sauer's project. The enormously influential conference held at Princeton in 1955, entitled *Man's Role in Changing the Face of the Earth*, that Sauer organized with urbanist Lewis Mumford and ecologist Marston Bates, represented in many respects an intellectual road map of the questions and concerns dear to the Berkeley School. Two contributions by Sauer and Glacken opened, and indeed anchored, the influential volume of the same name that subsequently appeared in 1956.[8]

Any account of the origins of a Berkeley School of Geography must acknowledge the sustained traffic in ideas between Geography and Anthropology: anthropologists Alfred Kroeber and Robert Lowie were colleagues of Carl Sauer at Berkeley and shared an interest in the relations between culture, land, and environment.[9] This cross-fertilization and inter-disciplinarity (similar exchanges were to be found later at Columbia and Michigan) was integral to the emergence of one of the most important legacies of Sauer and the Berkeley School, namely cultural ecology. Even though Sauer never deployed the term, cultural ecology is one of the red threads running through the Berkeley School's corpus (the others would be cultural landscape and historical morphology). Berkeley Geography had sister departments, institutions, and theoreticians of cultural ecology in the discipline. A trio of Geography departments at the University of Chicago, the University of Wisconsin, Madison, and the Australian National University (ANU) in Canberra, all of which had connections to Sauer in some way, were key sites in the emerging network of cultural ecological thought.

Cultural ecology was not, of course, the sole preserve of Geography or Sauer. It was anthropologist Julian Steward (1955) who first deployed the term – building upon, it must be said, a prior history of anthropological and geographical research on culture and environment – as a way of identifying how "adaptational strategies" led to multi-linear pathways of cultural evolution among Native American groups in the southwest. Steward's central claim was that environmental assemblages revealed functional and causal relationships to specific forms of social organization. Rather than thinking that all aspects of culture and nature are functionally interrelated, he identified a "culture core" in which functional ties with the natural setting are more explicit, and the interdependencies between cultural patterns and organism–environment relationship were most crucial. Subsistence and food activities were functionally related to the properties and structures of specific ecosystems and they provided the ground on which cultural groups and their culture evolved. Steward's ideas can be located on a larger canvas of environmentally oriented anthropologists including Raymond Murphy, Frederick Barth, Robert Edgerton, Clifford Geertz, Robert Netting, and Marshall Sahlins during the 1950s and 1960s.[10] For my purposes, the key point here is that any institutional and intellectual road map for cultural ecology as a theoretical field must necessarily reflect a resolutely dispersed, transnational and inter-disciplinary genealogy of ideas, institutions, research sites, and theoretical personalities.

Cultural ecology in both Anthropology and Geography was always (and remains) centrally concerned with processes of adaptation (Zimmerer 1996; Denevan 1983), a concept borrowed (along with others like niche and ecotope) from evolutionary biology. The early ethno-ecology of Harold Conklin (1975 [1957]) who studied indigenous Hanunóo agricultural practice and knowledge (in part through local taxonomic knowledge systems), and Clifford Geertz's (1963) famous concept of "involution" – inward directed intensification and elaboration in padi rice systems also referred to by Geertz as "overadaptation" – both focused on how internal system equilibrium or homeostasis in swiddening and padi rice systems were central organizing forces for society and culture in southeast Asia.[11] While each saw the defining properties of social systems as a stable ecology and the adaptation of the society to its local ecosystem, neither saw ecology as a totalizing explanatory framework as such. To quote Geertz (1963 [1974]: 6), cultural ecology entailed a "strict confinement of the application of ecological principles and concepts to explicitly delimited aspects of human social and cultural life for which they are particularly appropriate rather than extending them, broadly and grandly, to the whole of it". Ecology, culture, and adaptation were powerfully sutured nevertheless. Cultural ecology's epistemological position resonated in many disciplines across the social sciences as ecological thinking of various sorts entered social science discourse and especially the behavioral sciences.[12] Even Eric Wolf's (1966) foundational book on peasants – by an anthropologist most closely associated with political economy – devoted space to the ecological adaptation of peasant households seen through a system of material and energy transfers from the environment to human communities.[13]

Adaptation as a form of thought

Adaptation as a key concept and a form of thought was a consistent current flowing across several generations of the Berkeley School. It appeared in Karl Butzer's archaeological work on civilizations as adaptive systems (Butzer 1980a, 1980b), in the analyses of material flows in so-called primitive subsistence systems (for example, the work by David Harris (1971) on swiddening along the Orinoco, and William Denevan's (1970) reconstructions of pre-Columbian intensive cultivation systems) and in cultural geography's concern with how built material form, patterns of settlement and inheritance, and landscapes generally were functionally related to specific environmental settings. But perhaps the most rigorous and influential of these early forms of cultural ecological analysis in Geography lay in the extraordinary long-term research program directed by geographer Harold Brookfield in association with anthropologist Paula Brown and a bevy of Geography students (including William Clarke, Diana Howlett, and Eric Waddell) conducted through the School of Pacific Studies (now defunct) at the Australian National University in Canberra and largely focused on the highlands of Papua New Guinea.[14] At its core was the relation between population, carrying capacity, and territory on a larger canvas of regional agrarian intensification and social change since the 1930s. Human population and land relations were to be understood through "formal ecosystem analysis" (Clarke 1971: 183). Exceptionally detailed and granular community studies, beginning in the late 1950s, examined the cycles and fluctuations in segmentary social groups among the Chimbu and other groups shaped by what Paula Brown (1972) called "forces of change". While focusing on feasts, conflict, settlement, and forms of stratification, this body of research disclosed significant "geographical and historical variations in adaptive strategies" (Waddell 1972: 220).

From related but rather different vantage points, two other force-fields shaped the emergence of cultural ecology during the 1960s. One turned on the work of University of Chicago geographer Gilbert White and his students, most prominently Robert Kates, which focused on environmental hazards and "purposive adjustments" to them (this was in fact their definition of adaptation; see Burton et al. 1978). In his dissertation, *Human Adjustment to Floods: A Geographical Approach to the Flood Problem in the United States*, White (1945: 87–88) defined adjustment as the human process of occupying or living in an area and the transformations of the initial landscape which result. He went on to specify a raft of forms of human adjustment to floods: elevating land, abating floods by land treatment, protecting against floods by levees and dams, providing emergency warning and evacuation, making structural changes in buildings and transportation, changing land use to reduce vulnerability, distributing relief, and taking out insurance. In subsequent studies, geographers (and sociologists) extended the concept to a variety of natural and technological hazards, demonstrating the myriad adjustments people and societies had made to survive and even prosper while living with recurrent hazards (Ohio State University, Clark University, the University of Toronto, and especially the Natural Hazards Center at the University of Colorado, Boulder were important sites for this work).[15] It was these adjustments or adaptations that made possible what Kates (2001: 6) called "the fruitful occupance by human societies of an enormous range of environmental settings". North America provided the setting for much of the early hazards work, but there was an emerging body of cross-cultural research on the anthropology of environmental disasters and the relations between culture, behavior, and so-called extreme events such as typhoons in Micronesia or drought in East Africa (see Oliver-Smith 1996). There were close family resemblances between the hazards research and cultural ecology because each turned its gaze to the relations between human communities and environmental events and perturbations. Behavioralist in orientation, hazards research placed particular weight on environmental perception and cognitive dissonance (how, for example, individuals perceived or misperceived the risk of flood losses) and the menu of strategies (including the structure of decision-making trees) which corresponded to the local or emic understanding of the threat.[16]

The second field was in direct dialogue with cultural ecology but sought in some respects to distance itself from it. Adopting the moniker of ecological anthropology (see Vayda and Rappaport 1967; Vayda 1969), it drew explicitly on new theoretical developments from ecological theory, and the analysis of living systems (hence its being dubbed the "new ecology"). As a body of theoretically driven work it was the product of collaborations and cross-pollinations in the 1960s shaped by the famous Macy Foundation Conferences on cybernetics held between 1946 and 1953, and by developments in evolutionary biology and ecology (see Alland 1975). Ecological anthropology in its most elaborated form was the long-term research conducted in Melanesia that linked Roy Rappaport and Peter Vayda (both at Columbia University at the time[17]) with geographers from Berkeley and the Australian National University. It gave birth to a raft of empirically rich, deeply ethnographic and intensive local studies of subsistence and ecological dynamics – typically involving careful measures of labor and energy flows and other socio-demographic survey data – systems ecology associated with, and indices of, so-called adaptive processes (Rappaport 1984 [1968]; see Dove and Carpenter 2008).[18]

Ecological anthropology and the adaptive order

Ecological anthropology sought to identify how cultural institutions and relations functioned with respect to key environmental variables. More abstractly it addressed what Rappaport called the "orderly adaptive structure" – structured sets of processes and regulatory hierarchies – that inhered in all living, general purpose systems (i.e. systems in which the goal is simply survival sustained by an economics of evolutionary flexibility[19]). The canonical text – and by far the most theoretically elaborated, deploying the conceptual architecture of living systems theory – is Roy Rappaport's hugely influential *Pigs for the Ancestors* (1984 [1968]) which was expressly written as a "reaction against the special form of ecology that Steward thought necessary to accommodate the concept of culture" (Rappaport 1994: 167). Rappaport's approach, and in fact ecological anthropology in general, construed human society as analogous to any animal population when viewed as parts of ecosystems. More critically, ecological anthropology saw the relations between human populations (his own case study was among the Tsembaga Maring in Papua New Guinea) and the environment through *functions*: "statements of 'what it does' and 'how it does it' may well be among the most informative, important and interesting that can be made concerning an organ, an institution or a convention" (Rappaport 1984 [1968]: 34).[20]

For Rappaport one of the main purposes of ethnography was to see how systems work, and how rituals regulate social life. The "new ecology" was not the same as the old functionalism (of cultural ecology) because when the field of application had changed to focus on the rationality of institutions with respect to environments, it exposed "organic and adaptive considerations" (Rappaport 1984 [1968]: 352). As he put it, the regulatory function of ritual among the Tsembaga and other Maring helps to:

> maintain an undegraded environment, limits fighting to frequencies that do not endanger the existence of the regional population, adjusts man-land ratios, facilitates

> trade, distributes local surpluses of pig in the form of pork throughout the regional population, and assures people of high-quality protein when they most need it.
>
> *(Rappaport 1984 [1968]: 224)*

In his account, aspects of culture such as ritual should be grasped as parts of a transcendent adaptive structure, that is to say, an order or sequences of response to environmental variables or perturbations inherent in *all* living systems. A key theoretical reference point here was important research by biologists Roberto Frisancho, Lawrence Slobodkin, and others, but most especially the work of Gregory Bateson on the economics of somatic flexibility.[21] The Tsembaga ritual cycle accordingly was construed by Rappaport as a "first order" negative feedback mechanism, "operating to maintain the values of a number of variables within 'goal ranges' (ranges of values that permit the perpetuation of a system, as constituted, through indefinite periods of time)" (Rapapport 1984 [1968]: 224). Such analysis was conferred, according to Rappaport, a high degree of "objectivity" and empirical rigor rooted in the ecological and biological criteria deployed through detailed case studies.

Ecological anthropology, in sum, took populations as analytical units, examining the trophic relations (energy flows for example) within the ecosystem (Odum 1971). The notion of an ecosystem provided a convenient frame for the analysis of trophic exchanges between ecologically dissimilar populations occupying single localities. To overcome the problem that ecosystem analysis emphasized one trophic exchange between populations occupying different ecological niches within a local bounded area, Rappaport (1984 [1968]: 226) suggested the concept of "regional population" in which local populations of humans (as other species) participated in regional systems in which the Maring ritual of pig slaughter (*kaiko*) was of great importance in *articulating* the local and regional subsystems. Kaiko operated, in short, as a homeostat, maintaining a number of variables that comprise that total system within ranges of viability, but also functioned as a transducer (a term derived from cybernetics), translating changes in the state of one subsystem into information and energy capable of producing changes in the another subsystem:

> Like thermostats, rituals have a binary aspect. As the thermostat switches on and off, affecting the amount of heat produced by the furnace and the temperature of the medium, so the rituals of the Tsembaga are initiated and completed, affecting the size of the pig population, the amount of land under cultivation, the amount of labor expended, the frequency of warfare, and other components of the system. The programs that should be undertaken to correct the deviation of variables from their acceptable ranges are fixed.
>
> *(Rappaport 1984 [1968]: 234)*

In Maring communities, women begin to complain that the pig population is getting too large to manage which indicates there are sufficient numbers of animals for the performance of ritual sacrifice. A triggering of the *kaiko* ritual has a corrective effect on the animal–human–environmental balance; it functioned as a homeostatic operation, a sort of giant servo-mechanism involving a complex chain of interactions between the "cognized model" of the Maring, the ritual cycle, the regional system, and wider adaptive order of the ecosystem.

Thinking about maladaptation

One of the great strengths of Rappaport's program was the clarity and rigor with which adaptation (and indeed its counterpoint, maladaptation) was formulated. Adaptation referred to the processes by which living systems maintained homeostasis in the face of short-term environmental perturbations and long-term non-reversing changes in their environments. As he put it, "beneath specific and categorical differences, *adaptive systems are organized by a generally similar architecture*" (1984 [1968]: 412, my emphasis). The architecture or order reflected a temporal and economic structure. In the face of a perturbation of a key variable (say, the late onset of the rains) the adaptive responses (of, say, a peasant community in rural Africa) might begin with small, and flexible, responses that commit limited resources and in the event of their inadequacy in the face of continuing or deepening threats (severe drought for example), larger and higher order responses as it were "kick in", typically committing greater resources and exhibiting less flexibility (or reversibility).[22] Order and structure inhere in all systems exhibiting a raft of salient properties relevant to understanding how human populations adapt to their environments: lower order controls are flexible and shallow, higher order controls tend to correct not minor deviations but control relationships between lower regulators and their outputs.[23] Rappaport sees the hierarchy of the adaptive structure as "ordered along axes of specificity, concreteness, reversibility, authority, time, sanctity" (1977: 14). High order controls among human population are typically draped in sanctity and abstraction; they are the high order propositions or principles which pertain to the normative conditions prevailing among the critical subsystems or components of the ecosystem.[24] As he put it: "vagueness is not a flaw but an adaptive characteristic of the ultimate" (1979: 155). Orderly adaptive structure maintained organic function, meaning, and meaningfulness. Of course, under some circumstances the "first order" negative feedbacks that inhere in the adaptive order may be unable to dampen the perturbation and "second order" negative feedback will attempt to control the disequilibrium. At that point "second order" negative feedback produces what Wilden (1972: 209) calls a meta-system (the elaboration of new structures or morphogenesis) or alternatively the ecosystem is destroyed: as he puts it: "In social systems the first (meta-system) is known as revolution; the second (ecosystem destruction) is extinction" (1972: 209).

Maladaptation in social systems – what Rappaport called disorders of adaptive structure – occurs when the temporal and hierarchical order of adaptation is subverted. For example the goals appropriate to lower level responses might become those of higher level orders (for example what is good for General Motors is good for America, what advances the paramount chief is in the interest of the collective good). Rappaport (1984 [1968]) identifies a number of pathologies of the maladaptive order; that is to say conditions in which human populations do not respond to perturbations and perturbations consistent with homeostatic principles and self-regulation of the system of which they are part (see Figure 2.1). For example, the goals of a key variable can be mis-specified (maladaptation of the goal) or the structure of adaptation can be "misordered" (cybernetic malfunctions or hierarchical anomalies). One part of a social system might capture another in a way that compromises evolutionary flexibility or parsimony, or a social system might be overcentralized or too loosely coupled to its environment such that, in hierarchical or temporal terms, individuals, households, communities, or governments cannot respond appropriately and adequately.[25] For Rappaport, then, the essential problematic for ecological anthropology is "the relations of actions formulated in terms of meaning to the systems constituted by *natural law* within which they occur" (1984 [1968]: 402, my emphasis).

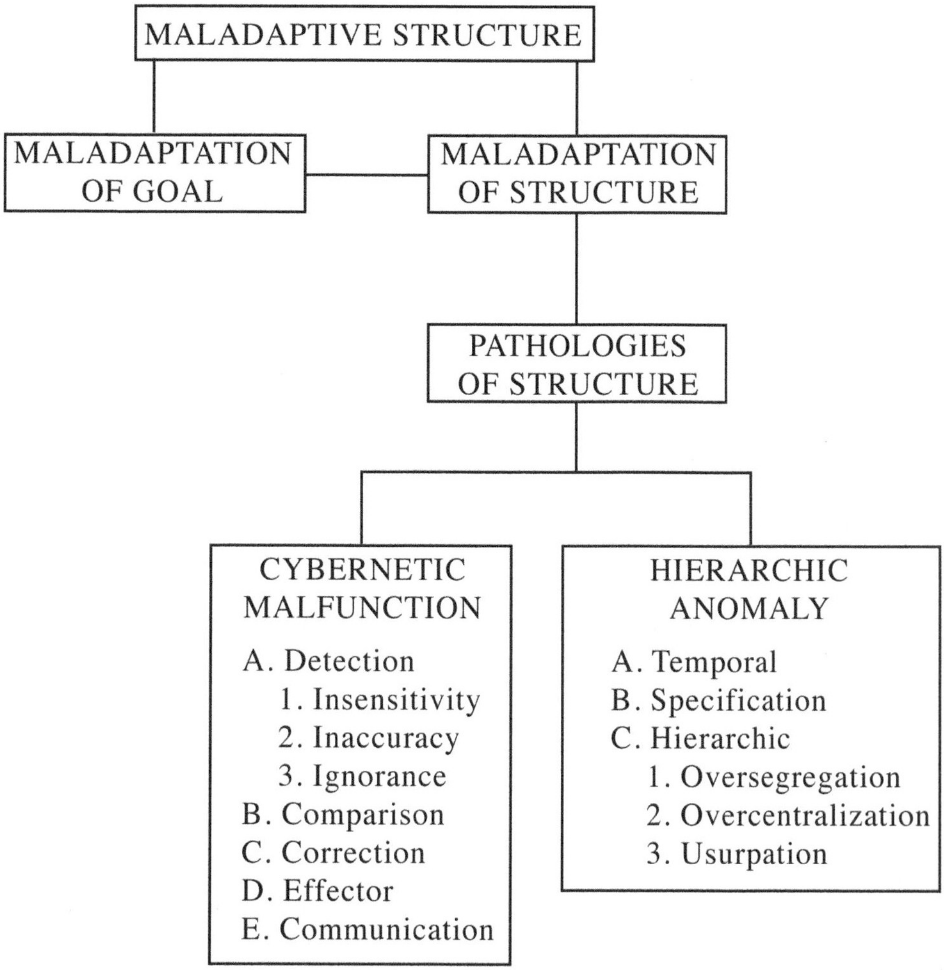

Figure 2.1 The structure of maladaptive orders (source: Rappaport 1974).

Adaptation in question

As ecological and ecosystemic thinking penetrated into the social sciences at large in the 1960s, the idea of adaptation was widely adopted and deployed, providing a powerful language for thinking not only human–environment relations but also behavior and communication in social systems. But there was suspicion too, even from within the Berkeley School itself, which helped give birth to a particular form of adaptive thinking. In the face of transformations wrought by global capital flows, the second great wave of marketization, and the dynamics of post-colonial development, talk of self-regulating systems, of Third World communities adapted to the ecological niches they occupied, and cognized models in the service of evolutionary flexibility all appeared increasingly problematic. Sauer himself was acutely aware of these tensions: not only the extent to which pre-capitalist and colonial practices could be deeply destructive of the environment, but the massive costs of a post-war rational, instrumental scientific hubris attached to American hegemony and growing global influence. In a letter to the Rockefeller Foundation (Sauer Papers, Bancroft Library, also available at http://rockefeller100.org/items/show/2590) in response to a request to reflect upon their early Green Revolution initiatives in Mexico, Sauer expressed a withering contempt for a program designed to "destroy the ecologic balance" of peasant communities, in which their "wisdom" is "greater than that of the scientific Deadaluses". It represented an intervention – a "crusade" as he put it – of an "alien group in another country" predicated upon an "overaccelerated economic and industrial growth". Many of us would call this empire. These ideas did not sit easily with certain notions of cultural ecological adaptation to the environment. Neither did they sit well with the Rockefeller Foundation. A minute on Sauer's report reads: "this must never be shown to anyone" (Sauer Papers, Bancroft Library, available at http://rockefeller100.org/items/show/2590).

What were the sorts of conceptual challenges that adaptation faced? To say that organisms adapt to their changing environments implies there are *processes* of adaptation and *end states* of being adapted. The concept of adaptation, with its inevitable affinity to evolutionary thinking, to Charles Darwin and to modern ecological theory, is a textbook example of what Raymond Williams (1977) calls a keyword. What distinguishes keywords like adaptation is that they are "binding" in relation to certain activities and their interpretations and "indicative" of certain forms of thought (Williams 1977: 15). Keywords, by definition, circulate widely and are deployed among heterogeneous communities in a variety of ways. Their historical semantics are typically complex and unstable. The language of adaptation is ubiquitous, if not promiscuous, traveling effortlessly across biological, cultural, social, and ideological boundaries. Whether deployed as a cover term for simply being in synch with a particular context or as an expression of evolutionary fit and fitness, by the 1970s the term could be found in research institutions, in the professions, in popular discourse: swimming animals have adapted by developing flattened appendages like fins; a lawyer is well adapted to her profession; therapists help us adapt to the stresses of urban living; rituals and cultural institutions among indigenous groups function as adaptive mechanisms for sustainable development; the Federal Reserve fulfills adaptive functions with respect to contemporary American capitalism.

Darwin's great materialist revolution was to show how evolutionary change is the result of variation among individuals converted into variation among species through dynamic and kinematic forces: the principles of variation, heredity, and natural selection. These principles were necessary and sufficient conditions for evolution by natural selection to occur. Adaptation was added by Darwin to explain the mechanical causes of differential reproduction and survival, namely the struggle for existence (which he took from his reading in 1838 of Malthus's *Essay on the Principle of Population*). Adaptation was to be understood in relation to an organism's niche – the "fit" – but also in relation to individuals' competition for the same resources, "the struggle for existence". In the classical Darwinian paradigm, variations within a species indexed different probabilities of success, while the external environment of the species or organism posed specific and concrete "problems" and challenges for which only certain morphological, physiological, and behavior traits were "solutions". As Richard Levins and Richard Lewontin put it, within the Darwinian frame: "organisms are the objects of the force of natural selection ... [which] sorts out the form that is the best solution to the problem" (1985: 97). Stanford human ecologist[26] Bill Durham extended this logic in his neo-Darwinian account proposing that "culture is generally adaptive in the biological sense" (1976: 91). Biological and cultural attributes are seen in terms of the selective retention of traits that enhance the inclusive fitnesses of individuals in their environments. To put it differently, adaptation derived from organic evolution is seen as simply a special case of a more general worldview, what Levins and Lewontin (1985) call "evolutionism".

Even within the confines of evolutionary biology, however, adaptation and adaptive processes have been, and remain, contested, freighted terms. Richard Levins and Richard Lewontin (1985) in their important book *The Dialectical Biologist* provide a sense of the controversy. In their view, the concept infers a problem or model that pre-exists to which organisms are fitted through dynamic processes. Even if there are assumed to be niches to which organisms adapt, evolution cannot be seen to be a process of becoming adapted. Furthermore the partitioning process of an organism into traits (which purportedly served adaptive ends) and the partitioning of environments (into niches or problems) raises the question of whether these are real or human constructions. Environmental problems are typically isolated and the fit to them is independent of other interactions with the environment. And not least, while the fact that adaptation has occurred seems self-evident, it is equally clear that many and perhaps most changes are *not* adaptive. The architecture of adaptation seems to rest on a billiard table view of organisms and their environments when in fact the organism is both subject and object. Organisms constrict every aspect of their environment: they are "creators and modulators of these objects of external forces" (Levins and Lewontin 1985: 104). To see the organism or the community as an agent, a subject active in the construction of its own environment, leads to a much more complex dialectical view, one in which the metaphor of adaptation must be substituted by a metaphor of construction. The organism, says Levins and Lewontin, organizes its own evolution by being an object of natural selection and the creator of the conditions of that selection (1985: 106). Inner and outer forces of which it is both subject and object must be held in tension. These broad sorts of challenges were central to the wider debates that linked biology, cognitive science, artificial intelligence, and so on to the emerging cybernetics and behavioral sciences which the Macy Conferences, and subsequently the Ford Foundation, promoted in the 1940s and 1950s.

Such concerns became absolutely central to the debates within cultural ecology and ecological anthropology from which political ecology was to emerge during the 1970s. One way to think about the conditions of possibility for the emergence of the first generation of political ecology – associated with Piers Blaikie, Harold Brookfield, Ben Wisner, Susanna Hecht, Larry Grossman, and so on – is that the very concepts of adaptation, of adaptive structures, adaptive orders, adaptive traits and functions, were put into question largely because they could not be made to speak to the conditions – often the realities of peasant communities in the throes of what Michael Burawoy (2014) calls the "second great wave of marketization". There were, in sum, all sorts of anomalies political ecologists confronted in their analyses of rural and agrarian communities in the post-colonial world: it was almost exclusively from studies of this world that political ecology emerged. To revisit some of these early and foundational political ecological works is to be immersed in the debates over functionalism and neo-functionalism, the uses and abuses of the organic analogy, the constraints imposed by "vulgar materialism", the limits of systems theories and of living systems, and of adaptation as ideology (the idea that Darwinian and evolutionary theory was an expression of nineteenth-century bourgeois sentiments and the flowering of an exuberant industrial capitalism). Whig history and Whig biology shared family resemblances if not affinities.

Political ecology and the crisis of adaptation

Clearly a number of advantages stemmed from seeing adaptation through the lenses of evolutionary theory, cybernetics, and systems theory – that is to say through the reigning conceptual and theoretical apparatuses within American social and behavioral sciences during the post-war period (see Nelson 2005). One was the sorts of quantitative data on material, energy, and information flows (in effect a sort of economism) required to calibrate the sorts of regulatory functions attributed to culture: detailed accounting of ecological relations and survey data on demographic change, labor flows, and land use was a hallmark of much of this work. Another was conceptual, namely that causality in biophysical systems was circular (the so-called feedback loop). Cybernetic principles highlighted what Luhmann (1993a, 1993b) called "recursivity", that is a process which uses its own outputs as inputs (this is key to the operations of negative feedback for example). Implicit here too is the notion of contingency of all observation: that A causes B and B causes A points to the fact that it is always possible to observe otherwise. In Gregory Bateson's (1972) language, the sort of knowledge you get depends upon the code or map that you use. In this sense systems theory contained an epistemological claim to the effect that the boundaries between system and environment or organism and environment were social constructions and arbitrary. Put this way, both cultural ecology and ecological anthropology, to the extent they framed adaptation as recursive, had unsettled the hard and fast boundaries between system and environment and the billiard board world of stressors and responses, and simple causes and effects.[27]

Yet, the functionalism and empiricism of systems and cybernetic theory, the strongly behaviorist thrust of the work, and the Cold War context out of which this science of control emerged (Heims 1991) all cast a long analytical shadow. Cybernetics was an instrument of technocratic management in which the angel of control was emphasized over the devil of disorder (Galison 1994: 266). Any sense of self-regulating equilibrium and balance or harmony seemed increasingly out of touch with the realities of communities marked by new patterns of social differentiation and inequality and what could only be called ecological destruction. In particular the withering critiques launched by Maurice Godelier (1972) and Jonathan Friedman (1974) exposed not just the mechanistic and often Hegelian character of much of what passed as adaptation theory (the idea that regulation of the environment was happening behind the backs of the actors through cultural thermostats), but also the difficulty of seeing how the adaptive structure of societies could be squared with not just the clear patterns of ecological destruction but the questions of power, class, property, and access which were central to other theoretical approaches, most especially Marxian political economy.[28] Central to this critique was a suspicion of organic analogies – subjecting social systems to an overriding logic of living systems or "ecological rationality" as Godelier (1972) called it – and the grave dangers of functionalism and a sort of inductive or crude materialism.[29] Social systems seemed to operate like giant servo-mechanisms. As Godelier put it:

> Here [in cultural ecology and ecological anthropology] we recognize empirical materialism, the "economism" that reduces all social structures to nothing but epiphenomena of the economy which is itself reduced, through technique, to a function of adaptation to the environment … a materialism like this is unable to explain the reasons why, the fundamental necessity of what exists, i.e. the reasons why the history of societies that are not always completely integrated totalities but totalities whose unity is the provisionally stable effect of a structural compatibility that enables different structures to reproduce themselves until they reach the point at which internal (and external) dynamics of these systems forbids this totality to go on existing as such.
>
> *(1972: xxiv–xxv)*

As a number of commentators noted, this form of ecological materialism was innocent of any form of contradiction. Sahlins said that ecologic rationalities "exchange … meaningful content for functional truth" (cited in Rappaport 1984 [1968]: 308). Society functions quite obviously, but to infer that everything in a society functions is an absurdity.

Economic and cultural anthropologists like Maurice Goldelier and Marshall Sahlins provided a powerful critique of what they saw as functionalist (in their view largely descriptive[30]) analysis but it was from geographical research struggling with these same misgivings over adaptation as a conceptual language that political ecology was born during the 1970s. It bears repeating that while the intellectual milieu was one in which peasant studies and a reconsideration of Marxist theory of development were on the rise, it was also a conjuncture in which a dominant Malthusianism and technological determinism (one thinks of the ubiquity of the tragedy of the commons and Club of Rome style thinking) confronted rising Third World nationalism and left-wing radicalism. Drawing inspiration from new research on the political economy of peasant societies (in part propelled by the Vietnam war and a reconsideration of peasant radicalism and politics) and on the role of the state in post-colonial development, a generation of political ecologists were shaped by a renewed interest in political economy of development writ large, and in agrarian political economy, or the so-called agrarian question in the Third World, in particular. Defined by Karl Kautsky (1988 [1899]), the agrarian question was concerned with the ways in which capital was taking hold of and transforming agriculture. Much of this political ecology addressed a post-colonial rural world in the throes of what Karl Polanyi called the "great transformation". Central to political ecology was not systems ecology as such but political economy, and how it shaped or even produced the environments which were, or were not, managed by differing sorts of "land managers".

Rather than examining the functional adequacy of culture or social structure, political ecology started with the relation of producers to the market, the commodification of land and labor, the forms of surplus extraction and the prismatic forms of social differentiation with peasant communities, the breakdown of the moral economy, emerging forms of class structure and the changing relations of production.[31] Rather than seeing environmental questions through the prism of society *and* nature or human response and biophysical trigger, political ecology, drawing on Marxist ideas of the labor process and notions of first and second nature, saw nature and society as dialectically constituted (Smith 1977; Sayer 1979). Environment was not some pre-given context, but was an object that could be construed in different ways by different communities and classes. Political ecology problematized what the environment meant and to whom – a central plank in Piers Blaikie's (1985) work on soil erosion for example.[32] What this meant was that the planetary ecosystem was constructed out of "the contradictory unity of capital and nature" (Harvey 2014: 248), that capital is a working and evolving "ecological system" in which "nature and capital are constantly being produced and reproduced". There is no transcendent adaptive or ecological order here, but an ecological system in which capital necessarily privatizes, commodifies, monetizes, and commercializes every aspect of nature.

Political ecology constructed a theory upon a more-or-less Marxist analysis of political economy in which the social relations of production, access to and control over resources, and power relations rooted in state and capital figured centrally. The dynamics of specific historical forms of capitalist accumulation – whether in the Brazilian Amazon (e.g. Susanna Hecht) or the Himalayan foothills (e.g. Piers Blaikie) – were its central starting points. Its object of critique was not only adaptation a such, but also a dominant Malthusianism ("population pressure" on the environment) which the rise of ecosystem thinking did little to change. The birth of political ecology was, not unlike its predecessor cultural ecology, a transnational, multi-sited, and trans-disciplinary enterprise (though Geography took pride of place). There were four geographically interconnected institutional settings each marked by the appearance of a sort of foundational text focused on field research in four different regions (Africa, Brazil, South Asia, and Melanesia).

What they all shared was a common engagement – in related but different ways – with the *political economy of development* and what Harvey (2014: 262) calls the "mindless extension of capital's ecology into our lifeworld". Systems of access to and control over resources, growing commodification of the resource base and social life, circuits of capital accumulation, and the role of the state were absolutely central. Each of these four sites and their founding figures – ANU (Harold Brookfield), Berkeley (Susanna Hecht and (immodestly) myself), Clark University (Ben Wisner), and the University of East Anglia (Piers Blaikie) – questioned not just functionalism and adaptation as a form of thought but also the cost–benefit and behavioralist assumptions of much of the hazards research construed as human responses and adjustments to threats and stressors. Political ecology turned the flashlight inward toward commercialization of agrarian societies, to how communities were being torn asunder and radically reshaped by the twin processes of globalization and to how the exercise of power was indispensable to the understanding of the institutions of property, resource control, and market dynamics (Watts 2012 [1983]).

The confluence of differing trajectories that merged to become political ecology not surprisingly contained a number of different points of intellectual departure, reflecting important analytical and institutional differences among its founding figures. The ANU-Melanesia group reflected the sense that Rappaport's account described adaptation without evolution, or as Harold Brookfield (1973: 155) put it, ecological function rather than sociological explanation. The sorts of adaptive functions imputed to pig cycles were not about the disposability of pigs but the reproduction of "a whole system of social relationships" rapidly being transformed by cattle, coffee, and the advancing frontier of capital. As Clarke put it describing the uplands in New Guinea, the communities were, in fact, at the "edge of a madhouse" (1977: 372). Brookfield himself was drawn to studying the costs of what he called inter-dependent development and the forms of specialization and risk associated with "the course of development" (Brookfield 1975).[33] In the ANU lineage, the death knell of adaptation was the appearance of Larry Grossman's book *Peasants, Subsistence Ecology and Development in the Highlands of Papua New Guinea* (1984, and originally a 1979 ANU PhD dissertation). Grossman identified his approach as cultural ecology yet his analysis saw the region through the lens of peasant theory and patterns of social differentiation, that is to say capital at work.

A similar set of developments were reflected in work that linked the universities of Michigan and Berkeley. Bernard Nietschmann's (1973) stimulating cultural ecological study of the Miskito communities on the Pacific Coast of Nicaragua, entitled *Between Land and Water*, proved to be a sort of limit case for cultural ecological analysis exposing the sorts of constrictions imposed by adaptation as a framework.[34] By making use of Marshall Sahlins's (1972) account of Marx's commodity circuit (and implicitly Karl Polanyi's (1944) work on markets), Nietschmann showed how the central dynamics of Misikito fishing and subsistence systems were increasingly driven by broader market changes, in large measure the commercialization of the turtle industry. My own 1979 dissertation at Michigan – which appeared as a book called *Silent Violence* (Watts 2012 [1983]) when I had relocated to Berkeley – certainly was influenced by these Polanyian insights into patterns of resource use and the politics of "fictitious commodities". But in examining the relations between drought and famine in West Africa I made use of structural Marxism and especially the so-called agrarian question. It was the intersection of markets (the role of merchant capital), patterns of social inequality, and climatic perturbations that shaped what sorts of decisions and realms of choice different classes of peasant households could make to manage risks like drought as well as why the systems of which they were part might collapse (i.e. famines as crises of social reproduction). Ben Wisner – who completed his PhD in 1977 working in eastern Africa – was exploring precisely these issues with students at Clark University, in a different part of the continent, as a way of upending the stimulus–response models of hazard research associated with the scholars at Chicago, Toronto, and Clark itself.[35] It was rigorous political economy analysis that demonstrated how vulnerability and marginality (both ecological and socio-economic) was being *produced* by particular sorts of social and economic exposure rooted in the circuits of capital and in the operations of what passed as state policy.

There were two other settings that proved to be foundational to an emerging political ecology: one was Berkeley Geography, in particular Susanna Hecht's[36] work on tropical deforestation in Brazil (1985), a frontier of land clearance and speculation propelled by a powerful logic of political alliances between landed elites and state-derived rents and subsidies. The other, located in the UK at the School of Development Studies at the University of East Anglia, centered on Piers Blaikie and his direct engagement with the political economy of development. Blaikie's hugely influential work (1985) emerged largely, but not exclusively, from South Asia on the subject of soil erosion and land management.[37] Again adaptation was not the central concern so much as the chains of inter-dependency linking farmers, households, and communities to the state and the world market which shaped – and often undermined – the capacity to manage the land and soil resources.

None of this should infer a common theoretical point of reference among those political ecologists for whom adaptation seems to confer a set of analytical blinders. For example, Blaikie and Brookfield were a mix of world systems theory, *dependencia*, and a very broadly defined (and not unequivocally Marxist) political economy. My own work drew heavily on Althusserian Marxism and the work of Karl Kautsky; Larry Grossman's book reflected the influence of peasant studies; Hecht's initial research on Brazil was shaped by Latin American theories of the state and rent seeking. Subsequent pathbreaking work – one thinks of Nancy Peluso's (1992) research on Indonesian forests – drew on social historians like E.P. Thompson or theories of property as much as accounts of the post-colonial state or of the peasantry. Both Wisner and Blaikie were directly engaged in policy and practical development work and consulting in a way that Grossman and I, for example, were not. What they all shared, I think, was common focus on patterns of accumulation, access to and control over resources, and changing class structure; political ecology could demonstrate that some individuals and households were rendered marginal (to their resource base) and made vulnerable to anticipated and unanticipated environmental processes in new ways. Small farmers might be degrading their environment because they had no choice (they were subject to a simple reproduction squeeze; see Watts 1983); forests were destroyed in a desperate attempt to establish property rights in areas where the rule of law was lacking; peasants worked harder and longer, often degrading their land, in order to ensure social reproduction in the face of price squeezes. In short, this political ecology had as its reference point what I would call *regimes of accumulation*, operating at a multiplicity of scales and through complex chains of causation, providing structures of opportunity and constraint – imposed by social relations of production and exchange and by property relations – that shaped how resources, environments, and perturbations might be managed and governed.

As a constellation of ideas and approaches, political ecology became, not surprisingly, something of a moving frontier. By the 1990s this first-generation political ecology had been broadened in two ways: to put it crudely, to Marx and regimes of accumulation were added Foucault and *regimes of truth* (Forsyth 2003; Li 2007), and Gramsci and *regimes of rule or hegemony* (Moore 2005). The new palette was partly a result of changing intellectual fashion (the growing influence of forms of post-structuralism), partly a function of cross-fertilization with other fields (science studies, race theory, environmental history, green justice), partly a function of the interest of deploying political ecology in First World, industrial, and advanced capitalist settings (rather than the world of peasants), and not least because of the blind spots and silences within

Marxian political economy. Political ecology had been relatively silent on the forms and dynamics of political contention surrounding the environment (see Peluso 1992). Environmental movements, the role of civil society, and later armed struggle (militant struggles over forests or oil) pushed political ecology to expand and deepen its understandings of the operations of power. No surprise then that the knowledge–power–institutions nexus, drawing especially from post-structural and discourse analysis, was taken up quickly. Careful examinations of forms of environmental expertise, such as how institutions like the World Bank were "greened", how conventional models of environmental degradation (e.g. the tragedy of the commons) constructed referent objects in particular ways with consequences for policy, and especially a focus on forms of green governance – for example understanding the effects of decentralized governance on forest regulation or common property institutions, and the politics of differing management regimes – all became central to political ecology in the 1990s (Ribot and Larson 2009; Goldman 2005; Leach and Mearns 1996; see Watts 2000 for review).[38]

There was a sort of intellectual traffic in other words between questions of rule, hegemony, discourse, and regimes of accumulation. Why did some environmental ideas and practices became dominant, and how might subalterns or oppressed groups contest and build alternatives to these practices and centers of power? Such questions once again brought Anthropology into conversation with Geography.[39] Tania Li's book, *The Will to Improve* (2007), explored what one might call environmental rule and green governmentality in Indonesia, exploring the power of a liberal "will to improve", understood as a two-century-long project to secure the welfare of populations, but rooted in a historically complex situation of government practice, operating within the jagged rhythms of capitalist accumulation. Government programmers draw boundaries around and "render technical" aspects of landscape, conservation, and livelihood. Simultaneously, Li demonstrates how these practices have limits, imposed by the contradictions between improvement and sovereign power, and between the rationalities and practices of government and their ability to actually regulate dynamic social relations. Jake Kosek's book *Understories* (2006) provides another illustration of how forests (in this case the US southwest as political ecology extended its domain both to the Global North and to the urban world) are classified, organized, and ruled in a way that is intended to produce particular sorts of subjects (including Smokey the Bear!) and property relations. At the very moment that forests are declining as local sources of revenue and employment, they become the basis for powerful (yet different) sorts of insurgent consciousness and practice among both Hispano and white rancher communities. These open up the terrain of "contestation and debate between people with different interests and claims" (2006: 270). Whatever else one may say about these post-structural approaches to political ecology – many of which retained, it should be said, a commitment to political economy in some way – the language of adaptation or adaptive order is entirely absent.

Adaptation rebooted: resilience, complex adaptive systems, and the market

Resilience, from the Latin *resilio* ("to spring back", "recoil", or "retreat"):

1. The mental ability to recover quickly from depression, illness or misfortune.
2. The physical property of material that can resume its shape after being stretched or deformed; elasticity.
3. The positive ability of a system or company to adapt itself to the consequences of a catastrophic failure caused by power outage, a fire, a bomb or similar.

(Merriam-Webster Dictionary)

Adaptation as a form of thought never really disappeared of course, any more than systems or evolutionary theory lost its appeal in the social sciences with the rise of political ecology. Perhaps during the 1970s and 1980s adaptation possessed less purchase but over the past two decades, the concept has returned arguably more robust than ever, attached now to the risks of global climate change and indeed to global threats of virtually all sorts (see Floyd and Matthew 2013). Re-tooled and re-purposed, adaptation and adaptive governance are put to the service of a new framework, designed to assist in the construction of resilient social systems[40] (Folke 2006). Adaptive capacity is, as some of its foremost theoreticians put it, "a core feature of resilient socio-ecological systems" (Nelson et al. 2007: 395). In place of the managed environmentalism or neo-Malthusian models of the 1960s and 1970s, is a new-fangled language of resiliency, adaptive community institutions, and market governance (see WRI 2008). Radical uncertainties about the effects of global climate change – global climate change models are robust on system dynamics but weaker on regional and local predictions – are the harsh realities to which adaptive and resilient systems are to be made to speak.

The resilience framework jettisons much (though by no means all) of the old modes of calculation – the insurance-based logic of calculable risks assessed through probabilities – and replaces them with modalities that can render an uncertain (and perhaps catastrophic) future thinkable, something that can be prepared for and remediated. It is at this point that governance and institutions meets up with theories of "complex adaptive systems" (CAS)[41] (see Holland 1995; Miller and Page 2007) designed to incorporate social and economic systems into an overarching science of "socio-ecological resilience" (Holling 1986, 2001; Folke et al. 2005). The scope and scale (and institutionalization) of resiliency thinking is now considerable, encompassing most fields of expertise which address security in the broadest sense from cybersecurity to global pandemics (Zolli 2012; Alexander 2013). Here is a blurb from a 2008 resiliency conference in the UK:

> The concept of resilience is now capturing high interest across academic, policy and popular debate. In a world where threats – whether linked to climate change, epidemic disease, or fluctuating financial markets – loom ever larger, resilience thinking valuably highlights the complex, open, path-dependent dynamics of coupled social-economic-environmental systems. Not only does it provide an increasingly vigorous and sophisticated body of analysis, resilience thinking also offers prospects for more integrated and effective policy making towards sustainability.
>
> *(Leach 2008: 1)*

Local knowledge and practice, notions of vulnerability and exposure – in other words the conceptual armory of political ecology – have been grafted onto this new turbo-charged systems theory, derived in particular from the work of ecologist C.S. Holling (1973), and brought together in a highly influential think tank called the Stockholm Resilience Center in 2006.[42] Resiliency is a risk-management tool for the sorts of communities – whether in New Orleans or Lagos or rural Papua – that political ecology saw as under threat or now confronting radical ecological change (or for that matter global terrorism, biosecurity, or financial crises).

Building resilient persons, communities, and institutions has become the *sine qua non* of contemporary forms of liberalism; that is to say a means by which all of us are purportedly able to anticipate and tolerate the disturbances, dangers, and radical contingencies of inhabiting a complex world in which, to quote the President of the Rockefeller Foundation, "we cannot predict where the next major shock to our well-being will manifest" (Rockefeller Foundation 2013: 1). Resilience occupies a common semantic space with a post-9/11 vocabulary: the keywords include risk, uncertainty, and security (O'Malley and Bougen 2009; Coaffee 2006). Each year sees another raft of books, policy documents, mission statements, and websites devoted to rearing resilient children, building resilient communities, constructing resilient states, critical infrastructures, cities, livelihoods and financial systems, and naturally to the pressing task of making a resilient military populated by men and women emboldened by "enhanced resilience training" (Bene et al. 2012; Neocleous 2013; Lentsoz and Rose 2009). A brief trawl through any internet search engine reveals many millions of resilience citations – roughly 20 million from a very cursory Google search – covering virtually every aspect of modern life, both ecological and social. So expansive is the prescriptive reach that it resembles what Paul Nasaday (2007) calls a "gospel". Resilience represents, in sum, the offspring of the meeting up of complex adaptive systems with what Amoore (2013: 12) calls the politics of possibility, that is to say a set of risk and security technologies "arraying and acting upon uncertain futures … [through] the fractionation of ever more finite categories of life" and degrees of safety and danger.

The origins of the resiliency framework lie in the 1970s within ecology. Holling attempted to locate the equilibrium-centered work of systems ecology on the larger landscape of the biosphere as a self-organizing and nonlinear complex system (Lindseth 2011). Complexity science – the hallmark of contemporary systems ecology – represents a meeting point of several multidisciplinary strands of science, including computational theory, non-equilibrium thermodynamics, evolutionary theory, and earth systems science.[43] In this account adaptation, organization and robustness transcend the disciplines and fields of expertise: the cell, the firm, and networks of machines also possess organizational properties in common (Miller and Page 2007). Viewed through this optic, adaptive social systems are "composed of interacting, thoughtful agents"; they stand between stasis and utter chaos (Miller and Page 2007: 12). Much of the focus is on how social agents, processing and deploying information, and engaged in complex interactions, lead to emergent phenomena.

At the heart of Holling's early work was the question of how systems retain cohesiveness under stress or radical perturbations (such as drought). Resilience determined the persistence of relations in a system. He explored the implications for management of ecosystems by emphasizing less stability than the unpredictable and unknowable nature of complex system interdependencies, which implied (in policy terms) a need to "keep options open, the need to view events in a regional rather than a local context, and the need to emphasize heterogeneity" (Holling 2001: 392). Through the Resilience Alliance and later the Resilience Center, resilience and adaptive systems thinking was pushed far beyond ecology as such to encompass a coevolutionary theory of societies (Gunderson and Holling 2002) and ecosystems as a single unified science ("panarchy"). Holling extended his view of resiliency by suggesting that all living systems evolved through disequilibrium, that instability was the source of creativity: crisis tendencies were constitutive of complex adaptive systems. What linked the social, economic, and ecological was, according to resilience theory, a theory of "capital accumulation" marked by episodic change, turbulence, and a lack of predictability.

Through complexity science and resiliency thinking, adaptation has had a remarkable new lease on life as my opening observations on the new IPCC report make clear. The overwhelming and self-evident need to adapt – "we already know that adaptation is necessary" (Adger et al. 2009: 2) – is anchored in a binary world composed of an external climate now capable of generating unprecedented threats and perturbations and a social systems characterized by uneven vulnerabilities (Taylor forthcoming).[44] But what, in conceptual terms, is entailed in adaptation when "adaptation to global environmental change" is now invoked by a small army of theorists, practitioners, and policy wonks? In what ways has it addressed or transcended the earlier political ecological critique? Pelling (2011) has properly pointed out that adaptation in this world of resiliency is "slippery". He distinguishes three forms: *resiliency* adaptation which he sees as technocratic and politically conservative; *transformational* adaptation that emphasizes the need for change and barriers to adaptation; and *transitional* adaptation which is situated between the two. This, of course, says very little about how and whether adaptation is old wine in new bottles. In more theoretically oriented work (Adger 2006b; Folke 2006) adaptation is a process of deliberate change or decision-making in anticipation of, or in reaction to, external stimuli and stress ("system adaptedness" is its outcome) while adaptive capacity refers to the preconditions for adaptation to occur (available resources and systemic attributes).

In so far as resilient systems embody adaptive capacity, then to the same extent resiliency is understood as the amount of change a system can undergo while retaining "the same controls on function and structure" (Nelson et al. 2007: 398) through self-organization, capacity for learning, and capacity to absorb change. In resiliency talk, adaptation is always coupled with a set of affinal terms – vulnerability, capacity, and exposure – while being embedded in larger socio-ecological living, self-organizing complex systems (see Ribot 2014). All of this is then harnessed to governance – adaptive governance is the moniker – which links self-organization to particular sorts of environmental problems (ecosystem restoration in the Everglades or water catchment systems in Kenya). Adaptation to drought in Nigeria would involve adjustments (switching occupations say) and resilience (social networks). The governance of drought related issues involve "building knowledge", "networking", and "leadership" (Olsson et al. 2006). Not unusually, much of this adaptation and resiliency is draped in the language of community empowerment, adaptive management, community regulation, and insurance using market mechanisms.[45]

The notion of adaptive capacity and adaptive management with respect to climate change, for example, relies upon a substantial body of research which demonstrates, for example, how rural communities in Africa (and elsewhere) adapt to climate change through mobility, storage, diversification, communal pooling, and exchange by drawing on social networks and their access to resources (Adger et al. 2009; Agrawal and Perrin 2009). But it is one thing to say that "vulnerability is driven by inadvertent or deliberate human action that reinforces self-interest and the distribution of power in addition to interacting with physical and ecological systems" (Adger 2006b: 270) and quite another to move from indicators of exposure to a causal structure of vulnerability and a robust theory of power and circuits of capital. Political ecology, after all, in its account of vulnerability emphasized structures of domination; the community was seen as a theater that had no simple unity, coherence, or equality, but was one in which power was contested and fought over, often violently. In these communities nature is internalized within the circulation and accumulation of capital. All of this – to say nothing of a broader grounding in social theory – is strikingly absent from the new adaptation studies. Rather what is on offer instead is a bland and bloodless shopping list of "conditions" for adaptive governance, including "policy will", "coordination of stakeholders", "science", "common goals", and "creativity". The same holds for accounts of the principles of resilient systems – a contested arena in any case – that typically enlist diversity, institutions, community, preparedness, equity, learning, social structure, non-equilibrium dynamics, and cross-scalar perspective as defining properties (see Bahadur et al. 2010: 14). It is not only that so much of this work – seen clearly in the analysis of Adger and his colleagues[46] – adopts the structure and function approach of earlier Rappaportian ecological anthropology, and in doing so obscures the recursive nature of structures and agency in human systems. It is also that the normative emphasis of so much of this work – getting the rules right as Cote and Nightingale put it (2011: 6) – is at the expense of the careful political economy that was the hallmark of political ecology from the outset.

A canonical policy statement like *Roots of Resilience* (WRI 2008) proposes to scale up "nature based income and culturing resilience", which require ownership, capacity, and connection. Ecosystem-based enterprises, rooted in community resource management, will entail local–state and private–civic partnerships and enterprise networking. Markets in ecosystem services, and delegation of responsibility to communities and households as self-organizing productive units, will constitute the basis for survival in biophysical, political, economic, and financial worlds defined by turbulence, risk, and unpredictability. Some will be resilient, but others will be too resilient or not resilient enough. Resiliency and adaptive capacity may confer in principle all manner of benefits: it focuses on poverty and vulnerability, it is "holistic", it purportedly brings nature and society together, it is forward-looking and dynamic and points to the saliency of complexity and disequilibrium (Cote and Nightingale 2011). But the value of these exhortations resides in the analysis offered, and it is here that the manifold strengths of political ecology – in which power, agency, struggles over access and control of property, labor and the disposition of surpluses – have been largely lost:

> Resilience is a cumbersome concept for social science ... It is difficult to avoid clashes with cornerstone concepts such as power, democracy and the right to self-determination when attempting to apply the concept of resilience to politics and governance. The reason for this is quite straightforward ... societies and ecosystems are ... fundamentally different.
>
> *(Duit et al. 2010: 365)*

In this sense, the normative qualities of resilience and adaption as a form of thought for prescribing policies and practices to address climate change can be thought of as a form of governing that resembles what Michel Foucault calls biopower (Foucault 2008): that is to say, the art of governing and the exercise of power in the form of the economy administered to and through populations.

How then can we understand the re-emergence of adaptation as a way of confronting not just global climate change but virtually any of the radical challenges currently the goal of global human security? This question is especially intriguing because the new wave of work on adaptation and adaptive capacity claims to have subsumed within it the insights of political ecology (see Adger 2006a; Smit and Wandel 2006; Janssen et al. 2006). The short answer is the development of "second order cybernetics" and the field of complexity science (Rasch and Wolfe 2000; Holland 2012; Varela and Maturana 1992). Far from invoking immanent determinism and the larger adaptive order, autopoietic organizations of second order cybernetics[47] are totally self-referential because they exist by virtue of what Varela et al. (1974) call operational closure. All living things are autopoietic in that they are continually self-reproducing according to their own internal rules and requirements; these systems are closed on the level of organization but open to perturbation on the levels of structure.[48] What is recognized as a trigger or perturbation is specified by the entity's organization and operational closure. All of this leads to unpredictable effects, emergent properties, and radical contingency. In his application of the new systems theory to human social systems, Luhmann (1990) sees modern society as a particular species of functionally differentiated social system composed of operationally closed self-referential function systems. Modern capitalism operates on a horizontal plane in which different autopoietic function systems exist side by side with no one system able to overdetermine the others (Rasch and Wolfe 2000: 24).

This account reproduces all of the problems of liberal technocratic functionalism. It is incapable of addressing the radical asymmetries of power which make some autopoietic systems function well for a few but badly for many others (Wolfe 2000 [1994]). Complexity seems irrelevant or at the very least beside the point. The epistemological insights drawn from second order cybernetics – contingency and the partiality of knowledge – run up against the principle of contradiction: that a sort of philosophical idealism and the formal equivalence of different observers in the social field confronts the real lack of equivalence in the world of political economy. Haraway's (1991) critique of systems theory, as a form of technological functionalism with an ideological appeal to the alleviation of stress that is crucial to the reproduction of capitalist social relations, seems as applicable to the second as to the first order cybernetics.

Climate adaptation is now embedded within a view of life understood as a living and complex adaptive system characterized by self-organization, non-linear, combinatorial transactions and radical contingency. Adaptation can only be meaningfully performed *through* contingency, which is to say through the conduct of shaping our exposure to, and creative exploitation of, contingent events and processes in nature and from the "independent actions and interventions of biological being itself" (Dillon 2008: 315). Contingency and transformation are the modalities of safety and survival, or more properly qualitative change in the nature of the living thing itself is the condition of possibility of security. Adaptation and resilience cannot be achieved solely by actuarial logic alone but is governed by an anticipatory logic: it seeks not to forestall through calculation but "to incorporate the very unknowability and profound uncertainty of the future into imminent decision" (Amoore 2013: 9).

Irrespective of its specific referent object (drought, youth, finance), the defining quality of virtually all resilience and adaptive thinking, at least in the social and socio-ecological sciences, is a robust relationship to systemic durability, flexibility and to a culture of preparedness, preemption, and precaution (Anderson 2010). But as Dillon and Reid note of contemporary liberal rule, resilience's reference point is *all of life itself*, and the practices required to "pre-empt the emergence of life forms in the life process that may prove toxic to life" (2009: 87). Resilience is in the business of forming governable subjects, a technology that, as Neocleous (2013: 4) observes, facilitates the connection between state bureaucracy and the political imagination. It is, he says, "nothing less than the attempted colonization of the political imagination by the state" (2013: 4). In sum, resilience provides a powerful anticipatory calculus, one of a flotilla of technologies associated with a security assemblage, rooted in a full-spectrum, and in some respects paranoid, social imaginary – a hyper-dangerous and threatening future. It is to this world of possibilistic logic that climate adaptation must speak.

Climate adaptation and resilience are textbook illustrations of biopolitics understood as the administration and regulation of life processes (Lemke 2011) drawing, however, upon a distinctively modern theory of life as a complex adaptive system. Resilience as it has emerged as a set of practices deployed by state and civil society groups, forms the basis for addressing the uncertainties and instabilities not simply of nature, but of contemporary capitalism, as well as the national security state, and it does so by endorsing a distinctive form of biopolitics and technologies of the self. Building resilient systems draws upon the adaptive and self-organizing capacities of the *market* above all else; resilience dissolves directly into neoliberalism understood as a way of life (Foucault 2008). At the time that Holling was laying out his first ideas, Friedrich Hayek delivered his Nobel Prize speech, which, as Walker and Cooper (2011) brilliantly show, has an elective affinity with Holling's ideas. Hayek was moving toward his mature theorization of capitalism as an exemplar of the biological sciences: the extended market order is "perfectly natural ... like biological phenomena, evolved in the course of natural selection" (Hayek 1988 cited in Walker and Cooper, 2011: 158). In his Nobel lecture, he returned to the epistemology of limited knowledge and uncertain future, a position which led him to explicitly reject and denounce the Club of Rome *Limits to Growth* report. It was to biological systems and complex, adaptive, and non-linear dynamics that he turned to provide the guide for his "spontaneous market order" of capitalism.

Conclusion

The world of climate adaptation is a world of shocks, contingencies, tipping points, thresholds, and risk (Collier 2008). Drawing from the putative self-organizing properties of social systems, climate adaptation, and resiliency constitutes a calculative metric for a brave new world of turbulent capitalism. The Global South's "bottom billion" provides a laboratory in which the poor will be tested as the impacts of change manifest (IPCC 2014). Resiliency has become a litmus test of the right to survive in the global order of things (Cooper 2006, 2008; Watts 2013). To return to Foucault and his notion of an expanded sense of biosecurity, resiliency-adaptation is an apparatus of security that will determine the process of "letting die". The Global South in particular has become the testing ground for a vision of security and care in which life is nothing more than permanent readiness and flexible adaptiveness. As such, it is a deeply Hayekian project – an expression of the neoliberal thought collective – in which the idea of a spontaneous market order has become, ironically, a form of sustainable development. Building resilient peasants and resilient communities in say the West African Sahel or north India turns on an unleashing their self-organizing potential by welding together indigenous capabilities and knowledge with the powers of the self-organizing market. Adaptation is self-organizing and self-engendering – consistent with life itself – entirely resonant with a sort of Hayekian economic subject and a spontaneous economic order (O'Malley 2010, 2011).

From the vantage point of political ecology, then, the new adaptation-resiliency theory raises at least two sorts of concerns. First, even if self-organization is taken as a powerful expression of the "laws of social systems", one can plausibly ask what are the analytical consequences of one form of self-organization (the market) dominating others; and why is it that the self-organizing capacities of the state seem correspondingly underplayed (Connolly 2013). At the very least self-organization must address why some forms of self-organization in a complex field of self-organization trump others. We are in short back in the world of how climate adaption analyses capture the fact that capital is a dominant working and evolving ecological system. And second, in normative terms adaptive capacity and resilience should be placed on the larger canvas of modern forms of biopower: theoretically they are consistent with a particular theory of life which is enrolled as a form of rule and governance. Adaptation entails embracing risk by linking the politics of the possible with "the drive for new areas of profit and the authorization of new actions and decisions" (Amoore 2013: 11). Adaptation, security, risk management, and resiliency are the contemporary hegemonic forms in which particular forms of life constitute the basis of neoliberal rule and governance. The challenges of adapting to the radical uncertainties and perturbations of global climate change invoke a new sense of *homo economicus*. A decision-maker in self-organizing, adaptive systems confronting catastrophic threats becomes "an entrepreneur of himself" (Foucault 2008: 241), a sort of hedge-fund manager for his contingent, turbulent, and unpredictable life.

Notes

1 I am grateful for the suggestions and advice provided by Nathan Sayre, Thomas Perreault, and James McCarthy. In addition I make considerable use of the excellent work of Michael Dillon on the relations between theories of life and human security, and Cary Wolfe's important overview of complexity theory, autopoeisis and second order cybernetics. I do justice to the sophistication

of neither. Some of the ground covered here is addressed in Ribot (2014), Bassett and Fogelman (2013), and Taylor (forthcoming), although my analysis differs from each. Much of my own research in political ecology has been struggling with, through and against the research program laid out 40 years ago by Roy Rappaport, a teacher of mine at the University of Michigan in the 1970s. I wish to emphasize that I am in this chapter almost wholly concerned with political ecology in the English-speaking world; other chapters in this volume address the Francophone and other lineages.

2 The *Impacts, Adaptation, and Vulnerability* assessment report is the second volume of a four-part assessment by the IPCC. It is the product of 243 lead authors and 66 review editors from 70 countries, and 436 additional contributing authors from 54 countries. The report cites more than 12,000 scientific references (see www.ipcc.ch/report/ar5/wg2/).

3 Popularized by atmospheric chemist Paul Crutzen, Anthropocene studies now has its own research journal of the same name.

4 According to the World Economic Forum, the constellation of four most likely risks in 2007 were chronic disease, oil price shocks, asset price collapse, and critical infrastructure breakdown; in 2014 the quartet included income disparity, extreme weather, climate change, and cyber attacks.

5 Writing in 2006, Smit and Wandel (2006: 289) say that there has been a "movement" to "mainstream" adaptation to climate change, a movement that has increasingly shifted from "adaptation analysis to practice". See Janssen et al. (2006) for an analysis of the growth of citations and references to adaptation in the top three climate journals over the 1977–2005 period. In their accounting adaptation's key figures are Roy Rappaport in the late 1960s, K. Butzer and P. Timmerman in the 1980s, R.T. Watson and B. Smits in the 1990s, and J.J. McCarthy in the 2000s.

6 Lance Gunderson and C.S. Holling, in their book *Panarchy: Understanding Transformations in Systems of Humans and Nature*, coined the term as "an antithesis to the word hierarchy (literally, sacred rules). Our view is that panarchy is a framework of nature's rules, hinted at by the name of the Greek god of nature, Pan" (2002: 21). Panarchy describes evolving hierarchical systems with multiple interrelated elements; it is the structure in which systems, including those of nature (e.g. forests) and of humans (e.g. capitalism), as well as combined human–natural systems (e.g. institutions that govern natural resource use), are interlinked in continual adaptive cycles of growth, accumulation, restructuring, and renewal.

7 Glacken's foundational book *Traces on the Rhodian Shore* (1967), and Paul Wheatley's *Pivot of the Four Quarters* (1972) are the definitive statements of their theoretical projects.

8 The conference included 70 scholars of wildly different theoretical persuasion (and the book was no less diverse).

9 See Philip Wagner (1960).

10 For a review of this anthropological research see Robert Nettings's *Cultural Ecology* published in 1977 and Peter Vayda's edited collection *Environment and Cultural Behavior* published in 1969.

11 Geertz (1974 [1963]: 6) considered that his research could be considered within cultural ecology while Conklin viewed his program in terms of the "ethnoecological" interpretations of key moments and conjunctures in the ordered sequence of the swidden cycle (Conklin 1975 [1957]: 154).

12 The Macy Conferences held during the 1940s on cybernetics and the influence of Gregory Bateson in particular were seminal influences on the incorporation of what one might call ecological and systems thinking into the social and behavioral sciences (see Heims 1991).

13 Much of this cultural ecological work was brought together in the canonical collection edited by Y. Cohen, *Man and Adaptation* (1971).

14 The geographer Marvin Mikesell identified a "New Guinea syndrome" centered on ANU comparable in importance to Sauer and his students' research focus in Mexico. The key figure was British geographer Harold Brookfield, appointed to the Research School of Pacific Studies, who actively recruited postgraduate students. As a result, the numbers of geographers (and anthropologists, economists, and historians) conducting research in PNG grew. What was often called the "Brookfield School" at ANU spawned a raft of hugely influential cultural ecologists such as Eric Waddell and Diana Howlett. Brookfield's program was in quite fundamental ways driven by an engagement with Esther Boserup, *The Conditions of Agricultural Growth* (1965). Much of this ANU research is drawn together in Harold Brookfield and Paul Brown, *The Struggle for Land* (1962) and Harold Brookfield and Doreen Hart, *Melanesia* (1971).

15 In addition to Gilbert White, the three major geographers in hazards research were Ian Burton, Robert Kates, and Martin Bowden. See also Kates and Burton (1986) and Burton et al. (1978).

16 Hazards analysis operated on quite naïve and culture-bound perception studies and rather pedestrian stressor-response models. The behavioral sciences (the Natural Hazards Center at Colorado was located within the Institute of Behavioral Science) tended to address decision processes and communication strategies within and between organisms in a social system while the sociological research on hazards typically examined the impacts of social organization on structural adjustment of the individual and of groups (Quarantelli and Dynes 1977). There was little understanding or interest in how power, class, status, or the webs of production relations shaped the capacity to "adjust" to environmental hazards (see Hewitt 1983 for a critique). This early work is quite different from some of the later critical sociological and anthropological analysis of so-called natural disasters like Hurricane Katrina (see for example the Social Science Research Council project on Katrina: http://understandingkatrina.ssrc.org/, or the extraordinary documentaries on New Orleans and the Katrina disaster by Spike Lee, especially *When the Levees Broke: A Requiem in Four Acts* (2006)).

17 Andrew (Peter) Vayda was Rappaport's doctoral advisor at Columbia and while they are typically seen as foundational figures (and holding similar views) in ecological anthropology, their theoretical opinions actually diverged quite sharply. Vayda believes, contra Rappaport, that ecosystems are not in fact self-regulating, self-organizing, and adaptive, and that Rappaport's program was a "chimerical undertaking" (1986: 298).

18 A hallmark of much (but by no means all) ecological anthropology was its attentiveness to material and energetic flows in ecosystems, an approach that built upon the foundational work of ecologist Howard Odum developed in the 1950s (see Odum 1971). Much of the best of this early work is collected in Dove and Carpenter (2008).

19 General purpose systems are, as Slobodkin and Rapoport (1974: 181), put it, forms of "existential games" in which there is no way of using the winnings for any other purpose than continuing to play the game for as long as possible. Successful evolution requires the maintenance of flexibility maintained in the most parsimonious way (organisms and populations must not make excessive or unnecessary commitments to perturbations in the system).

20 This overwhelming concern with structure and function and with humans as agents embedded within the logic of ecosystems dynamics is identical to the formulations of what I am calling the second phase adaptation research around global climate change (see Folke (2006) and Adger (in Leach 2008)).

21 Rappaport met Bateson in Hawaii in 1968 and noted that he was "the most profound of influences upon me" (Rappaport 1994: 167). See Bateson's hugely influential collection of essays, *Steps to an Ecology of Mind* (1972).

22 Bateson's (1972) economics of somatic flexibility, Frisancho's (1975) research on human functional adaptation to high altitude hypoxia, and the theoretical work of Slobodkin and Rapoport (1974: 198) on parsimonious evolutionary flexibility are cases in point.

23 Specificity in Rappaport's model refers to the nature of the goals of regulatory mechanisms at different levels. Low order controls regulate specific operations in accordance with reference values established by higher order operations. In social systems the higher order the goals or propositions are general "principles" which, in Rappaport's account, are often sacred and unverifiable or beyond reach.

24 A number of behavioral scientists explored, in the 1960s and 1970s, how these principles of living systems can be deployed in public policy or in the analysis of social systems more generally (the work of Ross Ashby, Gordon Pask, Emilio Lazlo, Edgar Dunn, and James Miller for example). But the person whose work – sadly neglected and largely unrecognized – followed Bateson and Rappaport most diligently, operating in the frontierlands of social theory and post-structuralism, was the brilliant Canadian communications theorist Anthony Wilden and his account of morphogenetic systems (1972; see Watts 1983 for a review).

25 The work of Flannery (1972) on "hypercoherent systems" extended this approach in his archaeological work, and Catalan biologist Ramón Margalef's very influential book *Perspectives in Ecological Theory* (1968) also explored similar lines of thinking.

26 Some anthropologists such as Durham identified as human ecologists which, for the purposes of this chapter, I am taking as roughly synonymous with ecological anthropology. In their concern with fitness and evolutionary theory this form of human ecology is quite different from the famous Chicago School of Sociology and its human ecology sub-field which emerged in the 1920s.

27 Wolfe (2000: 177) makes the important point that Bateson, for example, emphasized contingency in observation yet smuggled back in the notion of an overarching "pattern that connects" – this is in effect Rappaport's "adaptive order of living systems". What at first glance looks like contingent

observation is instead determined "from behind" by the total pattern of existence immanent in the total interconnected social and planetary ecologies.

28 It needs to be said that Rappaport in the second edition (1984) of *Pigs for the Ancestors* provides an extraordinary 200-page(!) response to his critics in which he concedes that he was not able to actually demonstrate that the *kaiko* pig ritual acted as a thermostat: "I probably did not emphasize sufficiently the role of conscious, pragmatic decision making in the affairs of the Maring. It did not occur to me that they would be noticed by readers in the course of the account even if they remained, in part, implicit … [Moreover] nowhere in the text is there any suggestion that it is otherwise" (Rappaport 1984 [1968]: 321). He spent much time defending his version of functionalism and responding to the critique of his putative negative feedback model by outlining his own theory of maladaptation, that is to say how social systems were not adapting, or were deviating from a normative adaptive order. Rappaport, however, made no systematic effort to shows how the dynamics of capitalism as a system was or was not maladaptive.

29 Cultural ecology was attacked as "vulgar materialism reinforcing rather than redoing the classical capitalist fetishization of 'things'", the domination of subjects by objects rather than by the social relations embodied in, and symbolized by, those objects (see especially Friedman 1974). Some of this critique was directed at the "cultural materialism" associated with anthropologist Marvin Harris (1968).

30 Sahlins (1976: 298) criticized the kind of "empiricism" expressed in *Pigs for the Ancestors* as a type of "ecology fetishism", because it practices "the idea that nothing is in fact what it appears, i.e. culturally, but is translated instead into natural coordinates or consequences".

31 For fuller accounts of the history and development of the field see Robbins (2005), Watts and Peet (2004).

32 While this concern with perception typically was expressed in terms of indigenous knowledge and subsequently green discourses of various sorts, there was nevertheless a powerful echo here of Rappaport's notion of a cognized environmental model.

33 Brookfield teamed up subsequently with Piers Blaikie to organize the important political ecological text *Land Degradation and Society* (1987).

34 Nietschmann, who had received his degree under William Denevan at Madison (a student of Sauer and James Parson at Berkeley) was part of the Michigan group working with Roy Rappaport, Kent Flannery and others, but left Michigan in 1977 to join the Berkeley Geography Department. His paper entitled "Cultural ecology: something old, something new, something borrowed, something blue" (sadly unpublished) delivered to the AAG meetings in Seattle in May 1974 made his grave misgivings about the adaptation framework explicit.

35 Wisner was involved in a major study of drought in eastern Kenya (and subsequently on water work in Tanzania) in the early 1970s which represented an important critique of the earlier natural hazards work of White and Gilbert that I described earlier. His work on "denaturalizing" natural disasters with Phil O'Keefe and Kirsten Haring was especially important. Much of the political ecology work on hazards and disasters was pulled together in Blaikie et al., *At Risk: Natural Hazards, People's Vulnerability, and Disasters* (1994). Wisner's critical approach was very much shaped by the Marxist inspired critical development debates in Dar es Salaam and Nairobi during the 1970s.

36 Hecht completed her PhD dissertation in 1980 and moved to the Planning Program at UCLA. A number of other persons – both faculty and students – on the Berkeley campus were working on political ecological questions including Louise Fortmann, Carolyn Merchant, myself, Nancy Peluso, and in a subsequent generation Judy Carney, Donald Moore, Nathan Sayre, and Jake Kosek.

37 Piers Blaikie received his PhD in 1971 and had worked on family planning, diffusion theory, and the spatial structure of agriculture in the 1970s before turning to land degradation and his regional political ecology approach. Again, this political ecology emerged from an engagement with the political economy of development, in his case through work in Zambia, north India, Morocco, and Nepal and subsequently he teamed up with Harold Brookfield to write *Land Degradation and Society* (1987), thereby bringing together two of the lineages I have outlined here.

38 Bruno Latour (2009) offers a rather different account of political ecology (including a political ecological critique of cybernetics, hierarchical and adaptational thinking) which turns on the inability or unwillingness of political ecology to grasp its own politics and ecology, not the least of which is the fact that the risk-free objects they study are increasingly becoming tangled, boundaryless, and invisible (indeterminate) (2009: 21–23).

39 It needs to be said, however, that much of the current writing from Anthropology on ecological questions is woefully ignorant of the earlier (and indeed much current) Geographical research on political ecology.

40 In their comprehensive theoretical overview, Martin-Breen and Anderies define resilience as "the ability to withstand, recover from, and reorganize in response to crises" (2011: 7). Drawing upon the lineage of ecological theory and the influential work of C.S. Holling, Nelson et al. (2007) define resilience as the amount of change a system can undergo while retaining (1) the same structure and function, and (2) the options to develop (2007: 396). Both of these definitions clarify the extent to which system, function, structure, and capacity to adapt – that is to say the conceptual hardware of General Systems Theory and the work on adaptive orders – are resilience's defining properties. The direct continuity with the 1960s adaptive orders of Roy Rappaport is striking.

41 A complex adaptive system is a system in which "large networks of components with no central control and simple rules of operation give rise to complex collective behavior, sophisticated information processing and adaptation via learning or evolution" (Mitchell 2009: 13).

42 In the mid-1990s the Beijer Institute in Stockholm and the University of Florida (where Holling was located) started the Resilience Network, a research program that later developed into the Resilience Alliance with its own journal, *Ecology and Society*.

43 Complex adaptive systems (CAS) entail (1) complex collective behavior, (2) signaling and information processing, and (3) adaptation which together *qua* system exhibit non-trivial emergent self-organizing behaviors.

44 Marcus Taylor's important forthcoming book *The Political Ecology of Climate Change Adaptation* covers much of the ground I survey here and is deeply critical of the climate adaptation framework (see also Bahadur et al. 2010). He argues that adaptation turns on a separation of climate and society that mirrors the modernist (Cartesian) distinction between natural and social world. While I agree with much of his critique, the theory of complex living systems draws this line very differently by folding all systems into an overarching theory of self-organization and functional differentiation.

45 Nelson et al. (2007: 404), for example, examine "adaptive actions" (selling assets, irrigation schemes, community based management, regulatory change) to resource stresses (drought, hurricane impact, coral reef damage) and the sources of resilience they index (social networks, security mechanisms, learning through consensus, self-efficacy). All of this is useful and well understood but it is completely devoid of any account of power, social relations, and political economy.

46 The cost of functionality is seen in the degree to which right and proper concern with inequality and legitimacy in, for example, Adger et al. (2005), is converted into a list of social factors that contribute to institutional function.

47 The key theoreticians are drawn from biology (Francisco Varela) and physics (Ilya Prigogine), and from the social sciences (Niklas Luhmann).

48 *Organization* denotes the relations necessary among components for a system to be a member of a class while *structure* denotes the actual relations and components that constitute a unity, making its organization real (see Wolfe 2000: 255). Structure and organization are continually self-producing.

References

Adger, W.N. 2006a. Vulnerability. *Global Environmental Change*, 16: 268–281.

Adger, W.N. (ed.), 2006b. *Fairness in Adaptation to Climate Change*. Cambridge, MA: MIT Press.

Adger, W.N. Arenell, N. and Tompkins, E. 2005. Adaptation to climate change. *Global Environmental Change*, 17: 77–86.

Adger, W.N. Eakin, H. and Winkela, A. 2010. Nested and teleconnected vulnerabilities to environmental change. *Frontiers in Ecology*, 7(3): 150–157.

Adger, W.N., Lorenzoni, I. and O'Brien, K. (eds.). 2009. *Adapting to Climate Change*. London: Cambridge University Press.

Agrawal, A. and Perrin, N. 2009. Climate adaptation, local institutions and livestock. In: Adger, N., Lorenzoni, I. and O'Brien, K. (eds.), *Adapting to Climate Change: Thresholds, Values, Governance*. Cambridge: Cambridge University Press, pp. 350–367.

Alexander, D. 2013. Resilience and disaster risk reduction: an etymological journal. *Natural Hazards Earth Systems Discussion Paper*, 1: 1257–1284.

Alland, A. 1975. Adaptation. *Annual Review of Anthropology*, 4: 59–73.

Amin, A. 2010. *Land of Strangers*. London: Polity.

Amoore, L. 2013. *Politics of Possibility: Risk and Security Beyond Probability*. Durham, NC: Duke University Press.

Anderson, B. 2010. Preemption, precaution, preparedness. *Progress in Human Geography*, 34(6): 777–798.
Bahadur, Aditya V., Ibrahim, Maggie and Tanner, Thomas. 2010. *The Resilience Renaissance? Unpacking of Resilience for Tackling Climate Change and Disasters*. Institute of Development Studies, University of Sussex, Brighton. Available at: http://community.eldis.org/.59e0d267/resilience-renaissance.pdf (accessed April 28, 2014).
Bassett, T. and Fogelman, C. 2013. Deja vue or something new? The adaptation concept in the climate change literature. *Geoforum*, 48: 42–53.
Bateson, G. 1972. *Steps to an Ecology of Mind*. New York: Ballantine.
Bene, C., Wood, R., Newsham, A. and Davies, M. 2012. Resilience: new utopia or new tyranny. Working Paper 405. Falmer: Sussex University, Institute of Development Studies.
Blaikie, Piers. 1985. *The Political Economy of Soil Erosion in Developing Countries*. London: Longman.
Blaikie, Piers and Brookfield, Harold. 1987. *Land Degradation and Society*. London: Methuen.
Blaikie, P.M., Cannon, T., Davis, I. and Wisner, B. 1994. *At Risk: Natural Hazards, People's Vulnerability, and Disasters*. London and New York: Routledge.
Boal, Iain. 2009. *Globe, Capital, Climate*. Berkeley: RETORT.
Boserup, Esther. 1965. *The Conditions of Agricultural Growth*. London: Allen & Unwin.
Brookfield, H. 1973. Explaining or understanding? The study of adaptation and change. In: Brookfield, H. (ed.), *The Pacific in Transition: Geographical Perspectives on Adaptation and Change*. New York: St. Martin's Press, pp. 3–23.
Brookfield, H. 1975. *Interdependent development*. London: Methuen.
Brookfield, H. and Brown, P. 1962. *The Struggle for Land: Agriculture and Group Territories among the Chimbu of the New Guinea Highlands*. London: Oxford University Press.
Brookfield, H. and Hart, D. 1971. *Melanesia: A Geographical Interpretation of an Island World*. London: Methuen.
Brown, Paula. 1972. *The Chimbu*. Morristown: General Learning Press.
Burawoy, M. 2014. Marxism after Polanyi, unpublished manuscript. Available at: http://burawoy.berkeley.edu/Marxism/Marxism%20after%20Polanyi.pdf.
Burton, I., Kates, R.W. and White, G.F. 1978. *The Environment as Hazard*. New York: Oxford University Press.
Butzer, K.W. 1980a. Adaptation to global environmental change. *Professional Geographer*, 32(3): 269–278.
Butzer, K.W. 1980b. Civilizations: organisms or systems? *American Scientist*, 69: 517–523.
Clarke, William C. 1971. *Place and People: An Ecology of a New Guinean Community*. Berkeley: University of California Press.
Clarke, W. 1977. The structure of permanence. In: Bayliss-Smith, T. (ed.), *Subsistence and Survival*. London: Academic Press, pp. 363–384.
Coaffee, J. 2006. From counter terrorism to resilience. *The European Legacy*, 11(4): 389–403.
Cohen, Y. (ed.). 1971. *Man and Adaptation*. Chiacgo: University of Chicago Press.
Collier, S. 2008. Enacting catastrophe: preparedness, insurance, budgetary rationalization. *Economy and Society*, 37: 224–250.
Conklin, C. Harold. 1975 [1957]. *Hanunóo Agriculture: A Report on an Integral System of Shifting Cultivation in the Philippines*. Food and Agriculture Organization of the United Nations (FAO). Reprinted by Elliot's Books, New Haven.
Connolly, William E. 2013. *The Fragility of Things: Self-Organizing Processes, Neoliberal Fantasies, and Democratic Activism*. Durham, NC: Duke University Press.
Cooper, M. 2006. Pre-empting emergence: the biological turn in the war on terror. *Theory, Culture and Society*, 23: 113–135.
Cooper, M. 2008. *Life as Surplus*. Seattle: University of Washington Press.
Cote, M. and Nightingale, A. 2011. Resilience thinking meets social theory. *Progress in Human Geography*, 36: 1–15.
Denevan, William M. 1970. Aboriginal drained-field cultivation in the Americas pre-Columbian reclamation of wet lands was widespread in the savannas and highlands of Latin America. *Science*, 169(3946): 647–654.
Denevan, W. 1983. Adaptation, variation, and cultural geography. *Professional Geographer*, 35(4): 399–406.
Dillon, M. 2008. Underwriting security. *Security Dialogue*, 39(2–3): 309–332.
Dillon, M. and Reid, J. 2009. *The Liberal Way of War*. London: Routledge.
Dove, M. and Carpenter, C. (eds). 2008. *Environmental Anthropology*. Oxford: Blackwell.
Duffield, M. 2011. Total war as environmental terror. *South Atlantic Quarterly*, 110(3): 757–769.

Duit, A., Galaz, V. and Eckerberg, K. 2010. Governance, complexity and resilience. *Global Environmental Change*, 20(3): 363–368.
Durham, William H. 1976. The adaptive significance of cultural behavior. *Human Ecology*, 4(2): 89–121.
Flannery, Kent V. 1972. The cultural evolution of civilizations. *Annual Review of Ecology and Systematics*, 3(1): 399–426.
Floyd, Rita and Matthew, R. (eds.). 2013. *Environmental Security: Approaches and Issues*. London: Routledge.
Folke, C. 2006. Resilience: the mergence of a perspective for socio-ecological systems analysis. *Global Environmental Change*, 16: 253–267.
Folke, C., Carpenter, S.R., Walker, B., Scheffer, M., Chapin, T. and Rockström, J. 2010. Resilience thinking: integrating resilience, adaptability and transformability. *Ecology and Society* 15(4): 20. Available at: www.ecologyandsociety.org/vol15/iss4/art20/.
Folke, C., Hahn, T., Olsson, P. and Norberg, J. 2005. Adaptive governance of socio-ecological systems. *Annual Review of Environment and Resources*, 30: 441–473.
Forsyth, Timothy. 2003. *Critical Political Ecology: The Politics of Environmental Science*. London: Routledge.
Foucault, M. 2008. *The Birth of Biopolitics*. London: Palgrave.
Friedman, Jonathan. 1974. Marxism, structuralism, and vulgar materialism. *Man*, 9: 44–69.
Frisancho, A. Roberto. 1975. Functional adaptation to high altitude hypoxia. *Science*, 187(4174): 313–319.
Galison, Peter. 1994. The ontology of the enemy: Norbert Wiener and the cybernetic vision. *Critical Inquiry*, 12: 228–266.
Geertz, Clifford. 1974 [1963]. *Agricultural Involution: The Process of Ecological Change in Indonesia*. Berkeley, Los Angeles and London: University of California Press.
Glacken, C. 1967. *Traces on the Rhodian Shore*. Berkeley: University of California Press.
Godelier, Maurice. 1972. *Rationality and Irrationality in Economics*, trans. Brian Pearce. London: NLB.
Goldman, M. 2005. *Imperial Nature*. New Haven: Yale University Press.
Grossman, Lawrence S. 1984. *Peasants, Subsistence Ecology, and Development in the Highlands of Papua New Guinea*. Princeton: Princeton University Press.
Gunderson, Lance and Holling, C.S. (eds.). 2002. *Panarchy: Understanding Transformations in Human and Natural Systems*. Washington, DC: Island Press.
Haraway, Donna. 1991. *Simians, Cyborgs, and Women: The Reinvention of Women*. London: Routledge.
Harris, D. 1971. The ecology of swidden cultivation in the upper Orinocco Rainforest. *Geographical Review*, LXI(4): 475–495.
Harris, Marvin. 1968. The cultural ecology of India's sacred cattle. *Current Anthropology*, 7(1): 51–64.
Harvey, David. 2014. *Seventeen Contradictions and the End of Capitalism*. London: Oxford University Press.
Hayek, F.A. 1974. The Pretence of Knowledge, Nobel Prize Speech, Salzburg, 11 December. Available at: http://nobelprize.org/nobel_prizes/economics/laureates/1974/hayek-lecture.html.
Hecht, Susanna B. 1985. Environment, development and politics: capital accumulation and the livestock sector in eastern Amazonia. *World Development*, 13(6): 663–684.
Heims, Steve J. 1991. *The Cybernetics Group*. Cambridge, MA: MIT Press.
Hewitt, K. 1983. *Interpretations of Calamity from the Viewpoint of Human Ecology*. Boston: Allen & Unwin.
Holland, John H. 1995. *Hidden Order: How Adaptation Builds Complexity*. New York: Basic Books.
Holland, John H. 2012. *Signals and Boundaries: Building Blocks for Complex Adaptive Systems*. Cambridge, MA: MIT Press.
Holling, C.S. 1973. Resilience and stability of ecological systems. *Annual Review of Ecology and Systematics*, 4: 1–23.
Holling, C.S. (ed.). 1978. *Adaptive Environmental Assessment and Management*. London: Wiley.
Holling, C.S. 1986. Resilience of ecosystems: local surprise and global change. In: Clark, W.C. and Munn, R.E. (eds.), *Sustainable Development of the Biosphere*. Cambridge: Cambridge University Press, pp. 292–317.
Holling, C.S. 2001. Understanding the complexity of economic, ecological and social systems. *Ecosystems*, 4: 390–405.
IPCC. 2007. *Intergovernmental Panel on Climate Change: Fourth Assessment Report*. Geneva. Available at: www.ipcc.ch/publications_and_data/publications_and_data_reports.shtml#1.
IPCC. 2014. *Fifth Assessment Report of the Intergovernmental Panel on Climate Change*. Geneva. Available at: www.ipcc.ch/report/ar5/wg2/.
Janssen, Marco A. and Ostrom, Elinor. 2006. Resilience, vulnerability, and adaptation: a cross-cutting theme of the International Human Dimensions Programme on Global Environmental Change. *Global Environmental Change*, 16(3): 237–239.

Janssen, M., Schoon, M., Weimao, L. and Borner, K. 2006. Scholarly networks on resilience, vulnerability and adaptation within the human dimensions of global environmental change. *Global Environmental Change*, 16(3): 240–252.

Kates, Robert W. 2001. Queries on the human use of the Earth. *Annual Review of Energy and the Environment*, 26(1): 1–26.

Kates, R.W. and Burton, I. (eds.). 1986. *Geography, Resources, and Environment*. Vol. 1. *Selected Writings of Gilbert F. White*. Chicago: University of Chicago Press.

Kauffman, Stuart. 2000. *Investigations*. New York: Oxford University Press.

Kautsky, Karl. 1988 [1899]. *The Agrarian Question: In Two Volumes*. London: Zwan Publications.

Kosek, Jake. 2006. *Understories*. Durham, NC: Duke University Press.

Latour, Bruno. 2009. *Politics of Nature*. Cambridge, MA: Harvard University Press.

Leach, M. (ed.). 2008. *Reframing Resilience*. STEPS Center, IDS, University of Sussex.

Leach, M. and Mearns, R. (eds.). 1996. *The Lie of the Land*. London: James Currey.

Lemke, T. 2011. *Biopolitics*. New York: New York University Press.

Lentsoz, F. and Rose, N. 2009. Governing insecurity. *Economy and Society*, 38(2): 230–254.

Levins, R. and Lewontin, R. 1985. *The Dialectical Biologist*. Cambridge, MA: Harvard University Press.

Li, Tania. 2007. *The Will to Improve*. Durham, NC: Duke University Press.

Lindseth, B. 2011. The prehistory of resiliency in ecological research. *LIMN*, 1. Available at: http://limn.it/the-pre-history-of-resilience-in-ecological-research/.

Luhmann, Niklas, 1990. *Essays on Self-Reference*. New York: Columbia University Press.

Luhmann, Niklas. 1993a. *Communication and Social Order: Risk: A Sociological Theory*. New Brunswick: Transaction Publishers.

Luhmann, Niklas. 1993b. Deconstruction as second-order observing. *New Literary History*: 763–782.

Margalef, Ramón. 1968. *Perspectives in Ecological Theory*. Chicago: University of Chicago Press.

Martin-Breen, P. and Anderies, J. 2011. *Resilience: A Literature Review*. Prepared for the Rockefeller Foundation, New York, unpublished manuscript.

Mathewson, K. 2009. Sauer and his critics. In: *Carl Sauer on Culture and Landscape: Readings and Commentaries*, co-edited with William M. Denevan. Baton Rouge: Louisiana State University Press, pp. 9–28.

Miller, J. and Page, S. 2007. *Complex Adaptive Systems*. Princeton: Princeton University Press.

Mitchell, D. 2001. *Cultural Geography*. London: Blackwell.

Mitchell, M. 2009. *Complexity: A Guided Tour*. London: Oxford University Press.

Moore, Donald S. 2005. *Suffering for Territory: Race, Place, and Power in Zimbabwe*. Durham, NC: Duke University Press.

Nasaday, P. 2007. Adaptive co-management and the gospel of resilience. In: Armitage, D., Berkes, F. and Doubleday, N. (eds.). *Adaptive Co-management*. Vancouver: University of British Columbia Press, pp. 208–227.

Nelson, D., Adger, N. and Brown, K. 2007. Adaptation to environmental change. *Annual Review of Environment and Resources*, 32: 395–419.

Nelson, R. 2005. Evolutionary social science and universal Darwinism. CCS Working Paper # 5, Columbia University, New York. Available at: http://capitalism.columbia.edu/files/ccs/CCSWP5_Nelson.pdf.

Neocleous, M. 2013. Resisting resilience. *Radical Philosophy*, 178: 3–7.

Nettings, R. 1977. *Cultural Ecology*. Berkeley: Waveland Press.

Nietschmann, B. 1973. *Between Land and Water: The Subsistence Ecology of the Miskito Indians, Eastern Nicaragua*. New York: Seminar Press.

Odum, Howard T. 1971. *Environment, Power, and Society*. New York: Wiley-Interscience.

OECD. 2003. *Emerging Risks in the 21st Century*. Paris: Organization for Economic Cooperation and Development. Available at: www.oecd.org/dataoecd/20/23/37944611.pdf.

O'Keefe, P., Westgate, K. and Wisner, B. 1976. Taking the naturalness out of natural disasters. *Nature*, 260: 566–567.

Oliver-Smith, Anthony. 1996. Anthropological research on hazards and disasters. *Annual Review of Anthropology*, 25: 303–328.

Olsson, Per, Gunderson, Lance H., Carpenter, Steve R., Ryan, Paul, Lebe, Louis, Folke, Carl and Holling, C.S. 2006. Shooting the rapids: navigating transitions to adaptive governance of social-ecological systems. *Ecology and Society*, 11(1): 18.

O'Malley, Pat (ed.). 2006. *Governing Risks*. Aldershot: Ashgate.

O'Malley, P. 2010. Resilient subjects. *Economy and Society*, 39(4): 488–509.
O'Malley, P. 2011. Security after risk: security strategies for governing extreme uncertainty. *Current Issues in Criminal Justice*, 23(1): 5–15.
O'Malley, P. and Bougen, M. 2009. Imaginable insecurities: imagination, routinisation and the government of uncertainty post 9/11. In: Carlen, P. (ed.), *Imaginary Penalities*. Devon: Willan, pp. 26–44.
Pelling, M. 2011. *Adaptation to Climate Change: From Resilience to Transformation*. London and New York: Routledge.
Peluso, Nancy Lee. 1992. *Rich Forests, Poor People: Resource Control and Resistance in Java*. Berkeley: University of California Press.
Polanyi, K. 1944. *The Great Transformation*. Boston: Beacon Press.
Quarantelli, Enrico L. and Dynes, Russell R. 1977. Response to social crisis and disaster. *Annual Review of Sociology*, 3(1): 23–49.
Rappaport, Roy. 1971. Nature, culture and ecological anthropology. In: Shapiro, H. (ed.), *Man, Culture and Society*. London: Oxford University Press, pp. 237–267.
Rappaport, Roy. 1974. Adaptation and maladaptation. Unpublished manuscript. Department of Anthropology, University of Michigan.
Rappaport, Roy A. 1977. Ecology, adaptation and the ills of functionalism (being, among other things, a response to Jonathan Friedman). Working Paper. Ann Arbor: Department of Anthropology, University of Michigan.
Rappaport, Roy. 1979. *Ecology, Meaning and Religion*. Berkeley: North Atlantic Book.
Rappaport, Roy. 1984 [1968]. *Pigs for the Ancestors: Ritual in the Ecology of a New Guinea People*. Enlarged edition. New Haven: Yale University Press.
Rappaport, Roy. 1994. Humanity's evolution and anthropology future. In: Borofsky, R. (ed.), *Assessing Cultural Anthropology*. McGraw-Hill, pp. 153–167.
Rasch, W. and Wolfe, C. (eds.). 2000. *Observing Complexity: Systems Theory and Postmodernity*. Minneapolis: University of Minnesota Press.
Ribot, J. 2011. Vulnerability before adaptation: toward transformative climate action. *Global Environmental Change*, 21(4): 1160–1162.
Ribot, J. 2014. Cause and response: vulnerability and climate in the Anthropocene. *Journal of Peasant Studies*, 41: 1–39.
Ribot, J. and Larson, A. 2009. *Democratic Decentralization through a Natural Resource Lens*. London: Routledge.
Robbins, Paul. 2005. *Political Ecology: A Critical Introduction*. New York: John Wiley & Sons.
Rockefeller Foundation. 2013. *Rebounding: Building a More Resilient World*. New York: Rockefeller Foundation.
Sahlins, Marshall. 1972. *Stone Age Economics*. Chicago: University of Chicago Press.
Sahlins, Marshall. 1976. *The Use and Abuse of Biology: An Anthropological Critique of Sociobiology*. Ann Arbor: University of Michigan Press.
Sauer, C. 1941. Foreword to historical geography. *Annals of the Association of American Geographers*, 31(1): 1–24.
Sayer, A. 1979. Epistemolgy and conceptions of people and nature in Geography. *Geoforum*, 10: 19–43.
Slobodkin, Lawrence B. and Rapoport, Anatol. 1974. An optimal strategy of evolution. *Quarterly Review of Biology*, 49: 181–200.
Smit, B. and Wandel, J. 2006. Adaptation, adaptive capacity and vulnerability. *Global Environmental Change*, 16: 282–292.
Smith, Neil. 1977. *Uneven Development: Nature, Capital, and the Production of Space*. Oxford: Blackwell.
Steward, Julian. 1955. *The Theory of Culture Change*. Urbana: University of Illinois Press.
Taylor, M. forthcoming. *The Political Ecology of Climate Change Adaptation*. London: Routledge.
Varela, Francisco and Maturana, Humberto. 1992. *The Tree of Knowledge*. Boston: Shambhala.
Varela, Francisco J., Maturana, Humberto R. and Uribe, R. 1974. Autopoiesis: the organization of living systems, its characterization and a model. *Biosystems*, 5: 187–196.
Vayda, A.P. (ed.). 1969. *Environment and Cultural Behavior*. Austin: University of Texas Press.
Vayda, A.P. 1986. Holism and individualism in ecological anthropology. *Reviews in Anthropology*, 13(4): 295–313.
Vayda, A.P. and Rappaport, Roy. 1967. Ecology, cultural and non-cultural. In: Clifton, J. (ed.), *Introduction to Cultural Anthropology*. Boston: Houghton-Mifflin, pp. 477–497.
Waddell, Eric. 1972. *The Mound Builders: Agricultural Practices, Environment, and Society in the Central Highlands of New Guinea*. Seattle: University of Washington Press.

Wagner, P.L. 1960. *The Human Use of the Earth*. Glencoe, IL: Free Press.

Walker, J. and Cooper, M. 2011. Genealogies of resilience: from systems ecology to the political economy of crisis adaptation. *Security Dialogue*, 43: 143–160.

Watts, M. 1983. On the poverty of theory: natural hazards research in context. In: Hewitt, K. (ed.), *Interpretations of Calamity from the Viewpoint of Human Ecology*. Boston: Allen & Unwin, pp. 231–262.

Watts, M. 2000. Political ecology. In: Barnes, T. and Sheppard, E. (eds.), *A Companion to Economic Geography*. Oxford: Blackwell, pp. 257–275.

Watts, M. 2012 [1983]. *Silent Violence*. Athens: University of Georgia Press.

Watts, M. 2013. Toward a political ecology of environmental security. In: Floyd, Rita and Matthew, Richard (eds.), *Environmental Security*. London: Routledge, pp. 82–102.

Watts, M. and Peet, R. 2004. Liberating political ecology. In: Peet, R. and Watts, M. (eds.), *Liberation Ecologies: Environment, Development, Social Movements*. London: Routledge, pp. 30–43.

Wheatley, P. 1972. *Pivot of the Four Quarters*. Edinburgh: University of Edinburgh Press.

White, G.F. 1945. *Human Adjustment to Floods. Res. Pap. No. 29*. Chicago: University of Chicago, Department of Geography.

Wilden, Anthony. 1972. *System and Structure: Essays in Communication and Exchange*. Second edition. London: Routledge.

Williams, R. 1977. *Keywords*. London: Oxford University Press.

Wisner, B. Jr. 1977. *The Human Ecology of Drought in Eastern Kenya*. PhD thesis. Clark University, Worcester, MA.

Wisner, B., Blaikie, P., Cannon, T. and Davis, I. 2004. *At Risk: Natural Hazards, People's Vulnerability and Disasters*. London and New York: Routledge.

Wolf, Eric Robert. 1966. *Peasants*. Vol. 10. Englewood Cliffs: Prentice-Hall.

Wolfe, C. 2000. In search of posthumanist theory. In: Rasch, William and Wolfe, Cary (eds.), *Observing Complexity: Systems Theory and Postmodernity*. Minneapolis: University of Minnesota Press, pp. 163–198.

Wolfe, Cary. 2000 [1994]. Making contingency safe for liberalism: the pragmatics of epistemology in Rorty and Luhmann. *New German Critique*, 61: 101–128 (reprinted in Rasch, William and Wolfe, Cary (eds.), *Observing Complexity: Systems Theory and Postmodernity*. Minneapolis: University of Minnesota Press, 2000, pp. 247–272).

World Bank. 2014. *Risk and Opportunity: Managing Risk for Development*. Wahsington, DC: World Bank.

World Economic Forum (WEF). 2014. *Global Risks*. Davos; World Economic Forum.

WRI. 2008. *The Roots of Resilience*. Washington, DC: The World Resources Institute.

Zimmerer, K. 1996. Ecology as cornerstone and chimera in human geography. In: Earle, C., Mathewson, K. and Kenzer, M. (eds.), *Concepts in Human Geography*. Lanham: Rowman & Littlefield, pp. 161–188.

Zolli, A. 2012. *Resilience: Why Things Bounce Back*. Boston: Free Press.

PART II

Origins, trajectories, and futures

The chapters in this section are *reflective chapters on the field of political ecology as a whole*: its origins, the reasons it has evolved in the particular ways it has, and what in the future it might become.

The first chapter, by Ben Wisner, one of the founders of the field and a continuous contributor since, offers a first-hand account of the formation of political ecology. Explicitly weaving together the personal and the professional, the academic and the activist, the chapter situates some of political ecology's key aspects and commitments at a pivotal political moment, yet also makes an argument regarding their continued salience for those newly drawn to the field under what may at first glance appear different circumstances.

The second chapter, by Enrique Leff, and the third, by Denis Gautier and Christian Kull, provide invaluable perspectives on Anglophone political ecology by looking at it from "outside" in a sense. Inasmuch as contemporary political ecology as represented in this volume is overwhelmingly an academic undertaking carried out by professionals in Anglophone universities and research traditions, it is necessarily shaped and limited (if also enabled) in many ways by those contexts: what questions it asks and where and how it asks them; where and in what language its products appear; what other groups outside of those institutions it engages with, and more. These two chapters on "political ecology," by important contributors to the field who are rooted in and work through other linguistic, national, and regional research traditions, help to make visible some of the situated aspects of the field that are so familiar to many practitioners as to often be invisible or unquestioned. Likewise, they make explicit the potential of the growing but still nascent trend towards productive interchanges between Anglophone political ecology and political ecology research traditions in other regions and languages: the growing interest in and explicit networks organized around "political ecology" in the EU and Latin America, for example.

Finally, the chapters by Paul Robbins and Bruce Braun raise linked questions regarding the "essential nature" and potential futures of political ecology. Specifically, they probe the relationship of political ecology to other fields, asking whether political ecology is an approach or undertaking that is *necessarily* oppositional, alternative, or one of critique, and if so, what it means for it to become widely accepted and institutionalized on the one hand, or what it might look like if turned more in the direction of constructivist experiments, rather than critique of what already exists, on the other.

3

SPEAKING TRUTH TO POWER

A personal account of activist political ecology

Ben Wisner

Introduction

This is a personal account of the development of what might be called 'radical geography', or 'activist' or 'engaged' political ecology (Akatiff, 2007; Wisner, 2012). It is personal in the simple epistemological sense that it's one person's experience and understanding. It's also personal in the somewhat more interesting sense that this 'take' on political ecology does not belong wholly to the streams that became rivers of research and advocacy within geography (Robbins, 2012; Castree, 2014), anthropology (Escobar, 2008; Goldman *et al.*, 2011) and development studies (Forsyth, 2003; Blaikie and Brookfield, 1987). Having dived in the deep end with two degrees in philosophy and paddled my way more or less blind (that is, without much theory or other flotation devises from these disciplines), I've ended up (washed up) on a shore whose name I do not know. I think of it as right livelihood or good work. I've been fortunate to have loved what I do, enjoying Schiller's dialectic of work and play. Dr Ack has his banjo. Jim Blaut sang calypso. I joke in Swahili with elders as we compare walking sticks and use them to 'map' the movement of livestock in the semi-circle of dirt between our three-legged stools. Maybe political ecology will continue to grow and eventually understand it has to do performance, with comedy and tragedy, and the aspirant will spend time in Clown School (www.clownswithoutborders.org/).

Looking and listening deeply

Looking back to 1959, when I worked for six weeks picking apricots and lived in a migrant labour camp in Hollister, California, I did more than stand every morning at 5 am in a line as farmers chose us to climb into their pick-up trucks. I learned to enjoy Mexican food and company. I absorbed with this food all the elements of what would become my political ecology: history, property, work and low pay, the company store, mobility and the Other; the land, irrigation water, the single golden fruit whose size and shape may be perfect today, these years later, for my grandson's hand, weighing down painfully on my shoulders then in their commodity form: 15 US cents per bucket hanging on my chest. I did not ask myself if this system was 'sustainable', and I didn't have words for 'exploitation' or 'surplus value'; I didn't even know about pesticides. Rachel Carson's book, *Silent Spring*, was not to be published for another five years. But as Thich Nhat Hanh would write, in the apricot there are non-apricot elements. The basis was laid down. It would be years before I understood these non-apricot elements – genocide against native people, water theft, land degradation, oppression and exploitation of migrant labour, global markets, union fight back and sometimes transformation (Jones, 2013; Pelling, 2010).

Fast forward to 2009, and I was two years into what has turned by this writing into a seven-year engagement with people in several Tanzanian villages and a group of schools that serve the villages. Here, too, as the reader will see below, I was made aware of many of the same processes that keep poor people in 'their place' in unjust societies, constructed risk, and literally push marginal people into marginal places. I was also reminded that oppression and injustice provoke resistance and that 'weapons of the weak' (Scott, 1987) and local knowledge can combine with specialist knowledge and result in a 'pedagogy' of both researcher and the oppressed (Freire, 1970).

The 'revolutionary' 1960s

War, huh, yeah
What is it good for
Absolutely nothing
Uh-huh
War, huh, yeah
What is it good for
Absolutely nothing
Say it again, y'all

War, huh, good God
What is it good for
Absolutely nothing
Listen to me.[1]

I was writing a master's thesis on the justifiability of civil disobedience at the University of Chicago when I first heard Thich Nhat Hanh speak. This was in March 1966. He was the founder of a group of more than 20,000 Vietnamese youth who cared for the dead and wounded, and rebuilt schools and clinics on both sides of that tragic war. Martin Luther King was to nominate him for the Nobel Peace Prize. I did not know at the time the important role this first of three teachers would have in my practice of political ecology.

In the protest march that followed Thay's speech[2] by a few days I saw a sign carried by four teenaged African American women: 'The Vietnamese never called me a nigger'. Aged 20-something, I inhabited a moral universe that I believed 'bends toward justice', in MLK's words.[3] The civil rights, anti-war movements and struggles within higher education over free speech framed and influenced my academic work. These, and also the environmental movement were behind the creation of *Antipode* when I was resident at Clark University from 1968–70 (Wisner, 2012). My mind was opened to non-violent civil disobedience and, increasingly, to Marxism. *Silent Spring* (Carson, 1964) met a very noisy spring in the streets of Paris in 1968, and also on campuses across the United States. Having been arrested for blocking the army recruitment centre in Worcester, MA, home of Clark University, I recall a memorable 'seminar' involving some Clark professors and a large number of geography grad students during our time together in a holding cell at the county jail.

I had lived in a Tanzanian 'African socialist' (*ujamaa*) village from 1966 to 1968, when I rejoined anti-war activists in the United States. The second of the great teachers who nourished the seeds of my off-beat political ecology was also called 'teacher'. He was *Mwalimu* (Swahili for 'teacher') Julius Nyerere, the first president of Tanzania. *Ujamaa* and self-reliance were core elements in his philosophy, honed while Nyerere studied at Edinburgh University, where he translated Shakespeare's *Julius Caesar* and Orwell's *Animal Farm* into Swahili. Daily life, villagers' confrontations with officialdom, environmental challenges and conflicts in the village taught me a great deal about 'habitat, economy and society' (Forde, 2010 [1934]). Although my BA had been at the great agricultural school, University of California at Davis, I did not know a pigeon pea from my elbow when I graduated in philosophy. Mbambara village was my finishing school, or, better, my kindergarten, where hands-on projects (construction of a dam, a bridge and large cooperative grain storage structure), the friendship and patience of the *wanakijiki* (villagers) made philosophy concrete and pushed me further in the direction of political ecology (Wisner *et al.*, 1975).

Back to Africa

I returned to East Africa in 1970, when my cohort of field workers was documenting the Biafra War in Nigeria, Sahel famine, and deadly cyclone in Bangladesh and hurricane-triggered mudslide flooding in Honduras (Blaikie *et al.*, 1994; Wisner *et al.*, 2004). Trained in the tradition of human ecology that led from Harlan Barrows to Gilbert White and his students Ian Burton and Robert Kates (Burton *et al.*, 1993), I began to question the purchase it gave on such events. I toyed with re-naming my work 'inhumane' ecology as a better way of focussing on power relations that trapped farmers in vicious cycles (Chambers, 1983; Blaikie and Brookfield, 1987; Wisner, 1988; Wisner *et al.*, 2012a; Gaillard *et al.*, 2014).

I had two concerns about the White–Burton–Kates approach. First of all, their macro analysis involved pairing of 'developed' and 'developing' country examples in their major work, *The Environment as Hazard* (Burton *et al.*, 1978). For instance, Australia and Tanzania as examples of drought management seemed to contain an uncritical, assumed belief in 'stages' of modernisation. Other geographers had used the idea of modernisation and later written autocritiques (e.g. Soja, 1979).[4] I rejected modernisation approaches erasing what Rodney (1972)[5] called 'the development of underdevelopment' (see also Gunder-Frank, 1966). Second, the approach was ethno-centric, specifically Euro-centric (Blaut, 1993).

My PhD uncovered the long history of marginalisation rather than modernisation. The post-colonial Kenyan elite took possession of the colonial settler core of Kenya's arable highlands and expanded this core. The politically unconnected had no alternative but to move farther into an expanded, semi-arid periphery. Adding insult to injury, the new elite neglected the periphery when it came to infrastructure and services. My field work revealed the value of local knowledge and provided evidence that people could resist political, economic and social power that was a root cause of their vulnerability to harm from natural hazards. My specific focus was drought in eastern Kenya, where the operation of triple marginalisation – ecological, economic and political – was clear in my early 1970s data and still present in a 30-year retrospective study carried out with Tom Smucker (Smucker and Wisner, 2008).

Two years at Universidade Eduardo Mondlane in the early Marxist-Leninist days of independent Mozambique (1978–80) exposed me to ideas that deepened my materialist understanding of power across spatial and temporal scales. An example is the magnificent group effort to study the history and condition of Mozambicans who migrate to the South African mines. This was underway while I was at the university and was coordinated by Ruth First (1983). She was later assassinated by the South African secret police.

It was also in FRELIMO's[6] single party ruled Mozambique that I was exposed to the problems created by top-down planning. I was nearly thrown out of the country by the rector of the university for criticising top-down imposition of plans for communal villages in a one-size-fits-all manner despite the huge agro-ecological, topographic and cultural-linguistic differences in this large country (Wisner, 1984). I didn't want Mozambique to make the same mistakes Tanzania had during the final phase of expanding *ujamaa* village programme, and I couldn't keep my mouth shut.[7]

In the geography department at UEM, I learned about state controlled rural and regional planning from colleagues from East Germany, Bulgaria and the USSR. Under this influence I suggested in an early *Antipode* article that one could theorise a 'socialist human ecology' (Wisner, 1978: 84). Of course, I soon found out that one could 'theorise' all sorts of things, but putting them into practice is quite different. In Mozambique I learned that when theory (that is, policy based on ideology) is applied from the 'top down' with no space for 'bottom up' agency, voice and spontaneity by ordinary people, trouble is the result (Wisner, 2010; Scott, 1998).

The result of these experiences and teachings from 1959 through roughly 1980 was the development of a view of political ecology that I continue to hold. It is an applied – even proactive – interdisciplinary study of society and the earth that focuses on political, economic and social power relations (as well as violence – structural and overt – and coercion) up and down a continuum of scales from global to local (Wisner, 1993). Activist political ecology also includes practice and study of resistance to these power relations: efforts to 'bend' the universe in the direction of justice that in the more insightful literature about 'climate justice' these days is called 'transformation' (Pelling, 2010).

Activist political ecology

What is an activist scholar?

J.M. (Jim) Blaut, my friend and mentor, was the third of the great teachers that shaped my own style of political ecology. He encouraged me to put together macro and micro as well as inside (emic) and outside (etic) perspectives that have defined my work. He also modelled activism in science. Reflecting on his life and work (Wisner *et al.*, 2006: 1046), two colleagues and I distinguished among 'application', 'advocacy' and 'activism':

> 'Application' is well known to geographers and planners, who very often are engaged in '*applied research*'. Here one works on a concrete problem – tropical soil erosion, child malnutrition, affordable housing, etc. – with a particular constituency or certain practitioners in mind. In applied research that constituency or stakeholder is often a governmental or official entity …
>
> In *advocacy research* the relationship is more often than not with non-governmental, unofficial groups of people – often disenfranchised or marginalized by society at large. The advocate takes the side of a group whose need, complaint, or demand is clarified and given strength and voice through one's scholarship.
>
> … Both applied and advocacy research aim to change things. They are 'active' in that sense. However, they both take prevailing and dominant social, political, and

> economic relations for granted. They constitute the framework within which research is applied and the positions advocated. The *activist scholar* questions these dominant power relations. [Emphasis mine]

Activist political ecology in Tanzania: case study

Since 2007 I have been involved with a team of Tanzanians and outsiders conducting research supported by the US National Science Foundation.[8] The goal is to comprehend how farmers, herders and fishers understand climate change in the broader context of many changes they had experienced over the past 20–30 years. These changes, people told us, include political, economic, environmental, demographic, technological, administrative, legal and social ones. Our team used mixed methods including 18 months of field scoping and site selection during 2007–2008, a large N household survey in all the village sites in 2008, and continuing with age and gender exclusive focus groups, numerous key informant interviews and larger community discussions (Wangui *et al.*, 2012; Wisner *et al.*, 2012b; Smucker *et al.*, 2015).

This work covered four of Kilimanjaro Regions' five districts and a series of 18 village study sites. In addition, the study team defined an altitude and agro-ecosystem gradient that ran from mountain and ridge top villages in the North Pare Mountains contained within one of these districts (Mwanga) down through middle slope sites to our driest study villages near the Ruvu River and the Nyumba ya Mungu hydroelectric reservoir. The sub-village of Emangulai B was the very driest of our study sites, and it was there that we began conventionally but ended up in a dialogue about land grabbing.

Along with others in our team, I worked hard over five years to build relationships with the predominantly Maasai population of Emangulai B, a sub-village of Kirya village belonging to Mwanga District in Kilimanjaro Region of northeastern Tanzania. This was a slow process that involved showing respect for the traditional leadership (the *laibon*), providing employment for secondary school graduates who worked as enumerators, providing lifts to people (and becoming a de facto taxi service), and giving small gifts such as solar powered lamps and a mobile phone charger.

In my own case, admiration for the efforts of a local secondary school head teacher who had painstakingly 'greened' his semi-arid, wind-swept school compound turned to friendship and solidarity. My wife and I co-funded with school and District authorities construction of a water cistern, intake and pump at a reservoir where twice a day students had drawn water in buckets they carried to the school to water seedlings and trees. This parallel solidarity work, not formally or financially part of our NSF research, expanded into a network of three secondary schools that now call themselves the Tanzania Green School Network.

We also co-funded construction of a roof rain water catchment system for the local primary school in Emangulai B. This assistance might be considered merely normal ethical or good neighbourly behaviour by long term resident-researchers, or seen as possibly small-scale development work.

Whatever our motivations (conscious and unconscious) the end result was to build trust that later provided the basis for the researchers to be welcomed deeply enough into the world of the residents to see their situation from the 'inside'. This allowed me and others in the team to grasp the historical continuity of exploitation, discrimination and marginalisation the Maasai have suffered and to use our expertise to 'stand with' the residents. The material expression of solidarity was a series of maps co-produced by the team and the community that highlighted community prioritised major problems they wanted addressed by district civil servants and political officials.

Brainstorming with the Maasai: 'We didn't kill the elephants!'

From 2008 until 2013 the Maasai residents of Emangulai B were experimenting with irrigation farming although not abandoning their herding activities. Herder-farmer tension was growing. Outside investors were positioning themselves to grab land and water as changes in Tanzanian law eroded the decentralised authority over water and land use of the elected village council.

In this context, very frank discussions of the past, present and possible future of this Maasai community ensued. In June 2013, the team held day long brain storming meetings (catered with breakfast and lunch). On this occasion Maasai women, elders and young warriors (*morani*) spoke passionately about their lack of control over outsiders who enter their sub-village to extract sand and burn charcoal illegally. Participatory mapping of the journey that women took to find desirable species of wood were used by the women themselves to emphasise the problem.

Resentment turned out to be long standing and focused not only on outside pressures on what these residents considered to be their own natural resources, but also decision making within the village of which Emangulai B is a sub-village. There had been a slow motion tug of war going on between irrigators (in adjacent sub-villages of Emangulai A and Kirya) and the Maasai over access of cattle and other livestock to the Ruvu River. Several years earlier the veterinary department built a cattle dip to protect against tick-borne disease. Non-Maasai irrigators made off with the pump, and the dip was never used. Later the veterinary department paid to fence the large seasonal cattle watering pond to the East of Kirya centre, in order to control Maasai animal movements. The cement pillars were destroyed and the wire turned up as useful material in various home compounds. None of this had risen to the level of violent conflict, but the number of land disputes was steadily increasing.

Earlier I cited Thich Nhat Hanh's idea that when one looks deeply at an apricot, one sees non-apricot elements. Activist political ecology depends on such 'seeing' and 'listening'. A break-through in our work came in 2012 when one of our senior Tanzanian researchers, Professor Adolfo Mascarenhas, noticed a very large tree in the background of a photo he took of some Maasai children. He knew that this tree was very old and a species that grows near rivers in areas currently inhabited by Tanzania's mega fauna, including hippo, rhino, lion and elephant. He asked Maasai elders, who confirmed that 75 to 80 years ago the zone now known as Emangulai B sub-village did have these animals. The elders hastened to add, 'We did not kill the elephants'. Rather, they were killed by the colonial White hunters and then the settlers from outside who came to farm along the river.

Why would the Maasai elders say a thing like that to Professor Mascarenhas? It is because the Maasai know they are seen by the government as second-class citizens and blamed for all sorts of environmental degradation. Indeed, the current President of Tanzania is on record as having said that pastoralism has no place in modern Tanzania (Pearce, 2012: 256). Suddenly it was obvious why the Maasai of Emangulai B worried so much about their land and water rights. At this point the team had already been facilitating participatory mapping as support for community-identified and controlled mini-research projects. Our perspective on the situation had begun to shift, and the master narrative of the long-dead elephants seemed to align and consolidate insights that had been coming from these mini-research projects.

These mini-research projects identified as priorities by the community led the team and the villagers working with us to formulate a number of questions about the root causes of the problems mapped. These, in turn, led to the passionate brainstorming in the meeting described in June 2013 and the formulation of questions and requests that Maasai representatives brought to the district meetings held in July of that same year. Questions to emerge from the joint community-team GPS based mapping included the following:

- Mapping showed serious waterlogging problems. Why was the formal irrigation scheme constructed with no return drainage to the river? Why had there been no rehabilitation of the irrigation scheme a year after money had been provided to a contractor both for rehabilitation and construction of a new intake and canal? Who would get the irrigable plots once the new intake was finished?
- What could be done to increase the productivity of farms served by the subsidiary irrigation canal dug by Maasai and revealed by participatory mapping of informal and formal irrigation systems? Would there be enough water for its continued use once the upstream intake began to divert Ruvu River water and the old intake closed?
- GPS based maps showed Maasai women having to go farther from their homes to find the sort of wood they preferred for domestic purposes. Why was this? The answer was that outsiders with lorries from town regularly harvested wood, excavated sand, made and carried away charcoal with no payment to Emangulai B sub-village, to Kirya village or to Kirya ward.
- What could be done to provide protection from crocodiles at critical sites along the river? On the basis of participatory mapping and discussion of river side domestic water points, there emerged a design concept that was costed out and presented to the district meeting in July 2013 by sub-village representatives.

These participatory mapping exercises and the conversations that followed caused the research team to see the landscape, local and regional economy and governance from the Maasai point of view. They were being exploited, stonewalled by non-transparent governance, inhibited in their attempts fully to integrate into an irrigation based village economy sited where, ironically, the ancestors of these Maasai pre-dated settlers who came from up in the North Pare Mountains. Thus, as an act of solidarity science, we 'stood with' these Maasai residents in June as we helped them articulate specific questions and requests for presentation to district civil servants and politicians.

In July 2013, a month after the sub-village brainstorming meeting with the Maasai, a delegation of three Maasai community members joined representatives from four other study site villages/sub-villages to present their concerns, together with supporting maps, to district political leaders and civil service professionals. The representatives were articulate and forceful. Working groups were also set up during the second day of district meetings, composed of civil servants responsible for water, education, livestock and farming and various community representatives. They undertook a process of working out the details of proposals in these areas.

Conclusion

I came late to geography as a discipline. I had two degrees in philosophy and two years learning about 'life, the universe and everything'[9] in a Tanzanian socialist village. Robert Kates introduced me to geography by inviting me to study that Tanzanian village from the 'outside' as a way of complementing the understanding I had gained from living 'inside'. Following Kates to Clark University, I soon found myself uneasy with the technocratic and modernist biases I detected in the up-and-coming sub-discipline of 'natural hazards geography'. I pushed against this paradigm, while I remained happy enough with the master narrative of a society–nature dialectic. Vietnam war protest and living in the Tanzania had immunised me against the cultural imperialism of the 'quick fix'. Living and working in Mozambique was also a caution against top-down 'solutions' to problems. All these experiences bumped up against, challenged or confirmed what I read in an eclectic and almost random way.[10] In the first edition of *Antipode* I explored advocacy planning. Who was to know that eventually unpacking and digesting the learning I had from Thich Nhat Hahn, Julius Nyerere and Jim Blaut, that I would stumble along a path that led me to participatory mapping as a tool for political empowerment.

Stepping back from my personal trajectory and acutely aware that many readers are graduate students who are much younger than I am, and whose lives are very different from mine, I wonder what I can say about activist political ecology that will be of use to you, the 20-something or 30-something reader. What messages do I want you to take away from my personal journey to political ecology and my past and present activist use of its tools? You are living through a rather different historical moment as you pursue your graduate studies and embark on field research. Much has changed about the world, but much has remained constant. What does activist political ecology allow us, as scholars and activists, to do?

I think the continuities outweigh the changes despite technological change, rise of social media, economic globalisation, the rise of the BRICS as economic and political powers, unravelling of the USSR and Yugoslavia, corporate concentration, financial system instability, growth of inequality (Piketty, 2014), decline of the post-war welfare state (Seabrook, 2013), escalating environmental assault on the world's oceans, biosphere and atmosphere, and spiralling cost of education. I do not underestimate the last mentioned and the fact that with heavy college and grad school debt, you have more at stake in 'doing the proper, expected, conventional thing' than I did. Nevertheless, the United States was at war when I 'came into this movie' and it still is at war. Various groups of people in every country in the world are still marginalised by systems of political, economic and social power and by coercive threat of violence. The way such marginalised people make use of natural resources, rural space and the built environment in urban contexts is still affected strongly by the power relations under which they struggle to survive and to raise their families. They still resist, they learn and they share local knowledge with one another. Therefore, despite all the changes, the lessons I learned from Thich Nhat Hahn, Julius Nyerere and Jim Blaut are still relevant and useful in the application of political ecology.

The take-home messages are these:

1 People matter. The most important research question is not about things and physical processes (soil moisture, climate, biodiversity) but the relationship between the researcher and the people who affect and are affected by these things and processes. Who are your interlocutors? What commitment do you make to their well-being? Will your research benefit them?
2 Context matters. Anywhere you work, there will be organised violence, call it 'war', recovery from war or preparation for war or coercion. This is the historical context of any and all field research. Researchers have to be aware of how a history of violence and displacement colours contemporary attitudes toward place, livelihood and the state (Mascarenhas and Wisner, 2012). Indeed, even absent overt violence at a particular moment in a village or country, one needs to be aware that economic and social life (some call it 'development') is not harmonising as modernisation theory assumes, but conflictual (Wisner, 1988). People have objectively different material interests, and these have to be negotiated.
3 Local knowledge matters. Deep listening, co-learning and problem solving with local people is a powerful mode of action research. Local knowledge exists everywhere. While much of my work has been in the rural, global South, I have also helped create and studied urban gardening in a low income, Hispanic area of Chicopee, Massachusetts and used participatory video as a tool for people's self-study of hazards, vulnerabilities and capacities in various parts of Mexico City and greater Los Angeles.

Some of the changes in the world make it easier to work in this manner. Increasingly the poor and marginalised are networked and organised, and most use social media. Consider, for example, international networks such as Via Campesina (http://viacampesina.org/en/) that connects many rural people or the International Slum Dwellers' Association (www.sdinet.org/).

All this is to be taken with a pupusa and a cold beer, and maybe a grain of salt! I am not preaching but trying to communicate across generations. Ultimately everyone comes to political ecology along a different path and will use differently its toolbox (containing, among other things, Robbins's (2012) hatchet and seed).

Notes

1 Excerpts from 'War', an anti-Vietnam war song written by Norman Whitfield and sung by Edwin Starr on the Motown label in 1970; hear it at: www.youtube.com/watch?v=bX7V6FAoTLc.
2 'Thay' is a term of respect used for senior Vietnamese Buddhist monks and means 'teacher'.
3 'The arc of the moral universe is long, but it bends towards justice' (King, 1958).
4 The 1993 revised edition of *The Environment as Hazard* is hardly a revision at all, and certainly contains no autocritique.
5 Walter Rodney, a brilliant economic historian from Guyana, taught at University of Dar es Salaam while I was there teaching in the department of community health at the medical school, 1972–4. Rodney was inspirational. Later he was assassinated when he returned and became politically active in Guyana.
6 Front for the Liberation of Mozambique.
7 From 1979 villagisation in Tanzania became compulsory. See Collier *et al.* (1986), Scott (1998) and Coulson (2013).
8 Local Knowledge and Climate Change Adaptation Project (LKCCAP), supported by US-NSF Grant No. 0921952. Any opinions, findings, and conclusions or recommendations expressed in chapter are those of the author and do not necessarily reflect the views of the National Science Foundation.
9 The famous phrase of Douglas Adams in his *Hitchhiker's Guide to the Galaxy* (1979).
10 I really do mean random. I stored books under the rafters that held up a grass roof over my mud and waddle house in the Tanzanian village. Over two years I read what the termites didn't eat.

References

Adams, D. 1979. *Hitchhiker's Guide to the Galaxy*. New York: Pan Books.

Akatiff, C. 2007. Then like now: The roots of radical geography. *Antipode Foundation*. Available at: http://antipodefoundation.org/2012/09/04/then-like-now-the-roots-of-radical-geography-a-personal-account/.

Blaikie, P. and Brookfield, H. 1987. *Land Degradation and Society*. London: Heinemann.

Blaikie, P., Cannon, T., Davis, I. and Wisner, B. 1994. *At Risk: Natural Hazards, People's Vulnerability and Disaster*. London: Routledge.

Blaut, J.M. 1993. *The Colonizer's Model of the World: Geographical Diffusionism and Eurocentric History*. New York: Guilford.

Burton, I., Kates, R. and White, G. 1978. *The Environment as Hazard*. New York: Oxford.

Burton, I., Kates, R. and White, G. 1993. *The Environment as Hazard*. 2nd edition. New York: Guilford.

Carson, R. 1964. *Silent Spring*. New York: Fawcett World Library.

Castree, N. 2014. *Making Sense of Nature*. New York: Routledge.

Chambers, R. 1983. *Rural Development: Putting the Last First*. London: Longman.

Collier, P., Radwan, R. and Wangwe, S. 1986. *Labour and Poverty in Rural Tanzania: Ujamaa and Rural Development in the United Republic of Tanzania*. Oxford: Clarendon Press.

Coulson, A. 2013. *Tanzania: A Political Economy*. 2nd edition. New York: Oxford.

Escobar, A. 2008. *Territories of Difference: Place, Movements, Life, Redes*. Durham, NC: Duke University Press.

First, R. 1983. *Black Gold: The Mozambican Miner, Proletarian and Peasant*. London: Palgrave.

Forde, D. 2010 [1934]. *Habitat, Economy and Society*. London: Routledge.

Forsyth, T. 2003. *Critical Political Ecology: The Politics of Environmental Science*. London: Routledge.

Freire, P. 1970. *Pedagogy of the Oppressed*. New York: Seabury Press [30th anniversary edition, available at: www.users.humboldt.edu/jwpowell/edreformFriere_pedagogy.pdf].

Gaillard, JC, Glantz, M., Kelman, I., Wisner, B., Delica-Willison, Z. and Keim, M. 2014. Taking the 'naturalness' out of natural disaster (again): An invited comment. *Natural Hazards Observer*, January. Available at: www.colorado.edu/hazards/o/archives/2014/jan14_observerweb.pdf.

Goldman, M., Nadasdy, P. and Turner, M. (eds). 2011. *Knowing Nature: Conversations at the Intersection of Political Ecology and Science Studies*. Chicago: University of Chicago Press.

Gunder-Frank, A. 1966. The development of underdevelopment. *Monthly Review*, 17 (September): 17–33.

Jones, L. 2013. Transformational change in climate policy: Triumph or travesty? London: Overseas Development Institute, 8 November. Available at: www.odi.org.uk/opinion/7991-transformational-change-ethiopia-climate-strategy.

King, M.L. 1958. Out of the long night. *The Gospel Messenger*, Official Organ of the Church of the Brethren, Elgin, IL: General Brotherhood Board, 8 February: 3 and 14. Available at: https://archive.org/details/gospelmessengerv107mors.

Mascarenhas, A. and Wisner, B. 2012. Politics: Power and disaster. In: B. Wisner, JC Gaillard and I. Kelman, eds. *The Routledge Handbook of Hazards and Disaster Risk Reduction*, pp. 48–60. London: Routledge.

Pearce, F. 2012. *The Landgrabbers: The New Fight Over Who Owns the Earth*. London: Transworld/Eden Project Books.

Pelling, M. 2010. *Climate Change Adaptation: From Resilience to Transformation*. London: Routledge.

Piketty, T. 2014. *Capital in the Twenty-first Century*. Cambridge, MA: Harvard University Press.

Robbins, P. 2012. *Political Ecology: A Critical Introduction*. 2nd edition. Malden: Wiley.

Rodney, W. 1972. *How Europe Underdeveloped Africa*. London: Bogle-L'Ouverture Publications.

Scott, J. 1987. *Weapons of the Weak: Everyday Forms of Peasant Resistance*. New Haven: Yale University Press.

Scott, J. 1998. *Seeing Like A State: How Certain Schemes to Improve the Human Condition have Failed*. New Haven: Yale University Press.

Seabrook, J. 2013. *Pauperland: Poverty and the Poor in Britain*. London: Hurst.

Smucker, T. and Wisner, B. 2008. Changing household responses to drought in Tharaka, Kenya: Persistence, change, and challenge, *Disasters*, 32(2): 190–215.

Smucker, T., Wisner, B., Mascarenhas, A., Munishi, P., Wangui, E., Sinha, G., Weiner, D., Bwenge, C. and Lovell, E. 2015. Differentiated livelihoods, local institutions, and the adaptation imperative: Assessing climate change adaptation policy in Tanzania. *Geoforum*, 59: 39–50.

Soja, E. 1979. Geography of modernization: A radical reinterpretation. In: R. Obudho and D. Taylor, eds. *The Spatial Structure of Development*, pp. 28–45. Boulder: Westview.

Wangui, E., Smucker, T., Wisner, B., Lovell, E., Mascarenhas, A., Solomon, M., Weiner, D., Munna, A., Sinha, C., Bwenge, H., Meena, P. and Munishi, P. 2012. Integrated development, risk management and community-based climate change adaptation in a mountain-plains system in Northern Tanzania. *Journal of Alpine Research*, 100(1). Available at: http://rga.revues.org/1701.

Wisner, B. 1978. Does radical geography lack an approach to environmental relations? *Antipode*, 10(1): 84–94.

Wisner, B. 1984. Eco-development and eco-farming in Mozambique. In: B. Glaeser, ed. *Ecodevelopment*, pp. 157–68. Oxford: Pergamon.

Wisner, B. 1988. *Power and Need in Africa*. London: Earthscan.

Wisner, B. 1993. Disaster vulnerability: Scale, power, and daily life. *GeoJournal*, 30(2): 127–40.

Wisner, B. 2006. J.M. Blaut. In: D. Simon, ed. *50 Great Thinkers on Development Studies*, pp. 40–5. London: Routledge.

Wisner, B. 2010. Climate change and cultural diversity. *International Social Science Journal*, 61: 131–40.

Wisner, B. 2012. Notes from underground: The beginning of *Antipode*. In: N. Theodore, S. Chari, V. Gidwani, K. McKittrick and J. Pickerill, eds. *Antipode: Past Editors' Reflections*. Available at: http://onlinelibrary.wiley.com/journal/10.1111/%28ISSN%291467-8330/homepage/editor_s_past_reflections.htm.

Wisner, B., Gaillard, JC and Kelman, I. 2012a. Framing disaster: Theories and stories seeking to understand hazards, vulnerability and risk. In: B. Wisner, JC Gaillard and I. Kelman, eds. *The Routledge Handbook of Hazards and Disaster Risk Reduction*, pp. 18–34. London: Routlege.

Wisner, B., Heiman, M. and Weiner, D. 2006. Afterword: Jim Blaut, scholar activist. *Antipode*, 37(5): 1045–50.

Wisner, B., Kasami, A. and Nuwagaba, M. 1975. Mbambara: The long load to *ujamaa*. In: L. Cliffe, ed. *Toward Rural Cooperation in Tanzania*, pp. 370–91. Dar es Salaam: Tanzania Publishing House.

Wisner, B., Blaikie, P., Cannon, T. and Davis, I. 2004. *At Risk: Natural Hazards, People's Vulnerability and Disasters*. 2nd edition. London: Routledge.

Wisner, B., Mascarehas, A., Bwenge, C., Smucker, T., Wargui, E., Weiner, D. and Munishi, P. 2012b. *Let Them Eat (Maize) Cake: Climate Change Discourse, Misinformation and Land Grabbing in Tanzania*. Contested Global Landscapes. A Multidisciplinary Initiative of the Cornell Institute for the Social Sciences. Ithaca, NY: Cornell University.

4

THE POWER-FULL DISTRIBUTION OF KNOWLEDGE IN POLITICAL ECOLOGY

A view from the South[1]

Enrique Leff

The critical epistemology of political ecology: the power in knowledge

The environmental crisis irrupted in contemporary history, in the 1960s and 1970s, as an expression of a crisis of civilization: a crisis of the hegemonic modes of understanding the world, of scientific knowledge and of techno-economic reason that had been institutionalized in the globalized world, stripping away the conditions for the sustainability of life. The construction of a sustainable world raises for political ecology an ontological and an epistemological question: it ponders the challenge of questioning the modes of thought, scientific paradigms, productive practices and social behavior that degrade life. The construction of a sustainable world implies the necessity of deconstructing[2] the power devices rooted in the hegemonic rationality of modernity – the *logocentrism* of science (Derrida 1982), the juridical norms (Foucault 1998), the modes of production and market logic (Marx 1965) – and erecting in their place theoretical and political strategies capable of conducting a process of social reconstruction in order with the ontological conditions of life. The environmental crisis calls us to think about something as yet unthought: the lack of knowledge and the unsustainability of life inadvertently produced by humankind.

Political ecology seeks to respond to the challenge posed by this civilizational crisis. As such, Anglophone political ecology has opened a critical space in the US and Anglo Saxon academy with the proposition of deconstructing theories that attempt to understand the relations between culture and nature – ecological geography and anthropology, political economy, agrarian and peasant studies – that have disregarded the epistemological causes of the environmental crisis: economic theory that drives economic decisions; evolutionary theory that normalizes human "adaptive" behavior. It also seeks to understand the social processes that affect, condition, determine and trigger the metabolism of ecosystems, global change – the entropic degradation of the planet and the unsustainability of life – that today afflict and challenge humanity, unraveling relations of power – and power in knowledge – that determine the modes of access, intervention, appropriation and degradation of nature.

Today, Anglophone political ecology is undergoing a self-critique, reflecting on its theoretical position in the terrain of science and politics regarding this global problematic. The construction of this domain of political ecology as a disciplinary field[3] of relations of power in knowledge (Foucault 1980) requires critical consideration of Anglophone literature, and in general the "Northern" understanding of the field of political ecology: its hegemonic position over this domain of knowledge, over the form in which it has shaped and woven its conceptual frameworks in the constitution of the emerging discipline – its schemes of intelligibility in theory, its research agendas, its political strategies in social activism and its incidence in public policy – in relation to other modes of understanding that emerge from other geographical latitudes and other economic, social, political and cultural contexts.

Today, diverse regional political ecologies are deployed across the entire planet, irrupting from the depths of the environmental crisis and from the roots of an ontology of difference, confronting the intention to seal them in a seamless unifying process of globalization. The focus on the geography of power in the field of political ecology is an invitation to engage a conversation with the purpose of enforcing discursive strategies to deconstruct the logics that drive the world toward socio-environmental degradation. In this perspective, political ecology unleashes the theoretical capacities and the sociological imagination and the social imaginaries that open ways toward sustainable futures in a diverse world. It is within this critical reflection that I will consider the geographies and the imaginaries of production of political ecological knowledge. More than a claim to reorder the puzzle of the environmental disciplines in the fields of social sciences, it is a call to sociological imagination to rethink the world from the perspective of the politics of knowledge in which political ecology is inscribed. As such, it opens a dialogue of knowledges that beyond any claim for universal truth, seeks the consistency of a mode of thought oriented toward the construction of a sustainable world, grounded in its geographical and cultural diversity.

In this chapter, I intend to discuss certain fundamental principles, ideas and proposals of political ecology from Latin America, contrasting these with the contributions of Anglophone political ecology. This endeavor will carry me beyond establishing a political socio-geography of environmental conflicts, to question the epistemological core of political ecology and stimulate a more cosmopolitan and critical reflection, which will strengthen the theoretical consistency and strategic efficiency in confronting the hegemonic powers that drive the world toward environmental degradation.

Origins and foundations of political ecology

Political ecology emerged as a new discipline in the terrain of social sciences in the 1960s–1970s, propelled by the environmental crisis. Political ecology opened an inquiry into socio-environmental conflicts generated by the capitalist appropriation of nature, strongly influenced by the Marxism that was much in vogue during that time of theoretical and cultural revolutions, with pioneering authors such as Murray Bookchin (1962), Eric Wolf (1972), Hans Magnus Enzensberger (1974) and André Gorz (1975). The field of political ecology has grown and spread to neighboring disciplines and fields, demarcating its positions from cultural ecology, ethno-ecological and geographical studies, overlapping with environmental sociology and ecological economics, expanding from political economy and fusing with post-development studies, eco-Marxism, social ecology and eco-feminism. Political ecology is forged in the melting pot of post-structural, deconstructionist, constructivist, post-colonial and after-nature studies (Escobar 2008). In a review of Anglophone political ecology, Peet and Watts (1993) saw its constitution from the perspective of a politicization of environmental sciences, cultural studies and agricultural practices. Political ecology was forged by the ecological critique of economic rationality (Gorz 1989) and the unravelling of the second contradiction of capital

(O'Connor 1998), in the critical margins of ecological economics (Martínez-Alier 1995). I will not attempt to synthesize a comprehensive mapping of its genealogy, its sources, its interdisciplinary core and relations with other disciplines, or its applications to diverse problematic processes and case studies. My intention is to think through the lines of demarcation of these emergent disciplines across regional political ecologies worldwide.

From its origins, the Anglophone school of political ecology emerged from a critique of the adaptationist theories derived from cultural ecology cultivated by authors such as Julian Steward (1972), Roy Rappaport (1968/1984, 1971), and Peter Vayda (1969, 1983), as well as other ecological paradigms – e.g. Bateson's ecology of the mind (Bateson 1972), the ecological systems of Holling (1973) and the socio-biology of Wilson (1975) – that constructed a biological, ecological and organic scheme for understanding society. From these organic views were derived the functionalist paradigms in sociology and anthropology, which assigned an adaptive response from the social order to diverse environmental problems, neglecting its political-epistemological character.

The problem of colonization of knowledge is not limited to masking the symbolic organization of cultures and obscuring of the relations of power through the naturalizing effects of biological theories in ecological anthropology, ecology, and cultural geography. This is also the problem of all interventions by anthropology and social sciences into the lifeworlds of traditional societies, with the intention of epistemological appropriation of their cultural organization through the concepts of science. This compels political ecologists to exercise a strategy of deconstructing knowledge, an epistemological vigilance and an ethic of otherness, in dialogue with, and through their interventions into, the cultural and environmental contexts with which they become involved in their scientific practices.

The disciplinary regionalization of knowledge of political ecology: *pater familias*

Under this critical deconstructionist approach to the theoretical and practical regionalization of political ecology, how can we understand theoretical differences, disciplinary domains and lines of demarcation in its discursive formations? If one of its objects of inquiry is the deterritorialization generated by capital, political ecology is constituted from a political geography of knowledge: of reason that gives meaning to ideas; of concepts and theories as devices of power. In this sense, political ecolog*ies* are regionalized from diverse perspectives and in different ways: from disciplinary schemes that converge in the field of political ecology, and from the various problematics triggered by the environmental crisis, manifested in socio-environmental conflicts in different regions of the planet, in different ecological and cultural contexts. The sensibilities, forms of reasoning and research practices that orient the theoretical interests of *Homo academicus* (Bourdieu 1984) lead to the organization of epistemic communities that drive political ecological thought through networks of alliances of intellectual and political actors.

This epistemological reflection about the field of political ecology opens up a research program for the sociology of knowledge, from which I can only outline some anchoring points. This domain has emerged from the influx of reasons and motivations that influenced authors from their academic formation in the fields of anthropology, geography, Marxism or political economy, building from theoretical lineages and affinities. Thus, before attempting to inscribe political ecology within an *episteme*, a paradigm or a theoretical scheme, it is possible to recognize its emergence within "disciplinary schools," academic niches, and epistemic communities, with research programs rooted in theoretical legacies, patriarchies and patrimonies of knowledge. Thus, the pioneer authors of Anglophone political ecology recognize the paternity of Carl Sauer's school of geography and of Julien Steward in cultural ecology.

In this way a political ecological framework was established, derived from Marxism, with diverse lines of argument in eco-Marxism, eco-anarchism and social ecology, emerging from the margins of economics and political economy. Together with this constellation of factors, it is possible to identify certain subjective conditions that have influenced the configuration of political ecology in the Anglophone academy. Here, the theoretical affinities and subjective sensibilities that result in some authors becoming theoretical referents, while others are ignored in spite of the similarity of their research – a question that is often elusive for the sociology of science – plays an important role. At play here are personal interests in recognition and repudiation by academic peers, theoretical seduction and disciplinary commitments that lead researchers to explore some areas and disdain others in the formation of their academic identity.

These considerations could perhaps explain the lack of dialogue of political ecology with related or neighboring fields, such as environmental sociology, that arose at the same time and are motivated by similar interests as political ecology.[4] Other examples include political ecology's distant relationship with ecological Marxism and ecological economics, the most consistent theoretical corpus from which it derives its philosophical sources of inspiration. The distance that Anglophone political ecology takes from authors such as Murray Bookchin, Barry Commoner and Michael Zimmerman is symptomatic of its self-enclosure, notwithstanding the close ties between political ecology and the eco-anarchism, social ecology and communitarianism of Bookchin (1982/1991, 1990/1996), with the critical analysis of Commoner's (1976) "poverty of power," and the essential conversation of radical ecology[5] with Heidegger's (1927/1962) existential ontology and postmodern thought (Zimmerman 1994). This lack of communication is symptomatic of the manner in which our theoretical preferences draw our gaze and imprint an identity on theoretical frameworks and discursive styles by which political ecology constructs its objects of study and strategies of action.

Anglophone political ecology has been constructed within an academic cloister insufficiently open to intercultural dialogue between regions: not only with other forms of knowledge and discursive practices, but with the protagonists of political ecology. If one of the core themes of political ecology is uneven access to resources among different populations, clearly there has been an asymmetry in the academic field in relation to access to the dissemination and distribution of political ecological thought. This has generated a concentration of ideas on the part of Anglophone authors that control and have privileged access to publications and the media of communication of the ideas at the global level. Certainly the barriers of language and difficulties of translation have limited opportunities for a more fluid dialogue and productive exchange between regional cultures of political ecology. This perhaps is most notorious in the communication with Latin American intellectuals and academics, and particularly in the case of authors of Latin American political ecology, who are less inclined to publish in the English language than their peers in Asia or Africa, where as a result of colonial history a majority of academics have been educated in Anglophone universities and are connected to its intellectual cultures. Nevertheless, the resonance of ideas does not depend solely on barriers of language and the control of the media of communication. Personal dispositions, intellectual cultures and theoretical affinities determine the interests and disinterests that promote or prevent academic interchange, the recognitions and rejections that generate the social environment in which a paradigmatic fence is built or an intercultural dialogue is opened between epistemic communities.

The politicization of ecology and the epistemic regionalization of political ecology

Before opening the interregional dialogue it is important to address the founding questions of political ecology: In what sense is ecology political; and what is the "regional" character of political ecology? Ecology – understood as the network of relationships of (non-human) populations with their environment, as the complex flows of matter, energy and information in the metabolism and organization of the biosphere, or as the relations of predation, trophic levels and ecosystem dynamics not driven by human activity – is not political in any sense. Ecology becomes political as an effect of human intervention in eco-*logical* transformations, which in this way cease to be governed by natural laws. Certainly, ecology understood as the Real that is enacted by human actions, presents itself as a diverse, heterogeneous and complex ontological order, where the effects of human intervention and diverse social rationalities become manifest in environmental change. Ecology becomes political as a result of the *will to power* that people exercise over nature, of the processes of appropriation guided by differentiated and often conflicting values and interests, and by the ways in which these are inscribed in rationalities that ascribe a meaning and an intensity to human transformative actions over nature. In this way, different strategies for appropriation of nature in different ecological contexts – be they cultural or capitalist – generate politicized ecological processes that are effects of power strategies. The political is the route by which the ontology of the Real becomes present in reality; political ecology is the transition from a global world ruled by the unifying power of the market to the construction of a diverse world. This process of politicization is constructed from an ontology of difference: by the deployment of the Real oriented by existential meanings and mobilized by cultural rights for the construction of diverse lifeworlds.

Political ecology is born of its detachment from two dominant theoretical paradigms of modernity: (1) economic theory that has configured the rationality of production that drives the modes of appropriation and transformation of nature that unleash the processes of entropic degradation; and (2) biological theory – in particular Darwinian evolutionary theory – that when transferred to the social sciences by way of cultural ecology, socio-biology and structural-functionalism became a normalizing social model. Political ecology was forged in the deconstruction of these two paradigms: against the "normality" of economics and political economy emerged eco-Marxism and ecological economics; political ecology positions itself in the margins, oriented toward the conflicts of ecological distribution that cannot be absorbed, adjudicated or resolved by ecological economics (Martínez-Alier 1995). Against the evolutionary theories, political ecology deconstructs the normalizing, ecologizing effects of social Darwinism on the ethnological order and anthropological practices (Watts 1983). From this line of demarcation, political ecology has sought to characterize and establish its identity in the context of scientific disciplines. Such an endeavor transcends the aim of designing a new specialty charged with unraveling the political character of ecological processes, or accounting for the political processes that leave their imprint in nature.

One of the keys to the deconstruction of the theories that dominate the field of environment – and political ecology – is the epistemological understanding of the concept of environment. While in economic and developmental theories the environment is thought of as an externality and cost, or the meaning of the environment of an organism and a cultural organization in the biological and ethnological sense, environmental epistemology thinks of the environment as the exteriority of the normal paradigms of science. The environment is the "other" of the *logocentrism* of science. Moreover, the concept of environment appears as productive potential, which opens perspectives for the construction of other possible worlds (Leff 2001). Thus, beyond policies intended to internalize environmental costs, it is possible to envision *other modes of production*, other ontologies and rationalities, and *other lifeworlds*, based on the ecological productivity and cultural creativity of the peoples of the earth (Leff 1986, 1995).

This epistemological concept of the environment allows us to understand the "regional" character of political ecology. Blaikie and Brookfield define their approach as "regional political ecology," in order to see "different geographical scales and hierarchies of socioeconomic organizations ... of environmental variability and the spatial variations in resilience and sensitivity of the land, as different demands are put on the land through time" (Blaikie and Brookfield 1987: 17). Thus, political ecology is regionalized by the impacts generated by the processes of techno-economic appropriation of nature in different geographical conditions, ecosystem resilience, and the resistance of affected social groups that together determine the levels of degradation from different land uses and demands on resources.

By contrast, the political ecology that emerges from the concept of the environment as conceived by an environmental rationality goes beyond the assessment of the complex social and ecological processes involved in land degradation to envision its potential for constructing alternative sustainable worlds: in rebuilding life-territories from the creativity of cultural knowledge and the ecological productivity of their environment. In this sense, geographical and anthropological conditions in political ecology acquire a more active ontological and political role, moving from social resistance to degradation processes, to the reconstruction of sustainable eco-cultural territories. Political ecology is not just focused on the economic asymmetries of a globalized world, the unequal distribution of economic benefits and environmental costs. Ecological distribution plays a positive role in the Southern regions, enacting the negentropic[6] productive potential of tropical ecosystems and the cultural creativity of their people.

The possibility of conceptualizing and putting into practice this productive ecological rationality emerges in the tropical ecosystems and in the ethnic territories of the South, where the planet's terrestrial ecological productivity is the highest and where creative cultural diversity is alive, embodied in the social imaginaries and the traditional practices that generated the ethno-ecological coevolution of their environments and the biocultural patrimony of their peoples (Boege 2008). The deconstruction of capitalist rationality from the ontological bases of environmental rationality – geographical, ecological, and cultural – is one of the main criteria that differentiates political ecology of the North from the political ecology of the South, and from Latin American environmental thought (Leff 2012).

In this view, political ecology does not constitute simply a field of research and social practices focused on socio-environmental conflict and the differentiated distribution of the costs and benefits of global change. Socio-environmental conflict drives new avenues to construct a sustainable world. If the deconstruction of the ecological principles that have colonized the social sciences – the critique of paradigms that have ecologized the cultural order and the agro-productive practices under the principle of biological adaptation – is one of the core orientations in the construction of Anglophone political ecology, then *environmental rationality* opens new theoretical strategies and guides social action for the cultural reappropriation of nature and the reconstruction of negentropic societies (Leff 2003, 2004, 2014a, 2014b).

Diversity and difference in the regional geography of political ecology

The Anglophone and the Latin American visions stem from similar understandings of the fundamental constitutive roots of political ecology and of the relations of power that organize its field. Both theoretical schemes are fed by shared theoretical traditions, which include Marx, Kropotkin, Thompson, Harvey, Deleuze and Foucault. They agree about the challenge of deconstructing the forms of power that subject the world, which degrade the biosphere, contaminate the environment and foreclose paths to ecological sustainability and cultural diversity. They see these effects in deforestation, soil erosion and the destruction of biodiversity in the Third World, in accumulation by dispossession and the deterritorialization of peoples, in the production of poverty and social inequality. The case studies in which political ecology has thrown its interest are located in the poor countries of the South (Blaikie and Brookfield 1987; Robbins 2012; Watts 1983). Authors such as Raymond Bryant (1992) have focused their interest in a Third World studies agenda and political ecology has oriented toward the differentiation of environmental movements in the North and South (Redclift 1987; Guha and Martínez-Alier 1997). In the context of the motivations of political ecologists from the North, political ecology of the South directs itself toward those processes that affect the socio-environmental conditions and movements that resist, defend and reconstruct their livelihoods and lifeworlds.

Nevertheless, beyond political ecology's interest in the asymmetries between North and South, in the socio-environmental impacts of the hegemonic power of globalization over the territories of the South; notwithstanding the variety of environmentalisms, of the theoretical sources, the disciplinary roots and the schools of thought that feed and shape the different frameworks and programs of political ecology, an unsolved question remains regarding the regional epistemological division in political ecology: is it possible to characterize theories, concepts and methods that permit us to establish regional typologies of environmental thought, delimit theoretical frameworks and methods of intervention in the field of political ecology? Since its inception, Anglophone political ecology has been concerned with the effects of power relations on environmental transformation (Bryant 1992); equally relevant is the inquiry of the effects of power relations on *knowledge* of environmental change.

Surely in one region or another, discursive strategies are configured from theoretical choices and conceptual frameworks to address environmental problems that shape different areas of analysis and drive social action. As such, Latin American political ecology is rooted in an emancipatory view, born of the colonization of knowledge, and is configured by a more direct and closer connection to the forms of knowledge and practices of the affected local peoples. Political ecology is rooted here in its proximity to the processes of ecological degradation and environmental conflict, existentially connected with its history, its cultures and its landscapes. This view emerges from indigenous peoples' knowledges, from the popular wisdom that expresses the "cry from the earth" (*grito de la tierra*) and the voice of Pachamama.

While Anglophone political ecology takes socio-environmental transformation of the Third World as critical themes and privileged objects of study, political ecology of the South is inserted in its processes of emancipation. If the former expresses itself discursively in its aim to deconstruct the dominant social rationality and to undertake case studies regarding the impacts of power on socio-environmental relations, the political ecology of the Third World is constructed as a discursive amalgam of academic and political actors, as a dialogue of knowledges between theoretical thought, participatory research and the social imaginaries of the people, in alliance with resistance movements and their political strategies for emancipation and reappropriation of their biocultural legacy (Guha 1989; Guha and Gagdil 1992; Arnold and Guha 1995; Shiva 1988; Escobar et al. 1998; Esteva and Prakash 1998; Acosta 2010; Quintero 2014). The political ecology of the Third World does not limit its purpose to the analysis of the processes of socio-environmental conflicts, or to the sociology of resistance movements. It has a theoretical and historical commitment to the construction of a sustainable future and of other possible worlds. Political ecology thus acquires a *strategic and prospective purpose*. In this sense, political ecology ceases to be just a new discipline or epistemic, interdisciplinary field. It understands the transition toward sustainability as a process of deconstructing the rationality of modernity and constructing in its place an environmental rationality in the confluence of cultural diversity and environmental complexity. Political ecology is the field where a dialogue of knowledges plays out – understood as the encountering of cultural beings constituted by their knowledge – the confrontation and alliances of diverse modes of being-in-the-world; in the re-encounter of nature and culture

(Leff 2001, 2004). The political ecology of the South is constructed through a dense discursive network of theoretical frameworks, social imaginaries and lifeworlds. It is the encounter between ontologies and rationalities, between modes of being-in-the-world and modes of appropriation of nature. It is the reinvention of identities and the movements of *rexistence*[7] of people with nature. In this perspective, the field of political ecology is moving toward the inquiry and practice of *political ontology* (Blaser 2009; Escobar 2013; Leff 2014a: chapter 3).

In the irruption of environmental complexity diverse epistemic communities have emerged, drawing from different theoretical sources and disciplinary identities that converge in the field of political ecology. Beyond comprising a new inter-scientific discipline, political ecology is forged as the field of forces where the dialectics of socio-environmental conflicts is expressed in a diversity of geographic, cultural and political contexts. However, the different modes of intelligibility and approaches to socio-environmental processes are not generated simply as a reflection in theory of different geographical and cultural conditions. The intellectual and academic environment has played an important role in constructing different analytical perspectives and theoretical developments in regional approaches to political ecology. Thus, in Latin America, the research of Eric Wolf and Angel Palerm (1972) regarding the ecological potential of agricultural systems in Mesoamerica, the analysis of John Murra (1956) regarding the organization of geographical space and vertical ecological zones[8] of the original people of Tawantiunsuyu, the *Geography of Hunger* of Josué de Castro (1975) and the studies of the *indigenato* of Darcy Ribeiro (1973), opened new avenues of anthropological, ethnographic, geographical and agrarian analysis tied to the knowledges and practices of traditional cultures in the territories of the South. The theories of dependency and internal colonialism (González Casanova 1965; Stavenhagen 1965), ecological liberation theory (Boff 1996), of decolonization and of the ethics of liberation (Dussel 1998; Quijano 2000; Lander 2000; Mignolo 2000, 2011; Mignolo and Escobar 2009), and the theories of agro-ecological practices, are inscribed in the call for a knowledge from/of the South (Sousa Santos 2008), and the emergence of a *Latin American environmental thought* (Leff 2012), that from the perspective of an ecology of difference and territorial conflict, puts its stamp on the political ecology of Latin America (Leff 2003, 2014a, 2014b).

Regional political ecologies are more a mosaic and a map of the diversity of foci regarding the relations of power that are exercised in different socio-ecological contexts and processes, from the forms of analysis, the tactics of negotiation and the strategies for resolving environmental conflicts. The field of political ecology is the meeting place of rationalities, logics of meaning and political practices in which a *will to power* is manifest in diverse and frequently opposing strategies in the struggle over life territories. It is the space of collision and resistance of the territorializing processes of hegemonic geopolitics of sustainable development (Leff 2002) that invade and foreclose possible alternative strategies for constructing sustainable worlds. In the field of political ecology socio-environmental movements unfold to resist the degradation of their livelihoods, reinventing collective identities and designing alternative strategies to reappropriate the biocultural heritage of the peoples of the Earth in order to construct a sustainable future.

Political ecology of the North and the winds from the South

Political ecology emerges from, and is manifested within, the rural problematic, the agrarian question and rural studies. It is within this space that environmental problems are radicalized as territorial conflicts, where disputes over modes and rights to appropriation of nature are played out. Surely today these processes are expanding to the oceans, the atmosphere and the cities. But it is in rural areas where these conflicts are the sharpest, and where possibilities open for constructing new, negentropic ways of (re)producing life. If, during the twentieth century, the countryside was the scene of agrarian revolutions and the struggle over land, the rural sphere is today the space where the processes of reappropriation and reconstruction of territory unfold. Beyond the struggles for land and the management of agrarian systems – of the division of lands as an economic development policy for agrarian subsistence and traditional ways of life – political ecology is the field in which the struggles over deterritorialization and re-territorialization are deployed. As such, political ecology has come to revolutionize studies of the traditional agrarian question of Lenin, Kautsky and Chayanov, proposing the reconstruction of the rural from new ontological bases. Struggles over agro-ecology, for cultural autonomy, and over territorial rights occupy a central place in Latin American political ecology. Within these practices arise the possibilities of constructing local sustainable economies founded on the ecological productivity and cultural creativity of peoples, in order to transition toward a sustainable global economy founded on the negentropic potential of the ecological organization of the biosphere (Leff 1995).

Political ecology emerges in the field of *externalities* of ecological economics. Ecological distribution refers to the unequal division of environmental costs and ecological potential, to these "ecological externalities" which are incommensurable with the values of the market, that appear as new entities to be internalized through economic instruments and ecological norms. These environmental costs generate social movements in response to ecological damage, and struggles for the social appropriation of nature. In this sense, ecological distribution refers to the power struggles that intercede in the social strategies for survival, to alternatives for sustainable production, and to the struggles for the social appropriation of nature. Ecological distribution focuses attention on the ways in which the imposition of economic rationality and the will to colonial domination have deterritorialized cultures as an effect of the modes of appropriation of nature according to the expansive requirements of capitalism and the geopolitics of sustainable development: modifying the climate, deforesting the biosphere and eroding biodiverse territories, over-exploiting sources of water and aquifers, exhausting subsurface resources. As a result, through unequal exchange, there is a transfer of resources from the poor countries of the South, rich in natural resources, to countries of the North, degrading the ecological potential of the territories of the South at an unprecedented scale (Houtart 2010).

If in the era of primitive accumulation capital expanded by colonizing the minds of indigenous peoples by the actions of missionaries in order to subject their consciences to the designs of their gods and kings, if they were dehumanized and dispossessed of their forms of knowledge in order to convert them into a pure labor force for the production of economic value, today peoples and their lands are deterritorialized in order to make them functional to the logic and geopolitical interests of "sustainable development." In reaction, people exercise their right to be, to reposition themselves in the world in the face of global change destined for the logic of the market and the rationality of modernity, through other social rationalities, constituted from their cultural identities, in the conflictive encounter and solidarity of their knowledge, their practices and their imaginaries, reopening the destinies of humanity toward a sustainable future, for the political repositioning of cultural being in the reconstruction of their lifeworlds.

Latin American political ecology is born of this perspective, striving to delink from global economic domination, demarcating itself from European and Anglo-American geography, anthropology and sociology. Latin American environmental thought drinks selectively from the fountains of sociological theory and philosophical thought, in order to attract critical thought to the life territories that are reborn from the roots of cultural diversity; to embed and embody the ontology of diversity in a politics of difference and an ethics of otherness; to hybridize "universal" thought with the thought that springs from other geographies – from the ecological and cultural conditions of Latin America and the South – with its political theories and traditional knowledges. Thus, a process of deconstruction, hybridization and reterritorialization of knowledge occurs, in the construction of other perspectives to construct a sustainable and diverse world order that emerges from the political ecology of the South (Leff 2014b).

Notes

1 Translated from the Spanish and edited by Tom Perreault.

2 *Deconstruction* designates the critical purpose of the philosophy of postmodernity in uncovering the conceptual framework and the social effects of metaphysical thought. The deconstructionist focus in the field of political ecology disentangles the ways in which the history of metaphysical thought has shaped the theoretical frameworks and paradigms that, legitimated by the dominant rules of the institutions of science, have generated hegemonic modes of understanding the world that determine the practices of intervention in nature. In this sense, theoretical deconstruction in the field of political ecology unravels and reveals the strategies of power in knowledge (Foucault 1980) that are interwoven in the logic and rhetoric of the discursive formations and the scientific disciplines against which political ecology seeks to demarcate itself, in order to analyze the power relations that cut across the field of socio-environmental processes.

3 Cf. Leff 2014a for my affinities and differences in the use of the concept of "field" (Bourdieu and Wacquant 1992) and of "strategic action field" (Fligstein and McAdam 2012), to comprehend both the power relations in the epistemic field of political ecology and the more general space of socio-environmental conflicts and struggles, as well as in relation to the practices and dispositions of social actors.

4 The dialogue between political ecology and the sociology of knowledge is fundamental to the establishment of differences, convergences and affinities in the approach to subjects of study, between realist and constructivist frameworks and causes (Leff 2014a).

5 Political ecology establishes its difference with deep and radical ecology not only in its epistemological scope of analysis of social conflict and power strategies that cross the field of ecological distribution, social inequality and sustainability, but also in its ontological and ethical commitments for emancipation from that conflict. While social ecology and ecofeminism search for emancipation in the suppression of dualities that support oppression, and in liberating the oppressed potentialities subjugated by such dualisms in patriarchal and modern social structures of domination, political ecology takes its stands in an ontology of difference between the Real and the Symbolic, being and entity, sexual difference, that unfold in hierarchical dualisms and socio-environmental conflict. Political ecology sees emancipation not so much as elimination of ontological differences, but in constructing an environmental rationality that can embrace them. Emancipation is not transcendence through ontological dialectics or subjective intentionality; it is not the reflexive restoration of modernity. Rather, emancipation is re-identification, the repositioning of being-in-the-world in the ontology of life (Leff 2004, 2014a, 2014b).

6 I adopt the concept of "negentropy" from Schrödinger (1944), as the original principle of life, the transformation of solar energy into biochemical energy through photosynthesis. Political ecology faces the challenge of extending this thermodynamic process toward an understanding of the social order founded in the immanence of life, in the ecological productivity of the biosphere and in cultural innovative practices under the thermodynamic, ecological and symbolic conditions of life in the planet.

7 *Rexistence* is the metaphorical concept invented by Carlos Walter Porto Gonçalves and adopted by this author that best expresses the ontological-epistemological-political turn from the movements of resistance to colonialism and to the impacts of the global economy on deterritorialization of original/traditional/local cultures, to the reconstruction of their livelihoods and world-lives rooted in the reinvention of their identities, in their cultural modes of existence and their social imaginaries for the sustainability of life (Porto Gonçalves and Leff 2014; Leff 2014a, 2014b).

8 Otherwise called "vertical archipelagos," a structure of exchange and access to the altitudinally separated resource zones (*pisos ecológicos*) of the Andes that were taken as fundamental to Andean civilizations.

References

Acosta, A. (2010), "El Buen Vivir en el camino del post-desarrollo. Una lectura desde la Constitución de Montecristi," Fundación Friedrich Ebert, policy paper 9, Quito.

Arnold, D. and Guha, R. (1995), *Nature, culture, imperialism: essays on the environmental history of South Asia*, Delhi: Oxford University Press.

Bateson, G. (1972), *Steps toward an ecology of the mind*, New York: Ballantine.

Blaikie, P.M. and H.C. Brookfield (eds.) (1987), *Land degradation and society*, London: Methuen.

Blaser, M. (2009), "Political ontology," *Cultural Studies*, 23 (5): 873–896.

Boege, E. (2008), "El patrimonio biocultural de los pueblos indígenas de México. Hacia la conservación *in situ* de la biodiversidad y agrobiodiversidad en los territorios indígenas," Mexico: INAH/Comisión Nacional para el Desarrollo de los Pueblos Indígenas.

Boff, L. (1996), *Ecologia, mundialização, espiritualidade*, São Paulo: Ática.

Bookchin, M. (1962), *Our synthetic environment*, New York: Knopf.

Bookchin, M. (1982/1991), *The ecology of freedom. The emergence and dissolution of hierarchy*, Montreal and New York: Black Rose Books.

Bookchin, M. (1990/1996), *The philosophy of social ecology. Essays on dialectical naturalism*, Montreal and New York: Black Rose Books.

Bourdieu, P. (1984), *Homo academicus*, Paris: Les Éditions de Minuit.

Bourdieu, P. and L. Wacquant (1992), *An invitation to reflexive sociology*, Chicago: Chicago University Press.

Bryant, R. (1992), "Political ecology: an emerging research agenda in Third-World studies," *Political Geography*, 11: 12–36.

Commoner, B. (1976), *The poverty of power*, New York: Alfred A. Knopf.

De Castro, J. (1975), *The geopolitics of hunger*, New York: Monthly Review Press.

Derrida, J. (1982), *Margins of philosophy*, Chicago: Chicago University Press.

Dussel, E. (1998), *Ética de la liberación en la edad de la globalización y de la exclusión*, Madrid: Trotta.

Enzensberger, H.M. (1974), "A critique of political ecology," *New Left Review*, 84: 3–31.

Escobar, A. (2008), *Territories of difference: place, movements, life, redes*, Durham and London: Duke University Press.

Escobar, A. (2013), "Territorios de diferencia: la ontología política de los derechos al territorio," Segundo Taller Internacional SOGIP, "Los pueblos indígenas y sus derechos a la tierra: política graria y usos, conservación, e industrias extractivas," available at: www.sogip.ehess.fr.

Escobar, A., L. Grueso and C. Rosero (1998), "The process of black community organizing in the Pacific coast of Colombia," in Álvarez, S., E. Dagnino and A. Escobar (eds.) *Cultures of politics/politics of cultures: revisioning Latin American social movements*, Boulder: Westview Press, pp. 196–219.

Esteva, G. and M.S. Prakash (1998), *Grassroots post-modernism: remaking the soil of cultures*, London: Zed Books.

Fligstein, N. and D. McAdam (2012), *A theory of fields*, Oxford: Oxford University Press.

Foucault, M. (1980), *Power/knowledge*, New York: Pantheon.

Foucault, M. (1998), *La verdad y las formas jurídicas*, Barcelona: Gedisa.

González Casanova, P. (1965), "Internal colonialism and national development," *Studies in Comparative International Development*, 1 (4): 27–37.

Gorz, A. (1975), *Écologie et politique*, Paris: Seuil.

Gorz, A. (1989), *Critique of economic reason*, London and New York: Verso.

Guha, R. (1989), *The unquiet woods: ecological change and peasant resistance in the Himalaya*, Delhi: Oxford University Press.

Guha, R. and M. Gagdil (1992), *This fissured land: an ecological history of India*, Delhi: Oxford University Press.

Guha, R. and J. Martínez-Alier (1997), *Varieties of environmentalism. Essays North and South*, London: Earthscan.

Heidegger, M. (1927/1962), *Being and time*, trans. John Macquarrie and Edward Robinson, San Francisco: HarperCollins.

Holling, C.S. (1973), "Resilience and stability of ecological systems," *An. Rev. Ecol. System.* 4: 1–23.

Houtart, F. (2010), *A agroenergia. Solução para o clima ou saída da crise para o capital?*, Petrópolis: Vozes.

Lander, E. (ed.) (2000), *La colonialidad del saber: eurocentrismo y ciencias sociales. Perspectivas latinoamericanas*, Buenos Aires: Clacso/Unesco.

Leff, E. (1986), "Eco-technological productivity: a conceptual basis for the integrated management of natural resources," *Social Science Information*, 25 (3): 681–702.

Leff, E. (1995), *Green production. Towards an environmental rationality*, New York: Guilford.

Leff, E. (2000), "Pensar la complejidad ambiental", in Leff, E. (Coord.), *La complejidad ambiental*, México: Siglo XXI Editores.
Leff, E. (2001), *Epistemología ambiental*, Sao Paulo: Cortez Editora.
Leff, E. (2002), "La geopolítica de la biodiversidad y del desarrollo sustentable: economización del mundo, racionalidad ambiental y reapropiación social de la naturaleza," in Ceceña, A.E. and E. Sader (eds.) *La guerra infinita: hegemonía y terror mundial*, Buenos Aires: CLACSO-ASDI, pp. 191–216.
Leff, E. (2003), "La ecología política en América Latina: un campo en construcción," *Polis*, Revista de la Universidad Bolivariana, II (5) 2003, Santiago de Chile, pp. 125–145.
Leff, E. (2004), *Racionalidad ambiental. La reapropiación social de la naturaleza*, Mexico: Siglo XXI Editores.
Leff, E. (2012), "Latin American environmental thinking: a heritage of knowledge for sustainability," *Environmental Ethics*, 34 (4): 431–450.
Leff, E. (2014a), *La apuesta por la vida: imaginarios sociales e imaginación sociológica en los territorios del Sur*, Mexico: Siglo XXI Editores.
Leff, E. (2014b), "Political ecology. A Latin American perspective," mimeo.
Martínez-Alier, J. (1995), "Political ecology, distributional conflicts and economic incommensurability," *New Left Review*, I/211.
Marx, C. (1965), *Œuvres, Économie I*, Paris: Gallimard.
Mignolo, W. (2000), *Local histories/global designs: coloniality, subaltern knowledges, and border thinking*, Princeton: Princeton University Press.
Mignolo, W. (2011), *Modernity and decoloniality*, Oxford bibliographies.
Mignolo, W. and A. Escobar (eds.) (2009), *Globalization and the decolonial option*, London: Routledge.
Murra, J. (1956), *The economic organization of the Inca state*, Chicago: University of Chicago.
O'Connor, J. (1998), *Natural causes. Essays in ecological Marxism*, New York: Guilford.
Peet, R. and M. Watts (1993), "Introduction: development theory and environment in an age of market triumphalism," *Economic Geography*, 69 (3): 227–253.
Porto-Gonçalves, C.W. and E. Leff (2014), "Political ecology in Latin America: the social re-appropriation of nature, the re-invention of territories and the construction of an environmental rationality," mimeo.
Quijano, A. (2000), "Colonialidad del poder, eurocentrismo y América Latina," in Lander, E. (ed.) *La colonialidad del saber: eurocentrismo y ciencias sociales. Perspectivas latinoamericanas*, Buenos Aires: Clacso/Unesco, pp. 201–246.
Quintero, P. (ed.) (2014), *Crisis civilizatoria, desarrollo y bien vivir*, Buenos Aires: Ediciones del Signo.
Rappaport, R.A. (1968/1984), *Pigs for the ancestors*, New Haven: Yale University Press.
Rappaport, R.A. (1971) "The flow of energy in an agricultural society," *Scientific American*, 224 (3): 116–132.
Redclift, M. (1987), *Sustainable development: exploring the contradictions*, London: Methuen.
Ribeiro, D. (1973), "Etnicidade, indigenato e campesinato," *Revista de Cultura Vozes*, Ano 73, Vol. LXXIII, outubro, No. 8, pp. 5–18, Rio de Janeiro.
Robbins, P. (2012), *Political ecology: a critical introduction*, Chichester: John Wiley & Sons, 2nd edition.
Schrödinger, E. (1944) *What is life? The physical aspect of the living cell*, London and New York: Cambridge University Press.
Shiva, V. (1988), *Staying alive: women, ecology and development*, London: Zed.
Sousa Santos, B. (2008), *Conocer desde el Sur. Para una cultura política emancipatoria*, Buenos Aires: CLACSO/CIDES-UMSA/Plural Editores.
Stavenhagen, R. (1965), "Classes, colonialism, and acculturation. Essay on the system of inter-ethnic relations in Mesoamerica," *Studies in Comparative International Development*, 1 (6): 53–77.
Steward, J. (1972), *Theory of cultural change: the methodology of multilinear evolution*, Urbana: University of Illinois Press.
Vayda, P. (ed.) (1969), *Environment and cultural behavior*, Garden City: The Natural History Press.
Vayda, P. (1983), "Progressive contextualization: methods for research in human ecology," *Human Ecology*, 11: 265–281.
Watts, M. (1983), *Silent violence: food, famine and peasantry in northern Nigeria*, Berkeley: University of California Press.
Wilson, E.O. (1975), *Sociobiology, the new synthesis*, Cambridge, MA and London: The Belknap Press of Harvard University Press.
Wolf, E. (1972), "Ownership and political ecology," *Anthropological Quarterly*, 45 (3): 201–205.
Wolf, E. and A. Palerm (1972), "Potencial ecológico y desarrollo cultural en Mesoamérica," in *Agricultura y civilización en Mesoamérica*, No. 32, Mexico: SepSetentas.
Zimmerman, M.E. (1994), *Contesting earth's future. Radical ecology and postmodernity*, Berkeley: University of California Press.

5

FRENCH RESEARCH TRADITIONS ON PEASANT AGRICULTURAL SYSTEMS

A convergence with political ecology?

Denis Gautier and Christian A. Kull

Introduction

Political ecology is largely an Anglophone research tradition. It has had, over the years, varying levels of contact and exchange with other linguistic, cultural, and regional research traditions outside its dominant centers in the United Kingdom and United States, via the literature as well as through personal contacts made in the field. Conversely, other national research traditions have been influenced by similar intellectual and contextual forces as those which led to political ecology, but have followed different trajectories. In France, for example, many of the key elements of a political ecological approach are present in the academy – including strong traditions of Marxist anthropology, post-structural inspirations (the names Foucault and Latour are hard to ignore), and field-based studies of agrarian systems – and yet they were never pulled together in the same way as political ecology: instead they produced alternative inspirations and communities of practice.

Knowledge production is geographically embedded, and the particular traditions that have emerged in France carry the imprint of that nation's own social, institutional, (post)colonial, and disciplinary history. Of relevance to typical political ecological themes, one might mention three strong Francophone traditions. First, tropical geography, with its focus on the *terroir* as the portion of land appropriated, managed, and used by the group that resides upon it (Sautter and Pélissier 1964; Blanc-Pamard and Cambrézy 1995; Bowd and Clayton 2005; Bassett et al. 2007; Gautier and Hautdidier forthcoming). Second, hydrogeographies, or integrated watershed studies, where critical considerations of political discourse, institutional structures, and power relations have been layered upon strong technical hydrological traditions (Molle 2008; Bouleau et al. 2009; Venot and Krishnan 2011; Blanchon and Graefe 2012; Bouleau 2014). Third, agrarian systems research – the focus of this chapter – which has always gone beyond straight agronomy to understanding farmers in a broader societal, political, economic, and developmental context.

In this chapter, we focus on the latter tradition of research on peasant agricultural systems, for its parallels and divergences with political ecology are instructive. We document its origins and character, and place it in socio-political context – for the kind of research that has emerged in this tradition is reflected in the mandates of its main institutional hosts, which themselves are shaped by France's own agrarian politics (at home) and postcolonial development legacy (overseas). We dissect the core assertions and concepts of the approach, and focus in particular on its approach to questions of land tenure and resource access – an area with obvious overlaps to political ecology. Then we conclude with a consideration of potential synergies despite the relatively limited exchange to date between the approaches.

Systèmes agraires: origins

"*Systèmes agraires*", or agrarian systems, is a set of approaches across French research institutions that studies rural farming communities at multiple scales and from both agro-technical and socio-economic perspectives. It encompasses detailed technical work on crop choices, rotations, tools, and practices; mid-scale analyses of the production system at the farm level (including land, labor, and capital); and – importantly – the higher order systems that emerge from relationships between farm systems and the overall economic, social, and bio-ecological worlds in which they are embedded. *Systèmes agraires* can and does take seriously the relations of production, questions of resource access, and the broad political, social, and economic contexts within which people seek to exploit and manage the environment, even if it is mainly with a technical and not actor-centered perspective.

The *systèmes agraires* approach emerged in the 1970s and has a history that parallels – with overlaps and disjunctures – the evolution of kindred approaches in the Anglophone world in the same time period. The most obvious overlap is with "farming systems research" (Norman 1980), which in the Anglophone world had an institutional base in natural resource management schools and agricultural faculties. While the two approaches appear similar on the surface, the overlap is largely constrained to the middle of the three scales of analysis common to *systèmes agraires* research, as we detail later. Indeed, farming systems research is illustrative of the "apolitical" approaches in response to which political ecology emerged. A second parallel would be with anthropologists and geographers working in "cultural ecology" and "human ecology", often in rural tropical landscapes, and who shared an interest in particular techniques and the local cultural systems in which they were embedded. At the boundaries between cultural ecology and farming systems research, scholars sought to compare rural farm societies (e.g. Turner and Brush 1987) in ways that approached what was happening in the French tradition, but with less emphasis on the broad political economic context. Finally, scholars from "agrarian studies" and "peasant studies" traditions approached rural societies with primary attention on labor and power dynamics and built a critique of rural marginalization.

The term *systèmes agraires* was first used in the 1940s by rural geographer Cholley (1946) in a way that emphasized the dynamic, evolving nature of agrarian societies and their systemic interactions. However, his dynamic vision of the concept was more often replaced by a more static concern with agrarian structures (*structures agraires*) – a more descriptive combination of analysis of spatial farm organization and tenure regimes (Cochet 2012).

With the growing popularity of "systems theory" in the 1960s and 1970s, the time was ripe for *systèmes agraires* to develop. There was a flourishing of Francophone writings based on a systemic perspectives (Piaget 1968; de Rosnay 1975; Crozier and Friedberg 1977; Le Moigne 1977; Morin 1977) inspired by, among others, the work of von Bertallanfy (1968) and also by the first "classic" works published at this time in the United States such as "Systems Approach" (Churchman 1979), "Systems Analysis" (Hare 1967), "System Theory" (Zadeh 1962), or "System Dynamics" (Forrester 1971). Popular French public intellectual Joël de Rosnay (1975), for instance, studied systems at MIT. In the realm of agrarian studies, Osty was one of the first researchers to promote the application of a systemic perspective. He wrote: "it is considering first the whole before studying deeply all the parts that we know how to analyze and that the farm is an organized whole that does not reply to simple criteria of optimization" (Osty 1978: 48). His work on "the farm viewed as a system" contributed to a new vision of agronomical studies that sought to better understand farmers' practices and choices, and thus to adapt extension efforts. His approach was influenced by Crozier and Friedberg (1977), for whom the actor does not exist outside the system that defines his liberty and the rationality that he can use for his actions, and conversely, for whom the system exists only as a construction of the actions and interests of different actors.

Institutional contexts

The story of how *systèmes agraires* developed in French research circles is tightly linked to the institutional context. A number of different, but related, approaches developed in the major French research institutions. Whereas much similar work in the Anglophone context takes place by individual researchers scattered across numerous universities, in the French context there is a large role played by government and parastatal research agencies. These agencies provide placements for PhD students being trained, and have established networks and facilities for field-based studies in France and overseas. In addition, the relatively centralized and hierarchical way in which academia functions in France gives quite some weight to dominant research programs. In this section we trace the context of the production of knowledge related to *systèmes agraires* across four main institutions.

ORSTOM (now IRD)

The French government created an organization dedicated to research in its colonies in 1943. It was called ORSTOM (*Office de la recherche scientifique et technique outre-mer*) reflecting its focus on overseas scientific and technical research. ORSTOM developed a tradition of interdisciplinary studies of village territories or small regions in developing countries. Paul Pélissier and Gilles Sautter (Sautter and Pélissier 1964; Pélissier 1966; Pélissier and Sautter 1970), as well as Augé (1970), Lericollais (1972), Boulet (1975) and others, published remarkable examples of a holistic and systemic approaches of agrarian societies. However, their studies were generally focused on the concept of "terroir", which refers to a portion of land appropriated, managed, and used by the group that resides upon it and draws from it their means of existence (Sautter and Pélissier 1964). The "terroir school" contributed to knowledge of peasant agriculture, nature–society relations in rural areas, and the efficiency of production systems, but at a village territory level (Pélissier 1979; Painter et al. 1994; Bassett et al. 2007).

ORSTOM's role in the genesis of French agrarian studies came via the activities of AMIRA ("*l'Amélioration des Méthodes d'Investigation en milieu Rural Africain*" [Improvement of Methods of Investigation in the African Rural Areas]). This was an informal group specializing in methodological research, active from 1975 to 1990, that linked ORSTOM, the French Ministry of Cooperation, and the national statistics agency INSEE (*Institut national de la statistique et des études économiques*). Building on the "terroir" studies described above, AMIRA aimed to contribute to renewing the methodological tools for investigating and analyzing the development process, as well as the importance of data, its gathering, processing, analysis, and use. Accordingly AMIRA had a prominent role in the genesis of French agrarian systems studies (Ancey 1975).

From the late 1970s, ORSTOM began to expand beyond its detailed "terroir" scale studies through two research units: "*Cadres spatiaux de l'indépendance alimentaire*" [Spatial frameworks of food independence] and "*Dynamique des Systèmes de production*" [Dynamic of production systems]. This latter group was focused on the circumstances and causes of the changes in rural societies at different scales: the plot, the farm, and the small region, bringing together multidisciplinary teams of agronomists, economists, geographers, and sociologists (Ancey 1977; Couty and Hallaire 1980; Hallaire and Savonnet 1985; Dubois et al. 1987). This research contributed to exploding outward the restricted framework of the "terroir" to promote the study of agrarian systems.

With decolonization the mission of ORSTOM mutated into more of a development cooperation role, symbolized (belatedly) by its name change in 1998 to IRD, or *Institut de recherche pour le développement* (Research institute for development). It is jointly overseen by the Ministry of Higher Education and Research and the Ministry of Foreign Affairs (which includes development cooperation). It has active missions and country offices in most French overseas territories, across the ex-French colonial world, and in other developing countries like Indonesia, Kenya, Brazil, and Peru. Some of its current research groups, notably GRED (*Gouvernance, Risque, Environnement, Développement*), have been keen to build bridges with political ecology.

L'Institut national agronomique Paris-Grignon *(INA P-G)*

The "Agro" is a venerable institution, an elite, competitive-entry university with a mission of training civil servants in agricultural fields. It recently merged with the water and forestry-focused ENGREF to form AgroParisTech, one of France's so-called "*grandes écoles*". It is here that a Chair of comparative agriculture and agricultural development was established and held by René Dumont from 1953 to 1973, followed by Marcel Mazoyer from 1974 to the early 2000s. From this position, the "French school of comparative agriculture" has been strongly promoted and has had a strong role in developing the concept of *système agraire* (Cochet 2012). This school of thought investigates the specificities of, and similarities between, the diverse forms of contemporary agriculture worldwide. It emphasizes the historical agricultural development of given societies and analyzes the linkages between the remnants of former agrarian systems and the elements of new ones. It seeks to draw overarching lessons to understand agricultural development, including those interpreted as "crises" or "revolutions", but avoiding broad generalizations or overly simplified modeling (Mazoyer and Roudart 1998; Dufumier 2006).

INRA

The *Institut national de la recherche agronomique* is a large research organization focused on agriculture, largely in France, and is jointly overseen by the Agriculture and Research ministries. During the post-war period – the so-called *Trente glorieuses* (or three decades of socio-economic prosperity) – French agriculture rapidly transformed and intensified, and INRA played an important role developing crop varieties and cultivation techniques, largely through a technical and micro-economic approach. In the later 1960s and 1970s, however, these sectoral technical approaches were integrated with a more global approach. A think tank on non-sectoral research was established, under the leadership of R. Gras. Simultaneously, J.-P. Deffontaines of the INRA's *Service d'Expérimentation et d'Information* (SEI, for Experimental and informational program), just after having defended a PhD in geography, developed an approach that set the farms and their dynamic in their geographical context in mountainous regions (Deffontaines 1977). It is worth noticing that in its initial phases, this kind of research often focused on marginal regions where the application of INRA's mainstream models of development and innovations transfer were causing serious problems. This kind of work by the SEI led to the establishment of a long-lasting department of *Systèmes agraires et développement* (SAD) within INRA in 1979 (Deffontaines 1980). Created to study the resistance of farmers to the adoption of innovation, the SAD from the beginning has brought together researchers from both the agronomic and social sciences, and centered its studies on practices, organizations, and perceptions linked to farming and natural resource management (Meynard 2010). It has been one of the pioneers of interdisciplinary research on the rural world (Deffontaines and Hubert 2004).

CIRAD

A fourth venue for *systèmes agraires* research is CIRAD (*Centre de coopération internationale en recherche agronomique pour le développement*, or the center for international cooperation in agricultural research for development). This major institution, with a presence across the tropical developing world, was created in 1984 out of the amalgamation of a number of technical institutes mainly dedicated to the cash crops of the former French colonial Empire. It is jointly answerable to the Ministries of Agriculture, Foreign Affairs, and Higher Education and Research. Generally focused on the industrial production of crops, such as palm oil, cotton, coffee, cocoa, and rubber, among others, these technical institutes tested crop improvements and techniques in laboratories or in experimental plots before seeking to transfer them to industrial groups, large modern farms, or peasants. However, it was realized that this transfer was more difficult in the case of food crops and small family farms. It was the technical institute in charge of the food crops, IRAT (*Institut de recherches agronomiques et des cultures vivrières* [Agricultural research institute for subsistence crops]) that first realized the necessity of escaping from the experimental domain to meet local people and understand their realities, in order to ensure better innovation transfer. The remarkable work of R. Tourte and his team in Sine Saloum, Senegal, which began in 1963, served as a foundation of agrarian studies within CIRAD (Kleene 1976; Tourte and Billaz 1982). This innovative approach led to the creation of an Agrarian Systems unit within IRAT in 1982, and then to the creation of CIRAD's Department of Agrarian Systems (DSA) in 1984.

Convergence

Despite the different backgrounds and institutional contexts, a systemic approach to agrarian studies spread quickly from the late 1970s across all these institutions, converging to become a distinct, consistent, and uniform approach (Brossier et al. 1990; Brossier et al. 1993; de Bonneval 1993) that sought to understand and integrate the complexity of rural activities (Conesa 1987). The uptake of this approach was motivated by an acknowledgment that technological advances in agronomy could not be implemented, or would not be adopted, without an understanding of farmers' behaviors and their broader socio-economic context. Often, the approach passed in France under the rather generic label "*Recherche-Développement*" (Pillot 1987). The pioneering groups founded in the 1970s and 1980s no longer exist under the same names except at INRA. But the theoretical basis of agrarian system studies continues to influence these research communities, even if they have been overtaken by more recent academic concerns such as local or territorial development (Caron 2005; Benoît et al. 2006). Research training in agrarian system studies remains particularly strong in the agronomic universities such as AgroParisTech (formerly INA-Paris Grignon) and SupAgro Montpellier as well as in some academic geography programs.

Agrarian systems at three scales

The *systèmes agraires* approach spans a number of different scales. Compared with political ecology and its residual influence from cultural ecology, there is a much stronger focus on agronomic practices at the smaller scale, and an orientation towards not just analyzing but also promoting change and adaptation. Here we present an overview of the three main scales of analysis, keeping in mind that the concepts have evolved over the decades.

The overall aim of *systèmes agraires* is to understand the agrarian system by studying its structure and character at different scales and the interactions between these scales. As the lowest level of scale, the study of *cropping and livestock technical systems* investigates in specific detail how farmers exploit and manage their environment, describing the particular patterns of activities for each type of crop or land use. It focuses on vegetal dynamics in cropfields and pastures, the "technical itineraries" (or combinations of tools, techniques, and practices that allow farmers to shape the environment for productive use (Sebillotte 1974)), and the plots or territories in which this occurs (Sebillotte 1982; Landais 1983; Lhoste 1984). The cropping and livestock technical systems are components of the farming system, but may be analyzed at a scale that is larger than the farm when they are related to practices of cropping or herding that are common to several farms.

The *système de production*, or production system, takes the analysis up one level of scale. It also moves away from strictly technical and agro-ecological analysis to consider the socio-economic system at the whole-farm scale. A production system has been defined as a combination, more or less coherent in space and time, of diverse means of production (labor, land, buildings, input, material, tools, livestock, etc.) in order to satisfy the farmer's socio-economic and cultural objectives at the farm level (Chombart de Lauwe et al. 1963; Tourte 1978; IRAM 1985). It should be noted that some authors use the term *système de production* somewhat differently, as a label for dominant aggregations of micro-economic systems at the regional level, such as the "cotton-sorghum system in the sudano-sahelian area". Such production systems are then used as a statistical unit for macro-economic analysis (Reboul 1976). However, in its most common sense at the micro-economic level, the definition of the French *système de production* and the Anglophone "farming system" are similar (Norman 1980; Pillot 1987). They both open the door to political economic analyses of farm labor, land tenure, and resource access. Thus, they both provide components for political ecology analyses. Yet conceptual differences appear when either scaling down or especially when scaling up.

The highest scale, the French notion of *système agraire*, acknowledges that interacting production systems are constitutive elements of a higher-order system, which emerges from the relationship between production systems and the overall economic and social structure. This is the original agrarian system idea – looking at an agrarian society and the lands that it uses through the twinned interactions of the bio-ecological system and the socio-cultural system (Deffontaines and Osty 1977; Vissac 1979). The idea was to go beyond the simple understanding of the internal functioning of the "production system" to take into consideration the overall conditions of production.

Marcel Mazoyer, of the comparative agriculture school at INA-PG, presented an important critique of this original conception of agrarian systems as interactions of the bio-ecologic and the socio-cultural. He argued that it was too static, and proposed a less structuralist, more dynamic definition:

> An agrarian system is a way of exploitation of the environment, evolved through time and lasting, a system with growth of production, adapted to bio-climatic conditions of

> a given area and answering the current conditions and needs. The internal coherence of the way of exploitation of the environment raises questions about the overall technical, economic and social conditions of production.
>
> *(Mazoyer 1987: 11)*

For Mazoyer, an agrarian system is a combination of the following essential variables in one form or another:

- The cultivated ecosystem: the original environment and its historical transformations to its present state.
- The production elements: tools, machines, plant cultivars, domesticated animals, and the social labor force (physical and intellectual) to manage them.
- The social division of labor between agriculture, craft industry, and industry which allow the reproduction of work tools, and then the agricultural surplus that allows the satisfaction of other social groups, beyond the needs of the farmers.
- The exchange relationships between these different but associated sectors, the relations of ownership and power which determine the share of the production work, of the production and consumer goods.
- Finally, the overall ideas and institutions, which allow social reproduction: production and exchange relationships and the sharing of production.

Agronomist Philippe Jouve (1988) added a more explicit consideration of space to the concept, and made the spatial and territorial aspects even more explicit in his development of a "rural systems" concept to replace "agrarian systems" (Jouve 1992). The framework of the "rural system" allows one to consider the increasing importance of off-farm activities and migration, while also enabling integration of all functioning elements of a rural society, such as health and religion.

The *systèmes agraires* approach in comparison with political ecology, then and now

Compared with contemporaneous Anglophone research traditions, if one may generalize, the *systèmes agraires* approach had no direct equivalent. It was more applied and geared towards effecting change – modernizing – than "cultural ecology" and "human ecology" approaches interested in how indigenous societies differed from modern ones. It brought much more technical and agro-ecological focus to the table than the traditions of "peasant studies" and "agrarian studies", while sharing the sharp focus on relations and modes of production. While it overlapped considerably with "farming systems" approaches in its intermediate analytical scale of *systèmes de production*, the *systèmes agraires* approach differed in giving a full account of the historical social and economic transformations at multiple scales that impact the processes of agricultural production. It also differed in better accommodating non-agricultural activities into its framework (Behnke and Kerven 1983; Tripp 1985). In a way, the Francophone agrarian systems approach was considering the historical transformations and political economic processes that political ecology sought to do in its own critique of "farming systems" approaches.

However, the challenge for *systèmes agraires* approaches has been to deal with current rapid transformations: globalizing trade, restructuring economies, and evolving world institutions. Few *systèmes agraires* scholars have paid much attention to the fact that agricultural modernization, for instance, has been leading to a marginalization and disappearance of small-scale farmers – the *petit paysannat*. As Cochet (2012: 133) notes:

> it is easier to analyze a relatively "stable" situation and *construct* an agrarian system ... than to analyze a system that is so dynamic that the various elements and their reciprocating interactions just barely have the time to stabilize before transforming again ... Perhaps the agrarian system concept is easier to wield when applied to history, to lay the groundwork of a system, than to rapidly changing modern agriculture.

These observations notwithstanding, both Dufumier (2007) and Cochet (2012) defend the *systèmes agraires* approach as profoundly relevant to analyzing such crises in agrarian systems and the processes that lead to their restructuring in different forms.

Undoubtedly, in its overall emphases, *systèmes agraires* focuses more attention on technical and agro-ecological aspects than political ecology. In turn, political ecology stresses not just production relations and global political economy (in concert with *systèmes agraires*), but also gives more attention to the institutional and discursive ways in which power relations play out. One area in which there has been significant overlap, if not interaction, is in the study of access rights and resource tenure. Researchers linked to ORSTOM/IRD and CIRAD have a long tradition of paying attention to the rights of access to land and resource and the conflicts around land tenure and use (Le Bris et al. 1982; Blanc-Pamard and Cambrézy 1995; Le Roy et al. 1996) and to theorizing the relationship of societies to land (Le Roy 1996). Even if political ecological research makes few explicit links via citation, it is likely that French research efforts on land tenure and access, particularly those focused on Africa, has had some influence on Anglophone political ecology (Bassett 1988; Kull 1998; Ribot 1999). Conversely, some Francophone work on land tenure issues and their relation to territorial policies presents strong similarities with research in political ecology, and has, since the year 2000, started to explicitly cite it (Chauveau 1997, 2000; Chauveau and Jacob 2006; Jacob 2007; Jacob and Le Meur 2010; Medernach and Burnod 2013).

These exceptions notwithstanding, until recent exchanges (see below), there have been few strong theoretical links made between political ecology and the French agrarian system approach. This is surprising given the strong similarities between, for instance, studies of land tenure and land use in the Francophone tradition of agrarian system research and the Anglophone corpus of political ecology (Bertrand et al. 2004). It appears to be more of a case of parallel or convergent evolution than a strong, direct connection. Beyond tenure issues, there are numerous divergences: for instance, the *systèmes agraires* approach does not pay much attention to the winners and the losers of rural development or environmental policies at the local level, nor do *systèmes agraires* studies frequently incorporate gender concerns.

It is only since the 2000s that a generation of French scholars who received a preliminary training in the *systèmes agraires* approach has been clearly inspired by some political ecology. This comes from personal interactions with Anglophone political ecologists through work in the field or academic exchanges, attending AAG conferences, or simply by the percolation of ideas from colleagues who have themselves been inspired by the political and social commitment which is often missing in agrarian system studies. One of the earliest French publications to recognize political ecology was an article by Blanc-Pamard and Boutrais (2003). This article cites political ecologists Piers Blaikie, Melissa Leach, and Raymond Bryant in a review of 50 years of development policy in the environment sector. Tom Bassett, who has long-standing relations with the *Centre d'Etudes Africaines* (home to the authors of that article), introduced geographers there to this approach in the 1990s.

The more forceful entry of political ecology on the French scene took place in the late 2000s, when the approach was embraced by a new generation of young researchers mainly trained in agrarian systems. Their enthusiasm for political ecology was championed and given institutional weight by established researchers like Bernard Hubert, President of Agropolis, a major multi-institutional hub of agricultural and development research in Montpellier, and Serge Bahuchet, director of the research group in eco-anthropology and ethnobiology at the Paris Natural History Museum (MNHN). In 2009, for example, CIRAD invited a number of prominent political ecological scholars (Paul Robbins, Tom Bassett, Nancy Peluso, Tor Benjaminsen) for a research school (www.politicalecology.fr), program evaluation workshops, and public lectures in the framework of the SETER project (Socio-Ecological Theories and Empirical Research) (www.seter.org/index.php/). The first articles in French on political ecology appeared in *Natures, Sciences, Sociétés* in 2009 (Benjaminsen and Svarstad 2009; Castro-Larrañaga 2009), and the first book – a result of the 2009 research school – in 2012 (Gautier and Benjaminsen 2012). A number of small informal research groups are now coalescing along political ecological themes in diverse institutions ranging from CIRAD, IRD, MNHN, to the National Institute for Research in Sciences and Technologies for Environment and Agriculture (IRSTEA).

Why this explosion of interest in Anglo-American political ecology? Two main reasons appear important. First, the dominance of the English language in an internationalizing academia has certainly helped. Young French researchers are encouraged to do post-docs in Anglophone countries, attend overseas conferences, and to publish in international, thus English-language, journals. This has increased exposure to Anglophone traditions, creating both inspiration as well as the need to "fit in". Second, we suggest that political ecology has grown because it helps researchers legitimate a more critical stance than hitherto possible under agrarian systems approaches (which were, after all, relatively technical and applied) and under their host institutions (which are part of the state's machinery and, in some cases, carry postcolonial legacies, albeit fading ones). Researchers seeking a more engaged approach that allows for the construction of tighter links between politics (broadly construed) and the environment can call on political ecology as a solid justificatory framework to do so.

Conclusion

With its opening towards political ecology in the past few years, French research has gained a legitimation for more critical approaches to work on agrarian change (Eloy 2005; Ducourtieux 2006), hydro-management (Bouleau 2014), and natural resource management (Gautier et al. 2011, 2013). There are some signs of convergence between the "*systèmes agraires*" tradition and Anglophone political ecology that allows a more critical approach within the French institutional context. Even if this convergence is still limited to individuals spread in different institutions, and even if political ecology is not taught as such in an academic program, an informal community is emerging that is building strong links between the agrarian systems and the political ecology traditions.

Yet the exchanges are largely unidirectional. It is relatively rare to find references to French agrarian systems literature, or its more modern manifestations, in Anglophone political ecology texts, except where researchers share field sites. There is surely much to be gained for political ecology from not just the very detailed systemic studies of the agricultural or livestock technical systems found in the French *systèmes agraires* tradition, but also from its structured, systemic approach to understanding the complex character of regional agrarian systems in all their bio-ecological and social splendor. For those seeking inspiration, some exemplary studies (such as Couty 1991; Aubert et al. 2003; Blanc-Pamard et al. 2005; Barnaud 2008) show how a *systèmes agraires* approach can provide strong field evidence to demonstrate the effects of development policies on rural societies and their environment.

References

Ancey, G. (1975) *Niveaux de décision et fonctions objectif en milieu rural africain. AMIRA, note de travail n°3.* Paris: ORSTOM.

Ancey, G. (1977) "Recensement et description des principaux systèmes ruraux sahéliens". *Cah. ORSTOM Sc. Hum.*, Vol. 14, pp. 3–19.

Aubert, S., S. Razafiarison, and A. Bertrand (2003). *Déforestation et systèmes agraires à Madagascar: les dynamiques des tavy sur la côte orientale.* Versailles: Editions Quae.

Augé, M. (1970) "Tradition et conservatisme Essai de lecture d'un terroir: Pays Alladian (Basse Côte-d'Ivoire)". *Etudes rurales*, Vol. 37–39, pp. 281–298.

Barnaud, C. (2008) *Equité, jeux de pouvoir et légitimité: les dilemmes d'une gestion concertée des ressources renouvelables. Mise à l'épreuve d'une posture d'accompagnement critique dans deux systèmes agraires des hautes terres du Nord de la Thaïlande.* PhD thesis, Université de Nanterre-Paris X.

Bassett, T.J. (1988) "The political ecology of peasant-herder conflicts in the northern Ivory Coast". *Annals of the Association of American Geographers*, Vol. 78, pp. 453–472.

Bassett, T.J., C. Blanc-Pamard, and J. Boutrais. (2007) "Constructing locality: the terroir approach in West Africa". *Africa*, Vol. 77, pp. 104–129.

Behnke, R. and C. Kerven (1983) "FSR and the attempt to understand the goals and motivations of farmers". *Culture and Agriculture*, Vol. 19, pp. 9–16.

Benjaminsen, T.A. and H. Svarstad (2009) "Qu'est-ce que la 'political ecology'?". *Natures Sciences Sociétés*, Vol. 17, pp. 3–11.

Benoît, M., J.-P. Deffontaines, and S. Lardon (2006) *Acteurs et territoires locaux: vers une géoagronomie de l'aménagement.* Versailles: Editions Quae.

Bertrand, A., J.C. Ribot, and P. Montagne (2004) "The historical origins of deforestation and forestry policy in French-speaking Africa: from superstition to reality?". In Babin, D. (ed.), *Beyond Tropical Deforestation: From Tropical Deforestation to Forest Cover Dynamics and Forest Development.* Paris: UNESCO, pp. 451–463.

Blanc-Pamard, C. and J. Boutrais (2003) "Les temps de l'environnement. D'un sauvetage technique à une gestion locale en Afrique et à Madagascar". *Historiens et Géographes, Regards sur l'Afrique*, No. 379, pp. 249–262.

Blanc-Pamard, C. and L. Cambrézy (eds) (1995) *Dynamique des systèmes agraires: terre, terroir, territoire: les tensions foncières.* Collection Colloques et Séminaires. Paris: ORSTOM.

Blanc-Pamard, C., P. Milleville, M. Grouzis, F. Lasry, and S. Razanaka (2005) "Une alliance de disciplines sur une question environnementale: la déforestation en forêt des Mikea (Sud-Ouest de Madagascar)". *Natures Sciences Sociétés*, Vol. 13, pp. 7–20.

Blanchon, D. and O. Graefe (2012) "La radical political ecology de l'eau à Khartoum. Une approche théorique au-delà de l'étude de cas". *L'Espace géographique*, Vol. 1, pp. 35–50.

Bouleau, G. (2014) "The co-production of science and waterscapes: the case of the Seine and the Rhône Rivers, France". *Geoforum*, Vol. 57, pp. 248–257.

Bouleau, G., C. Argillier, Y. Souchon, C. Barthélémy, and M. Babut (2009) "How ecological indicators construction reveals social changes—the case of lakes and rivers in France". *Ecological Indicators*, Vol. 9, pp. 1198–1205.

Boulet, J. (1975) *Magoumaz, pays Mafa (Nord Cameroun): Étude d'un Terroir de Montagne.* Paris: ORSTOM.

Bowd, G. and D. Clayton (2005) "French tropical geographies: editors' introduction". *Singapore Journal of Tropical Geography*, Vol. 26, pp. 271–288.

Brossier, J., L. de Bonneval, and E. Landais (1993) *Systems Studies in Agriculture and Rural Development.* Institut National de la Recherche Agronomique (INRA).

Brossier, J., B. Vissac, and J.-L. Le Moigne (1990) "Modélisation systémique et système agraire: décision et organisation". Actes du séminaire du Département de recherches sur les systèmes agraires et le développement (SAD), Saint-Maximin, France: Institut national de la recherche agronomique.

Caron, P. (2005) "À quels territoires s' intéressent les agronomes? Le point de vue d'un géographe tropicaliste". *Natures Sciences Sociétés*, Vol. 13, pp. 145–153.

Castro-Larrañaga, M.V. (2009) "Commentaire du texte de Benjaminsen et Svarstad: 'Qu'est-ce que la political ecology'?". *Natures Sciences Sociétés*, Vol. 17, pp. 12–17.

Chauveau, J.-P. (1997) "Jeu foncier, institutions d'accès à la ressource et usage de la ressource". In Contamin, B. and H. Memel-Fotê (eds.), *Le modèle ivoirien en questions: crises, ajustements, recompositions*. Paris: Karthala, pp. 325–360.

Chauveau, J.-P. (2000) "Question foncière et construction nationale en Côte d'Ivoire". *Politique africaine*, Vol. 2, pp. 94–125.

Chauveau, J.-P. and J.-P. Jacob (2006) *Modes d'accès à la terre, marchés fonciers, gouvernance et politiques foncières en Afrique de l'Ouest*. London: IIED.

Cholley, A. (1946) "Problèmes de structure agraire et d'économie rurale". *Annales de géographie*, Vol. LV, pp. 81–101.

Chombart de Lauwe, J., J. Poitevin and J.-C. Tirel (1963) *Nouvelle gestion des exploitations agricoles. Deuxième Edition*. Paris: Dunod.

Churchman, C.W. (1979) *The systems approach and its enemies*. New York: Basic Books.

Cochet, H. (2012) "The 'systeme agraire' concept in francophone peasant studies". *Geoforum*, Vol. 43, pp. 128–136.

Conesa, A.P. (1987) "Equipes francaises systemes agraires". *Cahiers de la recherche developpement*, No. 16, pp. 1–2.

Couty, P. (1991) "L'agriculture africaine en réserve. Réflexions sur l'innovation et l'intensification agricoles en Afrique tropicale". *Cahiers d'études africaines*, Vol. 31, pp. 65–81.

Couty, P. and A. Hallaire (1980) *De la carte aux systèmes. Vingt ans d'études agraires au Sud du Sahara. (ORSTOM, 1960 – 1980). Note AMIRA n° 29*. Paris: ORSTOM, p. 121.

Crozier, M. and E. Friedberg (1977) *L'acteur et le système*. Paris: Éditions du Seuil.

de Bonneval, L. (1993) *Systèmes Agraires. Systèmes de Production. Vocabulaire français-anglais avec index anglais [Agricultural systems. Production systems. French-English Dictionary with English index]*. Paris: INRA.

de Rosnay, J. (1975) *Le Macroscope: vers une vision globale*. Paris: Éditions du Seuil.

Deffontaines, J.-P. (1977) "Aménagement et espace: Evolution des exploitations agricoles et transformation de l'espace". *Économie rurale*, Vol. 119, p. 28.

Deffontaines, J.-P. (1980) "Systèmes de pratiques et terroirs. Pour une approche agronomique des systèmes agraires". Carribean Seminar on Farming Systems Research Methodology, IICA, INRA, Pointe à Pitre, pp. 163–177.

Deffontaines, J.-P. and B. Hubert (2004) "Un regard sur l'interdisciplinarité à l'Inra". *Natures Sciences Sociétés*, Vol. 12, pp. 186–190.

Deffontaines, J.-P. and P.-L. Osty. (1977) "Des systèmes de production agricole aux systèmes agraires". *Espace géographique*, Vol. 6, pp. 195–199.

Dubois, J.-P., A. Lericollais, P. Milleville, and G. Pontié (1987) "Terroirs anciens, approche renouvelée: analyse du changement dans les systèmes agraires au Sénégal". Communication au séminaire Dynamique des Systèmes Agraires, Paris, p. 38.

Ducourtieux, O. (2006) *Du riz et des arbres: L'élimination de l'agriculture d'abattis-brûlis, une constante politique au Laos*. Thèse de doctorat en agriculture comparée, INA P-G (AgroParisTech).

Dufumier, M. (2006) "Diversité des exploitations agricoles et pluriactivité des agriculteurs dans le Tiers Monde". *Cahiers Agricultures*, Vol. 15, pp. 584–588.

Dufumier, M. (2007) "Agriculture comparée et développement agricole". *Revue Tiers Monde*, No. 3, pp. 611–626.

Eloy, L. (2005) *Entre ville et forêt: le futur de l'agriculture amérindienne en question*. Thèse de Doctorat en géographie, Université de Paris III/ IHEAL.

Forrester, J.W. (1971) *World Dynamics*. Cambridge, MA: Wright-Allen.

Gautier, D. and T.A. Benjaminsen (eds.) (2012) *Environnement, discours et pouvoir: L'approche Political ecology*. Versailles: Editions Quae.

Gautier, D., T.A. Benjaminsen, L. Gazull, and M. Antona (2013) "Neoliberal forest reform in Mali: adverse effects of a World Bank 'success'". *Society & Natural Resources*, Vol. 26, pp. 702–716.

Gautier, D. and B. Hautdidier (forthcoming) "Connecting political ecology and French geography: on tropicality and radical thought". In Bryant, R.L. (ed.), *International Handbook of Political Ecology*. Cheltenham: Edward Elgar.

Gautier, D., B. Hautdidier, and L. Gazull (2011) "Woodcutting and territorial claims in Mali". *Geoforum*, Vol. 42, pp. 28–39.

Hallaire, A. and G. Savonnet (1985) "Le terroir, une formule rigide, ses transformations, ses éclatements". In Blanc-Pamard, C. and A. Lericollais (eds.), *A travers Champs, agronomes et géographes. Dynamique des systèmes agraires*. Paris: ORSTOM, Colloques et séminaires, pp. 31–56.

Hare, V.C. (1967) *Systems Analysis: A Diagnostic Approach*. New York: Harcourt, Brace and World.

IRAM (1985) *Recherche-Développement, compte-rendu des journées d'étude 12–13 septembre 1984*. Paris.

Jacob, J.-P. (2007). *Terres privées, terres communes. Gouvernement de la nature et des hommes en pays winye, Burkina Faso*. Paris: IRD éditions.

Jacob, J.-P. and P.-Y. Le Meur (2010) "Citoyenneté locale, foncier, appartenance et reconnaissance dans les sociétés du Sud". In Jacob, J.-P. and P.-Y. Le Meur (eds.), *Politique de la terre et de l'appartenance. Droits fonciers et citoyenneté locale dans les sociétés du Sud*. Paris: Karthala, pp. 5–57.

Jouve, P. (1988) "Quelques réflexions sur la spécificité et l'identification des systèmes agraires". *Les cahiers de la recherche développement*, Vol. 20, pp. 5–16.

Jouve, P. (1992) *Le Diagnostic du milieu rural de la region a la parcelle: approche systémique des modes d'exploitation agricole du milieu*. Montpellier: CNEARC.

Kleene, P. (1976) "Notion d'exploitation agricole et modernisation en milieu Wolof Saloum (Sénégal)". *L'Agronomie tropicale*, Vol. 31, pp. 63–81.

Kull, C.A. (1998) "Leimavo revisited: agrarian land-use change in the highlands of Madagascar". *The Professional Geographer*, Vol. 50, pp. 163–176.

Landais, E. (1983) *Analyse des systèmes d'élevage bovin sédentaire du Nord de la Côte d'Ivoire*. Maisons-Alfort: IEMVT.

Le Bris, E., E. Le Roy, and F. Leimdorfer (eds) (1982) *Enjeux fonciers en Afrique Noire*. Paris: ORSTOM – Karthala.

Le Moigne, J.-L. (1977) *La théorie du système général. Théorie de la modélisation*. Paris: PUF.

Le Roy, E. (1996) "La théorie des maîtrises foncières". In Le Roy, E., A. Karsenty, and A. Bertrand (eds.), *La sécurisation foncière en Afrique, pour une gestion viable des ressources renouvelables*. Paris: Karthala, pp. 59–76.

Le Roy, E., A. Karsenty, and A. Bertrand (eds) (1996) *La sécurisation foncière en Afrique. Pour une gestion viable des ressources renouvelables*. Paris: Karthala.

Lericollais, A. (1972) *Sob: étude géographique d'un terroir sérèr (Sénégal)*. IRD Editions.

Lhoste, P. (1984) "Le diagnostic sur le système d'élevage". *Les cahiers de la recherche et de développement*, No. 3–4, pp. 84–88.

Mazoyer, M. (1987) *Dynamique des Systèmes agraires. Rapport de synthèse présenté au Comité des systèmes agraires*. Paris: Ministère de la Recherche et de la technologie.

Mazoyer, M. and L. Roudart (1998). *Histoire des agricultures du monde: Du néolithique à la crise contemporaine*. Paris: Seuil.

Medernach, K. and P. Burnod (2013) "Recompositions inattendues d'un système agraire malgache par l'agrobusiness". *Etudes rurales*, Vol. 1, pp. 63–76.

Meynard, J.-M. (2010) "Avant-propos". In Muchnik, J. and C. de Sainte Marie (eds.), *Le temps des Syal*. Paris: Editions Quæ, pp. 7–11.

Molle, F. (2008) "Nirvana concepts, narratives and policy models: insights from the water sector". *Water Alternatives*, Vol. 1, pp. 131–156.

Morin, E. (1977) *La méthode. 1 : La Nature de la Nature*. Paris: Editions du Seuil, collection Points.

Norman, D.W. (1980) *The Farming Systems Approach: Relevance for the Small Farmer*. East Lansing: MSU Rural Development Paper No. 5. Department of Agricultural Economics, Michigan State University.

Osty, P.-L. (1978) "L'exploitation agricole vue comme un système. Diffusion de l'innovation et contribution au développement". *Technical Information Bulletin of the Ministry of Agriculture*, No. 326, pp. 43–49.

Painter, T., J. Sumberg, and T. Price. (1994) "Your terroir and my 'action space': implications of differentiation, mobility and diversification for the approche terroir in Sahelian West Africa". *Africa*, Vol. 60, pp. 447–463.

Pélissier, P. (1966) *Les paysans du Sénégal: les civilisations agraires du Cayor à la Casamance*. Saint-Yrieix: Imprimerie Fabrègue.

Pélissier, P. (1979) "Le paysan et le technicien: quelques aspects d'un difficile face-à-face". Actes du colloque de Ouagadougou 4–8 décembre 1978, "Maîtrise de l'espace agraire et développement en Afrique tropicale. Logique paysanne et rationalité technique," ORSTOM, Centre National de la Recherche Scientifique et Technologique de la Haute-Volta, pp. 1–8.

Pélissier, P. and G. Sautter (1970) "Bilan et perspectives d'une recherche sur les terroirs africains et malgaches (1962–1969)". *Etudes rurales*, Vols. 37-38-39, pp. 7–45.

Piaget, J. (1968) *Le Structuralisme*. Paris: coll. Que sais-je ?, P.U.F.

Pillot, D. (1987) *Recherche développement et farming system research. Concepts, approches et méthodes*. Paris: Réseau Recherche-Développement, p. 40.

Reboul, C. (1976) "Mode de production et systèmes de culture et d'élevage". *Economie rurale*, Vol. 112, pp. 55–65.

Ribot, J.C. (1999) "Decentralisation, participation and accountability in Sahelian forestry: legal instruments of political-administrative control". *Africa*, Vol. 69, pp. 23–64.

Sautter, G. and P. Pélissier (1964) "Pour un atlas des terroirs africains. Structure type d'une étude de terroir". *L'Homme*, Vol. IV, pp. 56–72.

Sebillotte, M. (1982) "Les systèmes de culture. Réflexions sur l'intérêt de cette notion à partir de l'expérience acquise en région de grande culture". Séminaire du département d'Agronomie INRA, 16–18 mars, Paris, pp. 63–80.

Sebillotte, M. (1974) "Agronomie et agriculture. Essai d'analyse des tâches de l'agronome". *Cahiers de l'ORSTOM*, Vol. 24, pp. 3–25.

Tourte, R. (1978) *Pour une étude régionalisée des systèmes techniques de production agricole en Côte d'Ivoire.* Montpellier: IRAT-GERDAT.

Tourte, R. and R. Billaz. (1982) "Approche des systèmes agraires et fonction recherche-développement. Contribution à la mise au point d'une démarche". *L'Agronomie tropicale*, Vol. 37, pp. 223–231.

Tripp, R. (1985) "Anthropology and on-farm research". *Human Organization*, Vol. 44, pp. 114–124.

Turner, B.L., II and S.B. Brush (1987) *Comparative Farming Systems.* New York: The Guilford Press.

Venot, J.-P. and J. Krishnan (2011) "Discursive framing: debates over small reservoirs in the rural South". *Water Alternatives*, Vol. 4, pp. 316–324.

Vissac, B. (1979) *Éléments pour une problématique de recherche sur les systèmes agraires et le développement.* Toulouse: INRA-SAD.

von Bertalanffy, L. (1968). *General System Theory: Foundations, Development, Applications.* New York: George Braziller.

Zadeh, L.A. (1962) "From circuit theory to system theory". *Proceedings of the IRE*, Vol. 50, pp. 856–865.

6
THE TRICKSTER SCIENCE

Paul Robbins

> During the moment of chaos, the world is reconstructed, reversed, and what was up is down, what was down is up. What was inside is outside, what was outside is inside; the powerful are weak, the weak are powerful; what was bad is good, what was good is bad.
>
> *(Scheub 2012: 31)*

Introduction

Political ecology is a field that engenders frustration. Despite its seemingly endless expansion and notorious lack of definition, it manages to survive, thrive, and maintain an identity (Blaikie 1999; Kepe et al. 2008). At the same time, a great deal of recent effort has been made to "synthesize" political ecology within other fields, including land change science and environmental health, yet it remains stubbornly hard to digest and metabolize (Brannstrom and Vadjunec 2013; King and Crews 2013). The field's most ardent critics suggest that it has no particular grasp on explanation and is instead agenda-driven narrative (Vayda 2009; Vayda and Walters 1999). Even its founding thinkers have difficulty in explaining its utility (Blaikie 2008). How does a field that continues to endlessly expand also manage to maintain coherence? If a field accepts almost any method, and a great many concepts, how does it sustain an identity? If it has no utility, why is it so often used?

This chapter lays out an argument for the longevity of political ecology, accounting for its ability to adapt and evolve, even while keeping true to its normative and conceptual underpinnings. It argues that political ecology's chief characteristic in this regard is its simultaneous ability to advance rigorous empirical assessment of socio-environmental conditions and change, freely adopting the methods and conceptual apparatus of related research traditions, while constantly critiquing and undermining the projects of these other fields. This insistence on borrowing and allying with neighboring approaches (e.g. land change science, resilience, vulnerability) even while undermining them, is both the engine of political ecology's survival and the key source of frustration for observers. Insofar as political ecology's role is dialogic, simultaneously advancing and undermining associated fields of study, the chapter advances a further argument, that political ecology is a kind of troublemaker, effectively adopting the role of Trickster. This folkloric figure is essential to the thriving of people but is also one who depends upon, and is made relevant by, other figures around it. For them, Trickster remains troublesome, conniving, and contradictory (Hyde 1998; Scheub 2012; Radin 1956).

Briefly reviewing the path-breaking work of James Fairhead and Melissa Leach in West Africa as a model, the chapter begins with a survey of political ecology's efforts to "have it both ways": advancing arguments by mobilizing diverse branches of nature/society research while simultaneously undermining these same branches of science. This section also introduces the folkloric figure of Trickster, and argues that the adoption of this mythic role is the key to political ecology's relevance, reproduction, and survival. Considering two examples from land change science and environmental health, the chapter then seeks to demonstrate how contemporary political ecological texts narrate this contradictory position. This section simultaneously emphasizes how political ecology maintains this posture even while embracing and engaging new and important research traditions. The chapter closes with a consideration of the limits and value of political ecology as Trickster, concluding that in a world hurtled forward by the forces of contradiction, a contradictory science like political ecology remains an essential field for explanation and action.

Having it both ways

At bottom, the definition given to political ecology so many years ago by Piers Blaikie and Harold Brookfield (1987: 17) remains perhaps the best and most inclusive operating framework. "The phrase 'political ecology'," they famously write (acknowledging a concept but not necessarily a formal field), "combines the concerns of ecology and a broadly defined political economy." The former term – "ecology" – we have to take to mean a systems perspective on relationships between land, soil, vegetation, and other organisms. The latter term – political economy – is a forgiving, broad, and generous one, but in this context it suggests a materialist perspective on capitalism, markets, social structure, and population.

Recent excellent work in political ecology appears to fit this definition with no needed alteration. Galt's extensive tracking of agro-pesticide flows seems to fit this bill, linking the household cropping scale (Galt 2008) to global economic circulation (Galt 2010). But Truelove's (2011) exploration of residential water use in Delhi fits too, in the way it depends both on an ecology of women's bodies/health in terms of water access, as well as the broader configuration of the city's political space economy and legal landscape. So too, Biehler's comprehensive analysis of urban pest ecologies can be recognized as political ecology in this vein, with its focus both on the detailed ecologies of the co-evolution of bedbugs, cockroaches, and rats with the pest control strategies (chemicals, traps, and screens) that accompanied them, all within the complex and racialized political economy of public housing, immigration, and struggles over the class-basis of public health. Rural or urban, productive or reproductive, global North or global South, food or health, contemporary or historical, the classical definition of political ecology is expansive enough to embrace a range of excellent research.

One might therefore recognize this kind of work everywhere and see the ongoing utility and power of explanation in this mold. This definition does little to account for, or explain, however, the full, diverse assemblage of approaches and research projects that fly under the banner of political ecology. On the one hand, the field increasingly includes work that is perhaps more familiar if described under the title of land change science, vulnerability research, or environmental health: examining regression residuals derived from a large set of cases to explain the success or failure of environmental cooperation (Agrawal and Chhatre 2011); using modeling to trace the socio-economic factors influencing species invasion (Brenner 2010);

using adaptive capacity as a framework for understanding differential vulnerability to climate change (Tschakert 2012); using socio-ecological systems as a lens into disease dynamics (Crews 2013). In recent years political ecology has adopted and evolved a set of scientific techniques and ideas that appear to exceed "political economy," both as a field and a conceptual framework. These approaches are heterogeneous, but share an insistence on the reliable materiality of measurable social and natural conditions and a growing set of rigorous techniques to assess them.

On the other hand, the traditional definition of political ecology does little to capture the revolution in rigorous historical and postcolonial environmental deconstruction that has also accompanied the evolution of political ecology. Wainwright's (2008) account of the Maya farm system is not dedicated to the ecological conditions of indigenous farming alone (or really at all), but instead to the invention of a "Maya farm system" through the deployment of colonial and developmentalist expertise, an invention resulting in the silencing of local people. Similarly, in Davis' compelling political ecological history of North African deforestation (2007), only a short few pages are given to the physical evidence for vegetation change in the region, with the remaining hundreds of pages dedicated to explaining the colonial and developmentalist roots of the unfounded arguments for desertification. Kull's *Isle of Fire* (2004) provides a fair bit of evidence on the ecological impacts of fire, but a great deal more ink is spilled in explaining the roots of problematic ideas about native fire use by state and international authorities. Lave's (2012) investigation of the context from which stream restoration theory and practice emerge shows how a specific school of technique and training comes to supersede others, owing to external economic shifts as well as the internal political economy of the field of restoration itself.

This kind of work is not merely an effort to "disprove" the weaknesses or fallacies of one or another competing explanation of environmental outcomes or change. Instead, it is an ambitious effort to precisely map the power-laden source and circulations of discourses and legitimate authority that forge ecological truth. So, political ecology seems also to have evolved and maintained a critical deconstructive apparatus that exceeds the "concerns of ecology" no matter how broadly defined. And this apparatus maintains an ambivalent, if not wary, relationship to the simple models of materiality and measurability that undergird expert practices and methods, most of which political ecologists themselves have adopted.

Forest, science, knowledge

What this suggests then is that, at least since 1987, political ecology has shown twin tendencies: evolution and ambivalence. The field has shown an uncanny ability to evolve, adopting a range of scientific techniques and concepts as it goes, often entering into complex, interdisciplinary team research programs. At the same time, however, the field has shown an ongoing tendency to oscillate between in-depth chains of empirical case explanation using these ever-evolving techniques, and a core effort at critical historical deconstruction of environmental science itself. Political ecology, in this sense, often seeks to have it both ways, as a rigorous participant in the advancement of scientific knowledge and a relentless critic of scientific ecological practice.

Consider the hallmark political ecology of James Fairhead and Melissa Leach, including a number of research papers, but especially their volumes: *Misreading the African Landscape* and *Reframing Deforestation* (1994, 1995, 1996, 1998). That work is notable for the careful use of time-series photography and satellite imagery (especially notable in the early 1990s) to examine rates of forest "island" change in West Africa. Their findings, now widely recognized across multiple academic fields and disciplines as well as in planning circles, are straightforward; forest cover loss is grossly overstated and many areas ham-fistedly regulated as losing forest cover, are actually *gaining* it, owing to local cultural practice and community decision-making. In a sense, this method and the conclusions associated with this work are effectively a form of land change science, albeit one predicated on postcolonial suspicions.

But Fairhead and Leach do not rest there. Not content to leave well enough alone, they go on to provide a rigorous historical analysis of the sources upon which contemporary estimations of deforestation in West Africa rest, carefully dissecting surveys, remotely sensed analyses, and historical records. Each of these, they demonstrate, rests upon previous sets of statistics and categories, all inherited from earlier researchers, who in turn borrowed from the researchers who came before them, back to shaky imperial foundations and before. Forest cover, and estimates of change that emerge therefrom, are merely projections reflected through a series of colonial funhouse mirrors. Traveling towards an epistemological vanishing point, Fairhead and Leach conclude that, to some degree, "it is turtles all the way down": all claims rest on the backs of previous representations, themselves resting on top of earlier texts, metaphors, and claims. As an inevitable result of such "intertextuality" (Barnes and Duncan 1992), current estimates of deforestation are, at best, grossly exaggerated and, at worst, rest on nothing more than colonial hot air.

To be sure, of course, these forest cover estimates are never *only* texts. The human and environmental stakes of claiming local communities either destroy or create forests are enormously high, since the control of local populations, the proliferation of rules that govern livelihoods, and the capacity of working people to reproduce themselves all sit in the crosshairs of forest-cover estimates. Such claims do very real and material work in the world, and Fairhead and Leach's work, by implication, represents an intervention into extremely urgent and practical politics.

Coyote, Raven, Spider

This research has been subject to reasonable critique over the years and is by no means the last word on West African forests. The reason it remains so widely recognized and cited, however, and has become such a point-of-pride for the scholarly community associated with political ecology, is that it elegantly weaves together a core contradictory project: to draw into doubt scientific accounts of environmental conditions or change while proliferating them. Fairhead and Leach engage in land change science even as they undermine it. Conversely, even while they challenge our ability to firmly establish accounts of environmental change outside of ideology and politics, they simultaneously seek to stabilize our understanding of landscapes through rigorous analysis, setting a tone of calm and familiar scientific explanation. In this way, they produce a "moment of chaos," in which the world and our understanding of it are reversed, inverted, and reconstructed. In short, Fairhead and Leach have it both ways.

Indeed, having it both ways is one of the principal characteristics of political ecology. The field pronounces explanatory rules that it sometimes violates, nervously cooperates in joint environmental projects while organizing arguments about their limits, and borrows methods prolifically while producing epistemological critiques of the capacity of these methods to explain.

In this sense, it may be best to think of political ecology as Trickster. As most folk traditions would have it, trickster figures are mischievous subversives, who act in contradictory ways. By turns, Trickster performs as a boastful clown or jester, occasionally acts as a thief, and constantly undermines haughty heroes. But Trickster also protects the weak, transports the seeds of culture, and provides gifts for humanity. Either way, Trickster rarely plays by the rules and is always an agent of often-revolutionary change. The Apache's Coyote tricks little children, Liberia's

Spider deceives a barren woman, and the Haida's Raven steals from the old man by the river. Coyote brings fire; Spider ends famine; Raven brings light.

Whether as the Native American Coyote or Raven, the Icelandic Saemundur, France's Reynert the Fox, Benin's Rabbit, or the Ashanti's Anansi the Spider, Trickster is a powerful mythic actor, both a hero and villain, but one whose activities are almost always activated in relationship to other figures, whether those are gods or men. Trickster is, in this sense, a relational figure, one always associated with transition or revolution, whose uncanny ability is to reverse roles with other, more powerful, actors. As folklorist Harold Scheub (2012: 31) explains: "This reversal of roles is at the heart of all trickster tales, and of all tales dealing with transitions. We are constantly destroying, being destroyed, and creating, being recreated, reborn. We are constantly in transition." Trickster adopts the forms and positions of others, if only temporarily, to upend the status quo. He schemes new forms to outwit those around him. *Misreading the African Landscape*, in this way, plays Trickster to other sciences.

Mimesis, infection, chaos

To a great degree, this core Trickster quality is one of the central reasons political ecology has persisted, as a familiar kind of approach, if not a field, despite more than 30 years of ontological and epistemological revolutions, fads, and innovations since the publication of *Land Degradation and Society*. This is because political ecology, like Trickster, fosters, and thrives within, a critical dialogue with those characters and actors who surround it.

For political ecology, these other actors represent allied research fields, foundationally including hazards research (Watts 1983b), conservation (Neumann 1998), and environment and development (Hecht and Cockburn 1989), and more recently including fields like land change science (Turner and Robbins 2008) and environment and health (King 2010). Wherever and whenever such fields establish important beachheads in explanation, political ecology emerges to undermine them, demonstrating the power-laden implications of any such foundational account of human/environment relationships. In that moment of engagement, however, political ecology also practices explanation, proliferating new and alternative environmental accounts, which ironically often mobilize the core elements and categories drawn from the very fields with which political ecology jousts. As political ecological arguments begin to enter the vernacular of these related fields, moreover, they begin to change these fields of study, sometimes in dramatic ways. As such, political ecology, as Trickster science, exhibits several core qualities:

1. Political ecologies are relational; their relationships to the other fields they engage are not contingent, but instead necessary and fundamental to political ecology's existence.
2. Political ecologies are mimetic; they tend to reproduce the forms and idioms of the fields with which they are engaged.
3. Political ecologies are infectious; their critical concepts commonly enter the taxonomies and variables of the sciences they engage.
4. Political ecologies are chaotic; founded on a contradictory mandate to advance and undermine explanation, they tend to encourage disorder in other fields.

The creation myth of political ecology itself reflects these very conditions. In most accounts, contemporary political ecology began to cohere in the 1980s as a response to other fields and approaches, including classical cultural ecology, a field dedicated to the study and explanation of cultural practices (especially agricultural practices) within an environmental context (Robbins

2012). Key works of political ecology at the time directed themselves to undermining, politicizing, contextualizing, and deconstructing the core conceptual apparatuses of cultural ecology: adaptation, function, and culture itself (Watts 1983a; Trimbur and Watts 1976).

But this did not stop researchers from precisely mimicking and extending cultural ecological methods and ideas in their work, appropriating them for new purpose. A generation of political ecologists performed research that in many ways reproduced and adopted many of the techniques and concepts of cultural ecology: tracking norms and practices, studying subsistence and survival, writing in ethnographic style, stressing community-level case studies (Gezon 1999; Muldavin 1996; Pelling 1999; Sheridan 1988). As the field matured, the borrowings would become more eclectic and wide-ranging, raiding techniques and technologies like remote sensing (Turner 2003), larger scope studies and surveys (Galt 2010), and so on. Through a kind of creative rearrangement of the fields of hazards and cultural ecology from which it emerged and upon which it turned, political ecology was propelled forward by mimicking the very fields it sought to undermine. In this way, political ecology, ever the Trickster, survives and replicates through a kind of *mimesis*.

At the same time, practitioners outside of political ecology, in more traditional areas of rural development science and cultural ecology, slowly softened to categories adopted from political ecology: class, power, gender, and even discourse. Criticisms from traditional fields began to give way to adoption, at least in limited ways, of core concepts from a broadly defined political economy. Vegetation and land use studies began to adopt critical lens on gender as a category of analysis (Potter 1996; Thapa et al. 1996). Even structure and agency experienced a resurrection in land change science (Chowdhury and Turner 2006).

In each case, an engaged field (like land change science), critiqued but also mimicked by political ecology, works to absorb critical components back into its corpus of research. Working to inoculate itself and to simultaneously improve the breadth of its explanatory capacity, core fields add political ecology to the mix. This, in turn, gives way to a proliferation of special forms of analytical exception, what might be called "*AND" political ecologies*. These include: land change science *and* political ecology (Aldrich et al. 2012; Walker et al. 2009); vulnerability analysis *and* political ecology (Eakin and Luers 2006); sustainability science *and* political ecology (Lawhon and Murphy 2012); environmental health *and* political ecology (Crews 2013). Political ecology advances, in this way, through a process of *infection by critique*.

But these efforts at "grafting" never fully take. Even while political ecology adopts and mimics the practices of its neighbors, lending them conceptual apparatus and perspectives, it remains steadfastly critical of these fields. Critiqued from the view of political ecology, for example, contemporary climate change and vulnerability research continue to mobilize backwards-looking approaches. According to Bassett and Fogelman (2013), notably, despite a serious effort to adopt and advance core concepts from political economy and related suites of critical concepts in human dimensions of climate change research, climate and vulnerability research clings stubbornly to an "adaptation as adjustment" perspective that is narrow and reminiscent of long-abandoned approaches from cultural ecology.

Similarly, land change science has come to embrace a range of core concepts from political ecology, as where Aldrich et al. (2012) stress the importance of adversarial and conflictive relationships between producers (rather than solely autonomous ones) in explaining Amazonian deforestation. Even so, critics continue to stress the way land change science lends itself too easily (and too problematically) to solutions borrowed from neoliberal economics (Lestrelin et al. 2013). The field's meta-technics and ambitions are therefore criticized for matching too closely those of "the territorial state, global investors, and first world consumers" (Robbins 2012: 148).

Likewise, sustainability science has made enormous headway by mobilizing concepts from political economy, including labor power (Yeh et al. 2014) and land security (Olabisi 2012). Even so, core and foundational concepts in the field, like carrying capacity (still a major trope in the field, see: Mihelcic et al. 2003; McMichael et al. 2003), receive ongoing critical attention in political ecology, which stresses the twisted metaphorical roots of such concepts (Sayre 2008).

Thus, ever-apolitical ecologies, no matter how well-dressed in the trappings of political economy, remain suspect and subject to withering critique. Still wedded to the "moment of chaos" that overturns the hegemony of environmental explanation, political ecology continues to play the role of Trickster, and keep its partners at arm's length.

Inside and outside

In sum, political ecology co-evolves with the fields around it, especially those fields in which it is most critically engaged. It mimics and mobilizes the insights of allied sciences and lends categories and concerns to other fields, even while confronting and subverting these fields. This awkwardly puts political ecology simultaneously both inside and outside many frontier areas of research. Chief amongst these are the fields of water resource management (Bakker 2004), urban ecology (Swyngedouw and Heynan 2006), ecological novelty (Robbins and Moore 2013), environmental health (King 2010) and land change science (Brannstrom and Vadjunec 2013). The congruence of these last two fields in particular for political ecology emerges both from their intertwined histories as well as their parallel concerns. Both of these fields make likely sites for engagement, therefore.

Land change science: flirting with objectivism

Land change science is a field that links remote sensing of landscape and modeling of human–environment interactions to explain and predict the trajectory of changes on the earth's surface, notably including deforestation and expansion of cultivated land (Turner et al. 2007). It shares a common history with political ecology, rooted in hazards analysis, adaptation, and an interest in land cover, land degradation, and local knowledge-based farming systems (Turner and Robbins 2008).

For many practitioners, the effort to link land change science and political ecology is therefore a natural and not overwhelming problematic one. In a notable example, Aldrich et al. (2012) forge this link by "fusing" the conceptual frameworks of political ecology (attention to power relationships, struggles over land, and so on) with the methodological approaches of land change science (remote sensing of property-scale land change, regression using multiple variables, and so on). The results of their research in the Amazon convincingly undermine traditional arguments, explaining land cover change, which stress microeconomic optimization, in favor of one that stresses complex and adversarial interactions. This certainly suggests that the linkage of political ecology with other fields is a merely practical one.

But, ever the Trickster, political ecology cannot rest there. Consider efforts by Wainwright et al. (2013) to quantify forest cover change in southern Belize. Their land cover change assessment is animated by explicitly political questions, since Mayan claims to land are challenged constantly by state authorities who question their ability to be stewards, stressing destruction of forests under Mayan control.

Their methodology, however, hews closely to traditional, objectivist, land change science. Using Landsat imagery in time-series analysis, they track the status of forest cover in Maya lands between 1975 and 2011, frequencies of cultivation, fallowing, and forest transitions. The analysis demonstrates that forests are not disappearing in Mayan areas, a fact which directly undermines the state's claims, though, as they point out, it doesn't tell the Maya anything they didn't already know. In this way, Wainwright et al. travel through the interior of a land change science question, applying sophisticated technological analysis to advance a key explanation with high political stakes.

Simultaneously, however, they sow chaos in their conclusions. Such studies, they argue, are extremely unlikely to create or foster the political outcomes that the authors seek, in part because the nature of the study itself – and its foundational objectivist mode of measurement – cannot, on its own, produce the creative questioning and reflection necessary to decolonize Mayan territories in Belize. Indeed, the very techniques and technologies employed act upon the researchers, the Maya, and the world in a way that may take us further from any such real confrontation:

> The essence of a study such as this one is to produce a specific representation of the world, one that is rigorous precisely because of its objectification of the land cover of Southern Belize and its categorical reduction into forest and non-forest. The existence of satellites and their capacities to objectify the world changes and reflects our being.
>
> *(Wainwright et al. 2013: 186)*

Such objectification, Wainwright and his co-authors insist, drawing upon Heidegger, in a final burst of classic political ecological mischief, actually reinforces the pernicious modern project of the "the conquest of the world as picture." They reject the otherwise-intuitive claim that because decision-makers and politicians understand the mechanics of models and maps, such approaches are politically more efficacious than other forms of confrontation. Wainwright and his co-authors insist instead that the urge to such forms of explanation move allies of the Maya further away from, rather than closer to, their political goals, and away from the possibility of more *just* outcomes. Reversing his role at the last minute, from land change hero back to political ecological Trickster, Wainwright and his colleagues end their journey into science once again on the outside, on the margin, where Trickster lives.

Environment and health: undermining and adopting

Much the same kind of fraught engagement can be seen at the borderlands between health research and political ecology. As Crews and King (2013) point out, political ecology has shown itself to be flexible enough to extend to explanations of unequal access to health services by class and gender, explicitly addressing power-laden patterns of disease exposure. Fusing the two can be, to some degree, a practical affair.

Hausermann et al.'s (2012) work on Burelli's ulcer is emblematic. A skin infection caused by the bacterium *Mycobacterium ulcerans*, the ulcer is rife across many parts of tropical and semi-tropical sub-Saharan Africa. By following a chain of causation familiar from political ecology, Hausermann shows how outbreaks of the disease are occurring where the surface hydrology and water flow have been changed by illegal and artisanal mining, which are in turn accelerated by poorly enforced regulatory and land ownership laws in countries like Ghana, and spurred onwards by the global runup in gold prices and the increasing incidence of foreign land ownership in Africa. As with land change science, political ecology and environmental health research can be merged.

But this interface is always problematic. Consider Guthman's (2011) efforts to assess the obesity epidemic in North America in her book *Weighing In*. A piece of classic political ecology, Guthman interrogates the claims made in the noisy field of environment and health. She demonstrates the way mortality/obesity relationships are massaged out of dubious studies, deconstructs arguments for obesogenic environments based on their blatant scalar inadequacies, and walks the reader through the mechanical simplifications of the energy-balance models used to explain weight gain. She further suggests that the allocation of responsibility and blame, which lie in the heart of such explanations, are fundamentally *unjust*. In this way, Guthman acts as the Fairhead and Leach of the human body, showing the largely taken-for-granted, common-sense global epidemic of obesity rests on shaky ground, poor science, and logics rooted in the reproduction of consumer economies. The work, in a sense, undermines environment and health research, at least in the area of obesity.

It does not do so, however, without advancing its own environmental and health arguments. Specifically Guthman raises the question as to whether Endocrine Disrupting Chemicals (EDCs) may play a key role in cell multiplication – and so, larger body mass in populations worldwide. EDCs are not rooted in the behavior of obese people, in the planning of towns or neighborhoods, or in any of the dietary fixations of the obesity research community. They are instead a product of a complex political economics of modern chemical processing, a ubiquitous component of the modern capitalist economy. In this way, Guthman uses EDCs as an erosive agent to challenge the other incoherent places one might lay blame for obesity (laziness, neighborhood sidewalks, and so on). Like a clever lawyer, Guthman brings these villains, subjects of a dawning frontier of objective, rigorously scientific investigation (Newbold et al. 2008), on stage only so long as is necessary to sow the seeds of doubt in the jury's mind concerning obesity. Even so, after dozens of pages that deconstruct and undermine environmental and health "truths," Guthman's arguments take a seditious turn towards foundational and objectivist science.

This is emblematic of political ecology's contradictory project. Having lurked outside on the margins of environmental health and played Trickster throughout her volume, Guthman still insists on having it both ways, introducing an "alternative theory of the crime" rooted directly inside frontier environmental toxicology. As Trickster science, this dual position is awkward but provocative, contradictory but productive.

Trickster Science: Living with Contradiction

The cases of land change science and environmental health underline both the important ongoing relationship of political ecology to other fields (arguably political ecology cannot exist without them), as well as the necessary impossibility of reconciling them into a single, stable, explanatory edifice.

This is because political ecology occurs at that moment when the human–environment accounts that we assemble become precarious and unstable. There is always a need to empirically explain important outcomes in land change, human vulnerability, environmental hazards/risk, and human health, especially relative to the role of power in causing these outcomes. But this need, or drive, is inevitably accompanied by the necessary counter-urge to advance skepticism about any such explanation, its implication in perverse systems of power, and its complicity in reproducing the very systems of power it seeks to unmask. The name we give to these simultaneous urges, and the Trickster stories we tell about them, is political ecology.

Now, political ecology is not an urge that scholars, researchers, activists, or students give in to all the time. As noted elsewhere, many who write political ecologies do so only *some* of the time, in between the practice of other forms of knowledge production. As a sort of double-move, political ecology is one that is often deployed strategically, and at key moments of intervention or mischief.

Still, we might conclude with the question this raises. Does indulging such contradictory urges, indeed erecting an interdisciplinary field of study upon them, have utility?

Certainly skeptics of political ecology think not (Vayda 2009). Tactically, they might argue, the language of political ecology is one without a broad enough audience in the halls of power to make a difference; bureaucrats speak positive models, not negative cases. Strategically, they might claim, any intellectual enterprise whose purview it is to undermine important scientific efforts (e.g. land change science), however intellectually interesting, is ultimately a liability; the National Academy of Sciences speaks objectivism, not deconstruction. Epistemologically, they might assert, the effort to "have it both ways" is foul play; researchers can't be allowed to make materially grounded assertions with one hand while pulling away the explanatory power of objectivism with the other. For all these reasons, trying to "have it both ways" might seem misguided.

Yet a thought experiment positing the absence of political ecology, however contradictory, suggests a terrible gap. Environmental deconstruction that cannot advance and recognize urgent material relationships has only limited utility in a world of very real environmental problems. Conversely, environmental research that cannot reflexively locate its relationship to power is self-evidently dangerous and has indeed proven truly violent in a world of urgent environmental justice challenges. The pursuit of one seems, almost inevitably, to give rise to the prodding of the other.

And so political ecology persists, despite calls for its reformation, its abandonment, and its absorption into other fields or approaches. It persists precisely, it would seem, because it repeatedly emerges from the instability of other fields of study, developing time-and-again from the internal contradictions and weaknesses that lie within these very fields. Political ecology is a symptom of the larger problem inherent in the rigorous pursuit of knowledge in a world filled with contradictions.

So, judging the momentum of changing material conditions of the Earth system itself, political ecology becomes more pressing all the time. Trickster stories are thought to do a certain kind of intellectual and emotional work on those who hear them, after all. Because Trickster is a benefactor and trouble-maker, he lays bare his own weaknesses as well as those around him, and shows the world in shades of gray. In a sense then, Trickster's central role is to make us sensitive to the inevitability and the power of contradiction. For environmental science and action in the Anthropocene, where what is "natural" is no longer easy to tell from what is "unnatural," and where what humans do both makes and destroys the world, the need to accept and interrogate such contradictions remains paramount. As with the cases of Maya land rights or the power over obese bodies, to behave ethically, we must explain. But any attention to ethics and justice necessarily brings with it the need to interrogate and undermine such explanation itself. And herein lies political ecology's utility. It abides, scheming new ways to extend, mimic, adopt, and upend the scientific fields around it in pursuit of more just outcomes.

References

Agrawal, A., and A. Chhatre. 2011. Strengthening causal inference through qualitative analysis of regression residuals: explaining forest governance in the Indian Himalaya. *Environment and Planning A* 43 (2): 328–346.

Aldrich, S., R. Walker, C. Simmons, M. Caldas and S. Perz. 2012. Contentious land change in the Amazon's arc of deforestation. *Annals of the Association of American Geographers* 102 (1): 103–128.

Bakker, K. 2004. *An Uncooperative Commodity: Privatizing Water in England and Wales*. Oxford: Oxford University Press.

Barnes, T.J., and J.S. Duncan. 1992. Introduction: writing worlds. In *Writing Worlds: Discourse, Text, and Metaphor in the Representation of Landscape*, eds. T.J. Barnes and J.S. Duncan, pp. 1–17. New York: Routledge.

Bassett, T.J., and C. Fogelman. 2013. Deja vu or something new? The adaptation concept in the climate change literature. *Geoforum* 48: 42–53.

Blaikie, P. 1999. A review of political ecology. *Zeitschrift fur Wirtschaftsgeographie* 43: 131–147.

Blaikie, P. 2008. Epilogue: towards a future for political ecology that works. *Geoforum* 39 (2): 765–772.

Blaikie, P., and H. Brookfield. 1987. *Land Degradation and Society*. London and New York: Methuen and Co. Ltd.

Brannstrom, C., and J. M. Vadjunec, eds. 2013. *Land Change Science, Political Ecology, and Sustainability: Synergies and Divergences*. New York: Routledge.

Brenner, J.C. 2010. What drives the conversion of native rangeland to buffelgrass (Pennisetum ciliare) pasture in Mexico's Sonoran Desert? The social dimensions of a biological invasion. *Human Ecology* 38 (4): 495–505.

Chowdhury, R.R., and B.L. Turner. 2006. Reconciling agency and structure in empirical analysis: smallholder land use in the southern Yucatan, Mexico. *Annals of the Association of American Geographers* 96 (2): 302–322.

Crews, K.A. 2013. *Positioning Health in a Socio-ecological Systems Framework*. London: Taylor & Francis.

Crews, K., and B. King. 2013. Human health at the nexus of ecologies and politics. In *Ecologies and Politics of Health*, eds. B. King and K. Crews, pp. 2–12. New York: Routledge.

Davis, D.K. 2007. *Resurrecting the Granary of Rome: Environmental History and French Colonial Expansion in North Africa*. Athens, OH: Ohio University Press.

Eakin, H., and A.L. Luers. 2006. Assessing the vulnerability of social-environmental systems. In *Annual Review of Environment and Resources*, pp. 365–394. Palo Alto: Annual Reviews.

Fairhead, J., and M. Leach. 1994. Contested forests. *African Affairs* 93: 481–512.

Fairhead, J., and M. Leach. 1995. False forest history, complicit social analysis: rethinking some West African environmental narratives. *World Development* 23 (6): 1023–1035.

Fairhead, J., and M. Leach. 1996. *Misreading the African Landscape: Society and Ecology in a Forest-Savanna Mosaic*. Cambridge: Cambridge University Press.

Fairhead, J., and M. Leach. 1998. *Reframing Deforestation: Global Analysis and Local Realities: Studies in West Africa*. New York: Routledge.

Galt, R.E. 2008. Toward an integrated understanding of pesticide use intensity in Costa Rican vegetable farming. *Human Ecology* 36 (5): 655–677.

Galt, R.E. 2010. Scaling up political ecology: the case of illegal pesticides on fresh vegetables imported into the United States, 1996–2006. *Annals of the Association of American Geographers* 100 (2): 327–355.

Gezon, L.L. 1999. Of shrimps and spirit possession: toward a political ecology of resource management in northern Madagascar. *American Anthropologist* 101 (1): 58–67.

Guthman, J. 2011. *Weighing In: Obesity, Food Justice, and the Limits of Capitalism*. Berkeley: University of California Press.

Hausermann, H., P. Tschakert, E.A.H. Smithwick, D. Ferring, R. Amankwah, E. Klutse, J. Hagarty, and L. Kromel. 2012. Contours of risk: spatializing human behaviors to understand disease dynamics in changing landscapes. *Ecohealth* 9 (3): 251–255.

Hecht, S., and A. Cockburn. 1989. *The Fate of the Forest: Developers, Destroyers and Defenders of the Amazon*. London: Verso.

Hyde, L. 1998. *Trickster Makes this World: Mischief, Myth, and Art*. New York: Ferrar, Straus and Giroux.

Kepe, T., J.F. Bissonnette, and D.J. Roberts. 2008. Why are students attracted to political ecology? *Environment and Planning A* 40 (11): 2539–2543.

King, B. 2010. Political ecologies of health. *Progress in Human Geography* 34 (1): 38–55.

King, B., and K.A. Crews, eds. 2013. *Ecologies and Politics of Health*. New York: Routledge.

Kull, C. 2004. *Isle of Fire: The Political Ecology of Landscape Burning in Madagascar*. Chicago: University of Chicago Press.

Lave, R. 2012. Bridging political ecology and STS: a field analysis of the Rosgen wars. *Annals of the Association of American Geographers* 102 (2): 366–382.

Lawhon, M., and J.T. Murphy. 2012. Socio-technical regimes and sustainability transitions: insights from political ecology. *Progress in Human Geography* 36 (3): 354–378.

Lestrelin, G., J.-C. Castella, and J. Fox. 2013. Forest transitions in Southeast Asia: synergies and shortcomings in land-change science and political ecology. In *Land-Change Science and Political Ecology for Sustainability. Synergies and Divergences*, eds. C. Brannstrom and J. Vadjunec, pp. 48–65. New York: Earthscan.

McMichael, A.J., C.D. Butler, and C. Folke. 2003. New visions for addressing sustainability. *Science* 302 (5652): 1919–1920.

Mihelcic, J.R., J.C. Crittenden, M.J. Small, D.R. Shonnard, D.R. Hokanson, Q. Zhang, H. Chen, S.A. Sorby, V.U. James, J.W. Sutherland, and J.L. Schnoor. 2003. Sustainability science and engineering: the emergence of a new metadiscipline. *Environmental Science & Technology* 37 (23): 5314–5324.

Muldavin, J.S.S. 1996. The political ecology of agrarian reform in China: the case of Helongjiang Province. In *Liberation Ecologies: Environment, Development, and Social Movements*, eds. R. Peet and M. Watts, pp. 227–259. New York: Routledge.

Neumann, R.P. 1998. *Imposing Wilderness: Struggles over Livelihood and Nature Preservation in Africa*. Berkeley: University of California Press.

Newbold, R.R., E. Padilla-Banks, W.N. Jefferson, and J.J. Heindel. 2008. Effects of endocrine disruptors on obesity. *International Journal of Andrology* 31 (2): 201–207.

Olabisi, L.S. 2012. Uncovering the root causes of soil erosion in the Philippines. *Society & Natural Resources* 25 (1–3): 37–51.

Pelling, M. 1999. The political ecology of flood hazard in urban Guyana. *Geoforum* 30 (3): 249–261.

Potter, L.M. 1996. The dynamics of Imperata: historical overview and current farmer perspectives, with special reference to South Kalimantan, Indonesia. *Agroforestry Systems* 36 (1–3): 31–51.

Radin, P. 1956. *The Trickster: A Study in Native American Mythology*. New York: Schocken Books.

Robbins, P. 2012. *Political Ecology: A Critical Introduction*. 2nd Edition. New York: Blackwell.

Robbins, P., and S. Moore. 2013. Ecological anxiety disorder: diagnosing the politics of the Anthropocene. *Cultural Geographies* 20 (1): 3–19.

Sayre, N.F. 2008. The genesis, history, and limits of carrying capacity. *Annals of the Association of American Geographers* 98 (1): 120–134.

Scheub, H. 2012. *Trickster and Hero: Two Characters in the Oral and Written Traditions of the World*. Madison: University of Wisconsin Press.

Sheridan, T.E. 1988. *Where the Dove Calls: The Political Ecology of a Peasant Corporate Community in Northwestern Mexico*. Tucson: University of Arizona Press.

Swyngedouw, E., and N. Heynan, eds. 2006. *In the Nature of Cities*. New York: Routledge.

Thapa, K.K., R.E. Bilsborrow, and L. Murphy. 1996. Deforestation, land use, and women's agricultural activities in the Ecuadorian Amazon. *World Development* 24 (8): 1317–1332.

Trimbur, T.J., and M. Watts. 1976. Are cultural ecologists well adapted? A review of the concept of adaptation. *Proceedings of the Association of American Geographers* 8: 179–183.

Truelove, Y. 2011. (Re-)conceptualizing water inequality in Delhi, India through a feminist political ecology framework. *Geoforum* 42 (2): 143–152.

Tschakert, P. 2012. From impacts to embodied experiences: tracing political ecology in climate change research. *Geografisk Tidsskrift-Danish Journal of Geography* 112 (2): 144–158.

Turner, B. L., and P. Robbins. 2008. Land-change science and political ecology: similarities, differences, and implications for sustainability science. *Annual Review of Environment and Resources* 33 (1): 295–316.

Turner, B.L., E.F. Lambin, and A. Reenberg. 2007. The emergence of land change science for global environmental change and sustainability. *Proceedings of the National Academy of Sciences of the United States of America* 104 (52): 20666–20671.

Turner, M.D. 2003. Methodological reflections on the use of remote sensing and geographic information science in human ecological research. *Human Ecology* 31 (2): 255–279.

Vayda, A.P. 2009. *Explaining Human Actions and Environmental Changes*. Lanham: Alta Mira.

Vayda, A.P., and B.B. Walters. 1999. Against political ecology. *Human Ecology* 27 (1): 167–179.

Wainwright, J. 2008. *Decolonizing Development: Colonial Power and the Maya*. New York: Blackwell.

Wainwright, J., S. Jiang, and D. Liu. 2013. Deforestation and the world as representation: the Maya forest of southern Belize. In *Land Change Science, Poltical Ecology, and Sustainability*, eds. C. Brannstrom and J.M. Vadjunec, pp. 169–190. New York: Routledge.

Walker, R., J. Browder, E. Arima, C. Simmons, R. Pereira, M. Caldas, R. Shirota, and S. de Zen. 2009. Ranching and the new global range: Amazonia in the 21st century. *Geoforum* 40 (5): 732–745.

Watts, M. 1983a. *Silent Violence: Food, Famine and Peasantry in Northern Nigeria*. Berkeley: University of California Press.
Watts, M.J. 1983b. On the poverty of theory: natural hazards research in context. In *Interpretations of Calamity*, ed. K. Hewitt, pp. 231–262. Boston: Allen & Unwin.
Yeh, E.T., Y. Nyima, K.A. Hopping, and J.A. Klein. 2014. Tibetan pastoralists' vulnerability to climate change: a political ecology analysis of snowstorm coping capacity. *Human Ecology* 42 (1): 61–74.

7

FROM CRITIQUE TO EXPERIMENT?

Rethinking political ecology for the Anthropocene

Bruce Braun

> Critique has everything – a tribunal of the peace, a registration room, a register – except the power of a new politics which would overturn the image of thought.
>
> *(Deleuze 1994: 137)*

> Are we not like those mechanical toys that endlessly make the same gesture when everything else has changed around them?
>
> *(Latour 2004: 225)*

Since its emergence in the late 1970s and early 1980s, political ecology has often been described as a 'critical' enterprise. While the story has been told in different ways (see the editors' introduction to this volume), a common narrative holds that in adopting a critical stance political ecology positioned itself against what were deemed 'apolitical' and 'uncritical' approaches to environmental crisis and environmental change.[1] The task that the nascent field set was nothing short of unveiling the political and economic causes of environmental change and mapping their uneven effects. It sought to show not only how environmental change was political through and through, but that theories of 'ecoscarcity' and 'modernization', and concepts and practices of 'bourgeois environmentalism', were themselves political insofar as they naturalized, and thus rendered invisible, a set of power relations and forces shaping environmental change.

For many political ecologists, Michael Watts' *Silent Violence* (1983) best embodied the critical turn, revealing drought and famine in West Africa to have causes that were social and political, rather than simply the outcome of inadequate knowledge, too many people, or 'backward' technologies. Indeed, a glance through publications in the 1980s reveals key themes from the period that have endured to the present: the problem of unequal access to resources and uneven distribution of the costs and benefits of environmental change, the ways in which practices of local land managers are shaped by larger forces and structures – market relations, property systems, state bureaucracies, legal systems, scientific paradigms, even state violence – and how these operated across a range of scales, from the local to the global. With the rise of neoliberalism, these themes expanded to include the construction of markets to solve environmental problems, the financialization of environmental governance, new modes of 'environmentality', and the openings and limits that accompanied each. If political ecology had a calling card, it was most certainly that it sought to 'get behind' the actions of land managers (and, at times, consumers) in order to reveal hidden power relations, and, as the field matured, to attend to how people contested them, or coped with the changes occurring around them.[2]

Critique is in political ecology's DNA. But might something have been lost along the way? Is critique as it has been practiced by political ecologists sufficient for the field's objectives and goals? Or have we become like those mechanical toys that endlessly go through the same motions even as everything around them has changed? In recent years it is critique itself that has come under scrutiny, and in its place calls made for more 'experimental', 'inventive', 'affirmative', 'creative' and 'playful' engagements with the socio-ecological worlds in which we live. Debates over the role of art and aesthetics, the place of imagination, speculation and wonder, even the need for utopian thinking, have garnered increased attention, to the point where a 'post-critical' political ecology is not impossible to imagine or at least to experiment with.[3] Perhaps most pronounced has been the shift toward experimentation as a new critical – or perhaps post-critical – practice.[4] What lies behind this impulse? To what problems, failures or urgencies does this 'experimental turn' respond, and what about existing critical practices is found wanting? What does the turn to experimentation promise, what might be its dangers, and what might it forget? This chapter seeks to understand and evaluate this turn, not in order to blindly endorse or dismiss it, but in order to locate new and effective ways of engaging with, inhabiting, and potentially transforming the complex socio-ecological worlds in which we today live.

In what follows I explore three propositions. First, that the experimental turn in political ecology must be understood in part as a response to the apparent exhaustion of the critical stance. This will require examining on what basis critique is said to have 'run out of steam', and why 'experimentation', 'play' and 'invention' have come into view as, if not solutions, at least necessary supplements. Second, I will propose that the experimental turn must also be seen as a response to the perplexities of the Anthropocene, and in particular, the challenges it poses to conservation, governance and environmentalism more generally.[5] This will require examining the ways in which the concept of the Anthropocene has introduced new ontological and epistemological questions that challenge divisions between 'nature' and 'society', destabilize the ground upon which critique stands, and place in question our ability to know and predict socio-ecological futures. With the Anthropocene it is not simply that critique is found wanting, but that in the absence of firm ontological and epistemological foundations socio-ecological assemblages are seen as experimental top to bottom – the outcome of compositional practices without aid of ecological baselines or certain knowledge. Third, I will suggest that the 'experimental turn' can also be seen as the expression of an emancipatory or at least democratic desire that in important respects extends political ecology's critical impulse by exploring and cultivating means by which we might know and live otherwise in the face of forces that shape, channel or restrict the composition of socio-ecological worlds. The experimental turn, I suggest, responds to all three, albeit in different ways. Less clear is whether the experimental turn signals the end of critique, or as I suggest in conclusion, requires its redefinition in order to attend to the limits and politics of experimentation itself.

Beyond critique: constructing cosmopolitical experiments

If critique is said to be in crisis, why is this so? What might a 'post-critical' political ecology look like? We can arrive at an initial answer to these questions by way of the recent work of Sarah Whatmore and colleagues on flood mitigation in a small village in Yorkshire, England (see Whatmore 2013; Lane et al. 2011; Whatmore and Landström 2011).

The story they tell starts in 2000, when a number of towns in the North York moors suffered serious flooding. After the flood, government authorities allocated funding to a flood protection scheme consisting of eight projects, of which one – a floodwall in Pickering – was never completed. As Whatmore and Landström (2011), explain, the UK Environment Agency (EA) had deemed floodwalls to be the most cost-effective way to protect the town, based on calculations made by flood control experts using software packages and models to estimate possible flood events, their effects, and the cost-efficiency of various mitigation schemes. However, the proposal encountered resistance from many residents who worried that the walls would impact the aesthetic values of the old town center and questioned the 'fast' process by which experts had come to their conclusions and recommendations. Delays followed, and before any further action was taken funding for the program was cancelled. Serious flooding occurred again in following years.

It is not difficult to imagine a critical analysis of the situation that revealed, variously, the production of vulnerable populations, the marginalization of particular actors, the retrenchment of the state, or even changing agricultural practices due to market forces or state policies that were changing the landscape and potentially increasing the frequency and scale of flooding incidents. Likewise, one could imagine a critique of the aesthetic values informing resistance to the plan, or a critical analysis of abstract models and expert knowledge, the conceptions of nature that lay behind them or the political interests reflected in them. Whatmore and her colleagues took a different approach: instead of adopting a critical stance, they used funds from the UK's Rural Economy and Land Use (RELU) Programme[6] to invent an 'experimental research apparatus'. This apparatus consisted of a 'competency group' arranged as a collaborative space in which the intermediate stages of expert knowledge production could be collectively explored by experts and town residents alike. Consisting of two flood modelers, three social scientists, eight volunteer residents from Pickering and the town's upstream catchment, a dedicated facilitator and a camcorder operator, the group met bi-monthly, held field-site visits, did archival research, and built, examined, prodded and ran mathematical flood models. Materials generated by members of the group were collected in a password-protected depository that all members could access and add to (see Whatmore 2013 for details).

At first glance this seems a remarkably prosaic project – a few people in a room, with maps, graphs, photographs and models, collectively exploring the movement of water through a landscape. On closer inspection, much more was going on. Perhaps the first thing to note was that the experimental research apparatus was not about gathering 'stakeholders' within a decision-making process aimed at finding a compromise between already established positions. Nor was it the gathering together of already competent knowers in which existing expertise was exchanged or combined. Instead, it was an experiment in knowledge production in which, borrowing the words of Isabelle Stengers, 'the citizens of whom scientific experts speak could be effectively present [and] participate in the invention' of knowledge (Stengers 2000: 160, as quoted in Whatmore and Landström 2011). By working together with materials, technologies and artifacts – photos, video montages, computer models, policy documents – the knowledge practices of all participants – flood modelers and social scientists included – were put at risk. The point of the competency group wasn't simply to empower residents, although that was one of its effects. Nor was it meant to enable resistance to powerful actors, or even to 'debunk' existing models and knowledges. Rather, the point was to empower *the situation* to force thought in those affected by it, and thereby slow down the reasoning of established experts (see also Gibson-Graham and Roelvink 2010). By empowering the situation to force thought, rather than simply empowering local residents to make decisions, the research apparatus had the effect of enabling the matter(s) at stake (the 'stuff' of water, soils, computers) to make a difference in the knowledge produced, building new scientific and political competencies among the people and communities concerned and enabling them to object or intervene in the matters that concerned them.

The result was something akin to a 'redistribution' of scientific and political capacity, achieved by enabling a situation to disrupt an established order of thought and produce *new* possibilities for knowing and acting.[7] Emerging from the competency group were a set of collective knowledge claims that previously had no place in flood planning, inscribed in visualization devises such as maps and computer models that could travel beyond the group, and concretized in the development of an alternative flood plan that emphasized water storage in multiple upstream sites, rather than flood walls through the center of town. The group's model was made public through an event at the local Civic Center, in which community members could examine the group's knowledge claims, get a sense of the working practices that produced the model, and try out the model for themselves on a computer set up for that purpose. From here the model traveled through various media and institutions, gathering a public, multiplying actors and amplifying the matter(s) at stake, in ways no longer controlled by the competency group (see Whatmore and Landström 2011). Eventually, well after Whatmore and her colleagues had left the scene, the model became the basis for a successful bid in a national competition to build new experimental flood control demonstration projects, potentially shifting the way that environmental risks are known and addressed across the UK.

Why stage a *cosmopolitical experiment* rather than merely an *exercise of critique*? For Isabelle Stengers, along with Bruno Latour, the problem with the critical attitude is that it subtracts reality from matters of fact, rather than adds reality to it. This is especially true where critique takes the form of 'social constructivism', in which scientific knowledge is continuously shown to reflect its social and political context, and objects are treated as fetishes, little more than empty screens on which is projected the power of society. What is lost is the difference that matter makes. Rather than seek to continuously show that scientific knowledge is constructed – a position that has been taken up enthusiastically by climate change deniers, the tobacco industry, and big oil – Stengers and Latour ask instead how knowledge production can occur in such a way that the issues or materials at stake can make a difference in the knowledge produced, and people can intervene in matters that concern them. Critique in the form of social constructivism can neither generate and detect difference nor inaugurate new political possibilities.

This concern with the limits and constraints of critique has affinities with other reservations about the critical stance. Eve Sedgwick (1997) famously worried that critique too often did more for the critic than the situation examined. This is especially the case when criticism becomes about revealing the complicity of the *other* critic – that pernicious game played all too often in the academy, in which we strive to show that everyone else is duped whereas we are not. The 'paranoid' reader, Sedgwick explained, seeks to never be taken by surprise. For others it is the political efficacy of critique that needs to be questioned, and in particular the assumption that a direct line can be drawn between critique and social transformation. For geographers Tara Woodyer and Hilary Geoghegan (2013), the problem lies equally in the affective dimensions of critique, which they complain can result in a 'dulling and deadening apprehension' of the world, and rather than inspire action, can leave us 'feeling helpless, depressed and defeated in the presence of unrelenting forces' (p. 196). Gibson-Graham (1996, 2006) argued something similar: more and more descriptions of power, and in particular the power of capital, can have the paradoxical effect of strengthening the very thing described, to the point of rendering it omnipotent. For Judith Butler, on the other hand, the problem with critique, and in particular those forms of Kantian critique that stand in judgment, is that it is unable to bring anything genuinely new or novel into the world. Critique-as-judgment not only separates the critic from the world, arrogating to the critic a transcendent perspective from which the world is judged, but is always at risk of 'subsuming a particular under an already constituted category' (Butler 2002: 213). In its inability to think its own occlusions and its own limits critique-as-judgment fails to grasp the singularity and potentiality of events, and thus too often forecloses on the very transformations for which it purportedly labors.

Despite their differences, a common set of concerns animates these writers: do our critical practices provide new openings, or might they paradoxically close down possibilities for imagining and composing the world otherwise? How might we cultivate new ways of knowing, and new ethical and political possibilities, rather than subsume the world under pre-given categories and political projects? For Sedgwick this meant developing forms of 'reparative reading' that sought to repair the damage of forms of prejudice and violence, rather than simply revealing abuse in more or unexpected places, while for Butler it meant forging new critical practices that did not simply assimilate thought back into the ordering function that it sought to question. For their part, Woodyer and Geoghagen, drawing upon Jane Bennett (2001) and the wider affective turn in political theory, propose the cultivation of 'enchantment', described as an open, ready-to-be-surprised disposition through which we might be struck by the extraordinary found in the ordinary.[8] For Woodyer and Geoghagen, as for Sedgwick and Butler, the point *is* to be taken by surprise.

Behind these efforts we can discern a shared concern with what Deleuze referred to as difference-in-itself and thus with warding off a form of critique or an image of thought in which 'difference becomes an object of representation always in relation to a conceived identity, a judged analogy, an imagined opposition or a perceived similitude' (Deleuze 1994: 138). But whether 'enchantment' or 'wonder' bear the heavy weight placed upon them is open for debate. They can readily appear too wistful, too wedded to the figure of the individual, too confident in the assertion that an 'open disposition' leads in some direct fashion to new ethical attunements to the things and people around us, and too grandiose in its claims about its founding of new political imaginations. In many accounts the individual who encounters the world as a place of wonder has an uncanny resemblance to the nineteenth-century 'flâneur', whose male privilege was left unremarked by Baudelaire but clearly recognized by feminist scholars (e.g. Wolff 1985). Wonder can also be practiced in a detached mode. In this respect the clear merit of Stengers' and Whatmore's 'experimental constructivism' is that it attends to the material arrangements in which situations can force thought, and the various effects that result. For both, it is not enough to simply have an 'open disposition' attuned to the affectations of matter. If experiments are to be 'a trial or venture into the unknown' they must be 'well constructed' (cf. Hans-Jörg Rheinberger 1997). This puts a premium on enacting situations in which actors can multiply, matter(s) can be amplified, and disturbances and openings can occur. In this sense, Whatmore and Landström's 'experimental research apparatus' can be seen as epistemological and political at the same time: expanding the ways in which a situation is able to affect its participants, and in so doing, generate new ideas, new powers, and perhaps new possibilities for composing socio-ecological assemblages otherwise.

The challenge of the Anthropocene

We will return to the experimental constructivism of Whatmore and colleagues, for it may be symptomatic of *other* turns within contemporary political and economic life and in its singular focus on generating difference it may have its own limits and occlusions. But there are other reasons why within political ecology 'experimentation', 'invention', and even 'play' have recently come to the fore. Not only is nature always more than what we imagine it to be, requiring experimental apparatuses that can detect difference, but nature is itself without essence, such that its past, present and future form is itself of the nature of an experiment. This appears ever more the case today, with the advent of the Anthropocene.

Since it was first proposed as the name for a new geologic epoch, the concept of the Anthropocene has had something of a meteoric career (Malm and Hornborg 2014). There are many reasons for this, not least the new life it has breathed into certain fields in the social sciences and humanities that have found in its questions and concerns renewed relevance and legitimacy. More significant for my purposes, it presents us with ontological and epistemological questions that present novel challenges to existing critical practices. The first, and perhaps most commented upon, is that it is no longer possible to imagine 'nature' and 'culture' as distinct ontological domains. With the Anthropocene there is no nature that stands outside human impact: earth systems are seen to be decidedly 'post-natural'. Indeed, the conceit here is that human activity has transformed earth systems to such an extent that a future intelligence, should it care to do so, will be able to read the presence of humanity in the geological record. There is considerable irony in this – for at the same time as it inflates humanity to the level of a geological agent acting on a global scale, the concept of the Anthropocene reminds us that our presence is but one strata in a much larger and longer archive: in what has now become a staple of apocalyptic novels and films, the 'age of humanity' names a moment in which humanity becomes aware (once again) of its future disappearance.[9]

For the purposes of this chapter, the significance of the Anthropocene is found in the added momentum and weight it gives to the experimental turn. This takes several forms. First, acknowledging anthropogenic transformation of earth systems at a global scale means that ecological 'baselines' no longer make much sense when it comes to knowing and inhabiting socio-ecological worlds. In the place of Nature we have the image of a fully worked-over world, and the proliferation of *natures* that are invariably marked by human activity. Today, this notion is increasingly commonplace, to the extent that major conservation organizations, such as the Nature Conservancy and the Breakthrough Institute, have turned their attention from preservation to composition (see Collard et al. 2015). We must 'love our monsters' (Latour 2011), aware that they are not corruptions of a pristine Nature but rather part of the construction of various and variable natures, for which responsibility must be taken and procedures and institutional spaces invented (Braun 2009).

To this is added another wrinkle: with the advent of the Anthropocene it is frequently – although not universally – assumed that anthropogenic changes to the environment have added volatility to earth systems, tipping existing systems out of a long period of relative stability (the Holocene) and pushing them into new and increasingly volatile states. This is an increasingly widespread view, popularized by institutions such as the Stockholm Resilience Center and the Resilience Alliance (cf. Rockström et al. 2009). According to these institutions, not only are socio-ecological systems complex and dynamic, they have been so transformed by humans that they now operate by new rules (cf. Robbins and Moore 2013). This emphasis on ecological novelty has been further reinforced by the ascendance of non-equilibrium thermodynamics and ecology, which have essentially pulled the rug out from under notions of timeless and ordered nature. In the place of 'baselines', we have, at best, multiple 'basins of attraction'; systems can exist in multiple quasi-stable states, but not all of these are necessarily amenable to familiar forms of life (if any life at all). Thus, not only is there no pure nature external to us, but the earth is increasingly seen as a 'fund' of potentiality that can be actualized variably, and earth systems seen to be continuously in flux.[10] Novelty is the new norm, both as a result of, and apart from, human activity. The implications are significant: not only does environmental governance shift from preserving diversity to sustaining nature's potentiality (often measured in terms of 'services'), but in the face of an increasingly volatile and unpredictable world, we are left without certain knowledge of the future. The past is no longer a reliable guide to the future, if it ever was.

As Paul Robbins and Sarah Moore (2013) note, these changes have significant implications for what they call the 'edenic sciences' – conservation biology, restoration ecology and invasion ecology – that 'share a tacit epistemological commitment to evaluating ecological relationships explicitly with regard to an a priori baseline' (p. 4). For these sciences, the issue of ecological novelty is at one and the same time an ontological, epistemological and political challenge, since it undermines the basis on which to make normative judgments or prescriptions about the composition of socio-ecological worlds.

These problems come to a head in practices like 'rewilding' which have attracted the attention of scholars interested not only in knowledge controversies but also in how, by whom, and at what scales future natures are to be invented and composed in a world without ecological foundations or certain ecological knowledge. Itself a response to anthropogenic environmental change, rewilding usually refers to efforts to return worked over or disturbed environments to their apparent 'natural' or 'wild' state. It can also refer to efforts to de-domesticate animals, especially those captured as part of the wild animal trade. Yet, what does it mean to *re*wild, when there is no wild nature that can stand as the place from which the authenticity of any such effort is to be judged? By definition, rewilding can only ever be an experiment in composing the wild, an experiment in making 'new' natures, in which the conservationist is always already implicated in the reality that he or she seeks to remake.

Rosemary Collard's (2014) superb ethnographies of the rewilding of animals captured as part of the wild animal trade demonstrates the irreducible nature of experimentation at the heart of rewilding practices (see also Chapter 9, this volume). Rewilding projects, after all, start with a set of perplexities, rather than certainties. If captured animals are to be released from captivity, what will keep them from being captured again? How can they be 'made wild' not just for a short period of time, but for good? As Collard shows, answering both questions has required the development and deployment of 'misanthropic' practices; that is, practices that teach animals to dislike humans, and to avoid them in the future. Rewilding, then, is not just artifice, it is experimental all the way down; one doesn't quite know what works, or even how to measure success. Rewilding is a journey into the unknown, a wager on what might work, for how long, and with which effects.

Jamie Lorimer and Clemens Driessen (2014) extend this point in the context of efforts to rewild or de-domesticate a polder next to the Dutch town of Lelystad. In this case, scientists and planners sought to determine what a wild landscape would have looked like prior to humans arriving on the scene, in an apparent search for 'baselines' in the absence of undisturbed nature. Were Western European landscapes characterized in their climax stage by continuous forest cover, or, as others have argued, were they characterized by a non-linear shifting mosaic of forest-pasture landscapes due to the influence of large mammals that kept forest growth in check? The rub, of course, is that in order to answer the question scientists had to create the very reality that they studied – from the outset the observer was part of the system observed.

The polder in question was in some senses an ideal candidate for such a rewilding experiment since it was abandoned shortly after its creation and then turned into a nature reserve. Why not simply let it go and see what would happen? Yet determining the 'original' nature of northern Europe was not so easy as introducing large mammals to an abandoned landscape and stepping back to observe what unfolds. As Lorimer and Driessen emphasize, real-world experiments are not so easily contained and controlled – there are always countless other, often unexpected, actors clamoring to be part of the experiment: invasive plant species and migrating animals and birds, for instance, who make a mess of experimental controls, and local residents – including animal rights activists and hunters, who in this case were able to walk up to the boundaries of the polder and witness the starvation and death of previously domesticated but newly 'wild' large mammals including cattle and horses that had been introduced to test the ecologists' theories. What started out as a simple experiment in what constitutes the 'wild', quickly became what Michel Callon (1998) has called a 'hot situation', increasingly entangled with politics, law and economy, and around which publics form, actors proliferate and the matter(s) at stake become amplified.

In the Anthropocene, Lorimer and Driessen explain, age-old distinctions between the 'made' and the 'found', 'artifice' and 'nature', 'wild' and 'tame', even 'order' and 'surprise', have little utility. The lesson to be learned is that in a world where nature is itself experimental all the way down, experiments in 'wilding' must themselves be 'wild experiments' – that is, they need to become *experiments in procedure* in a situation of ecological and political uncertainty, able to incorporate new and unexpected human and nonhuman actors, and making room for different ways of knowing by farmers, hunters, amateur naturalists and local residents. The process must be iterative and pragmatic, starting in the middle of things and learning from moment to moment (cf. Lehman and Nelson 2014). Where Lorimer and Driessen take us to is not a critique of 'wilderness' as an ideological concept, along the lines of Bill Cronon's now infamous 1995 essay. Their project isn't about 'debunking' nor does it take 'ideology critique' as its method. Nor even does it impose normative judgments about what is 'better' or 'worse' when it comes to the new natures constructed. Instead it takes us to a question, and a method, that begins to converge with that of our previous example from rural England: if in the Anthropocene the development of 'new' natures can only ever be experimental, how does one design experiments in nature's production that avoid closure – that is, experiments that are dialogical, open to new actors, both human and nonhuman, and designed to generate surprises rather than confirm what is already known or expected?

The end of critique?

The call for 'wild experiments' has considerable appeal, especially so if our goal is not to withdraw from the world, but actively participate in its composition. But might we also have reason to proceed with caution? Might the experimental turn have limits or occlusions of its own?

Responding directly to the compositionism of Bruno Latour, Paul Robbins and Sarah Moore (2013) suggest that what underlies much work in this vein is a trust in 'good liberal and communicative discussion'. To put it differently, the experimental turn frequently turns on the invocation of the 'dialogic', expanded now to include nonhuman actors who are present through spokespersons whose capacity to represent can always be refused or called into question. The possibility for new ethical and political openings is seen to emerge from staging experiments in such a way that actors are multiplied and the matter(s) at stake amplified.

This begs a series of questions. On the one hand, for all the attention paid to experimentation, invention and play, it is not clear that we have left critique behind. Indeed, it is difficult to avoid the sense that there are implicit yet unspoken norms that structure these accounts, namely the protocols of liberal democracy and its ideal of communicative rationality, and, following from this, an implied critique of an *inadequately democratic* constructivism in which not all affected parties are included. As such the 'post-critical' is itself founded on critique, albeit of a particular kind: on one hand, the *critique of critique*, which rightly finds in critique-as-judgment the closing off of new possibilities, and on the other what might be termed *democratic critique*, in which certain knowledge-making and earth-shaping processes are seen to limit the proliferation of actors and restrict possibilities for new ideas and competencies. In this light, then, the experimental turn actively seeks to invent and cultivate means by which we might know and live otherwise in the face of forces that shape, channel or restrict the composition of socio-ecological worlds, including the practice of critique itself.

Further, it may be that critique is not left behind so much as relocated, no longer *external* to the experimental apparatus but *immanent* to the experiment itself. Certainly this is the strength of Whatmore and colleagues' 'competency group', in which actants of all sorts (human and nonhuman) are empowered to refuse enrollment or demand to be accounted otherwise. It is precisely because the competency group was not a gathering of already given constituencies and competencies, but rather a group that came to its ideas and its competencies through collectively grappling with materials, experimenting with techniques and exploring the potential of various models and technologies, that it could be *generative* in the sense that Whatmore and Landström document. Critique, in this context, is located in the resistance of materials, the disagreements of individuals, and the affordances and constraints of technologies, rather than in a position external to the process that is occupied by the detached critic. Crucially, Whatmore and Landström also powerfully demonstrate that publics do not simply form organically or inevitably around issues, as is often assumed in Michel Callon's (1998) 'hot situations', Latour's (2005) 'making things public' and Marres' (2005) 'issues that spark publics'. Rather, they require what Whatmore calls 'the energetic business' of arousing, triggering and sparking connections between knowledge controversies and emergent publics, in her case by composing a competency group that generated new possibilities in relation to an existing issue, and then by staging an opportunity for publics to form around these possibilities.

This merits attention, for even as the 'dialogic' emphasizes an ethical openness in which one's knowledge is placed at risk, the process of producing situations in which this 'disposition' and these sorts of encounters are possible emerges in Whatmore's work as an explicitly *political* activity, even if this is not the language she or her collaborators use. The creation of a 'competency group' and the public event in which its proposals were made public were indeed fascinating experiments in which the situation was empowered to force thought. This is 'enchantment' understood in the most material of ways. Taking Whatmore's metaphor of 'energetic business' at face value, we can also understand the competency group as a relation of force. This is not a bad thing – indeed, it should be seen in the most positive light. As Wainwright (2005) has powerfully argued, the problem with approaches that emphasize democratic and procedural issues is that they often assume a politics without politics. That is, they posit forums in which disagreement is possible (including the refusal of objects to play the part assigned to them) but with no examination of barriers to participation, and no sense of the histories that brought the present situation into being with its particular controversies and various antagonisms. By emphasizing that experimental situations must be actively composed an irreducible political dimension of experimentation is brought into view and with this a train of further questions. How might we think about relations of power *within* the Pickering group? How might gender have figured or been reworked? What technologies were available and used and which were not, and with what effects on possible scenarios and proposals? And why bring together *these* actors rather than others?

Here we run up against a nagging question that haunts any turn to 'composition' and 'experimentation': is it enough to emphasize generative practices, or do we need also to attend to that which must be eliminated or negated in order that new or radical possibilities are possible? Insofar as it seeks transformative change, does something of the critical stance remain a necessary component of political ecology? Might the experimental turn have its own blindspots and occlusions that require reflection and critique? What is perhaps most striking about the 'experimental turn' in geography, science studies and political ecology is the spatially bounded nature of many of the experimental situations imagined, constructed or studied. In Whatmore and Landström's 'competency group', for instance, the problem is framed in terms of the fluvial dynamics of a *particular* watershed, with the implied goal being one of 'adaptation', even 'resilience', in the face of persistent and damaging floods. As innovative and constructive as the competency group was, must we not ask what gets bracketed out in such a framing? What larger patterns and relations were not examined, or not taken into account? Take, for instance, processes of global climate change, which may be exacerbating local and regional flooding: the experimental research apparatus may have been able to detect and generate difference within the local environs, but to delimit flooding as a 'local' problem in need of 'local' solutions, passes over a great deal, even if funds to mitigate the effects of floods are ultimately procured from the state. The same is true of political economic arrangements that may be transforming the decisions and actions of local farmers. It is often national economic and agricultural policy that finds expression in land use and land cover change; to frame this as a 'watershed' problem risks ignoring other actors and relations that exceed such a bounded sense of place.

For all its novelty in terms of 'knowledge controversies' the Pickering example may fit a bit too neatly within the terms of a now ubiquitous neoliberal understanding of resilience as 'local' adaptation, in which communities and individuals are charged with becoming resilient in the face of changes for which they bear little or no responsibility. Here a crucial question arises: how do we distinguish between experimentation as an opening for new ethical and political attachments and experimentation as a mode of government in which subjects and communities are required to permanently struggle to adapt to changing conditions?

Issues of scale have long been the bread and butter of political ecology in its 'critical' mode. It is precisely the scalar issues bracketed out in narrowly bounded notions of experimentation that lead Paul Robbins and Sarah Moore (2013) to emphasize that experimental ecologies are inherently political in the sense given by Wainwright. It may not be enough to stage experiments, even wild experiments; it may be equally necessary to attend to the histories and relations within which any experiment occurs. By moving to the global South and turning to efforts to 'restore' the native vegetation of Mauritius, Robbins and Moore locate experimentation within a longer geopolitical story of colonialism and capital.[11] Like Lorimer and Driessen, they show that restoration is necessarily an experiment in producing novel ecologies. But both the experiment and its urgency are understood in relation to unequal power relations between the global North and global South, and in the face of anthropogenic climate change that threatens to inundate almost the entire archipelago. The issue in Mauritius is most definitely local (after all, the future of the islands is at stake!), yet at the same time the issue's elements and dimensions stretch around the world in complex arrangements of animals, plants, climate, capital and institutions. While Western scientists proposed the ecological experiments, it could not be assumed in advance that they were desirable for those who inhabit the island, or for those from whose islands organisms were taken as part of the experiment. Nor could it be assumed what form the experiment should take. For Robbins and Moore, the Mauritius case reveals the need to see such experiments as themselves outcomes of struggle and sites of struggle, in which the fact that nature is a multi-species collaboration runs up against strategic and political considerations and where competing desires and relations of power must be negotiated. A similar issue faces indigenous communities in northern Canada, for whom the imperative to be 'experimental' in the face of melting sea ice and warming climate can only appear as the most cruel of demands in the face of a truly global experiment in climate change over which they have virtually no say (cf. Cameron 2012). It is not difficult to imagine staging a 'competency group' among these communities that would come up with plans for adapting to a changing climate, or even challenge the calculations and plans of experts. Such groups may be needed, and northern indigenous communities are certainly as 'experimental' as any other, but the implicit demand to 'adapt' or 'experiment' ignores crucial geopolitical and political economic questions of who is compelled to experiment with new ways of knowing and producing novel socio-ecological forms, and who is in the position to make the demand.

Coda: questioning regimes of truth

The turn to experimentation in political ecology shifts attention beyond critique to practices of composition through which difference can be identified and new knowledges and political possibilities generated. It frequently carries a radical democratic charge, but it also parallels and potentially embodies political and economic transformations that may not share the same radical intent. This may suggest the need to locate a different mode of critique, rather than announcing its end. For Michel Foucault (1997) critique was not a practice that stood in judgment of the world, assimilating difference into known categories. Indeed, critique-as-judgment was to be avoided at all costs. But this did not mean that critique was necessarily at an impasse. Instead, for Foucault critique named a *mode of questioning* that asked after the constitution of the categories themselves as part of an 'art of existence' or 'practice of freedom' that risked a mode of existence unsupported by a regime of truth. Critique was itself an experimental practice – a discipline, a mode of living, a self-making or desubjectivation – within an order that foreclosed on alternative possibilities of ordering. It was the thinking of what remained 'unthought' from within the terms of any particular order.

If, as Foucault insisted, the point of critique is not to object to this or that demand, but rather to ask about the *order* in which a demand becomes legible and possible, then a task for political ecology today may be precisely to ask after the order within which the demand – experiment! – *itself* becomes legible, possible and increasingly ubiquitous. Rather than simply embrace experimentation as the solution to the apparent exhaustion of critique, we might instead ask: under what conditions of knowledge and existence does experimentation become a necessary form for the management and administration of individual and collective life? Might we be witness today to the *experimentalization* of life as part of a mode of government proper to the Anthropocene? If so, is the experimental turn in political ecology and nature–society studies always the site of a hopeful 'post-critical' politics focused on the detection and generation of difference, or might it also be symptomatic of other turns that remain unthought? Perhaps we can imagine an experimental criticism – an 'art of existence' – that is at one and the same time a questioning of experimentation. As a more experimental political ecology emerges to grapple with the risks and potentials of the Anthropocene, this may be one of its most critical tasks.

Notes

1 These uncritical or apolitical approaches were, variously, the Malthusian language of ecoscarcity, the Limits to Growth paradigm of the Club of Rome, modernization theories that understood environmental degradation to result from inadequate knowledge, backward technologies or market inefficiencies, and what were deemed 'bourgeois' environmentalisms that placed a premium on 'wild' nature at the expense of local residents and land users, and were blind to forms of environmental injustice in everyday life.

2 These critiques were prosecuted through a powerful set of analytical and critical tools, borrowed from diverse sources: political economy (and Marxist political economy in particular); Weberian theories of the state and state rationality; science and technology studies; Foucauldian studies of 'power/ knowledge', 'governmentality' and 'biopolitics'; and from feminism, anti-racism and postcolonial

theory. Each added another layer of sophistication and often challenged the silences of earlier critical frameworks, to the point where the phrase 'critical political ecology' appeared in the literature, in an apparent attempt to distance recent work from the inadequately critical political ecology of the past.

3 See for example Laura Ogden's short paper at the 2014 Dimensions of Political Ecology Conference which creatively experimented with 'post-critical' political ecology (Ogden 2014).

4 The use of the term 'experimental' in human geography and political ecology refers less to the empirical testing of theory, such as in positivist science, and more to practices of composition that are meant as ventures into the unknown. For an excellent review of the increasingly frequent yet contentious use of the term in geography and related fields, see Last (2012).

5 The 'Anthropocene' has been proposed as a new geologic epoch in which humanity's transformation of earth systems will be readable in the geological record as a specific strata. Whether or not the category Anthropocene is adopted may be less important than the event in consciousness that it represents (on the 'event' of the Anthropocene, see Colebrook (2014)).

6 The RELU Programme was a collaboration between various UK funding bodies (for details, see Whatmore and Landström (2011)).

7 While Jacques Rancière's distribution of the sensible focuses primarily on politics and aesthetics, rather than science, and while his interest in *dissensus* is framed exclusively in terms of human actors, the affinities between Stengers and Rancière are notable: each is concerned with that 'part which has no part' and imagines how the order of the sensible might be transformed in order to account for them (see Rancière 2004).

8 To this we might add Donna Haraway's (2007) conception of 'play' as that which 'rearranges elements' and 'makes an opening'.

9 Claire Colebrook (2014) puts this nicely, along the way giving cybernetics a unique twist: the human is that organism that annihilates its own milieu. Whether blame for this suicidal tendency should be placed at the feet of capitalism, with its logic of accumulation, or must be seen as inherent to the evolved capacities of the human species, or both, is a question beyond the scope of this chapter.

10 This understanding is easily incorporated within neoliberal forms of conservation, and in particular the financialization of nature (see Nelson 2015).

11 The particular focus in this case was a slow-growing ebony hardwood tree that stabilized the island's soils and whose reproduction was historically dependent on a symbiotic relationship with the now extinct Saddle-backed Mauritius giant tortoise, leading researchers to experiment with introducing an 'alien' tortoise species from another social and ecological context.

References

Bennett, J. 2001. *The Enchantment of Modern Life: Attachment, Crossings and Ethics*. Princeton: Princeton University Press.

Braun, B. 2009. Nature. In N. Castree, D. Demeritt, D. Livermore and B. Rhoads, eds. *A Companion to Environmental Geography*. Oxford: Blackwell, 29–36.

Butler, J. 2002. What is critique? An essay on Foucault's virtue. In David Ingram, ed. *The Political: Readings in Continental Philosophy*. London: Blackwell, 212–228.

Callon, M. 1998. An essay on framing and overflowing: Economic externalities revisited by sociology. In M. Callon, ed. *The Laws of Markets*. M. Callon, ed. Oxford: Blackwell, 244–269.

Cameron, E. 2012. Securing Indigenous politics: A critique of the vulnerability and adaptation approach to the human dimensions of climate change in the Canadian Arctic. *Global Environmental Change* 22 (1): 103–114.

Colebrook, C. 2014. *The Death of the PostHuman: Essays on Extinction, Volume 1*. MPublishing, University of Michigan Library, available at: http://quod.lib.umich.edu/o/ohp/12329362.0001.001.

Collard, R. 2014. Putting animals back together, taking commodities apart. *Annals of the Association of American Geographers* 104 (1): 151–165.

Collard, R., Dempsey, J., and Sundberg, J. 2015. A manifesto for abundant futures. *Annals of the Association of American Geographers*.

Cronon, W. 1995. The trouble with wilderness; or, getting back to the wrong nature. In W. Cronon, ed. *Uncommon Ground: Rethinking the Human Place in Nature*. New York: W.W. Norton & Co., 69–90.

Deleuze, G. 1994 [1968]. *Difference and Repetition*. Trans. by Paul Patton. New York: Columbia University Press.

Foucault, M. 1997. What is critique? In M. Foucault, *The Politics of Truth*, edited by S. Lotringer. Los Angeles: Semiotext(e).

Gibson-Graham, J.K. 1996. *The End of Capitalism (As We Knew It): A Feminist Critique of Political Economy*. Oxford: Blackwell.

Gibson-Graham, J.K. 2006. *A Postcapitalist Politics*. Minneapolis: University of Minnesota Press.

Gibson-Graham, J.K. and Roelvink, G. 2010. An economic ethics for the Anthropocene. *Antipode* 41(S1): 320–346.

Haraway, D. 2007. *When Species Meet*. Minneapolis: University of Minnesota Press.

Lane, S., Odoni, N., Landström, C., Whatmore, S., Ward, N. and Bradley, S. 2011. Doing flood risk science differently: An experiment in radical scientific method. *Transactions of the Institute of British Geographers* 36 (1): 15–36.

Last, A. 2012. Experimental geographies. *Geography Compass* 6 (12): 706–724.

Latour, B. 2004. Why has critique run out of steam? From matters of fact to matters of concern. *Critical Inquiry* 30: 225–248.

Latour, B. 2005. From realpolitik to dingpolitik, or how to make things public. In B. Latour and P. Weibel, eds. *Making Things Public*. Cambridge, MA: MIT Press, 14–43.

Latour, B. 2011. Love your monsters: why we must care for our technologies as we do our children. In M. Shellenberger and T. Nordhaus, eds. *Love Your Monsters: Postenvironmentalism and the Anthropocene*. The Breakthrough Institute, 17–25.

Lehman, J. and Nelson, S. 2014. Experimental politics in the Anthropocene. *Progress in Human Geography* 28 (3): 444–447.

Lorimer, J. and Driessen, C. 2014. Wild experiments at the Oostvaardersplassen: rethinking environmentalism in the Anthropocene. *Transactions of the Institute of British Geographers* 39: 169–181.

Malm, A. and Hornborg, A. 2014. The geology of mankind? A critique of the Anthropocene narrative. *The Anthropocene Review*.

Marres, N. 2005. Issues spark a public into being. A key but often forgotten point of the Lippmann-Dewey debate. In B. Latour and P. Weibel, eds. *Making Things Public*. Cambridge, MA: MIT Press, 208–217.

Nelson, S. 2015. Beyond *The Limits to Growth*: ecology and the neoliberal counterrevolution. *Antipode*.

Ogden, L. 2014. Speculative wonder at the ends of the earth: an experiment in post-critical political ecology. Paper delivered at Dimensions of Political Ecology Conference, Lexington, Kentucky, February 2014.

Rancière, J. 2004. *The Politics of Aesthetics: The Distribution of the Sensible*. Ed. and trans. by Gabriel Rockhill. London: Continuum.

Rheinberger, H.-J. 1997. *Toward a History of Epistemic Things: Synthesizing Proteins in the Test Tube*. Stanford: Stanford University Press.

Robbins, P. and Moore, S. 2013. Ecological anxiety disorder: Diagnosing the politics of the Anthropocene. *Cultural Geographies* 20 (1): 3–19.

Rockström, J., Steffen, W., Noone, K., Persson, Å., Chapin III, F.S., Lambin, E., Lenton, T.M., Scheffer, M., Folke, C., Schellnhuber, H., Nykvist, B., De Wit, C.A., Hughes, T., van der Leeuw, S., Rodhe, H., Sörlin, S., Snyder, P.K., Costanza, R., Svedin, U., Falkenmark, M., Karlberg, L., Corell, R.W., Fabry, V.J., Hansen, J., Walker, B., Liverman, D., Richardson, K., Crutzen, P., and Foley, J. 2009. Planetary boundaries: exploring the safe operating space for humanity. *Economy and Society*, 14 (2).

Sedgwick, E. 1997. Paranoid reading and reparative reading; or, you're so paranoid you probably think this introduction is about you. In E. Sedgwick, ed. *Novel Gazing: Queer Readings in Fiction*. Durham: Duke University Press, 1–40.

Stengers, I. 2000. *The Invention of Modern Science*. Minneapolis: University of Minnesota Press.

Wainwright, J. 2005. Three recent works by Bruno Latour. *Capitalism Nature Socialism* 16 (1): 115–127.

Watts, M. 1983. *Silent Violence: Food, Famine and Peasantry in Northern Nigeria*. Berkeley: University of California Press.

Whatmore, S. 2013. Earthly powers and affective environments: An ontological politics of flood risk. *Theory, Culture and Society* 30 (7/8): 33–50.

Whatmore, S. and Landström, C. 2011. Flood apprentices: An exercise in making things public. *Economy and Society* 40 (4): 582–610.

Wolff, J. 1985. The invisible flâneuse: Women and the literature of modernity. *Theory, Culture and Society* 2 (3): 37–46.

Woodyer, T. and Geoghegan, H. 2013. (Re)enchanting geography? The nature of being critical and the character of critique in human geography. *Progress in Human Geography* 37 (2): 195–214.

PART III

Doing political ecology

The chapters in this section reflect on a set of *issues and questions that arise frequently in connection with the actual practice of conducting research in political ecology*. The chapters emphasize that such research does not take place in a vacuum or under laboratory conditions: it is often explicitly political, connected to current activism and policy. Likewise, the methodological challenges of understanding and representing complex socionatural relations as they actually exist in the world are formidable. The intention of this section is not to provide easy or prescriptive guidelines or formulae, but rather for a number of authors who have wrestled directly with these questions, which we see as perennially relevant and difficult ones for political ecologists, to convey some sense of how they have thought about and engaged with these issues over time in the course of their own work and thinking.

The section begins with two chapters on ethics, one by Juanita Sundberg and the other by Rosemary-Claire Collard, which work through the difficult questions of what it means to engage in "critical" research often focused on contextually embedded differential power relations, when researchers themselves are often products of, and thoroughly situated within, some of the very unequal relationships they are attempting to study. Key questions include who, or what, is considered within an ethical framework; how researchers position themselves and their work relative to those frameworks; and how to think through the perhaps inevitable ethical compromises involved in the conduct of much research. Next, two chapters on methods explore different parts of what is often referred to as the eclectic methodological "toolkit" of political ecology, engaging questions of both research design and implementation in the field. The chapter by Abigail Neely and Thokozile Nguse focuses on the crucial but under-examined roles of research assistants and relationships in much political ecological fieldwork and knowledge production, while Karl Zimmerer's chapter examines the complex role of environmental science in political ecology research.

The next two chapters, one by Nik Heynen and Levi Van Sant and the other by Alex Loftus, explore potential relationships between political ecology – a field that often studies and endorses calls for radical social change – and activism. They ask both how activism has been understood and studied within political ecology, and whether and how research in political ecology might count as or contribute to activism. Finally, the last two chapters in the section, one by Brent McCusker and the other by Tony Bebbington, reflect on their authors' experiences engaging directly in the realms of policy and development practice *as* political ecologists. They speak directly to a seemingly perennial debate in the field over whether such engagements are sorely lacking, as many have contended, or ultimately incompatible with political ecology's political commitments, as others have argued fervently.

8
ETHICS, ENTANGLEMENT, AND POLITICAL ECOLOGY

Juanita Sundberg

Introduction

The subject of research ethics is one that has provoked and aggravated me throughout my entire academic life. The difficulties I encountered doing fieldwork in Guatemala as a white woman and citizen of the United States brought the question of complicity with empire to the fore, forcing me to examine the politics and ethics of knowledge production (Sundberg 2003, 2005). In what ways do geopolitical relations condition research? What are the ethics of producing knowledge under imperial conditions? And, how do geopolitics inform what comes to count as research ethics? Years of struggling with these issues have convinced me there is no *dis*interested place from which to engage in research and, therefore, to practice research ethics. For me, the burning question now is what it means to start from a place of entanglement, as scholars situated in and often beneficiaries of the very politico-economic systems under consideration in our research. This chapter asks how the interpretation and practice of ethics are transformed when knowledge production is framed in terms of entanglement. I begin by examining the ethical dilemmas of research as I see them. I then elaborate the notion of entanglement and ask what this shift implies for ethics.

Before I begin, I wish to note that this reflection is the product of many years of thinking and reading about research ethics and the politics of producing knowledge. These issues are subject to lively debate, especially in feminist, anti-racist scholarship and activism. Hence, my modest goal here is to give the matter my own spin by weaving together threads of analysis spun by a variety of authors across the social sciences. In so doing, I write from my geopolitical and institutional position as a privileged feminist political ecologist who has US citizenship and works in a Canadian university. I address this reflection to students of political ecology and related fields of study who are most likely located in English-speaking countries in the Global North. I use the word "we" when referring to these readers.

Imperialism and research ethics

As a field of study, political ecology has long engaged ethical issues. Historically, normative questions about global inequalities and socio-ecological justice have dominated the field (Bryant and Bailey 1997; Forsyth 2008; Hecht and Cockburn 2010; Neumann 1998; Peluso 1993; Peet and Watts 1996; Rocheleau et al. 1996; Watts 1983). In Raymond Bryant and Lucy Jarosz's (2004: 808) words, political ecology's ethical stance is "one that privileges the rights and concerns (often livelihood-based) of the poor over those of powerful political and economic elites." Political ecologists, then, tend to share a concern with many geographers about "our obligations and responsibilities to 'distant strangers' near and far" (Jarosz 2004: 918). A hallmark of political ecology, fieldwork takes this responsibility seriously by seeking to understand the experiences of "distant strangers."

As a doctoral student and political ecologist ardently committed to socio-ecological justice, I identified with the concerns of marginalized actors in Guatemala whose livelihoods were threatened by the creation of a biosphere reserve and the subsequent imposition of management practices designed to protect an idealized vision of Nature as separate from Culture. While conducting fieldwork in Guatemala, however, I was more likely to be identified with the powerful actors responsible for implementing the biosphere reserve: educated, white 20-somethings from the United States. Close proximity to marginalized *and* elite actors brought the question of complicity to the fore; what was I doing in Guatemala, after all? Undoubtedly, I was more likely to benefit from the biosphere reserve than most Guatemalans. I was the beneficiary of a Fulbright fellowship funded by the US government, also funding the biosphere reserve; and, I would produce a dissertation in furtherance of my career goals. Not surprisingly, local people who gave time and energy to my research continuously asked me the same question posed to *gringos* and *gringas* (United States citizens) working with conservation organizations: how would our presence in the reserve benefit them? My answer was always insufficient.

Ongoing anxiety about conducting research under the sign of US imperialism compelled me to seek out reflections on the politics and ethics of knowledge production. My engagements and contributions to a lively debate on imperialism, epistemology, and ethics focus on objectivity as a taken-for-granted ideal in the social sciences (Sundberg 2003, 2005). Along with numerous feminist scholars, I have been particularly critical of the concept of objectivity, which depends on the idea that the researcher "is not influenced by her/his values, experiences or material conditions" (Staeheli and Lawson 1995: 328). Conventional notions of objectivity arrange the *observer* and the *observed* in a hierarchical spatial relationship: the observer is situated above the observed in order to attain perspective and this distance is assumed to guarantee neutrality, an unbiased view (Haraway 1991; Nightingale 2003; Rose 1993, 1997; Wolf 1995). While this spatial imaginary is problematic in various ways, I have been most interested in how it underpins ideas about fieldwork and invites particular constructions of selfhood.

As a student studying in the Area Studies tradition, Latin America was framed as a series of field sites from which I would choose the most appropriate site for my research interests. Travel between home and field dominated my life as I pursued my master's and doctoral degrees. From my home in Texas, I would travel to Guatemala to observe a particular group of people, a particular place, and a particular set of interactions. Then, I would return home with my fieldnotes, recorded interviews, and other documentary material I had collected. At home, I would analyze this material and engage in theoretical conversations about it with other Latin Americanists. For me, the problem with this framing of *home* and *field* as discrete, bounded locations emerged when I realized that I was not interested in Guatemala per se, meaning Guatemala as a site of distance and difference from home. Rather, I wanted to understand how the Maya Biosphere Reserve was constituted in and through asymmetrical relations between Guatemala and the United States. I wanted to know how *home* and *field*, *here* and *there*, *us* and *them* were bound together, constituted through imperial capitalist relations. These questions put me into conflict with my desire for objective observer status; how could I be an outsider looking in and on when I was entangled in and complicit with these relations?

I also struggled with the expectation that my objective observer status would allow me to produce authoritative knowledge about Guatemala (that somehow would be of relevance in the United States). Latin American Studies is a very *interested* mode of knowledge production, explicitly framed by the US government as instrumental to subjecting and managing distant and different populations (Berger 1995; Berman Santana 1996; Mignolo 2009; Morris-Suzuki 2000). Given the constant reminders of my complicity with empire, I could not comfortably claim to be "above and detached from the world under investigation," as this location assumes a "position of mastery" in relation to the observed (Bondi 1997: 247). Implicitly, I rejected the master-of-all-I-survey, a construction of self that recalls imperial tales of exploration and glorifies European men as the creators of scientific or cartographic knowledge by obscuring the contribution and participation of the many individuals involved, especially Indigenous peoples (Pratt 1992).

Struggles with these questions put me into conflict with political ecology, which provided few answers. For instance, in *Liberation Ecologies* (1996, 2004) Richard Peet and Michael Watts call for a renewed political ecology attentive to power and knowledge. Their discussion, however, does not extend to political ecologists as *producers of knowledge*, even though historically our gaze has been overwhelmingly focused on marginalized peoples primarily in what was once called the Third World. Even as we tend to position ourselves in opposition to imperial and other oppressive forms of (environmental) governance, epistemologies that enact un-located distance and detachment implicate political ecologists in modes of self-fashioning that run counter to this overt political stance (Sundberg 2005).

Taken-for-granted forms of objectivity have long been subject to critique – along with implicit assumptions of mastery. However, recent controversies in geography – like that generated by the Bowman Expedition undertaken in Oaxaca – oblige us to continue questioning how epistemological habits inherited from and operationalized in imperial relations inform conceptualizations of research ethics (Wainwright 2012). I suggest imperial epistemologies frame research ethics as a set of guidelines to direct behavior towards the observed. This is especially the case during fieldwork when the distance between the observer and the observed is breached and the researcher is in direct contact with the objects of research. In this framing, the primary goal of ethics is to minimize direct harm *to the observed* while continuing to pursue research (e.g. the AAG Statement on Professional Ethics). Unquestioned is the right of the researcher to pursue knowledge about whatever, wherever, whenever, and with whomever she chooses. Ultimately, this conception of research ethics safeguards the observer's position of distance and mastery in relation to the observed. And, home remains safely detached from the ethical dilemmas encountered in the field.

To be sure, feminist and anti-racist scholars have developed alternative framings of objectivity to situate knowledge – it comes from somewhere – and collaborative forms of research to shift power relations between researchers and the objects of research (Katz 1994; Nagar 2006; Nightingale 2003; Pratt 2008, 2003; Rocheleau 1995). Nonetheless, it would be naïve to suggest that such forms of research lie beyond the legacies of imperial research and are therefore free of ethical dilemmas. On the contrary, collaborative research generates new kinds of issues requiring further ethical consideration (Pain and Francis 2003). Such dilemmas led Daniel Mato (2000: 487) to suggest moving away from *studying with* Indigenous communities in Venezuela to studying "the practices of hegemonic agents and the global-local articulations of power" in order to make this knowledge available to marginalized communities through collaborative practices. Of course, the practice of studying elites is sure to produce its own ethical challenges (Gould 2010; Han 2010). Moreover, some questions simply do not lend themselves to collaborative research. I learned this the hard way as I tried to find collaborative ways of working on the environmental dimensions of boundary enforcement in the US–Mexico borderlands. To ask the questions that needed to be asked, I could not ally, align, or even ethically immerse myself with the US Border Patrol, the US Fish and Wildlife Service, or liberal environmental organizations. I did ally with organizations such as No More Deaths, not as research partners but political allies, which significantly enriched my research.

Years of reflection have pushed me to conclude there is no pure place from which to engage in research. As scholars trained in the Global North, we are always already marinated in and complicit with geopolitical relations and institutional knowledge that bear traces of imperial histories. And, our positions of privilege place us in asymmetrical relations with the majority of people on earth today. As a consequence, there is no way to conduct research that will absolve us of imperial complicity or ethical dilemmas. Does this mean that political ecologists should stop doing research? Arguably, ethical considerations do imply ending certain kinds of research, like when the researcher's objectives do not align with the interests, capacity, or political goals of marginalized communities with whom they wish to work (Smith 1999; Hodge and Lester 2006; Leeuw et al. 2012; Wainwright 2012). And yet, as Richa Nagar (2002: 181) suggests, the dilemmas of research do not justify the decision to remain silent on the struggles facing marginalized people; rather, she argues, scholars "should accept the challenge of figuring out how to productively engage with and participate in mutually beneficial knowledge production about those struggles." How might we engage with these dilemmas and challenges?

For me, the challenge is how to start from a place of entanglement or how to replace epistemologies that enact hierarchy and distance with those that assume interdependency and entanglement in asymmetrical conditions. Political ecologists are situated in, complicit with, and benefit from the very politico-economic systems that constitute our research subjects. Rather than *observers* who can extricate ourselves from imperial capitalist relations to look down on the practices of others, we are *participants in* these relations. What do ethics entail as a result of this shift? And, what kinds of methodological changes are needed to produce knowledge that reveals how we are intertwined in such systems? In what follows, I argue the concept of ethics must shift from being solely about our proximate relations with the objects of research; an ethics of entanglement requires a transformation of that very relationship at ontological, epistemological, and practical levels. My goal is not to suggest ways of eliminating ethical dilemmas but to invite more accountable responses.

Ethics of entanglement: from fieldwork to homework

To foster this transformation of ethics, I advocate *homework*, or the work one undertakes long before *fieldwork*. Gayatri Spivak (1990) uses the term homework to describe activities that identify the coordinates of one's location in a geopolitical world shaped by coloniality (see also Danius et al. 1993). For Spivak, homework entails a self-reflexive analysis of one's own epistemological and ontological assumptions; in other words, examining how these have been naturalized in academic practices in relation to geopolitical and institutional power relations. Homework is a key practice in *unlearning* that which one has learned; unlearning privilege, especially the privilege of sanctioned ignorance that allows the perpetuation of silence about ongoing colonial violence (Kuokkanen 2007, 2010). Homework starts from where we are but the goal is to move beyond identity politics to take responsibility for the epistemological and ontological worlds we enact through the everyday practices entailed in academic research. Next, I briefly identify some homework tasks at ontological, epistemological, and practical levels, even as I recognize that these are not separate but mutually constituting dimensions of knowledge production.

Ontological tasks

If we accept the argument that the contemporary university system is the outcome of European colonialism, with the social sciences initially institutionalized in Europe and European settler societies, then we are faced with Eurocentrism as its primary structuring feature (Amin 1989; Blaut 1993; Said 1979; Smith 1999; Wallerstein 1999). As James Blaut (1993) suggests, Eurocentric university systems offer "the colonizer's model of the world." Indeed, "colonial relations of power [have] left profound marked [*sic*] not only in the areas of authority, sexuality, knowledge and the economy, but on the general understanding of being as well" (Maldonado-Torres 2003: 243).

A principal task of homework, then, is to examine what Eurocentric models of the world imply for the basic categories used in the social sciences today. For instance, *humans, nature, race, history, progress, labor, property, capitalism*, and the *state* are all categories laden with very particular ontological assumptions. What does it mean when these categories are deployed in scholarship as if they are universally applicable, valid tools with which to analyze the world in all times and places (Chakrabarty 2000)? What forms of ontological violence are authorized when Eurocentric categories are the primary referents of analysis through which "to encode and represent" (Mohanty 1991: 55; Berman Santana 1996; Sundberg 2014)?

For non-Indigenous inhabitants of white supremacist settler societies like the United States and Canada, homework may involve *unsettling the settler* – "the colonizer who lurks within" – to account for how taken-for-granted settler geographies came into being (Regan 2010: 11). Joel Wainwright (2005: 1039), for instance, invites North American and European political ecologists to consider how we "speak of, understand, and read the spaces 'in' which environmental conflicts unfold" so we may better analyze how assumptions of geopolitical spatiality are produced *through* environmental conflicts. Likewise, Akhil Gupta and James Ferguson (1997: 5) suggest, "the prior conceptual segmentation of the world into different cultures, areas, and sites" make it possible "for the world to appear ... as an array of field sites." Taking these categories for granted risks naturalizing them as static, bounded units with particular sociopolitical and ecological traits, available for political ecologists to enter, observe, describe, map, and analyze (see also Coronil 1996). Our first homework task, then, is less about deciding whether or not we should do fieldwork but to determine how and why there is a "radical separation" between home and field, here and there (Gupta and Ferguson 1997: 15; Wainwright 2012).

Epistemological tasks

How does imperialism condition epistemology in Euro-American institutions? This question, posed by Edward Said (1989: 214) many years ago, has been only vaguely addressed and requires much more sustained treatment. Wainwright's (2012: xiii) analysis of the Bowman Expedition to Oaxaca takes on this task, gesturing to the "lingering relationship between empiricism and empire." Wainwright develops this point in relation to epistemologies historically and currently employed in geography that privilege observation and direct experience in the production of knowledge about the world. These epistemologies have their roots in empiricist enquiry, "which assumes that 'the facts' (observations) somehow speak for themselves and are independent of theory" (Barnes 2009: 191).

Doing our homework means questioning the epistemological practices through which knowledge is produced and legitimated in political ecology. Who counts as a legitimate producer of knowledge and why? How do we position ourselves in relation to the *objects of research*? Are we silent about the categories and technologies that make observation possible? Do we recognize the very geopolitical conditions that enable researchers to produce knowledge? To what extent are we involved in processes of extraction that mirror the imperial extraction of raw material (Wainwright 2012: 89)? Are our epistemologies designed to produce knowledge that universalizes particular categories? Or, do they account for partiality and the existence of knowledge grounded in other epistemologies? Whose interests are served?

Practical tasks

In addition to ontological and epistemological tasks, homework entails asking questions about our research interests and the affective attachments or investments driving them. Why do we want to do a particular kind of research, anyway? What is our purpose? Why is fieldwork in faraway places a rite of passage? How are particular places made to appear "available for research" (Gupta and Ferguson 1997: 9)? To what extent do we map "difference onto exotic sites," thereby presuming that "otherness means difference from an unmarked white Western self" (Gupta and Ferguson 1997: 14–15)? And, what makes us personally attached to the idea of going to such places? Is our decision-making driven by theoretical questions, personal investments, or career opportunities? Whose interests are we serving?

For Maori scholar Linda Tuhiwai Smith (1999: 2), these questions are crucial, as white or "Western" researchers often position themselves and their research as "serving a greater good 'for mankind'." Smith (1999: 3) points to "counter-stories" about the "absolute worthlessness [of research] to us, the indigenous world, and its absolute usefulness to those who wielded it as an instrument." As we listen to and reckon with these stories, the challenge is to avoid over-personalizing individual complicity in oppressive relations of power. I have taken heart from community and anti-racist educator Ann Bishop, who suggests that distinguishing between guilt and responsibility requires balancing an "understanding [of] oneself as an individual and as part of a collective reality." "Guilt means taking on all the weight of history as an individual; responsibility means accepting your share of the challenge of changing the situation" (Bishop 2002: 113, 115).

In sum, homework does not stand in for ethical responsibilities to those enrolled in our research; rather, it is meant to question and transform how we understand those very responsibilities. Homework implies *learning* about, so as to seriously consider and begin to *unlearn* imperial habits that produce the world as a series of peopled places available for our geographical expeditions and extractive practices. The aim of these tasks is not to achieve a position of purity from which to undertake research – as if that were possible (Rose 1997). Rather, the purpose of homework is to situate ourselves so as to enable ethical decisions about what to research, with whom, using which practices, and to what ends. Homework cultivates accountability for the epistemological and ontological worlds we enact through academic practices. What kind of world would we like to be involved in enacting?

Research as entangled engagements

This chapter reflects on how ethical concerns are framed in political ecology; I suggest conventional framings of ethics exist alongside research practices and modes of self-fashioning that may run counter to our stated political stance. Identifying and doing research with marginalized peoples involved in struggles against imperial capitalist relations, as I did in Guatemala, does little to question the epistemological position of the researcher in relation to the objects of research. And, it risks masking the ways in which we are entangled in and complicit with those very relations. Shifting our focus from ethical behavior in the field to homework obliges us to take a much more explicit stance regarding the why, where, when, and how of our political agendas, research engagements, and practices.

In thinking about how to do research that is informed by an ethics of entanglement, I often turn to the Zapatistas' *Sixth Declaration of the Selva Lacandona* (2005), which invites people engaged in struggles against neoliberalism and for democracy, liberty, and justice to *walk with* the Zapatista Army of National Liberation. This vision of *walking with* breaks with liberal notions of solidarity and political alliances; walking with does not mean helping the Zapatistas nor does it mean being like them. To walk with the Zapatistas means to be involved in the struggle for a just world from and in our own sites of entanglement and engagement. What would it be like to frame research in terms of walking with differently situated others in intersecting, yet distinct and unequally constituted struggles?

Certainly, there are many ways of walking with. In her reflections on radical theory and critical praxis, Perla Zusman (2004: 143) distinguishes between academically led research projects and analysis that evolves out of political engagements. Zusman's (2004: 144) experiences in Barcelona, doing politics in and through collective activism targeting city planning, lead her to conclude:

> Making activism the starting point for analysis makes it possible to incorporate knowledge produced in different temporal and spatial contexts into the academic sphere, in political and production of knowledge circumstances where a different intersubjectivity from the one that usually feeds academic practice takes precedence.

As I understand it, Zusman is suggesting that political commitments emerging from engagements to transform socio-ecological conditions at home, in places where we have a stake, will also transform our very conception of research subjects and the practice of research and ethics. For Zusman, the shifts result from experiencing knowledge production as an inter-subjective process, a collective rather than individual endeavor (see also Chapter 13, this volume).

My own efforts to transform my relations to home and field led me to research in the US–Mexico borderlands. Although I no longer *live* close to this region, US boundary enforcement plays a crucial role in constituting the meaning and material status ascribed to my citizenship and the ease with which I move through the world. Moreover, I am a direct beneficiary of the North American Free Trade Agreement (NAFTA) in a way that goes beyond the usual benefits to US and Canadian citizens (e.g. inexpensive products and labor); indeed, I was hired at the University of British Columbia because of provisions in NAFTA allowing for the easy movement of certain categories of professionals. Nonetheless, I feel the pull for lands further south of the US boundary with Mexico and continue to think about a form of research that will satisfy my own personal sense of ethics as walking with.

Along these lines, I am inspired by multi-sited research, especially the methodologies put forward by Cindi Katz (2001a: 720) to examine how people in distinct geographic locations are simultaneously connected through "the globalization of capitalist production, the prosecution of war, or the imposition of structural adjustment." Katz (2001b) advocates *critical topographies*, a strategy for analyzing the material effects of such processes in distinct yet linked locations, while *counter-topographies* entail connecting common struggles against such processes so as to build different kinds of (collective) responses. In my view, Katz's methodologies easily lend themselves to the usual forms of scholarship, wherein the researcher traces contour lines between places that (appear) disconnected from her world. Nonetheless, Katz's methodologies hold strong potential for research strategies to examine how we (our communities) are connected to other communities through particular policies. For instance, Roberta Hawkins (2012) elaborates a critique of ethical consumption campaigns that position Northern (female) consumers as saviors of (feminized) people and environments in the Global South. In short, the most radical stance we might take is to start where we are, from the entanglements that constitute our everyday lives.

Wherever we are located, entanglement as an ethical practice attends to interlocking power relations at multiple sites. Ultimately, an ethics of entanglement calls on political ecologists to be accountable for our political position by unlearning imperial epistemologies and making knowledge production a means of collective transformation.

References

Amin, S. 1989, *Eurocentrism* (trans. R Moore), Monthly Review Press, New York, NY.

Barnes, T. 2009, "Empiricism" in D. Gregory, R. Johnston, G. Pratt, M.J. Watts and S. Whatmore (eds.), *Dictionary of Human Geography* (5th Edition), pp. 190–191. Wiley-Blackwell, Hoboken, NJ.

Berger, M.T. 1995, *Under Northern Eyes: Latin American Studies and US Hegemony in the Americas 1898–1990*, Indiana University Press, Bloomington, IN.

Berman Santana, D. 1996, "Geographers, colonialism, and development strategies: the case of Puerto Rico," *Urban Geography*, vol. 17, pp. 456–474.

Bishop, A. 2002, *Becoming an Ally: Breaking the Cycle of Oppression in People*, Zed Books, New York, NY.

Blaut, J.M. 1993, *The Colonizers Model of the World: Geographical Diffusionism and Eurocentric History*, The Guilford Press, New York, NY.

Bondi, L. 1997, "In whose words. On gender identities, knowledge and writing practices," *Transactions Institute British Geographers*, vol. 22, pp. 245–258.

Bryant, R.L. and Bailey, S. 1997, *Third World Political Ecology*, Routledge, London.

Bryant, R.L. and Jarosz, L. 2004, "Ethics in political ecology: a special issue of *Political Geography*. Introduction: thinking about ethics in political ecology," *Political Geography*, vol. 23, pp. 807–812.

Chakrabarty, D. 2000, *Provincializing Europe: Postcolonial Thought and Historical Difference*, Princeton University Press, Princeton, NJ.

Coronil, F. 1996, "Beyond occidentalism: toward nonimperial geohistorical categories," *Cultural Anthropology*, vol. 11, pp. 51–87.

Danius, S., Jonsson, S., and Spivak, G. 1993, "An interview with Gayatri Chakravorty Spivak," *boundary 2*, vol. 20. no. 2, pp. 24–50.

Forsyth, T. 2008, "Political ecology and the epistemology of social justice," *Geoforum*, vol. 39, no. 2, pp. 756–764.

Gould, K. 2010, "Anxiety, epistemology, and policy research 'behind enemy lines'," *Geoforum*, vol. 41, no. 1, pp. 15–18.

Gupta, A. and Ferguson, J. (eds.) 1997, *Anthropological Locations: Boundaries and Grounds of a Field Science*, University of California Press, Berkeley, CA.

Han, J.H.J. 2010, "Neither friends nor foes: thoughts on ethnographic distance," *Geoforum*, vol. 41, no. 1, pp. 11–14.

Haraway, D. 1991, *Simians, Cyborgs and Women: The Reinvention of Nature*, Routledge, London.

Hawkins, R. 2012, "Shopping to save lives: gender and environment theories meet ethical consumption," *Geoforum*, vol. 43, no. 4, pp. 750–759.

Hecht, S.B. and Cockburn, A. 2010, *The Fate of the Forest: Developers, Destroyers, and Defenders of the Amazon*, University of Chicago Press, Chicago, IL.

Hodge, P. and Lester, J. 2006, "Indigenous research: whose priority? Journeys and possibilities of cross cultural research in geography," *Geographical Research*, vol. 44, pp. 41–51.

Jarosz, L. 2004, "Political ecology as ethical practice," *Political Geography*, vol. 23, no. 7, pp. 917–927.

Katz, C. 1994, "Playing the field: questions of fieldwork in geography," *The Professional Geographer*, vol. 46, no. 1, pp. 67–72.

Katz, C. 2001a, "Vagabond capitalism and the necessity of social reproduction," *Antipode*, vol. 33, no. 4, pp. 709–728.

Katz, C. 2001b, "On the grounds of globalization: a topography for feminist political engagement", *Signs*, vol. 26, no. 4, pp. 1213–1234.

Kuokkanen, R. 2007, *Reshaping the University: Responsibility, Indigenous Epistemes, and the Logic of the Gift*, UBC Press, Vancouver, BC.

Kuokkanen, R. 2010, "The responsibility of the academy: a call for doing homework," *Journal of Curriculum Theorizing*, vol. 26, no. 3, pp. 61–74.

Leeuw, S.D., Cameron, E.S. and Greenwood, M.L. 2012, "Participatory and community-based research, Indigenous geographies, and the spaces of friendship: a critical engagement," *The Canadian Geographer/ Le Géographe canadien*, vol. 56, pp. 180–194.

Maldonado-Torres, N. 2003, "On the coloniality of being," *Cultural Studies*, vol. 21, pp. 240–270.

Mato, D. 2000, "Not 'studying the subaltern,' but studying with 'subaltern' social groups, or, at least, studying the hegemonic articulations of power," *Nepantla: Views from South*, vol. 1, no. 3, pp. 479–502.

Mignolo, W. 2009, *The Idea of Latin America*, Blackwell, Malden.

Mohanty, C.T. 1991, "Under western eyes: feminist scholarship and colonial discourses" in C. Mohanty, A. Russo and L. Torres (eds.), *Third World Women and the Politics of Feminism*, pp. 51–80. Indiana University Press, Bloomington, IN.

Morris-Suzuki, T. 2000, "Anti-area studies," *Communal/Plural*, vol. 8, pp. 9–23.

Nagar, R. 2002, "Footloose researchers, 'traveling' theories, and the politics of transnational feminist praxis," *Gender, Place and Culture: A Journal of Feminist Geography*, vol. 9, no. 2, pp. 179–186.

Nagar, R. 2006, *Playing with Fire: Feminist Thought and Activism through Seven Lives in India*, University of Minnesota Press, Minneapolis, MN.

Neumann, R.P. 1998, *Imposing Wilderness*, University of California Press, Berkeley, CA.

Nightingale, A. 2003, "A feminist in the forest: situated knowledges and mixing methods in natural resource management," *ACME: An International E-Journal for Critical Geographies*, vol. 2, no. 1, pp. 77–90.

Pain, R. and Francis, P. 2003, "Reflections on participatory research," *Area*, vol. 35, no. 1, pp. 46–54.

Peet, R. and Watts, M. (eds.) 1996, *Liberation Ecologies*, Routledge, London.

Peet, R. and Watts, M. (eds.) 2004, *Liberation Ecologies*, 2nd edn, Routledge, London.

Peluso, N. 1993, "Coercing conservation?" *Global Environmental Change*, vol. 3, pp. 199–217.

Pratt, G. in collaboration with Ugnayan ng Kabataang Pilipino sa Canad 2003, "Between homes: displacement and belonging for second generation Filipino-Canadian youths," *B.C. Studies*, vol. 139, pp. 41–68.

Pratt, G. in collaboration with Philippines-Canada Task Force on Human Rights 2008, "International accompaniment and witnessing state violence in the Philippines," *Antipode*, vol. 40, no. 5, pp. 751–779.

Pratt, M.L. 1992, *Imperial Eyes Travel Writing and Transculturation*, Routledge, New York, NY.

Regan, P. 2010, *Unsettling the Settler Within: Indian Residential Schools, Truth Telling, and Reconciliation in Canada*, University of British Columbia Press, Vancouver, BC.

Rocheleau, D. 1995, "Maps, numbers, text, and context: mixing methods in feminist political ecology," *Professional Geographer*, vol. 47, no. 4, pp. 458–466.

Rocheleau, D.B., Thomas-Slayter, E. and Wangari, E. 1996, *Feminist Political Ecology: Global Issues and Local Experiences*, Routledge, New York, NY.

Rose, G. 1993, *Feminism and Geography: The Limits of Geographical Knowledge*, University of Minnesota Press, Minneapolis, MN.

Rose, G. 1997, "Situating knowledges: positionality, reflexivities and other tactics," *Progress in Human Geography*, vol. 21, pp. 305–320.

Said, E. 1979, *Orientalism*, Random House, New York, NY.

Said, E. 1989, "Representing the colonized: anthropology's interlocutors," *Critical Enquiry*, vol. 15, pp. 205–225.

Smith, L.T. 1999, *Decolonizing Methodologies: Research and Indigenous Peoples*, Zed Books, London.

Spivak, G. 1990, *The Post-Colonial Critic: Interviews, Strategies, Dialogues*, Routledge, New York, NY.

Staeheli, L. and Lawson, V. 1995, "Feminism, praxis, and human geography," *Geographical Analysis*, vol. 27, pp. 321–338.

Sundberg, J. 2003, "Masculinist epistemologies and the politics of fieldwork in Latin Americanist geography," *The Professional Geographer*, vol. 55, no. 2, pp. 181–191.

Sundberg, J. 2005, "Looking for the critical geographer, or why bodies and geographies matter to the emergence of critical geographies of Latin America," *Geoforum*, vol. 36, no. 1, pp. 17–28.

Sundberg, J. 2014, "Decolonizing posthumanist geographies," *Cultural Geographies*, vol. 21, no. 1, pp. 33–47.

Wainwright, J. 2005, "The geographies of political ecology: after Edward Said," *Environment and Planning A*, vol. 37, pp. 1033–1043.

Wainwright, J. 2012, *Geopiracy: Oaxaca, Militant Empiricism, and Geographical Thought*, Palgrave Macmillan, New York.

Wallerstein, I. 1999, *The End of the World as We Know it: Social Science for the Twenty-first Century*, University of Minnesota Press, Minneapolis.

Watts, M. 1983, *Silent Violence: Food, Famine, and Peasantry in Northern Nigeria*, University of California Press, Berkeley.

Wolf, D. 1995, "Situating feminist dilemmas in fieldwork," in D. Wolf (ed.), *Feminist Dilemmas in Fieldwork*, pp. 1–55, Westview Press, Boulder.

Zapatistas 2005, *Sixth Declaration of the Selva Lacandona,* viewed November 29, 2013, www.zcommunications.org/sixth-declaration-of-the-selva-lacandona-by-subcomandante-marcos.

Zusman, P. 2004, "Activism as a collective cultural praxis: challenging the Barcelona urban model," in D. Fuller and R Kitchin (eds.), *Radical Theory/Critical Praxis: Making a Difference Beyond the Academy?*, pp. 132–146. Praxis (e)Press, Vernon and Victoria, BC, Canada.

9

ETHICS IN RESEARCH BEYOND THE HUMAN

Rosemary-Claire Collard

Doing political ecology in multispecies contact zones

November 2011 was a comparatively quiet month at ARCAS Wildlife Rehabilitation Centre, which houses and attempts to rehabilitate and release formerly trafficked animals in El Petén, Guatemala. In thirty days, there was one animal release (a hawk, who rustled around for a full anticlimactic minute in its open cardboard box before bursting free and flapping up into a black, starry sky, tracked by the eager eyes of twenty staff and volunteers with flashlights); three deaths (a howler monkey, a red-lored Amazonian parrot, and a parakeet); two necropsies (which determined that the howler monkey, whose dissection filled the boot room beside the lab with a metallic stench, had died of unknown causes; and the parrot of respiratory disease, a common illness for high-flying birds whose lungs are not equipped for the dampness of a captive life at the forest floor); the arrival of a blind heron the volunteers named Harold (who refused to eat and so was force-fed pre-soaked dog biscuits and slippery little free-swimming fish caught daily from the lake); and two semi-escapes (a fierce kinkajou, who slipped into a neighboring spider monkeys' cage, where it was met with general wariness; and a spider monkey, Beatrice, who broke into her cage's anteroom, where she ate all the lunch leaves). As a newcomer, this seemed liked plenty of action to me. Between scrubbing floors and scribbling in my notebook, my daily participant-observation research that month at ARCAS, part of a larger multisited project on global live wildlife trade, kept me scrambling. But seasoned ARCAS workers assured me: it gets a lot crazier. Too bad you weren't here for something exciting, they said, like the time a box jammed with 470 iguanas arrived, more than half of whom were already dead of asphyxiation or later died because they wouldn't eat, or the time a government official escorted an upright-walking spider monkey here, hand in hand like a little kid, but with a leash so tight it rubbed off a band of fur around her neck.

November 2011 had no such drama. This was personally a relief and research-wise a disappointment. In thirty straight days, I partook in and saw only the routine, the mundane, nothing out of the ordinary. But this day-to-day labor provided plenty of stimulation. The immediate challenges were a matter of muscle and controlling revulsion. Caring for captive beings can be physically demanding and stomach-churning. Regular tasks include shoveling piles of steamy fecal compost and carrying fifty-pound sacks of corn down dirt trails from the delivery truck to the *cuarentena*, where the animals are kept. More quickly than I would have guessed, my body became accustomed to heavy lifting and to olfactory and tactile encounters with rot, feces, urine, and other metabolic realities. At that point, I began to register a discomfort that would not be so easy to overcome.

This discomfort had its root in a tension between the goals of the political ecology research I was doing – to identify and call into question the hierarchical power relations between humans and animals, particularly in the context of the global live wildlife trade – and my embodied experience working as a wildlife rehabilitator, where I was actively performing the very power relations I wanted to contest. I have written in more detail about wildlife rehabilitation elsewhere (Collard 2014), but here I want to briefly explain the aspects of rehabilitation practice that in particular generated ethical struggles for me, because they speak to a broader ethical dilemma at work in much participant-observation animal research. This dilemma is the focus of this first section of the chapter. The second part of the chapter steps back to consider how animals have tended to feature in political ecological research, and how there are signs that this may be changing so that animals are becoming more direct subjects of research. In light of this potential shift, the third part of the chapter reflects on the ethical review process – through my own experience and more broadly – as a currently inadequate but potentially promising site for preliminary discussion of the ethics of critical animal research. My own struggle at ARCAS serves as the driving thread through these reflections.

Three aspects of rehabilitation practice generated this struggle. First, the rehabilitation logic at ARCAS is that any animals who exhibit humanized behaviors – eating human foods, monkeys walking upright, or parrots uttering English or Spanish words – are un-releasable. Animals are only released if they behave "naturally," and an assumed part of that natural behavior is fear and distance from humans. For rehabilitators, any animal who does not avoid humans is at risk of death or re-commodification. Early on my trainer gave me strict instruction on how to conduct myself in cage two, home to five juvenile spider monkeys. Stand tall, he said. Don't turn your back. Shout at the monkeys if they come near you. Be aggressive to them. Spray them with water from the hose (which they hate). They should fear and dislike us. Our job is to make sure they stay away from humans. The assumption underpinning this logic is that contact with humans outside of ARCAS is bad because humans are dominant and cruel and will harm wild animals. Yet inside ARCAS's cages, volunteers and workers are called upon to perform that very conception of the human. Both the assumption and the performance of a particular human subject – one characterized by mastery and control – can be thought of as a sort of "misanthropic humanism," a suspicion or dislike of "the human" that simultaneously naturalizes and universalizes a specific figure of the human. This is not to say that some animals are not better served by living out their lives in relative isolation from humans (and likewise that there are some animals who humans are better served to avoid). It is rather to point out my own discomfort at performing the role of the aggressive, dominant human – a subject position that I centrally aim to unsettle.

Second, and related, ARCAS paradoxically attempts to make animals unencounterable to humans through intense and prolonged encounters. Parreñas (2012) describes this as a process of making independent through relations of dependence, what she calls "arrested autonomy." The rehabilitant animals rely on humans for basic necessities: food, water, shelter. The promise of intimate encounters with animals draws many international volunteers to pay to work at rehabilitation facilities (see Parreñas forthcoming). At the same time, rehabilitation aims to divest animals of their ties to humans. The most pronounced manifestation of this occurs with monkeys. Most monkeys arrive at ARCAS as babies, confiscated from a trapper who has shot a mother monkey out of the trees in order to capture the babies clinging to her body. These babies, who require a great deal of social interaction, care, and affection, enter into a relationship with a surrogate human parent, usually a female, at ARCAS. A "parent" is easy to spot around ARCAS, her monkey ward never far away – usually nestled close around her nape, peering over a shoulder, dark tail curled delicately around the parent's neck. Once the monkey is a juvenile, however, this bond is severed. The monkey is gradually exposed to the treatment outlined above: spraying, shouting, aggression. One of ARCAS's primatologists (2011, personal communication) admits that the process is an imperfect balancing act, requiring the "right level of human maternal care to give to orphan primates so that they are still wild and can still successfully reproduce in the wild, but also if you don't give them enough they won't survive as babies."

Finally, during my month at ARCAS I was continually reminded of the tension between working from and for an anti-captivity politics while acting in no uncertain terms as a captor. When I arrived at ARCAS, my first lessons were all in locking and unlocking. For example:

1. When entering the *cuarentena* hallway, which is lined with locked doors that open onto cage anterooms, always lock the door behind you.
2. When entering a cage from its anteroom, do not unlock the cage door until you have first locked the anteroom door.
3. To lock parrots in their own individual cages, which hang from the chain-link walls of a larger cage, elaborately wind multiple pieces of wire around their cage door, watching out for nips to your knuckles from beaks. Be prepared for bloody hands, and to catch escaped parrots regularly, especially one smaller blue parrot that promptly disentangles each wire after it is twisted.
4. Always apply double locks to the spider monkey cages, adding to the usual deadbolt a thin but strong contraption that requires delicately placing a wire into a hole and spinning it tight. Then stand back and watch a monkey, usually the tawny-furred one, Stevie, leap onto the door, weave his fingers through its cracks, and start deftly worrying the wire.

ARCAS is a world of locks and cages, of cages within cages, of captors and captive. Workers do their best to ensure the highest possible quality of life for the centre's animals. Over my month there, some volunteers complained that the animals ate better than we did, receiving more fruits, no beans. But it was clear who held the keys.

This chapter opens with these three dilemmas I experienced while acting as a researcher–captor–rehabilitator because in many ways they gesture to a broader ethical dilemma of "doing" political ecology in what can be thought of as a multispecies "contact zone," following Mary Louise Pratt (1991, 1992). For Pratt a contact zone is where two or more cultures intermingle, "clash, and grapple with each other, often in contexts of highly asymmetrical relations of power" (Pratt 1991, 34). As a thinking device and methodological frame, the contact zone "emphasizes how subjects are constituted in and by their relations to each other. It treats the relations among colonizers and colonized, or travelers and 'travelees,' not in terms of separateness or apartheid, but in terms of co-presence, interaction, interlocking understandings and practices" (Pratt 1992, 7). Unlike Pratt's, the contact zones I inhabited were multispecies spaces in which I came into contact with human and animal research subjects and where they came into contact with each other – spaces in which "a multitude of organisms' livelihoods shape and are shaped by political, economic, and cultural forces" (Kirksey and Helmreich 2010, 545). Like Pratt's, my contact zones were saturated with deeply asymmetrical relations of power and structured by histories, knowledge systems, and political economies that position animals as subordinate to humans.

Donna Haraway's work on contact zones is a helpful complement to Pratt's (also see Ogden 2011). For Haraway (2008, 244), humans' inextricable entanglement with human and nonhuman others means that "being" human is always a "becoming-with" a multitude of others. This occurs within contact zones that for Haraway, like Pratt, are always constituted by intersecting axes of power and exploitation. Gender, race, sexuality, and class are all at work in structuring human–animal relations, and vice versa (see Emel 1995; Haraway 1991; Plumwood 1993). An especially prominent power asymmetry involved in organizing human–animal relations is the human–animal dualism. In this dualism, a western, liberal figure of the human is positioned as superior and in opposition to a constellation of beings collected under the category "the animal," cast as a subordinate object. These are the categories that constitute western humanism, a dominant but by no means universal (see Sundberg 2013) apparatus relied on and perpetuated by capitalism and colonialism (Plumwood 1993; Wolfe 2003, 2010). The human–animal dualism takes difference, abstracts it, and constructs it as hierarchy (Plumwood 1993). Another term I use to describe this dualism is *speciesism*, or a disregard for animals on the basis of their species (i.e., namely, not being human) and a simultaneous belief in the superiority and mastery of the figure of the human, or what Haraway (2008) calls human exceptionalism.

For some posthumanist, feminist, and indigenous scholars, contesting the human–animal dualism requires not only *looking at* the animal in multispecies contact zones, but also recognizing that *the animal looks back* (see Danta and Vardoulakis 2008; Derrida 2008). This recognition disrupts the Descartian relegation of animals to the status of mere machines. The contact zone is an apt frame for this reciprocal looking. Host to "both the trouble and the vitality of … companion species" (Haraway 2010, 322), the contact zone acknowledges power and friction while staying open to possibility (Tsing 2005; Pirkey 2012). In this spirit, my research attempted to foreground the animals I encountered in contact zones as active, but not necessarily willing, participants in my research. ARCAS animals were "significantly unfree" research partners (Haraway 2008) – thinking, feeling beings with their own lookouts and their own ability to work for themselves.

This is all relatively comfortable theoretical territory. In practice, though, to look at animals and to be looked upon by animals often entails accessing an embodied proximity to them. Depending on the animal, this proximity may demand a degree of control over and manipulation of the animal. Many animals cannot be encountered in an uncontrolled manner without risk of harm to the researcher. At ARCAS, animals' captivity – power over the animal in the form of spatial confinement and relations of dependency – was a condition of my access to them. As Sundberg (Chapter 8, this volume) writes, "Political ecologists are situated in, complicit with, and benefit from the very politico-economic systems that constitute our research subjects." In particular, doing political ecology in multispecies contact zones (of mutually constituted and radically asymmetrical beings) can rely on particular power relations as a condition of possibility. As was my experience at ARCAS, research practices did not only rely on power asymmetry; they actually *performed*, or brought into being, these power relations. As hesitant as I was to brandish the hose and shout at monkeys, if these monkeys did not ultimately subordinate themselves to me and exhibit fear of me and other humans, they would not be released. So I sprayed and shouted.

Political ecologists have provided incisive reflections on the power dynamics at work in field research (e.g. Sundberg 2003; Chapter 8, this volume; Sultana 2007) and have productively engaged the notion of the contact zone (Sundberg 2006; Tschakert 2009; Ogden 2011; Pirkey 2012). As I argue in the next section, however, animals have not entered this discussion in earnest. Political ecologists have begun to acknowledge the mutual constitution of humans and animals, but power asymmetries – speciesism and the human–animal dualism – have been less recognized, especially in the actual *doing* of political ecology. Equally, the human ethics review process ignores or even perpetuates these asymmetries. Building from political ecologists' critical reading of ethical review, in the third part of the chapter I begin to assess the usefulness of the institutional ethical review process for other-than-human research. What I offer are not answers but reflections on my ongoing struggles to cultivate an academic practice (methodological as well as theoretical) that is sensitive to animal autonomy, by which I mean the full expression of animal life, including freedom of movement, social and familial association, work and play. My goal is to contribute to and prompt open conversation about institutional and theoretical shifts that may assist others beginning to negotiate the ethical dilemmas of doing political ecology in multispecies contact zones.

Political ecology and the human/animal dualism

You do not have to look hard to find animals in political ecology. From early investigations of deforestation and cattle ranching in the Amazon to political ecologists' long-standing interest in the politics of wildlife conservation, research in the discipline has frequently involved animals. But *how* are they involved? In answering this question, Hobson's (2007, 250; also Robbins 2005; Hinchliffe 2008) assessment that political ecology has largely treated animals as "static components of a thoroughly human sociality" still has relevance, although important exceptions are emerging and may indicate a shift underway in the discipline. Before turning to this, I summarize three dominant (and sometimes interconnected) ways that animals have appeared in political ecology: as resource, as object of conflict, and as social representation. These are, of course, only general tendencies, and my characterization is not intended as a critique but rather to point out some continuing gaps and opportunities for future research. Additionally, the following truncated review accounts for work firmly within the discipline of political ecology, bracketing out, for example, work in vitalist and new materialist theory, some of which can be loosely located within political ecology (see Whatmore 2002; Braun 2008; Bennett 2010).

The first principal way that animals have appeared in political ecology is as a "species of capital" (Sayre 2002). From fish (Mansfield 2011), to cattle (Sheridan 1988; Perramond 2010; Bobrow-Strain 2009), sheep and goats (Turner 1999), sea turtles (Campbell 2007), and horses (Rikoon 2006), political ecologists have examined animals as natural or physical resources around which swirl politics of access, capitalist logics, and contested property regimes. In this work, efforts are made to view human groups as heterogeneous. For example, in his study of ranching, Perramond (2010) finds and accounts for significant diversity within the ranching community. However, the animal groups – the cows – tend to remain abstract, standard, and uniform, mimetic of their treatment in capitalism. As Vandana Shiva (1992) argued over twenty years ago, the contemporary notion of "resource" strips the natural world of its own creativity and of any reciprocity between humans and nature. The logic of the resource is such that natures (or animals) cannot work for themselves; only the application of human knowledge can make anything of them. The animal rendered resource is passive. Much political ecological analysis of on commodified animals is silent on the animal, taking it as a backdrop to human affairs, politics, and sociality.

Recently some political ecologists have sought to understand how the specificity of biological nature shapes or resists commodification practices, particularly in a stream of work referred to as "neoliberal natures" (see Heynen et al. 2007). For example, Paul Robbins and April Luginbuhl (2005) trace physical and institutional efforts to enclose and privatize wildlife in the United States, looking specifically at elk in Montana, and how elk and disease resist enclosure. This specificity is generally limited to the *species* level. Commodification and its effects are considered at the level of the aggregate population. Difference *within* species, however, goes unrecognized, as does a sense of animals as sentient beings in their own right (see also Wilkie

2005). A key exception is Robbins's (2005) lecture, "We are the elk," which opens with an encounter between the author and an elk. Overall, the emphasis on populations rather than individual animals leads to a perception of animals as "without individual character, knowledge, subjectivity or experience" (Tovey 2003, 196). Political ecology is of course not alone in privileging the population scale. Population-level thinking also prevails in animal geography, as Bear (2011) argues, and it echoes the leading approach in conservation writ large, which is to treat wildlife as an aggregate population and neglect questions of animal welfare (Paquet and Darimont 2010).

A second way that political ecologists have tended to include animals in their analysis is as objects of conflict. As Peet et al. (2011, 27) identify, one of political ecology's key areas of concern has been "the impacts, logics, and operations of conservation and environmental protection," and so commonly, such studies address conflict in wildlife conservation (e.g. Wondrak 2002; Rikoon 2006; Gupta 2013). In this thread, much scrutiny has been directed at "global efforts for the protection of wildlife through the creation of national parks" (Peet et al. 2011, 27). As Brockington (2002) demonstrates, "saving animals," most often charismatic fauna like lions, pandas, and polar bears, has usually been attempted though a "fortress" model, in which urban elites' calls for protection lead to the enclosure of lands previously home to rural, indigenous, and local people. The importance of this work, in which conservation is shown to be a form of environmental control, or green governmentality, is undisputable. However, although this work necessarily mentions the animals whose "saving" is being attempted, these animals are typically rendered a mute background against which human social and political struggle unfolds. As Hobson (2007, 253) remarks, this work implicate animals "as co-oppressors of the world's marginalised and poor, rather than likewise subjects of and in spatially uneven practices." Political ecology's defining interest in questions of power and forms of rule are not often extended to consider animals as subjects of this rule.

Finally, animals are considered in political ecology as social representations, or products of knowledge. Political ecology is centrally interested in the categories, ideologies, and narratives – collectively, discourse – that people develop to understand and explain environmental change and to mobilize environmental politics. Questions about how nature is constructed, how people know nature, how expertise is formed, and how representations of nature circulate and compete all often involve animals as the part of nature being represented. Political ecologists have examined how representations of animals circulate and are drawn on by various actors in environmental politics and knowledges concerning, for example, endangered species (McGregor 2005), invasive species (Robbins 2004), bordering practices (Sundberg 2011), or wildlife management (Robbins 2006; Campbell 2011; Nadasdy 2011). In this work attention is increasingly given to how the specificity of certain species of animals is drawn on to construct especially potent and mobile representations – for example, charismatic megafauna like the panda (Lorimer 2007) or grizzly bear (Dempsey 2010). These are promising signs of political ecologists recognizing that animals have always been part of the political realm, and are key actors with their own stakes in a politics that is "also very much about reconfiguring our relationships with nonhumans" (Dempsey 2010, 1154).

The work in these three areas examines how animals are enrolled in projects of capital accumulation, uneven development and uneven distribution of power and wealth, building and maintaining expertise, directing environmental change, and so on. But on the whole, most of this work still considers power as a condition of *human* relationships, and so it is unsurprising that animals figure primarily as currency in these power relationships, as a means through which power and capital circulate. Rarely are animals considered as subjects who also bear the costs of environmental change; who can themselves be stripped of autonomy and self-determination;

who are excluded, exploited, and dispossessed (see Wolch and Emel 1998; Philo and Wilbert 2000). Animals largely appear *without making any ethical demands*, especially on the researcher. There are important exceptions, particularly Watts's essays on zoos and livestock (2000) and broiler chickens and hogs (2004), which link capitalist expansion to animal enclosure while highlighting the embodied effects for animals; Emel and Neo (2011), who also bring animal welfare into their consideration of the rise of factory farming; and Hobson (2007), who documents the conditions of life and death for sun bears in her case study of bear bile farming. This work, combined with the aforementioned emerging work that highlights animals' agency in environmental politics, is an indication that work in posthumanism, animal geography, and critical animal studies is making inroads into political ecology, which is not surprising. This all points to a great deal of opportunity and promise that lie ahead in political ecology for research that takes seriously the human–animal dualism, or speciesism, as a fundamental organizing and enabling axis of power.

But broadening the question of who "counts" in research, and conducting research in which animals can and do make ethical demands, not only requires different methodologies, as several scholars have noted (Whatmore 2006; Hobson 2007; Lorimer 2010), but also involves different ethical considerations in the research process. What to do, for example, about the thorny issue of consent? If strands of political ecology continue to move toward a more meaningful inclusion of animals in analysis, politics, and ethics, if the "multispecies ethnography" that is expanding in anthropology and other social sciences also finds fertile ground in political ecology, as it well should, some researchers will undoubtedly experience ethical struggles such as the one I outlined in the first section of this chapter. There are few resources for negotiating these struggles. Ostensibly, the chief institutional space for formal ethical deliberation is the ethical review process. But where in the ethical review process might we deliberate the ethical dilemmas associated with multispecies fieldwork?

Ethical review

About a year before my time at ARCAS, I sat in front of the computer at my desk on the University of British Columbia (UBC) campus, where I was a graduate student, about to create a new ethical review application through the Researcher Information Services (RISe) website. The ethical review process is a mandatory one for most university researchers today (see Martin 2007 for a history of the Institutional Review Board's [IRB] rise in the United States; and see Dyer and Demeritt 2009 for a discussion of Research Ethics Committees' later arrival in the UK). Political ecologists, among others, have well-founded concerns about ethical review as an institution and a process, as I will discuss shortly. Largely unaware of these at the time, I was simply focused on jumping through the ethics review hoop, as I perceived it. I read off the webpage: "I would like to create a new application for ..." and barely hesitated before clicking on the "human ethics" tab instead of "animal care." This may not sound significant. After all, I intended to conduct interviews with humans and I did not and do not do animal science or testing, although UBC does have a large and controversial collection of captive animals – 225,043 of them in 2011 (UBC 2012) – for experimentation. At the time, the decision was easy, and I gave it little thought.

Only once I was in the field and experiencing the challenges outlined above did my unthinking selection of "human ethics" begin to seem less obvious. Although I did not conduct experiments on or with animals (and this is what exempted me from requiring animal care ethics approval, according to later personal communication with UBC research ethics board staff), my research brought me into direct and sustained proximity with animals, many of whom were suffering profoundly in captivity, where it is impossible to meet the precise needs of each animal. Captive animals experience many negative effects: boredom, anxiety, emotional dysregulation, hypersensitivity to environmental change, uncontrollable aggression, self-inflicted wounding, post-traumatic stress disorder, malnutrition, disease, and death (see, for example, Clubb and Mason 2003; Weston and Memon 2007; Bradshaw et al. 2009; Yenkowsky et al. 2010; Lopresti-Goodman et al. 2012). My research involved actively enforcing these conditions, and even inflicting aggression and pain on animals. This felt, in no uncertain terms, unethical.

How would this research practice be deliberated within an ethical review? Animals were not mentioned once in my human ethics review. But the animal care review is equally unsuitable and irrelevant. At UBC, to be granted an animal care ethics application approval from the Animal Care Committee (ACC), applicants must pass a twelve-module "Experimental Animal User Training Program" offered by the Canadian Council for Animal Care (CCAC). Mandatory modules include familiarizing oneself with the "three Rs of Humane Animal Experimentation" (replacement, reduction, and refinement) and with practices of anesthesia and euthanasia (see ORS 2013). Elements of an animal care ethics application then involve categorizing, among other things, how painful the research will be for animals according to the CCAC's (2013) "Category of Invasiveness" (e.g. category B corresponds to "minor stress or pain of short duration" and category D to "severe pain near, at, or above the pain tolerance threshold of unanesthetized conscious animals"). Prospective researchers must describe the procedures they will employ, whether there are any alternatives to live animal use, what animals they will use, expected morbidity and mortality rates, and why they need the number of animals they request. This process would have had little relevance for the research I carried out. In her research on the dairy industry, for example, Kathryn Gillespie (forthcoming) was taken down a rabbit-hole of institutional ethics during her IRB process when she clicked "yes" to the question "does your research involve animals?" Gillespie (forthcoming) documents how she "learned the best methods for decapitating rats, but nothing about appropriate conduct relating to cows living on sanctuaries or farms." It deserves noting that in a recent review of animal ethics committees Varga (2013, 90) finds that "we know almost nothing about how well [they] work from the animal protection point of view."

Clearly, laboratory animal research is markedly different in terms of ethical regulation from human research, one of the key differences being human ethics' requirement for informed consent (Greenhough and Roe 2011), another obvious one being animals' killability. These differences aside, there is, I think, a more important similarity across both ethical review processes. From the brief description above, it is evident that the animal ethics review relies on the human–animal dualism, as animal experimentation does. Animals are, according to this dualism, killable and confinable. This power dynamic is assumed, not questioned, in the animal care ethical review. Critically, the human ethics process retains the same human–animal dualism in its bracketing out of any concerns beyond the individual researcher's interactions with other individual humans. Animals are considered as outside of the purview of "human" ethics. Both ethical review processes therefore rely on the human–animal dualism. Political ecologists and critical geographers have observed this, noting that ethical reviews "presuppose an individuated liberal humanist research subject that is incommensurate with the subjectivities of our actual research participants" (Butz 2008, 241). Dyer and Demeritt (2009, 49) write that "acknowledging the relationality of our place in a 'more-than-human world' unsettles the idea, absolutely central to the liberal principlism of conventional ethical review, of an autonomous individual subject who might give its informed consent to becoming the object of research by another." Butz and Dyer and Demeritt reach the heart of my own struggles with ethical review. As it currently exists, it cannot be used to meaningfully deliberate the ethical struggles associated with research that attempts to challenge liberal humanism and the human–animal dualism, because ethical review itself deploys these logics.

Does this mean researchers working with multispecies methodologies should just forget ethical review, be grateful that they fall through the institutional cracks? I think not. Yes, ethical reviews "bureaucratize ethics," potentially lulling "us into forgetting the need to take responsibility for thinking ethically on a day-to-day basis" (Valentine 2005, 485; also Martin 2007; Sultana 2007; Butz 2008; Dyer and Demeritt 2009). But most critics of ethical review admit that a complete absence of this institution is also problematic. This is perhaps especially the case, as Dyer and Demeritt (2009) note, when there is a "carer–cared for" dimension to the researcher–research subject relationship (as in medical research), or when there are risks of side effects, such as illness, for research participants. In research with animals – both lab research and multispecies research of the sort I carried out at ARCAS – researchers may indeed find themselves in such relationships of caring for their animal subjects. This suggests that multispecies research is possibly *more* in need of ethical review than strictly human research. In a broader sense, ethical review essentially determines what kind of research is done (Dyer and Demeritt 2009). The ethical review process is thus part of a broader institutional orientation. Of course, ethical review is but one site for ethical deliberation, and, as critics rightly point out, it is a limited one. But the lack of a relevant ethical review process for multispecies field research serves to further subordinate meaningful discussion about the ethics of animal research practice in non-experimental social science.

Moving forward, is it possible that a modified ethical review process could be made more relevant to the kind of research being undertaken in multispecies contact zones while acknowledging the relational and non-dualistic thinking that I and others have advocated? I think it is both possible and necessary, and that it could be a part of reorienting the academy to face animals in research not as dualistically opposed to humans, but as ethical subjects in their own right, entangled in multiple ways with human social and political life. Within such a shift, it will be important to develop a means of allowing animal research subjects *not to participate* (Greenhough and Roe 2011). The extent to which this choice is actually available is complicated by the conditions under which the choice is made – for example, can captive animals be given the choice to be captive or not? Perhaps, as Greenhough and Roe (2011, 60) suggest, shifting research ethics away from the capacity to consent and instead toward "the recognition of an affective capacity to respond" (60) would be important steps in moving toward a relational ethics, although what this would look like in practice is less clear. Regardless, a more appropriate or relevant ethics review would *not* have prevented the ethical dilemmas I experienced in the field. Ethics are dynamic, produced by ongoing friction and tensions, and their negotiation will vary across different contexts. But an ethics review enlivened to multispecies research might have prompted reflections and conversations that would have made me more attuned and better equipped to negotiate ethical dilemmas.

Toward doing anti-speciesist political ecology

Each evening at ARCAS it is as if someone slowly twists a knob that controls both volume and light. The day's incessant din trails off with the sun until you can stand beside the cages in the murky light and hear only muted murmurs and patters as hundreds of animals settle into sleep. In the middle of each spider monkey cage you can just make out the shadowy mass of what is known at ARCAS as a "monkey ball," a tight cluster of bodies huddled knees to chins. Watching them, I am reminded, as I am hundreds of times over the month at the centre, that these creatures are creatively independent and deeply social.

The words and research practices to account for this independence and sociality are still being developed. Constructing such a register is part of the broader critical task of articulating and mobilizing anti-speciesist thinking and practice. Political ecology can offer much to this effort. A core question across political ecology research is to ask what the costs and benefits of environmental change and decision-making are, and how they are distributed. This is a question that could account more fully for animals. An anti-speciesist political ecology would recognize animals as beings with their own complex social structures, affairs, material and spatial requirements, beings whose lives environmental change transforms, beings who can be dispossessed and displaced. Practically, research in this vein requires constant interrogation of gaps and silences, of assumptions about who matters, at all stages of a project – from research design (e.g. framing research questions that are open to registering animals as drivers of environmental change and bearers of change's effects), fieldwork (carrying out research in a way that recognizes dynamic animal subjectivity, capacities, and autonomy), analysis (constructing explanations that do not silence animals or collapse them into humanist frames), and writing (crafting narratives that enliven readers to the possibility of worlds beyond the strictly human).

I am not advocating that all political ecologists adopt this approach. My suggestion is decidedly more modest, and it is a contribution to work that is already being undertaken in the discipline. The suggestion is that we direct a political ecological lens – one that has proven itself well attuned to power's operation across gender, race, and class – onto the hierarchy positioning "the human" as separate from and superior to "the animal." Tackling this hierarchy – the human–animal dualism, or speciesism – will undoubtedly involve some research in multispecies contact zones, where political ecologists will look at animals, and animals will look back. But it will also involve choosing, at times, not to enter into proximate physical encounters, to let an animal be unencounterable. Ethical review is but one site for such deliberation, deliberation that ought to take place in a manner that balances enthusiasm for animal research with consideration of the power-laden subject formations and relations upon which research may depend.

Doing anti-speciesist political ecology is ethically fraught. Research itself has costs and benefits (see Chapter 8, this volume), and is performative (Pratt 2000), bringing worlds into being. I have outlined one dilemma I wrestle with in my research, and one that I think has bearing beyond my experience: the tension between conducting research with the objective of contesting power asymmetries between humans and animals, but at the same time, relying on or even re-performing these dynamics in the research process. Ultimately, much animal research relies on an *encounterable* animal. These encounters have ethical contours and costs and benefits. They are imbued with distinct power relations. They also formulate anew the political and ethical terrain on which subsequent encounters occur (Robbins 2005). The conditions and effects of animal encounterability must therefore be brought to the fore when deliberating the ethics of doing political ecology in multispecies contact zones.

References

Bear, C. (2011) "Being Angelica? Exploring individual animal geographies." *Area*, vol. 43, pp. 297–304.

Bennett, J. (2010) *Vibrant matter: A political ecology of things*. Durham: Duke University Press.

Bobrow-Strain, A. (2009) "Logics of cattle–capital." *Geoforum*, vol. 40, pp. 778–780.

Bradshaw, G., J. Yenkosky, and E. McCarthy. (2009) "Avian affective dysregulation: psychiatric models and treatment for parrots in captivity." Proceedings of the thirtieth annual association of Avian Veterinarians Conference with the Association of Exotic Mammal Veterinarians. August 8–15, 2009, Milwaukee, WI. www.kerulos.org/learn_more/parrots_files/Bradshaw_Yenkosky_McCarthy_2009.pdf.

Braun, B. (2008) "Environmental issues: inventive life." *Progress in Human Geography*, vol. 32, pp. 667–679.

Brockington, D. (2002) *Fortress conservation: The preservation of the Mkomazi Game Reserve, Tanzania.* Bloomington: Indiana University Press.

Butz, D. (2008) "Sidelined by the guidelines: reflections on the limitations of standard informed consent procedures for the conduct of ethical research." *ACME: An International E-journal for Critical Geographies*, vol. 8, pp. 239–259.

Campbell, L. (2007) "Local conservation practice and global discourse: a political ecology of sea turtle conservation." *Annals of the Association of American Geographers*, vol. 97, pp. 313–334.

Campbell, L. (2011) "Debating the science of using marine turtles: boundary work among species experts." In *Knowing nature: Conversations and the intersection of political ecology and science studies*, Goldman, M., P. Nadasdy, and M. Turner (eds.), Chicago: University of Chicago Press, pp. 47–64.

CCAC. (2013) "Categories of invasiveness." Ottawa: CCAC. Available at: www.ccac.ca/en_/standards/policies/policy-categories_of_invasiveness.

Clubb, R. and G. Mason. (2003) "Animal welfare: captivity effects on wide-ranging carnivores." *Nature*, vol. 425, pp. 473–474.

Collard, R.-C. (2014) "Putting animals back together, taking commodities apart." *Annals of the Association of American Geographers*, vol. 104, pp. 151–165.

Danta, C. and D. Vardoulakis. (2008) "The political animal." *SubStance*, vol. 37, pp. 3–6.

Dempsey, J. (2010) "Tracking grizzly bears in British Columbia's environmental politics." *Environment and Planning A*, vol. 42, pp. 1138–1156.

Derrida, J. (2008) *The animal that therefore I am.* New York: Fordham University Press.

Dyer, S. and D. Demeritt. (2009) "Un-ethical review? Why it is wrong to apply the medical model of research governance to human geography." *Progress in Human Geography*, vol. 33, pp. 46–64.

Emel, J. (1995) "Are you man enough, big and bad enough? Ecofeminism and wolf eradication in the USA." *Environment and Planning D: Society and Space*, vol. 13, pp. 707–734.

Emel, J. and H. Neo. (2011) "Killing for profit: global livestock industries and socio-ecological implications." In *Global political ecology*, Peet, R., P. Robbins and M. Watts (eds.), New York: Taylor & Francis, pp. 67–83.

Gillespie, K. (Forthcoming) *The cow with ear tag #1389.* Chicago: University of Chicago Press.

Greenhough, B. and E. Roe. (2011) "Ethics, space, and somatic sensibilities: comparing relationships between scientific researchers and their human and animal experimental subjects." *Environment and Planning D: Society and Space*, vol. 29, pp. 47–66.

Gupta, A.C. (2013) "Elephants, safety nets and agrarian culture: understanding human-wildlife conflict and rural livelihoods around Chobe National Park, Botswana." *Journal of Political Ecology*, vol. 20, pp. 239–254.

Haraway, D. (1991) *Simians, cyborgs, and women: The reinvention of nature.* New York: Routledge.

Haraway, D. (2008) *When species meet.* Minneapolis: University of Minnesota Press.

Haraway, D. (2010) "Interview with Annie Potts – Kiwi chicken advocate talks with Californian dog companion." *Feminist Psychology*, vol. 20, pp. 318–336.

Heynen, N., J. McCarthy, S. Prudham, and P. Robbins (eds.) (2007) *Neoliberal environments: False promises and unnatural consequences.* New York: Routledge.

Hinchliffe, S. (2008) "Reconstituting nature conservation: towards a careful political ecology." *Geoforum*, vol. 39, pp. 88–97.

Hobson, K. (2007) "Political animals? On animals as subjects in an enlarged political geography." *Political Geography*, vol. 26, pp. 250–267.

Kirksey, S. and S. Helmreich. (2010) "The emergence of multispecies ethnography." *Cultural Anthropology*, vol. 25, pp. 545–576.

Lopresti-Goodman, S., M. Kameka, and A. Dube. (2012) "Stereotypical behaviors in chimpanzees rescued from the African bushmeat and pet trade." *Behavioral Sciences*, vol. 3, pp. 1–20.

Lorimer, J. (2007) "Nonhuman charisma." *Environment and Planning D: Society and Space*, vol. 25, pp. 911–932.

Lorimer, J. (2010) "Moving image methodologies for more-than-human geographies." *Cultural Geographies*, vol. 17, pp. 237–258.

Mansfield, B. (2011) "'Modern' industrial fisheries and the crisis of overfishing." In *Global political ecology*, Peet, R., P. Robbins, and M. Watts (eds.), New York: Taylor & Francis, pp. 84–99.

Martin, D. (2007) "Bureaucratising ethics: institutional review boards and participatory research." *ACME: An International E-Journal for Critical Geographies*, vol. 6, pp. 319–328.

McGregor, J. (2005) "Crocodile crimes: people versus wildlife and the politics of postcolonial conservation on Lake Kariba, Zimbabwe." *Geoforum*, vol. 36, pp. 353–369.

Nadasdy, P. (2011) "'We don't harvest animals; we kill them': agricultural metaphors and the politics of wildlife management in the Yukon." In *Knowing nature: Conversations and the intersection of political ecology and science studies*, Goldman, M., P. Nadasdy, and M. Turner (eds.), Chicago: University of Chicago Press, pp. 135–151.

Ogden, L. (2011) *Swamplife: People, gators, and mangroves entangled in the Everglades*. Minneapolis: University of Minnesota Press.

ORS (Office of Research Services). (2013) "Animal care training." Vancouver: University of British Columbia. Available at: www.ors.ubc.ca/ors/animal-care-training.

Paquet, P. and Darimont, C. (2010) "Wildlife conservation and animal welfare: two sides of the same coin." *Animal Welfare*, vol. 19, pp. 177–190.

Parreñas, R. (2012) "Producing affect: Transnational volunteerism in a Malaysian orangutan rehabilitation center." *American Ethnologist*, vol. 39, pp. 673–687.

Parreñas, R. (Forthcoming) "The materiality of intimacy in wildlife rehabilitation: Rethinking ethical capitalism through embodied encounters with animals in Southeast Asia." *Positions*.

Peet, R., P. Robbins, and M. Watts. (2011) *Global political ecology*. New York: Taylor & Francis.

Perramond, E. (2010) *Political ecologies of cattle ranching in northern Mexico: Private revolutions*. Tuscon: University of Arizona Press.

Philo, C. and C. Wilbert (eds.) (2000) *Animal spaces, beastly places: New geographies of human-animal relations*. London: Routledge.

Pirkey, W. (2012) "Beyond militant particularisms: collaboration and hybridization in the 'contact zones' of environmentalism." *Capitalism Nature Socialism*, vol. 23, pp. 70–91.

Plumwood, V. (1993) *Feminism and the mastery of nature*. New York: Routledge.

Pratt, G. (2000) "Research performances." *Environment and Planning D: Society and Space*, vol. 18, pp. 639–651.

Pratt, M.L. (1991) "Arts of the contact zone." *Profession*, vol. 91, pp. 33–40.

Pratt, M.L. (1992) *Imperial eyes: Travel writing and transculturation*. New York: Routledge.

Rikoon, S. (2006) "Wild horses and the political ecology of nature restoration in the Missouri Ozarks." *Geoforum*, vol. 37, pp. 200–211.

Robbins, P. (2004) "Comparing invasive networks: cultural and political biographies of invasive species." *Geographical Review*, vol. 94, pp. 139–156.

Robbins, P. (2005) "We are the elk: ethics and explanation in political ecology." Plenary lecture on "Ethics, Justice and Political Ecology" at the Annual Meeting of the Association of American Geographers, Denver, April 6. Available at: www.uky.edu/~tmute2/geography_methods/readingPDFs/Robbins_Elk.htm.

Robbins, P. (2006) "The politics of barstool biology: environmental knowledge and power in greater Northern Yellowstone." *Geoforum*, vol. 37, pp. 185–199.

Robbins, P. and A. Luginbuhl. (2005) "The last enclosure: resisting privatization of wildlife in the western United States." *Capitalism Nature Socialism*, vol. 16, pp. 45–61.

Sayre, N. (2002) *Ranching, endangered species, and urbanization in the Southwest: Species of capital*. Tucson: University of Arizona Press.

Sheridan, T. (1988) *Where the dove calls: The political ecology of a peasant corporate community in northwestern Mexico*. Tucson: University of Arizona Press.

Shiva, V. (1992) "Resources." In Sachs, W. (ed.), *The development dictionary*, New York: Zed Books, pp. 206–218.

Sultana, F. (2007) "Reflexivity, positionality and participatory ethics: Negotiating fieldwork dilemmas in international research." *ACME: An International E-Journal for Critical Geographies*, vol. 6, pp. 374–385.

Sundberg, J. (2003) "Masculinist epistemologies and the politics of fieldwork in Latin Americanist geography." *The Professional Geographer*, vol. 55, pp. 180–190.

Sundberg, J. (2006) "Conservation encounters: transculturation in the 'contact zones' of empire." *Cultural Geographies*, vol. 13, pp. 239–265.

Sundberg, J. (2011) "Diabolic Caminos in the desert and cat fights on the Rio: a posthumanist political ecology of boundary enforcement in the United States–Mexico borderlands." *Annals of the Association of American Geographers*, vol. 101, pp. 318–336.

Sundberg, J. (2013) "Decolonizing posthumanist geographies." *Cultural Geographies*, vol. 21, pp. 33–47.

Tovey, H. (2003) "Theorising nature and society in sociology: the invisibility of animals." *Sociologia Ruralis*, vol. 43, pp. 196–215.

Tschakert, P. (2009) "Digging deep for justice: a radical re-imagination of the artisanal gold mining sector in Ghana." *Antipode*, vol. 41, pp. 706–740.

Tsing, A. (2005) *Friction: An ethnography of global connection*. Princeton: Princeton University Press.

Turner, M. (1999) "Merging local and regional analyses of land-use change: the case of livestock in the Sahel." *Annals of the Association of American Geographers*, vol. 89, pp. 192–219.

UBC. (2012) "Animals research at UBC – 2011 summary." Available at: www.animalresearch.ubc.ca/animal-statistics.html#2011.

Valentine, G. (2005) Geography and ethics: moral geographies? Ethical commitment in research and teaching. *Progress in Human Geography*, vol. 29, pp. 483–487.

Varga, O. (2013) "Critical analysis of assessment studies of the animal ethics review process." *Animals*, vol. 3, pp. 907–922.

Watts, M. (2000) "Afterword: enclosure." In Philo, C. and C. Wilbert (eds.), *Animal spaces, beastly places: New geographies of human-animal relations*, London: Routledge, pp. 291–301.

Watts, M. (2004) "Are hogs like chickens? Enclosure and mechanization in two 'white meat' filières." In Hughes, A. and S. Reimer (eds.), *Geographies of commodity chains*, London: Routledge, pp. 39–62.

Weston, M., and M. Memon. (2007) "The illegal parrot trade in Latin America and its consequences to parrot nutrition, health and conservation." *Bird Conservation*, vol. 9, pp. 76–83.

Whatmore, S. (2002) *Hybrid geographies: Natures cultures spaces*. London: Sage.

Whatmore, S. (2006) "Materialist returns: practising cultural geography in and for a more-than-human world." *Cultural Geographies*, vol. 13, pp. 600–609.

Wilkie, R. (2005) "Sentient commodities and productive paradoxes: the ambiguous nature of human–livestock relations in Northeast Scotland." *Journal of Rural Studies*, vol. 21, pp. 213–230.

Wolch, J. and J. Emel (eds.) (1998) *Animal geographies: Place, politics, and identity in the nature-culture borderlands*. New York: Verso.

Wolfe, C. (2003) *Animal rites: American culture, the discourse of species, and posthumanist theory*. Chicago: University of Chicago Press.

Wolfe, C. (2010) *What is posthumanism?* Minneapolis: University of Minnesota Press.

Wondrak, A. (2002) "Seen any wildlife? Community conflict and a struggle for the soul of Estes Park, Colorado." *Cultural Geographies*, vol. 9, pp. 68–94.

Yenkowsky, J., G. Bradshaw, and E. McCarthy. (2010) "Post-traumatic stress disorder among parrots in captivity: treatment considerations." Proceedings of the thirty-first annual association of Avian Veterinarians Conference with the Association of Exotic Mammal Veterinarians, August 1, San Diego, CA. Available at: www.aemv.org/members_only/SmallBook2010FINAL.pdf#page=27.

10
RELATIONSHIPS AND RESEARCH METHODS
Entanglements, intra-actions, and diffraction

Abigail H. Neely and Thokozile Nguse

Introduction

In the summer of 2009, we were sitting in the shade at the homestead of a *gogo* (*gogo* is the Zulu word for grandma) we knew well. We were in the middle of conducting a series of oral histories (for the years 1955, 1968, 1981, 1992, and 2009) in several households in Pholela, South Africa. In this particular community, the *gogos* from the two households with which we were working had decided we should conduct their oral histories together. (Following local custom, we use the term "community" to refer to different, named settlements in former African homeland areas.) They told us it was more "efficient," and besides, they had known each other for decades, so "they had no secrets." It was a hot afternoon, but the *gogos* insisted we have tea and biscuits. The gogo who lived at this homestead, whom we'll call Gogo Mbanjwa, sent Thoko (Thokozile) into the house to boil the water and bring out the tea. As Thoko prepared our snack, Gogo Mbanjwa and Gogo Sithole caught up on all the news and Abby (Abigail) did her best to follow and interject. Once Thoko returned, we got down to business, drinking tea – rooibos with lots of sugar – and asking questions about household demographics, livelihoods, agriculture, and health in 1968. As they answered, the *gogos* sprinkled in complaints about the fact that we had asked these questions last time (they were correct, though we were talking about 1955), how nothing had changed, and how they were getting bored. And then we got to a question about shopping – what they were buying, where, and with what money. Gogo Sithole answered that she bought food, soap, and other essentials with the money she earned selling food outside the local health center. This seemed fine, until we remembered that her husband was employed at the health center. When we asked if she used the money he earned to buy goods, she laughed and replied, "He was supposed to give me money to buy things for the house, but he spent it on other women instead – his girlfriends." She paused, and then followed up, "You're getting all of our secrets now."

That "secret" marked an important moment in our research for two interconnected reasons. First, it helped us understand the gendered dimensions of labor and livelihoods, leading to the realization that women's work – subsistence and wage – mattered more than men's for the story we would tell about Pholela. Second, and most important for this chapter, the very fact that we had uncovered this and other "secrets" – and that Gogo Sithole laughed as she told us so – reveals how important *relationships* were to the research we conducted, and by extension, to the stories we tell. In this chapter, we examine the importance of relationships to research in political ecology, with a particular focus on qualitative methods rooted in ethnography.

Geographers have long written about qualitative research methods (Crang 2002, 2003; Limb and Dwyer 2001), and since the 1980s, feminist scholars (including geographers and political ecologists) have written about the relationships that are so important to qualitative research, as well as the power and positionality that undergird them (Kobayashi 1994; Staeheli and Lawson 1995; Sultana 2007; Sundberg 2003; Crang 2003; Haraway 1988; Katz 1994; Rocheleau 1995; Rocheleau et al. 2013). As a political ecologist and an Africanist in graduate school in the mid-2000s, Abby's methodological training was steeped in these lessons. As a result, when she developed the methodology for the research we conducted, many of the insights you will read below were at the forefront of her mind. When she got to Pholela and once we started working together, we came to realize that those critiques were only a starting point; our relationship and the ones we had with people like Gogo Sithole offered yet another level of complexity. In this chapter, we use our experience to unpack the complexity of relationships – between us, between us and the people with whom we conduct our research, and between us and the institutions that govern higher education – and their importance for research methods in political ecology. It is worth noting that political ecology has a long, rich tradition of mixed methods and celebrates a plurality of approaches (Jackson and Neely 2015; Robbins 2012; Rocheleau 1995). Indeed, Abby employs more methods – surveys, focus groups, archival methods, participatory GIS, and the (critical) use of scientific work – than the ones we write about here. Nonetheless, we contend that regardless of the methods one employs, relationships are fundamental to the production of knowledge in political ecology.

We begin with an overview of the feminist literature on qualitative and ethnographic methods with a particular focus on questions around objectivity, positionality and reflexivity, and relationships. This literature teaches that researchers need to reflect on their positionality vis-à-vis the people with whom they conduct their research and the larger political-economic structures that shape life, livelihoods, and academic work (Kobayashi 1994; McDowell 1992; Sultana 2007). While some scholars see positionality as stable (at least on the broad terrain of power), others understand it as intersectional – as composed of multiple intersecting characteristics – and as in part shifting depending on circumstance (Kobayashi 1994; Sultana 2007). Rooting our analysis in Pholela, we build from this work to argue that a person's position is relational; it comes into being through relationships which are ever changing and constituted at multiple scales.

To do this work, we offer three related concepts: *entanglement*, *intra-action*, and *diffraction*. Drawing on feminist science studies scholars Donna Haraway (1991) and Karen Barad (2007), we use the term *entanglement* to highlight how everyone – us, the people with whom we conduct our research, our families, Abby's dissertation and tenure committees (as well as the non-human bodies that are always present) – is interconnected in ways that are large and small, clear and obscure, expected and unexpected. In other words, our research is entangled in a web of relationships. The idea of *intra-action* reminds us that we (the researchers) and the people with whom we conduct our research are and always have been relational beings. Whereas interaction relies on discrete preexisting bodies that interact when they come together, *intra-action* recognizes that all bodies are relational from the start; we (and our research subjects and our colleagues and so on) and our positionalities do not preexist our entanglements (Barad 2003). Therefore, research and positionality come out of intra-actions like the one we describe above. *Diffraction* offers a way of – a method for – reading different strands of research through each other. Whereas reflexivity, based on the visual metaphor of reflection, calls on the researcher to recognize and disclose her positionality by reflecting on her raced, classed, gendered, and geographic characteristics (among others), diffraction attends to the ways in which actors intra-act with, interfere with, and reinforce one another to produce difference in those bodies (and their positionalities) and in research. It is a relational method where process and change are constitutional. For example, reading a sick person's description of an illness through a doctor's diagnosis through an *isangoma's* (a healer who works in consultation with the ancestors) and through her father's explanation offers a rich, complex understanding of health and illness where difference comes to the fore. Attending to the ways in which those multiple strands are then read through the researcher(s) – attending to the diffraction pattern that stems from the researcher(s) – offers a way to think through how researchers' and research subjects' relational positionalities shape knowledge. By starting from a point of entanglement (see Chapter 8, this volume) and reading our research using the method of diffraction, we recognize the importance and pervasiveness of relationships to research methods, arguing that an attention to these relationships is central to political ecological methodologies and ethical entanglements in the field.

Objectivity, positionality, and situated knowledges

Since the 1980s, feminist scholars have questioned the idea of objectivity. They note that rather than a disembodied, distanced, and unbiased understanding of the world, objectivity is a white, male, privileged perspective (Cope 2002; Harding 1986, 1987). As such, objectivity fails to account for the full spectrum of perspectives, knowledges, and evidence (Kobayashi 1994; Sundberg 2003). Building from these insights, Donna Haraway offers a "feminist objectivity" that she terms situated knowledges (Haraway 1988: 581). Now axiomatic to feminist scholarship, situated knowledges calls for an accounting of the history, race, class, gender, and geography – the situation – of knowledge producers (Schumaker 2001). So doing, it celebrates the partiality of the perspective that each person has while being sensitive to the structures of power that shape them (Mcdowell 1992). Feminist scholars contend that through an attention to partial perspectives, scholars develop more objective and more responsible knowledge(s), and critical engagement, rather than critical distance, becomes the goal of research (Kobayashi 1994; Rose 1997).

To incorporate partial perspectives, Linda McDowell writes that "we must recognize and take account of our own position, as well as that of our research participants, and *write this into our research practice*" (McDowell 1992: 409, original emphasis). In response to provocations like this, scholars seek to account for their situated knowledges "by reflexively examining [their] *positionality*" (Rose 1997: 305, our emphasis; see also: Kobayashi 1994; Sultana 2007). A researcher's "positionality" typically includes her individual characteristics – race, class, gender, education, age – and her institutional privilege – job, institutional home, geography – all vis-à-vis the "multidimensional geography of power relations" that shape her work and life (Rose 1997: 308). Reflexivity refers to the next step, whereby the researcher *reflects* on her positionality and makes it "visible" (see Chapter 8, this volume) with the goal of rendering authorial transparency (Staeheli and Lawson 1995). As Gillian Rose explains, reflexivity "looks both 'inward' to the identity of the researcher, and 'outward' to her relation to her research and what is described as 'the wider world'" (Rose 1997: 309). In calling on researchers to be reflexive, these scholars assert that particular characteristics and their relationships to the broader terrain of power must be accounted for in all aspects of the research process, from research design to access to the field to data collected to analysis and to the scholarship produced (Mollett 2013; Sultana 2007; Sundberg 2003). Quite simply, knowledge is shaped by the people and places involved in its creation; in order to fully render that knowledge, we must reflect on our position and how it shapes our research.

In the interest of taking up the challenge of transparent reflexivity, we offer our positionalities: we are Abigail (Abby) Neely, a white, English- and Zulu-speaking (Zulu is a second language) American assistant professor of geography at Dartmouth College in New Hampshire, USA and Thokozile (Thoko) Nguse, a black, Zulu- and English-speaking (English is her second language, and is far better than Abby's Zulu) human resources intern in Durban, South Africa, and a life-long resident of Pholela. We are almost exactly the same age (Abby is three weeks older), and we are both single women. That said, our lives are quite different. Thoko has two children and Abby has none. In 2008 and 2009, when we conducted the bulk of our research, Abby was a PhD student at the University of Wisconsin and Thoko was a community health worker who had completed the first year of a tertiary degree but had stopped studying and returned home due to a lack of funds. We first met in early 2008 when Abby arrived in Thoko's community to conduct a household survey on health, agriculture, and livelihoods and then worked together on a daily basis for over a year. Throughout the research process, Abby received grants from the U.S. government and the University of Wisconsin, and from those funds she paid Thoko a salary. As will become clear, we conducted much of the fieldwork for Abby's dissertation and book manuscript together, as well as some of the analysis (Neely 2011).

While this description of our individual characteristics and institutional homes is surely illuminating, it is only partially so; we must also account for our positions in the context of the multidimensional landscape of power (Rose 1997). At first blush this seems pretty easy. Abby is from a capital-rich country, she is highly educated, she is white, and in 2008 and 2009 the stipend from her research grants allowed her to live very comfortably with electricity, running water, security, and a car. In other words, she is inscribed with a lot of power. During our research, Thoko lived in rural South Africa with no electricity or running water, she received a "stipend" of R1000 (US$100) per month for her work as a community health worker and helped to care for her children and grandmother. In other words, she is inscribed with very little power. So here we have a conundrum: two researchers working on the same project in radically different "positions" vis-à-vis the terrain of power. If we stopped our analysis here, what picture would you take away? Here's what we imagine: Abby calls the shots; after all, she's the one with the education and she is employing Thoko. In return, Thoko helps smooth Abby's – the white American's – entry into people's homes, translates, and guides her linguistically and culturally (Gade 2001; Marshall 2002; Schumaker 2001; Sharp 2005). From the reflexive description of our positionalities and what you know about the global terrain of power and how it intersects with race, class, gender, and geography (Mollett 2013; Sultana 2007; Sundberg 2003), this analysis makes sense. And yet, as we will see, this reading does not always fit our experience (see also Sultana 2007). After all, our relationship and the relationships we had with the people with whom we conducted our research – our *entanglements* – shaped our positionality in both small and big ways and changed that positionality over time. We too are relational beings.

Relationships and research methods

Researcher–research assistant relationships

A few scholars have written specifically about the relationship between academic researcher and research assistant (our relationship) (Rosaldo 1993; Sanjek 1993; Schumaker 2001). Much of this work notes the absence of attention to the research assistant in scholarly work. It also points to many of the hierarchies we detail above – researchers as "usually white and mostly male" and therefore in a position of power, and research assistants as "mainly persons of color" and therefore not in a position of power (Sanjek 1993: 13). A strand of this work seeks to show the impact of research assistants on scholarship, often seeing them as "cultural brokers" and as instrumental to research completed (Schumaker 2001; see also Bank 2008). Taken together, these works challenge the "myth of the lone ethnographer" (Rosaldo 1993), drawing attention to the important (if limited) role research assistants play in the sorts of community-based research that we conducted.

Our relationship and experience certainly fits with much of what scholars have written – Thoko not only helped Abby with language (Gade 2001; Sharp 2005), she also helped Abby fit in culturally (with greetings, dress, behavior, and so on), especially initially (Sultana 2007). But this was only half of the equation; Abby helped Thoko understand how a researcher puts together a project and what Donna Haraway and others write about situated knowledges. While these two forms of expertise carry different currency in the political-economy of higher education, in Pholela this was often reversed. Most importantly, it was the combination of these two forms of expertise – shared, discussed, and processed *together* – that undergirds the knowledge (and the diffraction pattern) we produce. In explaining the idea of diffraction, Karen Barad focuses on the importance of the apparatus – the equipment used in an experiment – for the production of difference in scientific research. It is the apparatus, she explains, that helps produce the diffraction pattern. In many senses, *we* are the apparatus through which the stories we collected are diffracted. As an apparatus, we offer a good example of how intra-action and entanglement are essential for diffraction, and by extension for knowledge; we are entangled and intra-acting with each other, just as we are entangled with the people with whom we conduct our research. The result of all this entanglement is a rich diffraction pattern where what we understand and what we write is read through our interpretations both together and separately.

In practice, this meant that we planned our weekly (and overall) research goals together – which homesteads to visit, what to ask, how to ask, and why it mattered. While these conversations began with Abby developing a plan and Thoko offering a few suggestions, they quickly evolved into more balanced conversations, as we came to better understand *our* research and our role in Pholela. For example, it was Thoko who, after months of interviews with various *isangoma* about illnesses and healing practices, suggested we see if one would take us on a trip to the forest to collect ingredients for *imithi* (medicines). This forest walk led us to think about resource use and landscape management in new ways and taught us much about the role of the local environment in health and healing.

As our collaboration developed, so did our disagreements. We often argued – sometimes loudly – about whether or not ancestors were real (with Abby taking the for position and Thoko the against) or whether or not Gogo Mzizi was telling the "truth" given what Gogo Dlamini said. These arguments were productive, helping us to think through and process – to diffract – what we were learning as we went, becoming integral to our research *method*. Organically, *our* conversations and disagreements emerged from the conversations we were having with Pholela's residents. They then fed back into the interviews we conducted as we built a diffraction pattern (see Kobayashi 1994). This iterative method yielded important insights, blurring the lines between our conversations and those we had with Pholela's residents. The more we worked together and taught one another, the more collaborative our research became, the better our understandings, and the richer our diffraction pattern.

As feminist scholars have long argued, and as we note above, the methodology devised, methods used, analysis completed, and scholarship produced are all part of the same process; these steps cannot be divided (Nast 1994; Staeheli and Lawson 1994, 1995; Mollett 2013; Moss 2002; Nightingale 2003; Rocheleau 2008; Sharp 2005). In many ways, the conversations we have had with each other throughout our research and after are as important to how we think about and write about the information we collected as were the months we spent visiting people's homesteads, and the years Abby has spent in academic libraries and seminars. This realization brings us to another insight of feminist scholars: theory and praxis, or theory and "empirics," are not so neatly divided (Staeheli and Lawson 1995; Englund and Yarrow 2013). Nowhere is this clearer than in our continuing conversations where our experiences in Pholela are ever present, even (or especially) as we talk through the insights of political ecology or feminist theory. This crucial point is what the scholars we cite above miss: the researcher–research assistant relationship is not only important because it facilitates the (often foreign) researcher's entry into the field enabling the collection of information, but because it is integral to the ways in which data is analyzed and scholarship produced. Research and relationships, theory and praxis, field and university, method and knowledge are all relational too.

Researcher(s)-researched relationships

Our relationship was not the only (and probably not the most important) relationship shaping our work. As we hint at above, we also had relationships with the people with whom we conducted our research. A number of scholars have written about these relationships, noting a binary between researcher(s) and "researched" (Cahill et al. 2007; Crang 2003; Nast 1994; Sultana 2007), and pointing to a hierarchy in which the (Western) researcher is in a position of power (Kobayashi 1994; Staeheli and Lawson 1995; Sultana 2007; Sundberg 2003). (Sultana offers an important variation to this story as a Western-trained, highly educated Bangladeshi scholar who conducts research in Bangladesh (Sultana 2007).) Following from lessons about objectivity and reflexivity, scholars explain that these hierarchies are important because they reveal a power imbalance where the researcher holds the power thanks to her position on its global terrain (Gilbert 1994; Kobayashi 1994; McDowell 1992). Significantly, scholars believe it essential to bridge the divides (Gibson-Graham 1994; Kobayashi 1994; Moss 2002; Crang 2003; Sultana 2007), advocating for self-conscious practices (beginning with reflexivity) that identify some common ground between researchers and research subjects in order to mitigate power differences and build a research project and relationship (Gibson-Graham 1994).

While certainly true on a broad level, our experience highlights two problems with understanding researcher–research subject relationships as a hierarchical binary. First, on a day-to-day basis, the power dynamics we experienced through our relationships were quite different from what one might expect knowing the broader terrain of power. For example, in the story with which we began, it was the *gogos*, not us, who decided that we would conduct our oral histories with them together. In another instance, we were holding a community meeting when a woman who was not attending walked into the crowd and pulled aside a local elected official. She told him that the *Inkosi* (Chief) wanted to see us immediately because we did not have permission to hold the meeting. Of course, we had received permission for our research, and so after some discussion with those in attendance, we decided to finish the meeting before going to see the *Inkosi*. As we were packing up, a couple of people told us that the woman who had interrupted was "jealous" of the people we had organized the meeting with, and so they thought that we would be able to clear everything up quickly. We then went to the *Inkosi*'s office where we learned that he had been gone all day and had never talked to the woman.

This story brings into sharp relief an experience that is familiar to many who conduct community-based research: when researchers enter communities, they enter places thick with preexisting power relationships (Marshall 2002; Sultana 2007). On a daily basis, the local terrain of power – always entangled with the global – shaped our relationships and the research we conducted. As our experience reveals, the accounting of our positions that we offer above –

white, American, educated and black, South African, less educated – and what that says about our authority and by extension our position(s) on the broad terrain of power is not sufficient for explaining how power works in Pholela. And power in Pholela had a profound influence on our day-to-day work and by extension on the scholarship we produce. In other words, while Abby's white skin and elite education paired with Thoko's black skin and "local knowledge" surely helped us access communities and homesteads, it did not determine how our interactions would unfold and what we would learn. As Rose writes, "Researchers are entangled in the research process in all sorts of ways, and the demand to situate knowledge is a demand to recognize that messiness" (Rose 1997: 314). We and the people with whom we conduct research are entangled in all sorts of (known and unknown) relationships from the local to the global, and as we intra-act those relationships shift and change, just as we shift and change. Research and relationships are messy business.

This insight brings us to our second point: we – researchers and research participants – are relational beings. The trouble with the idea that researcher and research subject relate to each other through a hierarchical binary is that it assumes a set of pre-given identities much like the ones we offered in our paragraph about our positionalities. If we take seriously the claim that feminism is a "relational ontology" (Gilbert 1994; Stacheli and Lawson 1995), then we must recognize that identities are relationally formed and further that they do not preexist relationships. If we take this lesson to heart, then we must account for the fact that researchers are *entangled* with the people with whom they conduct research, that research subjects are entangled with each other, and that everyone is entangled with the places in which they live and labor, just as those places – Pholela and Dartmouth College – are entangled with each other (McDowell 1992; Mollett 2013). If we start from a place of entanglement, rather than from a binary, proceed from research to analysis to writing, and allow the various information and perspectives we gather to intra-act (with each other and with our perspective), we find a rich diffraction pattern. This diffraction pattern, attuned to difference, interference, and reinforcement, developed out of an engaged and responsible (partial) objectivity, then offers a more complex knowledge which better represents the people and the place of Pholela.

More relationships still

As many scholars have pointed out, and as we hint at above, the relationships that affect research and scholarship are not bounded within a specific site (Rocheleau 2008; Chapter 8, this volume). For our project, Abby's doctoral dissertation committee had an important role to play in the methods we used, as well as in her dissertation and subsequent publications. Equally important have been the various interlocutors at universities – mainly in the Global North, but also in South Africa – who have listened to talks Abby has given, offered comments on pieces she and we have written, and chatted through various ideas in political ecology, feminist science studies, and other scholarship. For this piece alone, Ramah McKay, Matt Turner, Jess Krug, and the editors for this volume – all positioned at universities in the Global North – offered thoughtful critique and advice to strengthen our insights and contributions. These relationships offer indispensable opportunities to think *with* people who are "experts," but who were not with us in Pholela. And these opportunities are as important for the diffraction pattern and the scholarship we produce as our time with people like Gogo Sithole and Gogo Mbanjwa.

There are more relationships still: relationships between researchers and the funding agencies who make research and writing possible (for us, the NSF, the University of Wisconsin's Land Tenure Center, and Yale University's Agrarian Studies Program), not to mention relationships between those funding agencies and the governments or foundations through which they receive their money. There are also relationships between academic researchers and the committees that determine whether or not they get a degree, a job, an article or book published, and whether they keep their position (Rocheleau 2008). The scholarship we produce – written together or sole-authored by Abby – and the choices we make in that scholarship (and about its authorship) are always embedded in these relationships, just as they are embedded in the relationships we have in Pholela. And, of course, there are more relationships still. We both belong to various communities and our relationships with the people we care deeply for – our parents, siblings, children, cousins, neighbors, friends, colleagues, and students – all shape the decisions we make about where we live and work, what we do for work, how we think, who we are. And all of these relationships shape the research we do and the scholarship we produce, just as the relationships we have with our research participants shape who we are in the rest of our lives. In following all of these entanglements, one can't help think that the old adage, "it's turtles all the way down" got it wrong; it's relationships all the way down.

Conclusion

From Abby's earliest days as a master's student, she took the lessons from the scholarship on objectivity, reflexivity, and relationships to heart and then to Pholela, where they provided inspiration as we conducted our research. In turn, our experiences and our relationships with each other and with the people with whom we work led us to the concepts of entanglement, intra-action, and diffraction as a way to build on that scholarship. What we found in Pholela was that while an accounting of the characteristics that make up our positionalities on the broader terrain of power is necessary, it is not sufficient. Relationships make accounting for our role in the production of knowledge a bit messier. As some of the work on reflexivity has shown, we must simultaneously account for our positions in all of the relationships in which we find ourselves, from the households of Pholela to the broad, global networks of the discipline of geography. In other words, our positionalities, our diffraction patterns, and the knowledge we produce is always shaped by and entangled with scholarly review panels *and* the "jealousies" or the boredom of Pholela's residents. As a result, we must understand ourselves and our postitionalities as relational and as constituted at multiple scales. Doing so makes clear that a researcher's positionality is always changing as the research process unfolds over time and across space, and as she intra-acts with people near and far (Kobayashi 1994; Sultana 2007). For example, in the simple act of making and drinking tea with the *gogos*, our positionalities changed over time. At first Gogo Mbanjwa would make tea, then she started asking Thoko to do so, then Abby. Who made the tea depended on how much energy Gogo Mbanjwa had, how long we had known each other, and who else was visiting. In this simple example, it becomes clear that our positionality changed as a result of our entanglements. Our collaboration then adds yet another level of complexity: we have different positionalities, experiences, understandings, and expectations. As a result, even before we visited our first homestead, our research was the product of intra-actions and entanglement and our method was one of diffraction.

As the examples from our research reveal, power and positionality are complicated business. Gillian Rose writes that we might think about "the imperative to situate less in terms of surveying positions in a landscape of power and more in terms of seeing a view of power as punctured by gaps precariously bridged" (Rose 1997: 315; see also: Gibson-Graham 1994). In other words, there is a landscape of power that *matters* – Abby is an American with a PhD and a tenure-track job, Thoko is a South African with a steady job, and most of the people with whom we conducted our research are still pensioners or unemployed – but it is neither monolithic nor coherent. The landscape of power is punctured by gaps – local struggles for authority or two researchers asking questions about livelihoods and health – that are bridged; it shapes but does not determine the relations of power and day-to-day life. Relationships bridge the gaps. And these bridges reach beyond Pholela, to the halls of universities and to printing presses in the Global North. While the knowledge we produce in volumes like this is marked by what the editors are interested in and what other scholars have written, it is also marked by people like Gogo Mbanjwa and Gogo Sithole who not only share their life histories with us, but tell us how we should conduct our research (in pairs, asking certain questions, digging up secrets). In recognizing the entanglements through which our research is diffracted we can better account for the multiple positions, perspectives, and knowledges that shape our research and from that we can produce better scholarship.

References

Bank, A. (2008) "The 'intimate politics' of fieldwork: Monica Hunter and her African assistants, Pondoland and the Eastern Cape, 1931–1932." *Journal of Southern African Studies*, vol. 34, pp. 557–574.

Barad, K. (2003) "Posthumanist performativity: Toward an understanding of how matter comes to matter." *Signs*, vol. 28, pp. 801–831.

Barad, K.M. (2007) *Meeting the universe halfway: Quantum physics and the entanglement of matter and meaning.* Durham, NC: Duke University Press.

Cahill, C., Sultana, F., and Pain, R. (2007) "Participatory ethics: Politics, practices, institutions." *ACME: An International E-Journal for Critical Geographies*, vol. 6, pp. 304–318.

Cope, M. (2002) "Feminist epistemology in geography." In Moss, P. (ed.) *Feminist geography in practice: Research and methods.* Malden, MA: Blackwell Publishers, pp. 43–56.

Crang, M. (2002) "Qualitative methods: The new orthodoxy?" *Progress in Human Geography*, vol. 26, pp. 647–655.

Crang, M. (2003) "Qualitative methods: Touchy, feely, look-see?" *Progress in Human Geography*, vol. 27, pp. 494–504.

Englund, H. and Yarrow, T. (2013) "The place of theory: Rights, networks, and ethnographic comparison." *Social Analysis*, vol. 57, pp. 132–149.

Gade, D.W. (2001) "The languages of foreign fieldwork." *Geographical Review*, vol. 91, pp. 370–379.

Gibson-Graham, J.-K. (1994) "'Stuffed if I know!': Reflections on post-modern feminist social research." *Gender, Place and Culture*, vol. 1, pp. 205–224.

Gilbert, M.R. (1994) "The politics of location: Doing feminist research at 'home'." *The Professional Geographer*, vol. 46, pp. 90–96.

Haraway, D. (1988) "Situated knowledges: The science question in feminism and the privilege of partial perspective." *Feminist Studies*, vol. 14, pp. 575–599.

Haraway, D. (1991) *Simians, cyborgs, and women: The reinvention of nature.* New York: Routledge.

Harding, S.G. (1986) *The science question in feminism.* Ithaca: Cornell University Press.

Harding, S.G. (ed.) (1987) *Feminism and methodology: Social science issues.* Bloomington: Indiana University Press.

Jackson, P. and Neely, A.H. (2015) "Triangulating health toward a practice of a political ecology of health." *Progress in Human Geography*, vol. 39, no. 1, pp. 47–64.

Katz, C. (1994) "Playing the field: Questions of fieldwork in geography." *The Professional Geographer*, vol. 46, pp. 67–72.

Kobayashi, A. (1994) "Coloring the field: Gender, 'race,' and the politics of fieldwork." *The Professional Geographer*, vol. 46, pp. 73–80.

Limb, M. and Dwyer, C. (2001) *Qualitative methodologies for geographers: Issues and debates.* London: Arnold.

Marshall, J. (2002) "Borderlands and feminist ethnography." In Moss, P. (ed.) *Feminist geography in practice: Research and methods.* Malden, MA: Blackwell Publishers, p. xiii.

McDowell, L. (1992) "Doing gender: Feminism, feminists and research methods in human geography." *Transactions of the Institute of British Geographers*, pp. 399–416.

Mollett, S. (2013) "Mapping deception: The politics of mapping Miskito and Garifuna space in Honduras." *Annals of the Association of American Geographers*, vol. 103, pp. 1227–1241.

Moss, P. (2002) *Feminist geography in practice: Research and methods.* Malden, MA: Blackwell Publishers.

Nast, H.J. (1994) "Women in the field: Critical feminist methodologies and theoretical perspectives." *The Professional Geographer*, vol. 46, pp. 54–66.
Neely, A.H. (2011) *Reconfiguring Pholela: Understanding the relationships between health and environment from the 1930s to the 1980s*. PhD, University of Wisconsin-Madison.
Nightingale, A.J. (2003) "A feminist in the forest: Situated knowledges and mixing methods in natural resource management."
Robbins, P. (2012) *Political ecology: A critical introduction*, 2nd edn. Malden: John Wiley & Sons.
Rocheleau, D. (1995) "Maps, numbers, text, and context: Mixing methods in feminist political ecology." *The Professional Geographer*, vol. 47, pp. 458–466.
Rocheleau, D.E. (2008) "Political ecology in the key of policy: From chains of explanation to webs of relation." *Geoforum*, vol. 39, pp. 716–727.
Rocheleau, D., Thomas-Slayter, B., and Wangari, E. (2013) *Feminist political ecology: Global issues and local experience*. New York: Routledge.
Rosaldo, R. (1993) *Culture & truth: The remaking of social analysis*. Boston: Beacon Press.
Rose, G. (1997) "Situating knowledges: Positionality, reflexivities and other tactics." *Progress in Human Geography*, vol. 21, pp. 305–320.
Sanjek, R. (1993) "Anthropology's hidden colonialism: Assistants and their ethnographers." *Anthropology Today*, vol. 9, pp. 13–18.
Schumaker, L. (2001) *Africanizing anthropology: Fieldwork, networks, and the making of cultural knowledge in central Africa*. Durham, NC: Duke University Press.
Sharp, J. (2005) "Geography and gender: Feminist methodologies in collaboration and in the field." *Progress in Human Geography*, vol. 29, pp. 304–309.
Staeheli, L.A. and Lawson, V.A. (1994) "A discussion of 'women in the field': The politics of feminist fieldwork." *The Professional Geographer*, vol. 46, pp. 96–102.
Staeheli, L.A. and Lawson, V.A. (1995) "Feminism, praxis, and human geography." *Geographical Analysis*, vol. 27, pp. 321–338.
Sultana, F. (2007) "Reflexivity, positionality and participatory ethics: Negotiating fieldwork dilemmas in international research." *ACME: An International E-Journal for Critical Geographies*, vol. 6, pp. 374–385.
Sundberg, J. (2003) "Masculinist epistemologies and the politics of fieldwork in Latin Americanist geography." *The Professional Geographer*, vol. 55, pp. 180–190.

11
METHODS AND ENVIRONMENTAL SCIENCE IN POLITICAL ECOLOGY

Karl S. Zimmerer

The expanding roles of environmental science in political ecology amid major social-ecological changes

It is timely yet a formidable challenge to examine the roles of environmental science in political ecology with regard to methods and methodology. This topic is rarely singled out, notwithstanding general discussion in overviews of political ecology (Forsyth 2002; Neumann 2005; Paulson and Gezon 2005; Robbins 2012; Rocheleau et al. 1996; Stott and Sullivan 2000; Zimmerer and Bassett 2003). One of the few stand-alone treatments of methods remains a founding work of more than twenty-five years ago, namely a chapter on measurement in *Land Degradation and Society* (Blaikie and Brookfield 1987: 49–63). Then, as now, the attention to environmental scientific methods and techniques is recognized as embedded in the design of research and overall methodology, the research topic and its questions, and theory, concepts, and conceptual framework (Sharp et al. 2011). Political ecology today benefits from two-decades-plus of new directions and productive advances. My goal here is to focus on environmental scientific methods and methodology issues as they concern the multiple roles of science currently at work in political ecology. Thus, this chapter looks beyond and seeks to complement a "how-to" perspective. Its focus is designed to account for the mounting importance of environmental issues being interwoven with the concerns of sustainability and justice.

Political ecology and environmental science, including the important sub-area of ecological science, are defined broadly here to enable their full engagement. This perspective brings together a significant variety of environmental and social approaches along with wide-ranging terminology (Table 11.1). It reflects the basic recognition of specific methods and methodology as interwoven inextricably with corresponding concepts and theory. In addition, it underscores this chapter's pluralistic vision, intentionally eschewing the universal adoption of environmental scientific methods in political ecology. This vision is shared with similar studies (Zimmerer and Basset 2003: 2–3; see also Beymer-Farris 2013; Turner 2009; Walker 2005). It reflects awareness and support of the validity and strengths of the large number and diversity of methods and methodological designs in political ecology. While many political ecology methods and methodologies, such as discourse analysis, content analysis, and ethnography may not involve science, they can be increasingly combined, as discussed below.

Table 11.1 Terms to discuss the roles of science and methods in political ecology

	Term	*Synoptic definition and usage*	*Specific point(s) of relevance*	*Source examples*
a	Political Ecology	approach combining broad concerns of political economy, social power and ecology	integrated social science and environmental science approach to nature–society interactions	Blaikie and Brookfield 1987
b	Post-Structural Political Ecology	political ecology with emphasis on social power and knowledges, such as discourse	highly productive expansion of the core of political ecology	Escobar 1998; Peet and Watts 2004; Robbins 2003
c	Social Science	scientific approaches to human societies, cultures, and behavior	contain conceptual and methodological tools well-suited to political ecology	Singleton and Straits 2009
d	Nature-Society Geography, Human-Environment Geography	main sub-field of geography that engages both human/social and environmental analysis; these terms can be treated as synonymous, or can refer to distinct within-subfield areas	political ecology, as defined above, can be situated within this subfield, while other disciplines and interdisciplinary domains also are extensively engaged	Turner II 2002a, 2002b; Zimmerer 2010a
e	Cultural and Human Ecology	interdisciplinary approaches combining broad interests in cultural practices, economies, and human conditions together with ecology	precursors to political ecology that continue though not replaced, as sometimes claimed; contribute to new sciences (see below)	Turner II 2002a; Turner 2003a; Zimmerer 2004
f	Environmental Science (ESci), Environmental Studies (EStud). Development Studies (DStud)	wide-ranging interdisciplinary environmental realms with emphasis on science (ESci) and broad spectrum of environment- and development-related studies including humanities (EStud, DStud)	can be used to engage political ecology through such topics as environment-development and conservation-development	Turner 2003b; Zimmerer and Bassett 2003
g	Ecology, Biogeography Ecogeography	relations among organisms and environments; a concern for environmental quality and management	both meanings relevant to political ecology; first meaning is scientific usage referring to ecological sciences	Turner 2009; Zimmerer 1998
h	Land Change Science, Social Ecology	scientific approaches focused on deeply interdisciplinary interactions in human-modified environments	influenced through cultural and human ecology, political ecology, and other inter-disciplinary approaches during past twenty-five years	Turner 1991; Turner et al. 2001, 2007

Also includes Sustainability Science, Global Change Science, Biodiversity Science, Industrial Ecology, Urban Ecology, Resilience Ecology

Table 11.1 continued

	Term	*Synoptic definition and usage*	*Specific point(s) of relevance*	*Source examples*
i	Earth Systems Science	science of the interactions of the atmosphere, biosphere, and hydrosphere	includes focus on Anthropocene, the newly coined geologic time span of human influence	Turner 1991
j	Resource, Ecological, Environmental Economics and Sociology	approaches within social sciences that use both quantitative and qualitative techniques	concepts and methods include significant overlap with the use and roles of science in political ecology (includes focus on decision-making)	Blaikie and Brookfield 1987
k	Science Studies/ Science and Technology Studies	social science analysis of the production, use, influence, and communication of scientific knowledge	can be integrated with political ecology	Goldman et al. 2011
l	Environmental Politics	politics of the environment and meanings of environmental science, management, and policy	a foundation of political ecology	Bryant 1991; Forsyth 2002
m	Citizen Science, Public Sociology Public Science,	approaches linked to social science, as well as social movements and activism, that incorporate science, often termed alternative sciences	potentially synergistic with political ecology trends that include public political ecology	Burawoy 2005; Jansen 2009; Kalleberg 2005; Robertson and Hull 2001
Also includes Public Political Ecology, Public Ecology				
n	Causal Inference	determination of cause-effect relations based on principles of the knowledge system	differs significantly among the social sciences and the biogeophysical sciences; poses special challenges and opportunities in political ecology	Forsyth 2002; Singleton and Straits 2009; Turner 1999a; Zimmerer 2010a
o	Mixed Methods	integration of qualitative techniques and quantitative techniques	guides approaches, such as political ecology, using science in combination with other methods	Rocheleau 1995
p	"Science Wars" (ca. 1990–2005)	heated debate and animosity between pro- and anti-science proponents in academia and policy	restricted the integrated or selective use of science within sub-fields (e.g. political ecology)	Goldman et al. 2011

The use of environmental science qua science, which has occurred widely and is integral to political ecology (Table 11.2), is complemented through the latter's focused analysis of scientific narratives and discourses on central themes such as adaptation, vulnerability, resilience, deforestation, and community conservation. More generally, these twin roles have gained strength in political ecology and related fields (Taylor 2010). Indeed, the entwined engagements with environmental science have been productive in political ecology's use in numerous influential examples (Table 11.2). It leads to my chapter's call for expanding research, along

Table 11.2 Some examples of the use of environmental scientific methods within political ecology and closely related fields[a]

Source (example)	*Political ecology and/or related field(s)*	*Environmental scientific focus*	*Use of scientific methods*	*Contribution(s) of scientific method(s) in overall farming*
Hecht 1985	Environment-development interactions and environmental change narrative	soil nutrient status corresponding to pasture age	sampling and nutrient analysis of soils (pH, Ca, Mg, P, N, C)	revealed soil degradation worsening with time in widespread pastures being created in predominant tropical lowland development
Liverman 1990; Liverman and O'Brien 1991	Environmental hazards	climate variation, crop impacts, and region-based variation of farm size	statistical analyses of climate, crop yields, and regional farm types	revealed greater vulnerability of regions with peasant and indigenous communities of smallholders
Zimmerer 1992, 2014	Political ecology, environment-development interactions and change narrative	ecology and biodiversity of local food plants at multi-taxonomic levels corresponding to landscape	sampling and field experiments on adaptive capacities of local staple food plants	revealed varied adaptive capacities contributing to widespread spatio-temporal unevenness of agrobiodiversity change, rather than predominant Genetic Wipeout
Turner 1999b	Land use, local production systems, and land-use history	ecology and regrowth capacity of rangeland plants at patch level	field experiments in range ecology in semi-arid grasslands	demonstrated spatial and temporal variation in plant ecological dynamics of livestock grazing and influence on desertification
Bassett and Zuéli 2000	Environmental change narratives; regional discourses	plant succession ecology following fire and burning	sampling, identification, and height measurement of woody plants	revealed processes of tree establishment, rather than predominant deforestation explanation of Sahel desertification
Nightingale 2003	Feminist geography	forest ecology of community forests	image analysis and vegetation inventories	revealed how forest knowledge systems embedded in environmental politics
Galt 2008	Political ecology and resource economics	resource economics of pesticide use	household surveys and statistical modeling	reveals levels of pesticide use influenced through combined farm household, environment, and political-economic factors

Note: "Closely related fields" refer especially to those of the human-environmental sciences that are directly connected to the same author or research group also relying significantly on political ecology.

with corresponding study design and activities, integrating the dual instrumental and reflexive perspectives on environmental science in political ecology. This call for strategic pairing of instrumental-reflexive vantages reflects recent disciplinary-centered rapprochement following the so-called Science Wars (Table 11.1.p). Examples include nature-society geography and human-environment geography, social-ecological sciences, and public sociology (Table 11.1.d, 11.1.h, and 11.1.m, respectively). Such strategic pairing of instrumental-reflexive perspectives offers particular resonance and promise for the use of environmental science methods in political ecology. More generally, it serves as a guide to this chapter and the contributions being advanced.

Mounting evidence of major, global-scale social-ecological changes is propelling, to a large degree, the rapprochement of the instrumental/practitioner and reflexive/critique perspectives in environmental science and the social sciences, including political ecology. Scientific consensus now estimates the severely worsening impacts of major human-induced global environmental problems, such as climate change, biodiversity loss, and nitrogen-cycle alteration, already exceed the sustainability of planetary boundary conditions (Rockström et al. 2009). Other global social-ecological issues, such as changes in land and freshwater use, also pose serious threats. These global problems are composed of both biogeophysical environments per se and broadly defined societal interactions (Turner 1991; Vitousek 1992)—such as economic, politics, and cultural processes at various local, national, and regional scales—thus justifying the description of such multi-scale issues as political-ecological. These expanding problems are powerfully interwoven with major sustainability and justice concerns and potential capacities among virtually all human populations and sub-groups. Vulnerability and access to resilience and economic resources vary widely across class, race, gender, and other intersecting axes siting people and places (Eakin and Luers 2006; Ribot and Peluso 2003; Rocheleau et al. 1996; Swyngedouw and Heynen 2003). These broadly environmental concerns have reverberated also in the loftier realms of social analysis (Latour 2009, 2010).

Scientists themselves have become more actively aware of the political obstacles and social complexity of environmental problems, thus tending to bridge in the general direction of political ecology. Practitioners within environmental science, environmental studies, and earth system science (see Table 11.1), for example, can be seen as agreeing that it is not the lack of scientific knowledge and management interest, per se, but rather questions of politics and social power that often undermine and limit the capacity to address social-ecological problems and opportunities (McCarthy et al. 2014). For example, the Union of Concerned Scientists is committed to publicizing the policies, politics, and political economy surrounding environmental scientific findings on such issues as climate change, pollution, and health and safety issues (e.g., UCS 2013). While rooted in environmental science, much analysis of social-ecological issues has adopted a post-positive perspective that rejects naïve empiricism (Chowdhury 2013; Turner 2002a). It embraces the socially relevant and engaged formulation of research topics together with the commitment to inform and promote policy and management options (DeFries et al. 2012; Zimmerer 2014). Focus on problem framing, multi-scale interactions, and familiarity with the sway of political discourse and political economy are indicators of interdisciplinary openness and sociopolitical interest. Examples include recent and fast-growing approaches such as the global-change, sustainability, and biodiversity sciences, as well as such fields as industrial ecology, urban ecology, and resilience ecology (Table 11.1.h; see also next section). A specific case among many is the new concept of telecoupling. Telecoupling, premised on global-scale land-system changes involving long-distance couplings of highly diverse flows of environmental and information factors along with governance effects, is the sort of highly interdisciplinary concept distinguishing new environmental science.

At the same time, works in political ecology are expanding the range and depth of analysis based on the use of environmental scientific concepts and methods. Much of this political ecological engagement is focusing insight on environmental scientific concepts, such as resilience, adaptation, and vulnerability (Bassett and Fogelman 2013; Beymer-Farris 2013; Cote and Nightingale 2012; Eakin and Luers 2006; Turner 2013; Zimmerer 2010b, 2013). Resilience ecology, for example, is providing new insights into landscape change dynamics while political ecology is used to rethink core assumptions of social-ecological systems. One new example elucidates the use of resilience and disturbance ecological concepts amid the social power and development dynamics of conservation initiatives aimed at tropical coastal mangroves (Beymer-Farris 2013; Table 11.1.h). Another direction of political ecology's ongoing engagement with environmental science is to incorporate wildlife ecological analysis; for example, the latter is interwoven with the political ecology of marine conservation (Campbell 2007; St. Martin and Hall-Arber 2008). This expanded use of environmental scientific methods is being deployed to understand environmental governance, justice, and the social power and politically mediated access to resources that are increasingly of concern to conservation and sustainability while they are actively expanding pillars of political ecology (Blaikie and Brookfield 1987; Ribot and Peluso 2003; Turner 2009). Methodologically, the opportunities for political ecological use of environmental science have expanded as a result of increased, original uses of the techniques of remote sensing and GIScience (Robbins 2003; Turner 2003a, Turner and Taylor 2003), and, also, a specific engagement with land change/system science (Brannstrom and Vadjunec 2013; Chowdhury 2013; Turner and Robbins 2008; Zimmerer 2013).

Beyond binaries in the post-science wars? CCMs and the opportunities for rapprochement and fruitful frictions, or fateful impasses at the palace gates

The prospect is positive, to be sure, for political ecology to expand its use of environmental science concepts and methods in a context of urgent challenges and opportunities, entwining environmental change and justice issues. But it would be naïve to assume that fruitful engagement is ensured as a consequence of dilemma-driven needs combined with the post-Science Wars rapprochement. Indeed, fundamental philosophical distinctions are characteristic of the converging knowledge systems in political ecology that incorporate environmental science. A suite of post-structural, postmodern, and post-positivist ways-of-knowing characterize the epistemological stances of predominant social science approaches in current political ecology. These theoretical commitments anchor strong methodological tool kits that are widely used and applied in political ecology: in addition to discourse and content analysis, ethnography, and the use of detailed historical analysis are now central to much political ecology. This attribute of varied methods highlights the pivotal importance of careful, well thought-out integration in the combining of environmental science and political ecology. How, for example, can a political ecology framework integrate discourse analysis, common in post-structural approaches, with the use of environmental scientific methods?

Integration is enticing given the insights of productively crossing the divide of social and environmental-scientific analysis, together with the need to respond to the clarion calls for these kinds of inputs to policy, management, and public engagement arenas. The allure and potential rewards of this integration through its radiating engagement with multiple audiences—the enticing entry through "palace gates" referred to above—is reflective of the growing influence, importance, and popularity of social-ecological interdisciplinarity. It has become a defining goal of rapidly growing approaches. Their general claim is to integrate across the realms of human and non-human nature, on the one hand, and resource utilization along with the influential realms of economics, politics, culture, and society. This integration can be conceptualized as the coupling of multiple social-ecological interactions (Figure 11.1; see also Zimmerer and Young 1998). The promise of such integration depends on particular connecting concepts "doing this work" in the actual practice of environmental science in political ecology and related fields.

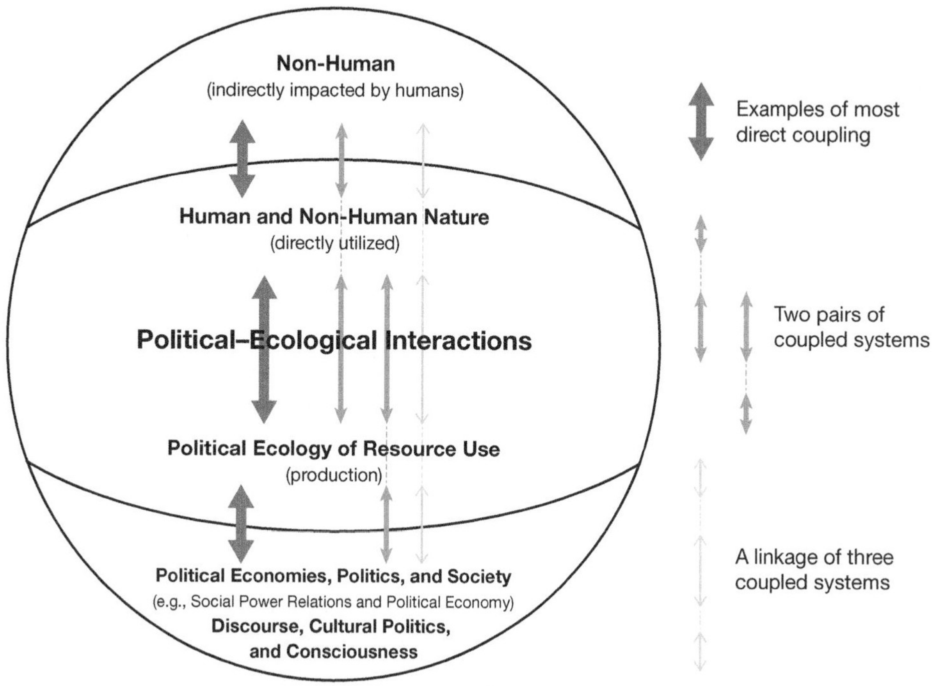

Figure 11.1 Diagram of general political-ecological integration based on coupled factors

I refer here to the expanding usage of certain connective conceptual modalities (Table 11.3). A connective conceptual modality (CCM) is the use of a concept to connect across the realms of the biogeophysical analysis of the environmental sciences and broadly human social understandings rooted in the social sciences and humanities. CCMs play a major role in the integrative social-environmental sciences, even if oftentimes only partly unacknowledged. As shown in Table 11.3, political ecology has tended to rely on a group of three CCMs in the integration of environmental scientific methods; other CCMs are important though somewhat less common. First, is a focus on resource properties, ranging from ecological relations to material flows, affecting socially differentiated access and impacts. One example concerns the ecological relations of essential food plants to their growing sites in which adaptive capacity is found to function through mutual interactions with the gendered resource access of smallholder households and communities, thus conferring either social-ecological resilience or vulnerability (Eakin 2006; Zimmerer 1992, 2004). A second CCM is critical pluralism (Table 11.3), including relevant insights from so-called methodological pluralism (Norgaard 1989). Here environmental scientific analysis is integrated via carefully designed comparisons (Sharp et al. 2011)—examples include the use of environmental scientific methods to compare with scientific narratives and discourses on degradation, conservation, and sustainability. Political ecology studies are often designed to use both these CCMs (see Table 11.3 for examples).

A third CCM is the idea of scientific boundary concepts that draws on a specific concept most widely associated with science studies. The scientific meaning of a "wildlife corridor," for example, crosses over between the ecological sciences and the social science of conservation that includes an explicitly spatial framework (Goldman et al. 2011; Zimmerer 2000). Numerous

Table 11.3 Examples of connective conceptual modalities (CCMs) in linking and bridging analysis in political ecology and related fields (emphasis on use of scientific methods and the roles of science in political ecology; see text for CCM definition and explanation)

Political ecology/related approach(es)	*Research topic*	*Use of scientific method(s)*	*Sources (example)*
Political ecology; human-modified or Anthropocene ecological biogeography	Resource properties, disturbance ecology and resilience	ecosystem analysis and ecological succession and disturbance theory and analysis	Beymer-Farris et al. 2012; Doolittle 2010; Turner 2009; Zimmerer and Young 1998
Political ecology; resource-use decision-making, valuation, and commodification	Pluralism to compare policy narratives and resource decision-making	econometric analysis, rapid ecological assessment	Blaikie and Brookfield 1987; Galt 2008, 2010, 2014; Robertson 2006
Political ecology; social studies of science and technology	Social analysis of scientists and scientific practice	focus on content of ecological models	Forsyth 2002; Lave et al. 2010; Goldman et al. 2011

examples of this third CCM build on the perspective of feminist standpoint theory while they also focus on the roles of environmental science in management and policy (Table 11.3). This third CCM in political ecology is also resonant with the many uses of environmental science in related fields such as anthropology and sociology (Brosius 1999; Dove 2006; Escobar 1998; Greenberg and Park 1994; Kaup 2008; Kinchy and Kleinman 2003; Paulson and Gezon 2005).

Elucidation of these CCMs, along with others, is crucial to political ecology in responding to calls for policy, management, and scientific research contributions. For example, they can clarify its value and complementarity relative to frameworks of broad social-ecological interdisciplinarity, such as sustainability science, that is relevant to geography and other disciplines (Turner II 2002a, 2002b; Zimmerer 2010a). Similarly these illustrations of the CCMs of political ecology are potentially of interest to the policy and environmental turns of recent social science. Such examples range from Burawoy's "public sociology" to Latour's "plea for earthly sciences" (Latour 2009). Valuable to note also are other influential CCMs, though they may engage differently with social-environmental science. Examples include health and disease of human bodies and psychology (Guthman and DuPuis 2006; King 2013; see also Chapters 26 and 43, this volume); associative critical pluralism (Sharp et al. 2011); and approaches based on assemblage theory and landscape technology (Bell 2013; Carney and Voeks 2003; Linton 2010; Zimmerer 2011). Elucidation of CCMs is a timely area for future research and discussion on combining the environmental sciences together with the social sciences and humanities.

Extending discussion from specific CCMs to a general prospectus requires incorporating the crucial interplay of current environmental policy and management issues. Insights hinge on the conjoined reflexive and instrumental uses of environmental science in political ecology. Examples of environmental scientific-derived insights include adaptation, resilience, and vulnerability in regard to climate variation and change (Eakin 2006; Liverman 1990; Chapter 23, this volume; Liverman and O'Brien 1991; Tschakert 2012); plant ecological, soil-plant nutrient, and spatial analysis of the social-ecological changes of tropical forests and woodlands (Duvall 2006; Hecht 1985; Rocheleau 1995), complex spatio-temporal ecological dynamics of rangelands and livestock-herding (Bassett and Zuéli 2000; Robbins 2003; Turner 1999a, 1999b,

2003b), landscape-level analysis of soil fertility and degradation (Blaikie 1985; Brannstrom and Oliveira 2000; Jansen 1999), wildlife ecology and biogeography related to conservation initiatives (Campbell 2007; Naughton-Treves et al. 2005); and the agrobiodiversity, pesticide-related toxicity, and human-health relations of agricultural landscapes and food systems (Carney and Voeks 2003; Coomes and Ban 2004; Duvall 2006; Galt 2010; Nabhan et al. 2011; Perreault 2005; Zimmerer 2014). In these examples, the use of scientific methods is integral to the careful examination of once predominant narratives of global environmental change—such as ignorant-colonist models of tropical deforestation, assumed large-scale processes of desertification, the so-called global circle of pesticide poisons, and the presumed spatio-temporal, wipe-out scenario in the "genetic erosion" and loss of agricultural biodiversity.

In these cases political ecology's use of environmental scientific methods enables new insights and narratives, including counter-narratives, with regard to powerful understandings and potentially mistaken interpretations of global change, resilience, and vulnerability (Adger et al. 2001; Beymer-Farris 2013). Such works are important to refining and sometimes countering socially and spatio-temporally uninformed scientific narratives of human-environmental change, collapse, and conservation (Zimmerer 2000). They can engage related works on themes such as environment–development interactions that examine the resource-related impacts on marginal social groups, such as rainforest-based peasant and indigenous peoples, of so-called poverty traps as well as environmental risks and hazards (Coomes et al. 2011; McSweeney 2005). In addition, new trends in political ecology offer clarion calls for critical, alternative scientific approaches, stemming from critiques of stream and wetland restoration (Lave 2012; Lave et al. 2010; Robertson 2006) and contributions of feminist theory (Cote and Nightingale 2012; Nightingale 2003; Rocheleau et al. 1996).

Professional risk and related obstacles are powerful factors, even if such consideration is largely missing to-date. Deployment of such methods can trigger unwelcome reaction by those members of the scientific community unsupportive or antagonistic toward integration with the concerns and concepts of political ecology. Conversely, it can elicit negative reactions in political ecology. In a personal experience some time ago, my conference presentation on political ecology selectively incorporating econometric techniques and a spatial externalities framework was characterized as inherently incompatible by a political ecologist discussant. Another risk stems from significant challenges and limitations in the degree of integration involving the use of environmental scientific methods in political ecology. Certain unnecessary limitations can result from the overly vague situating of scientific methods. Providing well-designed focus to the scientific inquiry is pivotal to successful integration (see the next section). Shortfalls are analogous to the uncritical use of field methods, where "everything is nothing" (see Bernard 2011: 23; Wolford 2006). Equally significant, the adoption of scientific inquiry in political ecology may be portrayed as highly promising, and perhaps politically desirable in parts of the academy, though it may not be clear how or why this combination is to occur (see discussion in Brannstrom and Vadjunec 2013). Finally, the risk of ineffective integration includes inadvertently reinforcing, rather than eclipsing, the epistemological binary of the human-social and environmental scientific realms.

Even the successful undertaking of social-ecological integration must counsel the commonness of selective integration in research practice and, by extension, raise the general phenomenon of productive tensions ("fruitful frictions"). These points echo the importance of reasonable expectation, rather than the sometimes common goal of the complete inclusivity of social-environmental integration. The latter inevitably falls short while it evokes an almost Holy Grail claim to all-encompassing comprehensiveness. Indeed, the successful roles of scientific methods in political ecology do not require, or even suggest, fully fusing or entirely eclipsing the distinctions of knowledge systems. Working with diverse ways-of-knowing is a vital and still-evolving strength of political ecology. As a result, the integrating of environmental scientific concepts and methods must be seen as most often linked and bridged selectively to other techniques, concepts, and philosophical perspectives located in political ecology.

Nor should the absence of a completely integrative approach be taken as a priori evidence of the shortcoming or failure of the use of environmental science in this type of research. In other words, the response to "beyond binaries?" posed in the heading to this section does not imply a negative answer in the case of anything short of sweeping, seamless synthesis. Instead, my point is to highlight that selective uses and productive tensions are often characteristic of the success of integrating environmental science and political ecology. In this regard, the expectation of fruitful frictions is meant to benchmark such uses that are rigorous, typically selective, and often creatively original. It should be anticipated they will help yield well-substantiated, important insights. Indeed, this selective integration of environmental science must be seen as vital to the analytic spectrum of knowledge systems engaged within political ecology and its various interdisciplinary borderlands.

Situating the use of scientific methods in political ecology research

Political ecology's engagement with environmental science is similar to other fields insofar as the development of methods and methodology are paramount to the practice of research. Specifically, they must take shape and be resolved in close coordination with the other major cornerstones of knowledge acquisition (Figure 11.2). This coordination is required with regard to both concepts and theory engagements ("purpose") and the identification of the research topic ("problem") (Figure 11.2). In other words, the design of methodology and choice of methods are undertaken in tandem, specifically with the choice of theoretical framework and conceptual issues to be examined (shown as (i) in Figure 11.2), and also with the identification of the research topic and questions or hypotheses (shown as (i) in Figure 11.2). These methodological cornerstones can also be referred to as: (i) the "purpose" of research and (ii) the object(s) of study (Sayer 1992). Separating the place of methods and methodology is necessarily abstract (iii in Figure 11.2), since each endeavor is embedded within the triad. The point of this illustration is to distinguish the place of methods and methodology in relation to other pivotal activities and to highlight its importance as a principal research endeavor.

In general, the use of environmental science requires the consideration of so-called mixed methods (Table 11.1), since most research in political ecology is unlikely to fit entirely within a scientific framework. Mixed methods, referring to the coordination of qualitative and quantitative techniques, is often necessary and well-suited to political ecology. It requires the careful design of research in accord with anticipated knowledge claims. A design heavy on scientific methods obviously would be ill-suited to a political ecology approach toward the analysis of social power relations. Conversely, other political ecology could fall short if the scientific methodology employed was insufficient for an adequate engagement with environmental evidence in the biogeophysical realm. Each of the examples given in Table 11.2 illustrates the careful matching of scientific methods to political ecology via mixed methods. Also helpful are key insights within the community of mixed methods research, per se. For example, the application of quantitative techniques does not necessarily constitute a scientific approach, as in the case of enumeration or statistical estimation that is purely descriptive. At the same time, it is possible that a scientific approach could involve sampling designs and addressing questions (or testing hypotheses) using primarily qualitative techniques.

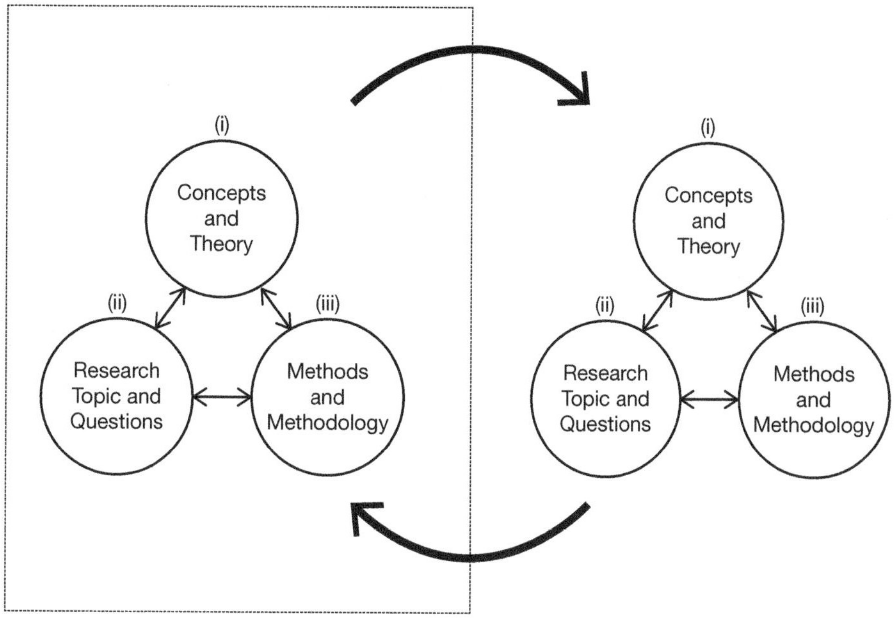

Figure 11.2 Triad of research cornerstones including methods and methodology

Figure 11.2 is intended to illustrate a pair of possible approaches that are helpful for the incorporation of environmental scientific methods in political ecology. One approach (Figure 11.2, left panel) assumes the multi-directional, inter-relational compatibility of the specifications in each of the three principal endeavors of research design (i–iii). The choice and use of scientific methods, like other techniques, must bear close relations to the concepts and theoretical framework being addressed. For example, environmental scientific methods can be related to social conditions via structuration processes in a theoretical framework of social science based on critical realism (Chowdhury and Turner 2006). Similarly, scientific methods must bear close relation to the research topic, question, and objects of study. In the case of social-ecological change, such methods must demonstrate well-suited compatibility with the processes, spatial scales, and temporal framing that distinguish the research focus. As illustrated, this approach can furnish a heuristic guide to research depending on the preliminary situating—involving interests and commitments—of the individual researcher. For example, a researcher who first identifies major interest in a certain topical area would need to think relationally with regards to theoretical-conceptual and methodological specifications, including the potential use of scientific methods.

The second approach (Figure 11.2, right panel) to research design is similarly applicable to planning the use of scientific methods within political ecology. Here scientific methods are sequenced as a distinct element within a multi-stage approach to methodological design. Several of the examples in Table 11.2 demonstrate this kind of design, though they do not draw explicit attention to the value of this research methodology in political ecology. Sampling and estimation of plant and wildlife populations, for example, may rely on the use of scientific methods on the part of the political ecologist or collaborators, who are also using other methods in order to study various social issues influencing resource access, power relations, governance, and institutions. Often these diverse activities are also sequenced in the timetable of research, including the possibility of phasing scientific methods at multiple stages in the overall design (e.g., re-sampling designs). This second approach recognizes that multiple, distinct triadic cores, each containing methods-topic-theory cornerstones, are in effect conjoined strategically within single political ecology-guided designs of research. In other words, it is possible to envision the design and accommodation of such multiple phases through the well thought-out interlinking of related research triads, each with a methodological component. Future research on scientific methods in political ecology would benefit from additional discussion and visualization of the multi-phase approach being described and illustrated here (Figure 11.2).

Recommendations: selective integration, novel interactions and surprise in the practice of environmental scientific methods in political ecology

A consequence of the above sections is to recommend a fuller awareness of the specific opportunities and challenges of environmental scientific methods in political ecology (Walker 2005). This awareness stems from recognition that the incorporation of scientific methods can generate notable insights in political ecology that potentially translate into significant interdisciplinary domains. At the same time such approaches can and do incur professional risks, as detailed above. Fuller awareness is also centered on recognition that scientific methods within political ecology, even when successful, are likely to be associated with selective use and strategic linkages, rather than a uniform or comprehensive unification across all the sub-components of research. Awareness of this partialness as a predictable, productive, and even desirable feature of successful works, rather than a sign of partly failed designs or ill-conceived methods, helps political ecology avoid a pair of pitfalls in its engagements with environmental science. First, such awareness enables the fuller appreciation of political ecology's existing and ongoing contributions to issues of sustainability, justice, and ethics associated with global change (Tables 11.1 and 11.2). Many of these political ecology contributions, rooted in the effective use of environmental and ecological science for purposes of social justice, have embraced an implicit ethics and philosophy of critical pluralism (Schlosberg 1999). The latter is also central to productive debates over philosophical and applied pragmatism in environmental approaches (Wescoàt 1992). Second, this awareness of selective applications offers grounds for reflecting productively on ongoing environmental scientific engagements along possible vanguardist impulses where new strands of scientific usages can be claimed, or at least seem, to arise *de novo* and transcend *en toto* other such engagements.

Selective use of science within political ecology is evident in productive research on issues of sustainability, justice, and ethics in the context of a rapidly changing world where interdisciplinary rigor, methodological innovation, and multi-scale integration are more critical than ever in social-ecological analysis. Here the specific environmental scientific methods deployed within political ecology yield insights into new global change including climate issues, environment–development interactions, land use and land cover change, and urbanization (Brannstrom and Vadjunec 2013; Zimmerer and Bassett 2003; see case studies in Table 11.2). Pressing social-ecological dimensions of these concerns—ranging from justice issues to technical management and administrative governance—are a powerful force in the potential expansion of environmental science in political ecology. At the same time, environmental science, including its use in political ecology, must grapple with the opportunities and challenges of rapidly amassing data, proliferating sensor-produced information including remote satellite imagery as well as inexpensive field-based sensors, data processing and analysis capacities including GIScience in particular, and multi-level data platforms well-suited to interdisciplinary environmental science. In all the preceding, it is the issues of ethics, as well as analytics, that leverage political ecology's advantage amid unprecedented needs and capacities for social-environmental scientific research.

The roles of environmental science in such opportunities and challenges are consistent with a view of scientific knowledge and scientific methods forming in relation to prevailing politics, economics, and social power. Science is envisioned as the practice of a system of knowledge that is subject to a continuum of influences operating both within and outside the demarcation of investigations. This multi-directional coherence of science is significant, while its culture and functioning are neither as idealized nor monolithic as once assumed (e.g., *The Economist* 2013). The social identities and context of scientists and their practices, including funding, lead to noteworthy influences of politics, economics, and social power, albeit without these forces overly determining, per se, the practice of science (Goldman et al. 2011; Kinchy and Kleinman 2003). This point is important in understanding environmental science and scientific methods, for its insight is a contrast to earlier assumption of these knowledge systems as stand-alone bastions or simple handmaidens derived from political and economic forces. Indeed the complexity of scientific practice, as described here, is integral to the agenda of social analysts and planners arguing for the bridging of the reflexive-instrumentalist components of knowledge systems (Burawoy 2005; Jansen 2009; Kalleberg 2005; Latour 2009, 2010).

Finally, awareness of the selective nature of the engagement of science with political ecology is a proposition that underscores the importance of new insights stemming from novel social-ecological interactions at multiple scales owing to mounting global change. Surprise in social-ecological interactions is increasingly characteristic of to the human dimensions of global change and such issues as sustainability, justice, and ethics due to the abundance of novel interactions. The theme of social-ecological surprise is being identified as having particular relevance in current and future environmental analysis, policy, and management. One of the take-home points of my chapter is that scientific approaches in political ecology are well suited and often strategically situated to provide fresh insights into novel social-ecological interactions tied to sustainability, justice, and ethics issues. The relative advantages of political ecology in this regard stem from its incorporation of environmental scientific methods with in-depth social, political, and economic analysis necessary to create a combined instrumental-and-reflexive perspectives (i.e., constructive-and-critical). For example, it enables political ecology to apply scientific methods in open-ended methodologies, which are the sort of approaches well-suited to research problems with lower levels of predictability and less existing information (Robertson and Hull 2001; see also Chowdhury 2013; Forsyth 2002; Goldman et al. 2011).

Ample opportunities for timely and insightful analysis abound in the application of scientific methods to innovative social research, such as political ecology, focused on persistent as well as novel issues of sustainability, justice, and the environment. The capacity of political ecology is notable for negotiating the realities of multiple data sources and diverse interpretive paradigms. As a result it can produce original findings that are both surprising in the context of research scholarship and practical in social-ecological management and policy. Explicit attention to scientific methods and analysis is integral to these practical realms in both political ecology and related fields. Additional importance of the role of surprise stems from the importance of communication with broader audiences and the general public. Novel insights from political ecology on major environmental issues gained through the incorporation of environmental scientific methods are likely to resonate with swelling numbers of scientific communication specialists. Strategically positioned to reach wide-ranging audiences of the general public and policymakers (Baron 2010), they increasingly cover environmental, sustainability, and justice issues related to climate change, biodiversity loss and conservation, and the roles of agriculture and food in society and global change.

Conclusion: productive tensions in the future of political ecology (Science is dead! Long may it live!)

A good part of political ecology has prospered and will continue to do so without the need for or use of environmental scientific methods. Broad emphasis on the politics and political analysis of environmental issues, especially environmental governance and power relations, has become a defining focus of political ecology that is quite skillfully and productively combined with local case studies, region-scale analytics, and global analysis based on a toolkit of rigorous, diverse methods other than the use of environmental science per se. These strengths could be taken to suggest a broad-based, academic division-of-labor in which ongoing trends are leading political ecology to adopt its identity as a distinct, mostly non-overlapping complement to environmental science. This prognosis of "Science is Dead" within political ecology would be quite unlike my analysis and recommendations in this chapter. That diametric prognosticating of a new complementarity of non-scientific political ecology would envision political ecology sans science occupying a niche as a partner of the enlarging social-ecological and global-change sciences. Ironically this prospectus would lead to a Science Wars-type conclusion by eschewing the practice of environmental science in political ecology.

Unlike such a prognosis, this chapter concludes that the use of environmental scientific methods confers valuable general capacities as well as specific strengths to political ecology. The latter are potentially unique in the universe of other human-environment and nature-society approaches, with this uniqueness depending in part on the use of environmental scientific methods. The latter's strengths are increasingly relevant amid mounting social-environmental issues involving management and policy as well as sustainability and justice issues. This relevance requires the continued advance of new research designs and the innovative use of environmental scientific methods within political ecology. Theoretical framing and pragmatic acumen in the selective use of environmental scientific methods will enable political ecology to undertake these innovations while building on the advance of productive contributions—thus this conclusion's rejoinder of "long may it live!" with reference to the multiple roles of science and environmental scientific methods in political ecology.

Similarly central to my conclusion is that the use of environmental scientific methods in political ecology offers potentially novel insights to urgent social-ecological issues with rapidly expanding impacts and multi-dimensionality. These capacities for original advances extend well beyond the borders of political ecology to spur communications that incorporate scientific analysis with public audiences and professional networks (Baron 2010). The latter span the environmental governance, management, policy, activist, and broadly institutional domains of social-ecological change, sustainability, and justice issues. These audiences tend to place particular value on the identification and analysis of unanticipated surprise findings in the interdisciplinary environmental sciences. Political ecology's continued and future engagements with environmental science and scientific methods can offer a particularly potent promise of being well-poised to address, communicate with, and influence such audiences.

Acknowledgements

I'm very grateful for the insightful comments and helpful suggestions of the editors and also for discussions in the "Political Ecology and Social-Ecological Change" seminar in spring 2014, with the members and participants in the new GeoSyntheSES lab group, and through ongoing conversations in the contexts of the Cultural and Political Ecology (CAPE) Specialty Group of the Association of American Geographers (AAG) that include the productive exchanges spurred by my Netting Award talk at the 2014 AAG meeting in Tampa, the 2014 Rappaport Prize Committee of the American Anthropological Association (AAA), and the National Socio-Environmental Synthesis Center (SESYNC) through my plenary talk there on "Surprise in Human-Environmental Change" (January 15, 2014) and subsequent SESYNC activities, and the science communication workshop organized in fall 2013 by Jennifer Balch, Sue Brantley and the Earth and Environmental Systems Institute (EESI) at Penn State.

References

Adger, W.N., Benjaminsen, T.A., Brown, K., and Svarstad, H. 2001, "Advancing a political ecology of global environmental discourses," *Development and Change*, vol. 32, no. 4, pp. 681–715.

Baron, N. 2010, *Escape from the ivory tower: a guide to making your science matter*, Island Press, Washington, DC.

Bassett, T.J. and Fogelman, C. 2013, "Déjà vu or something new? The adaptation concept in the climate change literature," *Geoforum*, vol. 48, pp. 42–53.

Bassett, T.J. and Zuéli, K.B. 2000, "Environmental discourses and the Ivorian Savanna," *Annals of the Association of American Geographers*, vol. 90, no. 1, pp. 67–95.

Bell, M.G. 2013, "The governance of food, technology and environmental resource flows: Connecting mills, water, wheat, and people in colonial Lima, Peru (1535–1700)." Doctoral dissertation, Pennsylvania State University, University Park, Pennsylvania.

Bernard, H.R. 2011, *Research methods in anthropology*, Altamira Press (Rowman & Littlefield), Lanham, Maryland.

Beymer-Farris, B.A. 2013, "Producing biodiversity in Tanzania's mangrove forests? A combined political ecology and ecological resilience approach to 'sustainably utilized landscapes'," in C. Brannstrom and J.M. Vadjunec (eds.), *Land change science, political ecology, and sustainability: synergies and divergences*, Routledge, London.

Beymer-Farris, B.A., Bassett, T.J., and Bryceson, I. 2012. "Promises and pitfalls of adaptive management in resilience thinking: the lens of political ecology," in T. Plieninger and C. Bieling (eds.), *Resilience and the cultural landscape: understanding and managing change in human-shaped environments*. Cambridge University Press, New York, pp. 283–289.

Blaikie, P. 1985, *The political economy of soil erosion in developing countries*, Longman, London.

Blaikie, P. and Brookfield, H. 1987, *Land degradation and society*, Methuen, London.

Brannstrom, C. and Oliveira, A.M. 2000, "Human modification of stream valleys in the western plateau of Sao Paulo, Brazil: implications for environmental narratives and management," *Land Degradation & Development*, vol. 11, no. 6, pp. 535–548.

Brannstrom, C. and Vadjunec, J.M. (eds.) 2013, *Land change science, political ecology, and sustainability: synergies and divergences*, Routledge, London.

Brosius, J.P. 1999, "Analyses and interventions: anthropological engagements with environmentalism," *Current Anthropology*, vol. 40, no. 3, pp. 277–310.

Bryant, R.L. 1991, "Putting politics first: the political ecology of sustainable development," *Global Ecology and Biogeography Letters*, pp. 164–166.

Burawoy, M. 2005, "For public sociology," *American Sociological Review*, vol. 70, no. 1, pp. 4–28.

Campbell, L.M. 2007, "Local conservation practice and global discourse: a political ecology of sea turtle conservation," *Annals of the Association of American Geographers*, vol. 97, no. 2, pp. 313–334.

Carney, J.A. and Voeks, R.A. 2003, "Landscape legacies of the African diaspora in Brazil," *Progress in Human Geography*, vol. 27, no. 2, pp. 139–152.

Chowdhury, R.R. 2013, "The intersection of independent lies: land change science and political ecology," in C. Brannstrom and J.M. Vadjunec (eds.), *Land change science, political ecology, and sustainability: synergies and divergences*, Routledge, London, pp. 224–240.

Chowdhury, R. and Turner, B.L. 2006, "Reconciling agency and structure in empirical analysis: smallholder land use in the southern Yucatán, Mexico," *Annals of the Association of American Geographers*, vol. 96, no. 2, pp. 302–322.

Coomes, O.T. and Ban, N. 2004, "Cultivated plant species diversity in home gardens of an Amazonian peasant village in northeastern Peru," *Economic Botany*, vol. 58, no. 3, pp. 420–434.

Coomes, O.T., Takasaki, Y., and Rhemtulla, J.M. 2011, "Land-use poverty traps identified in shifting cultivation systems shape long-term tropical forest cover," *Proceedings of the National Academy of Sciences*, vol. 108, no. 34, pp. 13925–13930.

Cote, M. and Nightingale, A.J. 2012, "Resilience thinking meets social theory situating social change in socio-ecological systems (SES) research," *Progress in Human Geography*, vol. 36, no. 4, pp. 475–489.

DeFries, R.S., Ellis, E.C., Chapin, F.S., Matson, P.A., Turner, B.L., Agrawal, A., ... and Syvitski, J. 2012, "Planetary opportunities: a social contract for global change science to contribute to a sustainable future," *BioScience*, vol. 62, no. 6, pp. 603–606.

Doolittle, A.A. 2010, "Stories and maps, images and archives: multimethod approach to the political ecology of native property rights and natural resource management in Sabah, Malaysia," *Environmental Management*, vol. 45, no. 1, pp. 67–81.

Dove, M.R. 2006, "Indigenous people and environmental politics," *Annual Review of Anthropology*, vol. 35, pp. 191–208.

Duvall, C.S. 2006, "On the origin of the tree *Spondias mombin* in Africa," *Journal of Historical Geography*, vol. 32, no. 2, pp. 249–266.

Eakin, H. 2006, *Weathering risk in rural Mexico: climatic, institutional, and economic change*, University of Arizona Press, Tucson.

Eakin, H. and Luers, A.L. 2006, "Assessing the vulnerability of social-environmental systems," *Annual Review of Environment and Resources*, vol. 31, no. 1, pp. 365–381.

The Economist 2013. How science goes wrong. October 19, 2013. Available at: www.economist.com/news/leaders/21588069-scientific-research-has-changed-world-now-it-needs-change-itself-how-science-goes-wrong (accessed November 11, 2013).

Escobar, A. 1998, "Whose knowledge, whose nature? Biodiversity, conservation, and the political ecology of social movements," *Journal of Political Ecology*, vol. 5, no. 1, pp. 53–82.

Forsyth, T. 2002, *Critical political ecology: the politics of environmental science*, Routledge, London.

Galt, R.E. 2008, "Beyond the circle of poison: significant shifts in the global pesticide complex, 1976–2008," *Global Environmental Change*, vol. 18, no. 4, pp. 786–799.

Galt, R.E. 2010, "Scaling up political ecology: the case of illegal pesticides on fresh vegetables imported into the United States, 1996–2006," *Annals of the Association of American Geographers*, vol. 100, no. 2, pp. 327–355.

Galt, R.E. 2014, *Food systems in an unequal world: pesticides, vegetables, and agrarian capitalism in Costa Rica*, University of Arizona Press, Tucson.

Goldman, M.J., Nadasdy, P., and Turner, M.D. (eds.) 2011, *Knowing nature: conversations at the intersection of political ecology and science studies*, University of Chicago Press, Chicago.

Greenberg, J.B. and Park, T.K. 1994, "Political ecology," *Journal of Political Ecology*, vol. 1, no. 1, pp. 1–12.

Guthman, J. and DuPuis, M. 2006, "Embodying neoliberalism: economy, culture, and the politics of fat," *Environment and Planning D*, vol. 24, no. 3, pp. 427–438.

Hecht, S.B. 1985, "Environment, development and politics: capital accumulation and the livestock sector in eastern Amazonia," *World Development*, vol. 13, no. 6, pp. 663–684.

Jansen, K. 1999, *Political ecology, mountain agriculture, and knowledge in Honduras*, Thela Publishers, Amsterdam.

Jansen, K. 2009, "Implicit sociology, interdisciplinarity and systems theories in agricultural science," *Sociologia Ruralis*, vol. 49, no. 2, pp. 172–188.

Kalleberg, R. 2005, "What is 'public sociology'? why and how should it be made stronger?" *The British Journal of Sociology*, vol. 56, no. 3, pp. 387–393.

Kaup, B.Z. 2008, "Negotiating through nature: the resistant materiality and materiality of resistance in Bolivia's natural gas sector," *Geoforum*, vol. 39, no. 5, pp. 1734–1742.

Kinchy, A.J. and Kleinman, D.L. 2003, "Organizing credibility: discursive and organizational orthodoxy on the borders of ecology and politics," *Social Studies of Science*, vol. 33, no. 6, pp. 869–896.

King, B. 2013, "Shifting spaces and hidden landscapes in rural South Africa," in C. Brannstrom and J.M. Vadjunec (eds), *Land change science, political ecology, and sustainability: synergies and divergences*, Routledge, London, pp. 191–206.

Latour, B. 2009, *Politics of nature*, Harvard University Press, Cambridge, MA.

Latour, B. 2010, "A plea for earthly sciences," *New Social Connections: Sociology's Subjects and Objects*, pp. 72–84.

Lave, R. 2012, "Neoliberalism and the production of environmental knowledge," *Environment and Society: Advances in Research*, vol. 3, no. 1, pp. 19–38.

Lave, R., Doyle, M., and Robertson, M. 2010, "Privatizing stream restoration in the US," *Social Studies of Science*, vol. 40, no. 5, pp. 677–703.

Linton, J. 2010, *What is water? The history of a modern abstraction*, University of British Columbia Press, Vancouver.

Liverman, D.M. 1990, "Drought impacts in Mexico: climate, agriculture, technology, and land tenure in Sonora and Puebla," *Annals of the Association of American Geographers*, vol. 80, no. 1, pp. 49–72.

Liverman, D.M. and O'Brien, K.L. 1991, "Global warming and climate change in Mexico," *Global Environmental Change*, vol. 1, no. 5, pp. 351–364.

McCarthy, J., Chen, C., López-Carr, D., and Endemaño Walker, B.L. 2014, "Socio-cultural dimensions of climate change: Charting the terrain," *GeoJournal*, vol. 79, no. 6, pp. 665–675.

McSweeney, K. 2005, "Natural insurance, forest access, and compounded misfortune: forest resources in smallholder coping strategies before and after Hurricane Mitch, northeastern Honduras," *World Development*, vol. 33, no. 9, pp. 1453–1471.

Nabhan, G.P., Chambers, K., Tecklin, D., Perramond, E., and Sheridan, T.E. 2011, "Ethnobiology for a diverse world defining new disciplinary trajectories: Mixing political ecology with ethnobiology," *Journal of Ethnobiology*, vol. 31, no. 1, pp. 1–3.

Naughton-Treves, L., Holland, M.B., and Brandon, K. 2005, "The role of protected areas in conserving biodiversity and sustaining local livelihoods," *Annual Review of Environment and Resources*, vol. 30, pp. 219–252.

Neumann, R.P. 2005, *Making political ecology*, Hodder Arnold, London.

Nightingale, A.J. 2003, "A feminist in the forest: situated knowledges and mixing methods in natural resource management," *ACME: An International E-Journal for Critical Geographies*, vol. 2, no. 1, 78.

Norgaard, R.B. 1989, "The case for methodological pluralism," *Ecological Economics*, vol. 1, no. 1, pp. 37–57.

Paulson, S. and Gezon, L.L. (eds.) 2005, *Political ecology across spaces, scales, and social groups*, Rutgers University Press, New Brunswick.

Peet, R. and Watts, M. (eds.) 2004, *Liberation ecologies: environment, development, social movements*, Second Edition. Routledge, Oxford.

Perreault, T. 2005, "Why *chacras* (swidden gardens) persist: agrobiodiversity, food security, and cultural identity in the Ecuadorian Amazon," *Human Organization*, vol. 64, no. 4, pp. 327–339.

Ribot, J.C. and Peluso, N.L. 2003, "A theory of acces," *Rural Sociology*, vol. 68, no. 2, pp. 153–181.

Robertson, D.P. and Hull, R.B. 2001, "Beyond biology: toward a more public ecology for conservation," *Conservation Biology*, vol. 15, no. 4, pp. 970–979.

Robertson, M.M. 2006, "The nature that capital can see: science, state, and market in the commodification of ecosystem services," *Environment and Planning D*, vol. 24, no. 3, p. 367.

Robbins, P. 2003, "Beyond ground truth: GIS and the environmental knowledge of herders, professional foresters, and other traditional communities," *Human Ecology*, vol. 31, no. 2, pp. 233–253.

Robbins, P. 2012, *Political ecology: a critical introduction*, Blackwell, Oxford.

Rocheleau, D. 1995, "Maps, numbers, text, and context: mixing methods in feminist political ecology," *The Professional Geographer*, vol. 47, no. 4, pp. 458–466.

Rocheleau, D., Thomas-Slayter, B., and Wangari, E. (eds.) 1996, *Feminist political ecology: global issues and local experience*, Routledge, London.

Rockström, J., Steffen, W., Noone, K., Persson, Å., Chapin, F.S., Lambin, E.F., Lenton, T.M. et al. 2009, "A safe operating space for humanity," *Nature*, vol. 461, no. 7263, pp. 472–475.

St. Martin, K. and Hall-Arber, M. 2008. "The missing layer: geo-technologies, communities, and implications for marine spatial planning," *Marine Policy*, vol. 32, no. 5, pp. 779–786.

Sayer, A. 1992, *Method in social science: a realist approach*, Routledge, London.

Schlosberg, D. 1999, *Environmental justice and the new pluralism: the challenge of difference for environmentalism*, Oxford University Press, Oxford.

Sharp, L., McDonald, A., Sim, P., Knamiller, C., Sefton, C., and Wong, S. 2011, "Positivism, post-positivism and domestic water demand: interrelating science across the paradigmatic divide," *Transactions of the Institute of British Geographers*, vol. 36, no. 4, pp. 501–515.

Singleton, Jr., R.A. and Straits, B.C. 2009, *Approaches to social research*, Fifth Edition. Oxford University Press, Oxford.

Stott, P.A. and Sullivan, S. 2000, *Political ecology: science, myth and power.* Arnold, London.

Swyngedouw, E. and Heynen, N.C. 2003, "Urban political ecology, justice and the politics of scale," *Antipode*, vol. 35, no. 5, pp. 898–918.

Taylor, P.J. 2010, *Unruly complexity: ecology, interpretation, engagement*, University of Chicago Press, Chicago.

Tschakert, P. 2012, "From impacts to embodied experiences: tracing political ecology in climate change research," *Geografisk Tidsskrift-Danish Journal of Geography*, vol. 112, no. 2, pp. 144–158.
Turner II, B.L. 1991, "Thoughts on linking the physical and human sciences in the study of global environmental change," *National Geographic Research and Exploration*, pp. 133–135.
Turner II, B.L. 2002a, "Contested identities: human-environment geography and disciplinary implications in a restructuring academy," *Annals of the Association of American Geographers*, vol. 92, no. 1, pp. 52–74.
Turner II, B.L. 2002b, "Response to Thrift's 'the future of geography'," *Geoforum*, vol. 33, no. 4, pp. 427–429.
Turner II, B.L. and Robbins, P. 2008, "Land-change science and political ecology: similarities, differences, and implications for sustainability science," *Annual Review of Environment and Resources*, vol. 33, pp. 295–316.
Turner II, B.L., Lambin, E.F., and Reenberg, A. 2007, "The emergence of land change science for global environmental change and sustainability," *Proceedings of the National Academy of Sciences*, vol. 104, no. 52, pp. 20666–20671.
Turner II, B.L., Cortina Villar, S., Foster, D., Geoghegan, J., Keys, E., Klepeis, P., Lawrence, D. et al. 2001, "Deforestation in the southern Yucatán peninsular region: an integrative approach," *Forest Ecology and Management*, vol. 154, no. 3, pp. 353–370.
Turner, M.D. 1999a, "Merging local and regional analyses of land-use change: the case of livestock in the Sahel," *Annals of the Association of American Geographers*, vol. 89, no. 2, pp. 192–219.
Turner, M.D. 1999b, "Spatial and temporal scaling of grazing impact on the species composition and productivity of Sahelian annual grasslands," *Journal of Arid Environments*, vol. 41, no. 3, pp. 277–297.
Turner, M.D. 2003a, "Methodological reflections on the use of remote sensing and geographic information science in human ecological research," *Human Ecology*, vol. 31, no. 2, pp. 255–279.
Turner, M.D. 2003b, "Environmental science and social causation in the analysis of Sahelian pastoralism," in K.S. Zimmerer and T.J. Bassett (eds.), *Political ecology: an integrative approach to geography and environment-development studies*, Guilford Press, New York, pp. 159–178.
Turner, M.D. 2009, "Ecology: natural and political," in N. Castree, D. Demeritt, D. Liverman, and B. Rhoads (eds.), *A companion to environmental geography*, Wiley, New York, pp. 181–197.
Turner, M.D. 2013, "Political ecology I An alliance with resilience?" *Progress in Human Geography*, doi: 0309132513502770.
Turner, M.D. and Taylor, P.J. 2003, "Critical reflections on the use of remote sensing and GIS technologies in human ecological research," *Human Ecology*, vol. 31, no. 2, pp. 177–182.
The Union of Concerned Scientists (UCS). 2013. "Groundbreaking new climate research: Since 1854 12.5% of all industrial carbon pollution has been produced by just 5 companies." UCS Newsletter. December. Available at: www.ucsusa.org/publications/pulse/december_2013.html?autologin=true (accessed December 10, 2013).
Vitousek, P.M. 1992, "Global environmental change: an introduction," *Annual Review of Ecology and Systematics*, vol. 23, pp. 1–14.
Walker, P.A. 2005, "Political ecology: where is the ecology," *Progress in Human Geography*, vol. 29, no. 1, pp. 73–82.
Wescoat, J. 1992, "Common themes in the work of Gilbert White and John Dewey: a pragmatic appraisal," *Annals of the Association of American Geographers*, vol. 82, no. 4, pp. 587–607.
Wolford, W. 2006, "The difference ethnography can make: understanding social mobilization and development in the Brazilian northeast," *Qualitative Sociology*, vol. 29, no. 3, pp. 335–352.
Zimmerer, K.S. 1992, "The loss and maintenance of native crops in mountain agriculture," *GeoJournal*, vol. 27, no. 1, pp. 61–72.
Zimmerer, K.S. 1998, "The ecogeography of Andean potatoes," *BioScience*, vol. 48, no. 6, pp. 445–454.
Zimmerer, K.S. 2000, "The reworking of conservation geographies: nonequilibrium landscapes and nature-society hybrids," *Annals of the Association of American Geographers*, vol. 90, no. 2, pp. 356–369.
Zimmerer, K.S. 2004, "Cultural ecology: placing households in human-environment studies-the cases of tropical forest transitions and agrobiodiversity change," *Progress in Human Geography*, vol. 28, no. 6, pp. 795–806.
Zimmerer, K.S. 2010a, "Retrospective on nature–society geography: tracing trajectories (1911–2010) and reflecting on translations," *Annals of the Association of American Geographers*, vol. 100, no. 5, pp. 1076–1094.
Zimmerer, K.S. 2010b, "Woodlands and agrobiodiversity in irrigation landscapes amidst global change: Bolivia, 1990–2002," *The Professional Geographer*, vol. 62, no. 3, pp. 335–356.

Zimmerer, K.S. 2011, "The landscape technology of spate irrigation amid development changes: assembling the links to resources, livelihoods, and agrobiodiversity-food in the Bolivian Andes," *Global Environmental Change*, vol. 21, no. 3, pp. 917–934.

Zimmerer, K.S. 2013, "The compatibility of agricultural intensification in a global hotspot of smallholder agrobiodiversity (Bolivia)," *Proceedings of the National Academy of Sciences*, vol. 110, no. 8, pp. 2769–2774.

Zimmerer, K.S. 2014, "Conserving agrobiodiversity amid global change, migration, and nontraditional livelihood networks: the dynamic uses of cultural landscape knowledge," *Ecology and Society*, vol. 19, no. 2, article 1.

Zimmerer, K.S. and Bassett, T.J. (eds.) 2003, *Political ecology: an integrative approach to geography and environment-development studies*, Guilford Press, New York.

Zimmerer, K.S. and Young, K.R. (eds.) 1998, *Nature's geography: new lessons for conservation in developing countries*, University of Wisconsin Press, Madison.

12

POLITICAL ECOLOGIES OF ACTIVISM AND DIRECT ACTION POLITICS

Nik Heynen and Levi Van Sant

Introduction

Political ecology has been informed by an engagement with activism since its beginning. We argue that the field should continue this trajectory, and suggest that there remain untapped lessons from historical, as well as ongoing, struggles that can inform the treatment of social mobilization within political ecology. Scholars in the field still have important intellectual work to do to benefit from decades of environmental activism, just as environmental activists have significant insights to glean from the rigorous analyses of political ecologists. In this chapter we explore the growing body of political ecological research around social movements, activism, and direct action environmental politics. We argue that engaging with direct action politics offers political ecologists the opportunity to further their contributions to broader forms of environmental activism. Because direct action often occurs when laws and other forms of governance are unable to, by themselves, intervene and mediate social claims on the state (think about for instance the suffragist movement, civil rights movement, gay rights movement, etc.), the political logics and solidarities that result through direct action offer a distinct way of understanding both the state from the outside as well as praxis more generally (see Chapter 35, this volume). Direct action offers a lens into individual activists' efforts working to convince the state of its duty, as opposed to working through the state, which is a more common theme within political ecological research to date. While direct action should only be thought of as one form of political expression, one that is interwoven with other forms of politics, focusing on it explicitly can open new, creative, and more articulate ways of thinking about the intersection of political ecology and activism.

This chapter will start by reviewing how political ecological work (broadly defined) has conceptualized political action motivated by environmental or ecological threats, problems, injustices, and other issues. Much of this evolution maps nicely onto the development of larger trends within geography, anthropology, and cognate fields, moving from a thoroughly political economic framing to one that more seriously engages with human and non-human agency. This trend helps open up possibilities of taking activism more seriously within political ecology, as well as creating bridges related to praxis, in a Gramscian sense (see Chapter 13, this volume), that can make the resources of political ecological insight more relevant to environmental activists. Indeed, thinking through Antonio Gramsci's (1971) distinction between "traditional" and "organic" intellectuals remains instructive, and inspires possibilities for cultivating more robust forms of political ecological praxis.

A brief review of the literature in political ecology related to "activism" prompts several provocative questions: What can political ecologists learn from environmental activists? Also, and as important, what can environmental activists learn from scholars working in political ecology? What are multiple ways of conceiving of "activism," and how have they been employed? And, finally, what might be gained by bringing a more differentiated understanding of activism, via direct action politics, into the political ecological orbit?

As others have noted, political ecology's firm grounding in critical theory has long fortified the field with a strong commitment to social justice (Peet and Watts 2004; Walker 2007). Thus, for many who identify with political ecology, the point is not just to understand the world – but to *change* it (see Chapter 13, this volume). These efforts shape individual research projects in many different ways, and there is no single, coherent framework for the elaboration of an ethical political ecology practice (Walker 2007; Mann 2009; though see Jarosz 2004 for a strong point of departure). One of the most common ways that political ecologists put critical theory to work is by searching for the contradictions in particular socio-ecological systems. In fact, it is arguable that political ecologists most often conceive of *politics* as resistance by social movements as part of these contradictions (Merchant 2005). Just as at other times in the development of political ecology when conceptual innovation has allowed scholars to better grasp the dynamics of their contemporary moment, we think that the language of direct action can push future efforts to think more creatively about environmental politics. One important contribution of focusing on direct action is the way that it can serve to decenter the theoretical dependence of working within the state to achieve emancipatory political ecological objectives. While we think the state is very important for understanding political ecology, necessary even, we also think that environmental direct action can offer other ways of thinking about the core questions political ecologists ask. We hope that doing so will help better embody political ecological theory as well as allow for important strategic connections within environmental praxis.

Moral economies and everyday resistance

Early efforts in political ecology to theorize social movements and other forms of political action were grounded in the field's concern with rural producers in the so-called "Third World" and drew heavily on Marxian peasant studies, specifically the concepts of "everyday resistance" and "moral economy" (Scott 1976, 1985). The moral economy approach highlights the ways that "small producers are faced with subsistence risks that help to create social systems of mutual assistance and tolerable exploitation" (Robbins 2011: 62). In this theoretical framework, political action is generally conceptualized as a response to changes in labor regimes and the related (re)distribution of risk (Peluso 1992). Together, this work suggests that the roots of many social movements centered on environmental issues in the so-called Third World can be found in the colonial era, when pre-existing systems of environmental management and social reciprocity were disrupted and replaced by liberal capitalist regimes.

The collapse or re-shaping of specific moral economies takes many different forms, of course, across time and space. One of the contributions of the political ecological literature is to catalog this multiplicity of experiences and to inquire into the particular conditions under which such movements cohere and organize. These matters are far from pre-determined, as changes in moral economies and intensification of exploitation do not always produce visible resistance, nor do these dynamics follow predictable linear paths. Political ecologists' initial interests in moral economies generally took the form of studies of enclosure that outlined in rich empirical detail the shifts in social relations and patterns of exploitation that were necessary to facilitate deeper integration into global capitalist systems (Neumann 1999). As grassroots actors were denied access to resources, the general pattern of response was one of "bitter covert resistance on a day-to-day basis in order to assert local rights" (Bryant and Bailey 1997: 171).

The broader literature on moral economy generally interprets the everyday resistance of marginalized producers as a practical strategy when open confrontation with power-holders carries the potential of a massive retaliatory response. In this sense everyday resistance is the opposite of "rebellion." As Bryant and Bailey (1997: 170) summarize the difference:

> Whereas peasant rebellion is overt and collective, everyday resistance is covert and often individual; while peasant rebellion directly challenges prevailing political and economic norms, everyday resistance does so indirectly and always on the sly. It is precisely the anonymity of everyday resistance which is, paradoxically, its greatest strength, and yet also its greatest weakness.

Perhaps the most important contribution political ecologists have made to the literature on moral economies is to highlight the forms of everyday resistance that are rooted in material practices of embodied knowledge (Carney 2002). As Robbins argues, this strain of work "attends to the material details that make up livelihoods (crop choices, labor rotations, seed storage, etc.) and the ways those specific material conditions present limits and opportunities for groups as they organize, struggle, and seek to define themselves and their ways of life" (2004: 191).

New social movements and poststructural theory

In the late 1980s and early 1990s, the increasing influence of poststructural theory and the growing prominence of a range of new social movements encouraged political ecologists to expand their field beyond its initial focus on the "traditional" male rural resource user in the so-called Third World. As Peet and Watts (2004: 4) describe them, these new social movements "typically link economic and ecological justice (the *politics of distribution*) with human rights and cultural identity (the *politics of recognition*)." Peet and Watts's edited collection, *Liberation Ecologies*,[1] published first in 1996 and then again in significantly revised form in 2004, embodies some of the earliest and most provocative efforts to bring poststructural theory to the study of political ecology and the "new" social movements. As the editors describe this task (which they conceptualize as "liberation ecologies"), poststructural theory demands more attention to the politics of meaning and the production of knowledge, while the contemporary conjuncture encourages "a practical political engagement with new social movements, organizations, and institutions of civil society challenging conventional notions of development, politics, democracy and sustainability" (Peet and Watts 2004: 6). For Peet and Watts, liberation ecology rests on the charge that scholars "listen to what social movements are saying without naively believing in the inherent wisdom of the 'traditional'" (2004: xiv). This is surely a difficult task but one that offers promising rewards.

These new conceptualizations of social movements and modes of intellectual engagement had several wide-ranging implications. Most broadly, this expanded treatment of social movements in political ecology re-shaped the ways that scholars in the field conceived of politics. According to Robbins (2011: 188–189) the new social movements represented "a new form of political action, since their ecological strands connect disparate groups, across class, ethnicity, and gender" (2011: 188–189). With new theoretical tools, political ecologists moved beyond the singular focus on class-based exploitation and resistance to more heterogeneous and diffuse understandings of social power.

The work of feminist political ecologists was especially productive in this regard, insisting that class exploitation never exists outside of unequal gender politics. Rocheleau et al. (1996) argue that prevailing gender relations (for instance, gendered production activities and norms relating to property rights) shape grassroots environmental movements around the globe. Similarly, they suggest that attention to gender in environmental movements can help bridge the urban/rural and First World/Third World divides that often prevail in conceptualizations of global environmental politics. This re-conceptualization of politics also entailed shifts in the sites of research and analysis. Feminist political ecology (see Chapter 40, this volume), for instance, shows that the household is a central domain for struggles over environmental resources and access (MacKenzie 1998). Investigations of environmental politics can no longer assume that these battles take place solely in the public realm.

A maturing field of study

As is the case with any successful analytical framing, political ecology has evolved and covered more geographic and theoretical terrain. In these continuing efforts many scholars suggest that the poststructural turn helped clarify and articulate these "new" social movements, yet in actuality, politics has always been broader than class struggle. Likewise, there has now been enough research produced about political ecological activism that we can more clearly see the progress made in closing the gap, to borrow from Gramsci again, between "traditional" and "organic" intellectuals. We echo the point, for instance, that "traditional" academics working within political ecology are indeed challenging more historically conservative roles by aligning with institutions working towards counterhegemonic goals, i.e. "solidarity work." It is not possible to review all of the research that increasingly connects political ecology with social movement dynamics; however, it is useful to point out some of the more recent research that has fertilized the overlapping terrain of political ecologists and activists.

Much of the recent political ecology research on land rights, indigenous rights, and resource extraction, for instance, has focused on social movement dynamics, and is often done in close collaboration with activists themselves (see Bebbington 2004; Bebbington et al. 2008a, 2008b; Perreault 2003; Perreault and Valdivia 2010; Perreault and Green 2013; Sundberg 2003, 2007; Valdivia and Benavides 2012; Wolford 2010). There are also important threads of social movement-oriented political ecology research in Sub-Saharan Africa, especially South Africa, working to more tightly partner with activists in an attempt to better embody political ecologies of activism (see Loftus 2012; Lawhon et al. 2014). There is also work that, while not explicitly self-identified as "political ecology," nonetheless contributes to this evolving tradition of blurring the lines between traditional and organic intellectual work (see Moore 2008; Holifield 2012). This research is not of a single thread, of course, yet these projects are broadly representative of the ongoing vibrancy of activist engagements within political ecology.

Much of this work effectively draws from the moral economy tradition and these scholars have, on occasion, explored politics which are clear examples of what many would call "direct action." One of the best examples of this type of analysis is Wendy Wolford's (2010; see also Chapter 44, this volume) influential study of the Brazilian Landless Workers' Movement (MST) in Brazil. Though Wolford does not explicitly conceptualize the politics of the MST as "direct action," or explore the intellectual history of this concept, she does pay significant analytical attention to MST's strategy of land occupation – a form of direct action politics par excellence. Wolford shows how, for instance, this strategy is employed as a sort of spectacle that both brings attention to the cause and serves to articulate participants' multiple identities together under the category of "landless."

Jumping off from these robust literatures, we think the concept and practice of direct action can aid efforts to broaden the understanding of politics within political ecology. Many political ecologists, for instance, concur with Goldman and Turner (2011: 22–23) that:

> One critique of political ecology is that politics are overly simplified—seemingly driven by the clash of interests between social groups over the natural resource in question. Alternatively, politics can be seen as a much more open, contingent interplay of divergent motivations, powers, and strategies of different actors within and across major social groups (and over meaning as much as materials).

Political ecological research is increasingly done in partnership with activist groups and holds the potential to engage this critique head on. To this end, we argue that activist-scholarship rooted in direct action can be brought into conversation with contemporary (and future) ecological struggles in productive ways for the sake of increasing the range and impact of political ecology. For example, doing so could help augment existing ways of describing these contradictions through more specialized language taken from activist experience. The concept of direct action, for instance, emphasizes forms of political action that differ significantly from some of the dominant understandings of activism in political ecology: much direct action rejects, for example, the anonymity central to everyday resistance. In the next section we will work to further conceptualize direct action, briefly examine this politics in practice, and then explore what insights and opportunities might come from more seriously engaging environmental direct action.

Defining environmental direct action politics

Voltarine de Cleyre (1866–1912), an anarcha-feminist active during the time of the Haymarket riot in Chicago, coined the term "direct action" in an effort to define a type of politics she witnessed that appeared to diverge from other dominant and more "traditional" forms of political action, despite the fact that similar sorts of political action had long existed. She argues that (1912):

> Every person who ever thought he [*sic*] had a right to assert, and went boldly and asserted it, himself, or jointly with others that shared his convictions, was a direct actionist ... Every person who ever had a plan to do anything, and went and did it, or who laid his plan before others, and won their co-operation to do it with him, without going to external authorities to please do the thing for them, was a direct actionist. All co-operative experiments are essentially direct action.

Because so much of political ecological work has been done with attention to operating within the organs of the state, thinking outside of the state as de Cleyre advocates brings to light other ways to do activism. Another way of thinking about this is that revolutions, either in thought or political practice, require fresh insights free from the slow pace of change consistent with the movement of state apparatuses. Not being beholden to the political tactics of the liberal state opens up new avenues for activists to think creatively about how to articulate and enact their demands.

While most frequently associated with anarchist praxis (see Franks 2003), direct action politics have in fact been central to significant historical and social change for more than a century and a half. Within the realm of US environmental politics we can see Henry David

Thoreau's abolitionist efforts while living at Walden Pond as a critical moment in bringing his direct action refusal to pay war tax into environmental discourse. More broadly, direct action politics were central to many important political moments of the nineteenth and twentieth centuries: Gandhi's non-violent politics and efforts to drive the British out of India; Martin Luther King's utilization of direct action within the US civil rights movement; and Nelson Mandela's embrace of direct action politics in the struggles against apartheid.

One of the most useful treatments of the relationship between direct action and democratic politics is April Carter's *Direct Action and Liberal Democracy*. Carter provides a conceptual foundation upon which political ecologists can build the logics of direct action politics into their research. She argues that (1973: 159):

> Direct action can be justified by constitutional, liberal, and democratic principles if the existing institutions cease to embody these principles ... If the radical implications of constitutionalism, liberalism, and democracy are extended direct action can be seen as an intrinsically valuable mode of expressing independence, practicing resistance and exercising popular sovereignty ... It creates a potential for social change by releasing new energy and determination and encouraging social imagination. But the direction it takes depends on the nature of the movement it is associated with.

Carter's definition is useful for thinking about political ecology because it opens up the possibilities for reaching toward creative new forms of politics that as of yet have not been brought into political ecology; for that matter have not necessarily been articulated anywhere. Furthermore, she offers an explicit tactical language that can help to expand the conceptual terrain across which political ecologists develop more imaginative and inventive ways of analyzing social movement action. Finally, Carter emphasizes the fact that direct action spans across the conventional left/right political tendencies. This is something to which political ecologists could pay more attention. In brief, we can use Carter's work to envision direct action as a more robust part of political ecological praxis.

Anderson (2004: 107) also characterizes environmental direct action as a political strategy that is amenable to a wide range of ideological and theoretical framings. As is clear from its history – and Anderson makes this point as well – individuals and groups choose to embrace direct action for a variety of reasons. Anderson makes this point by drawing on a central text in the ethos of environmental direct action politics, Edward Abbey's novel *The Monkey Wrench Gang* (1975). Anderson points out that the disparate range of people who engage in direct action can be seen in smaller cross-section in the different kinds of subjectivities inherent within the characters in the book. "The diversity of the movement," Anderson argues, "means that individuals operate under a variety of organizational (or dis-organizational) banners depending on the action concerned, and oscillate between the use of traditional spaces of political protest, more radical spaces of action, and the invention of new spaces of protest depending on the issue and activists involved" (2004: 107). While these different subjectivities should be noted for how they inform political moments, so too should the range of political affinities be noted for the way they can introduce the latent racism, xenophobia, and misogyny in some of Abbey's own sentiments, and in many of the founding moments of the radical environmentalism inspired by *The Monkey Wrench Gang*. One of the more interesting outcomes of this complexity and diversity, also mirrored in the increasing range and activity within left-to-right libertarian politics, is that direct action politics has been used by conservative groups (e.g., anti-abortion activists blocking or committing violence at clinics) as often as progressive left groups. Attention to the spectrum of left-to-right politics is already an important part of political ecological research (see McCarthy 2002), but could be brought more to the foreground.

Jonathan London's (1998) case study of California's environmental activist group, *North Coast Earth First!*, shows how environmental direct action politics can speak to the problematic (and prevalent) nature–culture dichotomy that so much political ecology research attempts to bridge. London shows this bridging, in part, through his account of how *North Coast EF!* worked through their direct action to cultivate both discursive and material politics that built synergies between forest workers and the redwood trees themselves. London's study shows how *North Coast EF!* activists worked to get away from the traditional discourse of "wilderness," which has often been a central environmental trope as well as a political wedge issue, by creating a new configuration of community that included both activists and timber workers who mobilized to challenge corporate claims on the redwood forest. In so doing they helped recast the ecological representation central to the political tensions that accumulated over time.

London's work also brings a spatial logic into the evolving story of environmental direct action politics as embodied through *Earth First!* and its various offshoots and regional groups. The formation of EF! in the early 1980s in the US Southwest was largely inspired by Abbey's *The Monkey Wrench Gang*, and the group was initially created by Dave Foreman, Mike Roselle, Bart Koehler, and others. This group is widely understood, and it seems accurately so, as having strong commitments to a narrow form of Deep Ecology that neglected many of the broader concerns of political ecology. As Joni Seager characterizes the early work of EF! in *Earth Follies: Coming to Feminist Terms with the Global Environmental Crisis* (1993: 227): "Deep ecology is saturated with male bravado and macho posturing. The American EF! Movement is particularly symptomatic of the masculinist ethos that suffused representations of deep ecology's philosophy." However, the direct action organizing of Judi Bari in the Pacific Northwest helped to bring in a feminist and class political sensibility to the struggles of EF! which have been well documented and are more akin to London's account. This brief overview suggests that the study of direct action politics opens up for analysis a rich terrain of contradictory ecological politics.

We want to add to this thread of discussion about environmental direct action politics and what it can offer political ecology by drawing on primary research on direct action conducted by one of us (Heynen), working with *Roadblock Earth First!* RBEF! commenced work in the fall of 2006 after a consensus decision to organize by six activists who lived in Bloomington, Indiana. It formed with the explicit objective of creating more radical resistance to the construction of Interstate 69 and doing so through more creative tactics than had been to date enacted by more liberal local activist groups. While there had long been social movement formation around the construction of the interstate, these politics had been pursued previously through letter writing campaigns, speaking at public meetings and hearings, and electoral politics throughout the early 2000s.

RBEF!'s official organizing statement signals their intentions as different from status quo liberal politics quite clearly and shows how organizers often internalize the politics of scale to complicate the more common local narratives:

> I-69 is a NAFTA superhighway, already constructed from Canada to Indianapolis and projected to extend down into Mexico. This highway is intended for the mass transportation of goods and resources, to further exploit workers and the land, and to lessen companies' accountability in terms of human and environmental rights. In 2008, they intend to begin construction of this road through southwestern Indiana, evicting hundreds of rural families, destroying hundreds of acres of land, and devastating

> the habitats of countless species of animals, many of them already endangered. We intend to stop them.

The initial liberal efforts to contest I-69 pushed for research on "the costs of the road," which helped to implicate the role of the state in the narrative of what they were organizing against. The section of I-69 that concerned RBEF! was approximately a 140-mile section of this transcontinental NAFTA superhighway. The majority opposition groups offered arguments based on estimates from several key studies done in 2002. They argued that the total financial cost would be approximately five billion in 2008 dollars, and they framed the issue as an environmental one, emphasizing several things: estimates suggested that approximately 4,546 acres of farmland would be taken both directly for highway lanes and for rights of way and frontage roads, as would over 2,100 acres of forestland and 103 acres of wetlands, totaling about 7,653 total acres. It was also estimated that 62 state endangered plant and animal species would be disrupted and/or destroyed through the construction. Critics of the project argued that at least 125 non-farm businesses would be forced to close or relocate if the interstate were built and that 400 homes would be taken through eminent domain.

RBEF!'s narrative, by contrast, often started with the 1994 signing of the North American Free Trade Agreement (NAFTA), a comprehensive trade agreement linking Canada, the United States, and Mexico in a set of free trade relations. Since NAFTA was signed, it has been difficult to analyze its macroeconomic effects due to the large number of other variables in the global economy, but NAFTA nonetheless was always central to RBEF!'s narrative. This narrative was both a way of showing the implications of the state as something they were organizing against as opposed to within, but also the ways the politics of scale through the sort of neoliberal regionalization NAFTA represented made traditional discussions of the state outdated. RBEF!'s direct action campaign stretched over three years and included numerous "tree-sits," banner drops, and other direct action demonstrations, including occupations. As is the point of direct action politics, the activists' symbolic and material tactics very quickly brought significant attention to their efforts, as witnessed both in the state's response through arrests but also the increasing numbers of activists who sought to engage the movement from outside the region. The use of their particular tactics demanded that the state and the press frame their political ecological campaign differently than other contestations, which helped bring much more attention to their efforts. RBEF!'s success can be measured by the number of activists from across the United States who came to Indiana to show solidarity and engage in direct action, as well as the fact that the many arrests made helped to raise awareness around the I-69 issue and thus spurred more creative ways of strategizing against how the state produces nature in the image of capitalist endeavor (see Smith [1984] 2008).

While the direct action of RBEF! brought a new energy and visibility to the construction of I-69, it would be a mistake to see these efforts in themselves as better, or more successful than the previously existing liberal politics. Instead, by focusing on the particularities of language and tactics used in the spirit of direct action, we see the ways they worked with, not against, other forms of activism, thus creating a more robustly mixed portfolio of tactics.

Direct action contributions to political ecology moving forward?

Direct action politics has raised considerable attention around the world about important environmental struggles, and in ways that other political efforts have not been able to. Direct action offers insight into new ways of imagining, framing and enacting political ecological praxis. As the innovations of the poststructural turn helped articulate a more responsive appreciation of "new" social movements, so too can paying more attention to the concept and practice of direct action environmental politics.

In response to Goldman and Turner's (2011: 22–23) critique that the politics within political ecology are overly simplified, a focus on direct action helps to differentiate many forms of politics that too often get lumped together, including (beyond direct action), direct service, self-help, Freirean-style popular education, and more liberally rooted forms of advocacy. Many of these tactics are common in actually existing forms of ecological activism can help political ecologists think about praxis in ways less dependent on state theory. Inversely, activists can glean important insights from in-depth studies of the historical-geographical processes, contradictions and possibilities for transformation increasingly produced by political ecologists.

In order to build on the existing connections between political ecology, social movements, and activism, more can be done to demonstrate how direct action as an analytical lens, not just a political tactic, complements the more familiar concepts of "social movements," "activism," "everyday resistance," and "moral economy" but does so by creating a distance from always being so focused on the state. In closing, it is important to note that scholarship on direct action is working to capture the very current experiments with radical democratic practice to inform political ecology moving forward (see Chapter 7, this volume). These experiments might of course lead to political "failures," and they are certainly not inherently "better" than other historical approaches. However, we believe that direct action as concept and practice offers a way to further expand the connections between political ecology and activism. Political ecologists have an opportunity to work toward better internalizing tactics within the current conjuncture that can inform future organizing logics and tactics that go beyond working through the state. Embodying these knowledges can be an important step in "doing political ecology," and working to close the gap between traditional and organic intellectuals of nature.

Note

1 *Liberation Ecologies* first appeared as a special double issue of *Economic Geography* in 1993, and most of those papers (with some new additions) were included in the 1996 edition.

References

Abbey, E. (1975). *The Monkey Wrench Gang*. New York: Perennial.

Anderson, J. (2004). "Spatial politics in practice: the style and substance of environmental direct action." *Antipode*, 36(1): 106–125.

Bebbington, A. (2004). "NGOs and uneven development: geographies of development intervention." *Progress in Human Geography*, 28(6): 725–745.

Bebbington, A., Hinojosa, L., Bebbington, D. H., Burneo, M.L., and Warnaars, X. (2008a). "Contention and ambiguity: mining and the possibilities of development." *Development and Change*, 39(6): 887–914.

Bebbington, A., Humphreys Bebbington, D., Bury, J., Lingan, J., Muñoz, J.P., and Scurrah, M. (2008b). "Mining and social movements: struggles over livelihood and rural territorial development in the Andes." *World Development*, 36(12): 2888–2905.

Bryant, R. and Bailey, S. (1997). *Third World Political Ecology*. London: Routledge.

Carney, J. (2002). *Black Rice: The African Origins of Rice Cultivation in the Americas*. Cambridge, MA: Harvard University Press.

Carter, A. (1973). *Direct Action and Liberal Democracy*. New York: Harper & Row.

de Cleyre, V. de (1912). *Direct Action*. Spunk Library. Available at: www.spunk.org/library/writers/decleyre/sp001334.html.

Franks, B. (2003). "Direct action ethic." *Anarchist Studies*, 1: 13–41.

Goldman, M.J. and Turner, M.D. (2011). "Introduction," in M.J. Goldman, P. Nadasdy, and M.D. Turner (eds.), *Knowing Nature: Conversations at The Intersection of Political Ecology and Science Studies*. Chicago: University of Chicago Press, pp. 1–23.

Gramsci, A. (1971). *Selections from the Prison Notebooks*. New York: Monthly Review Press.

Holifield, R. (2012). "Environmental justice as recognition and participation in risk assessment: negotiating and translating health risk at a superfund site in Indian Country." *Annals of the Association of American Geographers*, 102(3): 591–613.

Jarosz, L. (2004). "Political ecology as ethical practice." *Political Geography*, 23(7): 917–927.

Lawhon, M., Ernstson, H., and Silver, J. (2014). "Provincializing urban political ecology: towards a situated UPE through African urbanism." *Antipode*, 46(2): 497–516.

Loftus, A. (2012). *Everyday Environmentalism: Creating an Urban Political Ecology*. Minneapolis: University of Minnesota Press.

London, J.K. (1998). "Common roots and entangled limbs: Earth First! and the growth of post-wilderness environmentalism on California's North Coast." *Antipode*, 30(2): 155–176.

MacKenzie, F. (1998). *Land, Ecology, and Resistance in Kenya*. London: IAI.

Mann, G. (2009). "Should political ecology be Marxist? A case for Gramsci's historical materialism." *Geoforum*, 40(3): 335–344.

McCarthy, James (2002). "First World political ecology: lessons from the Wise Use movement." *Environment and Planning A*, 34(7): 1281–1302.

Merchant, C. (2005). *Radical Ecology: The Search for a Livable World*. New York: Routledge.

Moore, S.A. (2008). "The politics of garbage in Oaxaca, Mexico." *Society and Natural Resources*, 21(7): 597–610.

Neumann, R. (1999). *Imposing Wilderness*. Berkeley: University of California Press.

Peluso, N. (1992). *Rich Forests, Poor People*. Berkeley: University of California Press.

Peet, R. and Watts, M. (2004). *Liberation Ecologies: Environment, Development, Social Movements*. New York: Routledge.

Perreault, T. (2003). "Changing places: transnational networks, ethnic politics, and community development in the Ecuadorian Amazon." *Political Geography*, 22(1): 61–88.

Perreault, T. and Green, B. (2013). "Reworking the spaces of indigeneity: the Bolivian ayllu and lowland autonomy movements compared." *Environment and Planning D: Society and Space*, 31: 43–60.

Perreault, T. and Valdivia, G. (2010). "Hydrocarbons, popular protest and national imaginaries: Ecuador and Bolivia in comparative context." *Geoforum*, 41(5): 689–699.

Robbins, P. (2004). *Political Ecology*. Malden, MA: Blackwell Publishers.

Robbins, P. (2011). *Political Ecology*. Malden, MA: Blackwell Publishers.

Rocheleau, D., Thomas-Slayter, B., Asamba, A., and Jamba, M. (1996). *Feminist Political Ecology*. London: Routledge.

Scott, J. (1976). *The Moral Economy of the Peasant: Rebellion and Subsistence in Southeast Asia*. New Haven: Yale University Press.

Scott, J. (1985). *Weapons of the Weak: Everyday Forms of Peasant Resistance*. New Haven: Yale University Press.

Seager, J. (1993). *Earth Follies: Coming to Feminist Terms with the Global Environmental Crisis*. New York: Routledge.

Smith, N. ([1984] 2008). *Uneven Development: Nature, Capital, and the Production of Space*. Athens and London: University of Georgia Press.

Sundberg, J. (2003). "Conservation and democratization: constituting citizenship in the Maya Biosphere Reserve, Guatemala." *Political Geography*, 22(7): 715–740.

Sundberg, J. (2007). "Reconfiguring north–south solidarity: critical reflections on experiences of transnational resistance." *Antipode*, 39(1): 144–166.

Valdivia, G. and Benavides, M. (2012). "Mobilizing for the petro-nation: labor and petroleum in Ecuador." *Focaal*, 63: 69–82.

Walker, P. (2007). "Political ecology: where is the politics?" *Progress in Human Geography*, 31(3): 361–369.

Wolford, W. (2010). *This Land is Ours Now: Social Mobilization and the Meanings of Land in Brazil*. Durham, NC: Duke University Press.

13
POLITICAL ECOLOGY AS PRAXIS

Alex Loftus

"Doing political ecology" as "activism"?

What often draws both students and teachers to political ecological research is a sense that something is not quite right in the world. By studying and researching in the subfield one might, perhaps, begin to rectify those wrongs. It's what first drew me to research on water politics and I continue to draw sustenance from my students' own hopes of creating a space in which activist research might be possible. However, political ecology is not really a form of activism in any way. Writing a chapter on "activism" for a section of a handbook on "doing political ecology" could well be a frustrating venture that risks stretching the definition of political ecology or activism too far. Most of what goes by the name "political ecology" is written about and practised in universities. The research produced is nearly always published in academic journals and lengthy monographs. Only rarely do these writings serve as activist interventions of any kind. And the language spoken – laden with references to actants, anti-Malthusianisms, systems approaches and poststucturalism – is rarely the same as that spoken by activists. How then might we reconcile the fact that most of those who practise political ecology care deeply about intervening in the world with the difficulties that most also seem to encounter in practising activism?

The relationship can, of course, be approached from a different perspective. What if the claim is, instead, that "doing political ecology" requires *being attendant* to activist practices? If Hecht and Cockburn's (2010) *Fate of the Forest* is a work of political ecology, its success lies, in part, in listening to and giving voice to the activist practices of the "defenders of the forest". It learns from and is fundamentally shaped by the work of one particular activist, Chico Mendes. When done well, of course, this learning relationship can be mutually beneficial, with political ecology beginning to inform activist practices. Indeed the relationship can be, perhaps even *should* be, symbiotic, mutually reinforcing, and productive for both the subfield and active interventions in the world.

Again, however, I find myself falling back on a well-worn platitude. Ever since my undergraduate days I have argued that activism and scholarship should be symbiotic; but only rarely have I paused to think about what that claim really means and how my work might genuinely achieve such a symbiosis. In reality, the potential difficulties of "doing political ecology" as a form of engaged scholarly activism are many. Thus, political ecologists face the potential pitfalls of *either* an impoverished activism (disappearing into offices to write those tomes upon which jobs depend) *or* an impoverished scholarship (failing to step back, to historicise, spatialise, or trace the multiple determinants out of which political ecologies are produced). Getting the relationship right between informed scholarship and activist engagements seems crucial if we are to work towards theorising about, while simultaneously seeking to bring about change within, the world.

Taking some of these questions as my starting point, I will make a case for political ecology as a form of engaged praxis. I will do so through a series of (hopefully not too solipsistic) reflections on my own experiences of learning how to do political ecology through research in Durban. I will intersperse this excursus on (semi-) engaged research with a rereading of Marx's *Theses on Feuerbach* (Marx 1975; hereafter I will cite individual Theses), one of the most succinct meditations on the relationship between theory and practice. Rethinking the *Theses*, I will simultaneously argue that political ecology can be fruitfully understood as a terrain of debate over which the role of theory and practice come to be considered and contested. Working with activist or disadvantaged groups has often pushed political ecologists to question their own position within the production of knowledge. Moving between critical and normative positions, political ecologists are forced to confront the fixity of their conceptual models in relation to the situated knowledges from which they derive sustenance in the field. Drawing on my own research experiences, I will make the case for political ecology to be framed as a form of dialectical pedagogy. Moreover, I will argue that political ecology, when practised well, involves a search for a theory which is adequate to an existing practice, as well as an adequate practical form for the theories with which we work. Instead of having to choose between interpreting the world and changing it (or having to choose between activism and arm-chair theorising), political ecology can be framed as an effort to heighten the capacity to know and act within what Bailey and Bryant (1997) refer to as the politicised environment.

Prior to embarking on the chapter, I should add two prefatory notes that I've already alluded to. First, I have *not* found an adequate balance between theory and practice, *nor* have I been able to elegantly balance activism and engaged scholarship. Instead, I have spent far longer writing those academic tomes than I have working with and engaging with activists. Nevertheless, crucial moments in my own understandings of political ecology have been the result of an attempt to redress this balance. I clearly have a lot to learn; but I think learning from activism can be the most productive, inspiring and progressive way of "doing" political ecology. Second, I work with a somewhat loose and sometimes idiosyncratic understanding of what political ecology is. My own first experience of political ecology came through reading the brilliant politically committed essays that come together within *Liberation Ecologies* (Peet and Watts 1996) and from seeking to figure out what such work might mean in the field. From the introduction of *Liberation Ecologies*, I discovered a field of research that learnt from social movements, as well as authors who saw a role for theory in understanding how politics criss-crosses and produces specific environments. If still idiosyncratic, hopefully the result of the following meditation on praxis might speak to one or two others who know the canon far better than I.

Praxis

> If you ask me what is the object of my work, the object of the work is to always reproduce the concrete in thought—not to generate another good theory, but to give a better-theorized account of concrete historical reality. This is not an anti-theoretical

stance. I need theory in order to do this. But the goal is to understand the situation you started out with better than before.

(*Hall 1988, pp. 69–70*)

Praxis is a curious word. Of Greek origin, the expression was carried over into Latin and then European languages. Praxis was interpreted within Aristotelian thought as a form of "practical knowledge" distinct from both *theoria* and *poiesis*. I don't think that most political ecologists would have a problem with framing their approach as a form of "practical knowledge"; nevertheless, this isn't, perhaps, the dominant contemporary understanding. Thus, different philosophical traditions have understood the term in slightly different senses (and defined praxis against not *theoria* and *poiesis* but against self-alienating praxis). The continuum has been to emphasise the role of practice in embodying, reflecting and realising knowledge claims (for a thorough review see Petrovic 1991). Thus, for Freire, praxis is the reciprocal relationship between thought and action. Feminist scholarship and activism has built on such a conceptualisation to see praxis as "theory in action" (Nagar and Swarr 2010). Hartsock (1998, p. 87) draws more directly from Marx, writing that "the concept of praxis, or human work, is a definition of what it is to be human – a striving first to meet physical needs and later for the realization of all human potentialities. The concept of praxis refers to the idea that one can only know and appropriate the world (change it and be changed by it) through practical activity".

As elsewhere (Hartsock 1983), Hartsock builds her understanding of praxis from Marx's *Theses on Feuerbach*. These brief notes, not intended for publication, and composed of no more than 600 words in total, lay out a radically different way of thinking about and changing the world. At the time he jotted down the *Theses*, Marx was seeking to distance himself from two distinct schools of thought – idealism and crude materialism. In distancing himself, however, Marx provides a perfect synthesis of both; this synthesis ends up exceeding both idealism and materialism in its ambitions and brilliance. First, having fallen out with his once close friends amongst the idealist Young Hegelians, Marx had come to recognise that beneath their radical bluster there was a deep conservatism that remained antipathetic to active political interventions in the world. Having just spent time amongst the communist artisans of Paris, Marx saw the ways in which banding together as a committed political grouping could, in itself, begin to achieve change. Thus, he began to question means and ends, seeing the means of seeking change through working together as a crucial stage in the realisation of particular ends. Part of Marx's critique of the Young Hegelians' idealism came from Ludwig Feuerbach who was, at the time, infamous for his critique of religious thought. Feuerbach appeared to reverse the direction of idealist thought: ideas came not from on high but from the messy realities of the world. Although Marx had been increasingly drawn by the writings of the latter, he now saw a need to distance himself from some of the more crude bases of Feuerbachian thought. In the *Theses* Marx, in prototypical fashion, appears to call for plague on both their houses and, instead, emphasises "the significance of 'revolutionary', of 'practical-critical', activity".

Throughout the *Theses*, Marx sees practice – sensuous activity – as fundamental to the make-up of both reality and the ways in which human beings make sense of that reality. Relationally organised human practice constitutes both "society" and the "environment" of which those societies are a part. Against most naturalising and essentialist readings, the human essence, Marx argues, cannot be understood outside of the relational organisation of human society. The *Theses on Feuerbach* contain, Engels was later to claim, "the brilliant germ of the new world outlook". Praxis, or "practical-critical activity" is fundamental to this new world outlook. The *Theses* thus lays out brilliantly (a) the nature of reality ("sensuousness as practical activity"); and (b) a thesis on how everyday men and women make sense of that reality (again through their sensuous practices). Just as communist artisans had discovered a form of communism through their banding together, so that new communal identity became crucial to the modes of thought they then took forward in their struggles. The lessons for political ecologists are pretty clear: look at everyday practices and the ways in which they emerge out of specific forms of social organisation; and work with the forms of knowledge that emerge from such practices.

Such concerns are clearly central to most of what goes by the name of political ecology; however the manner in which praxis-based perspectives have been understood and put to work in different areas of the sub-discipline have varied greatly. Within the sub-discipline in which I work, urban political ecology, much has been written about the politics of water within the city: indeed some of the most innovative theoretical work has emerged through rethinking the relationship between the city and the resources that flow through and constitute that city. In quite brilliant ways, scholars have emphasised the newly emergent relations between water and social power, as well as the manner in which infrastructure and water come to be enrolled in wholly new ways to produce new choreographies of power. However, much less has been written about the ways in which change might be achieved. One of the criticisms of such work (and indeed of my own work) has been that a critical distance has emerged between theorising and learning from activists. This defies Robbins' (2004) claim that political ecology is motivated by both a critical and a normative agenda; instead, much of the work on the political ecology of water has eschewed normative research, focussing more singularly on critique. Even if following Marx's dictum of "ruthless criticism of all that exists", key questions clearly remain unanswered. Who, for example, might be the agents of change within these new choreographies of power? How, in turn, might these agents work with, or against, theoretical perspectives derived from scholarly practice? How might scholars and activists organise to seek change? And how might scholars most effectively seek to work with such activists? There is clearly no simple answer, if any answer at all, to such questions, but they remain crucial for forging a political ecology that is worthy of its claims of possessing a transformative edge. In what follows, I seek to build on the critical work that has been conducted within the sub-discipline as a way of understanding how a praxis-based political ecology might emerge.

Activist ontologies

Why might ontological questions matter to political ecology? And why, moreover, might they matter to one that is attendant to, and that seeks to learn from, *activist* practices? Ontological understandings matter because they lay open the nature of reality, demonstrating the types of relations that exist between humans and non-human others while throwing into question the immutability of particular historical and geographical forms and highlighting the different subjectivities in relation to these forms. In short, our ontological understandings strike to the heart of what can and can't be changed in the world. Influenced by the writings of David Harvey (1996), much of the work in urban political ecology and elsewhere in the subfield has developed a relational ontology which places particular emphasis on flows, processes and relationships between human and non-human (Heynen et al. 2006). Swyngedouw (2004, 2007) has thus shown how city and nation are historically constituted out of flows of water that are simultaneously flows of power. The waterscape is criss-crossed by relations of power: quasi-objects should be understood in relation to the discourses, ideologies and materialities out of which they emerge (Swyngedouw 1999). Crucial to this reality are historically and geographically specific practices, or sensuous human labour. Making what appears to be a perverse ontological claim, "reality", Marx writes in the first *Thesis on Feuerbach*, is "sensuous human activity, practice" (Thesis I).

Working with such an understanding within my own research, I have sought to better understand the ways in which pipes, taps, water tanks, meters and flow limiters are knitted together by changing sets of historically and geographically specific practices (Loftus 2006). In South Africa, the question of who has access to water and who doesn't is profoundly influenced by the raced and classed practices that emerged during the racist years of "separate development" or apartheid (a project described by David Smith (1982) as one of the most ambitious forms of applied geography). In the contemporary moment, such sedimented histories continue to shape the ability to access water, as do practices more closely associated with historically and geographically specific divisions of labour. The sensuous labouring practices of women thus remain absolutely central to how water is and isn't accessed within the country (Loftus 2007). It is primarily women's labour out of which water is carried from the standpipe to the home. It is women's labour that, when yard taps run dry, ensures a household has continued access to water. And it is largely women's labour that negotiates the complex discussions with the municipality when a supply has been disconnected and a household is plunged into socially produced forms of scarcity. If, as urban political ecologies (building on Marx) have repeatedly stressed, the ontological foundations of the waterscape rest on a metabolic relation in which water and humans come to be mutually transformed, this metabolic relation is forged predominantly through women's work (although not focussed on water, see Hartsock 1983), as well as the unacknowledged practical acts of myriad wage labourers, working under both apartheid and post-apartheid conditions, whose practical acts come to be congealed within the quasi objects making up the waterscape. The peculiar power that water meters seem to have achieved over low income households is a material expression of new divisions which have emerged as low income communities have been rescripted as consumers (and potential sources of profit) in the post-apartheid settlement. Throughout, practices matter and materialise the political ecology of the post-apartheid waterscape. Realising alternatives to such unjust ecologies requires being attendant to practices.

Activist epistemologies

What conceptions of the world might begin to emerge from such realities? And how are they related to the ontologies that they seek to make sense of? It is one thing for a political ecologist to claim that people make their own political ecologies (albeit not under conditions of their own choosing); it is another for those engaged in making such ecologies to foreground such a claim within their own practices ("Our practices are central to the ecology of this place; therefore we can make things differently"). Turning back to the *Theses on Feuerbach*, Marx makes the radical claim that objective truth is not simply a "*scholastic* question" but is rather "a *practical* question" (Thesis II). In short, "All mysteries which lead theory to mysticism find their rational solution in human practice and in the comprehension of this practice" (Thesis VIII). On the one hand this appears an anti-theoreticist stance; but it is also one that couldn't have been arrived at had it not been for the prior ontological claim. Reality is constituted out of "sensuous human activity, practice": epistemological frameworks can emerge out of the "practical knowledge" acquired from shaping those realities. Practice is thus central to both ontology and epistemology. In this manner, understanding the reality of political ecologies *necessitates* an activist practice, or, at least, learning from those actively involved in making that reality. A remarkably similar position forms the basis for Neil Smith's (2008) critique of bourgeois environmentalisms and his advocacy of an understanding rooted in the production of nature (and it is no surprise that Smith cites Thesis VIII with such relish).

To return to the case of South Africa, the labouring practices of those involved in the production and reproduction of the waterscape are of fundamental importance in seeking to understand, make sense of, and theorise about that waterscape. One clear example, which relates quite directly to the claim that I made at the outset of this chapter (that "doing political ecology" requires *being attendant* to activist practices) occurred in the informal settlement of Inanda when water ceased to flow to the impoverished sub-community of Amaoti (a tanker stopped bringing water to the locked standpipes in the settlement). Distinctly gendered responses emerged to this period of enforced thirst. Whereas men were, unusually, seen to be carrying water within the settlement (albeit for a fee and, some might cynically claim, to make the best of a bad situation through moneymaking) women began to organise a wildcat protest that descended on the offices of the local councillor and proved highly effective in bringing free water to this forgotten community. Women became insurgent architects of a fundamentally different set of social relations within the waterscape. Seeking to build on their *practical* knowledge of the power relations that enabled some to access water while excluding others, Amaoti's women wove this practical knowledge into a set of common demands and worked effectively to transform the situation. This was a form of political ecology in action, in which situated knowledges opened up conditions of possibility for the production of a fundamentally different waterscape. Making sense of the politics of water is, in large part, reliant on one's ability to draw from historically and geographically specific practices.

The political ecological activist

It is all too easy within such a framework to slip into crude essentialisms. The peasant's traditional ecological knowledge thus comes to be fetishised and the women of Amaoti crudely juxtaposed with their corrupted male brethren. Moving against such a position, Marx insists that "the human essence is no abstraction inherent in each single individual. In its reality it is the ensemble of social relations" (Thesis VI). There is no essential political ecological activist, nor should the practices associated with forging particular ecologies (always understood as a process of co-evolution between human and non-human) be assumed to lead directly to a standpoint or vantage point that can be derived *solely* by one's position within a division of labour. *Some* women in Amaoti chose to organise to change conditions for the better. *Some* men joined them in this struggle. Many others instead sought to accommodate themselves to the changed conditions and the enforced drought. The circumstances of working collectively to transform the conditions within Amaoti, similarly, left its imprint on the women who did participate in the struggle. Thus, drawing on the dialectical pedagogy to which he has already alluded, Marx makes the claim that "the changing of circumstances or self-changing can be conceived and rationally understood only as revolutionary practice" (Thesis III). Political ecological activists thus come into being in relation to the struggles of which they are a part.

Gramsci appears to take this anti-subjectivist reading a little further, viewing the human essence as shaped by both the ensemble of social relations and the historically and geographically specific ways in which humans relate to "nature". Thus, in response to the question "What is Man?", Gramsci claims the question should be framed as "what can man become?" He then goes on to claim that humanity should be conceptualised through the active relations forged between "1. the individual; 2. other men; 3. the natural world" (Gramsci 1971: 352). Humans shape environments, just as environments shape the humans who are a part of them. Activist practices come into being in relation to changing political ecological circumstances which transform human and non-human accordingly. Crucially, a praxis based understanding of political ecological activism must prioritise the conditions of possibility within practice, as opposed to assuming some linear path to an untainted vantage point (already arrived at in the mind of the detached academic who seeks to theorise *about* activists). Indeed "the materialist doctrine concerning the changing of circumstances and upbringing forgets that circumstances are changed by men and that it is essential to educate the educator himself" (Thesis III).

The point is to change it

Marx concludes the *Theses on Feuerbach* with the infamous claim that "The philosophers have merely interpreted the world, in various ways; the point is to change it". So far, engaging with the other theses, I have made an argument for the importance of learning through researching with activists who themselves are learning from their own efforts to reshape political ecologies. Environmental understandings, which emerge from historically and geographically specific practices of "metabolising nature", are transformed into political knowledges through the process of struggling to democratise those metabolisms. At specific conjunctures, theory and practice can come to be identified in a way that both heighten one another through an effort to interpret the world and simultaneously change it. For Peter Thomas, drawing on Gramsci's philosophy of praxis:

> The production of the identity of theory and practice then becomes the critical art of finding, in a Spinozist fashion, the adequate theoretical form of a practice, capable of increasing its capacity to act, on the one hand, or, on the other hand, the adequate practical form of a theory, capable of increasing its capacity to know … It attempts to act as the theoretical comprehension of actually existing practices, describing their tendencies and lines of potential development as concrete acts of organization and coordination rather than normatively prescribing their necessary forms from above.
>
> *(Thomas 2009, p. 33)*

In this penultimate section I will draw together some of the ways in which these different accumulated processes of knowledge making informed the activist practices of Inanda and shaped my own understandings of political ecology. In particular, I will draw on experiences of working with Thulani Ncwane, a long-time activist within Inanda. Thulani transformed the research that my partner and I were able to conduct, through being able to articulate a knowledge that emerged from the situated practices of the informal settlement (see also Chapter 10, this volume). This knowledge was inseparable from his role as an "activist" and it became inseparable from my own understandings of the political ecology of the settlement.

Thulani's knowledge of Inanda was largely shaped by his involvement in the Comrades movement in the late 1980s. As a young teenager, hawking on the commuter trains that plied the coast, Thulani came to meet United Democratic Front activists who spoke of their struggles to challenge the unjust conditions which the young man experienced so directly. Joining this struggle, which at the time was embroiled in a bitterly fought turf war with supporters of the Inkatha Freedom Party, Thulani was shot and hospitalised before returning to Inanda and working with the Inanda Marshalls to forge a viable governance structure for the settlement in the transition from apartheid to the post-apartheid period. Although the ANC sought to disable many of the structures formed during this period, Thulani remained active and became a key player within the Inanda Development Forum which sought to mediate between the different interests of the settlement. As the forum was later sidelined again, Thulani resumed struggles both within and outside of the ANC. Over a twenty-year period between 1986 and 2006, few were as well placed as Thulani to understand the politics of Inanda from the ground up. Over the time of our research, he worked as a translator, a source of continual inspiration, a friend and collaborator. Crucially, Thulani was already embedded in negotiating the relations out of which the politicised environment of the post-apartheid settlement is constituted. As an Inanda Marshall, he had been responsible for distributing areas of vacant land amongst needy squatters and knew, subsequently, the system of land-tenure in great depth. Thulani knew, both personally and professionally, the local councillors now being paid to do what he had always done for no financial gain. The political history of Inanda had been pieced together, learnt about, and indeed partly crafted by Thulani and his comrades from an early age. The politics of water – that Fiona and I set out to explore – was simply a new layer to add to that politicised reading of Inanda.

There are of course profound dangers to relying solely on one such perspective, and any budding political ecologist knows the importance of triangulating a grounded perspective with other perspectives. Thulani guided us to research participants who were likely to take us in some directions more than others, and was, through his very presence, likely to influence the responses that participants gave in interviews. Nevertheless, aware of these pitfalls we embarked on an exploratory journey through the situated knowledges of those profoundly implicated in shaping the post-apartheid politics of Inanda. Thulani permitted us to historicise, to spatialise and to learn from the intimate ways in which socio-ecologies had been co-produced within Inanda. From the start of the research, we were compelled to focus on the everyday practices through which people made the waterscape of which they are a part. We were forced to confront the ways in which such practical-critical knowledges, although always partial, translated into new conditions of possibility for thinking about and changing Inanda. Finally, on a somewhat pragmatic and banal level, Thulani insisted on us taking the research back to those from whom we had learnt.

Conclusions

There are no doubt traces of a long redundant humanism in my framing of political ecology as praxis. By placing the practical knowledge of *humans* at the centre of political ecological research, I appear to go against the decentring of human activities that has been so central a move within the subfield over the last decade or more. Nevertheless, throughout, I have sought to frame praxis as centrally concerned with the process of co-evolution between human and non-human other, as well as centrally concerned with how individual subjects and differentiated social groups seek to make sense of this process of co-evolution. If political ecology emerged as a framework that sought to radicalise what were perceived to be some of the more conservative foundations of cultural ecology (see Chapters 2 and 3, this volume), it was in part through seeking to put this practical knowledge to work.

Above all, I have sought to demonstrate that a praxis-based political ecology, framed in relation to Marx's most succinct formulation of praxis in the *Theses on Feuerbach*, poses crucial questions for the "doing" of political ecological research in relation to activism. First, I have argued that it prioritises the role of practice (along gendered, classed and raced lines) in making specific ecologies (always understood as a process of co-evolution). Second, and related, I have argued that conditions of possibility for rethinking political ecologies emerge from those very same practices. Third, I have made the obvious point that there is no guarantee in all this, just as there is no essential political ecological activist subject. Activists come into being through their political ecological struggles. These struggles can often be seen to emerge from the process of sense-making that comes from shaping ecologies in specific ways. Finally, I have sought to foreground the possibilities for a world-changing praxis to emerge in relation to this ontology and epistemology. Through the example of Thulani Ncwane's efforts to work as an insurgent architect of a transformed (and transformative) political ecology in Inanda, I have sought to emphasise the possibilities for making political ecologies profoundly differently.

References

Bailey, S. and Bryant, R. 1997, *Third World Political Ecology: An Introduction*, Routledge, Abingdon.

Gramsci, A. 1971, *Selections from the Prison Notebooks*, Lawrence and Wishart, London.

Hall, S. 1988, The toad in the garden: Thatcher among the theorists, in C. Nelson and L. Grossberg (eds), *Marxism and the Interpretation of Culture*, University of Illinois Press, Urbana, pp. 35–58.

Hartsock, N. 1983, *Money, Sex, and Power: Toward a Feminist Historical Materialism*, Longman, New York.

Hartsock, N. 1998, *The Feminist Standpoint Revisited and Other Essays*, Westview Press, Boulder.

Harvey, D. 1996, *Justice, Nature and the Geography of Difference*, Blackwell, Oxford.

Hecht, S.B. and Cockburn, A. 2010, *The Fate of the Forest: Developers, Destroyers, and Defenders of the Amazon*, University of Chicago Press.

Heynen, N., Kaika, M. and Swyngedouw, E. (eds) 2006, *In the Nature of Cities: Urban Political Ecology and the Politics of Urban Metabolism*, Routledge, Abingdon.

Loftus, A. 2006, Reification and the dictatorship of the water meter, *Antipode*, 38(5), 1023–1045.

Loftus, A. 2007, Working the socio-natural relations of the urban waterscape in South Africa, *International Journal of Urban and Regional Research*, 31(1), 41–59.

Marx, K. 1975, *Early Writings*, Pelican, London.

Nagar, R. and Swarr, A.L. 2010, Introduction: theorizing transnational feminist praxis, in A.L. Swarr and R. Nagar (eds), *Critical Transnational Feminist Praxis*, SUNY Press, Albany, pp. 1–20.

Peet, R. and Watts, M. (eds) 1996, *Liberation Ecologies: Environment, Development and Social Movements*, Routledge, Abingdon.

Petrovic, G. 1991, Praxis, in T. Bottomore, V. Kiernan and R. Miliband (eds), *A Dictionary of Marxist Thought*, Blackwell, Oxford, pp. 435–440.

Robbins, P. 2004, *Political Ecology: A Critical Introduction*, Wiley Blackwell, Oxford.

Smith, D. 1982, *Living Under Apartheid: Aspects of Urbanization and Social Change in South Africa*, George Allen & Unwin, London.

Smith, N. 2008, *Uneven Development: Nature, Capital, and the Production of Space*, University of Georgia Press, Athens.

Swyngedouw, E. 1999, Modernity and hybridity: nature, regeneracionismo, and the production of the Spanish waterscape, 1890–1930, *Annals of the Association of American Geographers*, 89(3), 443–465.

Swyngedouw, E. 2004, *Social Power and the Urbanization of Water: Flows of Power*, Oxford University Press, Oxford.

Swyngedouw, E. 2007, Technonatural revolutions: the scalar politics of Franco's hydro-social dream for Spain, 1939–1975, *Transactions of the Institute of British Geographers*, 32(1), 9–28.

Thomas, P. 2009, *The Gramscian Moment: Philosophy, Hegemony, and Marxism*, Brill, Leiden.

14

POLITICAL ECOLOGY AND POLICY

A case study in engagement[1]

Brent McCusker

Introduction

After several years of being a "practicing" political ecologist, I found myself sitting at the 2010 Annual Meeting of the Association of American Geographers in the Cultural and Political Ecology (CAPE) lecture being delivered by Pier Blaikie (Blaikie 2012). Well known for both his foundational writings on political ecology and his engagement with policy makers, I was intrigued by his thoughts on whether or not political ecologists could or should engage more closely with that community. As a geographer critical of international development, my first reaction had often been to reject engagement with policy makers, especially in the United States. I considered them greatly at fault for the profound unevenness in global development. "What good could come from engagement?" I asked myself. This view was reinforced by often raucous debates on the CAPE listserv that strongly challenged any good that could come out of increased interaction. At about the same time, the Development Geographies Specialty Group had been holding pre-conferences to the main AAG meeting posing very similar topics and questions. After encouragement from colleagues, I decided to explore closer interaction with policy makers during my sabbatical year via a fellowship from the American Association for the Advancement of Science. In 2011, I applied and was selected to serve as a "science advisor" to the United States Agency for International Development (USAID).

At the onset I decided to use this opportunity simply as a learning experience. Rather than starting from a critique of the agency, I decided to frame my daily work around three questions: first, is greater engagement possible and, if so, is it desirable; second, where are there areas of congruity (between this particular development donor and myself) in development epistemologies/approaches; and third, where are there divergences in epistemologies/approaches? Although I have viewed USAID's activities, particularly in southern Africa and South/Central America, as problematic, my goal was to learn why the agency works the way it does. Thus, I did not set out to perform a "participant observation" of the agency as Robertson (2010) had with the Environmental Protection Agency. While I was much more interested at the onset of my tenure at USAID in understanding the inner machinations of the agency than Robertson reported being in his piece, on reflection for the preparation of this chapter, we had remarkably similar experiences. I echo his finding that "the idea that such research involves going behind or inside something was eventually replaced by the complexity and internal divisions within the state, and the evident blurring of external and internal critique" (Robertson 2010: 7). The critiques I brought with me, but suppressed in a misguided notion that I simply needed to "learn," were often voiced by longtime staffers who had intricate knowledge of a particular project or program. They knew there were conceptual (design) or practical (implementation) problems, but their concerns were either never raised to decision makers or were just ignored. The notion that I would simply be a "detached observer" quickly proved naive as my analytical skills were valuable in addressing both conceptual issues with programs/projects but also in helping to demonstrate the value of geography or "spatial thinking" as it was termed. I quickly found myself in the role of an internal advocate for geography and political ecology.

Because I was never a direct employee of the agency, I never had the same level of accountability to it that Robertson had. However, his discussion of the "author-less" text struck an immediate chord with me. In my experience, this was initially frustrating. In academia, the measure of my performance is tied to the words I produce. Echoing Robertson, I gradually learned that the anonymity of authorship often helps push ideas forward that might not otherwise be considered. While I was always treated with respect, I was far too new, and temporary, for the bureaucracy to expect my ideas to be adopted *in toto*. What I did find is that by removing my name, some parts of my ideas could make it forward because they were attributed to my office as a whole.

A final similarity I share with Robertson is the feeling of being completely encompassed by neoliberal ideas while working in "my small part" of the state. My experience was too short to find the same degree of resistance or oppositional thinking that he uncovered, however, I was surprised at how thoroughly neoliberal principles permeated thinking about development policy at USAID. I had hoped at the beginning of my experience to be proven incorrect – a raving radical who finds capitalist conspiracy everywhere. While the halls of the Ronald Reagan Building are indeed *not* filled with capitalist conspirators, their ideas are strongly represented (it is, after all, in a building named after Ronald Reagan!). I did find allies who were critical of the "market-based" solution to any problem, but my tenure was too short to map the network.

I have chosen not to focus on a reflexive assessment of my time at USAID in this chapter, but rather to highlight some of the opportunities for engagement that I observed in my time there. This chapter is qualified by all the normal caveats one might expect: USAID is only one type of policy institution; it is part of the US government and carries with it a great deal of imperial baggage; these are only my very situated and biased observations (e.g. I *wanted* my time there to be productive); and I was only a very small part of a very large organization and not privy to high-level decision making or decision makers.

In the next section, I summarize the topic of engagement between political ecologists and policy makers in the literature. Given the paucity of critical reflection on the topic in the literature, I draw heavily from a few examples. I then move to my case study at USAID. For purposes of clarity, I distinguish between policy makers and policy implementers and then explore how USAID broadly conceptualizes development and its place in that field. I also problematize my own decision to work for a US government agency. In the penultimate section, I consider several areas of congruence and incongruence between USAID and a political ecological approach to development, noting that there is no singular path for either. My conclusion is that while there are many mechanisms for increased engagement, any relationship between international donors on one hand and political ecologists on the other will remain fraught with challenges and opportunities. My viewpoint, which changed as a result of my experience, is that it is worthwhile for political ecologists to engage with policy implementers.

Where is the policy?

Even a cursory review of major political ecology texts demonstrates a community of practice that is both incredibly varied and deep in analytical richness (Peet and Watts 2004; Peet et al. 2011; Neumann 2005; Brannstrom and Vadjunec 2013; Zimmerer and Bassett 2003; Bryant and Bailey 1997; Stott and Sullivan 2000; Blaikie and Brookfield 1987; Blaikie 1985). That same review would find serious consideration paid to the impact of development policies and projects on local people and their environment. It would, however, find less attention paid to policy *making* and policy *implementation*.

One of the few narratives to directly raise this issue was Peter Walker's 2006 article titled "Political ecology: where is the policy?" But Walker's article did not emerge out of the ether; rather, it was but one materialization of a long-running debate and dialogue between members of the Cultural and Political Ecology Specialty Group of the Association of American Geographers, amongst others. Walker cut straight to the crux of the issue in his paper when he wrote "despite its rich history and professed interest in engaging public debates, the actual engagement of political ecology with fields of research and public debate outside the academy has been limited" (Walker 2006: 383). To be precise, though, Walker was largely concerned in his text with *public* policy, usually embodied by academics who rise in stature high enough to shift *public* debate through their writing. He discussed the lack of effective counters to such dominant narratives as neo-Malthusianism and environmental determinism.

While he was certainly correct and his intervention long overdue, even it misses an altogether more important point: that political ecologists rarely engage with the people and institutions that create and implement the policies they so directly examine, deconstruct or challenge (Carr 2011). A good case is the work of Dick Peet (2007). His book *Geography of Power: The Making of Global Economic Policy* is a trenchant analysis of the ways in which international capitalist development policy is generated and reproduced through regimes of knowledge and power. It deftly describes economic, ideological and political power circuits in an approachable and persuasive language. I find the book so useful, in fact, that I teach it in my advanced development geography course. What Peet does not do, however, is report on any single interaction with a policy maker or implementer. There is no discussion of policy making and implementing challenges and no discussion of the very heterogeneous composition of such policy makers, their institutions or their implementing partners. Further, and perhaps most problematically, the book does nothing to identify and build alliances with like-minded policy makers and implementers. Such a book is a wonderful way to stir students' critical thinking skills, but it does little to find and reach out to potential allies inside what are presented as monolithic knowledge and power circuits. Similar strengths and weaknesses extend to a special edition of the journal *Human Geography*, edited by Peet (2013), and indeed to much political ecological work.

Political ecological analysis of policy and policy outcomes has also tended to homogenize policy making and policy implementation (notable exceptions are Chapter 15, this volume; Bebbington 2014; Robertson 2010). In the example that follows, one grounded in experience in Washington, DC, it is important to recognize the differences between policy makers, policy implementers and implementing partners. Policy *makers* are the legislators who enact policies, usually crafted directly or indirectly by special interest groups and think tanks, at least in the United States. They are the members of Congress and their staffs who enact legislation that has increasingly been accompanied by onerous limitations and stringent budget restrictions. Such limitations and restrictions are put in place to limit the flexibility of policy *implementers*, those members of the bureaucracy in the executive branch charged with carrying out (implementing) Congressional and Presidential directives and laws. Policy implementers still do retain some degree of latitude in exactly how they execute such laws and directives, but informal conversations indicate that it is much less than in the past.

In the 1990s, the Clinton administration's efforts at "reinventing government" led to significant reductions-in-force in the civil service. Skilled staff flowed out of executive agencies, replaced by or themselves hired into firms of *implementing partners*. These businesses, pejoratively referred to as "the Beltway Bandits" (or less pejoratively "the Highway Helpers") due to their location around the Washington, DC interstate highway system, perform many of the actual tasks of government, the day-to-day work of *doing*. Most development projects in developing countries are contracted out to either one of these large firms, or, more recently, to local implementing partners. What a political ecologist might see in a developing country is the action of one of these partners. What they might read as a "policy document" is largely written by an implementing partner, albeit with significant guidance from a policy implementer.

A limited engagement

Both Walker (2006) and Blaikie (2012) suggest many reasons for a limited engagement of policy in political ecology. Many political ecologists prefer to "speak to other political ecologists (and perhaps to others in closely related fields)" (Walker 2006). Blaikie (2012: 234) suggests one possible very good reason for this state of affairs. He wrote that "the content of a political ecology which claims to be emancipating, liberating and politically progressive will be critical of existing knowledges and power relations and maybe of specific actors in policy networks outside the academy altogether." This is a particularly important point. Many political ecologists will be critical of development programs and projects with just cause. However, he goes on to argue that, "deconstruction and other acts of aggressive epistemological attack ... may not appeal to any audience other than like-minded academics" (Blaikie 2012: 234).

A second major obstacle in the path of engagement is the structure of reward in academic institutions and increasing workloads. In many departments housing political ecologists, a strict and seemingly unchanging division of labor forces academics into a mix of teaching, research and service, with the former two roughly equal in importance. In those categories, there is often little reward for the creation and publication of policy documents or time spent engaging policy makers or implementers. Simply put, it is impact factor above actual impact in many academic institutions. As university budgets shrink and new sources for funding are pursued, this may well change, but may also result in even more overburdened faculty.

The policy side of this problem is no less inflexible, however. Many funders or policy institutions have very strict limitations on what can be written about them, either from staff, contractors or consultants working on the inside. Most have publication clearance processes that are cumbersome and time-consuming. New forms of communication, such as social media, are looked on with great hesitation – transgressing the boundaries in 140 characters or on a blog post can result in severe penalties. These are not hypothetical scenarios. During my time at USAID, a colleague was subjected to close oversight of all his external communications after a slight misstatement on a blog-post. Many actors within these institutions would rather avoid this additional burden.

A common refrain inside some policy institutions is that staff are already so overburdened with work and the institutions themselves so understaffed that engagement with academics is simply an unattainable luxury. While there may be some degree of hyperbole in this claim, and indeed there is great variation from institution to institution, it does evoke conversation about a very real disconnect between policy implementers and academics over timelines. Academic timelines stretch into multiple years to complete analysis, while policy makers and implementers often have much less time, in the range of a few days to a few weeks, to plan assessments of their work. Most academics are also limited in their flexibility by their teaching, with long field research trips confined to summer/winter breaks. The rhythm of the academic year simply does not match that of the quick turnaround needs of the policy world. Given that most political ecologists conduct some type of field research, such time inconsistencies can present a significant obstacle to engagement.

Finally, what analytical tools do political ecologists offer policy makers and implementers? As noted by Blaikie, discourse analysis and dialectics simply won't find an audience within most policy institutions. Some political ecologists have employed geospatial technologies critically in order to demonstrate their utility in answering key questions (see Chapter 19, this volume). I have previously published on the use of such technologies for political ecological research (McCusker and Weiner 2003) and other political ecologists have used geospatial technologies. Geospatial technologies are becoming increasingly widespread within even development institutions such as USAID. The agency's recently (mid-2011) established GeoCenter promotes spatial thinking and the use of GIS, GPS and remote sensing amongst USAID staff.

An example: policy implementation at USAID

Rather than remain at levels of abstraction that merely suggest what might be possible, I will discuss in this section an example of a policy implementing body, the US Agency for International Development (USAID) with which I have worked extensively beginning in 2011. USAID is the largest single bilateral development donor. It covers a variety of thematic areas related to development including but not limited to health, food security, the environment, science and technology, energy, climate change, gender, democracy, conflict and humanitarian assistance. Many of these topics are central to political ecology, thus this example seems most appropriate not only for USAID but also other policy implementers that touch on human-environment issues, such as NOAA, NASA or the EPA (Robertson 2010).

All of the agency's operating procedures and philosophies of practice, known as the Automated Directives System (ADS), are available to the public via its website (www.usaid.gov/who-we-are/agency-policy; accessed November 3, 2013). The ADS is divided into a series of six collections (Series 100–600), which are further broken down into more specific chapters. Of most direct relevance to this discussion is Series 200, which discusses policy programming, planning, assessment, environmental procedures and ensuring gender equality.

USAID promotes a set of operating principles, which upon cursory examination, do not seem to be all that divergent from a political ecological approach, including "promote gender equality and female empowerment; apply science, technology and innovation strategically, practice selectivity and focus, measure and evaluate impact, build in sustainability from the start, apply integrated approaches to development, and leverage 'solution-holders' and partner strategically" (USAID 2012: 14). Each of these is discussed in turn.

The document continues into a section titled "The Program Cycle Overview" that establishes how USAID projects are structured. All project implementation is informed in the first stage of the program cycle by reference to collected strategies and guidance documents. In the second stage of the cycle, country development cooperation strategies (CDCS) are developed in consultation with "host country governments, local civil societies, private sector organizations, the (US) State Department, and the broader USG [US government] interagency and other donors" (USAID 2012: 23). Project or program goals are structured into a hierarchical set of indicators that "graphically represents the development hypothesis and defines a CDCS goal, development objectives, intermediate results, sub-intermediate results and performance indicators (USAID 2012: 24). It is important to note here that while this may seem like an overly structuralist tool, USAID is required to account for each and every dollar spent with indicators of its impact. This leads to data such as "number of women trained" or "number of hectares of land under improved management" that at first glance appear trivial and meaningless but upon further review are essential to reporting and accountability. In my recent work with USAID, I have helped to develop a tool that measures impact more holistically. Called "stocktaking" (http://frameweb.org), this method encourages a retrospective, whole of landscape assessment that tries to understand how natural resource utilization/management changes actually occurred against the backdrop of obstacles such as lack of resources or resistance from others. While USAID staff are still required to quantify impact using reductionist indicators, qualitative and less reductionist methods exist and have been received well.

Specific projects are structured to fit into this program cycle and feed into each goal, objective and result. Each project is required to have a "logical framework" that maps against and contributes directly and clearly to the program. As the project is implemented, the agency stresses attention be paid to "what is learned, adapting project activities, revising work plans; and, if necessary modifying contracts, grants or other implementation modalities" (USAID 2012: 29).

All projects and programs must contain a rigorous and evidence-based evaluation and monitoring plan. One of the more refreshing changes that has happened in the agency recently is a strong push for decision making to be based on evidence rather than anecdotes or foreign policy goals (although it would be naive to assume the latter have been altogether excised as factors in decision making). As a result of monitoring and evaluation, managers and staff are expected to focus on key learning areas that include:

> facilitating coordination, collaboration and exchange of experimental knowledge internally and with external stakeholders; testing hypotheses, filing critical knowledge gaps and addressing uncertainties in the hypothesis with new research, evaluations or syntheses of existing analyses; ensuring new learning, innovations and performance information gained through monitoring and evaluation inform implementation, policy formulation, and strategy development; and identifying and monitoring game changers or broad conditions that ... could impede or improve implementation.
>
> *(2012: 31)*

Finally, all of this activity must help the agency demonstrate its budget efficacy. There is a constant struggle for resources, and given that much of the agency's funding comes from Congress as program funding (for instance to promote scientific literacy in country x or reduce deforestation in region y) rather than operation expenses, fiscal accountability is a top priority.

Points of overlap and potential engagement

The brief outline of basic principles above is meant only to give the reader an introduction to the program cycle and project design at USAID. In this section, I highlight areas of potential overlap between the work of USAID and political ecologists, and highlight the range of similarities and differences in both philosophy and approach. Sensitive to the fact that such reductionism might overgeneralize political ecology as a set of principles rather than a diverse community of practice and that it takes the ADS at face value and as the single set of principles for a quite diverse agency, I attempt to present a range of possible congruencies and incongruences between the two groups.

USAID and political ecologists share a focus on gender equity in development. The strong overlap includes issues such as women's empowerment, understanding roles and the effect of inequality on natural resource management, and women's role in economic growth. Where political ecologists might part company with USAID is over the implementation of programs designed to promote gender equity. Analysis regarding how USAID programming actually results in change is lacking. There are also deeply western cultural values that are often attached to gender programming that many political ecologists, especially those in countries that are the target of such programs, might find troubling.

A second area of overlap is the use of science and technology in development. The agency has dramatically increased the presence of all categories of trained scientists, from physicists to social scientists to ecologists, recently. These actors give considerable voice to science within the agency. As with gender equity, there will be a range of responses from political ecologists, from outright rejection of "science" as a label to the use of science as merely a tool of imperialism, to acceptance of its role in development (see Chapter 11, this volume). On this last point, political ecologists can bring a rich heritage of critical assessments of the impact of science and technology in development that is often lacking in a broad sense at USAID, but also in other donors such as the World Bank. The so-called "Green Revolution" is often referred to as an unmitigated success within the development/policy community. Political ecological studies strongly suggest otherwise.

The application of selectivity and focus yields more areas of potential overlap. The ADS refers to selectivity as "gaining a good understanding of the *conditions on the ground* that are needed to begin movement in a certain development objective and applying clear, measurable, and relevant criteria for selecting countries, sub-national regions, or sectors on the basis of those conditions" (USAID 2012: 15, emphasis added). The fact that the agency prioritizes what political ecologists might call "context" is a key area of congruence. How that context is assessed and for what purposes is an altogether different question, however, and one to which political ecologists can contribute.

Following from the understanding of "context," the agency promotes "making assistance investments where there is demonstrable local demand and ownership, and where a broad segment of the community has a stake in ensuring that the activity or service continues after the USAID program or project ends" (USAID 2012: 16). The participatory and consensual development phase of the 1980s and early 1990s did not pass USAID by. In addition, the agency mandates that projects should build local skills and foster the growth of local governmental and non-governmental institutions and be environmentally sustainable. Again, in principle and in broad terms this appears remarkably similar to much of the political ecological literature on local development.

"Stove-piping" is a common problem in development institutions. This refers to the isolation of programs and projects into their thematic areas (e.g. health, environment, energy) without sufficient recognition of the overlapping nature of development. The agency has recently rethought this issue and has tried to increase integrated approaches to development. This is a particular challenge given the nature of how funding is dispersed by Congress. Political ecology, if anything, is an integrated approach to human–environment relationships and as such could be employed to analyze exactly the interrelationships USAID seeks to better understand.

Throughout ADS, there are general statements that suggest philosophies in congruence with political ecology. As part of explanation of the project cycle, for instance, development is recognized as "not static and rarely linear" (USAID 2012: 19). Clearly a small bit of post-development has snuck into even the largest development institutions! Host country stakeholder opinions and guidance is referred to repeatedly throughout the document. As mentioned above, participation and local ownership is valued. Results are measured based on "evidence (citing specific assessments and evaluations) that illustrates why USAID should reasonably expect a specific investment will produce targeted development impacts" by utilizing "analytic rigor and the best available evidence" that "incorporate[s] continuous learning" (USAID 2012: 24–25).

On two general points, however, there is serious epistemological divergence. First, USAID promotes "development" in its most mainstream sense. The market and the private sector are continuously referenced in similarly uncritical terms as science and technology. Here, many political ecologists would take issue with USAID. Seeming areas of shared concern, such as focus on local context and integrated approaches described above, could become areas of disagreement given that the agency situates its solutions in both a neoliberal reading of development and within the context of promoting the interests of the US government *first*. It should be noted, however, that market limitations are recognized amongst many staff – the agency is simply not a homogeneous block that blindly follows prescribed solutions. The economism in the agency, is, however, pervasive.

In addition, the agency's "results framework" may strike many political ecologists as overly reductionist. The complexity of development, its actors and contexts, are often greatly simplified in order to demonstrate that some action has been taken or some goal achieved. Political ecologists might challenge the narratives that such frameworks normalize and treat without critique or deconstruction.

In reviewing these points of overlap, there is no suggestion that this is a comprehensive list nor will all readers agree in extent or tenor. This dialogue is relatively open for debate exactly because there has been so little engagement.

Conclusion

In this chapter, I have attempted to demonstrate that increased engagement between political ecologists and policy makers and implementers is possible. I have used my own experience at USAID to evoke and discuss areas of both convergence and divergence between the two groups. The areas that separate us may be fewer than we think; however, those that do remain are significant. I tried to avoid presenting this as a binary "either/or" case for or against engagement.

I have described a very specific experience in this chapter. However, a few general insights may apply to other contexts, such as sub-national (state/provincial) or non-governmental institutions. First, policy as written is almost always very different from policy as implemented. A bevy of competing agendas operate behind the public face of any institution. From an NGO perspective, this may result in confusing or contradictory messages. Given the fact that so many different individuals with vastly different agendas shape and reshape policy, largely anonymously, it can be difficult to discern what actually happens "on the inside." Taking time to engage with and understand some of the internal dynamics helps explain what appear to be contradictory or inexplicable policies. Second, building alliances with like-minded individuals within the policy community is critical to understanding how the given agency or department works. This helps identify not only needs, but also pressure points where policy changes can be attempted. Third, there will always be points of divergence between academics and policy makers/implementers over both the philosophies and impacts of policy and implementation, as well as many other areas. These disagreements should not be over-generalized and, building on the second point, may be more malleable that outward appearances might suggest. Finally, engaging policy makers and implementers of any type will require flexibility and persistence. Given divergent timelines and respective work pressures, for instance, significant effort and time will be necessary to forge any serious relationship.

Should political ecologists feel that increased engagement is desirable, I would suggest that several issues need further clarification and discussion. First, what should the role of a political ecologist, academic or otherwise, be when working in any type of policy institution? There are clearly a range of possibilities – what are the advantages and disadvantages of each? Second, how can academic political ecologists work together across their respective institutions to start to break down promotion and tenure requirements that either actively discourage such engagement or informally shun it? Third, how can a greater audience for such interaction be cultivated? What are the opportunities to show the value of such work and how is that value translated back into the work of political ecologists?

Of course, there will be many more issues and many more opinions than I can possibly capture here. Further work in this area will demand flexibility and hybridity from both academics and policy practitioners, however, we can at least begin to address the question "should some political ecology be useful?" (Blaikie 2012).

Note

1 The views set out in this chapter are in no way those of the US Agency for International Development (USAID), its employees, its contractors or anyone in any way affiliated the Agency. As such the chapter has not been cleared by or in any way approved or disapproved by anyone in the agency. As a result, no information about the agency's inner workings are described. I reference no internal documents (or conversations) here, other than those that are made available to the general public.

References

Bebbington, A. 2014. "The endogenous scholar: porous boundaries and travelling ideas in development." *Third World Quarterly*, 35(3): 520–523.

Blaikie, P. 1985. *The Political Economy of Soil Erosion in Developing Countries*. London: Longman.

Blaikie, P. 2012. "Should some political ecology be useful? The inaugural lecture for the Cultural and Political Ecology Specialty Group, Annual Meeting of the Association of American Geographers, April 2010." *Geoforum*, 43: 231–239.

Blaikie, P. and Brookfield, H. 1987. *Land Degradation and Society*. London: Routledge.

Brannstrom, C. and Vadjunec, J. 2013. *Land Change Science, Political Ecology and Sustainability: Synergies and Divergences*. London: Routledge.

Bryant, R. and Bailey, S. 1997. *Third World Political Ecology*. London: Routledge.

Carr, E. 2011. "If you are uncomfortable, you are probably doing it right," *Environment and Planning, A*, 43: 2797–2800.

McCusker, B. and Weiner, D. 2003. "GIS representations of nature, political ecology, and the study of land use and land cover change in South Africa," in K. Zimmerer and T. Bassett (eds.), *Political Ecology: An Integrative Approach to Geography and Environment-Development Studies*. New York: Guilford, pp. 201–218.

Neumann, R. 2005. *Making Political Ecology*. New York: Oxford University Press.

Peet, R. 2007. *Geography of Power: The Making of Global Economic Policy*. London: Zed Books.

Peet, R. 2013. "Comparative policy analysis." *Human Geography*, 6(2): 1–10.

Peet, R. and Watts, M. 2004. *Liberation Ecologies: Environment, Development, Social Movements*. London: Routledge.

Peet, R., Robbins, P. and Watts, M. 2011. *Global Political Ecology*. New York: Routledge.

Robertson, M. 2010. "Performing environmental governance." *Geoforum*, 41: 7–10.

Stott, P. and Sullivan, S. 2000. *Political Ecology: Science, Myth and Power*. New York: Oxford University Press.

USAID. 2012. ADS Chapter 200. Available at: www.usaid.gov/who-we-are/agency-policy, accessed November 3, 2013.

Walker, P. 2006. "Political ecology: where is the policy?" *Progress in Human Geography*, 30(3): 382–395.
Zimmerer, K. and Bassett, T. 2003. *Political Ecology: An Integrative Approach to Geography and Environment-Development Studies*. New York: Guilford.

* Full URL for Stock Taking Guide is http://frameweb.org/adl/en-US/11027/file/1687/A%20Guide%20to%20Taking%20Stock%20of%20Natural%20Resource%20Management.pdf.

15

AT THE BOUNDARIES OF *LA POLÍTICA*

Political ecology, policy networks and moments of government

Anthony Bebbington

I don't know if my colleagues in Latin America are political ecologists or not. They probably are, many of them at least. Most of them view natural resources and nature as a terrain of dispute. Most of them have been politically active, sometimes militant. Most of them bring some sort of political theory to bear on how they understand relationships between nature and society, environment and development. While few of them refer to themselves as political ecologists, and a good number of them are economists, all of them engage with *la política* in some sort of way.

This chapter draws on the work of these colleagues, and my own experiences in interacting with them. I suggest that this work and these experiences offer particular windows on intersections between political ecology and policy. The chapter is neither manifesto nor critique, and nor is it a full review of the literature – it is instead a personal reflection on a set of experiences, and an effort to draw out insights that these experiences might offer. My starting point is a comment on the nature of *la política* with which these colleagues engage. This then becomes a platform for reflecting on the career trajectories of a number of these colleagues, what the pathways they have followed might imply for how one thinks about the relationship between critical scholarship of the environment, policy and politics, and more generally for the ways in which we conceive of boundaries. I then discuss briefly how my own work has become, on and off, caught up in these different trajectories. I close by discussing the evolution of what is now a decade-long collaboration in El Salvador that might, in its own small way, constitute an engagement between political ecology, one ostensible political ecologist and policy.[1]

La política, policy, politics

The Spanish term *la política* is an interesting one, meaning both "policy" and "politics." While in conversation an Anglophone speaker could generally select the most fitting translation depending on how "*la política*" is being used, the fact that the same word refers to two terms that many political ecologists would be very careful to keep separate seems important to me. Recall that Peter Walker very deliberately dedicated two somewhat polemical *Progress in Human*

Geography essays on the state of political ecology to this distinction: "Political ecology: where is the policy?" (Walker, 2006), and "Political ecology: where is the politics?" (Walker, 2007). How would a translator handle this difference if they were reproducing the essays in Spanish? I pose the hypothetical question not to be cute, but to entertain a more serious possibility: namely that this peek into "*la política*" might suggest that we are working with a peculiar sort of boundary, one that is more porous, and perhaps even more imagined, than the policy/politics distinction might imply (Bebbington, 2014).

The politics/policy distinction is often invoked as a means of making other distinctions: critical/technocratic and radical/reformist among others. The implication, frequently, is that "policy" is a conservative arena that contaminates, demands compromises and ultimately leads those who engage with it down paths that take them away from initially progressive intentions (if they ever had them) and into a world that consistently seeks to "render technical" that which is rightly political (Li, 2007). What, though, is policy? In the pursuit of particular purposes, policies lay down commitments and rules intended to govern relations of authority, modes of interaction and the allocation of resources, opportunities and sanctions. All organizations have policies – policy is not just in the domain of the public sector. Some policies of course have far greater reach than do others – contrast the reach of a national mining policy with that of the policy of a legal defense non-profit organization, for instance. However, it remains the case that all organizations have policy governing their actions. And in no organizations are such policies immaculately conceived. They are the temporary product of discussions, contestations and/or authoritarian acts within these organizations and in the environments in which they operate.

As such, policies are clearly political – they are not socially neutral in their effects, and the processes that produce these rules always involve particular combinations of contention, conflict, negotiation, inclusion and exclusion. In some sense, then, a policy is the frozen product of politics, a product that is itself political and has effects that are also political. Indeed, it could be said that most of what passes as politics in political ecology is in some regard a struggle over the definition of policy at scales that run from community authority to international organization: land and resource tenure policy, forest policy, REDD, water law and on and on. Ultimately, social movements, NGOs, activists, lobby groups and the like seek to change the policies of other organizations: government policy, corporate policy, IFI policy, etc. Seen this way, it begins to make sense that Spanish uses the same word for *policy* and *politics*.

My charge in this chapter is to reflect on political ecology and policy, but I use this opening gambit to trouble any easy distinctions between policy and politics and to suggest that in very many regards they really are the same thing. This is also my segue back to the colleagues that I invoked in the opening sentences because their own lives and careers also challenge the separation between scholarship, policy and politics. As my own work on rural development, environmental governance, social conflicts and livelihoods in Latin America has unfolded over the years, I have had the pleasure and privilege to develop professional and personal relationships with professionals many of whom could easily have ended up in tenured academic positions and many of whom do, from time to time, publish in academic outlets and do university teaching in their spare time. In other institutional and political economic contexts, with different sorts of university environments, a number of these colleagues would quite probably have become scholars and been deemed political ecologists. In the following section I discuss elements of their career trajectories and suggest that they have implications for how one might think of the relationships between political ecology and policy.

Trajectories and boundaries

In a perceptive and personal account of democracy and development in twentieth-century Latin America, David Lehmann (1990) turned special attention to the nature of non-governmental organizations, social movements and the roles that they had played in the shifting relationships between politics, economics and religion. He suggested that NGOs could be understand as an "informal university," a network of institutions that performed social science research in a context in which the political economy of the academy made such work impossible or at least very difficult in most formal university settings in the region. While varying combinations of authoritarian government and lack of investment limited possibilities in the formal academy, the availability of financial resources from foundations, non-governmental bodies and social democratic governments in Europe and North America made it possible to build research capacity in the form of NGOs, all the more so when this capacity was combined with activism and different types of development intervention (see also Bebbington and Thiele, 1993). In some cases, these NGOs were also consciously supported as a "government in waiting" – they were vehicles to support professionals, to generate bodies of knowledge, and to experiment with models of development that would, once national political settlements shifted (cf. Khan, 2010), move into government. The consummation of this strategy was probably clearest during the transition to democracy in Chile in 1990 when many NGO leaders and ideas moved into the post-Pinochet state to be part of an interesting, difficult but historically momentous coalition of Christian Democrats, Social Democrats, Communists and Socialists committed to putting the heritage of military authoritarianism to rest forever. At different points, colleagues and friends, some of them very close, have been part of this pattern, themselves moving from NGOs and research centers to help build a post-authoritarian state, most of them working on the agrarian, rural and regional dimensions of this process (e.g. Julio Berdegué, Octavio Sotomayor, Eduardo Ramírez, Claudia Serrano, Cecilia Leiva, José Weinstein). While this phenomenon was especially striking in the Chilean transition, it has repeatedly been the case that politically committed technocrats and social scientists have trafficked to and fro between government and "civil society" as regimes have changed. Recent examples would include El Salvador following the election of the FMLN in 2009, Peru at the end of the Fujimori era in 2000, and again following the election of Ollanta Humala in 2012, Bolivia in 2006 when Evo Morales came to power, Ecuador when Rafael Correa became president in 2007 and now again Chile with the return of Michelle Bachelet.

Such movements are hardly surprising – elections and revolutions alike are, after all, about changing the composition of government and the socio-political networks that have privileged access to the institutions of the state. Nonetheless, these boundary crossings do raise questions for any more general understanding of the ways in which state and civil society, or public policy and political activism, are separated. They are a reminder that when government changes, people such as these seek to turn opposition politics into public policy through their newly acquired influence over the instruments and powers of the state. Thus, while there clearly *are* differences between the institutions of the state and civil society, the social composition of each of these domains is in many respects an artifact of the political moment (see also Chapter 35, this volume).

Very many of the people that I have been privileged to work with in different research projects have made these same journeys across boundaries – and several of them on more than one occasion. Some of these journeys are made easily – many jumped at the chance to work in building a post-Pinochet government in Chile. Other journeys are agonized over. I have had conversations with a good number of colleagues trying to decide whether they would be more effective (and also less compromised or exhausted) doing their work from their research centers, or whether they should accept the invitation to assume a senior position in government where they could take some of these research ideas with them. Some of the most anguished discussions relate to positions in Ministries of Environment – some of the weakest and youngest ministries in the region, some of the most subject to cooptation and evisceration, yet some of the most critical to a political ecology agenda concerned with fairer and less destructive modes of natural resource governance. Invariably those who opt to move into government find themselves in environments that are fascinating, frustrating, difficult, exhausting and compromising. Many, probably most of them, end up resigning after a number of years – though some have lasted the course until a change of government. Consistently, whether they leave or last, they appear to come to the view that the hardest thing to do, and to know how to do, is to *make* policy, especially policy with teeth, and then to *implement* it. Making policy requires an understanding of public administration, the politics of lobbying and the micro-politics of legislative change and cabinet dynamics that few if any political ecologists (indeed academics in general) are trained in at all. Becoming a Minister, Vice-Minister or Department Director is, by itself, simply not enough.

I consider these colleagues to be part of a very loose continent-wide social movement of broadly social democratic (some more radical, others more conservative) thinkers who are committed to deepening democracy (Fung and Olin Wright, 2003; Avritzer, 2002), thickening civil society (Fox, 1996; Watts, 2001), reducing the crassest asymmetries of power (cf. Acemoglu and Robinson, 2012), seeking more inclusive growth models (Kanbur, 2000) and holding back the dispossession and enclosure of natural resources. The more radical nodes in this network engage with the works of political ecologists such as Martínez-Alier (2002) and Escobar (2008), while the more reformist nodes are more inclined to seek inspiration in Acemoglu and Robinson (2012), Elinor Ostrom (1990, 2005) or that group of economists who sought to inject concerns for inequality and exclusion into the World Bank (World Bank, 2005). What this broad movement shares (even if from the inside it would, like many social movements, probably seem to have as many fractures and chasms as convergences) is a commitment to the absolute need for policy change as part of these changes. The paths to this policy change might be through changing public discourse, targeted policy engagement, occupation of positions in government or working with students who in later years will themselves become makers and framers of policy – but while the paths are varied the purpose is similar: to change *la política* in pursuit of these goals.

Not all of these colleagues have worked on substantive issues that one would consider bread-and-butter questions of political ecology. But many are working on rural and agricultural development, forest policy, environmental governance, extractive industries and indigenous peoples' issues. Nor are these people always operating in institutional settings that might seem like an obvious port of call for political ecologists. Just to take one example, since the mid-2000s I have come to know and share ideas with a group of lawyers working in the Human Rights Ombudsman's Office of Peru who have increasingly had to deal with human rights issues surrounding extractive industries disputes. The Ombudsman's office (*Defensoría del Pueblo*) is an interesting body in that it is part of the state, but not part of the governing administration. It answers to Congress who appoint the Ombudsperson, and remains independent of the Presidency. The Defensoría has few formal powers, though it does have significant moral power: it cannot set policy, but it intervenes in policy processes. One such set of processes has been the debate, and subsequent legislation, surrounding free, prior and informed consent for indigenous peoples. Peru had ratified the ILO Convention 169 in 1994 but never operationalized this commitment as national law. This group of lawyers, working in the Defensoría and then subsequently in venues such as the Ministry of Culture, played a critical role in pushing the issue of prior consent and consultation into the legislative process and public debate (using conflicts over resource extraction as the vehicle through which to do this). The process has been tortuous, with compromises and negotiations along the way, a number of disagreements with indigenous peoples' organizations and other activists, and more than a few resignations. *La política* is a contact sport, and the higher the stakes are for hegemonic interests and ideas, the more punishing the contact and the potentially more compromising the negotiations along the way. This clearly raises challenges, awkward decisions and potential litigation for academic political ecologists accompanying such processes, and there are colleagues among our number who have experience of these challenges.

If the colleagues I have been referring to *can* be seen as a loosely networked movement (and not all would agree with such a contention), one question that arises is how to characterize the place of academics who work closely with these colleagues? Are we part of this same movement? Do we at least share a conviction that an important part of movement processes is to shift the ways in which problems are framed and talked about in public and policy debates, and that our own work can be part of this politics of knowledge (sometimes whether we like it or not)? And how do we negotiate the boundaries between the personal and *la política* in these processes? If we are in some sense part of these networks, if we gain from them for elements of our own work and life, if colleagues are friends … then how far are these networks also mechanisms of accountability and reciprocal obligation in which we are somewhat embedded and that then influence our own decisions? Answers to such questions are personal, certainly not generalizable. It is however possible that one answer is that, even if "we" might at times comment critically on the work of these networks, the political ecologist can become involved in them to such a degree that these relationships carry us along into domains that we might never have anticipated and that might well be awkward and sometimes daunting. In the following section I discuss some of these issues through the lens of my own interaction with *la política* of environmental governance in El Salvador. This is a case where a layering of relationships – from the most intimate to the generally professional – have brought me into work situations that I would never have imagined a decade back and that have influenced how I think of *la política* and political ecology.

Political ecologies in post-conflict El Salvador

In the early 2000s, Denise Humphreys Bebbington (my wife) began to build a Latin American wide network of advisory committees for the work of Global Greengrants Fund (www.greengrants.org). Global Greengrants channels small, rapidly disbursed grants to activists and local groups involved in socio-environmental conflicts and is itself a network with which political ecologists have interacted from time to time (people such as Arturo Escobar, Serge Dedina, Ximena Warnaars) as advisors, brokers and fellow travelers. Through that work, Denise began collaborating with Deborah Barry, then Ford Foundation program officer and herself a Berkeley-trained political ecologist in most senses but name. That relationship in turn led indirectly to a collaboration of my own with Deborah in a project to work with eight Central American and Mexican non-governmental centers that combine environment-development research, policy advocacy and direct interventions. In that project we sought to theorize the role of non-governmental research in the politics of environment and development in the region and to use that framework to analyze the roles that each of these centers was playing (Bazán et al., 2007; Bebbington, 2007). These were studies that were at once analytical and psycho-analytical (in the sense that they involved a great deal of self-reflection and critique), and in more than one case they caused waves within the participating organizations. One of these centers was the Fundación PRISMA in El Salvador (Cuéllar and Gómez, 2007), a group that itself had collaborated with political ecologists such as Susana Hecht in the conduct of research aimed at affecting policy (in that case, the research dealt with forest resurgence in El Salvador: Hecht et al., 2006). I myself began to collaborate with PRISMA on a range of issues: research strategy, organizational design, guiding concepts, etc.

Though not formally linked to the FMLN, the political party that had grown out of the guerrilla group of the same name, PRISMA sympathized with the FMLN platform as it sought to wrest political power from the conservative ARENA government that had held the presidency since the civil war period and had been founded by the alleged intellectual author of the assassination of Archbishop Óscar Romero. After 20 years of ARENA government, the FMLN won the 2009 elections and as it began to build a cabinet, the new government appointed a member and former Director of PRISMA, Herman Rosa, to be Minister of Environment. After his nomination, but before his taking up the position, I was with Herman at a research meeting organized by one of the other policy-scholarly research networks in the region with which both he and I collaborate. He asked if I would have time to work with the new ministerial team in thinking through concepts and strategies of environmental governance. One of the more politically urgent environmental governance issues that the ministry had to address was what to do about hard rock mining (Spalding, 2013a, 2013b).

El Salvador currently has no large-scale mining activity, though it has had a limited amount in the past. Since the mid-1990s, exploration companies had become interested in El Salvador – both in areas of old mines and in areas with no mining history. Several of these exploration sites had become increasingly conflictive, and these conflicts had helped produce a national public debate on the advisability of promoting the mining sector or not. In the face of this contention, by 2007 even the pro-market ARENA government had frozen all administrative decisions related to environmental permits for mining companies. The practical effect of this was that companies could not progress with their projects because they were unable to gain environmental approvals to move to the next step in the project cycle (from prospection to exploration, or from exploration to construction and operation).

These debates, which continue through to the present, have been couched in familiar themes of political ecology. Let me note four of these. First has been the question of *risk*. Those who are skeptical of mining's role in El Salvador emphasize the risks that this would mean for the country's water resources in a context in which water quality has already been acutely compromised and in which water is scarce for reasons that are both material and socially constructed (see Perreault, 2013, for political ecological work on mining and water). Second is the *political economy of environmental transformation*. Here the argument has been that "nature" in El Salvador has been so transformed by prior rounds of environmentally unregulated capitalist expansion that it has been rendered acutely vulnerable – and that as a result, it would be unwise to allow the expansion of mining as a further source of transformation and vulnerability. These discussions tie risk and hazards to political economy in ways that resonate clearly with long-standing political ecological discussions of vulnerability and resilience (Wisner et al., 1994 [2003]; Watts, 1983). Third is the theme of *socio-environmental justice*. Movements, the Catholic Church, activists and others argue that the unequal geographies of risk and opportunity that mining would introduce would constitute new sources of injustice in ways that an already vulnerable society could ill afford (cf. Carruthers, 2008). Fourth, of course, is the issue of *the role of social movements in shaping the relationship between environment and development* (Peet and Watts, 1996). Here is an instance in which movements are quite self-consciously shifting the terms of national debate over the relationships between nature and society to which El Salvador should aspire.

The social movements involved in these conflicts have called for a national ban on mining in El Salvador (Spalding, 2013a, 2013b; Broad and Cavanagh, 2011) and in the 2008/9 electoral campaign, the FMLN candidate, Mauricio Funes, appeared to commit his party to such a ban. In parallel with this process, however, two mining companies (one US based, the other Canadian) opened litigation against the government of El Salvador demanding compensation for losses incurred by the *de facto* suspension of environmental permitting enacted by the Salvadoran government. Under both the Central America-Dominican Republic Free Trade Agreement (CAFTA-DR) and national legislation the companies were able to pursue these cases at the International Court for the Settlement of Investment Disputes (ICSID), part of the World Bank Group. Were El Salvador to lose these cases, the financial costs would be extremely high. As I have noted elsewhere (Bebbington, 2012), this placed the incoming FMLN government – itself historic, the first elected post-guerrilla government of Latin America – in an apparently impossible position. Align with movement demands and run the risk of fines so large that they would compromise government finances severely; or allow mining to proceed and so fall foul of popular demands and compromise the quality of the democratic transition by making clear the extent to which international trade agreements render the domestic politics/policy of environmental governance largely irrelevant (cf. McCarthy, 2004). The other dilemma facing the government was the sense that they really didn't have enough information at hand to know what was "the best" course of action, to understand the actual levels of environmental risk involved or the potential revenue that a mining sector might generate for the state.

Having never set foot in El Salvador until 2005 (barring one flight diversion), my personal and professional networks had now placed me into the mix of these debates. While I had initially been asked to share thoughts on environmental governance (and to draw on political ecological ideas in the process), I was soon being asked if I would work with the Ministry of Environment and the Ministry of Economy on the problem of mining policy. In line with the National Environment Law, a rarely enacted article of which requires national policies to be subject to strategic environmental assessments, the Ministries had decided to commission such an assessment for the mining sector. This assessment would generate the primary technical arguments on the basis of which the FMLN administration would decide how to govern mining. The contract for the assessment was ultimately won by a bid from a Spanish consulting company and I was asked by the government to chair a Technical Committee that would oversee the process and provide substantive commentary on draft chapters. The three other Technical Committee members were also people who had long worked on the boundaries of policy and science, in the worlds of consulting, the World Bank, and academia.

In some sense, this role of overseeing and commenting on research design, implementation and text was similar to that of seeing a doctoral thesis to completion. The text is somebody else's, and they have a better sense than do you of the raw data that underlies the arguments being made, while you the advisor are nonetheless implicated in the text, possibly have a greater sense of the broader context of which the text is a part, and ultimately have to pass judgment over it. The differences, of course, were that the stakes were of a completely distinct nature and that very many more players had an interest in the process. Managing relations with these players – governmental, corporate, activist – in ways that allowed access to information but retained "independence" was not straightforward.

At the end of the process, the Technical Committee was also asked to prepare a short report on our view of the issues at stake, of the advisability or not of promoting mining in El Salvador, and of the conditions under which mining *might* contribute to development while not introducing unmanageable risks into the already fragile relationships between the country and its environment. Subsequently I was also asked by the Minister of Environment to work with a team at the Ministries of Environment and Economy on the outline of a Legal Decree that would codify these implications. That decree was then sent to the National Assembly for debate. The decree, which calls for an indefinite suspension of all administrative procedures related to mining until a wide set of conditions have been met, is in many ways an exercise in applied political ecology and environmental management. It builds an argument regarding the current status of socio-environmental risk and vulnerability in El Salvador, relates this to an assessment of the resilience of El Salvador's environment and its institutions, and on that basis crafts a nuts and bolts outline of the sorts of legal and administrative frameworks and organizational capacities that would have to be in place before mining could proceed.

In the meantime, the legal case between one of the mining companies (Pacific Rim) and the Government of El Salvador was working its way through the system at ICSID (the other case collapsed). One particularly senior and accomplished member of the Technical Committee, Robert Goodland, had worked with the lawyers representing the Government of El Salvador as an expert witness. In December 2013 Robert died suddenly and I was asked to replace him in this role. I was initially hesitant. Asking colleagues whether this was a good idea or not, somebody in El Salvador said to me, "If you don't, you will be forever haunted by Robert's ghost." There was considerable truth to this – and so now the networks carrying me along into these spaces of *la política* were no longer just social and personal, they had become psychological and supernatural. As it turns out, the testimony I have prepared continues to draw on a mix of foundational ideas in political ecology, global change and more recent work on mining in Central America by authors whose work might be deemed political ecological regardless of how they label themselves (e.g. Dougherty, 2011, 2013; Pedersen, 2014; Spalding, 2013a; Laplante and Nolin, 2014).

What I have realized is that the lawyers find these ideas and studies dealing with socio-environmental conflict, participation, materiality, resilience and the like immensely useful (and interesting), though would never have had the time to find or process them alone. My function has been to make the link between the lawyers and this research, but also to interpret it in ways that they can cast as "expert." That label sits very awkwardly with me, as do the potential risks for my reputation and any future litigation. I am just not sure, however, that I could possibly have declined this invitation. To play off the title of an article of one close colleague (McCarthy, 2004), by that point the conditions of my own intellectual production had been anything but private, and I think I was in no position to erect a wall declaring that on the particular issue of mining in El Salvador, I was now going to exercise my own private control of that production. The networks that had made my own learning possible were asking that I put some of that learning to use for issues of real concern to those same networks.

Conclusions

James Ferguson (1997) has argued that underlying anthropology's ambivalent relationship with development is the constitution of anthropology itself as a "field that is always concerned with the 'less,' the 'under,' the 'not-yet' ... developed" (1997: 170). He goes on: "so intimately intertwined is the idea of development (and its lack) with the idea of anthropology itself, that to be critical of the concept of development requires, at the same time, a critical re-evaluation of the constitution of the discipline of anthropology itself" (1997: 170). As such, Development is Anthropology's evil twin, with Development being founded on a set of principles "that conflicts with the most basic theoretical and political commitments" (170) of the discipline of Anthropology, yet at the same time being so genetically related to that same discipline that it cannot simply be thrown out of the house.

Albeit less dramatically, something similar is going on in the relationship between political ecology and policy in ways that a reflection on "*la política*" helps make a little more apparent. Political ecology is founded on a political critique of arrangements that underlie some combination of ecological exclusion, injustice and destruction. These arrangements – existing across a range of scales and spheres – are nothing other than policies: the policies of communities, states, corporations, and organizations, policies that reflect the balance of power at any one time. Of course, some of these policies are far more formal (e.g. state policy on land tenure) and others are informal (e.g. social groups' ideas about the levels of inequality, prejudice and exclusion that should be allowed to exist). Likewise, some are explicit while others remain largely unspoken as collective doxa (Bourdieu, 1977): but in each instance these arrangements reflect how these different social bodies think that things should be. Analytically, political ecology is defined by a critique of policy writ large; normatively, political ecology is defined by some sort of commitment that policy should be otherwise. This commitment does not necessarily mean that political ecologists should have to engage in policy making; it does mean, though, that we cannot escape from policy. I suspect it probably also means that we cannot escape from some sort of theory of policy change. I also suspect that the escape clause that Ferguson offers anthropology is not available to political ecology. While Ferguson suggests that anthropology can rid of itself of its evil twin by redefining the discipline itself such that it is no longer constituted by the study of the cultural "other" existing in a condition of "lack," it is not at all clear that political ecology can avoid defining itself both analytically and normatively by its critique of "la política dominante" and its commitment to a "política alternativa."

The potential pathways to policy change are many (as, of course, are the resistances), and so to recognize the inescapability of policy does not imply the need to become involved in the supposedly tawdry worlds of World Banks, think tanks, technocracies and bureaucracies. That said, over the course of a career, during which the life of the political ecologist becomes enmeshed with the evolving careers of multiple other collaborators, friends and colleagues, it is quite possible that these more tawdry pathways to policy change might also present themselves. Even as they do, the choice is always open to not engage – but the personal and political reasons for not doing so may not always be clear-cut or easy to defend.

In his Inaugural Lecture for the Cultural and Political Ecology Group of the Association of American Geographers, Piers Blaikie asked "Should some political ecology be useful?" (Blaikie, 2012: 231). Posing the question this directly can be contentious, as demonstrated by what became an extensive and sometimes heated debate between Ed Carr and Ben Wisner on the listserve of the Cultural and Political Ecology Specialty Group in 2010 (Simon, 2011: 2789). Whatever the case, Blaikie's view was that, indeed, *some* political ecology should be useful. Others share similar views, not just about political ecology but also its close cousin (if not evil twin), development geography (see Simon, 2011; Simon and Carr, 2014). What those different contributions do, however, is to pose this question from the standpoint of the political ecologist deciding whether and how to make their work "useful" and "relevant" to others. The types of collaboration, collegiality and friendship developed over a career on which I have reflected in this chapter invert the way in which this question is posed. For regardless of what the scholar might decide, it may be that these networks end up finding the work of the scholar useful, and asking the scholar to play certain roles that they may not have been willing, or able, to choose for themselves. The making of these requests does not obligate the scholar to acquiesce, but when they are made in the context of overlapping personal and normative commitments they can be difficult to escape – not only because it is hard to let a friend and colleague down, but also because the very making of the request can suggest that the scholar has become part of something rather larger than they had initially planned. At that point, any existential angst generated by such invitations might derive less from a reflection on whether or not to engage with policy and rather more from the recognition that the political ecologist has become part of *la política*. This is, I think, a qualitatively (if not quantitatively) similar sort of existential angst that colleagues experience when they get the call asking them to become Minister of Environment and giving them 24 hours to decide to accept or not.

Note

1 From this introduction alone it will be clear that this chapter has innumerable debts and I cannot name them all here – that would be another chapter. I simply want to thank these friends and colleagues for the *camino recorrido* and for sharing so much with me, above all their trust even when I was sometimes an unreasonable critic. I am also grateful to James McCarthy and Tom Perreault for guidance and great comments (and for making me write the chapter) and to Herman Rosa and Julio Berdegué for their comments (and much else).

References

Acemoglu, D. and J. Robinson (2012) *Why Nations Fail: The Origins of Power, Prosperity and Poverty*. New York: Crown Business.
Avritzer, L. (2002) *Democracy and the Public Space in Latin America*. Princeton: Princeton University Press.
Bazán, C., N. Cuellar, I. Gómez, C. Illsley, I. Monterroso, J. Pardo, J.L. Rocha, P. Torres and A. Bebbington (2007) "Producing knowledge, generating alternatives? Challenges to research oriented NGOs in Central America and Mexico" pp. 175–195 in A. Bebbington, S. Hickey and D. Mitlin (eds.) *Can NGOs Make A Difference? The Challenge of Development Alternatives*. London: Zed Press.
Bebbington, A.J. (ed.) (2007) *Investigación y cambio social: desafíos para las ONG en Centroamérica y México*. Guatemala City: Editorial de Ciencias Sociales.
Bebbington, A.J. (2012) "Underground political ecologies." *Geoforum* 43(6): 1152–1162.
Bebbington, A.J. (2014) "The endogenous scholar: porous boundaries and travelling ideas in development." *Third World Quarterly* 35(3): 520–523.
Bebbington, A.J. and G. Thiele (1993) *NGOs and the State in Latin America: Rethinking Roles in Sustainable Agricultural Development*. London: Routledge.
Blaikie, P. (2012) "Should political ecology be useful?" *Geoforum* 43: 231–239.
Bourdieu, P. (1977) *Outline of a Theory of Practice*. Cambridge: Cambridge University Press.
Broad, R. and J. Cavanagh (2011) "Like water for gold in El Salvador." *The Nation*, August 1–8, 2011.
Carruthers, D. (ed.) (2008) *Environmental Justice in Latin America*. Cambridge, MA: MIT Press.
Cuéllar, N. and I. Gómez (2007) "Generación y Movilización de Conocimiento Sobre Desarrollo y Ambiente: El Caso de PRISMA" pp. 103–174 in A. Bebbington (ed.) *Investigación y cambio social: desafíos para las ONG en Centroamérica y México*. Guatemala City: Editorial de Ciencias Sociales.
Dougherty, M.L. (2011) "The global gold mining industry, junior firms, and civil society resistance in Guatemala." *Bulletin of Latin American Research* 30(4): 403–418.
Dougherty, M.L. (2013) "The global gold mining industry: materiality, rent-seeking, junior firms and Canadian corporate citizenship." *Competition & Change* 17(4): 339–354.
Escobar, A. (2008) *Territories of Difference. Place, Movements, Life, Redes*. Durham, NC: Duke University Press.
Ferguson, J. (1997) "Anthropology and its evil twin: 'development' in the constitution of a discipline" pp. 150–175 in F. Cooper and R. Packard (eds.) *International Development and the Social Sciences: Essays on the History and Politics of Knowledge*. Berkeley: University of California Press.
Fox, J. (1996) "How does civil society thicken? The political construction of social capital in rural Mexico." *World Development* 24(6): 1089–1103.
Fung, A. and E. Olin Wright (2003) *Deepening Democracy: Institutional Innovations in Empowered Participatory Democracy*. New York: Verso.
Hecht, S.B., S. Kandel, I. Gomes, N. Cuellar and H. Rosa (2006) "Globalization, forest resurgence, and environmental politics in El Salvador." *World Development* 34(2): 308–323.
Kanbur, R. (2000) "Economic policy, distribution and poverty: the nature of disagreements." *World Development* 29(6): 1083–1094.

Khan, M. (2010) *Political Settlements and the Governance of Growth Enhancing Institutions*. London: SOAS, mimeo.
Laplante, J.P. and C. Nolin (2014) "Consultas and socially responsible investing in Guatemala: a case study examining Maya perspectives on the indigenous right to free, prior, and informed consent." *Society and Natural Resources: An International Journal* 27(3): 231–248.
Lehmann, A.D. (1990) *Democracy and Development in Latin America: Economics, Politics and Religion in the Post-War Period*. Cambridge: Polity Press.
Li, T. (2007) *The Will to Improve: Governmentality, Development and the Practice of Politics*. Durham, NC: Duke University Press.
Martínez-Alier, J. (2002) *The Environmentalism of the Poor*. Cheltenham: Edward Elgar.
McCarthy, J. (2004) "Privatizing conditions of production: trade agreements and environmental governance." *Geoforum* 35(3): 327–341.
Ostrom, E. (1990) *Governing the Commons*. Cambridge: Cambridge University Press.
Ostrom, E. (2005) *Understanding Institutional Diversity*. Princeton: Princeton University Press.
Pedersen, A. (2014) "Landscapes of resistance: community opposition to Canadian mining operations in Guatemala." *Journal of Latin American Geography* 13(1): 187–214.
Peet, R. and M. Watts (eds.) (1996) *Liberation Ecologies: Environment, Development, Social Movements*. London: Routledge.
Perreault, T. (2013) "Dispossession by accumulation? Mining, water and the nature of enclosure on the Bolivian Altiplano." *Antipode* 45(5): 1050–1069.
Simon, D. (ed.) (2011) "Geographers and/in development." *Environment and Planning A* 43(12): 2788–2800.
Simon, D. and Carr, E. (2014) "Introduction: Engaging Critically from Theory to Policy and Implementation." *Third World Quarterly* 35(3): 505-506.
Spalding, Rose (2013a) *Contesting Trade in Central America: Market Reform and Resistance*. Austin: University of Texas Press.
Spalding, Rose (2013b) "Transnational networks and national action: El Salvador's antimining movement" pp. 23–55 in E. Silva (ed.) *Transnational Activism and National Movements in Latin America: Bridging the Divide*. London: Routledge.
Walker, P. (2006) "Political ecology: where is the policy?" *Progress in Human Geography* 30(3): 382–395.
Walker, P. (2007) "Political ecology: where is the politics?" *Progress in Human Geography* 31(3): 363–369.
Watts, M. (1983) "On the poverty of theory: natural hazards reviewed in context" pp. 231–262 in K. Hewitt (ed.) *Interpretations of Calamity: From the Viewpoint of Human Ecology*. Boston: Allen Unwin.
Watts, M. (2001) "1968 and All That ..." *Progress in Human Geography* 25(2): 157–188.
Wisner, B., P. Blaikie, T. Cannon and I. Davis (1994 [2003]) *At Risk. Natural Hazards, People's Vulnerability and Disasters*. London: Routledge.
World Bank (2005) *Equity and Development: World Development Report 2006*. Washington, DC: World Bank.

PART IV

Core questions in political ecology

The chapters in this section of the Handbook are organized around fundamental questions that political ecologists ask regarding nature and our relationship to it. These are:

- *How do we come to know nature, and what differences do forms of environmental knowledge make?* (Section A, on Environmental Knowledge)
- *In what ways are nature and society transformed through economic activity, and how does this metabolic relationship affect various social groups in different ways?* (Section B, on Environmental Change)
- *Through what sorts of social arrangements and forms of rule do people 'manage' nature, and to what effect?* (Section C, on Environmental Governance)
- *How are social subjectivities shaped through, and reflected by, differential access to and control over nature?* (Section D, on Environmental Identities)
- *In what ways and for what reasons do people mobilize politically around nature?* (Section E, on Environmental Politics)

PART IV, SECTION A

Environmental knowledge

This section of the Handbook considers the questions, *How do we come to know nature?* and *What differences do forms of environmental knowledge make?* Knowledge regarding nature and our relationship to it – from the root causes of soil erosion, famine, and deforestation, to the bodily ecologies of obesity and health – continues to be central to political ecology. Producing such knowledge unavoidably reflects and reproduces relations of social power as it involves questions about how, by and for whom, and to what effect knowledge is produced. Indeed, critique of dominant forms of environmental knowledge and the production of new, counter-hegemonic knowledges – Robbins' "hatchet and seed" – has been central to political ecology's method, theory, and politics since its inception. If knowledge is power, then a critical understanding of *how* we know *what* we know is a core element of any emancipatory project, academic, or otherwise.

Rather than examining knowledge production in relation to specific empirical problems in political ecology (say, urban water governance or desertification), the chapters in this section consider epistemological and methodological frameworks that political ecologists have used to understand the complex relations between society and non-human nature. The section begins with Rebecca Lave's critical consideration of Actor-Network Theory and its limitations as a conceptual framework in political ecology. This is followed by David Demeritt's analysis of popular participation in science, an effort at breaking down the distinctions between "expert" and lay understandings of scientific knowledge. The theme of popular or non-expert forms of knowledge production is continued in the next two chapters, by Leah Horowitz and Joe Bryan, who examine, respectively, local environmental knowledge and participatory mapping. Both chapters consider cross-cultural collaboration in the production of geographical knowledge, and the often contentious and highly unequal power relations this entails. The final chapter in this section, by Diana Davis, considers historical approaches – and the uses of history – in political ecology, and the occasionally complicated relationship between political ecology and allied fields such as environmental history and historical geography.

16

REASSEMBLING THE STRUCTURAL

Political ecology and Actor-Network Theory

Rebecca Lave

Introduction

Political ecologists have engaged with Actor-Network Theory (in both the matrimonial and military senses of the word) since the mid-1990s, and it is now a key theoretical strand of post-structural political ecology. I argue here that this centrality is both surprising and inappropriate given the deep incompatibilities between political ecology's core commitments and Actor-Network Theory's (ANT) conceptualization of actors, networks, and power relations.

I begin with an overview of ANT, including debates about what it actually claims, and the ways it has been incorporated into political ecology analyses. With this context in place, I lay out the central incompatibilities between ANT and political ecology, and the ways in which political ecologists and critical nature/society scholars more broadly have attempted to address them. Finally, I argue that much of what political ecology has borrowed from ANT could be more productively borrowed from other sources with more compatible theoretical commitments to situated subjects and analysis of inequality.

The world according to ANT

ANT arose in the early to mid-1980s in France. Initially, the central figures were Michel Callon, Bruno Latour, John Law, and Arie Rip; today, that list would also include Annmarie Mol. As political ecologists have focused almost exclusively on the work of Bruno Latour, I will do so as well, though it is worth noting that there is some variety within the ANT core.

Latour was a pioneer of post-Mertonian Science and Technology Studies (STS), which focuses on science as a social practice rather than as a superior form of knowledge production. Initially, he conducted ethnographic research describing the construction of "facts" within scientific laboratories (Latour and Woolgar 1986 [1979]), but he soon expanded outward to consider the broader networks that fortify scientific truth claims. These networks consist of humans and nonhumans (including microbes, scientific instruments, and transportation systems, among others), both of whom Latour refers to as *actants*.

Networks are built and extended via processes of *translation*, through which an individual *enrolls* other actants in his network by convincing them that they will achieve their goals more easily by supporting his work, or even that they should adopt his goals as their own. As

Latour describes it, in the translation process, scientists "*enroll others* so that they participate in the construction of the fact [by] ... tailor[ing] the object in such a way that it caters to these people's explicit interests" (1987: 108, emphasis original). That tailoring does not involve outright falsehood, according to Latour, merely distortion, as he describes in relation to Pasteur's work developing a vaccine for anthrax: "As for all translations it is possible and necessary to distort the meanings but not to betray them entirely. Groups that accepted to pass through Pasteur's hands in order to solve their problems nevertheless only go through him to their own ends" (1983: 263).

ANT researchers are enjoined to analyze all actants symmetrically rather than entering into research with *a priori* assumptions about the relative causal efficacy of, for example, humans versus animals versus computers in any given network. The directive to dissolve subject/object dualisms (2005: 76, fn 89) via symmetrical analysis is often described as a profound ontological leveling (Eden et al. 2000; Castree 2002; Kirsch and Mitchell 2004), and is routinely pointed to as one of ANT's key contributions to social theory, as Latour himself notes (2005: 70).

A number of scholars in political ecology and across the social sciences read Latour's insistence on symmetry as attributing agency, in the strong sense of intentional action, to nonhumans ranging from scallops to fax machines (e.g. Swyngedouw and Heynen 2003; Perkins 2007; Robbins 2007; Robbins and Marks 2009). For example, in the preface to his strongly ANT-influenced book, *Lawn People*, Robbins describes his primary goal as, "to provide a novel explanation of how daily life is ... controlled and disciplined by a nonhuman actor – the lawn itself" (2007: xvi). Or as he puts it later in the text, "to the degree that these 'objects' obey their own rules, as is so evident from ever-hungry turfgrass, it is their rules that set the pace and character of subjected lives. They do so tied to the exigencies of capitalist power, to be sure, but with independent, prior, and often ultimate authority" (2007: 135). ANT may not actually be taking that strong of an ontological stance, however, as geographers such as James Murdoch (1997) and Sarah Whatmore (1999, 2002) have argued. More recently, ANT scholar Edwin Sayes has written an admirably lucid analysis of its treatment of agency. Sayes makes a compelling case that, polemical declarations aside, ANT in fact espouses a quite weak view of agency for all actants, one that is "almost empty of meaning" (Sayes 2014: 144). Sayes writes that researchers,

> need only ask of an entity "[d]oes it make a difference in the course of some other agent's action or not? Is there some trial that allows someone to detect this difference?" [Latour 2005: 71] If we can answer yes to these two questions, then we have an actor that *is* exercising agency – whether it is nonhuman or otherwise. It is not the case ... that the human becomes the "standard measure" of agency, but that the "standard measure" of agency becomes dehumanized: the ability to make a difference.
>
> *(2014: 141, emphasis in original)*

Thus the interpretation of ANT as raising nonhumans to human status is inaccurate.

Another central claim of ANT is that "the social" consists only of networks of human and nonhuman actants. According to Latour, ideas such as society, culture, or structures of dominance cannot be used to explain particular outcomes because our social world consists not of these macro-level structures, but of undetermined agglomerations of star-shaped networks. Race, class, and gender are not important social structures; they are the truth claims of sociologists, who have effectively enrolled many into their networks. Latour writes that,

> In the translation model ... there are only heterogenous chains of association ... Analysts who use groups endowed with interests in order to explain how an idea

> spreads, a theory is accepted, or a machine rejected, are not aware that the very groups, the very interests that they use as *causes* in their explanations are the *consequence* of a ratification, extraction, and purification of a handful of links from these ideas, theories or machines ... [I]t is crucial ... to be immunized against the notion that there is a society and "social factors" able to shape, influence, direct or slow down the path of pure science and pure technics.
>
> *(1987: 140–411, emphasis in original. See also Latour 2005)*

Instead, Latour argues, power relations can and should be explained solely on the basis of network size: extensive networks are more powerful, smaller networks less so. Inequalities are thus not the result of structural forces but of the expansion or contraction of networks.

Latour consistently places scientists and their knowledge claims at the center of politics as the catalysts of network production. For example, Latour wraps up his analysis of Pasteur's spectacular network-building success by asserting that, "In our modern societies most of the really fresh power comes from sciences ... and not from the classical political process" (1983: 273). Or, more recently, he argues that the seemingly apolitical elements of the laboratory are the new elements of politics: "A vaccine, an incandescent lamp, an equation, a pollution standard, a building, a blood screening procedure: those were the new means through which politics was being carried out" (2007: 813). Latour argues that science is a particularly effective form of network building, and thus of politics, because of the special character of the laboratory as a testing ground for objects' characteristics, crucially enabled by scientists' ability to spatially displace their objects of study there from the outside world (1983: 272–273). He thus re-privileges science, giving it responsibility for the success of European colonialism (1987: 229), the power of the nation state (1983: 274), and even the possibility of our future collective existence through "political epistemology" (2005: 254).

In constructing this striking analytical framework, Latour develops some very useful vocabulary that is now in wide use in political ecology and the social sciences more broadly. In addition to terms such as actant and translation, defined above, Latour's analytical vocabulary includes:

- *obligatory passage points*, places or actants who have successfully established themselves as central points and/or arbiters in particular networks (an obligatory passage point is, "a routine black box in everyone's hands[;] ... whatever you do, and wherever you go, you have to pass through the contenders' position and help them further their interests" (1987: 120);
- *immutable mobiles*, objects (living or dead) that have been transformed in such a way that they can be transported without degradation or loss and then combined with other immutable mobiles; and
- *centers of calculation*, places where immutable mobiles are brought for processing, enabling the areas where they are located to expand their networks, and thus their influence and power.

On a final note, Latour is often viewed as a strong social constructivist who questions whether there is any material connection between science and the physical world. Other scholars, including Latour himself, challenge that view. There are certainly claims in Latour's work that support a constructivist reading; for example, he defines "facts" not as accurate reflections of nature, but as claims that are too expensive to challenge:

> As long as controversies are rife, Nature is never used as the final arbiter since no one knows what she is and says. But *once the controversy is settled, nature is the ultimate referee* ... [T]he scientists themselves ... from hardcore relativists, have turned into dyed-in-the-wool realists. Nature is now taken as the cause of accurate descriptions of herself. We cannot be more relativist than the scientists about these parts and keep on denying evidence when no else does. Why? Because the cost of the dispute is too high.
>
> *(1987: 97 and 100)*

Yet in his 2004 essay, "Why has critique run out of steam?" Latour argues that social constructivism has gone too far and that social science scholars need to employ more realist views. Many read this as Latour recanting his earlier views; Latour, by contrast, argues that he has never been a strong constructivist (e.g. 2005: 88–93), and there is evidence throughout his work to support the claim that he was analyzing the *process* of fact-making as social, but said nothing about whether the *content* of the claims themselves was constructed.

Political ecologists' use of Latour and ANT

Political ecologists have borrowed widely from Latour, in some cases skimming a key concept or two off the top (e.g. Braun 2002; Ogden 2011), and in others delving deeply into his epistemological and ontological foundations (Robbins 2007; Whatmore 1999). Rather than attempting to address this work comprehensively, I address three exemplars of how Latour's work has influenced research in political ecology and critical nature/society analysis: Eric Swyngedouw's hydro-social cycle (2004), Paul Robbins' analysis of American lawn culture (2007), and Sarah Whatmore's advocacy of hybrid analyses (1999, 2002, and with Lorraine Thorne 1998 and 2000).

Dipping only a toe or two into ANT waters, Swyngedouw's hydro-social cycle framework assimilates elements of ANT – hybridity and networks – into a profoundly political economic framework, putting them very much at the service of analyses of production, accumulation, and social justice. For example, in the Introduction to *Social Power and the Urbanization of Water*, Swyngedouw traces out some of the nodes in the network of urban water, employing ANT methods to an end it is difficult to imagine Latour approving:

> Every body and every thing is ... a mediator, part social, part natural, lacking discrete boundaries and internalizing the multiple contradictory relations that redefine and rework them. Take the example of urban water. Drinking tap water combines the circulation of productive, merchant and financial capital with the production of land rent and their associated class relations; the ecological transformation of hydrological complexes and the biochemical process of purification with the libidinous sensation and the physiological necessity of drinking fluids; the social regulation of access to water with images of clarity, cleanliness, health.
>
> *(2004: 18)*

Swyngedouw clearly recognizes that he is bending ANT against the joint: "Latour's networks and quasi-objects need to be historicized, as following Ariadne's thread through the Gordian knot of socio-nature's networks – as Latour suggests – is not good enough if this is stripped from the process of its historical geographical production" (2004: 21). Thus it is not too surprising that even five years later, Swyngedouw no longer felt the need to draw on Latour in explaining the hybrid character of hydro-social cycles (2009).

If Swyngedouw only tested the ANT waters, Paul Robbins wades right in in his exploration of American lawns and lawn people: the anxious subjects who tend them. Although *Lawn People* (2007) references a range of theorists, the primary scaffolding is a striking combination of Althusser's concept of interpellation and an ANT-derived valorization of nonhuman agency. Turf grass, in Robbins' telling, literally hails lawn people into being as the classic policeman's "Hey you there!" is transformed into plaintive demands for water, fertilizer, weed removal, pesticides, and endless, endless mowing. *Poaceae*, not people, are the primary sources of agency in this picture. Yet Robbins combines this exploration of the dominance of nonhuman agency with analysis of the political economy of lawn care. He also maintains a clear eye on the social justice and environmental implications of lawn culture, asserting that nonhuman agency is a key part of a political, and thus emancipatory, ecology. These more characteristic concerns of political ecology are married to a determinedly ANT stance, as in passages such as:

> The rhythms and behaviors of these neighborhoods, although enforced by human communities, are dictated by the pattern, pace, and specific ecological needs of other species. Lawn grass has at its disposal not merely the labor of individual homeowners (who might at any time neglect to mow or spray for grubs on an ad hoc basis) but instead an entire social machine, organized to enforce and make regular all of the practices necessary for turfgrass growth. Lawn people are, therefore, also perfectly enrolled participants in actor-networks.
>
> *(2007: 116)*

Robbins' feet may still be touching bottom on the political economic shore, but the ANT water has gotten pretty deep.

Critical nature/society scholar Sarah Whatmore, by contrast, dives in and swims decisively away from the political economic mainland in her germinal work on hybrid geographies. She calls out the inadequacies both of approaches that separate the natural from the social, and of Marxist dialectics:

> neither the "bracketing off" of an environmental sub-field common in other disciplines, nor the threadbare promise of a reintegration of physical and human geography, will suffice. Nor, in my view, does recourse to variants of dialectical reasoning centred on the ways in which nature and society interact provide a radical enough basis for critical enquiry ... Far from challenging this a priori categorization of the things in the world, dialectics can be seen to raise its binary logic to the level of a contradiction and engine of history.
>
> *(1999: 25)*

Instead, Whatmore argues for a *hybrid geography*: "an upheaval in the binary terms in which the question of nature has been posed and a recognition of the intimate, sensible and hectic bonds through which people, organisms, machines, and elements make and hold their shape in relation to each other in the business of everyday living" (1999: 26).

Whatmore's advocacy of ANT is by no means slavish, however. Even in her earliest ANT-inflected work she draws on feminist STS as well as Latour, and at times regards ANT with a critical eye, "I do not think that one can, or ought to, look to ANT to provide some sort of ready-made compass. None the less, there are useful beginnings here for journeys out of the impoverished wor(l)d of N/nature" (1999: 30).

These three examples demonstrate the range of political ecologists' engagements with ANT, and also that that engagement often involves supplementation from other theoretical frameworks. Even political ecologists who find ANT alluring seem to have an uneasy sense that it is not sufficient on its own. I turn now to the concerns that spark that discomfort.

Incompatibilities and attempts to address them

There has been a lot of discussion about the relationship (or lack thereof) between ANT and approaches more focused on structures of dominance (e.g. Castree 2002; Fine 2005; Gareau 2005; Hartwick 2000: 1180–1182; Holifeld 2009; Kirsch and Mitchell 2004; Latour 2005; Mann 2011: 80–83; Perkins 2007; Rudy 2005; Walker 1997; Whatmore 1999; Winner 1993). I address three primary incompatibilities here: (1) ANT's categorical denial of structural inequalities, (2) the uncertain political implications of approaching humans and nonhumans symmetrically, and (3) the neoliberal flavor of Latour's conceptualization of actors.

One core difference between the fields arises from Latour's denial of society and structural power. Instead, he describes society as an unstructured agglomeration of networks; the relations of dominance that political ecologists consistently analyze are explicitly excluded from ANT analyses. A Latour-inspired analysis of a development project would focus on the construction and extent of the network supporting the project, not on the legacies of colonialism, global political economy, racism, and sexism that shape who is enrolled and what level of influence they are accorded within that network. This presents a stark contrast with most political ecology analyses. As Noel Castree notes in relation to political economic analysis in particular, for many critical nature/society scholars, "to scrutinize society-nature relations in abstraction from processes of capital accumulation is to miss a vital aspect of their logic and consequences" (2002: 123). Political ecologists have, in Latour's words, an "infatuation with emancipation politics" (2005: 52).

A second stark difference stems from ANT's insistence on dissolution of subject/object dichotomies and the symmetrical treatment of humans and nonhumans, an ontologically radical position that is often described as one of ANT's core contributions to social theory. The key point I wish to make here is that there *is no necessary correlation between the ontologically radical and the politically radical,* as scholars such as Whatmore (1999), Castree (2002), and Robbins (2007) imply. Insisting on the ontological equivalence of nonhuman entities, such as ocean currents, does not obviously support emancipatory struggles for human beings suffering oppression. How, for example, would a class analysis of worker bees and colony collapse disorder help us develop deeper understanding of "political strategies and alliances possible for subaltern classes" (Robbins and Marks 2009: 188)? As Elaine Hartwick succinctly states, "I find the notion of nonhuman actants, such as fax machines, having as much of an active role as workers intensely, deeply troubling. What sort of 'radical' politics does this produce? A union for fax machines? 'Fax machines unite, you have nothing to lose but your electrical cords'?" (2000: 1181).

A third difference has received far less attention, but is to my mind equally confounding: the actors at the heart of Actor-Network Theory are quite different from the situated subjects that form the analytical core of political ecology. As historian of economics Chris McClellan notes, the "A" in ANT describes rationalizing, interest-maximizing actors strongly reminiscent of *homo economicus* and the core subjectivities espoused in neoliberal (and neoclassical) theory (1996: 203; see also Hayden 2003: 21).

Latour himself has explicitly denied this charge in *Reassembling the social,* his most recent introduction to ANT. There he describes subjects as star-shaped agglomerations of connections: actor-networks in and of themselves (e.g. 2005: 216–217). He does not explain the processes of translation and connection-building through which these actor-networks are formed, though, except perhaps obliquely in his use of puppeteers as a metaphor for the dynamics of actor-networks:

> what was wrong with the metaphor of the marionettes was not their activation by the many strings firmly held in the hands of their puppeteers, but the implausible argument that domination was simply transported through them without translation. Of course marionettes are bound! But the consequence is certainly not that to emancipate them, you have to cut all the strings. The only way to liberate puppets is for the puppeteer to be a *good* puppeteer.
>
> *(2005: 215)*

It seems that there is something acting at the center of the constellation (the puppeteer analogy), but in *Reassembling the social* Latour never spells out how that individual selects its connections and convinces them to be bound into its network. This is not an uncommon lacuna in ANT studies; the relative paucity of treatments of enrollment and translation in ANT has been noted not only by critics, but also by supporters, including his fellow ANT theorist John Law (1999).

The clearest and most detailed explanation of the process of translation is in Latour's initial introduction to ANT, *Science in action* (1987). I have quoted Latour's descriptions of the translation process above (pp. 213–214). McClellan summarizes it as follows, using Latour's vocabulary:

> At the center of a network is an actor who builds her network though the process of enrolment; that is, by translating the interests of other actors and actants to commensurate with her own in order to facilitate the exchange of interests that takes place as the link is generated. The actor at the center takes the position of spokesperson for the network, thereby controlling the network and the entities of which it is comprised.
>
> *(1996: 201)*

This description of interested actors enrolling others based on translations of their interests has clear resonance with the self-interested action of *homo economicus*. In the 1986 volume that contributors, including Latour, viewed as introducing the developed form of ANT, editors Callon, Law, and Rip called out this parallel clearly, claiming that, "the behavior of the scientists studied conforms in every way to that of the classical entrepreneur" (1986: 9).

With its denial of social structure and a suspiciously *homo economicus*-like subject at its core, ANT can be comfortably appropriated into neoliberal organizations and research agendas. And it has been, to the deep discomfort of critical nature/society scholars who regard ANT as a politically radical theoretical approach. When ANT's use as an instrument for organizing neoliberal voluntary governance structures came up during the keynote session at the 2013 Dimensions of Political Ecology conference, for example, there were shocked gasps from the audience. The very first question asked of the speaker, Ariel Salieh, could be sanitized as, "*We use* ANT; why are those neoliberal jerks picking it up?!?" The work of McClellan (1996), Mirowski and Nik-Khah (2007) and others in the STS community suggests strongly that the link between ANT and neoliberal policy Salieh described is not a one-time coincidence, but a core congruence that calls the radical *bona fides* of ANT sharply into question.

What then should we do with these stark differences in the core commitments and analytical frameworks of political ecology and ANT? Political ecologists and critical nature/society scholars more generally have taken three primary approaches. Some combine the two without attempting to modify either, grafting ANT branches onto trunks marked by political ecology's attention to political economy and injustice (Holifeld 2009; Robbins 2007; Swyngedouw 2004). Given the depths of the differences involved, I would argue that this type of grafting leads to a certain level of intellectual incoherence. A second group proclaims the profound incompatibility of ANT and political ecology's political economic roots. In a stirring pro-ANT polemic, for example, Sarah Whatmore (1999) flatly denies the potential for reconciling the two approaches, while political economists have made similarly Manichean arguments against rapprochement (Fine 2005; Hartwick 2000: 1180–1182; Mann 2011: 80–83; Rudy 2005; Walker 1997).

A third set of scholars attempt to smooth the differences between ANT and political economic approaches, filing the edges down from jagged incompatibilities to complementary strengths (Castree 2002; Gareau 2005; Kirsch and Mitchell 2004; Perkins 2007). Perkins (2007) and Kirsch and Mitchell (2004) both engage ANT through questions of labor; the latter in particular provides an elegantly argued treatment of the relations between ANT's ontological claims about nonhuman agency and Marx's conceptualization of dead labor. Noel Castree's 2002 article "False antitheses?" provides one of the most thorough and thoughtful attempts to fit ANT and political economic analyses together, though as Castree notes, this requires "weakening" ANT to a form that may no longer be recognizable as such.

To some scholars, then, an ANT/political ecology synthesis is within reach. By contrast, I would argue that even if the project of watering political ecology and ANT down to compatible forms is possible, it does neither side any favors. Sometimes different strengths are just that: different. Softening up ANT's conceptualization of the social world in order to make it compatible with political ecology's central values takes away some of its audacious, "look Ma, no hands!" analytic power. And I cannot imagine how political ecology's social justice agenda is advanced by placing a neoliberal subject at its core.

Implications for future research: reassembling the structural

While I do not see an easy or appropriate synthesis between political ecology and ANT, there are replacements for the most compelling aspects of ANT ready to hand. ANT is part of a larger shift in the social sciences which is often described as "the new materialism." In an excellent critical review Bakker and Bridge locate the attraction of new materialist approaches, "in a common desire to shift our frames of reference by saying something quite simple: things other than humans make a difference in the way social relations unfold" (2005: 17–18). As Robbins and Marks note, ANT approaches "are by no means the only way to re-embrace the material in Geography" (2009: 192). Bakker and Bridge (2005), Robbins and Marks (2009), and Whatmore (2006) highlight a range of potential guides to new materialist intellectual practice, from Timothy Mitchell to Judith Butler to Isabelle Stengers. I focus here on Donna Haraway and her writing on naturecultures, but there are clearly other useful materialist approaches.

Haraway's work is not cited in political ecology anywhere near as commonly as ANT. When present at all, Haraway references tend to serve merely as shorthand for constructivist approaches to science. For example, variations on Haraway's phrase "Biology is a discourse, not the living world itself" (1992: 298) can be found in Braun (2002: 225), Demeritt (1994: 177), and Forsyth (2003: 89), though unfortunately only Braun cites the original source. Less frequently, but more distressingly, Haraway is sometimes portrayed as a sort of Latour handmaiden either explicitly (Holifield 2009), or implicitly by referencing her work only in the same sentence as Latour. Her theoretical framework, however, is both complex and quite different from Latour's, in ways that align far better with political ecology's core commitments.

Haraway's work includes trenchant critiques of nature/culture dualisms as evidenced in her refusal to separate the two words, referring to them instead as *naturecultures*. This hybrid vision has had several avatars over the years, most prominently the cyborg (1991) and the companion species (2003). What sets Haraway's analysis of hybridity so clearly apart from Latour's is her insistence on the classed, raced, gendered, and deeply historical co-production of the human and nonhuman (2003, 2007; see also Robbins and Marks 2009: 185), adjectives that resonate with political ecologists' common concerns. In Haraway's analysis, a dog is not enrolled in a particular human's network through a successful translation process whereby one actant convinces another independent actant to join its cause; instead dogs and humans have changed each other, behaviorally and genetically, in ways indelibly marked by class privilege and histories of colonialism, racism, and sexism, so that there is no independent, pre-relational entity to enroll. Privilege and prejudice serve as touchstone and refrain, as humans and nonhumans co-produce each other over evolutionary time periods.

Haraway writes that companion species are engaged in "co-constitutive relationships in which none of the partners pre-exist the relating and the relating is never done once and for all. Historical specificity and contingent mutability rule all the way down, into nature and culture, into naturecultures" (2003: 12). This,

> multi-species, relentlessly complex legacy … crosses evolutionary, personal, and historical time scales of companions species. Every registered breed, indeed every dog, is immersed in practices and stories that can and should tie dog people into myriad histories of living labor, class formations, gender and sexual elaborations, racial categories, and other layers of locals and globals.
>
> *(2003: 96–97)*

In addition to reconceptualizing hybridity and nature/culure dualisms along post-structuralist lines, Haraway's work on companion species speaks to the ontological question of nonhuman agency. Unlike ANT, however, Haraway's take on the agency of nonhumans is unambiguous: dogs are not people, and the differences in power between us have profound ramifications (2003: 11–12, 21). Our relations with them may change us, physically and emotionally, but humans are the dominant partner. This clear recognition of the power dynamics inherent in significant othering strikes me as a much more productive basis for an emancipatory politics of the Anthropocene.

A final compelling component of ANT for many political ecologists is its methodological focus on networks. Yet there are many other ways of analyzing linked groups of humans and non-humans – from Arjun Appadurai's social life of things (1988) to classic commodity chain analyses – that can enable political ecologists to address connections between humans and nonhumans without abandoning our focus on injustice. Moving even closer to home, long-standing political ecology approaches, such as chains of explanation, can easily be expanded to include nonhuman links (e.g. Galt 2011; Guthman 2004); Blaikie and Brookfield (1987) were deeply concerned with ecology and the natural world, so it does no violence to their conceptual framework to align it with current materialist concerns.

Bluntly, it is time to retire ANT as a core element of the political ecology tool kit. ANT is incompatible with many of political ecology's core commitments, and the insights it does offer can be productively replaced from other sources. We have nothing to lose but our networks!

References

Appadurai, A. (1988) *The social life of things*, Cambridge: Cambridge University Press.

Bakker, K. and G. Bridge (2005) "Material worlds? Resource geographies and the 'matter of nature'," *Progress in Human Geography*, vol. 30, no. 1, pp. 5–27.

Blaikie, P. and H. Brookfield (1987) *Land degradation and society*, London: Methuen.

Braun, B. (2002) *The intemperate rainforest: Nature, culture and power on Canada's West Coast*, Minneapolis: University of Minnesota Press.

Callon, M., J. Law, and A. Rip (eds.) (1986) *Mapping the dynamics of science and technology*, London: Macmillan.

Castree, N. (2002) "False antitheses? Marxism, nature and actor-networks," *Antipode*, vol. 34, pp. 111–146.

Demeritt, D. (1994) "The nature of metaphors in cultural geography and environmental history," *Progress in Human Geography*, vol. 18, no. 2, pp. 163–185.

Eden, S., S. Tunstall, and S. Tapsell (2000) "Translating nature: river restoration as nature-culture," *Environment and Planning D: Society and Space*, vol. 18, no. 2, pp. 257–273.

Fine, B. (2005) "From actor-network theory to political economy," *Capitalism, Nature, Socialism*, vol. 16, no. 4, pp. 91–108.

Forsyth, T. (2003) *Critical political ecology: The politics of environmental science*, London: Routledge.

Galt, R. E. (2011) "Circulating science, incompletely regulating commodities: Governing from a distance in transnational agro-food networks," in Goldman, M., P. Nadasdy, and M. Turner (eds.) *Knowing nature: Conversations at the intersection of political ecology and science studies*, Chicago: University of Chicago Press, pp. 227–243.

Gareau, B.J. (2005) "We have never been human: Agential nature, ANT and Marxist political ecology," *Capitalism Nature, Socialism*, vol. 16, no. 4, pp. 127–140.

Guthman, Julie (2004) *Agrarian dreams: The paradox of organic farming in California*, Berkeley: University of California Press.

Haraway, D. (1991) *Simians, cyborgs, and women: The reinvention of nature*, New York: Routledge.

Haraway, D. (1992) "The promises of monsters: A regenerative politics for inappropriate/d others," in Grossberg, L., C. Nelson, and P. Treichler (eds.) *Cultural Studies*, New York: Routledge, pp. 295–337.

Haraway, D. (2003) *The companion species manifesto*, Chicago: Prickly Paradigm Press.

Haraway, D. (2007) *When species meet*, Minneapolis: University of Minnesota.

Hartwick, E. (2000) "Towards a geographical politics of consumptions," *Environment and Planning A*, vol. 32, pp. 117–192.

Hayden, C. (2003). *When nature goes public: the making and unmaking of bioprospecting in Mexico*, Princeton: Princeton University Press.

Holifeld, R. (2009) "Actor-Network Theory as a critical approach to environmental justice: A case against synthesis with urban political ecology," *Antipode*, vol. 41, no. 4, pp. 637–658.

Kirsch, S. and D. Mitchell (2004) "The nature of things: Dead labor, nonhuman actors, and the persistence of Marxism," *Antipode*, vol. 36, no. 4, pp. 687–705.

Latour, B. (1983) "Give me a laboratory and I will raise the world," in Biagioli, M. (ed.) *The science studies reader*, London: Routledge, pp. 258–275.

Latour, B. (1987) *Science in action: How to follow scientists and engineers through society*, Cambridge, MA: Harvard University Press.

Latour, B. (1993) *We have never been modern*, Cambridge, MA: Harvard University Press.

Latour, B. (2004) "Why has critique run out of steam? From matters of fact to matters of concern," *Critical Inquiry*, vol. 30, pp. 225–248.

Latour, B. (2005) *Reassembling the social: An introduction to Actor-Network Theory*, Oxford: Oxford University Press.

Latour, B. (2007) "Turning around politics: A note on Gerard de Vries' paper." *Social Studies of Science*, vol. 37, no. 5, pp. 811–820.

Latour, B. and S. Woolgar (1986 [1979]) *Laboratory life: The social construction of scientific facts*, Princeton: Princeton University Press.

Law, J. (1999) "After ANT: Complexity, naming and topology," in Law, J. and J. Hassard (eds.) *Actor Network Theory and after*, Oxford: Blackwell, pp. 1–14.

Mann, G. (2011) "Economie$," in Del Casino Jr., V., M. Thomas, P. Cloke, and R. Panelli (eds.) *A companion to social geography*, Oxford: Blackwell, pp. 72–91.

McClellan, C. (1996) "The economic consequences of Bruno Latour," *Social Epistemology*, vol. 10, no. 2, pp. 193–208.

Mirowski, P. and E. Nik-Khah (2007) "Markets made flesh: Performativity, and a problem in science studies, augmented with consideration of the FCC Auctions," in MacKenzie, D., F. Muniesa, and L. Siu (eds.) *Do economists make markets? On the performativity of economics*, Princeton: Princeton University Press, pp. 190–224.

Murdoch, J. (1997) "Inhuman/nonhuman/human: Actor-Network Theory and the prospects for a nondualistic and symmetrical perspective on nature and society," *Environment and Planning D: Society and Space*, vol. 15, no. 6, pp. 731–756.

Ogden, L. (2011) *Swamplife: people, gators, and mangroves entangled in the Everglades*, Minneapolis: University of Minnesota Press.

Perkins, H. (2007) "Ecologies of actor-networks and (non)social labor within the urban political economies of nature," *Geoforum*, vol. 38, pp. 1152–1162.

Robbins, P. (2007) *Lawn people: How grasses, weeds, and chemicals make us who we are*, Philadelphia: Temple University Press.

Robbins, P. and B. Marks (2009) "Assemblage geographies," in Smith, S., R. Pain, S. Marston, and J.P. Jones III (eds.) *The Sage handbook of social geographies*, London: Sage, pp. 176–194.

Rudy, A.P. (2005) "On ANT and relational materialisms," *Capitalism, Nature, Socialism*, vol. 16, no. 4, pp. 109–125.

Sayes, E. (2014) "Actor-Network Theory and methodology: just what does it mean to say that nonhumans have agency?" *Social Studies of Science*, vol. 44, no. 1, pp. 134–149.

Swyngedouw, E. (2004) *Social power and the urbanization of water: Flows of power*, Oxford: Oxford University Press.

Swyngedouw, E. (2009) "The political economy and political ecology of the hydro-social cycle," *Journal of Contemporary Water Research & Education*, no. 142, pp. 56–60.

Swyngedouw, E. and N. Heynen (2003) "Urban political ecology, justice and the politics of scale," *Antipode*, vol. 35, pp. 898–918.

Walker, R. (1997) "Field of dreams," in Goodman, D. and M. Watts (eds.) *Globalising food*, London: Routledge.

Whatmore, S. (1999) "Hybrid geographies: Rethinking the 'human' in human geography," in Massey, D., J. Allen, and P. Sarre (eds.) *Human geography today*, Cambridge: Polity Press, pp. 22–39.

Whatmore, S. (2002) *Hybrid geographies: Natures cultures spaces*, London: Sage.

Whatmore, S. (2006) "Materialist returns: Practising cultural geography in and for a more-than-human world," *Cultural Geographies*, vol. 13, pp. 600–609.

Whatmore, S. and L. Thorne (1998) "Wild(er)ness: Reconfiguring the geographies of wildlife," *Transactions of the Institute of British Geographers*, vol. N.S. 23, no. 4, pp. 435–454.

Whatmore, S. and L. Thorne (2000) "Elephants on the move: Spatial formations of wildlife exchange," *Environment and Planning D: Society and Space*, vol. 18, pp. 185–203.

Winner, L. (1993) "Upon opening the black box and finding it empty: Social constructivism and the philosophy of technology," *Science, Technology & Human Values*, vol. 18, pp. 362–378.

17

THE PROMISES OF PARTICIPATION IN SCIENCE AND POLITICAL ECOLOGY

David Demeritt

In our technological age, the sciences occupy an increasingly contested and contradictory position in environmental politics and policy. Consider the case of stratospheric ozone depletion. Here science is at once the means for knowing there is a problem and the source of potential solutions to it. At the same time, however, science, in the form of CFCs, is also the ultimate cause of the problem in the first place. Ulrich Beck is far from alone in seeing a paradox here. In the face of global environmental changes that seem to make them "more and more necessary," the sciences are "at the same time, less and less sufficient for the socially binding definition of truth" (Beck 1992b: 156).

This same ambivalence runs through the very heart of political ecology. The field emerged in the 1970s and 1980s to offer a more critically inflected and social scientifically based diagnosis of environmental change and the conflicts to which it often gives rise. In contrast to purely technical analyses of environmental change, political ecologists emphasized its "political sources, conditions and ramifications" (Bryant 1992: 13). To that end, they critiqued the way that science is often used to naturalize inequality and exploitative socio-natural arrangements. Drawing on currents in academic science studies and poststructural theory, they challenged the objectivity of science and its epistemic privilege to define what passes for nature (Demeritt 1998).

This gave political ecologists a very powerful way to critique established socio-natural arrangements, but it also left their preferred remedies vulnerable to the very same maneuver. Thus demands for reductions in greenhouse gas emissions have met with counter-charges that climate science is too unsettled, socially constructed, and politically tainted to justify any such action (Demeritt 2006). A number of political ecologists now worry that the pendulum of constructivist critique has swung too far and that there is too much politics and not enough science and ecology in political ecology (Walker 2005; Forsyth 2003). The fear is that without some authoritative basis for knowing about our environmental problems it will be impossible to devise appropriate and publicly acceptable measures to deal with them. This, of course, is the dilemma Ulrich Beck identified as the defining feature of an emergent risk society marked by far-reaching reflexivity and public debate about the grounds for truth and for political action.

One increasingly common response to this challenge is to call for more public engagement in science and science-based policy making. The political ecologist Tim Forsyth (2011: 44), for instance, recently declared, "environmental science must necessarily become more deliberative than commonly practiced … [T]he objectives and basic framings used to underpin scientific research will need to be opened to greater scrutiny." Such calls build on a long-standing tradition of participatory research methods in political ecology for tapping into so-called traditional ecological knowledge (Chapter 18, this volume), but they are also aligned to some broader claims in human geography and political theory about the need to democratize science through public participation and deliberation (Brown 2009; Whatmore 2009). Such calls are no longer restricted to the ivory towers of academe. From natural resource management to medicine, the rhetoric of public engagement, participation, and dialogue has become something of a mantra across a wide sweep of policy fields that were once the exclusive preserve of scientific experts.

However, as such calls have become more widespread, the reasons for them have become more diffuse and poorly defined (Demeritt 2005). The aim of this chapter, therefore, is to clarify the meaning and purposes of public engagement in science and political ecology. In so doing I will try to bring participatory currents in political ecology into conversation with a much wider body of work in academic science studies, political theory, and policy practice on the promises and perils of participation.

Why encourage public participation in science and political ecology?

Calls for public engagement and participation are underwritten by three distinct, if also often intertwined, rationales. First, there are normative arguments about participation as a fundamental democratic right (e.g. Brown 2009; Whatmore 2009). As the Royal Commission on Environmental Pollution (RCEP 1998: 102) put it, "Those directly affected by an environmental matter should always have an accepted right to make their views known before a decision is taken." Defending this right animates much work in political ecology (Peet and Watts 2004). For example, a rich vein of research on nature reserves and biodiversity conservation has exposed the continuity of "fortress-style" conservation with colonial practices of indigenous dispossession (e.g. Neumann 1998; Peluso 1992). In place of often oppressive systems of natural resource management by (and often *for*) scientific elites, political ecologists have promoted community-based resource management as a more just alternative (Brosius et al. 2005; Zimmerer and Young 1998). Participation gives local people a voice, and so is consistent with procedural ideas of environmental justice as recognition. It is also more attuned to local livelihoods and so is arguably also better placed to secure the just outcomes emphasized by consequentialist theories of environmental justice (Shrader-Frechette 2002).

Second, in response to these normative demands for public participation, many government agencies are themselves now trying to incorporate more participation by the public in their science and science-based policy making, albeit often for instrumental as much as normative reasons. Participation was a mainstay of Agenda 21, which was formulated at the 1992 UN Conference on Environment and Development (WCED 1987) and has since become a central plank for many international conservation initiatives (Adams 2001). According to the UK government's Council for Science and Technology (CST 2005: para 11), public engagement offers "a more efficient means of developing broadly acceptable policies for issues where the problem of public consent is real, and which cannot readily be sidestepped by a quick fix or political sleight of hand."

It should be clear what a departure this marks from long-standing traditions of technocratic policy making in which public opposition to any science-based proposals was attributed to public ignorance and a deficit of scientific knowledge (Sturgis and Allum 2004). The CST (2005: para 6) now acknowledges that "public concerns can rarely be reduced simply to scientific issues," even as it frames dialogue and public engagement as ways of "increasing ...

public acceptance of specific policy decisions" (CST 2005: para 15). As Alan Irwin (2006: 306) has observed, the participatory turn in science-based policy making has sometimes failed to acknowledge how public dialogue can "create further grounds for criticism and concern" rather than political consensus. While there is now a growing literature in science studies evaluating public engagement schemes and offering best practice recommendations (i.e. Chilvers 2009), political ecologists have tended to follow Foucault in critiquing these instrumental visions of participation as disciplinary mechanisms for molding individuals into self-regulating "environmental subjects ... for whom the environment constitutes a critical domain of thought and action" (Agrawal 2005: 17) In this guise, participation has sometimes been condemned as a new tyranny that coopts people into their own subjugation (Cooke and Kothari 2001).

Third, normative and instrumental rationales for public participation often find common ground in the seductive promise that it will also increase the quality of science and science-based policy. For instance, a recent US National Academy of Sciences report trumpets the importance of public participation in environmental assessment and decision-making for "getting the science right" (Dietz and Stern 2008: 50). Likewise, many political ecologists insist that community participation in natural resource management will also lead to more effective and ecologically sensitive forms of environmental conservation than coercive systems of technocratic management by scientific experts (Adams 2001; Brosius et al. 2005).

Such claims about the substantive contributions of public participation to science and science-based policy are beset by some fundamental ambiguities. To explore them further, I want to return for a moment to Ulrich Beck, both because his theory of reflexive modernization is influential in its own right and because it starkly illustrates the ambiguity about the wider claims made in science studies and political ecology about the value and purpose of public participation in science and science-based policy.

Beck (1992a: 119) writes:

> The public sphere, in co-operation with a kind of "public science" would be charged as a second centre of "discursive checking" of scientific laboratory results.

While superficially attractive, this vision of the public sphere's engagement with science begs some important questions. What kind of "discursive checking" does Beck hope the public will perform in his "upper house"? As I discuss below, Beck's vision of the substantive contributions of public participation can be understood in two quite different ways.

Participation as normative steering

One way to read Beck's vision of participation is as a kind of normative steering of science (see Figure 17.1). Here the role for the public upper house would be to apply the normative "standard 'How do we wish to live?' to scientific plans" (Beck 1992a: 119). In this role, the public or political sphere is responsible for regulating the techno-scientific innovation undertaken in the lower house of science. This vision depends on already established distinctions between the scientific work of discovery and the political work of agreeing on the values to regulate its development and application. Conventionally this normative steering has come after the fact (Figure 17.1a), in the form of restrictions on the socially acceptable use of technology. As such Beck may sound more like a description of the status quo than some new, more reflexive modernization, but this apparent contradiction might be resolved if it were read as a call for "upstream" public engagement in science itself (Figure 17.1b).

Upstream public engagement was popularized in an influential pamphlet from the London think-tank Demos (Wilsdon and Willis 2004). It called for engagement with the public to be moved "upstream" into the heart of the scientific research process where research agendas can be shaped and steered in publicly acceptable ways, rather than, as has been more typical of "downstream" public consultations, waiting until after the invention of new technologies before worrying about how to regulate them. For political ecologists, the idea of upstream public engagement might help to formalize and thereby strengthen the role for the public in their otherwise often rather vaguely articulated appeals to participatory action research as a research methodology.

Rather than dissolving entirely the distinctions between science and politics, this vision of participation as normative steering would make the institutional boundaries between them more porous while at the same time preserving the epistemic distinction between facts and values. The role for the public would be assessing "the values, visions and assumptions that usually lie hidden [i]n the theatre of science and technology" (Wilsdon and Willis 2004: 24). Public participation here serves a normative role, steering the direction science goes and deciding what goods science should serve, not the epistemic one of judging sound science or evaluating the truth of its epistemic claims. This is a reformist, rather than a radical agenda, and it is one that is already coming to fruition, in the form of ethical review by institutional review boards and research ethics committees (Dyer and Demeritt 2009), various "nanodialogues" and other participatory technology assessments (Chilvers 2009), and citizens' juries and other forms of public engagement in science-based policy making (Dietz and Stern 2008), to name just a few.

Two problems, at once of principle and of practice, plague the ideal of participation as normative steering. The first is about representation. How should participants be chosen to

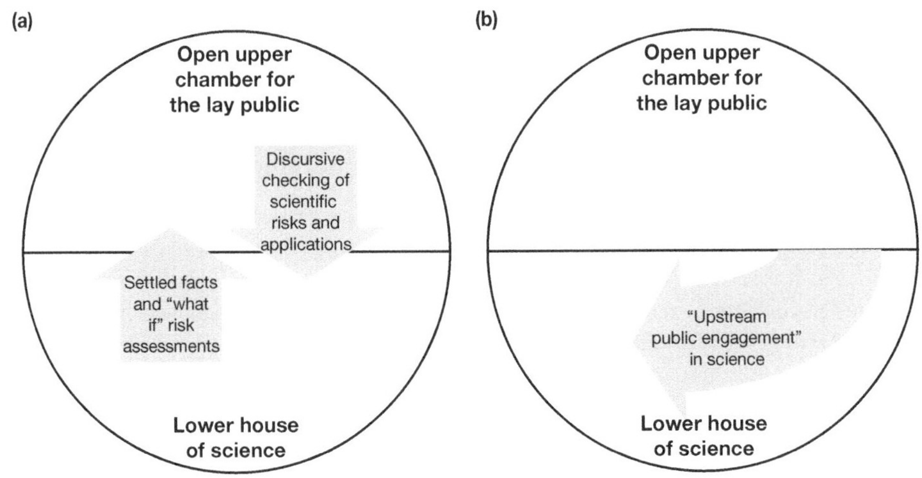

Figure 17.1 Beck's vision of participation as "normative steering" of science. (a) Conventionally this normative steering comes after the fact, in the form of regulations enacted by the upper public, or political, sphere on the socially acceptable use of technological innovations generated by the lower house of science (b) Moving public engagement farther "upstream" into the research process is intended to provide more effective normative steering of science, blurring the institutional boundaries between science and politics while preserving the metaphysical distinction between facts and values.

ensure that their normative judgments reflect those of the wider public they serve and represent? It is difficult to scale up from small scale deliberative fora, such as the citizen juries conducted in the UK on behalf of the Committee for Radioactive Waste Management (Chilvers 2005), to larger scale national decisions about where and how to store nuclear waste or whether to commission another generation of nuclear reactors. One persistent complaint about public engagement exercises is that they fail to represent the views of the "silent majority" (Irwin 2006). Similarly political ecologists have noted that participation in community resource management schemes is often skewed towards local elites and can reinforce existing inequalities based on caste, class, and gender (Agrawal 2005). But that same critique might also be turned inwards on political ecology itself, whose paternalistic tradition of radical vanguardism has not always encouraged reflexivity about the effects of its own interventions.

A second closely related problem is about democratic accountability. How can the public license the decisions taken by participants acting in its name but, unlike elected officials, not directly accountable to it through the ballot box? In its response to the CST (2005) report, the UK government enthusiastically endorsed the CST recommendation that the purpose of public dialogues on science "is not to determine but to inform policy … Government must retain responsibility for decision-making." In a representative democracy, governments are accountable in ways that focus group participants are not. In practice, however, the institutional imperative for participatory exercises is often precisely to create enough distance between elected officials and controversial regulatory decisions to allow for blame avoidance and political deniability. Rather than confronting this problem of political accountability, participation can serve to exacerbate it by adding a new layer of unelected and therefore unaccountable representatives from the lay public to already unelected and weakly accountable regulatory bodies like the HFEA. Indeed, as Rothstein (2007) notes, it is precisely among such unelected and weakly accountable arms of the regulatory state where the enthusiasm for public engagement has been greatest. Likewise in a development context, critics of the participatory turn worry that it tends to reinforce rather than resist hegemonic power (Cooke and Kothari 2001).

But as I have already noted, there is another, much more radical way to understand the substantive contributions of public engagement in science and science-based policymaking.

Participation as epistemic checking

Public participation can also be understood as offering an epistemic challenge to scientific authority (Figure 17.2). For instance, Beck writes:

> Only a strong competent public debate, "armed" with scientific arguments, is capable of separating the scientific wheat from the chaff.
>
> *(1992a: 119)*

Read in this context, Beck's (1992a: 119) claim about the role of the public sphere "as a second centre of 'discursive checking' of scientific laboratory results" can be understood as a truth determining one. Here the role for the public would be to double-check the factual claims made by the lower house of science. This amounts to a conventionalist theory of truth, in which, after Rorty (1991: 23), science is understood as a form of solidarity and rather than a method of objective inquiry, and "'truth' is simply a compliment paid to the beliefs we think so well justified that for the moment further justification is not needed." In such a world, scientific debate flows seamlessly into political debate. Indeed, the difference between them fades away altogether as epistemological and institutional divides between science and

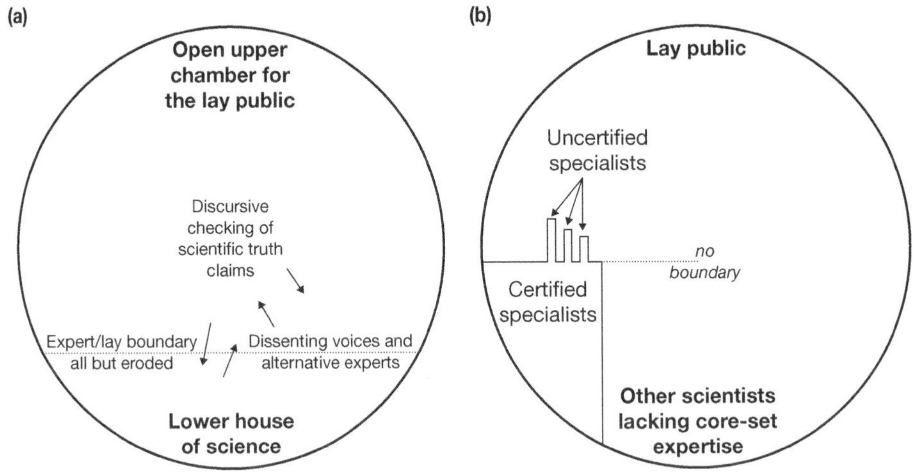

Figure 17.2 Beck's vision of participation as "epistemic checking" of science. (a) In the very strongest versions, the expert/lay boundary is all but eroded, as truth becomes a matter of convention and scientific experts lose their special epistemic warrant. (b) Collins and Evans (2002) preserve the epistemic authority conventionally granted to experts, by refining its basis in experience. They grant epistemic warrant to some uncertified specialists from among the lay public while at the same time denying it to credentialed scientists lacking expertise in the technical matters at hand.

politics, facts and values, are dissolved within an enlarged and invigorated public sphere (Figure 17.2a).

Beck, of course, is far from alone here in arguing that the boundaries between science and politics have been irreparably breached. This is a central theme for a generation of academic science studies (cf. Demeritt 1996) and of poststructural political ecology. Escobar (1998: 54–55), for example, rejected the idea that biodiversity is "a true object that science progressively uncovers," insisting instead that it is a social "construct around which a complex discourse of nature is being deployed … [to] anchor an entire apparatus for the dispersion of new truths throughout vast social domains." However, critics (i.e. Forsyth 2003) worry that constructivists leave political ecology with no way to distinguish warranted belief from mere opinion: debate about environmental degradation can always be extended by dissenters, however ignorant, ill-informed, or duplicitous their claims.

Harry Collins and Robert Evans (2002) call this the "problem of extension," and climate change provides a good example of the difficulties it creates. Notwithstanding the robust scientific consensus to the contrary, a host of conservative think tanks and industry-funded political action committees have spent millions in a slick public relations campaign to deny the risks posed by rising concentrations of greenhouse gases from fossil fuel consumption (Demeritt 2006; cf. Chapter 23, this volume). Pointing specifically to such special interest organized skepticism, Collins and Evans (2002: 280) ask:

> do we never want to say that the tobacco industry has for years falsified … epidemiological studies out of a concern for selling more cigarettes? Do we want to say, rather, that this was just [their] point of view and that the only fight there is to be had with them is a political fight, not a scientific fight?

To solve this problem of extension, Collins and Evans offer a more carefully differentiated definition of expertise, emphasizing experience rather than formal scientific qualifications, as the basis for warranting knowledge claims (Figure 17.2b). Their approach to expertise provides a basis for some public involvement in epistemic checking of scientific claims by valorizing the knowledge of uncertified specialists from among what had been previously regarded as a uniformly ignorant and unqualified lay public. At the same time Collins and Evans also insist that having a PhD in one specialist area does not qualify you as an expert in others. As a result the line in Figure 17.2b demarcating an expert-scientific realm from a public-political one is jagged to take in "the odd-shaped pockets of expertise found among the lay public" (251) and exclude scientists not possessing the particular expertise necessary to answer the scientific question at hand.

Their idea of uncertified expertise provides one justification for the claim, now widespread in political ecology and science studies, that "public engagement can be essential for 'getting the science right'" (Dietz and Stern 2008: 50). In political ecology, participatory mapping and GIS are now firmly established (Chapter 19, this volume), building on traditions of participatory rural appraisal for involving poor, often marginalized and non-literate groups as co-equal partners in development planning (Chambers 1994). Similar participatory approaches are now being developed to engage non-scientists in the design, testing, and validation of computer simulation models that are so important for many areas of environmental science and policy (e.g. Lane et al. 2011; Millington et al. 2011). Public engagement methods are an important addition to the political ecology toolbox of techniques for exploring the integration of nature and society (Chapter 11, this volume), but political ecologists have not always been as clear as Collins and Evans about what, how, and why public participation might be expected to contribute to science and science-based policy or to political ecology studies of them.

There are at least two distinct ways in which participation might make substantive, epistemic contributions to steer science and political ecology. First, members of the public might serve as sources of scientific data and thereby contribute to empirical discovery. Though the spatial imaginary of so-called "traditional ecological knowledge" typically locates it in less developed parts of the world (cf. Agrawal 1995), in both the United States and Europe there have been efforts to enroll the expertise of amateur naturalists in measuring wildlife populations through programs such as the Great Backyard Bird Count (Toogood 2013). Similarly user-generated data from flood victims is also being used to improve the measurement of extreme flood events, which are otherwise hard for scientists to measure (Demeritt and Nobert 2014; Lane et al. 2011; Parkes et al. 2013). Political ecologists have long celebrated the validity and value of traditional ecological knowledge (Chapter 18, this volume), but their embrace of participation has also been motivated by the emancipatory desire to move away from extractive relationships to ones of co-production in which research subjects are treated as partners and involved not just in generating data but also in deciding what counts as true and valid.

This points to a second way in the public might contribute to what Funtowicz and Ravetz (1993) call "extended peer review." Particularly for "post-normal" problems "where facts are uncertain, values in dispute, stakes high and decisions urgent" (744), Funtowicz and Ravetz (1993) insist that "public agreement and participation, deriving essentially from value commitments, will be decisive for the assessment of risks and the setting of policy" (751). In a similar way political ecologists have often championed the potential for such "citizen science" (Irwin 1995) to challenge the hegemony of expert claims about environmental problems such as deforestation (Fairhead and Leach 1996), hydraulic fracturing (Willow and Wylie 2014), and flooding (Whatmore 2009).

However, such claims about the substantive, epistemic contributions of public participation are beset by ambiguities about which members of the public might be qualified to make such contributions and on what basis. For instance Chambers (1994: 954) insists that "poor and exploited people can and should be enabled to conduct their own analysis of their own reality." In so doing he conflates a normative plea for participation as a fundamental democratic right with an epistemological claim that participatory methods involving such marginalized groups "come out better by criteria of cost-effectiveness, validity, and reliability ... compared with conventional methods" of expert-led science (956). These different aims imply different kinds of participation from differently defined constituencies.

Participation as democratic right would imply unrestricted participation by any member of the relevant political constituency. While there are important questions about how that constituency should be defined in any given case, it is clear that if participation is a right, then participation rests on identity and political standing, rather than possession of any substantive knowledge or experience beyond membership of the political community itself. But if this is the case, then how can universal participation avoid the problem of extension identified by Collins and Evans (2002)? Chambers provides no explicit answer. His reference to the poor having "their own reality" somehow different from that of "outsiders" whose "reality [often] blanketed that of local people" (963) hints at a poststructuralist sensibility that would regard the resolution of scientific disputes about knowledge as essentially and ultimately a question of power rather than of logical proof or independent empirical validation. While that anti-foundational instinct is something of a commonplace now in political ecology, it belies the more conventional truth claims Chambers makes elsewhere about participation contributing substantively to the epistemic checking of science and science-based policy. Here the claim is not about power effects, but about facts and truth, defined in rather conventional, naturalist terms.

This, in turn, would depend upon restricting participation to those with relevant knowledge to contribute. Such knowledge might be place-based, derived from traditional knowledge of some particular environment or process, or practical, emerging from first-hand experience of the environment. There are also hints here of a standpoint epistemology whereby marginality itself is imagined as the only valid basis for understanding marginalization (Demeritt 1996). Whatever its precise basis, participation in epistemic checking would seem to depend on knowledge and as such should be restricted to experts, whether certified or otherwise, actually possessing some relevant knowledge to contribute. Of course deciding what knowledge is relevant and thus who might be qualified to contribute to epistemic checking is contestable, but that is a political decision, and as such is, in principle at least, distinct from questions of truth and scientific validity, where participation is restricted to those with knowledge. This is very different from the universal participation in normative steering, which depends on identity and political status rather than knowledge.

Conclusion

Like the fields of environmental science and policy with which it is in critical conversation, political ecology has taken a participatory turn of late, but without always articulating the reasons for that move or the implications that follow from it. Thus, the aim of this chapter has been to clarify the various competing rationales for embracing public participation in science and political ecology. In particular, I showed how normative claims for participation as some kind of basic right of those affected sit uncomfortably alongside other more instrumental and substantive claims that public participation will also somehow increase the quality and legitimacy of scientific knowledge and of the policy decisions informed by that knowledge. These different rationales are in marked tension with one another. They imply engaging in different ways with differently constituted publics to different ends. Sharper distinctions are required here because without them the tendency will be both for the promises of participation to be oversold and for public engagement plans to be ill-suited to the specific contexts and contributions for which they are intended.

To that end, I distinguished two very different ways in which public participation might contribute substantively to science and science-based policy, which I termed normative steering and epistemic checking. Whereas the former promises to provide the moral compass needed to steer their development in democratically legitimate ways (and as such is continuous with some normative claims for participation as democratic right), the latter contributes new information and quality assurance procedures to ensure their truth and reliability and as such is restricted to those possessing relevant knowledge to contribute.

Arguably this is much too tidy a distinction. After all, political ecology was founded on the recognition that distinctions between science and politics are not self-evident and ontologically given but precariously and problematically *made* (Chapter 16, this volume). It was that realization that science already reflects certain tacit political values and power structures that gave rise to calls for participation as a way of challenging them.

However, those critiques are doomed to fail without greater clarity about the different kinds of public warrant involved in making different sorts of political ecology critique. Normative critiques of science and of the injustice of particular environmental policies are properly political questions for the public in general. Political ecologists have often been quick to unveil the values underpinning the science-based claims of others, but they have not always been terribly reflexive about whether their *own* research programs meet with the approval of the people in whose name they are acting (see Chapter 8, this volume). Political ecology's paternalistic tradition of radical vanguardism tends to close off the sorts of questions about the purposes, framing, and funding of political ecology research, which the idea of normative steering is supposed to open up to wider public scrutiny. Political ecologists have typically treated representativeness as an epistemological question about how truthfully they are reflecting the views and values of those they study, rather than a political one about their own accountability to the publics they purport to speak for and to. While institutional review boards and ethical review ensure the accountability of political ecology to a universalized coda of research ethics based on liberal ideas of individual informed consent (Holden and Demeritt 2008), upstream public engagement provides a way to formalize often vague claims about participatory action research methodologies by giving those involved a mechanism for playing a more active role in framing the aims and methods of political ecology. Opening up political ecology to normative steering by the public would thus involve relinquishing some of the autonomy and academic freedom traditionally enjoyed by academic researchers, but it might also provide new opportunities to articulate the wider public purposes of political ecology and in the process, perhaps, expand the breadth and depth of its public appreciation.

Opening up the substantive factual claims of political ecology to participatory challenge is potentially even more radical. In its very strongest form, it would dissolve any epistemic or institutional distinction between scientists and citizens into a vastly expanded public debate in which science carries no special epistemic status and extended public participation is required because truth is a matter of convention, determined through persuasion, popularity, and power, rather than by expert judgments about its correspondence to the independent reality it purports to represent. While the claims of poststructural political ecologists like Escobar (1998) about undecidability and the social construction of knowledge and nature can certainly be read in this very radical neo-Kantian way (Demeritt 1998), it is noteworthy that they are typically rendered in a fairly conventional register, with direct quotations, footnotes, and the whole architecture of academic referencing offering themselves up for evaluation not as imaginary, but realistic worlds, in the fashion of fiction, but rather as actual worlds, faithfully represented. Seen in this light it may be that some of the concerns about relativism are overblown and that poststructuralists are not nearly as radical as they claim to be. Nevertheless, for many political ecologists, this radical version of public participation as epistemic checking strays too close to relativism. They worry that the field needs some foundation for warranting belief and preventing the extension of debate by dissenters, however ignorant, ill-informed, or duplicitous their claims. While they embrace the potential for uncertified experts to use their experiential or traditional knowledge to contribute to political ecology, they limit such participation to those possessing some relevant to knowledge to contribute. This more modest vision of public participation as epistemic checking challenges the monopoly of experts on epistemic authority, but it is does not challenge the traditional grounds for evaluating the truth of those claims.

References

Adams, W.M. 2001 *Green development: Environment and sustainability in the third world*, London: Routledge.

Agrawal, A. 1995 Dismantling the divide between indigenous and scientific knowledge, *Development and Change* 26: 413–439.

Agrawal, A. 2005 *Environmentality: Technologies of government and the making of subjects*, Durham: Duke University Press.

Beck, U. 1992a From industrial society to the risk society: questions of survival, social structure, and ecological enlightenment, *Theory, Culture & Society* 9: 97–123.

Beck, U. 1992b *Risk society: Towards a new modernity*, trans. M. Ritter, London: Sage.

Brosius, P.J., Tsing, A.L., and Zerner, C., eds. 2005 *Communities and conservation: Histories and politics of community-based natural resource management*, Lanham: Rowman Altamira.

Brown, M.B. 2009 *Science in democracy: Expertise, institutions, and representation*, Cambridge, MA: MIT Press.

Bryant, R.L. 1992 Political ecology: an emerging research agenda in Third-World studies, *Political Geography* 11: 12–36.

Chambers, R. 1994 The origins and practice of participatory rural appraisal, *World Development* 22: 953–969.

Chilvers, J. 2005 Democratising science in the UK: the case of radioactive waste management. In Leach, M., Scoones, I. and Wynne, B. (eds.) *Science and citizens*, London: Zed Press, pp. 237–243.

Chilvers, J. 2009 Deliberative and participatory approaches in environmental geography. In Castree, N., Demeritt, D., Liverman, D., and Rhoads, B. (eds.) *Blackwell companion to environmental geography*, Oxford: Wiley-Blackwell, pp. 400–417.

Collins, H. and Evans, R. 2002 The third wave of science studies: studies of expertise and experience, *Social Studies of Science* 32: 235–296.

Cooke, B. and Kothari, U., eds. 2001 *Participation: The new tyranny?* London: Zed Books.

CST 2005 *Policy through dialogue: Informing policies based on science and technology*, report from Council for Science and Technology, available at: www2.cst.gov.uk/cst/reports/files/policy-through-dialogue/report.pdf, accessed February 22, 2009.

Demeritt, D. 1996 Social theory and the reconstruction of science and geography, *Transactions of the Institute of British Geographers* 21: 484–503.

Demeritt, D. 1998 Science, social constructivism, and nature. In Braun, B. and Castree, N. (eds.) *Remaking reality: Nature at the millennium*, New York: Routledge, pp. 177–197.

Demeritt, D. 2005 The promises of collaborative research, *Environment and Planning A* 37: 2075–2082.

Demeritt, D. 2006 Science studies, climate change, and the prospects for constructivist critique, *Economy and Society* 35: 453–479.

Demeritt, D. and Nobert, S. 2014 Models of best practice in flood risk communication and management, *Environmental Hazards* 13(4): 313–328.

Dietz, T. and Stern, P.C. 2008 *Public participation in environmental assessment and decision making*, Washington: National Academy of Sciences.

Dyer, S. and Demeritt, D. 2009 Un-ethical review? Why it is wrong to apply the medical model of research governance to human geography, *Progress in Human Geography* 33: 46–64.

Escobar, A. 1998 Whose knowledge, whose nature? Biodiversity, conservation, and the political ecology of social movements, *Journal of Political Ecology* 5: 53–82.

Fairhead, J. and Leach, M. 1996 *Misreading the African landscape: Society and ecology in a forest-savanna mosaic*, Cambridge: Cambridge University Press.

Forsyth, T. 2003 *Critical political ecology: The politics of environmental science*, London: Routledge.

Forsyth, T. 2011 Politicizing environmental explanations: what can political ecology learn from sociology and philosophy of science? In Goldman, M.J., Nadasdy, P., and Turner, M.D. (eds.) *Knowing nature: Conversations at the intersection of political ecology and science studies*, Chicago: University of Chicago Press, pp. 31–46.

Funtowicz, S.O. and Ravetz, J.R. 1993 Science for the post-normal age, *Futures* 25: 739–755.

Holden, K. and Demeritt, D. 2008 Democratising science? The politics of promoting biomedicine in Singapore's developmental state, *Environment and Planning D: Society and Space* 26: 68–86.

Irwin, A. 1995 *Citizen science: A study of people, expertise, and sustainable development*, New York: Routledge

Irwin, A. 2006 The politics of talk: coming to terms with the "new" scientific governance, *Social Studies of Science* 36: 299–320.

Lane, S., Odoni, N., Landström, C., Whatmore, S.J., Ward, N. and Bradley, S. 2011 Doing flood risk science differently: an experiment in radical scientific method, *Transactions of the Institute of British Geographers* 36(1): 15–36.

Millington, J.D.A., Demeritt, D., and Romero-Calcerrada, R. 2011 Participatory evaluation of agent-base land use models, *Journal of Land Use Science* 6: 195–201.

Neumann, R.P. 1998 *Imposing wilderness: Struggles over livelihood and nature preservation in Africa*, Berkeley: University of California Press

Parkes, B.L., Cloke, H.L., Pappenberger, F., Neal, J., and Demeritt, D. 2013 Reducing inconsistencies in point observations of maximum flood inundation level, *Earth Interactions* 17: 1–27.

Peet, R. and Watts, M., eds. 2004 *Liberation ecologies: Environment, development, and social movements*, 2nd edn, London: Routledge.

Peluso, N.L. 1992 *Rich forests, poor people: Forest access and control in Java*, Berkeley: University of California Press.

RCEP 1998 *21st Report of the Royal Commission on Environmental Pollution: setting environmental standards*, Cm 4053, The Stationery Office, Norwich, available at: www.rcep.org.uk/standardsreport.htm, accessed July 17, 2014.

Rorty, R. 1991 *Objectivity, relativism, and truth*, Cambridge: Cambridge University Press.

Rothstein, H. 2007 Talking shops or talking turkey? Institutional constraints to proceduralising consumer representation in risk regulation, *Science, Technology and Human Values* 32: 582–607.

Shrader-Frechette, K.S. 2002 *Environmental justice: Creating equity, reclaiming democracy*, New York: Oxford University Press.

Sturgis, P. and Allum, N. 2004 Science in society: re-evaluating the deficit model of public understandings, *Public Understanding of Science* 13: 55–74.

Toogood, M. 2013 Engaging publics: biodiversity data collection and the geographies of citizen science, *Geography Compass* 7: 611–623.

Walker, P.A. 2005 Political ecology: where is the ecology, *Progress in Human Geography* 29: 73–82.

WCED [World Commission on Environment and Development] 1987 *Our common future*, Oxford: Oxford University Press.

Whatmore, S.J. 2009 Mapping knowledge controversies: science, democracy and the redistribution of expertise, *Progress in Human Geography* 33(5): 587–598.

Willow, A. and Wylie, S. 2014 Politics, ecology, and the new anthropology of energy: exploring the emerging frontiers of hydraulic fracking, *Journal of Political Ecology* 21: 222–236.

Wilsdon, J. and Willis, R. 2004 *See-through science: Why public engagement needs to move upstream*, London: Demos, available at: www.demos.co.uk/files/Seethroughsciencefinal.pdf, accessed July 17, 2014.

Zimmerer, K.S. and Young, K.R., eds. 1998 *Nature's geography: New lessons for conservation in developing countries*, Madison: University of Wisconsin Press.

18
LOCAL ENVIRONMENTAL KNOWLEDGE

Leah S. Horowitz

Knowing the local environment

Over the past four decades, research has highlighted ways that local communities', and particularly indigenous peoples', close, historically deep relationships with their surroundings have often led to an intimate knowledge of ecosystem components and interactions. In the 1970s and 1980s, these understandings began to fall under the labels of "indigenous knowledge," "indigenous technical knowledge," "traditional ecological knowledge," or, more broadly, "local environmental knowledge," often abbreviated, respectively, IK, ITK, TEK, and LEK. (In the interest of inclusivity, this chapter will use LEK.) This has been defined as "a cumulative body of knowledge, practice, and belief, evolving by adaptive processes and handed down through generations by cultural transmission, about the relationship of living beings (including humans) with one another and with their environment" (Berkes 2012: 7). Such knowledge encompasses awareness of locally available natural resources such as foods, medicines, timber, and firewood – not only what is present, but how to harvest and prepare it, and equally importantly, how to avoid overharvesting. This knowledge often extends beyond the technical expertise appreciated by Western scientists, to encompass the location of taboo places and the spirits who reside there, and how these must be approached or placated (e.g. Scales 2012).

Anthropologists and geographers have observed and documented such ecological understandings as exhibited by a wide range of communities around the world, describing customary marine resource management in Asia and the Pacific (Hviding 1996; Johannes 1981; Ruddle and Johannes 1985) and Central America (Nietschmann 1973), as well as land-based "ethnoecologies" of communities throughout South America (Posey 2002), Asia (Ellen et al. 2000), Australia (Rose 1992; Rumsey and Weiner 2001), Africa (Anderson and Grove 1987), North America (Gordon and Krech 2012; Menzies 2006), and even Europe (von Glasenapp and Thornton 2011). Researchers have also often borne witness to the gradual loss of this knowledge as it ceases being transmitted intergenerationally, or loses accuracy or relevance in the face of large-scale environmental change. Nonetheless, they note that despite serious threats to their existence, these systems are often highly resilient and adaptable (Gómez-Baggethun and Reyes-García 2013).

It is well beyond the scope of this chapter to summarize, synthesize, or even provide an exhaustive list of the vast body of research on LEK, which has been thoroughly covered elsewhere (e.g. Berkes 2012; Ellen et al. 2000). Instead, I will examine how political ecology has gone beyond the cataloguing of indigenous or local environmental knowledges, to examine the reciprocal influence of broader politico-economic, socio-cultural, and biophysical forces on local people's engagements with nearby ecosystems. I will focus on the following key contributions, all of which explicitly or implicitly examine the political uses of knowledge and/ or perceptions of knowledge.

In the first section below, I briefly trace the history of political ecologists' interest in LEK and outline some debates that arose early in this history but still resonate today. In particular, political ecologists began cautioning in the early 1990s against an overly-romanticized and ultimately constraining view of local, particularly indigenous, knowledge. These insights continue to inform contemporary debates about LEK, scholar-activism, and indigenous/non-indigenous alliances. Next, I discuss political ecology's contributions to our understandings of ways that both the physical presence of outsiders and the influence of their belief systems have affected local people's relationships to their natural resources and surroundings. More dramatically, an increasingly global economic system has opened remote areas to resource extraction on a large scale, vastly changing local environments. In the next section, therefore, I examine political ecology's insights into tensions between large-scale development and LEK, and the micropolitical tensions that also arise as local knowledge systems articulate with global forces. However, political ecology does not simply analyze threats to LEK; from its inception, political ecology has had an explicitly emancipatory agenda, aimed at freeing Third World (Blaikie and Brookfield 1987; Bryant and Bailey 1997) and then also First World (McCarthy 2002, 2005) "land managers" from misplaced blame for environmental problems. In the subsequent section, I explore how research on LEK has helped to further that agenda by demonstrating that rather than causing environmental degradation through ignorance, poor rural people have often been forced into destructive practices through politico-economic marginalization. Political ecologists have also shown that in many other cases, rather than resulting in the overuse or degradation of natural resources, LEK has contributed to their long-term management. As I explain in the following section, this research has been instrumental in enabling and promoting co-management initiatives, in which local, experience-based knowledge is combined with information gathered through scientific methods. However, conservationist paradigms are beginning to swing back toward "fences and fines" approaches. In part, this is due to the failure of co-management efforts that were based on the flawed assumption that local communities shared conservationists' conceptual framings of local environments, as I discuss next. Meanwhile, however, as I note in the final section before the conclusion, despite sometimes stark differences in local and outsider worldviews, many conservationists create simplistic portrayals of "traditional ecological wisdom," which, as political ecology has shown, can be strategically useful in forging alliances but may also have negative repercussions for both conservationists and indigenous groups. I conclude with some suggestions for future research.

Early debates about the value of LEK

Political ecologists have long been interested in communities' understandings and management of their natural resources and environments. Indeed, political ecology itself emerged from a critique of earlier theories of "cultural ecology," which had developed rich insights into the role of culture in local resource management but had viewed indigenous societies as closed systems in which cultural practices served as homeostatic mechanisms, maintaining stable human–environment relationships (e.g. Rappaport 1984 [1968]). In contrast, political ecologists recognized the need to examine the ever-tightening linkages between small-scale land-use practices, inevitably steeped in local knowledge, and an increasingly global political economy. For instance, Michael Watts's classic tome, *Silent Violence* (1983), showed that famines in Nigeria did not result simply from climatic abnormalities or growing populations but were socio-politically created as capitalist modes of commodity production disrupted villagers' long-standing land management practices. A few years later, Piers Blaikie made a related argument in his pioneering book, *The Political Economy of Soil Erosion in Developing Countries* (1985; see also Blaikie and Brookfield 1987; Blaikie et al. 1980). Drawing upon fieldwork in Morocco, Zambia, and India, but mainly Nepal, he showed that small farmers did not undermine their natural resource base through destructive farming techniques due to ignorance, or even by choice, but because they were marginalized by national- and global-scale politico-economic forces. Although highly politicized and controversial, these insights became extremely influential within policy-making circles in Nepal and elsewhere, ultimately changing how decision-makers approached resource conservation (Walker 2006).

If peasant farmers were not destructive through ignorance, a logical corollary was that they might in fact possess useful information about local conditions, and viable strategies for coping with them. Drawing upon his studies of West African agricultural systems, Paul Richards (1985) argued that local farmers possessed valuable knowledge that Western scientists needed to incorporate, not dismiss as backward, within rural development projects. Along similar lines, Susanna Hecht (1990) argued that Kayapó farmers in the Amazon demonstrated soil management practices much better suited to local ecological conditions than techniques imported from temperate regions. Further studies revealed that women, previously dismissed within development projects as ignorant, actually possessed crucial farming knowledge not shared by men (Gururani 2002; Momsen 2007).

However, the purported superiority of LEK soon became a subject of debate within political ecology. Blaikie now pointed to the pitfalls of a "neo-populist developmentalism" that "reifies and idealizes indigenous knowledge" (1996: 84). He argued for a return to "basic and simple" research (1996: 85) that focused on ways that environmental knowledge and agendas at an international level could be, rather than dismissed, instead translated into greater human rights and environmental justice at the local scale. A few years earlier, Tony Bebbington (1990) had argued that the marginalization that indigenous Latin American farmers faced had resulted in levels of ecological degradation with which their knowledge could no longer cope, and they welcomed technical assistance. Indeed, contrary to academics' framings, Andean federations viewed "Green Revolution" technologies not as antithetical to traditional practices but rather as part of an "indigenous" strategy that helped farmers remain in their communities, thereby allowing for the maintenance of cultural practices and identities that were otherwise threatened with loss (Bebbington 1993). LEK did not, therefore, need to be conceptualized as fundamentally distinct from "Western" knowledge systems. Rather than touting LEK as an inherently virtuous, static body of knowledge, researchers might more constructively work toward enabling local resource users to play a role in deciding what forms of knowledge were most useful to them, and how.

Intercultural contact

Thus, a large part of political ecologists' analysis of LEK was to dismantle unhelpful dichotomies between "local" and "scientific" knowledge, and to acknowledge LEK's dynamism. Indeed, political ecologists showed that even "indigenous" knowledge systems have always already been evolving through intercultural communication, sometimes over centuries (Agrawal 1995;

Berkes and Berkes 2009; Ellen 1999; Nygren 1999). For instance, Michael Dove (2002) examined how indigenous farmers in Borneo readily adopted the cultivation of rubber, transplanted from the Amazon by European colonists, and used their knowledge of local conditions to innovate appropriate cultivation techniques. Similarly, smallholder farmers in Ethiopia adapt their practices through experimentation, communication with or observations of other farmers, and acquisition of external knowledge via extension packages (Dixon 2005).

While contact with members of other cultures clearly stimulates LEK's evolution, it can also lead to its loss, yet also – paradoxically – provides opportunities for its maintenance. Greater intercultural contact, through globalization processes, is a double-edged sword for traditional knowledge systems, although mostly a negative influence. Because environmental knowledge is transmitted and maintained intraculturally (within a culturally distinct group), the lifting of physical and politico-economic barriers to interactions between members of different cultures is reducing the transmission of locally based ecological understandings. This decline is accentuated where a stigma is attached to the use of certain natural resources, such as "traditional" medicines. Conversely, however, cross-cultural exchange can instead help to preserve ecological knowledge. For instance, school garden programs in urban areas of South Africa have helped to maintain phytomedicinal knowledge (Philander et al. 2011). Ethnobotanical recordings, through bioprospecting projects, can also help to preserve and valorize knowledge of medically useful plants. However, these activities are highly controversial, often seen as "bioimperialism" (Moran et al. 2001).

Indeed, the process of recording local knowledge for people outside the community, whether pharmaceutical companies or academic researchers, is infused with power relations. Providers of traditional ecological information may not always be adequately recognized or compensated, and may even be disempowered in the process, as has occurred in Madagascar where the international pharmaceutical industry has marginalized local scientists and resource users (Neimark 2012). However, local groups may be equally disempowered by outsiders' attempts to protect them. Anthropologists Brent and Elois Ann Berlin describe an ethnobotanical research project in Chiapas, Mexico that obtained communities' prior and informed consent and planned to distribute any benefits equally among communities, but that was shut down by NGOs categorically opposed to any form of bioprospecting (Berlin and Berlin 2004). Meanwhile, once these resources are discovered by outsiders, their commodification may result in overharvesting for urban markets and concomitant inaccessibility for those who have long relied upon them. This has been the case, for example, with herbal remedies that are essential for poor rural communities' health care in India, now being depleted and overpriced as a fashionable commodity for wealthy urban consumers (Shiva 2007).

LEK encompasses more than information about particular natural resources and their usefulness for humans. Spiritual ecological knowledge, such as the identities of taboo places, animals, and plants, is being eroded not only by increasing market accessibility but also by the spread of other belief systems such as Christianity. This has serious implications for resource management, as much of this knowledge refers to restrictions on harvesting resources or developing wild spaces (Anoliefo et al. 2003; Robbins 1995). Ultimately, the lifting of such restrictions has helped to facilitate the over-exploitation of natural resources, such as increased harvesting of wild animals and plants in Madagascar (Jones et al. 2008), and has limited resistance to the siting of industrial infrastructure in formerly off-limits areas, such as the construction of a nickel refinery on ancestral lands in New Caledonia (Horowitz 2008a). Intracommunity tensions may result as some community members seek to embrace economic development while others express concerns about angering ancestral spirits. Large-scale development projects in Australia, for example, often provoke arguments within Aboriginal communities about whether or not sites are sacred (Trigger and Robinson 2001). At a proposed mine site near a Jawoyn Aboriginal community, senior custodians of the area insisted that mining would risk angering a powerful ancestor while other community members, who hoped for local economic development through the mining project, claimed that they had been taught that the site in question held no spiritual danger (Ross 2001). Similarly, a proposal to build a bridge met with resistance from a group of Ngarrindjeri women who claimed that they possessed "secret and gender-restricted" knowledge that the bridge would destroy women's reproductive powers (Merlan 2001: 255–256). They were opposed by other Ngarrindjeri women who claimed that they had never received such information. Such concerns may reflect micropolitical struggles. For instance, a shrimp-farming project in Madagascar failed when the local ruler, fearing a loss of his personal influence, claimed that local spirits opposed the project and then banished the aquaculture company after a series of bizarre accidents (Gezon 1999).

Environmental change

Meanwhile, ecologically based cultural knowledge, such as ancestral myths inscribed in the landscape, may form a crucial component of indigenous groups' cultural heritage and identity (e.g. Graham et al. 2000; Horowitz 2001; Moore 1996, 1998; Rumsey and Weiner 2001). It is also a key element of the "lifeworld," a shared "stock of knowledge" that gives meaning to daily existence (Habermas 1987: 125). However, this cultural knowledge is challenged, and lifeworlds may be "colonized" (Habermas 1987; see also Crossley 2002), when large-scale development projects threaten to destroy landscape features and archaeological sites. A desire to protect their cultural heritage may play a role in motivating indigenous groups to resist this development, from South America (Bebbington et al. 2008) to Australia (O'Faircheallaigh 2008).

Development projects also conflict with LEK when scientists and local residents make very different predictions of environmental impacts. Each group's evaluations are based on their own "criteria of credibility" (Cash et al. 2003: 8088; see also Garvin 2001): While scientific "experts" collect data within a restricted timeframe, constructing "order" through normative methods (see Latour and Woolgar 1986), local residents may rely on observations collected incidentally to other activities, over several generations, that have been transmitted orally by respected elders (e.g. Birkenholtz 2008). This can lead citizens to predict, for example, the devastation of marine ecosystems from industrial pollution, contradicting industry scientists' assurance that currents will carry it away (Horowitz 2010). When scientists choose to ignore knowledge based in local experience, residents may feel that their identity is threatened by "ignorant but arrogant experts" (Wynne 1992: 295).

At a global scale, industry – at every stage along the commodity chain, from natural resource extraction to the consumption of manufactured products – is transforming the planet in startling ways. A dramatic example is climate change, which is resulting in significant alterations to local ecological conditions, forcing communities to adapt their long-standing knowledge systems and resource management practices. Arctic residents have experienced particularly rapid and drastic change. While these groups are known for their resilience and adaptive capacity, the plethora of extreme events in recent times has reduced opportunities for young people to safely go out and learn from their elders. It has also resulted in possibly "maladaptive" strategies which may function in the short term but may not be appropriate responses to ongoing and unpredictable environmental changes, and could potentially increase communities' vulnerability over the longer term (Ford et al. 2013).

Exculpating the victim

Political ecologists have also examined ways that scientists' and local residents' different environmental understandings inform conflicts over not only development, as discussed above, but also the conservation of natural resources and ecosystems. From colonial times to the present, official discourses have blamed local people for ecological degradation, attributing this degradation to allegedly harmful local practices based in ignorance or attachment to irrational traditions (Adams 1990; Coombes et al. 2012; Fairhead and Leach 1996; Scales 2012). These accusations have been used to justify harsh conservation measures: Following a conservation strategy based on models developed in Europe and America, authorities set aside large areas as national parks or forest reserves, making them off-limits to local people who had depended on them for subsistence (Brockington 2002; West et al. 2006). This has led to various forms of community-led resistance to the officially protected areas (e.g. Brenner and Job 2012; Orlove 2002), including illegal activities such as grazing cattle within reserve boundaries in Burma (Bryant 1996) or burning grasslands and woodlands in Madagascar (Kull 2002).

Following on Blaikie's ground-breaking work (discussed above), other researchers have shown that not only could many environmentally degrading activities ultimately be traced back to political and economic marginalization, as he and his colleagues had demonstrated; conversely, under the right conditions, local residents could rely upon their ecological knowledge to manage resources for long-term viability. In some cases, they had done so for millennia. This flew in the face of previous assumptions – popularized in the 1960s by the term "tragedy of the commons" (Hardin 1968), but expounded since the 1830s (Lloyd 1832) – that resources held in common would inevitably be overexploited unless they were either privatized or placed under state control. In contrast, political ecologists showed that close-knit communities with intimate understandings of local ecosystems could devise rules for resource use and informally, but highly effectively, monitor each other's behavior (e.g. McCay and Acheson 1987). Elinor Ostrom and colleagues identified several "design principles" for successful community-based management of "common pool resources" (Ostrom 1990; Ostrom et al. 1994, 2002). In documenting such management strategies based in local knowledge and customary rights, researchers offered an alternative to state control or privatization. They argued that where traditional resource management systems existed, there was potential for external institutions to help strengthen these systems, rather than weaken them through authoritarian control. This research coincided with growing disapproval of the human rights implications of coercive conservation measures, along with calls for "participatory" conservation that would benefit local communities (Adams and Hulme 2001).

Co-management

In light of these analyses and concerns, protected area management paradigms began, in the 1980s and 1990s, to move away from a "fences and fines" approach and toward co-management, also called community-based conservation (CBC) or community-based natural resource management (CBNRM), which entails the integration of local knowledge and practices with scientific data (see Agrawal and Gibson 1999). Co-management initiatives may be run by governments or non-governmental organizations (NGOs), and political ecologists have examined many such projects (e.g. Evans et al. 2011; Horowitz 1998, 2008b; Maliao et al. 2009; Stocks et al. 2007; Turner 1999; Worboys et al. 2001). In some cases, co-management processes have resulted in greater community empowerment and the strengthening of local rights to land or sea and natural resources, in conjunction with meaningful community participation in management decisions. In Maine, for instance, lobster fishermen have long managed their fishery through informal rules involving territorial boundaries and trap limits (Acheson 1988). Despite deep-seated disagreements over data quality and management strategies, these fishermen have been able to work with scientists and government regulators to create a successful co-management system (Acheson 2003). In other cases, however, official recognition of traditional resource management may actually be used as a means of increasing state control, disempowering local users. In Peru, for example, legislators formally recognized customary water use-rights in an attempt to boost the legitimacy of legal frameworks. However, this resulted in an over-simplified, rigid regulatory structure that in fact excluded many traditional practices, creating friction with local communities (Boelens and Seemann 2014).

Meanwhile, co-management paradigms have been challenged in recent times as conservationists have been disillusioned by the disappointing results of many well-intentioned projects (Brosius and Russell 2003; Hutton et al. 2005). Often, these results have been due to inadequate legislative frameworks and/or poor planning on the part of conservation practitioners (Mayaka 2002). In some cases, conservationists have misunderstood local people's aspirations, which may involve strong desires for autonomous resource control and/or empowerment (Campbell and Vainio-Mattila 2003). In other cases, conservationists have relied overly on market incentives for conservation, which may backfire dramatically. In The Gambia, for instance, a WWF project designed to encourage the conservation of local mangroves by commercializing women's traditional oyster gathering has resulted in a need for cooking fuel to preserve the oysters for sale, which in turn has placed severe pressure on mangrove forests (Crow and Carney 2013). Nonetheless, the perceived failures of co-management projects have resulted not in more careful and appropriate project design but rather in an ideological shift back toward a protectionist, "fortress" conservation model, despite the objections of many social scientists (Dressler et al. 2010).

Different ecological ontologies

As political ecologists have shown, the success of co-management projects relies on the pre-existence of strong local institutions with rules for resource use that coincide, or at least overlap, with the project organizers' conservation objectives, even if the reasons behind the restrictions are based in different worldviews (Berkes 2004). Often, however, local residents have ecological ontologies that differ so greatly from those of conservationists as to proscribe effective collaboration. Community members may not view ecosystems and their components as possessing intrinsic, aesthetic, and/or recreational values that must be protected from any exploitation, as conservationists may insist, but rather as resources "to be respected and used responsibly" in accordance with locally developed management practices and/or ancestral taboos (Scales 2012: 74; see also Campbell and Vainio-Mattila 2003). When resources are abundant, local residents may not see any reason to conserve them; even when resources grow scarce, however, ecological knowledge has sometimes led to overexploitation rather than conservation (Alvard 1993; Dore 1997; Headland 1997). This has particularly been the case where markets for these natural resources exist and are rapidly expanding (Cinner et al. 2012; Godoy et al. 2005). Meanwhile, in adapting their knowledge systems to changing environments, local people may grow to depend upon the altered conditions that conservationists view as environmental hazards. For instance, the Jawoyn, an Australian Aboriginal group, have come to view Kakadu National Park's water buffalo as an important food source and the horses living there as pets, in contrast to government officials' views of these animal populations as composed of feral species that need to be reduced (Robinson et al. 2004).

Alternatively, local communities may share conservationists' visions of wild species and spaces as valuable and vulnerable, yet may have more pressing concerns about the direct or indirect threats that this very wildlife poses to themselves, their neighbors, and their children. These concerns can lead community members to oppose conservation efforts that they perceive as creating dangers for humans. Local residents may perceive these threats as stemming directly from protected wildlife, as reflected in a fear of large carnivores expressed by neighbors of Tanzania's Serengeti National Park (Kaltenborn et al. 2006). Alternatively, community members may view the conservation efforts themselves as constituting a major threat to human well-being, as exemplified by New Jersey residents' opposition to environmentalists' attempts to preserve an urban wetland, which would prevent construction of a truck bypass that townspeople saw as the only way to protect themselves from the pollution and risk of accidents that the trucks represented (Horowitz 2013).

Strategic alliances

As discussed above, local (particularly indigenous) communities' understandings of the value of natural spaces and resources, and how they should be managed, often differ greatly from those of Northern conservationists. Nonetheless, non-indigenous activists may feel powerless to combat environmentally destructive development without an alliance with indigenous groups, which confers the legitimacy that these groups represent. In some cases, this strategy may work to mutual advantage. In Montana and South Dakota, for instance, Native Americans and white ranchers were able to overcome long-standing quarrels over grazing lands and water rights in order to band together in opposition to exogenous threats including coal and uranium mines and bombing ranges (Grossman 2003). The success of these struggles was partly attributable to the Native Americans' treaty rights but also to public sympathy for their arguments that these lands were sacred to them. In Cochabamaba, Bolivia, too, indigenous farmers were able to draw upon perceptions of their legitimacy, based in customary water management systems, in joining with urban activists to force the government to reverse its decision to privatize drinking water and sewerage services (Perreault 2008). In other cases, however, environmentalists may be deeply disappointed to discover that their indigenous allies have more multifaceted concerns, including community members' material well-being, which can ultimately work against environmentalist efforts. In New Caledonia, for example, urban grassroots environmentalists tried to "translate" an indigenous protest group as being, like the environmentalists, completely opposed to a multinational mining project, and were stunned when the group's leaders secretly signed a "pact" with the mining company that allowed the project to proceed in exchange for benefits for the community (Horowitz 2012b).

In their efforts at forging such strategic alliances, environmentalists often portray indigenous people as exhibiting a conservation ethic that corresponds to Euro-American sensibilities. Simplifying, exotifying, and misinterpreting ethnographic accounts, environmentalists attempt to transform indigenous ecological "'knowledge' into wisdom, spiritual insight, or some other such quality" (Brosius 1997: 54). As "indigeneity" has become an internationally recognized source of political and moral legitimacy (Merlan 2009), these discourses have gained traction with international institutions, civil society, and the popular press. Political ecologists have discussed the deceptive nature of such Western environmentalist depictions of "traditional ecological wisdom" (e.g. Baviskar 1995; Krech 1999). They have also recognized the need to be careful with what they write, however, as indigenous groups themselves have often adopted academic, activist, or official interpretations of their own knowledge systems and have reflected these interpretations back, in support of political aims such as greater autonomy, authority, rights, or territorial claims (Berkes 2004). In Guinea, for instance, hunter "brotherhoods" have been able to mobilize conservation organizations' and the government's simplistic definitions of "indigenous" hunters' knowledge and social position, in order to secure greater control over "outside" gun hunters (Leach and Fairhead 2002). An excess of "ethnographic scrutiny" could, then, become "reactionary," contrary to "emancipatory" goals (Brosius 1999: 288). Nonetheless, valorizing their own indigenous ecological knowledge through a Western lens can also involve pitfalls for indigenous groups. Communities seeking to improve their living standards through economic development may cease to be considered "indigenous," and may thereby lose the popular support associated with this concept (Conklin and Graham 1995; Li 2002; Neumann 1997). An example of this risk comes from Papua New Guinea, where indigenous communities sought not to close down the highly polluting Ok Tedi mine but rather to claim compensation for its damages and to ensure its continued operation, albeit with additional environmental safeguards, as the mine provided them with economic benefits. This led to accusations that community members were "greedy rather than green" (Kirsch 2007: 314).

Conclusions and agendas for future research

In summary, political ecology has expanded our understandings of LEK systems by analyzing them not as localized, isolated, fixed sets of knowledge but as components of wider networks, both influenced by, and influencing, broader political, economic, and social forces such as the global spread of market economies, industrial development, and Euro-American belief systems. For instance, political ecologists have examined how politico-economic circumstances can marginalize small farmers and force them into unsustainable practices, but researchers have also uncovered ways that, under the right conditions, local knowledge has contributed to natural resource management. Such work has helped pave the way for co-management paradigms; however, scholars have also analyzed the often contentious relationships between conservationists' and local residents' environmental understandings when these paradigms are put into action. Political ecologists have also examined how "traditional ecological knowledge" has been used strategically, by both community members and environmentalists, to counter industrial development projects – or has been turned on its head through assertions that communities seeking higher living standards are no longer "indigenous."

Much scope remains ahead for political ecology to contribute to our understandings of LEK systems and their evolutions in contemporary politico-economic contexts. For example, more relevant than ever is the question of whether and how, despite differences with Western environmentalist views and accepted scientific methodologies, LEK can be mobilized in support of goals that benefit both ecosystems (at a global and/or local scale) and human societies. Efforts to set aside and protect areas of forest, particularly tropical, have been given a boost in recent times by UN- and World Bank-backed initiatives to "Reduce [carbon] Emissions from Deforestation and forest Degradation" (REDD+). These projects have sometimes involved the forcible displacement of communities and other abuses by corrupt governments eager for financial compensation (e.g. Beymer-Farris and Bassett 2012). Further research could help expose the broader politico-economic context of such human rights violations, but could also bring international attention to traditional resource management systems and explore the possibility of strengthening rather than weakening these. Another avenue for further research is the question of whether and how indigenous groups can still mobilize LEK, and their relationships to local ecosystems, as a source of legitimacy in the eyes of an increasingly cynical international community. On a related note, scholars could explore the ways LEK (both indigenous and non-indigenous) may become both an asset and a liability within environmental protest movements (see Horowitz 2011, 2012a). They might also examine ways that LEK, as a knowledge system but also as a component of cultural identity, informs relationships between stakeholder groups with different backgrounds and aims who are nonetheless collaborating toward overlapping environmental goals. Clearly, LEK continues to play an important role in contemporary environmental debates, and further research could productively elucidate its many dimensions, contexts, and relationships.

References

Acheson, J.M. 1988. *Lobster gangs of Maine*. Hanover, NH: New England University Press.

Acheson, J.M. 2003. *Capturing the commons: Devising institutions to manage the Maine lobster industry*. Hanover, NH: University Press of New England.

Adams, W.M. 1990. *Green development: Environment and sustainability in the Third World*. London and New York: Routledge.

Adams, W.M., and D. Hulme. 2001. If community conservation is the answer in Africa, what is the question? *Oryx* 35 (3): 193–200.

Agrawal, A. 1995. Dismantling the divide between indigenous and scientific knowledge. *Development and Change* 26: 413–439.

Agrawal, A., and C.C. Gibson. 1999. Enchantment and disenchantment: The role of community in natural resource conservation. *World Development* 27 (4): 629–649.

Alvard, M.S. 1993. Testing the "ecologically noble savage" hypothesis: Interspecific prey choice by Piro hunters of Amazonian Peru. *Human Ecology* 21 (4): 355–387.

Anderson, D., and R. Grove, eds. 1987. *Conservation in Africa: People, policies and practice*. Cambridge: Cambridge University Press.

Anoliefo, G.O., O.S. Isikhuemhen, and N.R. Ochije. 2003. Environmental implications of the erosion of cultural taboo practices in Awka-South local government area of Anambra State, Nigeria: 1. Forests, trees, and water resource preservation. *Journal of Agricultural and Environmental Ethics* 16: 281–296.

Baviskar, A. 1995. *In the belly of the river: Tribal conflicts over development in the Narmada Valley*. Delhi and New York: Oxford University Press.

Bebbington, A. 1990. Farmer knowledge, institutional resources and sustainable agricultural strategies: A case study from the eastern slopes of the Peruvian Andes. *Bulletin of Latin American Research* 9 (2): 203–228.

Bebbington, A. 1993. Modernization from below: An alternative indigenous development? *Economic Geography* 69 (3): 274–292.

Bebbington, A., D.H. Bebbington, J. Bury, J. Lingan, J.P. Muñoz, and M. Scurrah. 2008. Mining and social movements: Struggles over livelihood and rural territorial development in the Andes. *World Development* 36 (12): 2888–2905.

Berkes, F. 2004. Rethinking community-based conservation. *Conservation Biology* 18 (3): 621–630.

Berkes, F. 2012. *Sacred ecology*. Third edition. London and New York: Routledge.

Berkes, F., and M.K. Berkes. 2009. Ecological complexity, fuzzy logic, and holism in indigenous knowledge. *Futures* 41 (1): 6–12.

Berlin, B., and E.A. Berlin. 2004. Community autonomy and the Maya ICBG project in Chiapas, Mexico: How a bioprospecting project that should have succeeded failed. *Human Organization* 63 (4): 472–486.

Beymer-Farris, B.A., and T.J. Bassett. 2012. The REDD menace: Resurgent protectionism in Tanzania's mangrove forests. *Global Environmental Change* 22: 332–341.

Birkenholtz, T. 2008. Contesting expertise: The politics of environmental knowledge in northern Indian groundwater practices. *Geoforum* 39: 466–482.

Blaikie, P. 1985. *The political economy of soil erosion in developing countries*. London: Longman Development Series.

Blaikie, P. 1996. Post-modernism and global environmental change. *Global Environmental Change* 6 (2): 81–85.

Blaikie, P., and H. Brookfield. 1987. *Land degradation and society*. London: Methuen.

Blaikie, P., J. Cameron, and J.D. Seddon. 1980. *Nepal in crisis: Growth and stagnation at the periphery*. Oxford: Oxford University Press.

Boelens, R., and M. Seemann. 2014. Forced engagements: Water security and local rights formalization in Yanque, Colca Valley, Peru. *Human Organization* 73 (1): 1–12.

Brenner, L., and H. Job. 2012. Challenges to actor-oriented environmental governance: Examples from three Mexican biosphere reserves. *Tijdschrift voor Economische en Sociale Geografie* 103: 1–19.
Brockington, D. 2002. *Fortress conservation: The preservation of the Mkomazi Game Reserve, Tanzania*. Oxford: James Currey.
Brosius, J.P. 1997. Endangered forest, endangered people: Environmentalist representations of indigenous knowledge. *Human Ecology* 25 (1): 47–69.
Brosius, J.P. 1999. Anthropological engagements with environmentalism. *Current Anthropology* 40 (3): 277–288.
Brosius, J.P., and D. Russell. 2003. Conservation from above: An anthropological perspective on transboundary protected areas and ecoregional planning. *Journal of Sustainable Forestry* 17 (1–2): 39–65.
Bryant, R.L. 1996. Romancing colonial forestry: The discourse of "forestry as progress" in British Burma. *The Geographical Journal* 162: 169–178.
Bryant, R.L., and S. Bailey. 1997. *Third World political ecology*. London and New York: Routledge.
Campbell, L.M., and A. Vainio-Mattila. 2003. Participatory development and community-based conservation: Opportunities missed for lessons learned? *Human Ecology* 31 (3): 417–437.
Cash, D.W., W.C. Clark, F. Alcock, N.M. Dickson, N. Eckley, D.H. Guston, J. Jäger, and R.B. Mitchell. 2003. Knowledge systems for sustainable development. *Proceedings of the National Academy of Sciences* 100 (14): 8086–8091.
Cinner, J.E., T.R. McClanahan, M.A. MacNeil, N.A.J. Graham, T.M. Daw, A. Mukminin, D.A. Feary, A.L. Rabearisoa, A. Wamukota, N. Jiddawi, S.J. Campbell, A.H. Baird, F.A. Januchowski-Hartley, S. Hamed, R. Lahari, T. Morove, and J. Kuange. 2012. Comanagement of coral reef social-ecological systems. *Proceedings of the National Academy of Sciences* 109 (14): 5219–5222.
Conklin, B.A., and L.R. Graham. 1995. The shifting middle ground: Amazonian Indians and eco-politics. *American Anthropologist* 97 (4): 695–710.
Coombes, B., J.T. Johnson, and R. Howitt. 2012. Indigenous geographies I: Mere resource conflicts? The complexities in Indigenous land and environmental claims. *Progress in Human Geography* 36 (6): 810–821.
Crossley, N. 2002. *Making sense of social movements*. Buckingham and Philadelphia: Open University Press.
Crow, B., and J. Carney. 2013. Commercializing nature: Mangrove conservation and female oyster collectors in The Gambia. *Antipode* 45 (2): 275–293.
Dixon, A.B. 2005. Wetland sustainability and the evolution of indigenous knowledge in Ethiopia. *Geographical Journal* 171 (4): 306–323.
Dore, E. 1997. How sustainable were pre-Columbian civilizations? In *Green Guerrillas: Environmental Conflicts and Initiatives in Latin America and the Caribbean*, ed. H. Collinson, 47–51. London: Latin America Bureau.
Dove, M.R. 2002. Hybrid histories and indigenous knowledge among Asian rubber smallholders. *International Social Science Journal* 54 (173): 349–359.
Dressler, W., B. Büscher, M. Schoon, D. Brockington, T. Hayes, C.A. Kull, J. McCarthy, and K. Shrestha. 2010. From hope to crisis and back again? A critical history of the global CBNRM narrative. *Environmental Conservation* 37 (1): 5–15.
Ellen, R.F. 1999. Forest knowledge, forest transformation: Political contingency, historical ecology and the renegotiation of nature in Central Seram. In *Transforming the Indonesian Uplands*, ed. T.M. Li, 131–157. Amsterdam: Harwood.
Ellen, R.F., P. Parkes, and A. Bicker, eds. 2000. *Indigenous environmental knowledge and its transformations: Critical anthropological perspectives*. Amsterdam: Overseas Publishers Association.
Evans, L., N. Cherrett, and D. Pemsl. 2011. Assessing the impact of fisheries co-management interventions in developing countries: A meta-analysis. *Journal of Environmental Management* 92: 1938–1949.
Fairhead, J., and M. Leach. 1996. *Misreading the African landscape: Society and ecology in a forest-savanna mosaic*. Cambridge: Cambridge University Press.
Ford, J.D., G. McDowell, J. Shirley, M. Pitre, R. Siewierski, W. Gough, F. Duerden, T. Pearce, P. Adams, and S. Statham. 2013. The dynamic multiscale nature of climate change vulnerability: An Inuit harvesting example. *Annals of the Association of American Geographers* 103 (5): 1193–1211.
Garvin, T. 2001. Analytical paradigms: The epistemological distances between scientists, policy makers, and the public. *Risk Analysis* 21 (3): 443–455.
Gezon, L.L. 1999. Of shrimps and spirit possession: Toward a political ecology of resource management in northern Madagascar. *American Anthropologist* 101 (1): 58–67.
Godoy, R., V. Reyes-García, E. Byron, W.R. Leonard, and V. Vadez. 2005. The effect of market economies on the well-being of indigenous peoples and on their use of renewable natural resources. *Annual Review of Anthropology* 34 (1): 121–138.

Gómez-Baggethun, E., and V. Reyes-García. 2013. Reinterpreting change in traditional ecological knowledge. *Human Ecology* 41: 643–647.
Gordon, D.M., and S. Krech, eds. 2012. *Indigenous knowledge and the environment in Africa and North America.* Athens, OH: Ohio University Press.
Graham, B., G.J. Ashworth, and J.E. Tunbridge. 2000. *A geography of heritage: Power, culture and economy.* London: Arnold.
Grossman, Z. 2003. Cowboy and Indian alliances in the northern plains. *Agricultural History* 77 (2): 355–389.
Gururani, S. 2002. Construction of Third World women's knowledge in the development discourse. *International Social Science Journal* 54 (173): 313–323.
Habermas, J. 1987. *The theory of communicative action, vol. 2. Lifeworld and system: A critique of functionalist reason.* Boston: Beacon Press.
Hardin, G. 1968. The tragedy of the commons. *Science* 162 (3859): 1243–1248.
Headland, T.N. 1997. Revisionism in ecological anthropology. *Current Anthropology* 38 (4): 605–630.
Hecht, S.B. 1990. Indigenous soil management in the Latin American tropics: Neglected knowledge of native peoples. In *Agroecology and small farm development*, eds. M.A. Altieri and S.B. Hecht, 151–158. Boca Raton: CRC Press.
Horowitz, L.S. 1998. Integrating indigenous resource management with wildlife conservation: A case study of Batang Ai National Park, Sarawak, Malaysia. *Human Ecology* 26 (3): 371–403.
Horowitz, L.S. 2001. Perceptions of nature and responses to environmental degradation in New Caledonia. *Ethnology* 40 (3): 237–250.
Horowitz, L.S. 2008a. Destroying God's creation or using what He provided? Cultural models of a mining project in New Caledonia. *Human Organization* 67 (3): 292–306.
Horowitz, L.S. 2008b. "It's up to the clan to protect": Cultural heritage and the micropolitical ecology of conservation in New Caledonia. *The Social Science Journal* 45 (2): 258–278.
Horowitz, L.S. 2010. "Twenty years is yesterday": Science, multinational mining, and the political ecology of trust in New Caledonia. *Geoforum* 41 (4): 617–626.
Horowitz, L.S. 2011. Interpreting industry's impacts: Micropolitical ecologies of divergent community responses. *Development and Change* 42 (6): 1379–1391.
Horowitz, L.S. 2012a. Power, profit, protest: Grassroots resistance to industry in the global North. *Capitalism Nature Socialism* 23 (3): 20–34.
Horowitz, L.S. 2012b. Translation alignment: Actor-Network Theory, resistance, and the power dynamics of alliance in New Caledonia. *Antipode* 44 (3): 806–827.
Horowitz, L.S. 2013. Toward empathic agonism: Conflicting vulnerabilities in urban wetland governance. *Environment and Planning A* 45 (10): 2344–2361.
Hutton, J., W.M. Adams, and J.C. Murombedzi. 2005. Back to the barriers? Changing narratives in biodiversity conservation. *Forum for Development Studies* 2: 341–370.
Hviding, E. 1996. *Guardians of Marovo Lagoon: Practice, place, and politics in maritime Melanesia.* Honolulu: University of Hawai'i Press.
Johannes, R.E. 1981. *Words of the lagoon: Fishing and marine lore in the Palau district of Micronesia.* Berkeley and Los Angeles: University of California Press.
Jones, J.P.G., M.M. Andriamarovololona, and N. Hockley. 2008. The importance of taboos and social norms to conservation in Madagascar. *Conservation Biology* 22 (4): 976–986.
Kaltenborn, B.P., T. Bjerke, and J. Nyahongo. 2006. Living with problem animals: Self-reported fear of potentially dangerous species in the Serengeti region, Tanzania. *Human Dimensions of Wildlife* 11 (6): 397–409.
Kirsch, S. 2007. Indigenous movements and the risks of counterglobalization: tracking the campaign against Papua New Guinea's Ok Tedi mine. *American Ethnologist* 34 (2): 303–321.
Krech, S. 1999. *The ecological Indian: Myth and history.* New York and London: W.W. Norton and Company.
Kull, C.A. 2002. Madagascar aflame: Landscape burning as peasant protest, resistance, or a resource management tool? *Political Geography* 21 (7): 927–953.
Latour, B., and S. Woolgar. 1986. *Laboratory life: The construction of scientific facts.* Princeton: Princeton University Press.
Leach, M., and J. Fairhead. 2002. Manners of contestation: "Citizen science" and "indigenous knowledge" in West Africa and the Caribbean. *International Social Science Journal* 54 (173): 299–311.
Li, T.M. 2002. Engaging simplifications: Community-based resource management, market processes and state agendas in upland Southeast Asia. *World Development* 30 (2): 265–283.

Lloyd, W.F. 1832. Two lectures on the checks to population, delivered before the University of Oxford, in Michaelmas Term 1832. Oxford: University of Oxford.

Maliao, R.J., R.S. Pomeroy, and R.G. Turingan. 2009. Performance of community-based coastal resource management (CBCRM) programs in the Philippines: A meta-analysis. *Marine Policy* 33 (5): 818–825.

Mayaka, T.B. 2002. Wildlife co-management in the Bénoué National Park-Complex, Cameroon: A bumpy road to institutional development. *World Development* 30 (11): 2001–2016.

McCarthy, J. 2002. First World political ecology: Lessons from the Wise Use movement. *Environment and Planning A* 34 (7): 1281–1302.

McCarthy, J. 2005. First World political ecology: Directions and challenges. *Environment and Planning A* 37 (6): 953–958.

McCay, B.J., and J.M. Acheson, eds. 1987. *The question of the commons: The culture and ecology of communal resources*. Tucson: The University of Arizona Press.

Menzies, C.R., ed. 2006. *Traditional ecological knowledge and natural resource management*. Lincoln: University of Nebraska Press.

Merlan, F. 2001. Development, rationalisation and sacred sites: Comparative perspectives on Papua New Guinea and Australia. In *Mining and indigenous lifeworlds in Australia and Papua New Guinea*, eds. A. Rumsey and J. Weiner, 244–269. Adelaide: Crawford House Publishing.

Merlan, F. 2009. Indigeneity: Global and local. *Current Anthropology* 50 (3): 303–333.

Momsen, J.H. 2007. Gender and agrobiodiversity: Introduction to the Special Issue. *Singapore Journal of Tropical Geography* 28 (1): 1–6.

Moore, D.S. 1996. Marxism, culture and political ecology: Environmental struggles in Zimbabwe's Eastern Highlands. In *Liberation ecologies: Environment, development, social movements*, eds. R. Peet and M. Watts, 125–147. London and New York: Routledge.

Moore, D.S. 1998. Subaltern struggles and the politics of place: Remapping resistance in Zimbabwe's Eastern Highlands. *Cultural Anthropology* 13 (3): 344–381.

Moran, K., S.R. King, and T.J. Carlson. 2001. Biodiversity prospecting: Lessons and prospects. *Annual Review of Anthropology* 30 (1): 505–526.

Neimark, B.D. 2012. Industrializing nature, knowledge, and labour: The political economy of bioprospecting in Madagascar. *Geoforum* 43: 980–990.

Neumann, R.P. 1997. Primitive ideas: Protected area buffer zones and the politics of land in Africa. *Development and Change* 28 (3): 559–582.

Nietschmann, B. 1973. *Between land and water: The subsistence ecology of the Miskito Indians, Eastern Nicaragua*. New York and London: Seminar Press.

Nygren, A. 1999. Local knowledge in the environment-development discourse: From dichotomies to situated knowledges. *Critique of Anthropology* 19 (3): 267–288.

O'Faircheallaigh, C. 2008. Negotiating cultural heritage? Aboriginal–mining company agreements in Australia. *Development and Change* 39 (1): 25–51.

Orlove, B. 2002. *Lines in the water: Nature and culture at Lake Titicaca*. Berkeley: University of California Press.

Ostrom, E. 1990. *Governing the commons: The evolution of institutions for collective action*. Cambridge: Cambridge University Press.

Ostrom, E., R. Gardner, and J. Walker. 1994. *Rules, games, and common pool resources*. Ann Arbor: University of Michigan Press.

Ostrom, E., T. Dietz, N. Dolsak, P.C. Stern, S. Stonich, and E.U. Weber. 2002. *The drama of the commons*. Washington, DC: National Academy Press.

Perreault, T. 2008. Custom and contradiction: Rural water governance and the politics of *usos y costumbres* in Bolivia's irrigators' movement. *Annals of the Association of American Geographers* 98 (4): 834–854.

Philander, L.E.A., N.P. Makunga, and S.J. Platten. 2011. Local medicinal plant knowledge in South Africa preserved by apartheid. *Human Ecology* 39: 203–216.

Posey, D. 2002. *Kayapó ethnoecology and culture*. London: Routledge.

Rappaport, R.A. 1984 [1968]. *Pigs for the ancestors: Ritual in the ecology of a New Guinea people*. New Haven and London: Yale University Press.

Richards, P. 1985. *Indigenous agricultural revolution: Ecology and food production in West Africa*. London: Hutchinson; Boulder: Westview Press.

Robbins, J. 1995. Dispossessing the spirits: Christian transformations of desire and ecology among the Urapmin of Papua New Guinea. *Ethnology* 34 (3): 211–224.

Robinson, C.J., D. Smyth, and P.J. Whitehead. 2004. Bush tucker, bush pets, and bush threats: Cooperative management of feral animals in Australia's Kakadu National Park. *Conservation Biology* 19 (5): 1385–1391.

Rose, D.B. 1992. *Dingo makes us human: Life and land in an Australian Aboriginal culture*. Cambridge: Cambridge University Press.

Ross, H. 2001. Social impact assessment: Coronation Hill. In *Working on country: contemporary Indigenous management of Australia's lands and coastal regions*, eds. R. Baker, J. Davies, and E. Young, 320–336. Melbourne: Oxford University Press.

Ruddle, K., and R.E. Johannes, eds. 1985. *The traditional knowledge and management of coastal systems in Asia and the Pacific*. Jakarta: UNESCO.

Rumsey, A., and J.F. Weiner, eds. 2001. *Emplaced myth: Space, narrative, and knowledge in Aboriginal Australia and Papua New Guinea*. Honolulu: University of Hawai'i Press.

Scales, I. 2012. Lost in translation: Conflicting views of deforestation, land use and identity in western Madagascar. *The Geographical Journal* 178 (1): 67–79.

Shiva, V. 2007. Bioprospecting as sophisticated biopiracy. *Signs: Journal of Women in Culture and Society* 32 (2): 307–313.

Stocks, A., B. McMahan, and P. Taber. 2007. Indigenous, colonist, and government impacts on Nicaragua's Bosawas Reserve. *Conservation Biology* 21 (6): 1495–1505.

Trigger, D., and M. Robinson. 2001. Mining, land claims and the negotiation of indigenous interests: Research from the Queensland Gulf country and the Pilbara region of Western Australia. In *Mining and Indigenous lifeworlds in Australia and Papua New Guinea*, eds. A. Rumsey and J. Weiner, 226–243. Adelaide: Crawford House Publishing.

Turner, M.D. 1999. Conflict, environmental change, and social institutions in dryland Africa: Limitations of the community resource management approach. *Society & Natural Resources* 12 (7): 643–657.

von Glasenapp, M., and T.F. Thornton. 2011. Traditional ecological knowledge of Swiss Alpine farmers and their resilience to socioecological change. *Human Ecology* 39: 769–781.

Walker, P.A. 2006. Political ecology: Where is the policy? *Progress in Human Geography* 30 (3): 382–395.

Watts, M. 1983. *Silent violence: Food, famine and peasantry in northern Nigeria*. Berkeley: University of California Press.

West, P., J. Igoe, and D. Brockington. 2006. Parks and peoples: The social impact of protected areas. *Annual Review of Anthropology* 35: 251–277.

Worboys, G., M. Lockwood, and T. De Lacy. 2001. *Protected area management: Principles and practice*. Oxford and New York: Oxford University Press.

Wynne, B. 1992. Misunderstood misunderstanding: Social identities and public uptake of science. *Public Understanding of Science* 1: 281–304.

19
PARTICIPATORY MAPPING

Joe Bryan

Introduction

Open any political ecology book and one of the first things you will find is a map. Nestled in the flyleaf or included in the opening chapter, the map acts like a geographical preface locating the study within our understandings of nature and the world. Dots locate villages, lines mark boundaries, and beneath it all lies nature. Our very understandings of "nature" and "the world" are so deeply informed by maps that we are often at a loss without them (Crampton 2009a; Wood and Fels 2008; Wood 2010). The indispensability of maps raises any number of questions. Can maps be used to think of nature otherwise? What sort of map might be adequate for that task? Who would make it?

As a field, political ecology offers multiple answers to these questions. One response has been to read multiple maps of the same area against each other, showing how nature, much like space, is a contested category (e.g. Braun 2002; Raffles 2002). Other maps show multiple or conflicting boundaries in a single frame, using their tangle of lines to link struggles over resources with struggles over meaning (e.g. Moore 2005). Still others use mapping as a method, showing how resource struggles produce their own geographies through techniques variously known as "participatory mapping," "ethnocartography," or "counter-mapping" (Brody 1981; Bryan and Wood 2015; Chapin et al. 2005; Herlihy and Knapp 2003; Nietschmann 1995). Each of these responses helps clarify the problematic role that maps play in shaping understandings of the world that in turn shapes how maps are made and used. Despite their creativity, they do not resolve this problem so much as they sharpen an understanding of how power works through practices of reading, using, and making maps.

Of these approaches, participatory mapping has proven to be the most controversial. Advocates of participatory mapping have made broad claims about its potential for empowering local communities, challenging the "top-down" view found on official maps with "bottom-up" perspectives on land and resources. Despite the implied creativity in approach, participatory mapping still has to adhere to certain cartographic conventions that make maps recognizable to others. This is more than a matter of making maps that look like maps. Mapping is shot through with power relations that inform what can be mapped, their visual style and content defined by the forms of power and economy that they otherwise contest. Refusal to engage with mapping is no more intellectually satisfying than adhering to the status quo. Indeed, participatory mapping's inability to free itself from dominant understandings of the world foregrounds unique possibilities for research in political ecology, affording insight into how mapping is used to shape understandings of nature, differences, and justice.

Putting mapping in its place

Like the history of cartography more generally, the emergence of participatory mapping is often explained in terms of technological progress. Participatory mapping was initially conceived as a method for compiling local knowledge and presenting it as data. Access to GPS units, GIS software, and satellite imagery made it possible to simultaneously expand the number of participants and improve the accuracy of the data collected, and with it their ability to work as maps. New online mapping software has expanded these possibilities further through projects like OpenStreetMap's effort to create a free, user-edited online map of the world (Haklay et al. 2008; Perkins 2014). Advocates of participatory mapping have hailed these changes as "democratizing" cartography, providing a vehicle for social change (Herlihy and Knapp 2003; Nietschmann 1995).

Others have taken a more critical approach, challenging the importance of technology by focusing on how maps work with regard to power (Crampton and Krygier 2005; Crampton 2009a; Wood 2010). Often referred to as "critical cartography," this approach goes beyond questions of who makes maps and the technology used, focusing instead on the crises and problems that call for the production and use of maps. Rather than treating maps as products of exploration, critical approaches tie mapping to colonial efforts to extending authority over people and lands that remained largely unknown to colonizers. Cobbling together knowledge from explorers' accounts and information extracted from local populations allowed colonial authorities to produce authoritative knowledge that bolstered their claims to sovereign control over people and places they had often never seen (Harley 1988; Thongchai 1994). State officials and private interests followed suit, producing "official" maps to proclaim their authority (Craib 2004; Scott 1998; Wood 2010).

The materiality of the world overwhelmed the ability of maps to definitively represent reality (Mitchell 2002; Turnbull 1989, 2000). Every map was out-of-date the moment it was finished. People moved, rivers changed course, areas under cultivation expanded, and wars shifted boundaries. The only resolution to this problem was to produce another map. Nor could maps show everything. In order to work, maps had to be selective, their contents reduced to the most essential information needed to convey a pattern or perspective (Wood 2010; see also Monmonier 1996). Their incompleteness – the information they omitted or excluded – afforded states and other official entities the ability to bring the materiality of the world into accordance with what was shown on the map. At the same time, it created the grounds for challenging the authority of the map, to say nothing of the perspective it conveyed. Through it all, maps came to dominate understandings of the world that in turn shaped the production and use of maps.

Participatory mapping techniques follow suit. Like political ecology, the advent of participatory mapping techniques is inextricably linked to political economic factors related to the spatial dynamics of capitalism, anti-colonial movements, and environmental change in the post-Cold War era. These dynamics present a host of economic and environmental problems that define contemporary capitalism, driving the need to expand and extend markets while contending with a growing range of environmental and social crises (e.g. Harvey 1996; Pred and Watts 1992). These crises have, in turn, marked the fracturing and fragmentation of "old" spatial categories while creating new ones. While these crises may not have created participatory mapping technologies per se, they define its applications and techniques.

In particular, digital technologies make it possible to continually update and revise maps in "real time" with ever increasing accuracy. The goal of producing authoritative knowledge of the world remains, but their incompleteness is now less a threat to their authority than an invitation to add to it by making a "better" map. Google Earth's mission to "map everything" is paradigmatic of this approach (Rushe 2012; see also Dalton 2013). In its spinning image of continually updated satellite data and user-generated place-marks, Jorge Luis Borges' (1999, p. 325) musings about the imperial desire for a map "of the same Scale as the Empire and that coincided with it point to point" find their contemporary form. The measure of a map's authority has shifted from the accuracy of its representation to the frequency and intensity of its use. At every turn amateurs and "volunteers" overrun the idea of cartography as a profession with their eagerness to contribute new data, correct errors, and adapt maps to new problems (Crampton 2009b; Wood 2003). Armed with GPS units and access to online mapping programs, the volunteers become "sensors" that convert daily habits and knowledge into data (Elwood et al. 2012; Goodchild 2007). This endless mapping affirms their indispensability for imagining the world, to say nothing of changing it.

Cultural ecology: maps as data

Participatory mapping techniques pre-date political ecology. Since the colonial era administrators, military officials, and speculators relied on "native" guides to compile knowledge of the terrain, guide survey crews, and produce maps (Craib 2004; Edney 1997; Mundy 1996; Turnbull 2000; Warhus 1997). Participation was a means to an end, its messiness obscured by the neat lines drawn by expert surveyors and cartographers (Mitchell 2002; Thongchai 1994). Closer inspection always revealed a more complicated state of affairs. Cultural ecologists were among the first to use maps as evidence of this complexity. Their use of maps was perhaps not surprising giving the field's origins. Conceived of in opposition to environmental determinists' efforts to read social differences off of locations on a map, cultural ecologists used maps to show how cultures produced their own geographies through recursive modification of the environment. That approach was first developed in the late nineteenth century by spatially minded anthropologists like Franz Boas and Alfred Kroeber. Kroeber in particular used maps to compile and present evidence of Native Americans' ability to adapt to and modify the environment as the basis for their unique cultures (Kroeber 1947). Along with other evidence they collected, Kroeber and his students at the University of California, Berkeley used maps to develop "cultural ecology" as theory of evolutionary change (Steward 1955). Kroeber's student Julian Steward used maps of Native American societies to dissolve the "frontier" in the western United States into a patchwork of cultural landscapes. His work further broke down abstract linguistic and cultural categories, converting colonial tribal categories such as "Shoshone" into discrete, geographically defined societies of "rabbit eaters," "pine nut eaters," and so forth (Steward 1937). Omer Stewart, another of Kroeber's students, used maps to interpret the environment of North America as a socio-natural hybrid, its natural features inseparable from Native American practices (Stewart 2002). Kroeber's cross-campus colleagues in Geography at Berkeley also contributed to the field. Led by Carl Sauer, they mapped the extent of indigenous peoples' modifications of the environment in the Americas (Denevan 2001; Doolittle 2000; Sauer 1969; Whitmore and Turner 2001). Across the board, the maps allowed researchers to use their data to make visible hidden landscapes obliterated by environmental change and overlooked due to cultural and political biases.

As much as cultural ecologists treated these maps as data, their political significance was unavoidable. Part of that significance stemmed from the fact that what they were mapping –

indigenous peoples, traditional lifestyles, and customary use – constituted the very basis for challenging colonial authority. Maps allowed them to make sense of these challenges, evaluating their claims, and managing the political challenges they posed. Both Kroeber and Stewart testified before the Indian Claims Commission convened by the US government for the purposes of settling land disputes with tribes in the 1950s (Heizer and Kroeber 1976; Stewart 1961; Sutton 1986). In the 1960s, the Canadian government relied extensively on land use and occupancy studies done by anthropologists and Native organizations to evaluate the impacts of oil and gas development on subsistence practices and manage opposition (Freeman 2011). Geographer William Denevan dramatically altered estimates of the pre-Columbian population of the Americas, challenging widely held perceptions of the continent as environmentally pristine prior to 1492 (Denevan 1992). In Central America, geographer Bernard Nietschmann used maps to compile and analyze his data documenting Miskito subsistence strategies, drawing attention to the political and ecological precariousness of their position with respect to the Nicaraguan state (Nietschmann 1973, 1979). Harold Conklin, a student of Kroeber's, used linguistics to study indigenous concepts of space in the Philippines that culminated in his detailed *Ethnographic atlas of Ifugao* (Conklin et al. 1980). Anthropologist Keith Basso took a similar approach to mapping Western Apache landscapes in the southwestern United States, depicting an "ethnography of lived topography" through analysis of place names (Basso 1996). As divergent as these approaches were, they helped transform cultural landscapes into "habitats" and "territories" foundational to the collective ways of life they mapped.

Participation followed a similar trajectory. From the very beginning, cultural ecology was impossible without native informants whose knowledge and practices were the objects of research. The more scientific methods in cultural ecology became, the more indispensable participation was. Nietschmann relied extensively on Miskito families' willingness to weigh and measure the food they produced before consuming it, constructing detailed caloric budgets that linked bodily health with geographical space (Nietschmann 1973). Researchers in Canada added detailed information about hunting and trapping to calorie counts to compile physical evidence of use of and occupancy (Brody 1981; O'Malley 1976; Watkins 1977; Weinstein 1976). In both instances, participation allowed researchers to substantiate their claims to representative understanding. In turn, their efforts produced maps and related studies that could be used to represent the people they were studying. The political salience of their work was thus a function of methodological rigor rather than explicit intent.

Maps as tools: counter-mapping

Broader engagement with anti-colonial movements inverted cultural ecology's focus on method, insisting instead on a more explicitly political approach to mapping as a vehicle for social change. It was no longer enough to show cultural modification of the environment in order to secure a better position for indigenous and traditional peoples within state societies. As anti-colonial movements made clear, the goal was to change society itself.

Aboriginal groups in Canada were among the first to use maps to challenge the state authority, transforming "cultural landscapes" into "territories" controlled by community authorities. First commissioned by the Canadian government as means of mitigating social impacts of resource development, Native communities began using the mapping process to support demands for legal recognition of their rights. Dene efforts to counter Canadian plans to develop oil and natural gas development in the Mackenzie River basin in the Yukon captured this approach. Summing up the findings of their mapping work, Dene scholar Phoebe Nahanni wrote "through evidence of our land use and occupancy, we are showing that we have tolerated at great cost to our culture the kind of development thrust upon us, and from here on it is our right to control and direct the changes that affect our survival as people" (Nahanni 1977, p. 27). Nahanni's assertion fused land use and occupancy mapping with a growing sense of Native peoples as nations, transforming the twist of trap lines used to map land use and occupancy into a territory vital to the Dene nation that countered Canadian claims to sovereignty.

Less than a decade later, Miskito communities in Nicaragua pursued a similar strategy. Using historical information they collected themselves and informed by maps made by Nietschmann, they produced a map in 1981 depicting their claims to territorial sovereignty over the eastern half of Nicaragua. The map fueled Sandinista fears of ethnic conflicts being manipulated by the US as part of the Reagan Administration's efforts to remove them from power. Within a year, the Sandinistas launched a counter-insurgency campaign aimed at stamping out that threat, opening a new front in the Contra War (Hale 1994; Nietschmann 1989; Ohland and Schneider 1983). In 1987, the Sandinistas controversially established autonomous regions in the territory identified by the Miskitos' map. Though the Sandinistas' move was aimed at integrating the region's pluri-ethnic population within the state, others interpreted their action as a concession to indigenous political authority. Maps made by indigenous peoples could push this agenda further, an assertion that led Bernard Nietschmann to proclaim that "more indigenous territory can be reclaimed and defended with maps than by guns" (Nietschmann 1995, p. 37). The line has since become a mantra of "counter-mappers," underscoring the importance that maps play in imagining forms of common identification and linking them with claims to political authority.

Throughout those projects, participation shifted from methodological to a political necessity. The more participatory the process, the more legitimacy the maps had when it came to representing a community or populations. One of the first mapping projects to make those political commitments explicit was the Detroit Geographical Expedition and Institute. The Expedition was launched in the late 1960s by geographer William Bunge as a joint project of Michigan State University and community organizations in Detroit, with the goal of making "educational and planning services available to inner city Blacks" (Horvath 1971). Under the leadership of an 18-year-old high school "push out," Gwendolyn Warren, the expedition produced a number of maps that made the city's stark racial inequalities visible. A map of pedestrian deaths downtown was thus titled "Where commuters run over black children on the Pointes-Downtown Track" (The Detroit Geographical Expedition and Institute 1971; see also Wood 2010). Another showed the "Region of Rat-bitten Babies" overlaying Detroit's "Black Ghetto." These maps were not produced as data for research done by "geographers [who] know less about geography than anyone else about their respective trades" (Bunge 1977). Instead, they were maps made to effect social change by expanding understandings of the environment and formulating new approaches to justice.

Participatory mapping at an impasse

No matter how much participation lent legitimacy to the maps, their validity often remained subject to evaluation by judges, state officials, and other outside experts (Rambaldi et al. 2006; Sparke 1998; Wood 2010). The maps could challenge disciplinary boundaries and advance political causes but they still had to be readable as maps. That requirement was more than just a matter of adhering to cartographic conventions. It meant fitting claims and participation into dominant understandings of the world, re-inscribing problematic distinctions between nature and culture, tradition and modernity.

Indigenous peoples' claims to territory were particularly vulnerable to this predicament. In order to assert their legitimacy as indigenous peoples, community claims to land and resources hinged on establishing historical and geographical continuity with a pre-colonial past consistent with generic understandings of indigeneity (see Chapter 41, this volume). As problematic as that stereotype is, it became the basis for defining indigenous peoples as political subjects. To take but one example, the Convention 169 on the Rights of Indigenous and Tribal Peoples adopted by the International Labor Organization in 1989 recognized indigenous peoples' right to control the "total environment" they customarily used (Anaya 1996; Brysk 2000; Rodríguez-Piñero 2005). That definition of territory picked up the work previously done by cultural ecologists, a point underscored in the translation of "total environment" as "*hábitat*" in the Spanish version of the Convention. It also reinforced widespread stereotypes of indigenous peoples as living close to nature that made their claims intelligible to a broader audience by equating indigenous practices with sustainable use of land and resources (Center for Native Lands 1992; Chapin 1992; Chapin et al. 2005). Over the course of the 1990s, conservation organizations and development agencies supported participatory mapping projects to promote sustainability and resolve conflicts over land and resources. The Ford Foundation played a prominent role, making participatory mapping a key component of community forestry projects in Latin America, the Caribbean, Southeast Asia, and the United States (Fox et al. 2005; Poole 1995). Early figures in the field of political ecology, including Bernard Nietschmann, Nancy Peluso, and Diane Rocheleau, figured prominently in all aspects of the work, acting alternately as promoters and critics of mapping (Nietschmann 1995; Peluso 1995; Rocheleau et al. 1995). The inclusion of mapping within development projects, particularly those aimed at property reforms supported by the World Bank, further employed methods for documenting use and occupancy as the basis for allocating rights to property (Gordon et al. 2003; Hale 2005). That effort was bolstered by the growing use of maps to formulate claims to land, modifying mapping methods further in order to adhere to legal criteria for adjudicating claims (Povinelli 1999; Sparke 1998; Wainwright and Bryan 2009).

Having experts back in control of mapping was not enough to ensure their validity. Conservationists remained skeptical about the sustainability of indigenous practices. State officials continued to be hostile to the very idea of recognizing indigenous claims. To address that persistent problem, mapping experts led efforts to make their participatory methods more scientific. Access to GPS units and GIS helped address the question of validity. So too did efforts to make participatory mapping methods more rigorous through sampling and spatial analysis (Tobias 2000, 2009). The Humanitarian OpenStreetMap Team (HOT) has fused both to use participatory mapping to guide disaster relief efforts, demonstrating the technique's relevance as a tool for governing (Soden and Palen 2014; Zook et al. 2010). As with critiques of participatory approaches more generally, the emphasis on accuracy and technology obscures the political economic factors that make community "participation" desirable and even necessary to expanding the reach of state authority and markets (Cooke and Kothari 2001; Mohan and Stokke 2000). Emphasis on technology also obscured difficult questions about whose knowledge mattered when it came to mapping, as well as what to map (Peluso 1995; Rocheleau 2005; Tsing 1999).

Gender differences brought those concerns into sharp relief, with mapping projects continually struggling to do anything other than prioritize male participation (Rocheleau 2005). Gender dynamics within communities only partially accounted for this problem, reinforced by broader codings of space outside the home as male that mirror similarly gendered divisions of public and private space (Haraway 1991; McClintock 1995). Male knowledge of spatially dispersed activities such as hunting, logging, and travel between communities was thus favored over more spatially compact practices of gardening, food provisioning, and household maintenance, particularly when it came to staking claims to land (Rocheleau et al. 1995). In a similar vein the legitimacy of "indigenous knowledge" was often conditioned upon the ability to document traditions, often at the expense of more contemporary understandings of resource use and conflict (Bryan 2009; Roth 2007; Tsing 1999).

As states turned lines on paper into property boundaries, the dilemmas of participatory mapping became more pronounced. There were immediate concerns about how mapping transformed customary practices and community identities in ways that radically altered access to land and resources, elaborating on questions first raised by Nancy Peluso in her 1995 article, "Whose woods are these? Counter-mapping forest territories in Kalimantan, Indonesia." Commenting on the proliferation of community mapping projects funded by the Ford Foundation, Peluso noted three key theoretical questions: How did new maps reflect prior spatial configurations of power produced under colonial administration? How did mapping "reinvent" traditions relating to customary access to land and resources? And, how did the role of NGOs – to say nothing of academics and development agencies – alter community access to land and resources in all aspects of the mapping process, from their production to dissemination?

Critiques of mapping projects in Southeast Asia, Africa, and Latin America elaborated on the importance of Peluso's questions (Hale 2005; Hodgson and Schroeder 2002; Rambaldi et al. 2006; Roth 2007; Walker and Peters 2001). Instead of bringing communities under the rule of law, mapping projects everywhere seemed to be contributing to the expansion of state control and assimilation of community land and resources into markets (Roth 2007; Wainwright 2008; Wainwright and Bryan 2009). Hopelessly compromised, mapping seemed destined to reproduce Cartesian understandings of space configured into nested scales of state sovereignty and private property (Bryan 2011; Rocheleau 2005; Roth 2009; Sparke 1998). There was very little that was "counter" about this project at all. Worse, participatory mapping accentuated conflicts between communities that further undermined the very rights to land and resources that they were intended to protect (Fox et al. 2008; Mollett 2013; Wainwright and Bryan 2009).

The controversy that erupted in 2009 over the American Geographical Society's (AGS) Bowman Expeditions brought together theoretically oriented critiques of mapping with practical concerns about their political application. The AGS launched the project in 2007 as the first of the Society's "Bowman Expeditions" aimed at using geographical research to produce foreign intelligence for US policymakers (Dobson 2012). Funded by the US Army, the initial Expeditions to Mexico, the Antilles, and Colombia made extensive use of participatory mapping to record property rights and assess social unrest. Information collected by participatory mapping projects could further be used to build a digital "human terrain" that could tell Army officials and others not only where people lived, but also who they were and what they did. Conversely, Army officials read the lack of that information as a threat to security, equating the absence of a map with violence (Demarest 2011). Over time, the AGS hoped to send an Expedition to every country in the world.

The AGS used its inaugural Bowman Expedition to Mexico, also known as the México Indígena project, to provide a proof of concept for this approach. Led by University of Kansas geographer Peter Herlihy, the Expedition sought to use participatory mapping to evaluate the impact of privatizing land collectively owned by indigenous communities. The project targeted two areas in particular, the Huasteca Potosina in Central Mexico and the Sierra Juaréz in the southern state of Oaxaca (Herlihy et al. 2008; Smith et al. 2009). Communities initially agreed to participate in the project, hoping to use the maps produced by the project for their own purposes. However, that relationship soured when communities in Oaxaca raised concerns about the US Army's role in the project (Bryan 2010; Wainwright 2013). Their concerns went beyond the funding for the project, accusing the researchers of turning over all of the data collected by the project to the US Army (Cruz 2010). The communities also demanded that the Expedition leaders return all data, and charged Herlihy and AGS President Jerome Dobson with violating ethical standards. In spite of support from numerous geographers, neither the AGS nor the Association of American Geographers responded to calls for a substantive response to the communities' allegations (Agnew 2010). Nor did either organization revise their standards with respect to military-funded research. Emboldened by the lack of response, the AGS has expanded its Bowman Expeditions program with support from the United States Department of Defense, launching a new Central America Indígena project with the same goals as the Mexico project in 2013. The newest Expedition is further designed to complete the AGS's effort to build a "digital regional geography" using GIS of the "US Borderlands" region that includes "all Latin American countries bordering the Gulf of Mexico and the Caribbean."

The implications of the Bowman Expedition for participatory mapping are extensive. It brings to light the role that militaries themselves have played in shaping participatory mapping techniques through their efforts to gain knowledge of the human terrain vital to counter-insurgency (Boyce and Cash 2013; Bryan and Wood 2015). It also mirrors efforts by militaries to use the social sciences more generally to compile data on cultural practices as a means of waging war (González 2009; Kelly et al. 2010; Price 2011). The problems that the Bowman Expedition controversy raises are thus not limited to maps alone any more than they are specific to the disciplines of geography or anthropology. Instead, maps reveal the grip that colonial categories have on understanding the present, drawing distinctions between nature and culture, tradition and modernity (Wainwright 2005). The Army's active support for projects like the Bowman Expeditions shows just how tight that grip is, aided and abetted – even if only passively – by the lack of any formal response from professional organizations like the AAG. Nor have critiques of participatory mapping done much to slow its growth, in spite of their ability to raise pointed questions.

New methods? From the power of maps to cartographies of power

A better map will not resolve the problems of participatory mapping. Instead what is required is a more careful consideration of the problems that such maps have been used to identify and resolve. In both cultural ecology and political ecology, the goal of participatory mapping has been to make cultural differences intelligible to people at some remove from the places mapped. While this may be politically necessary, it sidesteps questions about how maps might be used to create and sustain the forms of collective life they are often used to defend. Rather than focusing on questions of representation or technological accuracy, participatory mapping foregrounds key questions about the political and intellectual consequences of mapping itself. Maps cannot resolve the impasse that participatory mapping faces, but practices of mapping can offer a means of moving through that condition.

Some of the strongest responses to these challenges come from communities directly involved in mapping. In Latin America, a growing number of participatory mapping projects have shifted their attention away from the maps themselves to the practices used to make and use them. Sometimes referred to as "social cartography," these efforts foreground the politics of everything from decisions about what to map to questions of who participates in their production, and how that information is shared and controlled (Acselrad and Régis Coli 2008; de Almeida 2011; Bryan 2011; Sletto 2009). As varied as this work is, it represents a departure from the instrumental reading of maps. The resulting maps are neither evidence of a struggle nor definitive of cultural differences. Instead they mobilize both as resources for struggle, showing how space is shaped by power (Kitchin and Dodge 2007; Pickles 2004). Racialized differences are recast as an outcome of dispossession and displacement by settlement rather than expressions of culture (Mollett 2011). Village locations are not organic expressions of cultural integrity, but artifacts of prior attempts by colonial administrators and missionaries, among others, to settle people for the purposes of extracting resource wealth (Gordillo 2004; Wainwright 2008). Even the need to map itself is understood as an expression of political economic change, organizing "empty" or unclaimed land on state frontiers into orderly systems of ownership of land and resources (Asher 2009; Bryan 2011; Hale 2005).

Viewing mapping as a social practice can be used to rethink the role of maps in political ecology. Attention to the production and use of maps dispenses with empty calls for participation, and instead asks whose knowledge counts when it comes to questions of space and power (Bryan 2009; Roth 2009; Turnbull 2005). Those efforts can also deform our very understanding of what counts as a map, producing images that reflect their historically and geographically specific understandings of space generative of new forms of politics and collective forms of life (de Almeida 2011; Enote and Mclerran 2011; Johnson et al. 2006; Povinelli 2011; Sletto 2014; see also Counter Cartographies Collective et al. 2012). Put differently, it begins to address questions of how understandings of the "environment" or "nature" shape understandings of "justice," and vice versa, pressing forward with political ecology's contention that struggles over resources are invariably struggles over meaning (Braun 2002; Kosek 2006; Moore 2005; Sparke 1998). Another aspect of this approach focuses on how mapping itself, through the reading, production, and use of maps, informs decisions about what to map and how to go about doing it (Bryan 2011; Caquard 2014). These practices are inevitably embodied, drawing out the performative aspects of how mapping simultaneously references spatial norms while reworking them in light of new challenges and goals (Sletto 2009; Sletto et al. 2013).

Conclusion

Treating mapping as a social process brings its methodological contributions full circle. If maps cannot be used instrumentally to show what is there but instead alter understandings of the world and our ways of being in it, their importance as a site for struggle over space becomes immediately apparent. So too is their importance to practices of knowledge production more generally. Process oriented approaches have the potential to draw this out by shifting attention to the kinds of situations that call for a map, how mapping practices are organized, and how these shape the use and production of maps themselves. Much in the way previous iterations of political ecology have struggled to liberate nature from ecology (Peet and Watts 1996, 2004), social cartography will need to free space from the disciplinary constraints of geography. The potential is there for a critically informed, collaborative approach to mapping, one that draws from a critical understanding of how mapping has been used to reinforce and deepen relations of power and their attendant inequalities.

References

Acselrad, H., and Régis Coli, L., 2008, Disputas Cartográficas e disputas territoriais, in H. Acselrad (ed.), *Cartografias Sociais e Território*, IPPUR/UFRJ, Rio de Janeiro, pp. 13–44.

Agnew, J., 2010, Ethics or militarism? The role of the AAG in what was originally a dispute over informed consent, *Political Geography*, 29(8), pp. 422–423.

de Almeida, A.W.B., 2011, Terras tradicionalmente ocupadas: processos de territorialização e movimentos sociais, *Revista Brasileira de Estudos Urbanos e Regionais*, 6(1), pp. 9–32.

Anaya, S.J., 1996, *Indigenous peoples in international law*, Oxford University Press, Oxford and New York.

Asher, K., 2009, *Black and green: Afro-Colombians, development, and nature in the Pacific lowlands*, Duke University Press, Durham, NC.

Basso, K.H., 1996, *Wisdom sits in places: Landscape and language among the Western Apache*, University of New Mexico Press, Albuquerque, NM.

Borges, J.L., 1999, Of exactitude in science, in J.L. Borges, *Collected fictions*, Penguin Books, New York, p. 325.

Boyce, G., and Cash, C., 2013, Geography, counterinsurgency, and the "G-Bomb": the case of México Indígena, in K. Williams, L. Messersmith-Glavin and W. Munger (eds.), *Life during wartime: Resisting counterinsurgency*, AK Press, Oakland, pp. 245–257.

Braun, B., 2002, *The intemperate rainforest*, University of Minnesota Press, Minneapolis.

Brody, H., 1981, *Maps and dreams: Indians and the British Columbia frontier*, Waveland Press, Prospect Heights.

Bryan, J., 2009, Where would we be without them? Knowledge, space and power in indigenous politics, *Futures*, 41(1), pp. 24–32.

Bryan, J., 2010, Force multipliers: Geography, militarism, and the Bowman Expeditions, *Political Geography*, 29(8), pp. 414–416.

Bryan, J., 2011, Walking the line: Participatory mapping, indigenous rights, and neoliberalism, *Geoforum*, 42(1), pp. 40–50.

Bryan, J., and Wood, D., 2015, *Weaponizing maps: Indigenous peoples and counterinsurgency in the Americas*, Guilford Press, New York.

Brysk, A., 2000, *From tribal village to global village: Indian rights and international relations in Latin America*, Stanford University Press, Stanford, CA.

Bunge, W., 1977. The Canadian-American geographical expedition. Report (mimeograph).

Caquard, S., 2014, Cartography II: Collective cartographies in the social media era, *Progress in Human Geography*, 38(1), pp. 141–150.

Center for Native Lands, 1992, The coexistence of indigenous peoples and the natural environment in Central America, Research & Exploration, *The National Geographic Society*, 8(2), map insert.

Chapin, M., 1992, Disappearing forests; disappearing peoples, *Cultural Survival Quarterly*, 16(3), pp. 63–66.

Chapin, M., Lamb, Z., and Threlkeld, B., 2005, Mapping indigenous lands, *Annual Review of Anthropology*, 34, pp. 619–638.

Conklin, H.C., Lupaih, P., and Pinther, M., 1980, *Ethnographic atlas of Ifugao: A study of environment, culture, and society in northern Luzon*, Yale University Press, New Haven.

Cooke, B., and Kothari, U. (eds.), 2001, *Participation: The new tyranny?* Zed Books, London and New York.

Counter Cartographies Collective, Dalton, C., and Mason-Deese, L., 2012, Counter (mapping) actions: Mapping as militant research, *ACME: An International E-Journal for Critical Geographies*, 11(3), pp. 439–466.

Craib, R., 2004, *Cartographic Mexico: A history of state fixations and fugitive landscapes*, Duke University Press, Durham, NC.

Crampton, J.W., 2009a, *Mapping: a critical introduction to cartography and GIS*, Blackwell Publishing, Malden, MA.

Crampton, J.W., 2009b, Cartography: maps 2.0, *Progress in Human Geography*, 33(1), p. 91.

Crampton, J.W., and Krygier, J., 2005, An introduction to critical cartography, *ACME: An International e-journal for Critical Geographies*, 4(1), pp. 11–33.

Cruz, M.K., 2010, A living space: The relationship between land and property in the community, *Political Geography*, 29(8), pp. 420–421.

Dalton, C.M., 2013, Sovereigns, spooks, and hackers: An early gistory of Google Geo Services and map mashups, *Cartographica: The International Journal for Geographic Information and Geovisualization*, 48(4), pp. 261–274.

Demarest, G., 2011, *Winning insurgent war: Back to basics*, Foreign Military Studies Office, Fort Leavenworth, Kansas.

Denevan, W.M., 1992, The pristine myth: The landscape of the Americas in 1492, *Annals of the Association of American Geographers*, 82(3), pp. 369–385.

Denevan, W.M., 2001, *Cultivated landscapes of Native Amazonia and the Andes*, Oxford University Press, Oxford.

The Detroit Geographical Expedition and Institute, 1971, *Field notes III.*

Dobson, J.E., 2012. The why, what, and where of Bowman Expeditions, *FOCUS on Geography*, 55(4), pp. 117–118.

Doolittle, W.E., 2000, *Cultivated landscapes of Native North America*, Oxford University Press, Oxford.

Edney, M.H., 1997, *Mapping an empire: The geographical construction of British India, 1765–1843*, University of Chicago Press, Chicago.

Elwood, S., Goodchild, M.F., and Sui, D.Z., 2012, Researching volunteered geographic information: Spatial data, geographic research, and new social practice, *Annals of the Association of American Geographers*, 102(3), pp. 571–590.

Enote, J., and Mclerran, J., 2011, *A:shiwi A:wan Ulohnanne—The Zuni World*, A:shiwi A:wan Museum and Heritage Center and the Museum of Northern Arizona, Zuni, NM and Flagstaff, AZ.

Fox, J., Suryanata, K., and Hershock, P. (eds.), 2005, *Mapping communities: ethics, values, practice*, East-West Center, Honolulu, HI.

Fox, J., Suryanata, K., Hershock, P., and Pramono, A.H. 2008, Mapping boundaries, shifting power: The socio-ethical dimensions of participatory mapping, in M.K. Goodman, M. Boykoff, and K. Evered (eds.), *Contentious geographies: Environmental knowledge, meaning, scale*, Ashgate, Hampshire, England and Burlington, VT, pp. 203–18.

Freeman, M.M.R., 2011, Looking back-and looking ahead – 35 years after the Inuit land use and occupancy project, *Canadian Geographer/Le Géographe canadien*, 55(1), pp. 20–31.

González, R.J., 2009, *American counterinsurgency: Human science and the human terrain*, Prickly Paradigm Press, Chicago.

Goodchild, M.F., 2007, Citizens as sensors: The world of volunteered geography, *GeoJournal*, 69(4), pp. 211–221.

Gordillo, G.R., 2004, *Landscapes of devils: tensions of place and memory in the Argentinean Chaco*, Duke University Press, Durham and London.

Gordon, E.T., Gurdián, G.C., and Hale, C.R., 2003, Rights, resources, and the social memory of struggle: Reflections on a study of indigenous and black community land rights on Nicaragua's Atlantic coast, *Human Organization*, 62(4), pp. 369–381.

Haklay, M., Singleton, A., and Parker, C., 2008, Web mapping 2.0: The neogeography of the geoweb, *Geography Compass*, 2(6), pp. 2011–2039.

Hale, C.R., 1994, *Resistance and contradiction; Miskitu Indians and the Nicaraguan State, 1894–1987*, Stanford University Press, Stanford.

Hale, C.R., 2005, Neoliberal multiculturalism: The remaking of cultural rights and racial dominance in Central America, *Political and Legal Anthropology Review*, 28(1), pp. 10–28.

Haraway, D., 1991, *Simians, cyborgs, and women: The reinvention of nature*, Routledge, New York.

Harley, J.B. 1988, Maps, knowledge, and power, in D. Cosgrove and S. Daniels (eds.), *The iconography of landscape: Essays on the symbolic representation, design and use of past environments*, Cambridge University Press, Cambridge, pp. 277–312.

Harvey, D., 1996, *Justice, nature, and the geography of difference*, Blackwell, Malden, MA.

Heizer, R.F., and Kroeber, A.L., 1976, For sale: California at 47 cents per acre, *Journal of California Anthropology*, 3(2), pp. 38–65.

Herlihy, P.H., and Knapp, G., 2003, Maps of, by, and for the peoples of Latin America, *Human Organization*, 62(4), pp. 303–314.

Herlihy, P.H., Dobson, J.E., Aguilar Robledo, M., Smith, D.A., Kelly, J.H., and Ramos Viera, A., 2008, A digital geography of indigenous Mexico: Prototype for the American Geographical Society's Bowman Expeditions, *The Geographical Review*, 98(3),pp. 395–415.

Hodgson, D.L., and Schroeder, R.A., 2002, Dilemmas of counter-mapping community resources in Tanzania, *Development and Change*, 33(1), pp. 79–100.

Horvath, R.J., 1971, The "Detroit geographical expedition and institute" experience, *Antipode*, 3(1), pp. 73–85.

Johnson, J.T., Louis, R.P., and Pramono, A.H., 2006, Facing the future: Encouraging critical cartographic literacies in indigenous communities, *ACME: An International E-Journal for Critical Geographies*, 4(1), pp. 80–98.

Kelly, J.D., Jauregui, B., Mitchell, S.T., and Walton, J., (eds.), 2010, *Anthropology and global counterinsurgency*, University of Chicago Press, Chicago.

Kitchin, R., and Dodge, M., 2007, Rethinking maps, *Progress in Human Geography*, 31(3), pp. 331–344.

Kosek, J., 2006, *Understories: The political life of forests in northern New Mexico*, Duke University Press, Durham, NC.

Kroeber, A.L., 1947, *Cultural and natural areas of Native North America*, University of California Press, Berkeley, CA.

McClintock, A., 1995, *Imperial leather: Race, gender and sexuality in the colonial contest*, Routledge, New York.

Mitchell, T., 2002, *Rule of experts*, University of California Press, Berkeley and Los Angeles, CA.

Mohan, G., and Stokke, K., 2000, Participatory development and empowerment: The dangers of localism, *Third World Quarterly*, 21(2), pp. 247–268.

Mollett, S., 2011, Racial narratives: Miskito and colono land struggles in the Honduran Mosquitia, *Cultural Geographies*, 18(1), pp. 43–62.

Mollett, S., 2013, Mapping deception: The politics of mapping Miskito and Garifuna space in Honduras, *Annals of the Association of American Geographers*, 103(5), pp. 1227–1241.

Monmonier, M., 1996 (1991), *How to lie with maps*, 2nd edn, The University of Chicago Press, Chicago.

Moore, D.S., 2005, *Suffering for territory: Race, place and power in Zimbabwe*, Duke University Press, Durham, NC.

Mundy, B., 1996, *The mapping of New Spain; indigenous cartography and the maps of the Relaciones Geográficas*, University of Chicago Press, Chicago.

Nahanni, P. 1977, The mapping project, in M. Watkins (ed.), *Dene nation: The colony within*, University of Toronto Press, Toronto and Buffalo, pp. 21–27.

Nietschmann, B.Q., 1973, *Between land and water; the subsistence ecology of the Miskito Indians, eastern Nicaragua*, Seminar Press, New York.

Nietschmann, B.Q., 1979, *Caribbean edge: The coming of modern times to isolated people and wildlife*, Bobbs-Merrill Company, Indianapolis.

Nietschmann, B.Q., 1989, *The unknown war: The Miskito nation, Nicaragua and the United States*, Freedom House, New York.

Nietschmann, B.Q., 1995, Defending the Miskito reefs with maps and GPS: Mapping with sail, scuba and satellite, *Cultural Survival Quarterly*, 18(4), pp. 34–37.

Ohland, K., and Schneider, R., 1983, *National revolution and indigenous identity: The conflict between Sandinistas and Miskito Indians on Nicaragua's Atlantic Coast*, International Work Group for Indigenous Affairs, Copenhagen.

O'Malley, M., 1976, *The past and future land: An account of the Berger Inquiry into the Mackenzie Valley pipeline*, P. Martin Associates, Toronto.

Peet, R. and Watts, M. (eds.), 1996, *Liberation ecologies; environment, development, social movements*, Routledge, London and New York.

Peet, R. and Watts, M. (eds.), 2004, *Liberation ecologies; environment, development, social movements*, 2nd edn, Routledge, London and New York.

Peluso, N.L., 1995, Whose woods are these? Counter-mapping forest territories in Kalimantan, Indonesia, *Antipode*, 27(4), pp. 383–406.

Perkins, C., 2014, Plotting practices and politics: (im)mutable narratives in OpenStreetMap, *Transactions of the Institute of British Geographers*, 39(2), pp. 304–317.

Pickles, J., 2004, *A history of spaces; cartographic reason, mapping and the geo-coded world*, Routledge, London and New York.

Poole, P., 1995, Indigenous peoples, mapping, and biodiversity conservation: an analysis of current activities and opportunities for applying geomatics technologies, Report prepared for The Biodiveristy Support Program, US Agency for International Development, Washington, DC.

Povinelli, E., 1999, Settler modernity and the quest for an indigenous tradition, *Public Culture*, 11(1), pp. 19–48.

Povinelli, E., 2011, Digital futures, *Vectors Journal*, 3(2), online at: http://vectors.usc.edu/projects/index.php?project=90 (accessed August 1, 2014).

Pred, A., and Watts, M., 1992, *Reworking modernity: Capitalisms and symbolic discontent*, Rutgers University Press, New Brunswick.

Price, D.H., 2011, *Weaponizing anthropology: Social science in service of the militarized state*, CounterPunch, Petrolia, CA.

Raffles, H., 2002, *In Amazonia: A natural history*, Princeton University Press, Princeton.

Rambaldi, G., Chambers, R., McCall, M., and Fox, J., 2006, Practical ethics for PGIS practitioners, facilitators, technology intermediaries and researchers, *Participatory Learning and Action*, pp. 106–113.

Rocheleau, D., 2005, Maps as power tools: Locating communities in space or situating people and ecologies in place?, in P. Brosius, A. Tsing, and C. Zerner (eds.), *Communities and conservation: Histories and politics of community-based natural resource management*, Altamira Press, Walnut Creek, pp. 327–362.

Rocheleau, D., Thomas-Slayter, B., and Edmunds, D., 1995, Gendered resource mapping: Focusing on women's spaces in the landscape, *Cultural Survival Quarterly*, 18(4), pp. 62–68.

Rodríguez-Piñero, L., 2005, *Indigenous peoples, postcolonialism, and international law: The ILO regime, 1919–1989*, Oxford University Press, Oxford and New York.

Roth, R., 2007, Two-dimensional maps in multi-dimensional worlds: A case of community-based mapping in Northern Thailand, *Geoforum*, 38(1), pp. 49–59.

Roth, R., 2009, The challenges of mapping complex indigenous spatiality: From abstract space to dwelling space, *Cultural Geographies*, 16(2), pp. 207–227.

Rushe, D. 2012, Google's Mr. Maps sets his sights on world delineation, *Guardian*, December 7, 2012, online at: www.guardian.co.uk/technology/2012/dec/07/google-maps-street-view-world (accessed August 15, 2014).

Sauer, C.O., 1969, The morphology of landscape, in J. Leighly (ed.), *Land and life: A selection from the writings of Carl Ortwin Sauer*, University of California Press, Berkeley, pp. 315–350.

Scott, J.C., 1998, *Seeing like a state: How certain schemes to improve the human condition have failed*, Yale University Press, New Haven and London.

Sletto, B.I., 2009, We drew what we imagined, *Current Anthropology*, 50(4), pp. 443–476.

Sletto, B.I., 2014, Cartographies of remembrance and becoming in the Sierra de Perijá, Venezuela, *Transactions of the Institute of British Geographers*, 39(3), pp. 360–372.

Sletto, B.I., Bryan, J., Torrado, M., Hale, C., and Barry, D., 2013, Territorialidad, mapeo participativo y política sobre los recursos naturales: la experiencia de América Latina, *Cuadernos de Geografía: Revista Colombiana de Geografía*, 22(2), pp. 193–209.

Smith, D.A., Herlihy, P.H., Kelly, J.H., and Ramos Viera, A., 2009, The certification and privatization of indigenous lands in Mexico, *Journal of Latin American Geography*, 8(2), pp. 175–207.

Soden, R., and Palen, L., 2014, From crowdsourced mapping to community mapping: The post-earthquake work of OpenStreetMap Haiti, in C. Rossito, L. Ciolfi, D. Martin, and B. Conein (eds.), *COOP 2014 – Proceedings of the 11th International Conference on the Design of Cooperative Systems, 27-30 May 2014, Nice (France)*, Springer International Publishing, Cham, pp. 311–326.

Sparke, M., 1998, A map that roared and an original atlas: Canada, cartography, and the narration of nation, *Annals of the Association of American Geographers*, 88(3), pp. 463–495.

Steward, J., 1937, Linguistic distributions and political groups of the Great Basin Shoshone, *American Anthropologist*, 39(4), pp. 625–643.

Steward, J., 1955, *Theory of culture change: The method of multilinear evolution*, University of Illinois Press, Urbana.

Stewart, O.C., 1961, Kroeber and the Indian claims commission cases, *Kroeber Anthropological Society Papers*, 25, pp. 181–190.

Stewart, O.C., 2002, *Forgotten fires: Native Americans and the transient wilderness*, University of Oklahoma Press, Norman.

Sutton, I. (ed.), 1986, *Irredeemable America: The Indians' estate and land claims*, Olympic Marketing Corp, Minnetonka.

Thongchai Winichakul, 1994, *Siam mapped: A history of the geo-body of a nation*, University of Hawaii Press, Honolulu.

Tobias, T., 2000, *Chief Kerry's moose: A guidebook to land use and occupancy mapping, research design, and data collection*, Union of British Columbia Indian Chiefs and Ecotrust Canada, Vancouver.

Tobias, T.N., 2009, *Living proof: The essential data-collection guide for indigenous use-and-occupancy map surveys*, Ecotrust Canada, Vancouver.

Tsing, A.L. 1999, Becoming a tribal elder, and other green development fantasies, in T. Li (ed.), *Transforming the Indonesian uplands: Marginality, power and production*, Harwood Academic Publishers, Amsterdam, pp. 157–200.

Turnbull, D., 1989, *Maps are territories: Science is an atlas*, University of Chicago Press, Chicago.

Turnbull, D., 2000, *Masons, tricksters and cartographers: Comparative studies in the sociology of scientific and indigenous knowledge*, Harwood Academic Publishers, Amsterdam.

Turnbull, D., 2005, Locating, negotiating, and crossing boundaries: A Western Desert land claim, the Tordesillas line, and the West Australian border, *Environment and Planning D: Society and Space*, 23, pp. 757–770.

Wainwright, J., 2005, The geographies of political ecology: After Edward Said, *Environment and Planning A*, 37(6), pp. 1033–1043.

Wainwright, J., 2008, *Decolonizing development: Colonial power and the Maya*, Blackwell, Malden, MA.

Wainwright, J., 2013, *Geopiracy: Oaxaca, militant empiricism, and geographical thought*, Palgrave Pivot, New York.

Wainwright, J. and Bryan, J., 2009, Cartography, territory, property: Postcolonial reflections on indigenous counter-mapping in Nicaragua and Belize, *Cultural Geographies*, 16(2), pp. 153–178.

Walker, P.A., and Peters, P.E., 2001, Maps, metaphors, and meanings: Boundary struggles and village forest use on private and state land in Malawi, *Society and Natural Resources*, 14, pp. 411–424.

Warhus, M., 1997, *Another America: Native American maps and the history of our land*, St. Martin's Press, New York.

Watkins, M. (ed.), 1977, *Dene Nation: The colony within*, University of Toronto Press, Toronto.

Whitmore, T.M., and Turner, B.L., 2001, *Cultivated landscapes of Middle America on the eve of conquest*, Oxford University Press, Oxford.

Weinstein, M.S., 1976, *What the land provides: An examination of the Fort George subsistence economy and the possible consequences on it of the James Bay hydroelectric project*, Grand Council of the Crees of Quebec, Montreal.

Wood, D., 2003, Cartography is dead (thank God!), *Cartographic Perspectives*, 45(45), pp. 4–7.

Wood, D., 2010, *Rethinking the power of maps*, The Guilford Press, New York.

Wood, D., and Fels, J., 2008, *The natures of maps: Constructions of the natural world*, The University of Chicago Press, Chicago.

Zook, M., Graham, M., Shelton, T., and Gorman, S., 2010, Volunteered geographic information and crowdsourcing disaster relief: A case study of the Haitian earthquake, *World Medical & Health Policy*, 2(2), pp. 6–32.

20

HISTORICAL APPROACHES TO POLITICAL ECOLOGY

Diana K. Davis

Introduction

Much work in political ecology has been historically focused since it appeared as a subdiscipline in geography in the early 1980s. The subsequent three decades have witnessed varying degrees of emphasis on historical depth and context in political ecological research but historical approaches remain both theoretically and empirically important to political ecology generally. More deeply historical work in political ecology is often termed "historical political ecology." Historical approaches to political ecology are as varied and multifaceted as political ecology itself but they share the common political ecological core of being informed by various forms of critical social theory (from Marx, Foucault, and Said to Polanyi, Gramsci and Haraway), utilizing multiple, mixed research methods (archival, survey/interview, ethnographic, biophysical), being committed to social and environmental justice, and aiming to have contemporary policy relevance. They take seriously the political and economic forces of environmental change, in addition to social and cultural forces, and directly address questions of various kinds of power in their analyses.

Just as there is no single theoretical or methodological approach to political ecology (Peet *et al.* 2011; Peet and Watts 2004; Robbins 2012), there is no single definition of or approach to "historical political ecology." Rather than asking "what is historical political ecology?" which has been explored elsewhere (Davis 2009; Offen 2004; Peluso *et al.* 2014), in this chapter I would like to examine the ways in which historical approaches to political ecology strengthen our analyses. What is to be gained from taking more historical approaches to political ecology, especially when so much political ecology research deals with very contemporary questions of socio-environmental change and the conflicts that such change often engenders? I suggest here that it is precisely the historical components of political ecology research that help to guard against "apolitical" analyses. This is so because it is only with a sophisticated and critical understanding of the historical development of landscapes/environments, of social relations, and of knowledge and the privilege that attends it, that we can reveal the hidden relations of power often at play in the questions studied by political ecologists.

This chapter analyzes why and how history has been important to political ecology and overviews many of the classic works in "historical political ecology" to demonstrate the benefits of critical historical approaches to political ecology. It then outlines and discusses some of the areas of similarity and difference with the cognate disciplines of historical geography and environmental history. Finally, it examines some of the newer work in political ecology and the ways in which historical approaches strengthen such research. Rather than argue that all political ecology should somehow be "historical political ecology," I conclude that critical historical approaches, whether a lengthy examination of several hundred years or a short surgical strike on only a few decades, remain crucial to political ecology. That is to say that critical historical analysis is at the heart of political ecology and it should become more clearly articulated in more of our work.

Why history?

During the early years of political ecology, history was important to a great deal of research because the focus was on Marxist or neo-Marxist political economy in the formerly colonial parts of the world. An engagement with peasant studies, post-colonialism, feminism, and discourse analysis infused political ecology work in the 1980s and 1990s as well, bringing more reasons to take history seriously in political ecology. Much of the more recent work in political ecology, such as that interfacing with science and technology studies (STS), theories of governmentality and post-humanism likewise highlights the importance of history in producing sophisticated, critical analyses. None of this kind of work is possible without careful diachronic research that reveals how the current (often dialectical) situation came to exist. This begs the question, though, of what kind of history is being utilized and/or written in such historically informed research.

A reliance on mainstream, often Euro-triumphalist histories is not usually adequate for studies in political ecology precisely because they so frequently privilege existing, exploitative power structures. Critical historical analyses that move beyond such historicism are particularly adept at revealing what is often the very political nature of issues represented as apolitical environmental problems with simple technical fixes. Sometimes this critical historical approach requires historical (re)construction and sometimes it requires deconstruction of mainstream histories, or both. This is vitally important because the ways in which history is told can mask and/or naturalize inequitable and exploitative relations at the center of contemporary nature–society problems.

Historically informed research in political ecology can take more than one approach to incorporate such critical historical analysis. In longer publications, usually of monograph length, time can be spent on deconstructing and/or (re)constructing histories to create new sophisticated political ecological analyses. In shorter publications, of chapter or article length, though, it is crucial to take adequate time deciphering which historical sources of those available take a critical perspective, that is, finding those sources that don't simply repeat biased, triumphalist stories of the past. This can represent a substantial investment of time on the part of the researcher/author since geographers are not often terribly well-trained in history and vetting historical sources.

This also raises the question of "what counts as history?" Even professional historians tend to disagree over this question with some, for example, denying the importance or validity of "recent history" while others may discount histories that utilize sources other than archival (for example, oral histories). It is important to keep in mind that historical approaches to political ecology don't need to "go back" a minimum number of years, decades, or centuries. Each research question for which an historical approach is beneficial will have its own key time period(s), some quite recent, for which detailed and critical historical analysis is necessary. Historical approaches to political ecology are not limited to monograph length works, although sometimes a series of two or more articles can be very productive.

The strengths of critical historical approaches are significant for many of the socio-environmental problems tackled by political ecologists, from understanding changing social relations and resource use under capitalism and colonialism, to revealing the hidden inequalities in nature conservation and environmental development programs, to understanding environmental conflict and governance. The historically situated and contingent production of knowledge, scientific and otherwise, is also crucial to many political ecology analyses. It is not simply a question, though, of examining the political economy of knowledge production but also a question of exposing other related facets of power like ideology and hegemony in our understanding of knowledge production, circulation, and operationalization, which usually requires an historical approach (Mann 2009; Peet *et al.* 2011, p. 15). And as Nancy Peluso has recently argued, this requires excavating historically situated knowledges *and* historically situated practices, particularly with respect to studies of socio-natural commodities (Peluso 2012). These are some of the greatest strengths of historical approaches to political ecology. They have invigorated our research for decades and are expanding and enriching our avenues of research to move it in new and productive directions.

Classics in historical political ecology

Some of the earliest work in political ecology has great historical depth and shows successfully why and how history matters to political ecology (Blaikie 1985; Blaikie and Brookfield 1987; Hecht and Cockburn 1989; Jarosz 1993; Peluso 1992; Watts 2013). One of the most influential early works, Michael Watts' 1983 *Silent Violence*, is a profoundly historical account firmly grounded in five months of archival research as well as over a year of meticulous field work that included interviews, participant observation, surveys, and documentary analysis at the village level. Utilizing a Marxist approach to the political economy of famine in Nigeria, Watts demonstrates that only by understanding the historical intricacies of how African peasants were enfolded into the machinations of global capitalism, during the colonial period and since, can we understand the roots and also, importantly, the contemporary reality, of marginalization and famine.

Moreover, this work shows that only through such detailed historical understanding, combined with contemporary fieldwork, can we fully realize the difference between the apolitical explanations of the "natural disaster" of drought and famine and the silent violence of historically specific changes in political economy and social relations that replaced a moral economy with a market economy at great social and environmental cost (Watts 1983, 2013). Watts was also careful to point out the potential weaknesses of his methods. Since archival documents are essentially class products, he warned that "the distorted optic provided by a wholesale dependence on archival sources can, and I would argue must, be complemented by oral fieldwork" (Watts 2013, p. 34). Many subsequent political ecologists have heeded this warning and some have further elaborated on it (Robbins 2012, p. 67). This theoretically informed triangulation of multiple methods is one of the hallmarks of political ecology.

Nancy Peluso's *Rich Forests, Poor People* appeared a decade after *Silent Violence* and likewise revealed the violence inherent in the restructuring of nature–society relations under capitalism, from the colonial period to the present, but in a different part of the world: Java, Indonesia. Focused on forests and forest livelihoods rather than the agrarian sector, Peluso's book incorporated long-term ethnography with documentary and historical research. Although not "archival" in the sense of being conducted in formal archives, the historical depth of this research is rich and compelling and shows clearly how crucial are the history of struggles for political-economic power and the history of resistance to understanding conservation's past and present. Inspired by peasant studies, she demonstrates vividly that scientific forestry can also be used for environmental and social control. Her argument is strongly convincing in large part because she so carefully delineates the historical transformations of nature–society relations in the Javanese forest over 150 years.

These two works are emblematic of much of what has been so productive about political ecology research that incorporates a carefully historical approach. These two foci of the problems of marginalization in the agrarian sector and the problems of social and environmental control in the conservation sector have provided inspiration and direction for a great deal of subsequent research in political ecology. It is also significant that both of these books deal with socio-environmental change in formerly colonized countries, from the pre-colonial period to the present, a common theme in much political ecology research that takes an historical approach (Bryant 1997; Carney 2001; Carney and Rosomoff 2012; Hecht and Cockburn 1989; Schroeder 1999; Sluyter 2012; Zimmerer 1996).

A second generation of political ecology work has built on the insights of this early research and further engaged with history in significant and productive ways. One of these has been the development of research in (post)colonial settings concerned with nature conservation and resource use that takes environmental representation and the power of discourse and narrative seriously (Bassett and Zuéli 2003; Davis 2007; Fairhead and Leach 1996, 1998; Goldman 2005; Hecht *et al.* 2014; Kull 2004; Leach and Mearns 1996; Neumann 1998; Showers 2005; Wainwright 2008). Much of this research has been very influential in this more genealogical approach to the political ecology of nature conservation. By tracing current, frequently erroneous, "received wisdoms" about ecology and historical landscape change to what are often their colonial roots, these authors have shown in multiple settings that contemporary conservation and agricultural laws and policies frequently replicate colonial policies to varying degrees. Moreover, they show that such policies and the narratives that support and justify them are often maintained because they provide state and/or elite actors with various forms of power and control over natural resources and human populations. This historically informed political ecology research has been particularly good at interrogating the relationships between discourses or narratives and material changes in landscapes and social relations over time. Important to many of these works is the idea that such powerful environmental narratives are often informed by long-standing notions of what particular types of landscape should look like and how they should be used – environmental imaginaries – that traveled with the colonial actors.

Rod Neumann's research on Arusha National Park in Tanzania is a particularly good example of how historical approaches that consider various forms of environmental representation strengthen political ecology (Neumann 1998). This analysis of the development of a major African national park and its attendant problems of "poaching" and the marginalization of proximate population groups reveals a long and complicated history dating back to the colonial period. In detailing the rise of faunal protection for big game hunting by European elites and how it led in many ways to the kind of environmental protection embodied today by national parks that aim to protect individual species and entire ecosystems, Neumann explains the profound changes in property rights, livelihood strategies, and moral and political economies in the region. The political-economic and social changes detailed here were driven less by global capitalist agriculture or markets for natural resources, though, than by international nature tourism and the commodification of the landscapes being "sold." Neumann explains that how the colonial elite conceived of "wilderness," and the ideas of what "nature" should look like and how it should be represented and used, derived from their historically specific social, economic, and political context of nineteenth-century England. As he details, these ideas themselves had a significant history in ideas of landscape norms/ideals in English culture.

Neumann deftly traces the history of the English wilderness imaginary from its early roots to its current incarnation as part of the global discourse that drives mainstream nature protection today.

Working to uncover the history and politics of environmental knowledge production, its sometimes contested spread to become a dominant narrative, and its use in a variety of development interventions is especially important to political ecology research in post-colonial settings because so many "baseline" data sets and other environmental understandings are derived from the colonial era and frequently contain biases and errors. Such baseline data sets include deforestation statistics, estimates of desertification, rates of erosion, and natural vegetation maps among others. These data sets, occasionally lightly revised, continue to inform contemporary environment and development plans, laws, and policies in a great many parts of the world, sometimes with very negative social and ecological effects (Hecht *et al.* 2014; Kull 2004; Leach *et al.* 1996; Showers 2005). Moreover, development policies based on data sets from a particular location are sometimes applied to different locations with very different ecological conditions resulting in questionable outcomes (Forsyth 2011). By problematizing such data sets, historical approaches to political ecology have much to offer land-use change studies, restoration ecology, and landscape ecology globally since baselines are usually crucial to their work.

For instance, in trying to understand the great divergences in opinion regarding desertification and land degradation between nomads in southern Morocco on the one hand and government officials and NGO personnel on the other, part of my research in the Maghreb sought to trace the evolution of the official history of the Maghreb's allegedly ruined environment. The development in the 1930s of potential vegetation maps by an influential French ecologist was a key moment in the long development and deployment of this colonial environmental narrative (Davis 2007, chapter 5). Informed by a century of speculation derived primarily from literary sources, and based on the subjective French phytosociology tradition, these vegetation maps effectively transformed a biased, incorrect colonial story of environmental destruction by the "natives" into an apolitical, established scientific fact. The maps were then used by many, including the powerful French colonial forester Paul Boudy, to calculate rates of deforestation and erosion as well as estimates of desertification that formed the basis of many environmental policies that marginalized North Africans in the Maghreb during the colonial period and since. These maps, however, simply served to reinforce and institutionalize earlier and largely incorrect assumptions about the nature and extent of environmental damage blamed on the indigenous populations.

Although the colonial environmental narrative had been contested (unsuccessfully) several times, the production of the potential vegetation maps gave the narrative great power and authority (Harley 1988). This power and authority has remained, helping to perpetuate the colonial environmental history and marginalizing indigenous populations, despite paleoecological and contemporary arid lands ecology research demonstrating that the Maghreb countries have not been as severely deforested or desertified as colonial and post-colonial texts have claimed. Furthermore, this research suggests that indigenous understandings and uses of the environment may in fact be better suited to its stochastic and non-equilibrium ecological dynamics much of the time (Davis 2007; Sayre 2006 documents a similar situation in the US Southwest). The colonial environmental history has remained the dominant story of environmental change in the Maghreb in part because it provides significant political and economic benefits to the state and other powerful actors when it is operationalized (Davis 2006).

Another reason for its success, however, is that the colonial narrative was part of a dominant ideology of understanding the environment that was based on notions of equilibrium that had become nearly universal. As such it had very similar counterparts in other parts of the French, British, and other European as well as American imperial territories that facilitated a hegemonic environmental scientific understanding in the Euro-American community. Moreover, given how the problems were framed, the solutions were relatively simple to conceive and apply (reforestation, strict forest laws, and curtailing pastoral grazing, for example). That is, the specific problematization of the Maghreb environment under French colonialism led to the alleged problems being rendered technical in certain ways that were simple to understand and operationalize which in turn led to their widespread adoption and application (Lave 2011; Li 2007). My research on the Maghreb, then, tried to show, in addition to the evolution and effects of the colonial narrative, the conditions of production of some of the environmental knowledge (potential vegetation maps and the data sets derived from them) used to drive that narrative from the early twentieth century in the hopes of enabling better environmental policy formulation.

Having been dominated by research in the "third world" for a long time, political ecology welcomed new research in "first world" and urban political ecology as the subdiscipline matured. Some of this work demonstrates how historical approaches can also strengthen research and analysis in these relatively new subfields (Biehler 2009; Freidberg 2009; Guthman 2004; Hollander 2008; McCarthy 2001; Sayre 2006; Swyngedouw 1997, 2004). Bruce Braun's *The Intemperate Rainforest* is at one and the same time "first world" and post-colonial political ecology. In this book, Braun highlights the power of nature's representation, textual, oral, and visual, as well as the importance of the social construction of nature (and the often old and complex history of such construction) for understanding contemporary conflicts over the rainforest of Canada's west coast. In common with much "third world" political ecology, Braun successfully explains with detailed historical analysis that the indigenous populations, the First Nations, have essentially been written out of a history of forest landscape use by a (neo) colonial elite, a process begun in the eighteenth century. Braun draws on documents from scientific and exploration expeditions, photography and artistic representation to trace the genealogy of representation of the forest and its indigenous populations. The largely unwitting use of these very old tropes today by environmentalists, foresters, the state, and others has perpetuated a dominant view of the forest which reinforces inegalitarian power structures and denies the people of the First Nations any substantial voice in decision making about the forest. Braun thus provides a sophisticated analysis of contemporary conflicts over the rainforest that has the potential to show what a "postcolonial environmentalism" might look like, one in which nature is understood as part of, rather than separate from, "the messy world of history and politics" (Braun 2002, p. x).

While they are very different books based on US experiences, both Paul Robbins' *Lawn People* (Robbins 2007) and Jake Kosek's *Understories* (Kosek 2006) use historical context strategically, as does Braun, to make key points throughout their political ecological analyses. This use of history allows both authors to make convincing arguments about environmental subjectivities, how such subjectivities were created and changed over time, and how and why they are so important today. In both cases, one treating the "good citizen's" use of toxic chemicals to maintain the suburban lawn ideal, and the other dissecting the insidious relationship between race and nature that has helped to form subjectivities and conflicts of a different kind, this historical depth provides much more nuanced and much more political research. Similarly, new research in the Camargue, southern France, demonstrates the effectiveness of in-depth analysis of particular historical moments in the evolution of the Scamandre marshes and their uses to successfully explain the current, very political, "crisis" in the region (Peluso *et al.* 2014).

Historical political ecology and allied fields

The strategic use of historical analysis is one aspect of political ecology research that frequently differentiates it from a majority of research in either historical geography or environmental history, two related subdisciplines. Historical geographers and environmental historians are very skilled at producing detailed histories of places that read smoothly (and usually beautifully) as coherent stories. As historian David Biggs has recently explained, "the field of environmental history is largely oriented to the phenomenological flow of stories" (Biggs 2014). Political ecologists, on the other hand are frequently more interested in providing historical detail(s) of certain aspects of their analysis that they consider important for understanding the research problem. As Jake Kosek explains, his book about Northern New Mexico is "not meant to be a detailed history of place but rather a place-based history of the articulations and politics of nature and difference" (Kosek 2006, p. 28). This is a subtle difference but it highlights that political ecologists tend to be problem-oriented while environmental historians and historical geographers tend to conceive of their research chronologically.

The political ecologist's concern with contemporary issues of environmental and social justice, and the goal of relevance of their research to contemporary problems and policy development, also sets their work apart from environmental historians and historical geographers who have eschewed "activist" research on the whole (Offen 2004, see p. 22). Much work in critical social geography, including political ecology, aims to make tangible improvements in the "real world" as is evident from the recent volume titled: *The Point is to Change It: Geographies of Hope and Survival in an Age of Crisis* (Castree *et al.* 2010). A majority of environmental history has also been declensionist (McNeill 2003), something that historically minded political ecologists have worked long and hard to critically deconstruct (Davis 2007; Kull 2004; Leach *et al.* 1996). The utilization of ethnographic and biophysical research and data also often differentiates the research of these three subdisciplines since political ecologists nearly always use multiple methods and data sources that include these in addition to critical social theory.

The dearth of social theory in environmental history has attracted attention and criticism from within the field itself (Sorlin and Warde 2007) as has its dominant focus on North America (Sutter 2003). As explained by geographer Robert Wilson, one influential environmental historian "has gone so far as to claim that many historians, especially environmental historians, were 'refugees' from theory" (Wilson 2013, p. 356). The work of environmental historians of the Global South, by contrast, often embraces various forms of social theory and has been much more strongly informed by attention to political economy, policy, and narrative in its analyses (Arnold 1996; Arnold and Guha 1998; Beinart 1984; Grove *et al.* 1998; Sutter 2003).

There are of course many exceptions to the general differences laid out above and the research in all three subdisciplines, which are constantly changing, has much to offer as several recent articles show (Colten 1998; Endfield 2009b; Naylor 2006; Offen 2012; Weiner 2005; Wynn *et al.* 2014). These three subdisciplines are becoming increasingly synergistic. There is a growing body of environmental history work that is directly relevant to contemporary nature–society problems (Carey 2010; Cronon 1983, 1992; Jacoby 2001; Langston 2010; Nash 2007; Safier 2008; Schiebinger 2004; Warren 1997). Likewise, there is a growing body of historical geography that has much to offer contemporary questions of interest to political ecologists (Cosgrove and Daniels 1989; Driver and Martins 2005; Endfield 2009a; Endfield and Nash 2002; Williams 2003; Wynn 2007). In particular, the exciting body of work on "geographies of knowledge" within historical geography is engaging in very productive ways with environmental knowledge constructions and their uses (Livingstone 2011; Livingstone and

Withers 2011; Naylor 2005). This interface between geography and science studies and histories of science is proving particularly fruitful.

New directions

One productive direction in which some historically informed work in political ecology has been heading in the last decade is to try to engage with insights from science and technology studies (STS) in order to analyze more carefully the creation, circulation, and operationalization of environmental knowledge (Lave 2012). As has been pointed out, some research in political ecology, while very strong in uncovering the "history and political economy behind scientific truth claims in certain places," has been less successful in explaining "the success of certain knowledge claims over others" in detail (Goldman *et al.* 2011, p. 20). To be sure, there are many debates over the strengths and weaknesses of incorporating an STS approach, in part because STS is itself multifaceted with competing approaches. Actor network theory (ANT), for example, one popular branch of STS, has been criticized by political ecologists for being apolitical and thus a poor fit for the subfield's focus on justice (Goldman *et al.* 2011, especially the introduction; Lave 2011 and Chapter 16, this volume; Robbins 2012). Several of these scholars have instead advocated the concepts of hybridity, co-production, and Bordieu's field concept (Lave 2012) as more compatible with political ecology.

Some very interesting historically informed work has been conducted, for example, on the creation, circulation, and application of ecological knowledge as it is mediated by and through technologies like remote sensing, GIS, and systems biology models. Both Paul Robbins and Matthew Turner have argued, in India and Niger respectively, that without acknowledgment and critical interrogation of the many assumptions of what constitutes a "forest" or a "healthy" rangeland when interpreting remotely sensed pixels and building a GIS, old mistakes may be carried forward and unintended consequences can occur. Robbins deftly demonstrates that in India, the use of remotely sensed data has shown tree cover expanding whereas the reality on the ground is that many valuable trees have been disappearing (Robbins 1998). In this case, an exotic invasive tree has been expanding, to the state foresters' delight in seeing increased forest cover, but at the expense of other local tree species long valued by the indigenous populations. It is by critically investigating the history of what "forest" and land degradation mean to various groups and how categories of landscape cover are assigned that Robbins is able to demonstrate that the apparently neutral and objective nature of satellite data is actually very political and that its application can easily have ecologically and socially negative outcomes (Robbins 2001a, 2001b). Turner's work in the Sahel likewise analyzes the application of remote sensing technologies and the use of GIS in evaluating arid rangelands (Turner 2003). By critically studying the history of how satellite data has been used and comparing it with a longer history of environmental scientific traditions in the Sahel, in particular the French phytogeographical tradition, he is able to demonstrate that what appears to many to be new, objective data actually replicate to a large degree the known errors of interpretation that have occurred in ecological research since the early twentieth century to the detriment of development in the region.

More recently, Joan Fugimura has revealed how important examination of the historical production of "technobiological imaginaries" is to understanding some of the pitfalls of the mechanistic models being developed in post-genomic research in systems biology (Fujimura 2011). Her research shows how simplistic adoption of mechanistic models may not only limit understanding of complex systems but also have questionable regulatory implications. The historical perspective in these three examples also highlights the flexibility of historical analysis for political ecology in terms of historical depth. The history they trace in terms of the recent use of these technologies is not very long, a few decades at most. Although fruitfully compared with much older notions of nature and earlier ecological and biological research dating to the colonial period or earlier in the twentieth century, the time period covered in detail in these short pieces is not terribly deep. Despite this, the historical analysis is detailed and critical and therefore adds key components to the main arguments of each author; indeed, they could not have made their main points without this critical historical analysis or with the simple repetition of mainstream (often triumphalist) histories.

A successful example of succinct critical historical analysis is Julie Guthman's book *Weighing In* (Guthman 2011), in which she examines, problematizes, and dismantles the "obesity epidemic" in the US and its related problems with critical historical analysis that reaches back only a few decades for the most part. While many might not consider this book to be "historical political ecology," it provides critical historical analysis of the biomedical concepts of BMI and obesity, of food and agricultural policy, of the capitalist expansion and transformation of food production, and of labor and social justice issues, among others. That is, critical historical analysis is a crucial part of the methodology and argumentation of this important work.

Further examples may be found in some of the newest work in political ecology, such as the recent work on management of public health and mosquitoes, or the emergence of mysterious new forms of kidney disease, from the perspective of assemblage theory and co-production that are informed by critical historical analysis derived from institutional archives (Robbins *et al.* 2008; Senanayake 2014; Shaw *et al.* 2010). Other new research utilizing similar theoretical tools has used historical analysis as a kind of surgical strike, focusing on the narrow period of World War II, to argue that desert locusts in French colonial North Africa were managed in certain ways that operationalized techno-politics to facilitate imperial rule across much of Africa and to legitimize *Free France* during the Nazi occupation (Peloquin 2013).

Conclusion

Such historical approaches to political ecology strengthen our work in the vast majority of cases. An understanding of the nexuses of social relations, of knowledge and power and their constitution, spread and operationalization, so necessary in political ecology, is provided only with careful, critical historical analysis. Whether the time period is short and recent or long and in the deep past, matters less than conducting critical historical research that interrogates environmental knowledge production, power relations, and their representation, application, and the ways these have spread, changed, and been internalized over time under global capitalism. The strengths of historical approaches to political ecology are more obvious, perhaps, in lengthier works that deal with the *longue durée* and/or the colonial period. They are, however, equally valuable to research more focused on recent problems for the very reason that a lack of critical analysis of situated histories will likely lead to apolitical analyses which do not fulfill the aims of political ecology. All political ecology research should include critical historical analysis as a key component.

References

Arnold, D. (1996) *The Problem of Nature: Environment, Culture and European Expansion.* Oxford: Blackwell.

Arnold, D., and R. Guha, eds. (1998) *Nature, Culture and Imperialism: Essays on the Environmental History of South Asia.* Delhi: Oxford University Press.

Bassett, T.J., and K.B. Zuéli. (2003) The Ivorian Savanna: Global Narratives and Local Knowledge of Environmental Change. In *Political Ecology: An Integrative Approach to Geography and Environment-Development Studies*, edited by K.S. Zimmerer and T.J. Bassett. New York: Guilford, 115–136.

Beinart, W. (1984) Soil Erosion, Conservation and Ideas about Development: A Southern African Exploration. *Journal of Southern African Studies*, 11: 52–84.

Biehler, D.d. (2009) Permeable Homes: A Historical Political Ecology of Insects and Pesticides in US Public Housing. *Geoforum*, 40 (6): 1014–1023.

Biggs, D. (2014) New Spaces for Stories: Technical and Conceptual Challenges to Using Spatial Imagery in Environmental History. *Environmental History*, 19 (2): 271–280.

Blaikie, P. (1985) *The Political Economy of Soil Erosion in Developing Countries*. London: Longman.

Blaikie, P., and H. Brookfield. (1987) *Land Degradation and Society*. London: Methuen.

Braun, B. (2002) *The Intemperate Rainforest: Nature, Culture, and Power on Canada's West Coast*. Minneapolis: University of Minnesota Press.

Bryant, R. (1997) *The Political Ecology of Forestry in Burma, 1824–1994*. Honolulu: University of Hawaii Press.

Carey, M. (2010) *In the Shadow of Melting Glaciers: Climate Change and Andean Society*. Oxford: Oxford University Press.

Carney, J. (2001) *Black Rice: The African Origins of Rice Cultivation in the Americas*. Cambridge, MA: Harvard University Press.

Carney, J., and R. Rosomoff. (2012) *In the Shadow of Slavery: Africa's Botanical Legacy in the Atlantic World*. Berkeley: University of California Press.

Castree, N., P. Chatterton, N. Heynen, W. Larner, and M.W. Wright, eds. (2010) *The Point is to Change It: Geographies of Hope and Survival in an Age of Crisis*. Oxford: Wiley-Blackwell.

Colten, C. (1998) Historical Geography and Environmental History. *Geographical Review*, 88 (1): iii–iv.

Cosgrove, D., and S. Daniels. (1989) *The Iconography of Landscape: Essays on Symbolic Representation, Design, and Use of Past Environments*. Cambridge: Cambridge University Press.

Cronon, W. (1983) *Changes in the Land: Indians, Colonists, and the Ecology of New England*. New Haven: Yale University Press.

Cronon, W. (1992) A Place for Stories: Nature, History and Narrative. *The Journal of American History*, 78 (March): 1347–1376.

Davis, D.K. (2006) Neoliberalism, Environmentalism and Agricultural Restructuring Morocco. *The Geographical Journal*, 172 (2): 88–105.

Davis, D.K. (2007) *Resurrecting the Granary of Rome: Environmental History and French Colonial Expansion in North Africa*. Athens: Ohio University Press.

Davis, D.K. (2009) Historical Political Ecology: On the Importance of Looking Back to Move Forward. *Geoforum*, 40 (3): 285–286.

Davis, M. (2001) *Late Victorian Holocausts: El Nino and the Making of the Third World*. London: Verso.

Driver, F., and L. Martins, eds. (2005) *Tropical Visions in an Age of Empire*. Chicago: The University of Chicago Press.

Endfield, G. (2009a) Reconsidering Climate and Causality: Case Studies from Colonial Mexico. In *Nature's End: History and the Environment*, edited by S. Sorlin and P. Warde. Houndsmills: Palgrave Macmillan, 304–328.

Endfield, G.H. (2009b) Environmental History. In *A Companion to Environmental Geography*, edited by N. Castree. Oxford: Wiley-Blackwell, 223–237.

Endfield, G.H., and D.J. Nash. (2002) Missionaries and Morals: Climatic Discourse in Nineteenth-Century Central Southern Africa. *Annals of the Association of American Geographers*, 92 (4): 727–742.

Fairhead, J., and M. Leach. (1996) *Misreading the African Landscape: Society and Ecology in a Forest-Savanna Mosaic*. Cambridge: Cambridge University Press.

Fairhead, J., and M. Leach. (1998) *Reframing Deforestation: Global Analyses and Local Realities: Studies in West Africa*. London: Routledge.

Forsyth, T. (2011) Politicizing Environmental Explanations: What can Political Ecology Learn from Sociology and Philosophy of Science? In *Knowing Nature: Conversations at the Intersection of Political Ecology and Science Studies*, edited by M.J. Goldman, P. Nadasdy and M.D. Turner. Chicago: The University of Chicago Press, 31–46.

Freidberg, S. (2009) *Fresh: A Perishable History*. Cambridge, MA: Harvard University Press.

Fujimura, J.H. (2011) Technobiological Imaginaries: How do Systems Biologists know Nature? In *Knowing Nature: Conversations at the Intersection of Political Ecology and Science Studies*, edited by M.J. Goldman, P. Nadasdy and M.D. Turner. Chicago: The University of Chicago Press, 65–80.

Goldman, M. (2005) *Imperial Nature: The World Bank and Struggles for Social Justice in the Age of Globalization*. New Haven: Yale University Press.

Goldman, M.J., P. Nadasdy, and M.D. Turner, eds. (2011) *Knowing Nature: Conversations at the Intersection of Political Ecology and Science Studies*. Chicago: The University of Chicago Press.

Grove, R., V. Damodaran, and S. Sangwan, eds. (1998) *Nature and the Orient: The Environmental History of South and Southeast Asia*. Delhi: Oxford University Press.

Guthman, J. (2004) *Agrarian Dreams: The Paradox of Organic Farming in California*. Berkeley: University of California Press.

Guthman, J. (2011) *Weighing In: Obesity, Food Justice, and the Limits of Capitalism*. Berkeley: University of California Press.

Harley, J.B. (1988) Maps, Knowledge, and Power. In *The Iconography of Landscape: Essays on the Symbolic Representation, Design and Use of Past Environments*, edited by D. Cosgrove and S. Daniels. Cambridge: Cambridge University Press, 277–312.

Hecht, S., and A. Cockburn. (1989) *The Fate of the Forest: Developers, Destroyers, and Defenders of the Amazon*. Chicago: The University of Chicago Press.

Hecht, S.B., K.D. Morrison, and C. Padoch, eds. (2014) *The Social Lives of Forests: Past, Present and Future of Woodland Resurgence*. Chicago: The University of Chicago Press.

Hollander, G.M. (2008) *Raising Cane in the 'Glades: The Global Sugar Trade and the Transformation of Florida*. Chicago: The University of Chicago Press.

Jacoby, K. (2001) *Crimes Against Nature: Squatters, Poachers, Thieves, and the Hidden History of American Conservation*. Berkeley: University of California Press.

Jarosz, L. (1993) Defining and Explaining Tropical Deforestation: Shifting Cultivation and Population Growth in Colonial Madagascar (1896–1940). *Economic Geography*, 69 (4): 366–379.

Kosek, J. (2006) *Understories: The Political Life of Forests in Northern New Mexico*. Durham: Duke University Press.

Kull, C.A. (2004) *Isle of Fire: The Political Ecology of Landscape Burning in Madagascar*. Chicago: The University of Chicago Press.

Langston, N. (2010) *Toxic Bodies: Hormone Disruptors and the Legacy of DES*. New Haven: Yale University Press.

Lave, R. (2011) Circulating Knowledge, Constructing Expertise. In *Knowing Nature: Conversations at the Intersection of Political Ecology and Science Studies*, edited by M.J. Goldman, P. Nadasdy and M.D. Turner. Chicago: The University of Chicago Press, 263–279.

Lave, R. (2012) Bridging Political Ecology and STS: A Field Analysis of the Rosgen Wars. *Annals of the Association of American Geographers*, 102 (2): 366–382.

Leach, M., and R. Mearns, eds. (1996) *The Lie of the Land: Challenging Received Wisdom on the African Environment*. London: The International African Institute.

Li, T.M. (2007) *The Will to Improve: Governmentality, Development, and the Practice of Politics*. Durham: Duke University Press.

Livingstone, D. (2011) Darwinian Landscapes. In *Envisioning Landscapes, Making Worlds: Geography and the Humanities*, edited by S. Daniels, D. DeLyser, J.N. Entrikin and D. Richardson. London: Routledge, 106–117.

Livingstone, D., and C. Withers. (2011) *Geographies of Nineteenth-Century Science*. Chicago: The University of Chicago Press.

Mann, G. (2009) Should Political Ecology be Marxist? A Case for Gramsci's Historical Materialism. *Geoforum*, 40 (3): 335–344.

McCarthy, J. (2001) States of Nature and Environmental Enclosures in the American West. In *Violent Environments*, edited by N.L. Peluso and M. Watts. Ithaca: Cornell University Press, 117–145.

McNeill, J.R. (2003) Observations on the Nature and Culture of Environmental History. *History and Theory*, 42 (December): 5–43.

Nash, L. (2007) *Inescapable Ecologies: A History of Environment, Disease, and Knowledge*. Berkeley: University of California Press.

Naylor, S. (2005) Introduction: Historical Geographies of Science – Places, Contexts, Cartographies. *The British Journal for the History of Science*, 38 (1): 1–12.

Naylor, S. (2006) Historical Geography: Natures, Landscapes, Environments. *Progress in Human Geography*, 30 (6): 792–802.

Neumann, R.P. (1998) *Imposing Wilderness: Struggles over Livelihood and Nature Preservation in Africa*. Berkeley and Los Angeles: University of California Press.

Offen, K. H. (2004) Historical Political Ecology: An Introduction. *Historical Geography*, 32: 19–42.

Offen, K. (2012) Historical Geography I: Vital Traditions. *Progress in Human Geography*, 36 (4): 527–540.

Peet, R., and M. Watts, eds. (2004) *Liberation Ecologies: Environment, Development, Social Movements*. Second edn. New York: Routledge.
Peet, R., P. Robbins, and M.J. Watts, eds. (2011) *Global Political Ecology*. London: Routledge.
Peloquin, C. (2013) Locust Swarms and the Spatial Techno-Politics of the French Resistance in World War II. *Geoforum*, 49 (1): 103–113.
Peluso, N.L. (1992) *Rich Forests, Poor People: Resource Control and Resistance in Java*. Berkeley: University of California Press.
Peluso, N. L. (2012) What's Nature Got to Do with It? A Situated Historical Perspective on Socio-Natural Commodities. *Development and Change*, 43 (1): 79–104.
Peluso, N., P. Robbins, and R. Mathavet. (2014) *Water, Reeds and Biodiversity in the French Western Camargue: Why Historical Political Ecology Matters*. Unpublished manuscript.
Robbins, P. (1998) Paper Forests: Imagining and Deploying Exogenous Ecologies in Arid India. *Geoforum*, 29 (1): 69–86.
Robbins, P. (2001a) Fixed Categories in a Portable Landscape: The Causes and Consequences of Land-Cover Categorization. *Environment and Planning A*, 33 (1): 161–179.
Robbins, P. (2001b) Tracking Invasive Land Covers in India, Or Why Our Landscapes Have Never Been Modern. *Annals of the Association of American Geographers*, 91 (4): 637–659.
Robbins, P. (2007) *Lawn People: How Grasses, Weeds, and Chemicals Make Us Who We Are*. Philadelphia: Temple University Press.
Robbins, P. (2012) *Political Ecology: A Critical Introduction*. Oxford: Wiley-Blackwell.
Robbins, P., R. Farnsworth, and J.P. Jones. (2008) Insects and Institutions: Managing Emergent Hazards in the US Southwest. *Journal of Environmental Policy & Planning*, 10 (1): 95–112.
Safier, N. (2008) *Measuring the New World: Enlightenment Science and South America*. Chicago: The University of Chicago Press.
Sayre, N.F. (2006) *Ranching, Endangered Species, and Urbanization in the Southwest: Species of Capital*. Tucson: University of Arizona Press.
Schiebinger, L. (2004) *Plants and Empire: Colonial Bioprospecting in the Atlantic World*. Cambridge, MA: Harvard University Press.
Schroeder, R. (1999) *Shady Practices: Agroforestry and Gender Politics in the Gambia*. Berkeley: University of California Press.
Senanayake, N. (2014) *From Biochemical Flows to the Disease-Poverty Nexus: Tracing the "Emergence" of Chronic Kidney Disease of Unknown Etiology in Sri Lanka's North Central Dry Zone*. MA Thesis, Geography Graduate Group, University of California, Davis.
Shaw, I., P. Robbins, and J.P. Jones. (2010) A Bug's Life and the Spatial Ontologies of Mosquito Management. *Annals of the Association of American Geographers*, 100 (2): 373–392.
Showers, K.B. (2005) *Imperial Gullies: Soil Erosion and Conservation in Lesotho*. Athens: Ohio University Press.
Sluyter, A. (2012) *Black Ranching Frontiers: African Cattle Herders of the Atlantic World, 1500–1900*. New Haven: Yale University Press.
Sorlin, S., and P. Warde. (2007) The Problem of the Problem of Environmental History: A Re-Reading of the Field. *Environmental History*, 12 (1): 107–130.
Sutter, P. (2003) What can US Environmental Historians Learn from Non-US Environmental Historiography? *Environmental History*, 8 (1): 109–129.
Swyngedouw, E. (1997) Power, Nature, and the City: The Conquest of Water and the Political Ecology of Urbanization in Guayaquil, Ecuador, 1880–1990. *Environment and Planning A*, 29: 311–332.
Swyngedouw, E. (2004) *Social Power and the Urbanization of Water: Flows of Power*. Oxford: Oxford University Press.
Turner, M.D. (2003) Methodological Reflections on the Use of Remote Sensing and Geographic Information Science in Human Ecological Research. *Human Ecology*, 31 (2): 255–279.
Wainwright, J. (2008) *Decolonizing Development: Colonial Power and the Maya*. Oxford: Wiley-Blackwell.
Warren, L. (1997) *The Hunter's Game: Poachers and Conservationists in Twentieth-Century America*. New Haven: Yale University Press.
Watts, M.J. (1983) On the Poverty of Theory: Natural Hazards Research in Context. In *Interpretations of Calamity*, edited by K. Hewitt. Boston: Allen & Unwin, 140–162.
Watts, M.J. (2013) *Silent Violence: Food, Famine and Peasantry in Northern Nigeria*. Athens, GA: University of Georgia Press.
Weiner, D.R. (2005) A Death-Defying Attempt to Articulate a Coherent Definition of Environmental History. *Environmental History*, 10 (3): 404–420.

Williams, M. (2003) *Deforesting the Earth: From Prehistory to Global Crisis*. Chicago: The University of Chicago Press.

Wilson, R. (2013) Environmental Histories. In *The Wiley-Blackwell Companion to Cultural Geography*, edited by N. Johnson, R. Schein and J. Winders. Oxford: John Wiley & Sons, 355–370.

Wynn, G. (2007) *Canada and Arctic North America: An Environmental History*. Santa Barbara: ABC-CLIO.

Wynn, G., C. Colten, R. Wilson, M. Melosi, M. Fiege, and D. Davis. (2014) Reflections on the American Environment. *Journal of Historical Geography*, 43 (1): 152–168.

Zimmerer, K.S. (1996) *Changing Fortunes: Biodiversity and Peasant Livelihood in the Peruvian Andes*. Berkeley: University of California Press.

PART IV, SECTION B

Environmental change

The chapters in this section of the Handbook ask, *In what ways are nature and society transformed through economic activity, and how does this metabolic relationship affect various social groups in different ways?* Dynamics of environmental transformation – from deforestation, soil erosion, and the enclosure of forests and game reserves, to the planting of chemical-intensive lawns, and the rapidly changing ecologies of health – are central foci of research in political ecology. Such processes may usefully be considered expressions of the metabolic relationship between nature and (capitalist) society. In the section's first chapter, Noel Castree presents an explicitly Marxist reading of political ecology, through the work of the late Neil Smith. Never a self-described political ecologist, Smith's writings on the capitalist production of nature and the politics of scale have nonetheless provided a theoretical foundation for political ecology. The next chapter, by James Wescoat, considers the political ecologies of risk, hazard and vulnerability, one of the oldest streams of political ecological thought. Emerging from the human ecology tradition, research in this area remains vital to political ecology as the following chapter attests. Diana Liverman critically reviews recent work in climate change which, given its fundamental linking of human and physical processes, as well as its clear implications for social justice, is a rapidly growing area of political ecological research,

The next two chapters in the section consider the dynamics of economic development and social reproduction, both core themes in political ecology. In the first of these, Astrid Ulloa examines the relationship between environment and development, and the various ways development, and alternatives to development, have been conceptualized and enacted in Latin America and the global South more broadly. The second of these two chapters, by Edward Carr, examines the concept of livelihood – a core analytical category in political ecological work on (mostly rural) development. Following this, Brian King examines the political ecologies of disease and health, placing these medical, scientific questions squarely in social, political, and environmental context.

The section ends with three chapters that directly engage with ways in which environments are transformed through metabolic relationship with capitalist production. In the first of these, Tor Benjaminsen examines environmental degradation, and how it has been understood in the political ecology literature. Next, Stefania Barca and Gavin Bridge examine the dynamics of industrialization as a motor of environmental transformation. The final chapter in the section, by Alf Hornborg, employs Marxist political economy and ecological economics to examine the ecologically unequal ramifications of trade.

21

CAPITALISM AND THE MARXIST CRITIQUE OF POLITICAL ECOLOGY

Noel Castree

> Marx nowhere talked explicitly about the production of nature. But in his work there is an implied understanding ... which leads firmly in this direction.
>
> *(Smith 1984: 50)*

> In all forms of society there is one specific kind of production which predominates over the rest, whose relations ... assign rank and influence to all others. It is a general illumination which bathes all colours and modifies their particularity.
>
> *(Marx 1973: 106)*

In a world where corporations create new life-forms and may soon geo-engineer the skies, does what we call 'nature' any longer possess autonomy and agency? In what ways, and to what degree, can the capitalist mode of delivering goods and services be said to 'produce' something that is, by definition, thought to be given rather than made? Is 'nature', in its various forms (large and small), something that can and should found a politics devoted to reforming contemporary capitalism or, perhaps, to superceding it? If not, how can the biophysical dimensions of capitalist accumulation be factored-in to a critique of political economy? This chapter will address these analytical and normative questions by reviewing over 40 years of Marxist scholarship focussed on the relationships between capitalism and what we by convention call nature – human and non-human.

Since my questions are large ones and the scholarship voluminous, I want to place the writings of the late geographer Neil Smith (1954–2012) at the heart of my review. Though Smith was only one of many Marxists who wrote about the capitalism–nature nexus, his several publications on the subject warrant especial attention for three reasons. First, since the mid-1990s they have been a key reference point for Marxist and Marxisant geographers seeking to more expansively 'materialise' Marx's theory of capital accumulation. Second, even those – for the most part based outside the geography discipline – who did not engage with Smith's writings in this quest, can usefully have their contributions interpreted in light of his striking insistence that capitalism makes even nature in its own image. The reverse applies too, of course: how defensible is this insistence when Smith's work is read in light of some other Marxists' belief that nature, and those who speak for it, can and do resist capitalism's entreaties *sui generis*? Third, notwithstanding his subsequent attempts to update and clarify key claims about nature advanced in *Uneven development* (1984), a certain interpretive openness attaches to them. By glossing this in the search for what Smith 'really intended', several critics – myself included some years ago (e.g. Castree 1995) – risk overlooking important aspects of his thinking. We might regard these aspects as 'productive ambiguities'.

In sum, this chapter foregrounds Smith's writings about capitalism and nature as a means of exploring key themes and insights in a larger corpus of Marxist work on the subject (what is sometimes called 'ecoMarxism'). Though I thereby intend to pay tribute to one of geography's most inspirational thinkers, my aim here is not to suggest that Smith's contributions trumped those of other Marxists similarly interested in nature. Instead, by reading the latter's arguments in relation to Smith's, I hope to shed light on important ideas and persistent points of analytical and normative disagreement. Given that the literature is now large, I will necessarily be quite selective in my coverage of writings beyond Smith's own. I aim to identify contributions that are representative of key themes and issues. My interpretation of them, while hardly definitive, emerges from over 20 years of immersion in the debates about Marxism, capitalism and nature as a sometime contributor sympathetic to Smith's project.

The chapter is organised chronologically. I begin by introducing Smith's thesis that nature is produced, first put forward back in 1980 and fully articulated in his book *Uneven development*, paying attention to the wider intellectual and political context in which he was writing. After this long section, I then summarise succinctly a set of other contributions to a then still-nascent Marxist political ecology that were published subsequent to – and usually in ignorance of – Smith's thesis (roughly 1987–2000). I read these in light of Smith, and Smith in light of them, making retrospective connections where none (or few) were made at the time. I then, again telegraphically, focus on the recent ('neoliberal') period, in both an intellectual and political economic sense. I place Smith's original thesis and his subsequent commentaries on it in relation to newer Marxist theories of capitalist–nature relationships and wider currents of thinking and politics. As I will explain, my continuous attention to context is of more than merely historic interest. Throughout, my overarching concern is whether and how a Marxist critique of political ecology can function without 'nature' as an ontological reference point for analysis and evaluation. I apologise to knowledgeable readers at the outset: lack of space prevents me from exploring the work I summarise here in anything like the detail it deserves.

Before I get down to business, a point about terminology, one about literature omitted from this review, and one about the consequences of knowledge. As far as I know Smith never called himself a 'political ecologist' (even after the term caught-on in geography and anthropology from the early 1980s), but his disquisitions on nature clearly make a certain definition of the term applicable. After all, Marx's middle and later writings – upon which Smith drew heavily and creatively his entire career – were billed as an exploration of the social definition, creation, distribution, regulation, effects and politicisation of wealth in a capitalist world ('political economy'). To draw-out what Marx left largely implicit, as Smith and fellow travellers have done, is to show that his political economy always necessarily had an ecological dimension. How the biophysical realm both enables and hinders the creation, growth and capture of wealth in its capitalist form is a question of prime analytical and political importance. Indeed, though Smith rarely referred to their publications, it should come as no surprise that some of the pioneers of a self-designated 'political ecology' – such as Piers Blaikie (1985) – were inspired by Marx's germinal account of the capitalist mode of production (see Chapter 2, this volume).

This reference to Blaikie's influential work suggests strong links should exist, via Marx, between the writings of Smith and the field of political ecology ('third world' and 'first world) showcased in this volume. However, these links are – surprisingly – few and far between.

Smith's highly theoretical treatment of Marx contrasted with the concrete, empirical preoccupations of Blaikie, Michael Watts and others who pioneered political ecology. Though his work has been widely read by political ecologists it has rarely been used in their research. A full review of the way Marx's ideas were operationalised by self-styled 'political ecologists' and elaborated by Marxists like Smith – who explored the nature question in a different way – remains to be written. Accordingly, when I refer to 'political ecology' in this chapter I'm referring not to the field of that name but to theoretical work like Smith's that examines how wealth in capitalist societies has a constitutively biophysical dimension. It is interesting to speculate whether and how the work I review here would have developed differently had its authors better engaged with Marxist political ecologists like Watts.

As we shall see, the concepts we ultimately favour in addressing the question of 'capitalist nature' significantly affect the answers Marxists like Smith have offered. Since capitalism is far, far more pervasive than ever it was in Marx's day, getting these answers 'right' – and inserting them into the discourses and programmes of political movements – recalls Marx's stirring injunction that critics must change the world, not merely aim to understand it. Universities are far more subject to the invisible hand of the market and the visible hand of the state than in decades past, but they remain crucial sites for the creation of oppositional thinking. Without them, contemporary Marxist thought would be smaller and less sophisticated than it is – notwithstanding the fact that fewer academics and students are drawn to it than a generation ago. For Neil Smith, and I hope for readers of this chapter, Marxism has some vital things to tell us about the 'nature' of our capitalist world that other approaches ignore, wilfully or otherwise. But these will never be verities for more than a few unless progress can be made in the perennial battle to win the minds and hearts of enough people in universities and beyond. As I will show, Marxist analysis is as vibrant and incisive as ever, but it appears politically impotent – notwithstanding the powerful anti-capitalist sentiments expressed worldwide in the late 1990s and a decade later when the global financial crisis erupted.

Denaturalising nature: Neil Smith thinking against the grain

A distinctive perspective

Thirty years ago Western capitalists were hunting for a cure to the wide and deep economic crisis that began in the early 1970s. As part of this crisis Left political organisations lost their former ascendancy in many countries. Meanwhile, knowledgeable observers had long realised that actually existing 'communism' fell far short of the ideals of any credible Marxist revolutionary. Yet radical dreams were hardly dead: the environmental, feminist and anti-racist movements had built up a head of steam in many Western countries through the 1970s, with the 'events' of 1968 a still inspirational memory. Though internally diverse, elements of these movements provocatively went against the grain of current thinking and practice. Universities afforded these New Leftists the time and space to match their political ambitions with powerful philosophies and theories. They also became a redoubt for Marxists trying to make sense of a more turbulent world – one in which Marxism would, outside the universities, become something of a dirty word once the 'eastern bloc' collapsed almost overnight (1989–91) and 'free market capitalism' seemed to reign triumphant in a world on the cusp of 'globalisation'. It was in this contradictory, febrile context that the then young Marxist geographer Neil Smith published a sophisticated work of abstract theory – *Uneven development* – in 1984. As he explained on the very first page, it was an 'exploration and critique of *concepts* as a means to interrogate more sharply the *reality* we live in' (1984: xv, emphasis added).

At the time of publication, the book was distinctive for a number of reasons, of which I want to highlight four. First, it had a lot to say about what we call 'nature' – even though Marx's own comments on the subject had been most fragmentary. In fact, it was among the first systematic attempts to integrate biophysical phenomena into Marx's political economy, linking them to space, scale and geographical inequality in the process. Prior to Smith, the main Marxian authors to consider Marx's view on nature (Friedrich Engels aside) were Alfred Schmidt, Sebastiano Timparano, Raymond Williams and Norman Geras (see Castree 2000). Second, what Smith said was – to use his own words – 'jarring' and 'quixotic' because he claimed that nature is *produced* not given. To quote him at some length:

> [T]his idea ... defies the conventional, sacrosanct separation of nature and society ... We are used to conceiving of nature as external to society, pristine and pre-human, or else as a grand universal in which human beings are but small and simple cogs. But here ... our concepts have not caught up with reality. It is capitalism which ardently defies the inherited separation of nature and society, and with pride rather than shame. In its constant drive to accumulate larger and larger quantities of social wealth ... capital[ism] transforms the shape of the entire world. No God-given stone is left unturned, no original relation with nature unaltered, no living thing unaffected.
>
> *(1984: 7–8)*

Third, Smith's emphasis on capitalism's transgressive powers called into question the deep-seated ontological assumptions underpinning both radical and more mainstream thought in the 1980s. His insistence that there is no nature intelligible outside contingent social discourses, relations and practices posed a challenge to much 'environmentalist' thinking, to politics and policies predicated on ideas of 'human nature' (mental and/or physical), and to the idea that 'natural science' (including physical geography) studies an intrinsically asocial world (leaving social scientists and humanists – including human geographers – to study everything else). As Smith explained in chapter 1 of his book, analytical and normative references to nature (without the scare quotes) are ideological, both in the sense of misleading and actively reproductive of capitalist society. I will say more about why presently. Finally, while this claim about ideology presaged later writings by post- or non-Marxists about the 'discursive construction' and 'cultural constitution' of 'nature', Smith refused to limit nature's social character to linguistic frames or semiotic sieves.

Understanding 'production'

What exactly did Smith mean by 'production'? At the heart of his conception were the terms *metabolism* and *labour* (1984: 33–34). The former, far more than a word like 'interaction', posits what we call 'people and environment' or 'society and nature' as *unities* not dualities. As Smith argued, 'Society is internal to nature' (1984: 33). For him, and for Marx, the motor of this internalisation is humans' propensity to make the material world into things of use and to thereby alter their own physical and mental 'nature'. Chapter 2 of *Uneven development* explores this in some conceptual detail. In a section on 'Production in general' he makes the key point that all work involves not just a relation with what we call nature (e.g. with water, soil or cattle), but with other people. The latter condition how work is performed, what it is in 'external nature' that is deemed useful, and how 'human nature' is thereby altered by the collective results of work. This immediately alerts us to the idea that all 'production' of goods and services extends beyond the physical act of individuals wresting useful items out of the non-human world in particular locations. In this light, we might say that metabolism alerts us to *flows* (of energy, ideas and materials), and labour to the key *relations* determining the specific pattern of those flows (metaphorical pipes or wires, if you will).

How, then, is production organised in capitalism? The answer culminates in chapter 2 of Smith's book. Like all modes of production capitalism proliferates use values (qualitatively specific entities designed to be of practical or symbolic utility). But since it is not a subsistence economy, useful items are produced in order to be exchanged. Exchanged for what? After Marx, Smith argues that the answer is money. For workers this is essential because in capitalism they are wage-workers, i.e. they must sell their capacity to work in exchange for a salary they can use to purchase the goods and services to reproduce themselves physically and psychologically. For capitalists – who own the 'forces of production' (e.g. factories) – it is essential for a different reason. Yes, they must sell enough products to pay for their own socio-physical reproduction; but they must also accumulate (or borrow) enough money to ensure future rounds of commodity production. This is more than a question of covering their production costs: capitalists do not go to the trouble of employing workers, and paying for material inputs, premises and equipment, with no expectation of a return on investment. Instead, they aim to accumulate more money than they laid out at the start of each production round. And since they must compete with other capitalists for market share they are compelled to innovate in any number of ways (e.g. inventing new use values, reducing production costs, expanding into overseas markets, or creating new demand niches in existing markets). In short, the ensemble of social relations specific to capitalism – relations of ownership, exchange and competition – ensure that 'accumulation for accumulation's sake' is, as Smith wrote, 'a socially imposed necessity' (1984: 70). These relations make expanding circulation – the entry of entities, goods and people into, along and out of various commodity chains – the economic norm.

Quite aside from the fact that it confronts workers, capitalists and everyone else as an impersonal force eluding control, and quite aside from its contradictory character (e.g. tending towards boom and bust periods), there is for Smith (after Marx) something else peculiar about capitalist production in this expanded sense. It is that one commodity (wage labour) is, in fact, the source of the wealth represented by the money that capitalists devote their energies to accumulating. *Contra* mainstream economic thinking, commodities do not have 'intrinsic value', value is not merely 'conferred' by consumers' preferences, and nor does profit originate from the skill or efficiency of specific capitalists. As Marx explained in his 'labour theory of value' – still controversial among analysts to this day – workers collectively create and unconsciously alienate social wealth by way of a process that conceals the fact and operates 'behind their backs'. His concepts of fetishism, concrete labour, abstract labour, socially necessary labour time, and surplus value were key to this theory of how commodities (pre-eminently money) are the material form assumed by social relations and the cloak hiding the transfer of social wealth between classes. In this light, capitalism's *differentia specifica* is that tendentially growing (and empirically changing) *flows* of energy, ideas and materials are both compelled by, and a displaced form of, particular inter- and intra-*relations* among two social classes. Here metabolism is unique because specific acts of work are profoundly conditioned by real, but abstract, social forces 'stretched-out' over time and terrestrial space. It is special too because at one level capitalism is tediously changeless yet, at another, extraordinarily dynamic.

In light of all this, it might seem perfectly reasonable to argue that capitalism 'utilises', 'relies upon' and often 'destroys' what we call 'nature' on an expanding scale – but not that that it *produces* it. After all, as Smith conceded in *Uneven development*, 'Nature is generally seen as precisely that which cannot be produced' (1984: 49). Furthermore, the just-mentioned concept of metabolism and Marx's/Smith's emphasis on relations apparently point us towards a process whereby various *different and discrete entities* (non-human and human) *connect with and co-constitute each other in historically and geographically specific ways*. Yet Smith held fast to the idea of production his entire career, and in a seemingly *literal* not metaphorical sense. 'Where capitalism is unique', he wrote 30 years ago, 'is that for the first time human beings produce nature at a world scale' (1984: 77). What is more, he considered – and then dismissed – the argument that because some parts of 'nature' are not socially produced (e.g. lava, our brains or gravity) the idea of production must be carefully circumscribed:

> these rather extreme examples hardly testify to the falsity of the 'production of nature' thesis, especially when one looks at more down-to-earth examples of supposedly unproduced nature, such as Yellowstone Park or Yosemite.
>
> *(1984: 80)*

Even though key natural resources appeared suddenly scarce (again) after 2000 (e.g. oil), and humans powerless to arrest the future effects of past greenhouse gas emissions, Smith did not back-track: '[T]he production of nature', he wrote in his final major essay on the subject, 'is being dramatically intensified and its dimensions multiplied' (2007: 21). As Julie Guthman recently noted, 'In effect, the production of nature thesis flipped materialism on its head, by repositioning nature as an outcome of social relations rather than an asocial input to the economy' (2011: 235).

Why disavow human and non-human nature?

Why was Smith such a fierce critic of the idea that 'nature' has an autonomous existence, agency or moral–ethical considerability – especially given how prominent 'environmentalism' in its various forms had become in the years when Smith was researching and writing *Uneven Development*? We can only speculate, but I would point to two aspects of the context in which he crystallised his ideas. Both, in part, were reactions to the nature–society dualism that had long organised thinking in Smith's own discipline of geography.

First, Smith's doctoral thesis advisor was David Harvey, who undoubtedly exerted a huge influence on his thinking. In 1974 Harvey published what, in time, became a germinal Marxist critique of the neo-Malthusian thinking. Such thinking formed a key strand of the just-mentioned environmentalism that arose because of perceived resource scarcities and anthropogenic destruction of species and ecosystems. Harvey took strong issue with the idea, popularised by the likes of American biologist Paul Ehrlich, that the world was 'over-populated'. Instead of highlighting 'natural limits' to economic growth, Harvey 'denaturalised' and relativized the question of how the biophysical world affects the social one. For him, problems of human poverty and scarcity reflected the systematic maldistribution of material wealth (e.g. food) because of unequal transfers of social wealth (represented by money) – such that 'limits to growth' were *internal* to capitalism.

Second, while Harvey's attention was directed at 'natural resources', he did not focus on 'natural hazards', such as hurricanes or tsunamis. These periodic threats to people were surely independent of any social conditioning. Yet in a 1976 *Nature* paper, the geographers Phil O'Keefe, Ken Westgate and Ben Wisner sought, as per their title, to take 'the naturalness out of natural disasters' (again, see Chapter 2, this volume). Noting that more people than ever were being badly affected by extreme biophysical events, they pointed to the socio-economic and political processes that rendered some vulnerable but not others. This led them to suggest that avoiding settling in hazardous areas or spending more money on technical solutions (e.g. flood barriers) was not necessarily the best response. Instead, they argued that attempts to address poverty and social marginalisation would render the worst affected groups more resilient to biophysical extremes. These extremes were thus experienced *contingently*, not as absolutes. Smith was undoubtedly aware of this argument: four years later he and O'Keefe (1980) together authored the very first presentation of the 'production of nature' thesis in *Antipode*.

In this light, it is not hard to see why Smith – to use the words of a later associate echoing Raymond Williams – believed that 'ideas that draw upon the authority of nature nearly always have their origin in ideas about society' (Ross 1994: 15). Here I return to the subject of 'ideology', mentioned in passing earlier. For Smith, references to a supposedly society-free nature not only served to anchor and legitimate all manner of capitalist projects, such as cures for 'genetic diseases' in humans marketed by biotechnology corporations. More than this, they were – and remain – *the conceptual mirror of the everyday forms in which capitalism presents itself* (see chapter 1 of *Uneven Development* for more on this). For the mode of production that takes hold of 'nature' in all its forms does not make plain the ramified and complex flows and relations that comprise it. Instead, Smith argued, it manifests as a world of entities – things, people, and so on – that may be conjoined but appear to exist regardless of any particular connections established between them. For Smith the job of Marxism is to contest the appearance and show that 'the question of nature', whatever else it may be, is really a question of how any society defines, creates and distributes the wealth that sustains it. The normative up-shot is to ask not what nature prevents or enables, but to consider how 'nature' might be produced in ways more democratic, more just and more subject to collective control (see Biro 2005).

Placing analytical limits on 'the production of nature'?

As we have seen, Smith appeared to believe that capitalist nature was produced 'all the way down'. This testified to how powerfully Hegelian holism, materialised by Marx, permeated his thinking. In this he presaged neo-Marxist Steven Vogel's plenary argument in *Against nature* that what 'we take for granted as "natural" turns out on investigation to be the product of human labor and hence *literally* socially constructed' (1996: 7). Yet in *Uneven development*, and subsequent essays, close readers could (can) spot some signs of equivocation. For instance, Smith observed that 'Unlike gravity, there is nothing natural about the "law" of value' (1984: 82) – a statement which posits the very distinction his book was intended to challenge. Similarly, in a chapter published the following decade, he said of his 'thesis' that 'If it indulges a certain anthropomorphism ... it expresses the extent to which advanced capitalist societies have intruded human activity at the centre of nature' (1996: 50). Here the image of 'intrusion' and the concept of the 'anthropos' both suggest/ed a residual Kantianism (or, if one prefers, Cartesianism) that Smith otherwise dissented from.

Was Smith simply inconsistent, or did his apparent vacillation reflect something important about late twentieth-century capitalism? After all, by the mid-1990s biotechnology firms were routinely crossing species barriers at the genetic level with considerable precision. Meanwhile, a new and thoroughly global regime of capital accumulation seemed both to exhaust nature's bounty (e.g. oceanic fish stocks) and to overestimate the environment's capacity to absorb waste (witness ozone layer thinning and 'global warming', both of which were headline news not long after *Uneven Development* was published). Perhaps capitalism produced some natures but not others, meaning that Smith (and other Marxists) needed a more differentiated sense of the natural according to (1) its malleability and (2) which capitalists were seeking to profit from its use. A number of analysts, though working separately, together addressed this need. None paid particular attention to Smith's thesis, but their use of Marxian concepts makes it easy enough to establish the connections with hindsight. I will focus briefly on seven published contributions, organising them into two clusters.

The uneven internalisation of 'nature' by capitalism: biophysical barriers and opportunities

The first five pertain to the analysis of agriculture and the question of how capitalists explore new frontiers in their desire for profit. Like mining, fisheries, forestry and other natural resource industries, agriculture must 'confront nature directly'. In what became a classic intervention, the rural sociologists Susan Mann and James Dickinson (1978) argued that agriculture's economic 'exceptionalism' – that is, its historic resistance to capitalist social relationships – had something to do with its biophysical basis. For instance, the naturally determined gap between investment and work (e.g. buying a tractor and sowing seeds), and return on investment (because foodstuffs take time to grow) can make agriculture unattractive to capitalist entrepreneurs. Following Marx, and his epigones Karl Kautsky and V. I. Lenin, Mann codified and elaborated this argument in *Agrarian Capitalism in Theory and Practice* (1990). The 'obstacle' of nature, she showed, helped explain why agriculture remained dominated by rentiers, families and various smallholders.

Yet these obstacles, others showed, were not all of a piece. In their book *From Farming to Bio-technology*, neo-Marxist agro-food analysts Goodman, Sorj and Wilkinson (1987) focussed on how capitalist firms had 'taken hold' of some aspects of agriculture. They focussed on 'appropriation' – manufacturing things farmers needed (e.g. combine harvesters) by altering their sense of what precisely they need – and 'substitution' – replacing on-farm inputs to farming (e.g. cow manure) with manufactured ones (e.g. chemical fertiliser). The same year, Marxist rural sociologist Jack Kloppenberg showed how both processes had unfolded historically in the United States in his monograph *First the Seed* (1987). In effect, his account of how agricultural science and democratically elected government had indirectly founded a new set of private firms supplying genetically altered seeds (and other inputs) to farmers year-on-year was an illustration of 'the production of nature' in all but name. At both a discursive and physical level, Kloppenberg showed, these firms created new commodities that circumvented previous biological obstacles to agrarian accumulation (see Castree 2001). Capitalist production in an expanded sense here produced 'nature' in a concrete sense.

All this suggested that capital literally circulated *through* some elements of nature but had to circulate *around* others, depending on prevailing technology. In his magisterial book about the growth of large-scale agriculture in California, geographer George Henderson (1999) evidenced the latter in compelling detail. He showed how 'finance capital' (banks, in this case) made money by extending credit to aspiring commercial farmers confronting the barriers to accumulation Mann and Dickinson had identified. The banks thereby enabled the intrusion of 'productive capital' into farming, notwithstanding the obstacles, and made money in the process. As Henderson argued in a trailer essay for his book, the point 'is that nature repels and attracts capital in different ways according the historical … contingen[cies]' (1998: 76). This enjoins us to attend to different circuits and sectors of capital, and – if we broaden the point beyond agriculture and money-lending – the differential affordances nature presents entrepreneurs.

Capitalism's biophysical outsides

Few, if any, of the contributions just mentioned paid attention to the concerns expressed by environmentalists from the early 1970s onwards. Additionally, their analyses were focussed on aspects of nature deemed directly 'useful' by capitalists and others. However, what about all those elements of nature, from fresh water to oxygen to human intelligence, that capitalism treated as 'free inputs' or else as 'sinks' for the release of the by-products of production? This question preoccupied a group of what became known as 'eco-Marxists'. Their aim, achieved differently in the detail, was two-fold: first, to explain how and why capitalism was systematically degrading the biophysical basis of its own existence; second, to thereby explain to Marxists and Left-leaning environmentalists alike that they needed to make common political cause.

It will suffice to point to two authors who, like Smith, drew directly on Marx's original writings in presenting their late twentieth-century critique of capitalism–nature relationships. In *The Future of the Market* German Elmar Altvater attended to the 'largely neglected dimension of economic processes [whereby] ... transformations [are] undergone by raw materials and energy in the course of production, consumption and distribution' (1993: 5) – ignored, that is, by mainstream economists and Marxists alike (in his view). He placed capitalist labour – in both its concrete and abstract senses – at the heart of these transformations, highlighting the partial way 'nature' registers as use values:

> Nothing can be defined as a use-value ... without regard to the ... biotic and abiotic environment. But this is *precisely what happens* if it becomes a bearer of value and acquires the properties of a commodity in the capitalist social formation.
>
> *(1993: 193, emphasis added)*

On this basis Altvater identified 'five dimensions of the contradiction between ecology and economic' (p. 198) arising from a clash of the 'ordering principles' (p. 204) governing capital accumulation, on the one hand, and a nature not designed for use by capitalism on the other.

Independently of Altvater, American Marxist James O'Connor was making similar arguments. Assembled in his 1998 book *Natural Causes,* O'Connor's many essays presented the concepts of 'conditions of production', 'under-production' and 'second contradiction of capitalism' – all of which have since become influential in certain Marxist circles. The first pointed to all those things (biophysical and social) upon which capitalism relies at any one moment but which it had no hand in (re)producing. The second pointed to these things' scarcity once capitalists utilise them as if they are limitless. This scarcity results in rising costs, new regulatory requirements (imposed by governments) and other burdens that are not shouldered by capitalists alone and may become politicised. The third concept pointed to an 'ecological dialectic' arising from capitalism's engagement with nature. O'Connor regarded this as just as important as Marx's 'first contradiction' between the 'relations and forces of production'. Accordingly, he suggested that radical environmentalists should join trades unions, communist organisations and others in any revolt against capitalism. 'Green' politics needs to be 'red', and vice versa.

From Altvater's and O'Connor's perspective, Smith's thesis – had they engaged with it at the time – would doubtless have appeared more metaphorical than literal: for them, we might say, capitalism treats all nature *as if* it is (or can be) 'produced', yet eventually runs-up against the physical contradictions and political backlash this creates. This arguably reflects the influence of neo-Marxist historian Karl Polanyi (1944) on their work. Polanyi's concept of 'fictitious commodities' pointed to all those things – values, relations, institutions, norms and physical entities – whose characteristics *exceeded* those 'demanded' by capitalism at any one time.

An imagined Smithian response

How might Smith have responded to the slew of Marxian work summarised in the two subsections above? I say 'might' because in neither of his two substantial 1990s essays about nature (Smith 1996, 1998) did he refer much to any of this scholarship. The seemingly equivocal Smith, with which I introduced this section, might have appreciated the qualifiers everyone from Mann to O'Connor introduced to the question of capitalist nature. He might have applauded the way a politics of class was thus shown to be wedded to a politics of nature, especially given that trades unions and socialist political parties were no longer in the vanguard after the economic crises of the 1970s and 1980s.

However, the apparently unequivocal Smith of the previous section would surely have argued something else: namely, that these authors (re)imported ideological thinking into Marxism by implying that much of 'nature' can be understood as possessing 'independent' qualities that capital either cannot profit from, circulates around or ignores at its peril. He had criticised Alfred Schmidt (1971), one of the first to systematically theorise nature as a Marxist, on just these grounds (see chapter 1 of *Uneven development*). Against any 'neo-Kantian revival' (1998: 266), as he called it, we need, he argued, to resist 'the fetishism of nature' (1996: 51) evident in both the environmental movement *and* attempts to 'green' capitalism from the early 1990s (witness The Nature Company: see Smith 1996: 36–39, 51–52).

It is with the unequivocal Smith that I wish now to end this chapter. As I will explain, a critique of capitalist nature can proceed under the sign of 'production' without falling prey to a *tabula rasa* argument ('capitalism can produce nature willy-nilly') and without, on the other hand, appealing to aspects of 'nature' that supposedly exist outside the production process. But this 'both/and' view necessitates understanding production in a broader sense than the literal fabrication of things like genetically modified organisms. Misinterpreting this breadth as covering nature 'internal' and 'external' to capitalism is, I believe, why Neil Smith can be misread as a 'hyper-constructionist' who accords 'nature' neither agency nor moral worth.

Capitalism, nature and radical politics in the Anthropocene

Contextual considerations

The early twenty-first century has been exceedingly eventful. First, earlier warnings about the magnitude of 'global environmental change' have been repeated more loudly. Recent IPCC reports foresee 'dangerous climate change' if runaway atmospheric pollution is not abated, while a network of environmental scientists proclaim the recent period of Earth history (the Holocene) over: people are now, they argue, equivalent to the 'great forces of nature', such is 'the human impact'. Second, this has given environmental politics a boost of sorts, and comes after the wave of anti-capitalist protests that helped to re-politicise this mode of production from the late 1990s. Third, the global financial crisis of 2008–9 reignited these protests and offered widespread opportunities to think about a more humane, eco-friendly capitalism (if not its outright replacement). 'Decarbonising' capitalism has become a seeming priority, and many identify the massive stored energy of fossil fuels as the motor that has carelessly driven it forward this last 150 years (see Huber 2013).

However, and fourth, critics suggest that the sting of radical thinking has been drawn, or simply ignored, by the powers that be. Despite the 2012 Rio+20 Earth Summit, 'environmental issues' have slid down the agendas of most major governments (witness the dismal international attempts to curb greenhouse gas emissions). Meanwhile, environmental economics and broader neoliberal policies now seemingly dominate attempts to give the invisible hand a 'green thumb'. Everything, even a 'nature' recognised as in need of better 'management', must conform to the imperative of economic growth in these times of recession and austerity. What's more, some capitalists remain determined to physically remake nature, rather than 'adapt' to it, as they harness the latest science to their profit-seeking ventures. Noted American biotechologist and businessman Craig Venter is, one might say, the poster child of this attempt to denaturalise ever more matter. Fifth, despite the evident problems with contemporary capitalism (environmental and otherwise), the relevance of Marxism is not widely appreciated outside (or often within) universities where people like me ply my trade. Furthermore, because 'environmental issues' are widely understood to be serious ones, it is likely that future attempts to popularise the Marxian critique will fare best if they speak to these issues and the political responses they have inspired.

In this context, Smith's idea of the production of nature may, as it approaches middle age, appear 'jarring' and 'quixotic' for all the wrong reasons. Unless carefully qualified, it seems unable to speak the plethora of analytical and normative questions pertaining to a world of dwindling oil supplies and higher ambient temperatures. The context has changed and so, perhaps, should our assessment of what value Smith's thesis (any longer) possesses. Even if one appreciates its analytical thrust, its refusal to grant the category of 'nature' political potency appears to make it irrelevant to the many radicals who want to protect, defend, restore and preserve whales or ice sheets.

A renaturalised Marxism?

Given all this we can see the appeal of more recent work by certain Marxist political ecologists. Chief among them is John Bellamy Foster and his associates. Their idea of a 'metabolic rift' between capitalism and the biophysical world has become well-known and influential in some Marxist academic circles. Though they acknowledge the specific and contingent character of all human–environment relationships, in *The Ecological Rift: Capitalism's War on the Earth* (Foster *et al.* 2010) they suggest that capitalism is pushing the Earth's biophysical systems beyond their capacity to function. In a recent essay Foster (2012) refers to the new scientific idea of 'planetary boundaries' (of which there are said to be nine) to describe this capacity. This argument echoes those made by Altvater and O'Connor. However, Foster and his associates focus more on 'high impact planetary ecological crises' (2010: 16), thus mirroring the Earth-system focus of many campaigning environmental scientists like James Hansen and Manchester University's Kevin Anderson.

This focus seems apropos. It serves as a corrective to the 'Prometheanism' and 'utopianism' some critics detect in earlier strands of Marxist political ecology (see Soper 1991 and Benton 1991). Yes, we live in a world where capitalist firms can remake nature forensically, but it is also one where capitalism's unintended 'environmental externalities' must be acknowledged and arrested for fear of massive and uncontrollable Earth-system changes. Indeed, high-level discussions of 'green capitalism' show that even (some) capitalists realise that 'the second contradiction' is no figment of the Marxist imagination. So how can Smith's critique of neo-Kantianism remain intact given the 'fact' of our Anthropocene condition? Surely Marxist analysis and politics must reckon with nature's agency at the largest scale?

Since Smith himself provided an answer to these questions we do not, fortunately, have to speculate. In *Socialist Register 2007* (Smith 2007) and the 'Afterword' to the third edition of *Uneven Development* (Smith 2008) he addressed the new way 'nature' was being mobilised by capitalist elites, by environmentalists and by certain Marxists. Two arguments stand out. First, he identified 'nature-washing' as 'the process in which social transformations of nature are well enough acknowledged, but in which that socially changed nature becomes a new super-determinant of our social fate' (2008: 245). Nature (external and/or universal) here becomes an ontological reference point justifying arguments for 'carbon offsetting' or, more radically, 'the revenge of Gaia'. Either way, Smith argued, its invocation fails to properly politicise capitalism, since the problem is not 'technology', 'over-consumption' or over seven billion human mouths to feed. Equally, the solution is not simply 'clean technology', less consumption and fewer babies.

Second, Smith argued that – notwithstanding the widespread recognition that nature needs better looking after – capitalism today

> absorbs nature more fully and completely ... For all that capitalism is more voracious than ever in vacuuming a supposedly external nature in search of commodifiable use values, we can also glimpse the starts of a new ... regime whereby the task of producing a useable nature begins to pass from so-called external to social nature.
>
> *(2007: 26)*

How, we might ask, are capitalist attempts to price (and profit from) environmental 'bads' and 'goods' (aka 'services') a form of 'production'? Smith's answer is that these goods and bads are not, in the end, 'natural' – though they are clearly anchored in real biophysical phenomena. Instead, they are phenomena framed discursively and practically by capitalists, usually working hand-in-hand with various field scientists. To bear 'value' in the Marxist sense, Smith argued, what we call nature becomes visible in circumscribed ways that are governed by capital's 'laws of motion' – *even when it is nature's 'real qualities' that are supposedly being valued for their own sake or for non-economic reasons* (see Robertson [2012] for more on this). A critique of capitalist political ecology cannot thus fall prey to its own kind of 'nature-washing', even as it objects to the way capitalism virtually usurps the power to determine our relation with the non-human world and our own corporeality (see Bakker and Bridge 2006). Much of the new 'critical resource geography' aims to strike this balance (see Chapters 13, 33, 34, 35 and 43, this volume).

Conclusion

This chapter has explored some big questions all-too-briefly. I have ignored relevant literatures about environmental in/justice and ecological economics. I have also ignored the writings of many talented (neo-)Marxists writing about nature, among them Karen Bakker, Ted Benton, Uli Brand, Gavin Bridge, Dan Buck, Stephen Bunker, Paul Burkett, Bram Buscher, Esteve Corbera, Gareth Dale, Peter Dickens, Michael Ekers, Vinay Gidwani, Matt Huber, Ray Hudson, Maria Kaika, Joel Kovel, Mazan Labban, Richard Levins, Richard Lewontin, Alain Lipietz, Alex Loftus, Minqi Li, James McCarthy, Philip McMichael, Jason Moore, Sandra Moog, Martin O'Connor, Tom Perreault, Scott Prudham, Morgan Robertson, Allan Schnaiberg, Richard Smith, Erik Swyngedouw, Richard Walker and Michael Watts (the list goes on). But, by focussing on Neil Smith's notion of nature's 'production' in relation to a selection of other Marxist writings, I have gone some way to addressing key analytical and normative issues subtending the work of these authors. Is 'nature' internal or external to capitalism (or both), a constraint or opportunity (or both); is it the 'enemy of nature' and if so what sort of 'nature'; and what should a 'politics of nature' look like in the critique of capital? As is now plain, I believe Smith's notion has analytical merit, while its political message is hopeful: for him, we can (and should) change our collective relation with what we call nature, but not because of any 'objective' imperatives emanating from the biophysical world. Analytically, the challenge is to find a way of registering 'the difference that nature' makes to all our lives while avoiding recourse to all those dualisms that have organised Western thought for centuries. Arguably, that challenge is being met in much of the recent research reviewed in this Handbook, even if – as I said at the outset – the sort of 'political ecology' practised by Smith largely ran parallel to the evolution of political ecology as a field.

Finally, what of politics and action? After all, Marxism famously aspires to change the world not only to understand it. Smith's arguments arguably remain too counter-intuitive to be of service in politics outside the academy. 'Production', conventionally understood, seems a strange concept – a peculiar *metaphor* even – to organise an ecologically aware anti-capitalist discourse. Moreover, like many academic Marxists writing about nature, Smith wrote virtually nothing about *realpolitik*: how, practically, might the social relation with 'nature' be changed for the better? Interestingly, certain strands of environmental and body-politics operative outside universities are now dispensing with 'nature' as an ontological referent (see, for example, Shellenberger and Nordhaus 2007). In a generic sense, this mirrors Smith's insistence that we need new terms of radical political discourse. The challenge, though, is to find a lexicon that resonates in everyday life without becoming assimilated to a soft reformism that does nothing to reign-in capitalism's appetite to commodity everything. Whatever happens, biophysical questions – questions of ice sheets, sea levels, atmospheric, temperature, genes, fresh water, and much else besides – will be absolutely central to politics (mainstream and radical) in the twenty-first century. How those questions are answered discursively and practically may, literally, determine the future of life on this planet. Can Marxists provide solutions that have mass appeal without invoking 'common sense' terms like 'nature'? Notwithstanding the bad name capitalism currently has in many quarters, the prospects are not terribly promising.

References

Altvater, E. (1993) *The future of the market*, trans. P. Camiller (London: Verso).

Bakker, K. and Bridge, G. (2006) 'Material worlds?'. *Progress in Human Geography* 30, 1: 5–27.

Benton, T. (1991) 'The Malthusian challenge: ecology, natural limits and human emancipation', in P. Osborne (ed.) *Socialism and the limits of liberalism* (London: Verso) pp. 241–269.

Biro, A. (2005) *Denaturalizing ecological politics* (Toronto: University of Toronto Press).

Blaikie, P. (1985) *The political economy of soil erosion in developing countries* (London: Methuen).

Castree, N. (1995) 'The nature of produced nature', *Antipode* 27, 1: 12–47.

Castree, N. (2000) 'Marxism and the production of nature', *Capital & Class* 72: 5–36.

Castree, N. (2001) 'Marxism, capitalism and the production of nature', in N. Castree and B. Braun (eds) *Social nature* (Oxford: Blackwell) pp. 189–207.

Foster, J.B. (2012) 'The planetary rift and the New Human Exemptionalism', *Organization & Environment* DOI: 10/1177/1086026612459964.

Foster, J.B., Clark, B. and York, R. (2010) *The ecological rift: capitalism's war on the Earth* (New York: Monthly Review Press).

Goodman, D., Sorj, B. and Wilkinson, J. (1987) *From farming to bio-technology* (Oxford: Blackwell).

Guthman, J. (2011) 'Bodies and accumulation: revisiting labour in the "production of nature"', *New Political Economy* 16, 2: 233–238.

Harvey, D. (1974) 'Population, resources and the ideology of science', *Economic Geography* 50, 3: 256–277.

Henderson, G. (1998) 'Nature and fictitious capital', *Antipode* 30, 2: 73–118.

Henderson, G. (1999) *California and the fictions of capital* (Oxford: Oxford University Press).

Huber, M. (2013) *Lifeblood* (Minneapolis: University of Minnesota Press).

Kloppenberg, J. (1987) *First the seed* (Cambridge: Cambridge University Press).

Mann, S. (1990) *Agrarian capitalism in theory and practice* (Chapel Hill: University of North Carolina Press).

Mann, S. and Dickinson, J. (1978) 'Obstacles to the development of a capitalist agriculture', *Journal of Peasant Studies* 5, 4: 466–481.

Marx, K. (1973) *Grundrisse*, trans. M. Nicolaus (Harmondsworth: Penguin).
O'Connor, J. (1998) *Natural causes* (New York: Guilford).
O'Keefe, P. and Smith, N. (1980) 'Geography, Marx and the concept of nature', *Antipode* 12, 2: 30–39.
O'Keefe, P., Westgate, K. and Wisner, B. (1976) 'Taking the naturalness out of natural disasters', *Nature* 260, April 15: 566–567.
Polanyi, K. (1944) *The great transformation* (New York: Rinehart).
Robertson, M. (2012) 'Measurement and alienation: making a world of ecosystem services', *Transactions of the Institute of British Geographers* 37, 3: 386–401.
Ross, A. (1994) *The Chicago gangster theory of life* (London: Verso).
Schmidt, A. (1971) *The concept of nature in Marx* (London: New Left Books).
Shellenberger, M. and Nordhaus, T. (2007) *Break through* (New York: Houghton Mifflin).
Smith, N. (1984) *Uneven development* (Oxford: Blackwell).
Smith, N. (1996) 'The production of nature', in J. Bird, B. Curtis, M. Mash, T. Putnam, G. Robertson and L. Tickner (eds) *FutureNatural* (London: Routledge) pp. 35–54.
Smith, N. (1998) 'Nature at the millennium', in B. Braun and N. Castree (eds) *Remaking reality* (London: Routledge) pp. 271–285.
Smith, N. (2007) 'Nature as an accumulation strategy', in L. Panitch and C. Leys (eds) *Socialist register 2007* (Monmouth: The Merlin Press) pp. 16–36.
Smith, N. (2008) 'Afterword to the third edition', in N. Smith *Uneven development* (Athens: University of Georgia Press) pp. 239–266.
Soper, K. (1991) 'Greening Prometheus', in P. Osborne (ed.) *Socialism and the limits of liberalism* (London: Verso) pp. 271–293.
Vogel, S. (1996) *Against nature* (Albany: State University of New York).

22
POLITICAL ECOLOGY OF RISK, HAZARDS, VULNERABILITY, AND CAPACITIES

James L. Wescoat Jr.

Key concepts, criticisms, and chapter outline

This chapter presents hazards research as a leading rather than lagging theme in political ecology. This reframing is warranted by the increasingly powerful alliance between theory and practice in hazards research, along with its progressive adjustment over the past decade to criticism from all quarters, including political ecology. Risk and vulnerability have been central concerns in political ecology. At the same time it may be asked whether advances in political ecology and hazards research are diverging or converging in promising directions.

Hazards research is sometimes portrayed as one of the antecedent fields of political ecology (Robbins 2011; see also Chapter 2, this volume). It followed human ecology research of the 1930s and arose somewhat before and then concurrently with cultural ecology in the 1960s. As will be elaborated below, early political ecologists criticized hazards research for its behavioral emphasis on risk perception and decision-making to the neglect of structural power relations and differential vulnerability of social groups. Vulnerability research thrived as a corrective to these issues but was later questioned as to whether it was disempowering, as compared with research on resilience that gives greater weight to human agency and capacity (in the terminology of current hazards research; Wisner et al. 2012). An additional challenge for political ecology and hazards research are critiques of "risk societies," characterized by ever-increasing anxiety to manage the symptoms but not the underlying social driving forces of environmental danger (Beck 1992, 2009). As if this were not enough, there are at least two incompatible definitions of risk within hazards research that muddy the waters.

This chapter strives to succinctly review these conceptual and historiographical issues, though it does not treat them as at all settled. Substantive research on risk, hazards, and vulnerability is increasing at a rapid pace. It is sometimes attentive to political ecology but is more commonly developing along parallel lines. The next section introduces key concepts in hazards research and their inter-relationships. The following section retraces the unfolding relationships between hazards research and political ecology leading up to recent research contributions. Based on the review of recent studies, the final section identifies emerging topics for research.

Conceptual frameworks

It is useful to begin with key concepts in risk and hazards research as they are employed by leading practitioners at the start of the twenty-first century. Wisner et al. (2012) offer the following conceptual model:

$$R = H \times (V/C - M)$$

In this formulation, Risk [R] is understood as the joint product of low frequency, high magnitude hazardous events [H] *and* the differential vulnerability of human and natural systems to them [V]. Vulnerabilities may be offset by an array of human capacities [C] that can take the form of preparedness, response, recovery, reconstruction, and more broadly resilience. Aggregate risk may be further reduced by public mitigation measures [M] (Wisner et al. 2004). It is useful to elaborate each of these variables:

H = hazards are defined in this framework as low frequency, high magnitude environmental events. This is a problematic definition for two reasons. A large body of early geographic research argued that hazards are the joint product of risk (low frequency, high magnitude events) and human vulnerability (Burton et al. 1993). Some branches of risk analysis such as engineering statistics likewise define risk in probabilistic terms. However, the related field of disaster research has been moving in the opposite direction by designating risk as the more comprehensive term than hazard in this formula. Those involved in the field employ both definitions, depending upon the audience and context; this chapter retains the broader conception of hazards developed by environmental geographers.

V = vulnerability has a complex history in late twentieth-century hazards research. It refers most commonly to the actual and potential suffering of marginalized social groups and individuals, i.e., those marginalized by economic class, gender, age, ethnicity, caste, (dis)ability, and so on. Wisner et al. (2012) refer to a triangle formed by the political, economic, and social structures of vulnerability. A significant body of vulnerability research is rigorously qualitative and reflexive. However, a broad analytical approach has also been developed in a Social Vulnerability Index in the United States (SoVI; Cutter et al. 2003). Its 32 social and environmental variables have been statistically analyzed and mapped at county, state, and regional levels.

C = capacity is a relatively new term in the risk framework. It complements vulnerability analysis by focusing on forms of human agency and social action that reduce losses. In this respect, it is comparable to research on capitals and capabilities (Bebbington 1999; Sen 1989). While capacity per se is less widely pursued in risk, hazards, and political ecology literatures to date, it hearkens back to Gilbert White's emphasis on human adjustment as actions that strive to "broaden range of choice" among alternative courses of action (Wescoat 1987).

M = mitigation became one of the most widely adopted concepts in natural hazards policy in the late twentieth century. It emphasizes actions that reduce risk in advance of disaster, rather than concentrating on recovery and relief. It is analogous to, although not the same as, the concept of mitigation in climate change research and policy, which refers to pollution prevention vis-à-vis adaptation to climate change. Mitigation came into prominence as the primary focus of natural hazards policy in the United States in the mid-1990s (US FEMA 1995).

This type of risk framework is associated with various types of uncertainty analysis. Whereas risk (or hazard) in probabilistic terms refers to the frequency, duration, and magnitude of damaging events, uncertainty refers to a lack of knowledge about the probability of such events. Uncertainty can take at least four forms (US National Research Council, Water Science and Technology Board 2000):

1 Aleatory uncertainty (i.e., unknown variability in the natural and social phenomena themselves).
2 Data uncertainty (i.e., missing data and unknown data quality).
3 Model uncertainty (i.e., missing variables and unknown fit between the model and reality).
4 Epistemic uncertainty (i.e., unknown approaches to problem framing and ways of thinking about the hazard).

However, as White et al. (2001) point out, increasing knowledge and reducing uncertainty do not necessarily lead to the mitigation of risk. Social and behavioral change do not follow from enhanced knowledge alone. While behavioral scientists have concentrated on psychological aspects of this problem through research on judgment under uncertainty (Kahneman et al. 1982), political ecology has broadened research into social forces and political dynamics that explain the persistence of risk and hazards even in the context of increasing knowledge and aggregate economic growth.

Research on risk is increasingly related to research on resilience, which is itself a multi-dimensional concept whose meanings range from physical mechanics to population ecology, ecosystem dynamics, and various forms of social response to damaging events. While physical resilience is recognized as inadequate for human-environmental systems, it is still incorporated in some systems models. Debates between political ecological and ecosystem resilience approaches are more challenging (see Walker and Cooper 2011 for a genealogy of ideas). Turner (2014) contrasts the explicitly normative commitments of political ecology with more positivist conceptions of resilience in systems thinking, though further excavation of the values embedded in contemporary systems approaches is warranted. A recent US National Research Council (2012) report on *Disaster Resilience: A National Imperative* argued for strengthening the "culture" and "practice" of disaster resilience. Although that report does not mention political ecology per se, it frequently mentions political, economic, and environmental aspects of resilience jointly in ways that resonate with political ecology approaches. The annual Natural Hazards Workshop at the University of Colorado, Boulder has convened scholars and practitioners and scholars from the United States and internationally to establish an expanding community of practice for more than three decades, just as political ecology has recently been described as a "community of practice" (Robbins 2011).

In summary, the concepts of hazards, disasters, and risk research have been perennially debated in ways that will likely continue. These debates include the denotation of the concepts, e.g., do disasters include chronic or slow onset events like drought and climate change? Do hazards include technological failures, economic depressions, violence, crime, wars, terrorism, or disease (Glantz 1999; Herrick 2012; Hewitt 1983; Mitchell 2003; Mustafa 2013; Wisner and Gaillard 2009)? Some aspects of the political ecology of hazards are less well developed than others. For example, the "ecological" in political ecology has referred more to social relations than to ecosystem processes that include human action (e.g., what is sometimes called coupled natural and human systems; Dove and Hudayana 2008; Walker 2005). However, this contrast may be more applicable to review articles than to substantive research, as important exceptions date back at least to Hewitt's (1983) *Interpretations of calamity: From the viewpoint of human ecology* (the human ecology antecedents of political ecology warrant further historical research). The politics of vulnerability and mitigation are not yet fully mainstreamed in the field of hazards research. The conceptual framework outlined above is also more synchronic than historical in its methods (though see Alexander 2002). The next section offers an historical perspective on the relations between political ecology and hazards research that helps shed light on recent inquiry.

Unfolding affinities between political ecology and hazards research

Political ecology has strong yet often implicit affinities with hazards research. This section starts with antecedent fields of inquiry, followed by implicit and explicit historical connections, and then moves toward recent patterns of research.

Some common antecedents

Political ecology and hazards research share at least three interesting antecedents. Cultural ecology is perhaps the closest antecedent field in time and substance, as it focuses on subsistence strategies as the "core" around which issues of risk, scarcity, and adaptation revolve (Zimmerer 2006). Hazards research and cultural ecology actually developed concurrently and to some extent perhaps in dialogue with one another (Turner 2008). For example, less well remembered today is the influence of Sprout and Sprout's (1979) research on environmental politics at multiple levels of inquiry including international relations, though it is less clear how that work might have influenced hazards research or whether it influenced political ecology. A thorough review of the historical links between political science, political geography, hazards, and political ecology is warranted (e.g., in research by Kasperson [1994] and others). In the early twentieth century, the field of resource geography provided a broad platform for human ecology and hazards research. Zimmerman's (1933) functional theory of resources operated at multiple scales, and linked the agency of resource creation with the structures and dynamics of resource appropriation, allocation, trade, depletion, and substitution. The mid-twentieth century witnessed a strong renewal of natural resources research at institutions like Resources for the Future, although that research agenda was framed in terms of applied microeconomics, in contrast with the more strongly political and Marxian strands of critical resource and hazards geography (e.g., on forests, water, and mining), which developed in leading geography programs such as Berkeley and Clark (e.g., Peet et al. 2011, for a sample of contributions to that prodigious body of work).

The reframing of hazards research and emergence of political ecology

In historical terms, hazards research began with an agenda of progressive reform in the pragmatist sense of the term from the 1940s onward (Robbins 2011; Wescoat 1992). Gilbert White's (1945) flood hazards research challenged US federal agencies with data showing that the more the government spent on flood control the greater the losses. White explained how the construction of levees reduced flood risk perception in ways that induced increased floodplain occupance, investment, and long-term losses. He argued consistently over a half-century that while extreme events are natural ("acts of God"), hazards losses were "acts of man" (White 1945, p. 2). Now even extreme geophysical events are shown to have human dimensions, e.g., in anthropogenic processes of desertification and climate change.

In the meantime, however, hazards research came under heated criticism as atheoretical, as uncritically accepting the politics of vulnerability, as overemphasizing the power of science and technology, and having an inadequate framework for explaining processes of social change (e.g., Hewitt 1983; Neumann 2005; Watts 1983). As there was little political ecology per se at that point, the main lines of criticism came from human ecology and Marxist political economy, though many of these critics and their students would become leaders in political ecology.

While these criticisms triggered new lines of research, they also included some biases. Research on risk perception, behavior, and communication were proscribed as "scientism"

(Waddell 1977). Research on environmental and resource economics that advances the ideas and methods of neoclassical or applied welfare economics receives less attention in political ecology, as compared with radical political economic approaches. Nevertheless, at least three profound lines of research have emerged from these critiques:

- vulnerability research
- social power relations
- politics of resource access and conflict.

Hazards researchers as well as their critics have turned their attention to these issues. Political ecology offered new approaches for addressing vulnerability, inequality, and the politics of hazards; but it is not clear whether the two fields were developing in dialogue or in parallel with one another. The next section shows that the linkages to date have been limited with perhaps slightly more attention to political ecology by hazards researchers than vice versa (cf. Peet et al. 2011 with Wisner et al. 2012).

A review of recent research

A systematic bibliographic search was conducted to assess the relationships between political ecology and various aspects of hazards research (for "bibliographic mapping" methods see Collaboration for Environmental Evidence 2013; Wescoat 2014). The search terms linked political ecology with key words in hazards research:

"political ecology" AND "risk"
"political ecology" AND "hazard"
"political ecology" AND "vulnerability"
"political ecology" AND "disaster"
"political ecology" AND "mitigation"
"political ecology" AND "resilience"

A second key consideration was the online digital libraries to be searched. Each library has different strengths, and the following were chosen to capture a broad range of work:

- WorldCat – books published from 2000 to 2013
- Proquest – dissertations and theses all dates
- Scopus – scholarly articles published from 2000 to 2013
- Web of Science – scientific articles published from 2000 to 2013
- EICompendex – engineering articles published from 2000 to 2013.

Results were screened for relevance, mainly by confirming whether the keywords were used as research terms or as ordinary language (the latter were excluded). In cases where more than one hundred hits were returned from a keyword search, the term was re-entered as a title word. As shown in Table 22.1 that was rare, for the number of hits were relatively few.

Several search results stand out. First, "risk" is the keyword most frequently associated with political ecology, which reflects its use across many disciplines. "Vulnerability" is the second most commonly associated keyword, which reflects the critical turn in both fields towards the issues of equity, injustice, and marginalization. Subsequently hazards research shifted toward studies of "resilience" to re-balance the emphasis on structural inequalities suffered by victims

Table 22.1 Bibliographic search results for the keywords "Political ecology" AND "…"

Search term	*WorldCat*	*Scopus*	*Web of Knowledge*	*Compendex Kta*	*Proquest Diss*
Risk	88 Kw/19 Ti	26	23	7	29
Hazard	7 Ti	4	5	2	10
Disaster	8	9	6	3	14
Vulnerability	10	17	15	4	16
Mitigation	9	0	0	0	10
Resilience	5	5	3	6	9

Note: Kw = keyword; Ti = title; Kta = keyword, title, abstract.

with increased attention to human agency and capacity (Cutter and Corendea 2013). By comparison, political ecology citations still appear to focus more on vulnerability than resilience (Miller et al. 2010).

One surprising search result was the high frequency of research on the political ecology of hazards in doctoral dissertations (i.e., the right-hand column of Table 22.1). Eight of the ten books on political ecology and vulnerability identified in WorldCat were theses or dissertations. Dissertation research ranges from agrarian sustainability (e.g., Murphy 2011) to urban environmental justice (Seo 2010; and Sicotte 2003). Interestingly, as in earlier generations the most frequent research topic involves water-related hazards (e.g., Miller 2003; Nethengwe 2007; Nijbroek 2012; Pelling 1997). The legacy of flood hazards research thus continues in a new frame.

Interestingly, few dissertation writers continued to write about the political ecology of hazards in subsequent journal articles. This observation may support Paul Robbins' (2011) point about political ecology as a community of practice that works within a broad conceptual framework vis-à-vis a sustained project of theory building. Pelling (1997, 1999) is an important exception whose doctoral research led to book chapters and articles on the political ecology of flood hazards in peri-urban Georgetown, Guyana. This research historicized social power relations and vulnerability. It shifted from agrarian toward urbanizing environments, and toward local community development rather than government institutions. However, Pelling's (2003b) book-length treatment of the subject also shifted toward social resilience, emphasizing citizen agency and adaptive potential in ways consonant with trends in hazards research. This observation should not be exaggerated as leading critical water hazards researchers employ ideas and methods associated with political ecology (e.g., Mustafa 2013, on flood and irrigation hazards in Pakistan; and Sultana 2006, on the political ecology of arsenic exposure in Bangladesh).

Of course political ecology research extends far beyond flood hazards (Collins 2008). A substantial and increasing body of research examines climate impacts and adaptation (Crate 2011). Insofar as it addresses mitigation, however, it deals with climate change impacts rather than risk reduction (cf. Birkenholtz 2012). There is a rapidly growing body of research on the relationships between climate adaptation and disaster risk reduction, a limited proportion of it appears to be framed in terms of political ecology (Shearer 2011). An increasing number of studies examine the political ecology of environmental health risks (Bidwell 2010; Little 2013; Mayer 1996; see also Chapter 26, this volume). In a study of the political ecology of tornado disasters, Donner (2007) developed a political model of human ecology (POET) in which population, organization, environment, and technology are examined to assess the social distribution of tornado disasters. This conceptual framework is not unlike the IPAT model (Impact = Population x Affluence x Technology) that is sometimes used, and criticized, in hazards research (Forsyth 2002).

Prospects

Studies at the intersection of political ecology and hazards research are interesting and important, yet relatively few in number. While hazards research was criticized in the late twentieth century for its limited attention to politics, especially at larger scales, the field subsequently addressed and moved beyond many of those criticisms. It offers increasing precision on issues of risk, vulnerability, mitigation, and resilience, and it could thus become a key theme in the field of political ecology. Emerging research topics include: environmental risks in nonstationary climates; multiple-hazards and all-hazards preparedness; political ecologies of insurance and finance; convergence of climate adaptation and disaster risk reduction; intersections of environmental risk and terror; and efforts to envision what if, anything, might lie beyond a global risk society (Beck 2009).

To give a sense of possible intersections between hazards research and political ecology, Wisner and Walker (2005) presented 12 research topics at a major conference in Kobe, Japan in 2005, which marked the tenth anniversary of the devastating earthquake in that city. The first five topics involved a broader scope of hazards:

1 governance and respect for human rights
2 globalization and disasters
3 war and disasters
4 climate change and disaster risk reduction
5 urbanization and hazards.

Another seven topics involved a selection of "ways forward":

6 local initiatives and innovations
7 meaningful and effective local participation
8 knowledge and risk communication
9 merging risk reduction and development
10 global alliances of disaster-affected peoples
11 women's crucial role in disaster reduction
12 full accountability and transparency.

Although the authors presented these topics under the banner of political ecology, they did not discuss how they draw upon or contribute to political ecology. A decade later, there have been dramatic advances in both hazards research and political ecology, but arguably less so in the rigorous development of their intersection. Political ecology arose in part to pose critical alternatives to physical and behavioral science approaches in hazards research, and it helped stimulate sustained research on hazards vulnerability. Its association with development studies deepened the geographic perspective of hazards research. While political ecology has drawn insight and inspiration from hazards researchers from Gilbert White onwards, it is less clear from the evidence compiled here what hazards research has bequeathed to political ecology, or to put it more critically, whether political ecology draws as intensively as it could upon advances across the enormous community of practice in risk, hazards, and disaster research.

The U.N. Hyogo Framework for Action on disaster resilience (2005–2015) that Wisner and Walker addressed in 2005 comes to a close, and the Third World Conference on Disaster Risk Reduction to replace it approaches in March 2015. In this context, political ecology has the opportunity to focus its advanced capabilities on unfolding processes of risk, vulnerability, resilience, and structural change, and to increase its engagement with the annual community of practice convened by the University of Colorado Natural Hazards Center. Similar opportunities abound at the intersection of climate adaptation and disaster risk reduction. Political ecology can once again provide a powerful conceptual framework for expanding the community of practice in both fields. Hazards research can help political ecology focus on specific dimensions and dynamics of environmental risk in sustained theory-building ways. These are just a few of the prospects and challenges for researchers on the political ecology of risk, hazards, and disasters.

Acknowledgments

I would like to thank the editors for inviting this chapter and offering excellent advice on how to strengthen it, as well as former colleagues and graduate students, now colleagues, from the Department of Geography and Natural Hazards Center at the University of Colorado-Boulder for advancing my understanding of both hazards research and political ecology.

References

Alexander, D. 2002, "Nature's impartiality, man's inhumanity: Reflections on terrorism and world crisis in a context of historical disaster," *Disasters* vol. 26, no. 1, pp. 1–9.

Bebbington, A. 1999, "Capitals and capabilities: A framework for analyzing peasant viability, rural livelihoods and poverty," *World Development* vol. 27, no. 12, pp. 2021–2044.

Beck, U. 1992, *The risk society: Towards a new modernity*, Sage, London.

Beck, U. 2009, *World at risk*, Polity, Cambridge.

Bidwell, D. 2010, "Bison, boundaries, and brucellosis: Risk perception and political ecology at Yellowstone," *Society & Natural Resources* vol. 23, no. 1, pp. 14–30.

Birkenholtz, T. 2012, "Network political ecology: Method and theory in climate change vulnerability and adaptation research," *Progress in Human Geography* vol. 36, no. 3, pp. 295–315.

Burton, I., Kates, R. and White, G. 1993, *The environment as hazard*, 2nd edn, Guilford Press, New York.

Collaboration for Environmental Evidence (CEE) 2013, *Guidelines for systematic review and evidence synthesis in environmental management*, Version 4.2. Environmental Evidence, www.environmentalevidence.org/Authors.htm/Guidelines 4.2.pdf (accessed November 17, 2013).

Collins T. 2008, "The political ecology of hazard vulnerability: Marginalization, facilitation and the production of differential risk to urban wildfires in Arizona's White Mountains," *Journal of Political Ecology* vol. 15, pp. 21–43.

Crate, S.A. 2011, "Climate and culture: Anthropology in the era of contemporary climate change," *Annual Review of Anthropology* vol. 40, pp. 175–194.

Cutter, S. and Corendea, C., eds. 2013, *From social vulnerability to resilience: Measuring progress toward disaster risk reduction*, SOURCE No. 17/2013, United Nations University Institute for Environment and Human Security (UNU-EHS), Bonn.

Cutter, S., Boruff, B. and Shirley, L. 2003, "Social vulnerability to environmental hazards," *Social Science Quarterly* vol. 84, no. 2, pp. 242–261.

Donner, W. 2007, "The political ecology of disaster: An analysis of factors influencing U.S. tornado fatalities and injuries, 1998–2000," *Demography* vol. 44, no. 3, pp. 669–685.

Dove, M. and Hudayana, B. 2008, "Conversations across the divide: The time and place for political ecology: The life-work of Piers Blaikie. Biocomplexity in coupled human-natural systems: The study of population-environment interactions," *Geoforum* vol. 39, no. 2, pp. 736–746.

Forsyth, T. 2002, *Critical political ecology: Environmental knowledge, policy and risk*, Routledge, London.

Glantz, M., ed. 1999, *Creeping environmental problems and sustainable development in the Aral Sea Basin*, Cambridge University Press, Cambridge.

Herrick, C. 2012, "The political ecology of alcohol as 'disaster' in South Africa's Western Cape," *Geoforum* vol. 43, no. 6, pp. 1045–1056.

Hewitt, K., ed. 1983, *Interpretations of calamity: From the viewpoint of human ecology*, Unwin Hyman, London.

Kahnemann, D., Slovic, D., and Tversky, A. 1982, *Judgment under uncertainty: Heuristics and biases*, Cambridge University Press, Cambridge.

Kasperson, R. 1994, "Global environmental hazards: Political issues in societal responses," in G. Demko and W. Wood (eds.), *Reordering the world: Geopolitical perspectives on the 21st century*, Westview Press, Boulder, pp. 141–166.

Little, P. 2013, "Vapor intrusion: The political ecology of an emerging environmental health concern," *Human Organization* vol. 72, no. 2, pp. 145–155.

Mayer, J. 1996, "The political ecology of disease as one new focus for medical geography," *Progress in Human Geography* vol. 20, no. 4, pp. 441–456.

Miller, F. 2003, "Society-water relations in the Mekong Delta: A political ecology of risk," Ph.D. dissertation, Department of Geography, University of Sydney, Sydney.

Miller, F., Osbahr, H., Boyd, E., Thomalla, F., Bharwani, S., Ziervogel, G., Walker, B., Birkmann, J., Van der Leeuw, S., Rockström, J., Hinkel, J., Downing, T., Folke, C., and Nelson, D. 2010, "Resilience and vulnerability: Complementary or conflicting concepts?" *Ecology and Society* vol. 15, no. 3, p. 11. www.ecologyandsociety.org/vol15/iss3/art11/.

Mitchell, J. 2003, "The fox and the hedgehog: Myopia about homeland security in U.S. policies on terrorism," in L. Clarke (ed.), *Terrorism and disaster: New threats, new ideas (Research in Social Problems and Public Policy, Volume 11)*, Emerald Group Publishing Limited, Bingley, pp. 53–72.

Murphy, D. 2011, "Going on *otor*: Disaster, mobility, and the political ecology of vulnerability in Uguumur, Mongolia," Ph.D. dissertation, Department of Geography, University of Kentucky, Lexington.

Mustafa, D. 2013, *Water resource management in a vulnerable world: The hydro-hazardscapes of climate change*, I.B. Tauris, London.

Nethengwe, N. 2007, "Integrating participatory GIS and political ecology to study flood vulnerability in the Limpopo Province of South Africa," Ph.D. dissertation, Department of Geography, West Virginia University, Morgantown, WV.

Neumann, R. 2005, *Making political ecology*, Routledge, London.

Nijbroek, R. 2012, "Mangroves, mudbanks and seawalls: political ecology of adaptation to sea level rise in Suriname," Ph.D. dissertation, Department of Geography, University of South Florida.

Peet, R., Robbins, P., and Watts, M. 2011, *Global political ecology*, Routledge, London.

Pelling, M. 1997, "A political ecology of urban flood hazard and social vulnerability in Guyana," Ph.D. dissertation, Department of Geography, University of Liverpool, Liverpool, UK.

Pelling, M. 1999, "The political ecology of flood hazard in urban Guyana," *Geoforum* vol. 30, no. 3, pp. 249–261.

Pelling, M. 2003a, "Toward a political ecology of urban environmental risk: The case of Guyana," in K. Zimmerer and T. Bassett (eds.), *Political ecology: An integrative approach to geography and environment-development studies*, Guilford Press, New York, pp. 73–94.

Pelling, M. 2003b, *The vulnerability of cities: Natural disasters and social resilience*, Earthscan, London.

Robbins, P. 2011, *Political ecology: A critical introduction*, 2nd edition, Wiley-Blackwell, Hoboken.

Sen, A. 1989, "Development as capability expansion," *Journal of Development Planning* vol. 19, pp. 41–58.

Seo, W. 2010, "Political ecology and environmental justice analysis of information and communication technology," Ph.D. dissertation, School of Urban Affairs and Public Policy, University of Delaware.

Shearer, C. 2011, "The political ecology of climate adaptation assistance: Alaska Natives, displacement, and relocation," *Journal of Political Ecology* vol. 19, pp 174–183.

Sicotte, D. 2003, "Race, class and chemicals: The political ecology of environmental injustice in Arizona," Ph.D. dissertation, Department of Sociology, Arizona State University, Scottsdale.

Sprout, H. and Sprout, M. 1979, *The ecological perspective on human affairs: With special reference to international politics*, Greenwood Press, Westport.

Sultana, F. 2006, "Gendered waters, poisoned wells: Political ecology of the arsenic crisis in Bangladesh," in Kuntala Lahiri-Dutt (ed.), *Fluid bonds: Views on gender and water*, Stree Publishers, India, with Australian National University, Canberra, pp. 362–386.

Turner, B. 2008, "Land-change science and political ecology: Similarities, differences, and implications for sustainability science," *Annual Review of Environment and Resources* vol. 33, pp. 295–316.

Turner, M. 2014, "Political ecology I: An alliance with resilience?" *Progress in Human Geography*, vol. 38, pp. 616–623.

U.S. Federal Emergency Management Agency (FEMA) 1995, *National mitigation strategy*, FEMA, Washington, DC.

U.S. National Research Council 2012, *Disaster resilience: A national imperative*, National Academies Press, Washington, DC.

U.S. National Research Council, Water Science and Technology Board 2000, *Risk analysis and uncertainty in flood damage reduction studies*, National Academies Press, Washington, DC.

Waddell, E. 1977, "The hazards of scientism: A review article," *Human Ecology* vol. 5, no. 1, pp. 69–76.

Walker, P. 2005, "Political ecology: Where is the ecology?" *Progress in Human Geography* vol. 29, no. 1, pp. 73–82.

Walker, J. and Cooper, M. 2011, "Genealogies of resilience: From systems ecology to the political economy of crisis adaptation," *Security Dialogue* vol. 42, no. 2, pp. 143–160.

Watts, M. 1983, "On the poverty of theory: Natural hazards research in context," in K. Hewitt (ed.), *Interpretations of calamity for the viewpoint of human ecology*, Allen & Unwin, Boston, pp. 231–262.

Wescoat, J. 1987, "The practical range of choice in water-resources geography," *Progress In Human Geography* vol. 11, no. 1, pp. 41–59.

Wescoat, J. 1992, "Common themes in the work of Gilbert White and John Dewey: A pragmatic appraisal," *Annals of the Association of American Geographers* vol. 82, no. 4, pp. 587–607.

Wescoat, J. 2014, "Searching for comparative international water research: Urban and rural water conservation research in India and the United States," *Water Alternatives* vol. 7, no. 1, pp. 199–219.

White, G. 1945, *Human adjustment to floods*, Chicago: University of Chicago, Department of Geography Research Paper, Chicago.

White, G., Kates, R., and Burton, I. 2001, "Knowing better and losing even more: The use of knowledge in hazards management," *Environmental Hazards* vol. 3, no. 3–4, pp. 81–92.

Wisner, B. and Gaillard, P. 2009, "An introduction to neglected disasters," *JMB: Journal of Disaster Risk Studies* vol. 2, no. 3, pp. 151–158.

Wisner, B. and Walker, P. 2005, "The World Conference on disaster viewed through the lens of political ecology: A dozen big questions for Kobe and Beyond," *Capitalism Nature Socialism* vol. 16, no. 2, pp. 89–95.

Wisner, B., Blaikie, P., Cannon, T. and Davis, I. 2004, *At risk: Natural hazards, people's vulnerability and disasters*, Routledge, London.

Wisner, B., Gaillard, P. and Kelman, I. 2012, *Routledge handbook of hazards and disaster risk reduction*, Routledge, London.

Zimmerer, K. 2006, "Cultural ecology: At the interface with political ecology – the new geographies of environmental conservation and globalization," *Progress in Human Geography* vol. 30, no. 1, pp. 63–78.

Zimmerman, E. 1933, *World resources and industries: A functional appraisal of the availability of agricultural and industrial resources*, Harper & Brothers, New York.

23

READING CLIMATE CHANGE AND CLIMATE GOVERNANCE AS POLITICAL ECOLOGIES

Diana Liverman

Introduction

Political ecology provides powerful insights into understanding the causes, consequences, and responses to climate change from local to global scales – interweaving material nature with the structural drivers of emissions and vulnerabilities, as well as with the agency of individual and institutional actors and the narratives they embrace to describe, explain, and debate what is happening to climate and what should be done about it. Climate change is an issue that highlights questions of environmental and social justice and can connect many political ecologists to activism and policy. It is distinguished by its global reach through the atmospheric commons, by the existential scope of the threats and solutions, and by a research community that is institutionally well connected to international policy.

This chapter explores how political ecology can analyze the geographical causes of climate change, including the differential responsibilities of countries for fossil fuel consumption and land use change. Second, the chapter focuses on climate change vulnerability where political ecology has provided valuable insights into the political economy of drought and other climate extremes in different regions, and the ways in which a changing climate intersects with other stresses, such as those of neoliberalism on peasant farmers. Third, it considers responses to climate change and how political ecology can be used to analyze mitigation – through energy policy and carbon offsets for example – and adaptation.

I will argue that climate change offers a wealth of questions for political ecology in a diverse critical literature. I suggest that climate change connects political ecology back to some of its origins in efforts to understand hazard vulnerability and the intersections between poverty and environmental degradation – but also takes us forward into highly politicized debates about the future of development, energy and land use. A political ecology perspective can counter an over-emphasis on the political economy of climate that can erase the agency of individuals and communities or fail to take science and nature seriously. Political ecology can also provide insights into changing attitudes to climate, including climate skepticism, through understandings of culture, discourse, and science studies. However, most critical research on climate is not self-identified as political ecology and often overlooks the spatiality, materiality, and embodiment of climate change causes and consequences.

Although there are several scholars of climate change who explicitly embrace political ecology as the frame for their work, in many cases a political ecology approach has to be read from the ways in which key dimensions of political ecology – political economy, human agency, material nature and discourse; critical analyses of environmental degradation, conservation, and conflict; governmentality and the creation of environmental subjects, networks, and political actors (Robbins 2012) – are reflected in climate research. Political ecology has much to gain from the broader critical approaches to climate change and especially in analysis of climate governance. Critical work on climate change shares historical origins with political ecology in the emergence of political economy approaches to hazards and famine and an early focus on marginalized populations and social justice (Hewitt 1983, Blaikie et al. 1994, Bohle et al. 1994). There is a growing community of practice that collaborates and communicates about critical approaches to climate change, with scholars connecting to practitioners from environmental organizations and governments to uncover and challenge the politics and discourses of climate politics. And as a counterpoint to a gender imbalance in much of academia, many of those showing a long-term commitment to understanding the political ecologies of climate change are women – such as Karin Bäckstrand, Harriett Bulkeley, Hallie Eakin, Karen O'Brien, Petra Tschakert, or Coleen Vogel.[1]

The international consensus on climate change research is commonly associated with the regular reports of the Intergovernmental Panel on Climate Change, with Working Group II focusing on impacts vulnerability, and adaptation (IPCC 2014a) and Working Group III reviewing research on responsibility for greenhouse gas emissions and mitigation (IPCC 2014b). The most recent reports include several chapters that draw extensively on critical work in political economy and political ecology on vulnerability, climate and development, equity and carbon markets.[2] Other political ecologists have kept their distance from mainstream climate research and policy, delivering intense critiques of climate models and politics (Demeritt 2001, Swyngedouw 2010).

The causes of anthropogenic climate change

While climate has changed and varied over the millennia in response to the geophysical influences that initiated ice ages and warmer periods, the human influence on climate is relatively recent, and is mostly associated with the development of fossil fuels, agriculture, and deforestation since about 1850 – a period now termed the Anthropocene. The greenhouse gases released by these activities are causing the planet to warm and the continuing upward trend in emissions will lead to significant changes and warming in climate, including greater extremes, within the next few decades (IPCC et al. 2013).

Political economy of emissions

The political ecologies underlying emission trends are clear – a global political economy dependent on oil, coal, and gas, on cement construction and livestock, and on converting forests to cropland, fuel, settlement, and commercial lumber. An overarching political economy of emissions is associated with capitalism and colonialism exploiting forest and then fossil fuels across the globe for accumulation, and to a multinational fossil fuel industry supported by states through subsidy, warfare, and special interests. The link between political economy and greenhouse gases is evident in Figure 23.1 when the worldwide recession hit in 2008. The signal of global economic slowdown was reflected in a temporary drop in atmospheric emissions, which then however returned to an even higher growth rate. The overall upward trend demonstrates the failure to control emissions since 1990.

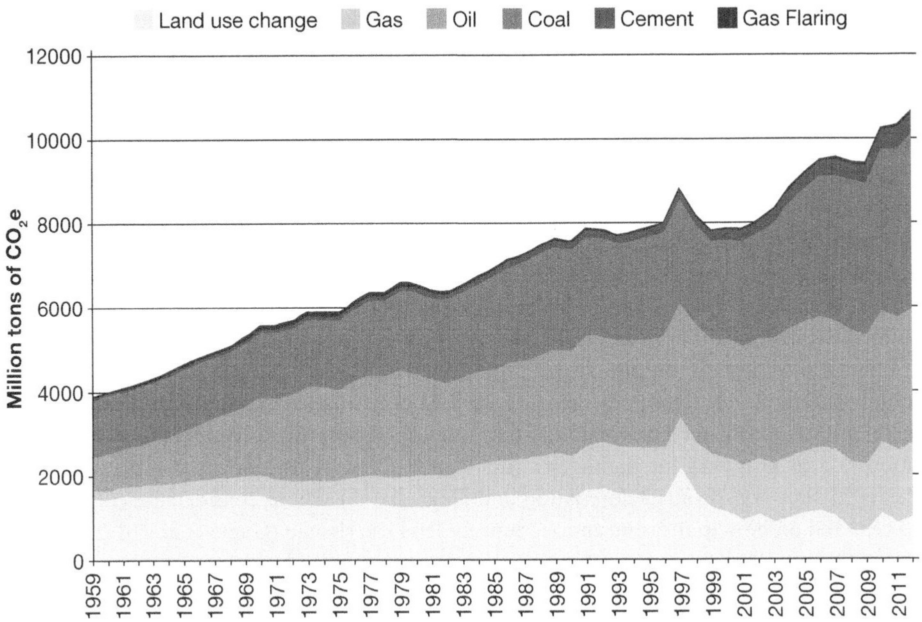

Figure 23.1 Trends in greenhouse gas emissions, 1959–2012

Nature and materialities

Although greenhouse gases mix throughout the atmosphere, sources have varied across time and space, with patterns of responsibility linked through international trade to production and consumption. Responsibility is also associated with the differing radiative potentials (heat trapping) of different greenhouse gases, with CO_2 trapping less energy per unit than methane, but remaining in the atmosphere for many more years. The processes of radiation, precipitation, and carbon cycling are all material and active components of the earth system, interacting with and constraining human activities. Political ecology draws attention to such materialities; showing how the biophysical characteristics of water, wetlands, and biodiversity can make them difficult to manage, commodify, and trade (Castree 2003, Robertson 2000, Bakker and Bridge 2006).

The materialities of greenhouse gases manifest in their differing impacts on climate, and in their spatial patterns and origins in nature – in the fossilized and living plants that are burned as fossil fuels or woodfuel and produce carbon dioxide, the decomposing wastes and rice paddies that produce methane, the millions of cows and sheep that generate methane in their digestive systems, and the chemicals and geological materials that release a range of greenhouse gases in the manufacture of cement, petrochemicals, and other products. The characters of these various greenhouse gases and their origins pose scientific and technical governance challenges for measurement and monitoring and when climate change response includes sequestering them or converting them into tradable commodities.

An initial accounting step in climate mitigation was to create a common unit of CO_2e – which converts some (but not all) greenhouse gases into a single unit or currency of carbon expressed as 'carbon dioxide equivalents.' As Bumpus (2011) argues, a ton of carbon is very much a virtual and abstract commodity – estimated as the difference between emissions that might have occurred (a counterfactual) and lower emissions as a result of a carbon offset project. He and others note the challenges in measuring baselines and additionality of carbon reductions, for communities trying to demonstrate their curation of material carbon reductions, and in the complex accounting and trading systems for invisible gases (Lansing 2012, Lovell et al. 2013).

Global carbon budget studies show major uncertainties about whether, on balance, forestry, grasslands, and agriculture release or sequester carbon in different regions (Dilling et al. 2003). This uncertainty over land use as a source or sink of greenhouse gases means that emissions accounting often separates overall national or per capita emissions into emissions with and without land use change. This has political implications for countries with significant forest cover, rapid deforestation, or reforestation.

International agreements also struggle with accounting for land use change. For example, carbon trading under the Kyoto Protocol initially only included credits for afforestation and reforestation and not for protecting existing forests. Pressure from conservationists and forest nations led to proposals for trading and funding of forest protection in the form of REDD (Reduced Emissions from Deforestation and Degradation) where considerable critical debate has centered on how to measure and account for land use change (Gupta et al. 2012, Leach and Scoones 2013). The recognition that other land uses – grassland, peat lands, agriculture – also influence the carbon budget, brings even more complexity, trading options, and governance challenges into the climate regime.

The scientific challenges of understanding the carbon budget have brought many physical scientists into debates about the management of the carbon cycles, and into partnerships with social scientists in studies to explore the political ecologies of land use emissions (cf. Chapter 11, this volume). For example, geographers Petra Tschakert and Susanna Hecht engaged with earth science to understand soil carbon in West Africa and forestry in El Salvador (Hecht and Saatchi 2007, Tschakert and Tappan 2004). Emerging questions include how to measure and manage the important role of the oceans in the carbon cycle and whether 'blue carbon' can be sequestered or released through the geoengineering, management, or protection of coasts and oceans (Dilling and Hauser 2013, Locatelli et al. 2014).

Responsibility for emissions

Debates about responsibilities for climate change contrast the role of north and south, rich and poor, and business, individuals, and governments based on a variety of calculations and claims. For example, different patterns of national emissions and blame are associated with 'historical emissions' (accumulated emissions), 'per capita emissions' (average emissions per person), and 'current emissions.' The 1991 benchmark paper 'Global warming in an unequal world: a case of environmental colonialism' argued for a distinction between 'survival' and 'luxury' emissions which would provide the poor with an entitlement of emissions necessary for basic human security but would penalize high per capita emissions of excess consumption of richer people and countries (Agarwal and Narain 1991). The United States and Europe bear greater responsibility historically, and India and China rank much lower on both historical and per capita emissions while playing a large role in current and future shares (Table 23.1). Taking an environmental justice perspective, Paul Baer and colleagues argue for 'Greenhouse Development Rights,' which combine estimates of 'climate debt' (historical responsibility) and countries' capacity to act (national income) to allocate emissions responsibility above a baseline development threshold that provides for the needs of the poor (Baer 2013).

Table 23.1 Calculating responsibility for climate change

	Total GHG emissions 2011 $mTCO_2e$	*GHG per capita 2011 Tons*	*Cumulative historical (1850–2011) $mTCO_2e$*	*Net traded emissions $mTCO_2e$ (negative is imports)*
Europe	7,631	10.3	467,628	
North America	7,266	21.0	389,076	
Latin America and Caribbean	3,310	5.5	50,587	
Asia	21,425	5.2	338,916	
Sub-Saharan Africa	2,391	2.7	21,608	
USA	6,550	21.0	361,300	-479
China	10,552	7.8	140,860	1329
India	1,861	1.5	35,581	94

The globalization of trade and the possibilities of carbon trading further complicate this debate as responsibility for emissions alters with 'embodied' emissions in the import and export of goods or with carbon offsetting. Recalculations of emissions responsibility estimate the role of trade and show that 10–25 percent of China's emissions are associated with exports to countries like the United States that are, in turn, net importers of embodied emissions (Peters et al. 2011).

An alternative to the national basis for responsibility in a globalized world is to assign blame to the corporations who extract, burn, or distribute energy and other products that release greenhouse gases. A study that examined the role of major multinational corporations calculated that 90 major corporations – mostly oil and coal – are responsible for two-thirds of carbon emitted in the last 150 years (Heede 2014). Some responsible corporations use a greenhouse gas protocol to distinguish and reduce their onsite emissions and emissions embodied in electricity use from their supply chains (Green 2010).

The scientific community is struggling to marshal more accurate data on emissions and biogeochemical cycles and to associate these with activities in different regions and sectors. Meanwhile governments, nongovernmental organizations, and corporations are selecting those data, images, and reports that represent and advance their interests and, perhaps, reduce their responsibilities (Lovell et al. 2009). Emissions information appears in corporate reporting, the media, international negotiations, and NGO campaigns and, as with other aspects of contemporary environmental policy, one of the more problematic discourses is that which makes individuals bear primary responsibility for their own emissions and carbon 'footprints' (Hobson 2013, Lorenzoni et al. 2011).

Climate change impacts and vulnerability

Political ecology is central to understanding how climate changes affect people and places, with approaches rooted in a cultural ecology which reacted to climate determinism by showing how people adapted to extreme environments, and a political economy of hazards that showed the importance of colonial legacies and inequality in creating vulnerability to drought and other hazards (cf. Chapters 2 and 3, this volume).

Political ecology of climate vulnerability

Political ecologist Piers Blaikie's coauthored book *At Risk: Natural Hazards, People's Vulnerability, and Disasters* (Blaikie et al. 1994) analyzes several climate hazards including drought, floods, and severe storms from a political economy perspective and defines vulnerability as 'the characteristics of a person or group and their situation that influence their capacity to anticipate, cope with, resist and recover from the impact of a natural hazard.' Explanatory variables include class, occupation, caste, ethnicity, gender, health, age, land tenure, immigration status, and social networks with vulnerability measured in terms of loss of life, property and livelihoods (pp. 11–12) (see also Liverman 1990b, Ribot 2014; Chapter 22, this volume).

Vulnerability to climate change became pivotal in the work of the Intergovernmental Panel on Climate Change and in the campaigns of many climate activists. The IPCC defined vulnerability as 'the degree to which a system is susceptible to, or unable to cope with, adverse effects of climate change, including climate variability and extremes' and identified three components of vulnerability as exposure, sensitivity, and adaptive capacity.

While this definition has framed much subsequent work on vulnerability, some political ecologists worry that the concept of vulnerability has been plagued with contested definitions and a lack of conceptual clarity, especially with regard to how to measure the relative role of political economic, individual, and biophysical factors (Eakin and Luers 2006, O'Brien et al. 2007, Smit and Wandel 2006). Partly in response to this concern, the concept of resilience was introduced as an alternative and more affirmative measure of people's ability to cope with environmental stress (Turner 2013, Cote and Nightingale 2012). Participatory research has provided local knowledge on vulnerability (Roncoli 2006, Tschakert 2007).

My own view is that we spent too much time on academic debates on vulnerability and resilience and too little time on listening to how people define their own experience of climate change. I also believe that using simple proxy measures of vulnerability (such as income or food security) and empirically evaluating their connection to losses and suffering may provide more immediate insights and solutions.

The most vulnerable populations often include the poor, women, people of color, and indigenous groups because of historical structures and discourses that have marginalized their rights. Gendered vulnerability analyses have documented the ways in which women are differentially harmed by climate – because they do not have access to land and other resources, suffer discrimination within households and communities, or do not receive warnings (Alston and Whittenbury 2013, Sultana 2013). But care must be taken not to erase the agency of the poor, women, or indigenous peoples by seeing them as passive victims without knowledge or capacity to respond to climate change (Arora-Jonsson 2011, Tschakert and Machado 2012, Whyte 2014) and it is important to recognize the multiple identities and intersectionality of people who may be poor but are also leaders, networkers, and key sources of community knowledge (Kaijser and Kronsell 2013, Nightingale 2011). Thinking of vulnerability as an embodied experience prompts more nuanced, feminist, post-development methodologies that are an important way forward in research on the political ecology of climate change.

Political ecology perspectives on changing human–environment relations as a result of neoliberalism and globalization (Castree 2008, Liverman and Vilas 2006) influenced work on climate change. For instance, Hallie Eakin has explained climate vulnerability in Mexico in terms of the neoliberal processes of free trade, structural adjustment, and changes in land tenure, showing how they have affected peasant farmers in central Mexican communities (Eakin 2006). Karen O'Brien and Robin Leichenko captured the intersecting risks of global environmental change and economic globalization with their influential concept of 'double exposure'

(Leichenko and O'Brien 2008). Others have drawn on Sen's theory of entitlement to goods and services to see vulnerability as a lack or failure of entitlements to, for example, food, land, or disaster relief (Ribot 2014).

Empirical analyses of vulnerability

Empirical analyses of vulnerability range from global and regional to local studies and from GIS indices and maps to qualitative ethnographies. The numerous case studies of local vulnerability are often difficult to compare because of very different contexts and survey questions and hence thwarted attempts at meta-analysis.

While household studies can acquire a wide range of information, climate vulnerability researchers can find that climate change and variability are less salient than other risks or that some questions are culturally inappropriate or invasive. For example, in our fieldwork in the Southwestern United States and Mexico, we understood that American Indian communities see weather and climate as private and connected to the spiritual realm, and that ranchers do not want to reveal the number of cattle they have or have lost because it is too close to revealing their incomes (Austin et al. 2000, Vasquez-Leon et al. 2003).

At broader scales, empirical analysis of vulnerability is often limited by the type, frequency, and spatial detail of the data that is collected by governments and others. For example, demographic and agricultural censuses are only conducted every ten years and rarely measure climate losses, making it difficult to trace the dynamics of changing vulnerability along with a lack of information below the state or county level. Researchers use yields as a proxy for climate impacts, and irrigation as a proxy for technology that reduces vulnerability including in my own work comparing climate vulnerability at the local level in Mexico where climate impacts were only expressed in terms of area lost to various hazards (Liverman 1990a). In the United States, Susan Cutter used census data in an index of social vulnerability to hazards that includes information on poverty, education, income, age, employment, race, and gender (Cutter and Finch 2008). The index shows pockets of vulnerability in the Deep South and Southwest, with race and ethnicity a strong influence on vulnerability in the Mississippi Valley and Texas–Mexico border and age in the Great Plains and Florida.

Preston et al. (2011) provide a helpful review of vulnerability mapping and its challenges. They note the benefits of GIS and maps in supporting land use planning and public education through visualization, but note that the maps will reify particular understandings and snapshots of vulnerability, create a false sense of confidence and accuracy, provide little information on dynamics or the future, or overlook key social, cultural, or contextual factors. However, they and others involved in vulnerability mapping have not addressed the concerns of other political ecologists about the potential use of the maps by powerful others, the appropriation of local knowledge, and the rights of surveillance (Bryan 2011).

Some of the more apocalyptic visions of future climate change suggest that it will trigger mass migrations and violent conflict, supported by correlation analyses that link drought to outmigration and to civil or interstate conflict and rooted in a form of climate determinism (Feng et al. 2010, Homer-Dixon 1994, Hsiang et al. 2013). Political ecology offers a more nuanced vision of these connections, especially when we talk to migrants, take account of historical influences, and take careful measure of the spatial patterns and impacts of environmental changes (Hartmann 2010, Dalby 2013, Piguet 2013). It can show that migration is a temporary and logical adaptation to extremes or is driven mostly by economic factors, and that tensions over water are more likely to drive cooperation than conflict (Farbotko and Lazrus 2012, Wolf 2007).

Discourses of vulnerability

The understanding of how people view and discuss climate vulnerability has ranged from more positivist and behavioral survey work to post-structural analyses of climate discourse, performance, art, and representation. A large body of work on how climate change is perceived, communicated, and narrated uses surveys of public opinion and media coverage to track and compare views of climate change over time and geographies. It links to political ecology in the analysis of environmental narratives and the structural influences on perception and news. For example, Max Boykoff uses empirical analysis of news and digital media to show discursive biases and their political and cultural implications (Boykoff 2011). Tony Leiserowitz's influential work on changing American attitudes to climate change pays explicit attention to the role of political affiliation, income, ethnicity, religion, and region. Saffron O'Neill provides some provocative analyses and reviews of climate change imagery and its influences, noting the influence of NGOS, the prevalence of arctic imagery – especially the polar bear – and of melodramatic climate impacts that promote fear such as flooding or fire (O'Neill and Smith 2014). And, O'Brien and colleagues argue for the policy impacts of vulnerability discourses, contrasting a scientific discourse focused on climate model results and recommending emissions mitigation and technical adaptation measures with a human security narrative about differential access to assistance, markets, and irrigation recommending addressing inequality in order to reduce vulnerability (O'Brien et al. 2007).

Post-structural approaches to understanding vulnerability include discourse and actor network based analyses that show how countries and people are constructed as vulnerable and studies that highlight how some groups and countries are 'performing' their vulnerability in order to justify access to international financial assistance (Cannon and Muller-Mahn 2010). For example, Yamane shows how data, stories, and maps in Sri Lanka were used to demonstrate vulnerability to the international climate community in the hope of financial assistance (Yamane 2009). Webber examines how government in Kiribati enacts vulnerability to secure international financing through an assemblage of facts, experts, and objects (Webber 2013).

Mike Hulme is interesting for his shift from climate scientist to cultural theorist of science and technology studies. He now sees social construction of knowledge and deep cultural influences in how we see climate change (Hulme 2010). He raises concerns about the use of climate models, the framing of climate change, and the ways in which climate change is used politically, recently arguing against technosolutions of geoengineering (Hulme 2014). Finally, the artistic response to climate change has been critically evaluated for its representation, activism, and role as a site for engagement between art and science (Miles 2010, Dixon 2013, Liverman 2009b).

Responses to climate change

The last decade has seen an expansion in the critical literature on responses to climate change, paralleling a proliferation of general research on the topics of climate mitigation/emissions reductions, climate adaptation, and geoengineering. While political ecology had its origins in critiques of environmental policies relating to forests, water, agriculture, fisheries, biodiversity, conservation, and land there was little initial attention by political ecologists to climate policy in foundational texts or elsewhere.[3] More recently political ecology has turned its attention to climate mitigation and adaptation, with sessions at the Association of American Geographers and contributions to journal special issues (Leichenko et al. 2010, Boykoff et al. 2009, Ford and Furgal 2009).

Mitigation policies - governing greenhouse gas emissions

Several important critical texts on climate mitigation and governance have been written from theoretical perspectives that include global governance, political economy – drawing on Marx, Gramsci, and Polayni – and post-structuralism – drawing on Foucault and Derrida (Okereke et al. 2009, Stripple and Bulkeley 2013). There are strong parallels with political ecology – especially in the analysis of power, discourse, agency, the state, and governmentality.

The history of climate mitigation is usually traced to the international negotiations that led to the signing of the UN Framework Convention on Climate Change (UNFCCC) at the Rio conference in 1992 (Gupta 2010, Liverman 2009a). The UNFCCC – an intergovernmental treaty – had the goal of preventing dangerous anthropogenic interference with the climate system and established principles of common but differentiated responsibility, a focus on the most vulnerable countries, the precautionary principle, and the right to sustainable development and an open economic system. The 1997 Kyoto Protocol created the framework for implementation, committing signatories from industrial nations to emission reductions, setting up mechanisms for carbon trading and offsetting, and creating a modest adaptation fund. Meanwhile non-nation state actors – states, cities, private firms, international banks, NGOs, and individuals – were making commitments to reduce emissions or convincing others to do so. Because most commitments were far less than the 50 percent emission reductions needed to reduce warming, and because emerging economies such as China were rapidly increasing emissions, by the time of the 2009 UNFCCC Copenhagen climate negotiations, there was a renewed sense of urgency for further emission reductions, especially from small islands at risk from sea level rise, from NGOs, and from the scientific community. New proposals for emission reductions were emerging including efforts to protect forest carbon through REDD (Reduced Emissions from Deforestation and Degradation) and for more substantial financial transfers to poorer countries for both mitigation and adaptation.

All of these developments provided a rich landscape for critical scholars to analyze the structures of power and influence of state and non-state actors in the international negotiations, the discursive and material emission pledges, and the curious mechanisms of trading carbon. Of the scholars who have engaged the political ecology of climate mitigation Harriet Bulkeley and Peter Newell stand out for their sustained critical attention, together with their collaborators. Bulkeley began with a focus on the politics of climate change in Australia using concepts of discourse coalitions and the risk society to explain links between government and industry in Australia (Bulkeley 2000). She then initiated the work for which she is best known – on how cities were governing climate change through local actions, energy policy, transnational networks, rescaling, and public–private partnerships – using methods that include interviews, discourse analysis of texts, and empirical data on policy outcomes (Bulkeley and Betsill 2003, Bulkeley and Broto 2013, Broto and Bulkeley 2013). Peter Newell's work is rooted in political economy and has examined, in particular, the intersection of climate policy with NGOs, business, and capital (Newell 2000, Levy and Newell 2005, Newell and Paterson 2010) and the political economy of the carbon markets, especially Kyoto's Clean Development Mechanism (CDM) that creates carbon credits from investments in greenhouse gas reductions in developing countries. His particular contributions are to explain the logics of business action and inaction on climate change, and to document the powerful interests involved in the CDM that often overlook local needs, energy access, and sustainability. The European version of carbon markets – the Emissions Trading System (ETS) – has been similarly criticized for serving powerful business interests (Bailey 2007). Swedish scholars have employed concepts of governmentality to uncover the discourses, technologies, and rationalities of forest carbon, carbon sinks, carbon markets, REDD, and carbon offsets (Lövbrand and Stripple 2011, Paterson and Stripple 2010).

I have argued that carbon offsets are a fascinating topic for political ecology, created 'as a new commodity that links north and south through a complex set of technologies, institutions and discourses' (Bumpus and Liverman 2011). We studied the political ecologies of carbon offsets from a variety of perspectives and in different places on the ground, exploring the global geographies of the CDM, recommending policies that would provide a greater advantage to poorer regions and local technologies. We proposed the notion of 'Accumulation by Decarbonization,' tracing how both the CDM and the voluntary carbon offset markets produce carbon credits through a commodity chain that links offset consumers in the rich world to project developers and local communities in poorer regions. We show how carbon offers opportunities to profit from offsets, governed beyond the state by supranational and private actors (Bumpus and Liverman 2008). The construction of carbon subjectivities for offset consumers and the agency of different carbon reduction technologies also links climate mitigation to new developments in political ecology (Lovell et al. 2009, Lovell and Liverman 2010). Our work on the greenhouse gas reductions and social impacts of local carbon projects paints a complex picture in which offsets are not always detrimental where renewable and efficiency projects provide income, jobs, and reduced air pollution within communities.

Carbon offsets provide critical fodder for others who see them as a form of neocolonial and unequal exchange that privatizes the atmosphere and damages local communities. Larry Lohmann and Patrick Bond provide vigorous arguments against offsetting, engaging with communities and activist groups (e.g. www.carbontradewatch.org/) who were opposing offset projects (Lohmann 2005, 2012, Bond 2008). Kathy McAfee and Elizabeth Shapiro have been especially critical of offsets in their work on payments for environmental services in Mexico (McAfee and Shapiro 2010; Shapiro-Garza 2013).

Forest offsets are especially interesting to political ecologists, given the sizeable literature on the political ecology of forestry around the world. Emily Boyd, Esteve Corbera, Tracey Osborne, and David Lansing began their studies of forest offsets with PhD dissertation case studies in Latin America informed by political ecology (including one case in Chiapas financed by Formula 1 racing to offset its emissions!). They identify issues with the distribution of funds, property rights, labor allocation, monitoring, and state institutions and have continued the debate around REDD (Boyd 2002, Corbera and Brown 2008, Corbera and Schroeder 2011, Osborne 2011, Lansing 2011). Arun Agrawal and Ashwini Chhatre raise important questions about centralization and governance of REDD (Agrawal et al. 2011, Phelps et al. 2010) with Connie McDermott and Heike Schroeder focusing on REDD governance and implementation (McDermott et al. 2012, Schroeder and McDermott 2014). A controversial article on REDD projects displacing local residents in Tanzania prompted a strong reaction from the NGO community (Beymer-Farris and Bassett 2012, Burgess et al. 2013)

Political ecologies of adaptation

Climate adaptation has been defined as adjustments to reduce vulnerability and increase resilience, through technologies such as irrigation and coastal defense, social networks of sharing and disaster warning and relief, and financial strategies such as insurance. Climate adaptation has been linked to the concept of adaptive capacity – the potential to respond to climate risks. Many studies of vulnerability also include a component of adaptation. While some see climate adaptation as a new field, others have pointed out the legacies from cultural ecology in geography and anthropology and the risks of reinventing the wheel (Bassett and Fogelman 2013, Head 2010, Tschakert 2012).

Studies of community and agricultural adaptation dominate; usually based on observations and interviews with local decision makers and residents and often focusing on adaptation to current climate variability rather than future climate changes. Bassett and Fogelman (2013) claim that the majority of adaptation studies focus on technical adjustments or modest development reform and do not pay attention to transformative adaptation that tackles the social roots of vulnerability. I would suggest, however, that there are studies that pay considerable attention to the deep-rooted institutional barriers to successful adaptation and to the differential ability to adapt within society by women and the poor. For example, Mark Pelling studies the political ecology of flooding in Guyana in historical and political context, arguing that development programs have undermined grassroots agency and options for adaptation (Pelling 1999). On the Mexican Caribbean coast he shows how structures of legitimation and domination have produced rigid governance that impedes transformative adaptation in communities (Manuel-Navarrete et al. 2011).

Neil Adger, who sometimes identifies as a political ecologist, is perhaps the best-known scholar of adaptation. He tends to approach adaptation from a political economy perspective, with attention to issues of equity and justice, and his empirical work – often with colleagues from development studies at the UK Tyndall Center – includes studies in Vietnam and the UK (Adger et al. 2009, Conway and Schipper 2011, Few et al. 2007, Osbahr et al. 2008).

Indigenous adaptations are another emerging area of inquiry using concepts from political ecology – including regional work in Australia (Leonard et al. 2013, Petheram et al. 2010), South Africa (Ziervogel et al. 2014) and the Arctic (Ford et al. 2010).

Climate adaptation is one topic where insights from scholars have had a significant impact on public policy and on international climate governance. For example, Saleemul Huq has published many articles where he argues for the value of local knowledge and adaptation strategies developed in the global South and for equity in climate financing (Ayers et al. 2014, Schipper et al. 2014). But he also participates in the climate negotiations – helping to draft text and advising negotiators and activists – and started a grassroots adaptation program in Bangladesh. Geographer Richard Klein has been influential at European and international levels in pushing for just climate adaptation and helped start 'Adaptation Watch' to promote transparency in the governance of adaptation funds (Klein et al. 2005, Smith et al. 2011). In the United States, Susi Moser is well known for her commitment to co-producing her research with communities and is a voice for the need to clearly communicate adaptation research to citizens and policy makers making it relevant on the ground (Moser and Dilling 2007, Moser 2010).

The onset of climate change, and the risks that warming will exceed 4 degrees C, means that adaptation is joining mitigation and carbon markets in the world of international development, finance, and local, national, and international politics. There are many unresolved questions for the political ecology of climate adaptation (Liverman and Billett 2010). Who should be eligible for adaptation assistance and who will pay for it? How can we ensure that high technology and large-scale options (such as sea walls or large dams) do not swamp effective small-scale local actions (such as crop diversity or ecosystem protection)? Should we meet what has been called the 'adaptation deficit' – from vulnerability to current climate variability – before focusing on climate change? And how can we ensure that aid for climate adaptation is appropriate, equitable, accessible to women and people of color, and does not divert from other human development priorities? How can we connect our critical scholarship to action on the ground or to influence policy?

There is a movement that suggests perhaps reducing emissions will be so difficult, and the risk of warming so great, that we should consider options for geoengineering the planet as a response to climate change. This might involve managing incoming solar radiation (e.g. through altering atmospheric composition to create clouds or changing the surface to reflect sunlight) or carbon capture (e.g. through new technology, fertilizing the oceans to take up more CO_2). As yet political ecology has hardly begun to engage with questions about the governance and socio-ecological impacts of geoengineering although there is a growing literature from science studies and critical international relations (Lövbrand et al. 2009, Hulme 2014).

Conclusion

This chapter set out to show that political ecology has a lot of offer in how we understand the human dimensions of climate change – whether explicitly named, or reflected in research that takes political economy, human agency, and nature seriously, revealing narratives that can create or oppose injustice. Although one might view political ecology and critical climate change research as following parallel paths there are important examples of cross-fertilization and future opportunities. The influence of political ecology on vulnerability is perhaps the most significant, not only in the early impact of critical hazards scholars but also in the flow of ideas from political ecology work on governmentality, feminist political ecology, and neoliberalism.

In the case of emissions and of climate policy the flow of ideas has been more from critical international relations and political economy to political ecology in the critique of emission responsibilities and of carbon markets. Political ecology has potential when it comes to carbon offsets and proposals for REDD in the way it examines the interaction between institutions across scales, the critique of practices and discourses of market environmentalism and development, and in the importance of material nature as an actor and explanation. The debate about carbon offsets continues – although the market has not grown as fast as anticipated because of carbon market uncertainties, low prices, and a backlash against offsetting. Moving forward I would suggest that there is need for well-designed comparative case studies, more space in academic papers to the voices of those affected by mitigation policy, attention to whether material carbon savings are truly additional, and more rigor in assessing offsets embraced or opposed by NGOs and the private sector. REDD+, in particular, has been popular with the conservation community as one of the only strategies they think might save tropical forests, and is strongly opposed by some indigenous rights movements.

Students in a recent graduate seminar – looking to design their own studies – were frustrated at how many articles in political ecology and environmental governance argued mostly from theory and were thin empirically, and how few critical articles discussed research design, methods, or positionality, provided summary statistics on fieldwork, analyzed biophysical data, or used quotes from interviews as evidence or to give voice to local people. In discussion we identified reasons that included a desire to demonstrate theoretical sophistication, innovation and anti-positivist stance, normative commitments to telling a convincing story and making a strong argument, lack of time and resources to extend case studies to comparative cases and baselines or to analysis of material nature, protecting individual informants, or poor record-keeping. In order to address these frustrations and provide clearer guides to others, political ecologists could do a better job of discussing their methodology in their publications and include more quotes and references to field data.

What political ecology gains from a focus on climate change is an engagement with one of the most existential and political environmental issues of our time – with serious implications for global geographies, social and environmental justice – and with a growing community of scholars and publics who wish to understand and act in small and larger ways to influence the future.

Notes

1 Space constraints mean that I am only able to cite a small sample of the dozens of articles by some of the key scholars but their other relevant work can easily be found through online literature searches.
2 For example, Neil Adger, Arun Agrawal, Kirstin Dow, Karen O'Brien, Mark Pelling, Petra Tschakert, Coleen Vogel, Julian Agyeman, Paul Baer, Emily Boyd, Esteve Corbera, Richard Klein, Larry Lohmann, Matt Paterson, Chuks Okereke, and Timmons Roberts.
3 With the exception of Tim Forsyth who did some work on forest offsets (2003).

References

Adger, W.N., S. Dessai, M. Goulden, M. Hulme, I. Lorenzoni, D.R. Nelson, L.O. Naess, J. Wolf and A. Wreford (2009) Are there social limits to adaptation to climate change? *Climatic Change*, 93, 335–354.

Agarwal, A. and S. Narain (1991) *Global Warming in an Unequal World: A Case of Environmental Colonialism*. New Delhi: Centre for Science and Environment.

Agrawal, A., D. Nepstad and A. Chhatre (2011) Reducing emissions from deforestation and forest degradation. *Annual Review of Environment and Resources*, 36, 373–396.

Alston, M. and K. Whittenbury (2013) *Research, action and policy: Addressing the gendered impacts of climate change*. Netherlands: Springer.

Arora-Jonsson, S. (2011) Virtue and vulnerability: discourses on women, gender and climate change. *Global Environmental Change*, 21, 744–751.

Austin, D.E., S. Gerlak and C. Smith (2000) *Building partnerships with Native Americans in climate-related research and outreach*. Tucson: Institute for the Study of Planet Earth.

Ayers, J.M., S. Huq, A.M. Faisal and S.T. Hussain (2014) Mainstreaming climate change adaptation into development: a case study of Bangladesh. *Wiley Interdisciplinary Reviews: Climate Change*, 5, 37–51.

Baer, P. (2013) The greenhouse development rights framework for global burden sharing: reflection on principles and prospects. *Wiley Interdisciplinary Reviews: Climate Change*, 4, 61–71.

Bailey, I. (2007) Neoliberalism, climate governance and the scalar politics of EU emissions trading. *Area*, 39, 431–442.

Bakker, K. and G. Bridge (2006) Material worlds? Resource geographies and the matter of nature. *Progress in Human Geography*, 30, 5–27.

Bassett, T.J. and C. Fogelman (2013) Déjà vu or something new? The adaptation concept in the climate change literature. *Geoforum*, 48, 42–53.

Beymer-Farris, B.A. and T.J. Bassett (2012) The REDD menace: resurgent protectionism in Tanzania's mangrove forests. *Global Environmental Change*, 22, 332–341.

Blaikie, P., T. Cannon, I. Davis and B. Wisner (1994). *At risk: Natural hazards, people's vulnerability and disasters*. London: Routledge.

Bohle, H.-G., T.E. Downing and M. Watts (1994) Climate change and social vulnerability: the sociology and geography of food insecurity. *Global Environmental Change*, 4, 37–48.

Bond, P. (2008) The state of the global carbon trade debate. *Capitalism Nature Socialism*, 19, 89–106.

Boyd, E. (2002) The Noel Kempff project in Bolivia: gender, power, and decision-making in climate mitigation. *Gender & Development*, 10, 70–77.

Boykoff, M.T. (2011) *Who speaks for the climate? Making sense of media reporting on climate change*. New York: Cambridge University Press.

Boykoff, M.T., A. Bumpus, D. Liverman and S. Randalls (2009) Theorizing the carbon economy: introduction to the special issue. *Environment and Planning A*, 41, 2299–2304.

Broto, V.C. and H. Bulkeley (2013) Maintaining climate change experiments: urban political ecology and the everyday reconfiguration of urban infrastructure. *International Journal of Urban and Regional Research*, 37, 1934–1948.

Bryan, J. (2011) Walking the line: participatory mapping, indigenous rights, and neoliberalism. *Geoforum*, 42, 40–50.

Bulkeley, H. (2000) Discourse coalitions and the Australian climate change policy network. *Environment and Planning C-Government and Policy*, 18, 727–748.

Bulkeley, H. and M.M. Betsill (2003) *Cities and climate change: Urban sustainability and global environmental governance*. Abingdon: Routledge.

Bulkeley, H. and V.C. Broto (2013) Government by experiment? Global cities and the governing of climate change. *Transactions of the Institute of British Geographers*, 38, 361–375.

Bumpus, A.G. (2011) The matter of carbon: understanding the materiality of tCO2e in carbon offsets. *Antipode*, 43, 612–638.

Bumpus, A.G. and D.M. Liverman (2008) Accumulation by decarbonization and the governance of carbon offsets. *Economic Geography*, 84, 127–155.

Bumpus, A. and D. Liverman (2011) Carbon colonialism? Offsets, greenhouse gas reductions, and sustainable development. *Global Political Ecology*, 203–224.

Burgess, N.D., S. Mwakalila, P. Munishi, M. Pfeifer, S. Willcock, D. Shirima, S. Hamidu, G.B. Bulenga, J. Rubens and H. Machano (2013) REDD herrings or REDD menace: response to Beymer-Farris and Bassett. *Global Environmental Change*, 23, 1349–1354.

Cannon, T. and D. Muller-Mahn (2010) Vulnerability, resilience and development discourses in context of climate change. *Natural Hazards*, 55, 621–635.

Castree, N. (2003) Commodifying what nature? *Progress in Physical Geography*, 27, 273–297.

Castree, N. (2008) Neoliberalising nature: processes, effects, and evaluations. *Environment and Planning, A*, 40, 153.

Conway, D. and E.L.F. Schipper (2011) Adaptation to climate change in Africa: challenges and opportunities identified from Ethiopia. *Global Environmental Change*, 21, 227–237.

Corbera, E. and K. Brown (2008) Building institutions to trade ecosystem services: marketing forest carbon in Mexico. *World Development*, 36, 1956–1979.

Corbera, E. and H. Schroeder (2011) Governing and implementing REDD+. *Environmental Science & Policy*, 14, 89–99.

Cote, M. and A.J. Nightingale (2012) Resilience thinking meets social theory: situating social change in socio-ecological systems (SES) research. *Progress in Human Geography*, 36, 475–489.

Cutter, S.L. and C. Finch (2008) Temporal and spatial changes in social vulnerability to natural hazards. *Proceedings of the National Academy of Sciences*, 105, 2301–2306.

Dalby, S. (2013) The geopolitics of climate change. *Political Geography*, 37, 38–47.

Demeritt, D. (2001) The construction of global warming and the politics of science. *Annals of the Association of American Geographers*, 91, 307–337.

Dilling, L. and R. Hauser (2013) Governing geoengineering research: why, when and how? *Climatic Change*, 121, 553–565.

Dilling, L., S.C. Doney, J. Edmonds, K.R. Gurney, R. Harriss, D. Schimel, B. Stephens and G. Stokes (2003) The role of carbon cycle observations and knowledge in carbon management. *Annual Review of Environment and Resources*, 28, 521–558.

Dixon, D. (2013) Placing climate change. *Science*, 342, 1171–1172.

Eakin, H. (2006) *Weathering risk in rural Mexico: Climatic, institutional and economic change*. Tucson: University of Arizona Press.

Eakin, H. and A.L. Luers (2006) Assessing the vulnerability of social-environmental systems. *Annu. Rev. Environ. Resour.*, 31, 365–394.

Farbotko, C. and H. Lazrus (2012) The first climate refugees? Contesting global narratives of climate change in Tuvalu. *Global Environmental Change*, 22, 382–390.

Feng, S.Z., A.B. Krueger and M. Oppenheimer (2010) Linkages among climate change, crop yields and Mexico-US cross-border migration. *Proceedings of the National Academy of Sciences of the United States of America*, 107, 14257–14262.

Few, R., K. Brown and E.L. Tompkins (2007) Public participation and climate change adaptation: avoiding the illusion of inclusion. *Climate Policy*, 7, 46–59.

Ford, J.D. and C. Furgal (2009) Foreword to the special issue: climate change impacts, adaptation and vulnerability in the Arctic. *Polar Research*, 28, 1–9.

Ford, J.D., T. Pearce, F. Duerden, C. Furgal and B. Smit (2010) Climate change policy responses for Canada's Inuit population: the importance of and opportunities for adaptation. *Global Environmental Change*, 20, 177–191.

Forsyth, T. (2003) *Critical political ecology: The politics of environmental science*. London: Routledge.

Green, J.F. (2010) Private standards in the climate regime: the greenhouse gas protocol. *Business and Politics*, 12, 1469–1569.

Gupta, A., E. Lövbrand, E. Turnhout and M.J. Vijge (2012) In pursuit of carbon accountability: the politics of REDD+ measuring, reporting and verification systems. *Current Opinion in Environmental Sustainability*, 4, 726–731.

Gupta, J. (2010) A history of international climate change policy. *Wiley Interdisciplinary Reviews: Climate Change*, 1, 636–653.

Hartmann, B. (2010) Rethinking climate refugees and climate conflict: rhetoric, reality and the politics of policy discourse. *Journal of International Development*, 22, 233–246.
Head, L. (2010) Cultural ecology: adaptation-retrofitting a concept? *Progress in Human Geography*, 34, 234–242.
Hecht, S.B. and S.S. Saatchi (2007) Globalization and forest resurgence: changes in forest cover in El Salvador. *BioScience*, 57, 663–672.
Heede, R. (2014) Tracing anthropogenic carbon dioxide and methane emissions to fossil fuel and cement producers, 1854-2010. *Climatic Change*, 1–13.
Hewitt, K. (1983) *Interpretations of calamity: From the viewpoint of human ecology*. Boston: Allen & Unwin.
Hobson, K. (2013) On the making of the environmental citizen. *Environmental Politics*, 22, 56–72.
Homer-Dixon, T.F. (1994) Environmental scarcities and violent conflict: evidence from cases. *International Security*, 19, 5–40.
Hsiang, S.M., M. Burke and E. Miguel (2013) Quantifying the influence of climate on human conflict. *Science*, 341, 1235367.
Hulme, M. (2010) *Why we disagree about climate change*. Cambridge: Cambridge University Press.
Hulme, M. (2014) *Can science fix climate change? The case against climate engineering*. Cambridge: Polity Press.
IPCC (2014a). *Climate Change 2014: Impacts, Adaptation, and Vulnerability. Part A: Global and Sectoral Aspects. Contribution of Working Group II to the Fifth Assessment Report of the Intergovernmental Panel on Climate Change [Field, C.B., V.R. Barros, D.J. Dokken, K.J. Mach, M.D. Mastrandrea, T.E. Bilir, M. Chatterjee, K.L. Ebi, Y.O. Estrada, R.C. Genova, B. Girma, E.S. Kissel, A.N. Levy, S. MacCracken, P.R. Mastrandrea, and L.L. White (eds.)]*. Cambridge, UK and New York: Cambridge University Press.
IPCC (2014b) *Climate Change 2014: Mitigation of Climate Change. Contribution of Working Group III to the Fifth Assessment Report of the Intergovernmental Panel on Climate Change*. Cambridge University Press.
IPCC, T. Stocker, Q. Dahe and G. Plattner (2013) Climate change 2013: the physical science basis. In *Working Group I Contribution to the Fifth Assessment Report of the Intergovernmental Panel on Climate Change. Summary for Policymakers (IPCC, 2013)*.
Kaijser, A. and A. Kronsell (2013) Climate change through the lens of intersectionality. *Environmental Politics*, 1–17.
Klein, R.J.T., E.L.F. Schipper and S. Dessai (2005) Integrating mitigation and adaptation into climate and development policy: three research questions. *Environmental Science & Policy*, 8, 579–588.
Lansing, D.M. (2011) Realizing carbon's value: discourse and calculation in the production of carbon forestry offsets in Costa Rica. *Antipode*, 43, 731–753.
Lansing, D.M. (2012) Performing carbon: materiality: the production of carbon offsets and the framing of exchange. *Environment and Planning-Part A*, 44, 204.
Leach, M. and I. Scoones (2013) Carbon forestry in West Africa: The politics of models, measures and verification processes. *Global Environmental Change*, 23, 957–967.
Leichenko, R. and K. O'Brien (2008) *Environmental change and globalization: Double exposures*. New York: Oxford University Press.
Leichenko, R.M., K.L. O'Brien and W.D. Solecki (2010) Climate change and the global financial crisis: A case of double exposure. *Annals of the Association of American Geographers*, 100, 963–972.
Leonard, S., M. Parsons, K. Olawsky and F. Kofod (2013) The role of culture and traditional knowledge in climate change adaptation: Insights from East Kimberley, Australia. *Global Environmental Change*, 23, 623–632.
Levy, D.L. and P. Newell. (2005) *The business of global environmental governance*. Cambridge, MA and London: MIT Press.
Liverman, D.M. (1990a) Drought impacts in Mexico: Climate, agriculture, technology, and land tenure in Sonora and Puebla. *Annals of the Association of American Geographers*, 80, 49–72.
Liverman, D.M. (1990b) Vulnerability to global environmental change. In *Understanding global environmental change: The contributions of risk analysis and management*, eds. J.X. Kasperson and R.E. Kasperson, 27–44. Tokyo: UNU Press.
Liverman, D.M. (2009a) Conventions of climate change: constructions of danger and the dispossession of the atmosphere. *Journal of Historical Geography*, 35, 279–296.
Liverman, D.M. (2009b) Seeking inspiration: a scientist turns to the cultural sector. In *Long horizons: An exploration of art and climate change*. London: Julie's Bicycle and British Council.
Liverman, D. and S. Billett (2010) Copenhagen and the governance of adaptation. *Environment*, 52, 28–36.
Liverman, D.M. and S. Vilas (2006) Neoliberalism and the environment in Latin America. *Annual Review of Environment and Resources*, 31, 327–363.

Locatelli, T., T. Binet, J.G. Kairo, L. King, S. Madden, G. Patenaude, C. Upton and M. Huxham (2014) Turning the tide: How blue carbon and payments for ecosystem services (PES) might help save mangrove forests. *Ambio*, 1–15.

Lohmann, L. (2005) Making and marketing carbon dumps: commodification, calculation and counterfactuals in climate change mitigation. *Science as Culture*, 14, 203–235.

Lohmann, L. (2012) Financialization, commodification and carbon: the contradictions of neoliberal climate policy. *Socialist Register*, 48, 85–107.

Lorenzoni, I., G. Seyfang, M. Nye, I. Whitmarsh and S. O'Neill (2011) Carbon budgets and carbon capability: lessons from personal carbon trading. In *Engaging the public with climate change: Behaviour change and communication*, eds. L. Whitmarsh, S.J. O'Neill and I. Lorenzoni, 31–46. London: Routledge and Earthscan.

Lövbrand, E. and J. Stripple (2011) Making climate change governable: accounting for carbon as sinks, credits and personal budgets. *Critical Policy Studies*, 5, 187–200.

Lövbrand, E., J. Stripple and B. Wiman (2009) Earth system governmentality: reflections on science in the Anthropocene. *Global Environmental Change*, 19, 7–13.

Lovell, H. and D. Liverman (2010) Understanding carbon offset technologies. *New Political Economy*, 15, 255–273.

Lovell, H., H. Bulkeley and D. Liverman (2009) Carbon offsetting: sustaining consumption? *Environment and Planning A*, 41, 2357–2379.

Lovell, H., J. Bebbington, C. Larrinaga and T.R.S. de Aguiar (2013) Putting carbon markets into practice: a case study of financial accounting in Europe. *Environment and Planning C: Government and Policy*, 31, 741–757.

Manuel-Navarrete, D., M. Pelling and M. Redclift (2011) Critical adaptation to hurricanes in the Mexican Caribbean: development visions, governance structures, and coping strategies. *Global Environmental Change-Human and Policy Dimensions*, 21, 249–258.

McAfee, K. and E.N. Shapiro (2010) Payments for ecosystem services in Mexico: nature, neoliberalism, social movements, and the state. *Annals of the Association of American Geographers*, 100, 579–599.

McDermott, C.L., L. Coad, A. Helfgott and H. Schroeder (2012) Operationalizing social safeguards in REDD+: actors, interests and ideas. *Environmental Science & Policy*, 21, 63–72.

Miles, M. (2010) Representing nature: art and climate change. *Cultural Geographies*, 17, 19–35.

Moser, S.C. (2010) Now more than ever: the need for more societally relevant research on vulnerability and adaptation to climate change. *Applied Geography*, 30, 464–474.

Moser, S.C. and L. Dilling (2007) *Creating a climate for change: Communicating climate change and facilitating social change*. Cambridge: Cambridge University Press.

Newell, P. (2000) *Climate for change: Non-state actors and the global politics of the greenhouse*. Cambridge: Cambridge University Press.

Newell, P. and M. Paterson (2010) *Climate capitalism: Global warming and the transformation of the global economy*. Cambridge: Cambridge University Press.

Nightingale, A.J. (2011) Bounding difference: intersectionality and the material production of gender, caste, class and environment in Nepal. *Geoforum*, 42, 153–162.

O'Brien, K., S. Eriksen, L.P. Nygaard and A. Schjolden (2007) Why different interpretations of vulnerability matter in climate change discourses. *Climate Policy*, 7, 73–88.

Okereke, C., H. Bulkeley and H. Schroeder (2009) Conceptualizing climate governance beyond the international regime. *Global Environmental Politics*, 9, 58–78.

O'Neill, S.J. and N. Smith (2014) Climate change and visual imagery. *Wiley Interdisciplinary Reviews: Climate Change*, 5, 73–87.

Osbahr, H., C. Twyman, W. Neil Adger and D.S. Thomas (2008) Effective livelihood adaptation to climate change disturbance: scale dimensions of practice in Mozambique. *Geoforum*, 39, 1951–1964.

Osborne, T.M. (2011) Carbon forestry and agrarian change: access and land control in a Mexican rainforest. *Journal of Peasant Studies*, 38, 859–883.

Paterson, M. and J. Stripple (2010) My space: governing individuals and carbon emissions. *Environment and Planning. D, Society and Space*, 28, 341.

Pelling, M. (1999) The political ecology of flood hazard in urban Guyana. *Geoforum*, 30, 249–261.

Peters, G.P., J.C. Minx, C.L. Weber and O. Edenhofer (2011) Growth in emission transfers via international trade from 1990 to 2008. *Proceedings of the National Academy of Sciences*, 108, 8903–8908.

Petheram, L., K. Zander, B. Campbell, C. High and N. Stacey (2010) 'Strange changes': indigenous perspectives of climate change and adaptation in NE Arnhem Land (Australia). *Global Environmental Change*, 20, 681–692.

Phelps, J., E.L. Webb and A. Agrawal (2010) Does REDD+ threaten to recentralize forest governance. *Science*, 328, 312–313.
Piguet, E. (2013) From 'primitive migration' to 'climate refugees': the curious fate of the natural environment in migration studies. *Annals of the Association of American Geographers*, 103, 148–162.
Preston, B.L., E.J. Yuen and R.M. Westaway (2011) Putting vulnerability to climate change on the map: a review of approaches, benefits, and risks. *Sustainability Science*, 6, 177–202.
Ribot, J. (2014) Cause and response: vulnerability and climate in the Anthropocene. *Journal of Peasant Studies*, Online.
Robbins, P. (2012) *Political ecology: A critical introduction*. Chichester: Wiley Blackwell.
Robertson, M.M. (2000) No net loss: Wetland restoration and the incomplete capitalization of nature. *Antipode*, 32, 463–493.
Roncoli, C. (2006) Ethnographic and participatory approaches to research on farmers' responses to climate predictions. *Climate Research*, 33, 81.
Schipper, E.L.F., J. Ayers, H. Reid, S. Huq and A. Rahman (2014) *Community based adaptation to climate change: Scaling it up*. New York: Routledge.
Schroeder, H. and C. McDermott (2014) Beyond carbon: Enabling justice and equity in REDD+ across levels of governance. *Ecology and Society*, 19, 31.
Shapiro-Garza, E. (2013) Contesting the market-based nature of Mexico's national payments for ecosystem services programs: four sites of articulation and hybridization. *Geoforum*, 46, 5–15.
Smit, B. and J. Wandel (2006) Adaptation, adaptive capacity and vulnerability. *Global Environmental Change*, 16, 282–292.
Smith, J.B., T. Dickinson, J.D. Donahue, I. Burton, E. Haites, R.J. Klein and A. Patwardhan (2011) Development and climate change adaptation funding: coordination and integration. *Climate Policy*, 11, 987–1000.
Stripple, J. and H. Bulkeley (2013) *Governing the Climate: New Approaches to Rationality, Power and Politics*. Cambridge: Cambridge University Press.
Sultana, F. (2013) Gendering climate change: geographical insights. *The Professional Geographer*, 1–10.
Swyngedouw, E. (2010) Apocalypse forever? Post-political populism and the spectre of climate change. *Theory, Culture & Society*, 27, 213–232.
Tschakert, P. (2007) Views from the vulnerable: understanding climatic and other stressors in the Sahel. *Global Environmental Change-Human and Policy Dimensions*, 17, 381–396.
Tschakert, P. (2012) From impacts to embodied experiences: tracing political ecology in climate change research. *Geografisk Tidsskrift-Danish Journal of Geography*, 112, 144–158.
Tschakert, P. and M. Machado (2012) Gender justice and rights in climate change adaptation: opportunities and pitfalls. *Ethics and Social Welfare*, 6, 275–289.
Tschakert, P. and G. Tappan (2004) The social context of carbon sequestration: considerations from a multi-scale environmental history of the Old Peanut Basin of Senegal. *Journal of Arid Environments*, 59, 535–564.
Turner, M.D. (2013) Political ecology I: an alliance with resilience? *Progress in Human Geography*, 38, 616–623.
Vasquez-Leon, M., C.T. West and T.J. Finan (2003) A comparative assessment of climate vulnerability: agriculture and ranching on both sides of the US-Mexico border. *Global Environmental Change*, 13, 159–173.
Webber, S. (2013) Performative vulnerability: climate change adaptation policies and financing in Kiribati. *Environment and Planning A*, 45, 2717–2733.
Whyte, K.P. (2014) Indigenous women, climate change impacts, and collective action. *Hypatia*, 29, 599–616.
Wolf, A.T. (2007) Shared waters: conflict and cooperation. *Annu. Rev. Environ. Resour.*, 32, 241–269.
Yamane, A. (2009) Climate change and hazardscape of Sri Lanka. *Environment and Planning, A*, 41, 2396.
Ziervogel, G., M. New, E. Archer van Garderen, G. Midgley, A. Taylor, R. Hamann, S. Stuart-Hill, J. Myers and M. Warburton (2014) Climate change impacts and adaptation in South Africa. *Wiley Interdisciplinary Reviews: Climate Change*, 5, 605–620.

24

ENVIRONMENT AND DEVELOPMENT

Reflections from Latin America[1]

Astrid Ulloa

Introduction

In Latin America, development seen as a discourse of modernization, as well as the environmental degradation that it generates, has been analyzed by academic approaches from political ecology and other development critics, as well as by social movements. Similarly, sustainable development and its new vision of the green economy has been analyzed, which has led to a rethinking of this discourse in order to consider the ways and means in which cultures construct alternative worlds, based on relationships between humans, and between humans and natures. From this starting point, proposals are being created whose main arguments are focused on rethinking development, including environmental issues, and reconceptualizing unequal relations with the biophysical context. These proposals are referred to in different ways: "alternative development," "self-development," "counter-hegemonic globalizations," "postdevelopments," "counter-developments" and "alternatives to development," among others. How has political ecology, focused as it is on development, analyzed these new processes? How alternative are these new proposals, what interests are involved, who are the actors and how do they take shape in the local context?

Political ecology (PE) has generated diverse perspectives in its critical analysis of the unequal power relations that underlie development-environment in the context of local-global articulations (Escobar 1998, 1999; Peet and Hartwick 1999; Radcliffe 2012; Zimmerer and Bassett 2004). Discussions from political ecology have influenced the analytical perspectives of the social sciences in Latin America (LA). Similarly, the conceptual proposals from LA have provided critical insights into, and a rethinking of, the most recognized approaches of political ecology (Alimonda 2009; Escobar 1999, 2010; Escobar and Pedrosa 1996; Leff 2003; Porto 2001). Currently, in the research field of environment and development relationships, there are several works that are based on the contributions of PE regarding access, use, control, rights and decision-making, and power relations that critically explore the new dynamics of extractivism (Alimonda 2011; Moncada et al. 2013). Highlighting all the discussions here would be pointless, since previous works have already done this (see for example, Alimonda 2009; Escobar 2010; Leff 2012).

I want to highlight here that LA has become a new area for processes of creation, appropriation and globalization of natures associated with resource extractivism, which has generated local political dynamics that rework and contribute to PE not only in LA but in the global context. I consider the contributions of LA to be focused on reconsidering the natural, spatial and scalar processes that are generated from the dynamics of various actors, specifically from local people, because they involve proposals for alternatives to development and the traditional relationship with the environment, based on other views of nature and spatiality, and other ways of being and living in a territory. This perspective is necessary, given the demands of indigenous peoples, Afro-descendants and peasants seeking to establish other ways of understanding development and its relationship with the environment. I believe that these local relationships with the nonhuman in specific spatial contexts allow us to concretely reconceptualize processes associated with capitalist nature, and identify unprecedented proposals for alternatives to development. Such proposals, however, must arise from specific local contexts and not generalize or essentialize these options.

This chapter is organized as follows: the first part presents the approaches to the natural and the spatial and its linkage with ways of life from the perspectives of local people in LA. In the second part, local proposals and the dynamics that have been generated around the construction of alternatives to development are analyzed, highlighting the politics of place and difference, biocultural memory, "living well" (*vivir bien or buen vivir*) and indigenous alter-geopolitics. The third part focuses on a critical analysis of alternatives to development, highlighting their implications in a historical context of inequalities between environment and development, and current contexts of resource extraction in LA. Finally, the chapter contributes to the environment/development debate by reflecting on considerations necessary for generating proposals situated in social, political, environmental and historical contexts.

Natural, spatial and local ways of life: a view from Latin America

Modernity and modernization, nature and territory (through development programs) are concepts that have different meanings and interpretations. This is especially the case for LA, given its particular socio-historical processes, which in turn allow local movements to propose alternatives to development or reframe the analysis of modernity and postdevelopment. Whereas in PE and environmental justice analyses the causes and effects of environmental inequalities respond to broader structural processes, enrolled in the vision of modern nature which considers stakeholders, power dynamics, access, use, control, and decision-making processes about nature, and in which different analytical categories intersect, it is necessary to account for discussions around natural and spatial relationships that examine power relations from the perspective of the dual categories of nature/culture. In this sense, for example, socio-ecological inequalities are not a dimension of social inequalities. In fact, when the dynamics of environmental disputes are analyzed, the claims and resistances, as well as proposed development alternatives, involve other views about nature and spatiality, since they are not only about access, control and effects, but about other ways of being and living in a specific *territory*.

Society–nature relationships allow an understanding of the positioning of the human, and interactions with the nonhuman. In this relationship, the material aspect has different connotations, for example, to different cultures a mineral is part of a living body, actants that are transmuted and transformed. Everything happens in a space, but this space changes due to the fact that there is a vertical and horizontal geopolitics in what is conceived of as territory.

In the current processes of the globalization of nature, indigenous peoples' proposals arise, for example, from the local control of natural resources to the rethinking of what is meant by *resource*. That is, control of the subsurface according to alternative understandings and visions of territory: a *vertical indigenous territoriality*, including various notions and scales of the territory, both horizontally and vertically. For example, the Embera people of Colombia and Panama believe that their Universe (*tatsira trua*) contains different worlds, scales and relationships between them, which are inhabited by various beings (humans, animals, stones, plants, water, half-human/half-other beings, among others) that can transform from one category to the other. There are nine worlds, three above, three in the middle (the human) and three below, and horizontally there are different areas of appropriation, which may be contiguous or in different locations, but are interconnected. Beginning with their understandings of their relationships with the nonhuman, the Embera conceive the territory as being beyond land and soil, relating it to territorial projections that consider the fixed and the mobile (extensions, discontinuities and continuities), reflected in various scales (body-territory-nationalities), including vertical and horizontal scales, which involves control in several dimensions: spiritual, material and political (Ulloa et al. 2004; Ulloa 2014b).

Global demands for resources such as gold, biodiversity and land, have challenged this territorial logic, which in turn has generated proposals for resistance on the part of the Embera, based on their concept of territory and their relations with the nonhuman, as well as their environmental, territorial and political self-determination, as a way to confront the pressures of encroachment and appropriation (Ulloa 2014a).

PE focuses centrally on the causes of unequal relations with nature and with local contexts, articulating these with other notions: gender, race or ethnicity, interlocking social and environmental inequalities. Likewise, it analyzes how cultures and social groups are located in specific territories. Therefore, it analyzes how environmental inequalities have arisen not only from those articulations but also how inequalities are reinforced. Under this logic inequalities would respond to a social construction of nature, which was consolidated in the nineteenth century, and which consequently implies inequalities in access, use, control and decision-making. Inequalities are in constant transformation, due to the changes in the valuations of nature (genetic material or minerals, for example), which in turn implies the presence of new national and transnational actors in specific territories. This process is transforming boundaries, rights, presences and absences, deterritorializations/territorializations of ideas, people and social dynamics.

However, considering other visions of nature and interactions with it implies a broader perspective, which contemplates new relationships between development and environment. Escobar (1999) presents these processes by stating that there is a permanent interaction between organic nature, capitalist nature and techno-nature, and by locating other territorialities and relationships among Afro-descendants in Colombia. Moreover, Escobar (2010: 2) argues that in so-called third generation PE, "Some recent trends discuss the multiplicity of socio-natural worlds or culture-nature, relational ontologies versus dualistic ones, structural forms of analysis versus network analysis, and even a renewal of the question of what constitutes life." Nevertheless, it is necessary to extend these discussions given the diversity of indigenous peoples, Afro-descendants and peasants in LA.

On the one hand, we must understand spatiality from other perspectives, where a space is constructed by places filled of meaning and identities, and territoriality is exercised, not only by political but by symbolic control, related to other beings, entities or actants that embrace it. On the other hand, other spatial notions are presented in territorial struggles among indigenous peoples where the recognition of rights over their territories has been a basic claim in their political demands and protests. Territories are living entities with memories where the geographies of relations with nature are inscribed, territoriality is exercised, and various symbolic, political, economic and social relations are intertwined (Ulloa 2012).

In that sense I focus on the importance of rethinking the relationship between society and nature (which would require looking at other ideas), and whether under other understandings of this relationship it makes sense to speak of the materiality of nature. This would imply the need for distinct theoretical and methodological approaches in order to analyze such interrelationships.

Part of the analytical and methodological problems currently existing in analyses of LA, stems from the fact that a single notion of nature is assumed. This vision of nature is analyzed under similar approaches, which are productive, but can only address one vision of nature, the "capitalist" one (analyzing the following elements: global capitalism and its location, the role of disciplinary knowledge production and the hegemonic imposition of that vision, as proposed by Escobar (1999)). In this sense it is necessary to combine different analytical perspectives that account for local processes (indigenous and Afro-descendants), and analyze their ontologies and territorial dynamics, which in turn requires us to focus on more than one approach. Likewise, other notions of nature permanently coexist, which struggle to position themselves in a world of capitalist nature. However, their analysis requires other approaches, by responding to relational ontologies in which the notions of subjects and human/nonhuman change.

Alternatives to development: rethinking the relationship human/nonhuman and spatiality

Several authors (Escobar 1999; Leff 2003; Ulloa 2005) have criticized development as a discourse of modernization. Similarly, these authors have argued that the logic of hegemonic globalization and economic development have generated environmental degradation, giving rise to new proposals for relationships with nature and space, mobilized from the culture of local actors. Local people demand more pluralistic and democratic rules of social coexistence (Leff 2003), which has led to a rethinking of development, to consider alternative forms, for example the ways local cultures construct possible worlds based on various relationships between humans and between humans and nature. In this respect, different perspectives on political ecology and critiques of development have been articulated. In the Colombian context, Eschenhagen and Maldonado (2014) state that alternatives to development have to be contrary to "the existing system of free market, production and relationship with nature" (2014: x). The chapters in this edited volume take a critical stance toward development and the construction of alternatives and their epistemological foundations and practices, in order to build real alternatives and not merely generate "alternative" proposals within the same logic of development (Ulloa 2014a). From this perspective, social movements, through their interaction with territory, are formulating relationships with nature that allow for the emergence of alternatives to dominant economic globalization and challenge modern notions of development and sustainable development (Leff 2002, 2012; Martínez-Alier 1997, 2004, 2011; Ulloa 2014b).

These approaches have entered into dialogue with perspectives from Latin American PE regarding environment and development in order to redefine the approaches centered on a specific form of knowledge production. This has resulted in a critique of the way knowledge is produced about LA and local people in their relations with the environment. These critics reposition the production of local knowledge and allow the emergence of other ontologies that have new perspectives toward the "natural" (see also Chapter 4, this volume).

From this perspective, and in conjunction with perspectives on the historical relations between humans and nonhumans in a context of unequal power relations, it is necessary to reclaim a historical dimension to concepts of *ecological debt* and *ecological distribution* (Alimonda 2011; Escobar 2011; Martínez-Alier 1997, 2004). This implies the recognition of the epistemic impacts, ethno- and ecocide caused by the conquest and how these are still caused under the logics of development and appropriation of nature by global processes. It also implies other ways of relating to development that arise from local peoples (Alimonda 2002, 2006, 2009, 2011; Leff 2003; Escobar 2005, 2009, 2011; Ulloa 2011b). The Latin American perspective proposes a view that focuses both on critiques of development and on proposals from local interaction with the environment, which are particular in terms of spatial, territorial and historical processes.

Moreover, rethinking what we mean by natures, especially in contexts where materiality is co-produced: land, forests and species all imply new analytical approaches. In this way, the role of the production of natures is what allows us to rethink relationships with natures and spatialities, where the starting point is a different conception of nature and spatiality (the territory is alive and control can be vertical but also horizontal, where beings are constantly interacting in a changing temporal dimension). This implies a reconceptualization and dislocation of the frameworks of local nature.

Escobar (2005) argues that the critiques of development involve the need to consider poverty and capitalism as embedded development issues, and that processes of resistance to development should be analyzed critically, and that local processes should not be romanticized. Elsewhere, Escobar (2009: 30) argues that post-development, "aims at the creation of a collective space/time where 'development' ceases to be the central organizing principle of economic and social life." In general, the proposals for alternatives to development involve a reworking of the relationship with nature, searching for options for individual and collective responsibility, rethinking global-local economic dynamics of capitalism and the state, and recuperating the philosophical principles through which indigenous peoples relate with their environments. Several trends and positions on alternative to development exist, and I will highlight the following: the politics of place and difference, biocultural memory, living well, and indigenous alter-geopolitics.

The politics of place and difference

Since the 1990s, the critiques of development led by Arturo Escobar and Enrique Leff highlighted the politics of place and recognition of ethnic identities. For example, Leff (2012) argues that critiquing development implies an understanding of where such alternatives originate:

> The perspectives of political ecology are not only to understand the ontological and political nature of socio-environmental conflicts and the power strategies involved in social struggles over ecological distribution, but to envision new potentials arising from "other" knowledge – from social imaginaries, the reinvention of identities and renewal of traditional productive practices – through the rights of being of cultural diversity, a politics of difference and a dialogue of knowledge, to open new paths toward sustainability; to analyze the organization of emergent social movements for the reappropriation of nature and to construct a political ethics and juridical procedures for the peaceful solution of such conflicts.
>
> *(Leff 2012: 35)*

Similarly, in the Colombian context, Escobar (1999) has argued that the practices and relationships located in specific places that have been led by the Process of Black Communities in the Colombian Pacific are alternatives to development, whose constitutive elements are identity, culture and relationships with territory, biodiversity and alternative development:

> For the organizations that make up the Process of Black Communities, development must be guided by principles that reflect the aspirations and rights of these communities,

> which foster and maintain the values of the ancient culture and natural wealth of the region. In this regard, development plans should be channels to empower decision, creativity, solidarity, mutual respect, recognition of the self, dignity and awareness of rights and duties, ethnic identity and sense of belonging to the territory. Development plans should be based on an overall consideration of the people of the Pacific, must have a vision of the present and the future, allow a collective rather than an individual view of themselves, and facilitate decision-making processes within the region. A plan is not only the creation of infrastructure and material conditions; it must respect the local languages and nurture the traditions and cultures.
>
> *(Escobar 1999: 197)*

These proposals position local perspectives in specific places, encompassing everyday dynamics of confrontation and transformation as a possible alternative to top-down, hegemonic processes, and providing a political dimension about culture to political ecology. This in turn leads to the recognition of local production of knowledge and how they are located.

Biocultural memory

The proposals based on local knowledge associated with nature-related practices have informed alternatives to development. There are based on the principle that the knowledges of indigenous peoples, Afro-descendants and peasants respond to ancestral memory based on relationships with nature that are historically constructed (Escobar 1999; Leff 2004; Ulloa 2005). Such collective memories, residing on the frontiers of modernity, are what Toledo and Barrera-Bassols (2008) called "biocultural memories," as spaces of resistance. In the words of Toledo (2014):

> These territories are the setting for the battles of resistance that impede the implementation of the models of modernity or development, led by agro-industrial forms of production: small-scale production against large properties; diversified production systems (polyculture farming and forestry) against the monotonous factories of specialized agriculture; organic agriculture against the use of agrochemicals; solidarity and a just economy against market logic; food self-sufficiency against the use of land for biofuels; and finally, community life and responsible consumption against individualism and consumerism.
>
> *(Toledo 2014: 154)*

While the concept of biocultural memory may be less well known, this proposal echoes the recognition of local knowledges and allows us to analyze the interrelationship of various forms of knowledge about development and environment. Moreover, it fosters discussion on how certain ways of producing knowledge have been articulated with specific powers of production, circulation and distribution of knowledge. Similarly, local knowledges provide options in the face of contemporary environmental transformations (such as climate change, see Ulloa 2011a) by including practices and strategies already implemented at the local level (see also Chapter 18, this volume). That is, local knowledges generate cultural and political strategies concerning environmental change, in order to propose local ways of well-being and living well.

Living well

In much of Latin America, most of the debates relating to alternatives to developments and environmental options are focused on proposals and notions of indigenous peoples, specifically around the idea of "living well" (*sumak kawsay* in Quechua or *suma qamaña* in Aymara). These proposals are reflected in the Constitutions of Ecuador (2008) and Bolivia (2009). There are various interpretations of the notion of living well: the proposals that arise from indigenous peoples from Ecuador and Bolivia, indigenous and non-indigenous intellectuals, and the use of this concept among conventional academic arenas. Xavier Albó (2009) presents an historical account of the concept of *suma qamaña* and how it arises from indigenous conceptions, practices and cultural dynamics, which have been systematized in the papers of Aymara intellectuals Javier Medina, Simón Yampara and Mario Torrez since 2001. Javier Medina (2001) states in his text the idea of "sweet life":

> The sweet life is not a world of knowledge, because the Amerindian people do not claim or want to change the world (as with revolutionaries) but rather love it as it is (as with mystics). Nor does the sweet life aspire to perfection, but rather to the mutual fostering of all types of living beings, from stars, to the plants and animals. It wants every being to live, even those considered to be weeds or pests, because they are all persons, that is to say, beings with whom it is possible to dialogue and to communicate, and they all have the right to live.
>
> *(Medina 2001: 34)*

In a similar way, Medina (2001), using Aymara concepts proposed by Mario Torrez, states:

> In order to explain what the Aymaras understand by **Qamaña**, Torrez started defining their concept of "well-being" as a complementary duality: *Jakañaes* "the well-being of home in the house" and ***Qamaña*** is "the well-being of community in the *ayllu*."[2] Note that in neither of the concepts does the "individual" appears as a subject of well-being; the Jaqi: men/women is always inside of a network: the family and, then, within a broader network: the community.
>
> *(Medina 2001: 662, highlighted in the original)*

In the Ecuadorian context, Acosta (2008) states the importance of living well (*sumak kawsay*) arising from indigenous peoples' proposals, conceptions and political statements. As he puts it:

> Living well ultimately has to do with another way of life, with a series of rights and social, economic and environmental safeguards. It is also reflected in the guiding principles of the economic system, characterized by promoting a harmonious relationship between human beings individually and collectively, as well as with nature. In essence it seeks to build an economy based on solidarity, while recuperating diverse sovereignties as a central concept of the country's political life.
>
> *(Acosta 2008: 38–39)*

Later this concept has been used and disseminated by academic authors (Acosta 2012; Gudynas 2009), who use the idea of "living well" as an alternative to confront extractive processes. On the other hand, the Permanent Working Group on Alternatives to Development (contributions systematized by Lang and Mokrani 2013) is looking to rethink development and its association with ideals of progress under a capitalist model, in order to propose life alternatives centered on the concept of "living well." These authors tend to break with Western rationality and foster alternatives from the perspective of cultural diversity and relationships with nature within multiple historical contexts. Such proposals are connected to post-neoliberal developments, and aim to rethink development, reversing material and social exclusions generated by capitalism (Radcliffe 2012; see also Artaraz and Calestani 2013).

Indigenous alter-geopolitics

Social relations and daily symbolic and spatial practices that are inscribed in the territories of indigenous peoples help consolidate defensive strategies in the context of threats posed by dominant economic and political actors (Oslender 2010, see also Chapter 41, this volume). The spatial practices connected to a place and its importance in building multi-scale relationships, are new strategies for consolidating alternative spatialities. Everyday cultural activities enrolled in specific relationships with territory are an option to rethink relations with other places. This means forming alliances with other social networks and movements that permit the defense of territories and position specific locations (Preciado and Uc 2010; Ulloa 2012).

In these processes, alternative ways of relating to the nonhuman or relational ontologies (Escobar 2010) must be set in place, in order to consolidate what I call an indigenous *alter-geopolitics of territory and knowledge* (Ulloa 2012, 2013). This *alter-geopolitics* involves indigenous peoples' territorial control and planning, and the continuity of ancestral relationships with the nonhuman through the transmission of knowledges, in order to renew species diversity and nature management, consistent with their own environmental practices, and revitalize strategies for the defense of territory, autonomy and political and environmental self-determination. These are related to cultural governability and "symbolic relationships and everyday spatial practices that are part of the territories of indigenous peoples, which may strengthen defense strategies against the intervention of economic and political actors that confronts them" (Ulloa 2012: 13).

This is the case of indigenous peoples in Colombia, who face the new geopolitics linked to the extraction of resources and whose rights are not recognized, and who have generated local forms of resistance and nonviolent strategies to defend their territories. Cultural and environmental political objectives of indigenous peoples' actions are directed primarily at the defense of their territories, always referring to them in terms of Mother Earth, Sacred Territories or Ancestral Territories. Indigenous people view territory as the central axis around which their lives revolve. Their claims focus on the struggle for reclaiming territory, and on self-determination and autonomy in managing their relationships with nature or the nonhuman.

One example is the organizational processes and public mobilizations of indigenous peoples in Cauca, Colombia, through what became known as the "Indigenous Minga"[3] of 2008. In particular, the *National Minga of Indigenous and Popular Resistance*, led by the indigenous peoples of Cauca, which sought to re-position indigenous claims in public arenas, by making apparent the specific, territorial relations between indigenous peoples and their environments, as well as the impacts of various megaprojects. Moreover, this movement also invigorated alliances with other processes, and articulated a series of demands of other social movements led by students and peasants. In this sense a spatial practice in defense of place was consolidated, which was mobilized through regional and national scales, with global impact, as it was disseminated through indigenous media and national and international coverage. It also had international support and solidarity that were channeled through the websites of the Minga (Ulloa 2012).

The political control of indigenous territories must be reconsidered in national contexts, as a territorial strategy with cultural control (Zambrano 2001). Cultural governability – understood as a political, defensive strategy – is evident in the proposals for autonomy on the part of the Kogui people in Colombia. These proposals are related to territorial control, environmental management and food sovereignty. This form of cultural governability arises from six cultural concepts: an ancestral vision of territory and the relationship between nature and society (*senúnulang*); the Law of Origin in their territory and the connection among all the sacred sites (*ezwama*); the political places from where governance arises (*nujwákala*); processes related to the production and harvesting of seeds, plants, animals and humans or food sovereignty (*kualamas*); family relations and lineages (*tuke*); and calendars of ritual activities. In sum, for the Kogui, autonomy is closely related to cultural governability of territory, according to their own understandings, which correspond to ancestral laws and revolves around the responsibilities and activities that every member of society has in relation to the sacred sites, within their lineages, and in specific places and times (Conchacala et al. 2010).

These are examples of the ways that indigenous peoples' specific places and spatial practices, territories and natures, and collective actions in defense of their knowledges are made visible. These local proposals become important as alternatives to development for other social movements and further the dialogue among different scales (local, regional, national and global) (Ulloa 2010, 2012).

Environmental alternatives to development?

The above proposals vary from country to country and within each country, depending on the political, social and environmental conditions in each area. However, one wonders if all these alternatives are actually based on indigenous perspectives, and if they respond to their contemporary social, cultural and political views. In criticizing economic development we run the risk of homogenizing various notions of space and nature. While proposals from various perspectives of alternative environmental thinking have permeated the approaches of development critics and political ecology, sometimes they do so schematically and are based on idealized associations with the local (Medina 2011), and do not consider conflicts or previous inequalities in which local people are involved, especially in contexts of legal and illegal property, resource extraction and the globalization of nature. This, in turn, brings several consequences for local peoples because of the ideals implicit in the recognition of these processes (essentialized images of local people, articulation among identity, territory and nature as harmonious, self-contained worlds, absence of conflict, etc.).

Since the 1990s, in the context of global environmentalism, indigenous peoples have represented the desire to return to a "primitive" and environmentally sustainable world, and to a preindustrial lifestyle. These representations have permeated mass media. Thus, within the "global village," indigenous peoples and their knowledge (as homogeneous categories) have been recognized as environmentally conscious, community driven, wise, healers, etc. In the collective imagination of the media, the return to indigenous traditions is presented as a hope for urban people. These understandings of the indigenous are necessary in order to criticize the modern world (Ulloa 2005). However, all these representations of the ecological native, associated to Mother Nature, show that indigenous peoples, now more than ever, are thought to be part of nature. Furthermore, the images of the "traditional native" or "noble savage" match the modern relationship with nature: colonization (human control of nature through the process of domestication-civilization) and protection (care of endangered species) (Ulloa 2005).

Moreover, certain representations of local peoples associated with environmental ideals legitimate some actors who meet these criteria, while ignoring others. For example, cultural recognition of many Latin American indigenous peoples was consolidated in the 1990s, particularly associated with the environment. However, these processes generate the consolidation of a collective idea of what indigenous peoples should be, as ecological natives (Ulloa 2005). In subsequent decades, the very arguments that helped position indigenous peoples in the global eco-politics, paradoxically served as tools to challenge the demands of indigenous peoples in "re-ethnicization" processes at odds with ecological ideals. Such is the case of indigenous peoples who sell petroleum products like gasoline, or who negotiate their land rights or market their cultural heritage. These actions do not correspond to the ideals of the ecological native, nor to those that frequently lie at the heart of alternatives to development. In addition, gender relations and inequalities between local people are blurred in the interest of collective political processes, without taking into account the differential effects of environmental and development policies for men and women.

Finally, contemporary market dynamics have produced an opening for new patterns of consumption. Economic and environmental processes have generated new forms of territorial appropriation and overlapping of territorialities, due to interests linked to conservation, production or extraction, as well as land-grabbing practices that impose territorial logics that ignore local people's rights and their relationships with the nonhuman. In these contexts, extractive processes have increased even in countries – such as Ecuador and Bolivia – that have promoted alternative ideas to development, and proposed (neo)extractivist models focused on the social distribution. This calls attention to the need to critically consider the link between post-neoliberal ideas of development and its relationship with redistributive "neo-extractivist" processes (Burchardt and Dietz 2014; see also Acosta 2012).

Conclusions

Considering the contradictions that may arise with utopian or idealized alternatives to development, I argue that in order to rethink "development" from a political ecology perspective, one must begin from the analysis of concrete situations of local contexts, inequalities, the lack of recognition, and territorial fragmentation that have been generated by logics of development and its historical relationship with environmental issues.

It is difficult to generalize for all political ecological and critical development approaches, but I argue that these perspectives have some ideas in common. Some of these would be the desire to understand how local groups establish a political space to act in a collective way, in order to generate their own relationships between humans and with nature, through networks of reciprocity and solidarity, based on their territorial, political and environmental autonomy. This would allow them to define their own agendas and ways of life, while recognizing that they are linked to national and global development and environmental policies. We also need to differentiate the effects and processes that trigger the economic, political, social and environmental interdependencies between local, regional, national and global processes according to gender differences.

Notes

1 Translated from the Spanish by Naira Bonilla; edited by Tom Perreault.
2 The *ayllu* is a form of cultural and territorial organization found among Aymara and some Quechua-speaking indigenous peoples in the Andes of Bolivia and Peru. Ayllus pre-date the Inca empire and today form a basis for indigenous political, territorial and cultural mobilization in the Andean region (ed.).
3 *Minga* is a Quechua term used to describe forms of collective labor performed for the common good (ed.).

References

Acosta, A. (2008) "El buen vivir, una oportunidad por construir." *Revista Ecuador Debate*, no. 75, pp. 33–48.

Acosta, A. (2012) "El buen vivir en la senda del posdesarrollo," in Massuh, G. (ed.) *Renunciar al bien común. Extractivismo y (pos)desarrollo en América Latina*, Buenos Aires: Mardulce, pp. 283–305.

Albó, X. (2009) "Suma qamaña = el buen convivir." *Revista OBETS*, no. 4, December, pp. 25–40.

Alimonda, H. (2002) "Introducción: política, utopía, naturaleza," in *Ecología política. Naturaleza, sociedad y utopia*. Buenos Aires: CLACSO, pp. 7–14.

Alimonda, H. (2006) "Una nueva herencia en Comala. Apuntes sobre la ecología política latinoamericana y la tradición marxista," in *Los tormentos de la materia. Aportes para una ecología política latinoamericana/* Buenos Aires: CLACSO. Available at: http://bibliotecavirtual.clacso.org.ar/ar/libros/grupos/hali/C3HAlimonda.pdf.

Alimonda, H. (2009) "Sobre la insostenibilidad colonial de la naturaleza latinoamericana," in Palacio, G. (ed.) *Ecología política de la Amazonia*. UNAL-Bogota: ILSA-Ecofondo, pp. 61–96.

Alimonda, H. (2011) "La colonialidad de la naturaleza. Una aproximación a la Ecología Política Latinoamericana," in Alimonda, H (ed.) *La Naturaleza colonizada. Ecología política y minería en América Latina*. Buenos Aires: CLACSO, pp. 21–58.

Artaraz, K. and Calestani, M. (2013) "Vivir bien, entre utopía y realidad," *Tabula Rasa*, no. 18, pp. 105–123.

Burchardt, H.J. and Dietz, K. (2014) "(Neo-)Extractivism – A New Challenge for Development Theory from Latin America." *Third World Quarterly*, vol. 35, no. 3, pp. 468–486.

Conchacala, A., Chimonquero, M., Zarabata, J., Garavito, P., Barragán, J. and Los Mama. (2010) Kwalzhumun Kwalama. *Caracterización cultural de la soberanía alimentaria en territorio kággaba. Cuenca del río Palomino, Sierra Nevada de Santa Marta, Colombia*. Bogotá: UNODC-OGT.

Eschenhagen, M.L. and Maldonado, C. (eds.) (2014) "Un viaje por las alternativas al desarrollo," in Eschenhagen, M.L. and Maldonado, C. (eds.) *Un viaje por las alternativas al desarrollo. Perspectivas y propuestas teóricas*. Bogotá: Universidad Pontificia Bolivariana, Universidad del Rosario, pp. ix–xv.

Escobar, A. (1998) *La invención del tercer Mundo: Construcción y reconstrucción del desarrollo*. Bogota: Norma.

Escobar, A. (1999) *El final del salvaje. Naturaleza, cultura y política en la antropología contemporánea*. Bogotá: ICANH-Cerec.

Escobar, A. (2005) "El 'posdesarrollo' como concepto y práctica social," in Mato, D. (coord.) *Políticas de Economía, Ambiente y Sociedad en tiempos de la globalización*. Caracas: Universidad Central de Venezuela, pp. 17–31.

Escobar, A. (2009) "Una minga para el postdesarrollo." *América Latina en Movimiento*, no. 445, pp. 26–30.

Escobar, A. (2010) "Ecologías políticas postconstructivistas," available at: www.sustentabilidades.org/revista/publicacion-02/ecologias-politicas-postconstructivistas (accessed February 1, 2011).

Escobar, A. (2011) "Ecología Política de la globalidad y la diferencia," in Alimonda, H. (ed.) *La Naturaleza colonizada. Ecología política y minería en América Latina*. Buenos Aires: CLACSO, pp. 61–92.

Escobar, A. and Pedrosa, A. (eds.) (1996) *Pacífico, ¿desarrollo o diversidad?* Bogotá: Cerec, Ecofondo.

Gudynas, E. (2009) "La dimensión ecológica del Buen Vivir: Entre el fantasma de la Modernidad y El Desafío Biocéntrico." *Revista OBETS*, no. 4, pp. 49–53.

Lang, M. and Mokrani, D. (2013) *Beyond Development: Alternative Visions from Latin America*. Quito-Ámsterdam: Fundación Rosa de Luxemburgo-Transnational Institute.

Leff, E. (2002) *La geopolítica de la biodiversidad y el desarrollo sustentable: econo- mización del mundo, racionalidad ambiental y reapropiación social de la natu- raleza*. Bogotá: Manuscript.

Leff, E. (2003) *La Ecología Política en América Latina. Un campo en construcción*. Buenos Aires: (Grupo de Ecología Política de CLACSO). CLACSO.

Leff, E. (2004) Racionalidad ambiental y diálogo de saberes: significancia y sentido en la construcción de un futuro sustentable. Revista: Polis, Revista de la Universidad Bolivariana 2004 2(7). Available at: www.revistapolis.cl/7/leff.htm.

Leff, E. (2012) "Political Ecology – A Latin American Perspective," in *Culture, Civilization and Human Society* (Eds. UNESCO-EOLSS Joint Committee), in Encyclopedia ofLife Support Systems(EOLSS), Developed under the Auspices of the UNESCO, Eolss Publishers, Oxford, UK. Available at: www.eolss.net.

Martínez-Alier, J. (1997) "Deuda ecológica y deuda externa." *Ecología política*, vol. 14, pp. 157–193.

Martínez-Alier, J. (2004) "Justicia ambiental," in *El ecologismo de los pobres. Conflictos ambientales y lenguajes de valoración*. Barcelona: Icaria.

Martínez-Alier, J. (2011) "Conflictos ecológicos y justicia social," in Álvarez Cantalapiedra, S. (coord.) *Convivir para perdurar*. Barcelona: Icaria-Antrazyt, pp. 51–70.

Medina, J. (2001) *La comprensión indígena de la buena vida*. La Paz: PADEPGTZ.

Medina, J. (2011) "Suma qamaña, vivir bien y de vita beata. Una cartografía boliviana." Available at: http://lareciprocidad.blogspot.com/search/label/Javier%20Medina, consultado 9 de agosto de 2014.

Moncada, E., Hernández, C., de Diego Correa, L., Delgado, G.C., Moncada, M., D'Amico, P., Carvajal, F., Sabogal, J., Torroba, R., Méndez, O.L., Ponce, A., Berardi, A., Bucio, C. and Ángeles, L. (2013) *Ecología política del extractivismo en América latina: casos de resistencia y justicia socioambiental.* Buenos Aires: CLACSO.

Oslender, U. (2010) "La búsqueda de un contra-espacio: ¿hacia territorialidades alternativas o cooptación por el poder dominante?" *Geopolítica(s): revista de estudios sobre espacio y poder*, vol. 1, no. 1, pp. 95–114.

Peet, R. and Hartwick, E. (1999) *Theories of Development.* New York: The Guilford Press.

Porto, C. (2001) "Del desarrollo al desarrollo sustentable y la geografía política del desarrollo sustentable," in *Geo-grafías. Movimientos sociales, nuevas territorialidades y sustentabilidad.* México: Siglo veintiuno editores, pp. 31–70.

Preciado Coronado, J. and Uc, P. (2010) "La construcción de una geopolítica crítica desde América Latina y el Caribe. Hacia una agenda de investigación regional," *Geopolítica(s). Revista de estudios sobre espacio y poder*, vol. 1, no. 1, pp. 65–94.

Radcliffe, S. (2012) "Development for a Postneoliberal Era? Sumak Kawsay, Living Well and the Limits to Decolonisation in Ecuador." *Geoforum*, no. 43, pp. 240–249.

Toledo, V. (2014) "Latinoamérica como laboratorio socioambiental: pueblos indígenas, memoria biocultural y cambio civilizatorio," in Eschenhagen M.L. and Maldonado, C. (eds.) *Un viaje por las alternativas al desarrollo. Perspectivas y propuestas teóricas.* Bogotá: Universidad Pontificia Bolivariana, Universidad del Rosario, pp. 143–162.

Toledo, V. and Barrera-Bassols, N. (2008) *La memoria biocultural. La importancia ecológica de las sabidurías tradicionales.* Barcelona: Icaria.

Ulloa, A. (2005) *The Ecological Native: Indigenous Movements and Eco-governmentality in Colombia.* New York: Routledge.

Ulloa, A. (2010) "Reconfiguraciones conceptuales, políticas y territoriales en las demandas de autonomía de los pueblos indígenas en Colombia." *Tabula Rasa*, no. 13, pp. 73–92.

Ulloa, A. (2011a) "Autonomie indigène et politiques globales du changement climatique: repenser la relation avec la nature dans la Sierra Nevada de Santa Marta, Colombie," in Gros, C. and Dumoulin Kervran, D. (eds.) *Le multiculturalisme "au concret". Un modèle latino-américain.* Paris : Presses Sorbonne Nouvelle, pp. 361–375.

Ulloa, A. (2011b) "The Politics of Autonomy of Indigenous Peoples of the Sierra Nevada de Santa Marta, Colombia: A Process of Relational Indigenous Autonomy." *LACES*, vol. 6, no. 1, pp. 79–107.

Ulloa, A. (2012) "Los territorios indígenas en Colombia: de escenarios de apropiación transnacional a territorialidades alternativas." *Scripta Nova. Revista Electrónica de Geografía y Ciencias Sociales*, vol. XVI, no. 418 (65), Barcelona: Universidad de Barcelona. Available at: www.ub.es/geocrit/sn/sn-418/sn-418-65.htm.

Ulloa, A. (2013) "Controlando la naturaleza: ambientalismo transnacional y negociaciones locales en torno al cambio climático en territorios indígenas, Colombia." *Revista Iberoamericana.* Instituto Ibero-Americano-Berlín. No. 49, pp. 117–133.

Ulloa, A. (2014a) "Conocimientos, naturalezas y territorios: Repensando las alternativas al desarrollo a partir de prácticas y estrategias de los pueblos indígenas en Colombia," in Eschenhagen, María Luisa and Maldonado, Carlos Eduardo (eds.) *Un viaje por las alternativas al desarrollo. Perspectivas y propuestas teóricas.* Bogotá: Universidad Pontificia Bolivariana, Universidad del Rosario, pp. 119–142.

Ulloa, A. (2014b) "Territorio, conflicto y participación democrática en el (neo)extractivismo," lecture in Rohstoffausbeutung und die Zukunft der Demokratie in Lateinamerika. Befunde und Herausforderungen, Heinrich Böll Stiftung, Berlin, 13–14.

Ulloa, A., Rubio, H. and Campos, C. (2004) "Conceptual Basis for the Selection of Wildlife Management Strategies by the Embera People in Utria National Park, Chocó, Colombia," in Silvius, K.M., Bodmer, R.E. and Fragoso, J.M.V. (eds.) *People in Nature: Wildlife Conservation in South and Central America.* New York: Columbia University Press, pp. 11–36.

Zambrano, C. (2001) "Territorios plurales, cambio sociopolítico y gobernabilidad cultural," in Nates, B. (ed.) *Territorio y cultura: Territorios de conflicto y cambio socio cultura.* Manizales: Grupo de investigación territorialidades (Departamento de antropología y sociología, Universidad de Caldas), pp. 19–67.

Zimmerer, K.S. and Bassett, T. (eds.) (2004) *Political Ecology: An Integrative Approach to Geography and Environment-Development Studies.* New York: The Guilford Press.

25
POLITICAL ECOLOGY AND LIVELIHOODS

Edward R. Carr

Introduction

Political ecology and livelihoods studies are natural partners. Whether examining the influence of multinational capital on local agricultural decision-making, or the community-level impacts of conservation projects that materialize global discourses of environmental management, exploring how people live in particular places has long been at the center of political ecological inquiry. At the same time, livelihoods research is fundamentally integrative, focused on how particular people in particular places mobilize environmental, economic, and social resources to meet challenges to well-being and achieve various goals (Hussein, 2002; Valdés-Rodríguez and Pérez-Vázquez, 2011). Many ostensibly local livelihoods decisions and outcomes are linked to extra-local processes and structures (e.g. Bagchi et al., 1998; Bebbington, 1999; Carr, 2013; Hussein, 2002; King, 2011; Murray, 2001). There is little in political ecological inquiry that might not be approached through livelihoods, for as Scoones (2009: 172) notes, this concept can be attached to "locales (rural or urban livelihoods), occupations (farming, pastoral or fishing livelihoods), social difference (gendered, age-defined livelihoods), directions (livelihood pathways, trajectories), dynamic patterns (sustainable or resilient livelihoods)" and other ideas and foci.

Any discussion of livelihoods, however, risks conflating disparate discussions about ways of thinking, sets of principles, and frameworks for analysis (Farrington, 2001). Conversations about ways of thinking about livelihoods and principles for livelihoods analysis tend to speak of livelihoods in holistic, locally focused terms. The application of these ideas to the investigation of how people live in particular places via livelihoods *approaches* is substantially narrower (see Prowse, 2008, 2010 for a discussion of the disjoints between the ontological and epistemological assumptions of livelihoods approaches), framing both the motivations behind particular livelihoods decisions and the evaluation of the outcomes and sustainability of particular livelihoods activities in economistic ways, often without any explicit theorization of these decisions or outcomes (Bebbington, 1999; Carr, 2013; de Haan and Zoomers, 2005; Jakimow, 2012; Kaag et al., 2004; Scoones, 2009; Small, 2007). Despite wide awareness of the limitations of livelihoods approaches that rely on such framings of behavior, efforts to reframe livelihoods decision-making and outcomes (e.g. Bebbington, 1999; Carr, 2013; Jakimow, 2012, 2013; King, 2011; Prowse, 2010) have generated little impact on practice.

In this chapter, I argue that the persistence of narrow economism is enabled and perpetuated, at least in part, by the fact that most livelihoods *frameworks* are deeply *cultural* ecological in their assumptions about how people live in particular places. Such frameworks generally treat livelihoods as local systemic relationships between people and their environment through which individuals and households access needed assets, relegating broader economic, environmental, and political processes to the role of shocks and pressures that "arrive" in particular places. Thus, in the broad field of livelihoods inquiry, livelihoods are treated as systems of local resources and networks intermittently connected to social, economic, political, and environmental relations that cross scales.

This narrowly economic, cultural ecological construction of livelihoods represents a lost opportunity to better understand how people live in particular places, the choices they make in their day-to-day lives, the outcomes of those choices for their quality of life, and the sustainability of their way of life over time. This chapter argues for greater alignment between political ecology and livelihoods studies. When compared with their implicitly cultural ecological counterparts, livelihoods frameworks built on political ecological understandings of the nature–society relationships at the core of rural livelihoods produce more robust representations of how people live in particular places, and more comprehensive assessments of the future sustainability and trajectories of those ways of living.

I begin with a brief review of livelihoods as an object of research, and the livelihoods approaches that emerged in the late 1990s as various development actors attempted to better focus aid on the needs of the global poor. I briefly review well-rehearsed critiques of these approaches as being too focused on economic factors in both framing livelihoods decisions and evaluating livelihoods outcomes. I then examine the (relatively unremarked) influences of cultural ecology implicitly embedded in these approaches, and demonstrate how they align with economizing frameworks to form a coherent whole. The result is an analytic process that treats behaviors that are inefficient with regard to local economic outcomes as outliers or problems to be addressed, instead of intentional efforts to make a living in a particular place. This process pushes critical factors shaping livelihoods decisions outside the analytic frame, limiting our ability to understand livelihoods decision-making and evaluate the efficacy and sustainability of those decisions. I then discuss efforts to engage livelihoods decision-making and outcomes through a political ecological lens. I demonstrate that these promising, if incomplete, efforts to consider factors and forces that play out across multiple scales provide us with greater analytic purchase on the decisions and outcomes of those we work with and for in the Global South. I close with a brief discussion of new frontiers at the intersection of political ecological and livelihoods research.

Livelihoods: a brief history of a concept and its frameworks

In his detailed review of the livelihoods concept and approach, Scoones (2009: 174) notes that the topics often grouped under livelihoods or livelihoods studies have been conducted for some time under any number of disciplinary headings, including "village studies, household economics and gender analyses, farming systems research, agro-ecosystem analysis, rapid and participatory appraisal, studies of socio-environmental change, political ecology, sustainability science and resilience studies (and many other strands and variants)." Cultural and political ecologists have long contributed to these bodies of knowledge. For example, work conducted under the broad heading of cultural ecology has promoted the holistic, local-scale investigation of particular social groups' efforts to make a living in particular places, for example through the importation and integration of ecological concepts like the community with anthropological concepts such as the household (Brookfield, 1964; Grossman, 1981; Netting, 1993; Reenberg et al., 2008; Steward, 1977; Vayda, 1969). More recent political ecological work extends this initial interest in how people live in particular places beyond a consideration of local relations to understand how that which is experienced as "the local" is constituted by forces, pressures, and shocks operating at many scales, and over which communities and individuals have varying degrees of control and influence. For example, political ecological work on conservation and development (e.g. Brockington, 2002; Goldman, 2011; Hanson, 2007; Horta, 2000; Jones, 2006; King, 2010; Ramutsindela, 2007; Schroeder, 1999; Selfa and Endter-Wada, 2008; Wilkie et al., 2006) demonstrates how local livelihoods are impacted by the enactment of conservation efforts driven by discourses that mobilize various forms of global capital. Another long-standing political ecological theme is that of agrarian change, where political ecologists have examined how discourses of development and "improvement" mobilize international capital and investments, and drive policy changes, in ways that produce sometimes dramatic livelihoods impacts for agrarian and pastoral communities (e.g. Bassett, 2006; Batterbury, 2001; Carney, 2004; Carr, 2011; Geoghegan et al., 2001; Kea, 2013; Little, 1994; McCusker and Carr, 2006; Perreault, 2003; Schroeder, 1997; Turner, 2004).

With the rise of modernization approaches in development, a broad body of cultural-ecological knowledge was condensed into the precursors of contemporary livelihoods approaches. As Scoones (2009) notes, these approaches focused on technical solutions for development challenges, and contributed to the rise of monodisciplinary social science perspectives, especially economic perspectives, in development thinking (see also Prowse, 2008: 15). This perspective moved integrative, locally focused work to the margins of development thought and implementation, at best used to inform the selection and design of projects aimed at achieving national-level goals set by economists. Once so situated, this proto-livelihoods work continued in a proscribed manner. As Scoones (2009) notes, in the 1980s especially there was a proliferation of interesting work that could easily be labeled as livelihoods studies, such as that on farming systems and agro-ecological systems, that while marginal, was still integrated into development efforts.

Livelihoods became more formalized as an area of inquiry with the emergence of the idea of sustainable livelihoods. This concept, coined by an advisory panel to the World Commission on Environment and Development (1987), and popularized by Chambers and Conway (1992) in what became a very influential paper, formalized both the definition of livelihood, and the basis upon which a livelihood might be seen as sustainable. In their now-famous framing:

> A livelihood comprises the capabilities, assets (including both material and social resources) and activities for a means of living. A livelihood is sustainable when it can cope with and recover from stresses and shocks, maintain or enhance its capabilities and assets, while not undermining the natural resource base.
>
> *(Chambers and Conway, 1992: 6)*

This framing of livelihoods did little by itself to shift livelihoods research from the margins of development policy and implementation. Scoones (2009: 176–181) traces the rise of livelihoods approaches in development to the growing frustration with the Washington Consensus in the late 1990s. In 1997, the new Labour government in Great Britain articulated a livelihoods and poverty focus for development. This created a space into which long-peripheral livelihoods studies moved (see also Small, 2007: 28). This move, however, was not without friction. Integrative, multidisciplinary, grounded livelihoods work had to be translated into the language of an economics that thought only a little about institutions, and even less about the contexts in which markets and institutions took shape. It is from this pressure, Scoones (2009) argues, that a framing of livelihoods emerged which focused on how assets (livelihood capitals) were accessed through social networks to address particular challenges (such as seasonality of livelihoods activities, or shocks that might upset those activities). Under contemporary livelihoods frameworks, the measurement of the sustainability of a given livelihood is framed in terms of these capitals and their drawdown. For example, any livelihood that continually draws down a livelihoods asset (such as a farmer who draws down the natural capital in his/her land through extensification) without the replacement of that capital in some form (such as through investment in children's education, which might be expected to enhance the human capital available to the household over time) is usually deemed unsustainable. As Carr (2013: 80) has argued, this presumes an economizing logic on the part of the farmer, where "livelihoods are principally about the maintenance and improvement of the material conditions of life."

As others have noted (Hussein, 2002; Valdés-Rodríguez and Pérez-Vázquez, 2011), livelihoods frameworks generally continue to follow this asset-based approach (see Carney, 1998; Ellis, 2000; Scoones, 1998). The impact of these approaches in both development policy/implementation and academia has been profound. Scoones (2009: 181) notes that the use of livelihoods approaches meant "Aid money was spent in different ways, new people with different values and skills were hired, and, for once, even if grossly inadequately, local contexts were better understood and poor, marginalised people were involved in plans and decisions." The world of development studies and development implementation responded accordingly, and today, livelihoods approaches are at the center of vulnerability analyses, project designs, and monitoring and evaluation efforts (see Hussein, 2002: 54; Scoones, 2009: 178–179 for discussion).

Livelihoods approaches and livelihoods analysis: criticisms

Despite their widespread adoption in development studies and implementation, asset-based livelihoods approaches have come under considerable critique. Scoones (2009: 181) focuses on four "recurrent failings of livelihoods perspectives": A lack of engagement with the processes of economic globalization, a lack of attention to power and politics, a failure to appropriately engage with climate change and its impacts, and the limited engagement of livelihoods approaches with ongoing agrarian transformation in many parts of the Global South (see Knutsson and Ostwald, 2006; Prowse, 2010; Small, 2007 for other, largely commensurate, typologies of problems in asset-based livelihoods frameworks). Here, I argue that these four recurrent failings are all linked by an inherent, uninterrogated scale of analysis that privileges local social, economic, political, and environmental contexts in the explanation and evaluation of livelihoods.

Generally speaking, livelihoods approaches pay attention to extralocal processes, and indeed the extralocal components of otherwise local processes, when they impinge upon the vulnerability context in the form of shocks and pressures. These processes and institutions, such as the state or global commodities markets, are otherwise vaguely sited "out there." As Scoones (2009: 181) notes,

> Livelihoods approaches, coming as they did from a complex disciplinary parentage that emphasised the local, have not been very good at dealing with big shifts in the state of global markets and politics. In the frameworks, these were dumped in a box labelled "contexts". But what happens when contexts are the most important factor, over-riding the micro-negotiations around access to assets and the finely-tuned strategies of differentiated actors?

The implicit scaling of livelihoods under asset-based approaches either reduces extralocal processes, and those aspects of local processes that transcend the local, to pressures and shocks felt through the vulnerability context, or it removes them from the analytic frame entirely. Livelihoods approaches tend to construct communities and individuals as receiving and responding to the influences and impacts of events and processes in other places, or operating at other scales, as opposed to participating actively in those processes. Thus, under the sustainable livelihoods approach, the assessment of environmental sustainability is largely local, and refers "to coping with immediate shocks and stresses, where local capacities and knowledge, if effectively supported, might be enough [to achieve sustainability]" (Scoones 2009: 182). At least in terms of their scale of analysis and framing of social-ecological relationships, contemporary livelihoods frameworks are much more cultural ecological than political ecological.

Understanding contemporary livelihoods approaches as cultural ecological helps to explain the most commonly voiced criticism of livelihoods approaches, those that either explicitly or implicitly argue that livelihoods approaches tend to underplay or overlook social processes, especially power relations, in their analyses (see, for example, Arce, 2003; Bebbington, 1999; Carr, 2008, 2013; de Haan and Zoomers, 2005; Jakimow, 2012, 2013; Kaag et al., 2004; King, 2011; McSweeney, 2004). The implicit scale of analysis in these approaches enables a narrowly materialist framing of individual and community motivations under contemporary livelihoods frameworks by producing an oddly stilted view of the social relations and processes that shape local outcomes (Carr, 2013). For example, land tenure rules whose local exercise are often used as a means of coercing particular livelihoods decisions or behaviors frequently derive legitimacy through ethnic ties or laws and regulations established and (unevenly) enforced by the state. Livelihoods approaches generally do not create the space for analysis of such translocal decision-making, pushing issues of complex identity politics and translocal social networks out of the analytic frame. Without consideration of the ways in which these larger considerations might enable, constrain, or otherwise shape the motivations that mold their livelihoods decisions, it becomes possible to rely on narrowly conceived framings of material self-interest in livelihoods analysis. Carr (2013: 80) argues that such reductionism is now inherent to contemporary livelihoods approaches, as they rest on an unstated and largely uninterrogated assumption that livelihoods are, principally, "about the maintenance and improvement of the material conditions of life" (see also Bebbington, 1999; de Haan and Zoomers, 2005; Jakimow, 2012; King, 2011; McSweeney, 2004).

Asserting the political ecological in livelihoods

Recovering the political ecological component of livelihoods and reinserting it into livelihoods analysis is one means of addressing the trend toward narrow analysis and interpretation that contemporary livelihoods frameworks have promoted. A political ecological approach to livelihoods analysis explains local livelihoods decisions and their sustainability through locally specific materializations of translocal economic, political, and environmental processes and structures. In this way, livelihoods become a political ecological lens, an ordering principle for making sense out of the complex, often-messy negotiation of these processes and forces by particular people in particular places.

There have been a few attempts to bring this political ecological framing to questions of scale and economism in livelihoods analysis. Bebbington's (1999) efforts to reframe livelihoods around capitals and capabilities rests fundamentally on the idea that "people's assets are not merely means through which they make a living: they also give *meaning* to the person's world ... This meaning will then be one of several influences in subsequent decisions people make about their livelihood strategies" (Bebbington, 1999: 2022). For example, in accessing water, a woman is doing more than merely gathering a resource for her household that is needed for sanitation and sustenance. In many places, she is also enacting "women's work," thus defining her gender (and, by deferral, other genders) while also likely performing the role of a "good wife" (with such status comes access to different social resources and networks). Similarly, a woman or group of women might use the gathering of water to challenge these same categories. The gathering of water may, in conforming to or contesting the roles and responsibilities associated with her gender or household status, serve as an act of definition that is as important to these women as the actual acquisition of the resource. These meanings, made and remade through this and other livelihoods activities, are critical to decision-making, but cannot be evaluated through relatively simple assessments of total availability or consumption as under contemporary livelihoods frameworks. Bebbington's framework thus "widens the lens" of analyses that examine the sustainability of livelihoods, noting that such sustainability is not only a biophysical question, but also a social question, a question of "tradeoffs between economic growth, human development, social integration and environmental integrity that are implied by different development options" (Bebbington, 1999: 2031; see also Batterbury, 2001; Bebbington and Batterbury, 2001)

Bebbington's challenge to the economism of livelihoods frameworks is political ecological in that the assets engaged in people's livelihoods should not be seen as either purely natural resources, or purely local resources, especially as many livelihoods engage both local natural resource exploitation and various labor and commodities markets that extend beyond the local. As Bebbington notes,

> This conceptualization [of livelihoods] has a related benefit, perhaps more potential than so far real, of conceiving livelihood sustainability within a framework that could also be used for thinking of regional and national sustainability...thus suggesting elements of a framework that could link levels of analysis in research and practice addressing the relationship between environment, society and development.
>
> *(Bebbington, 1999: 2022)*

For example, changing land uses can alter the availability of commodities commonly exchanged in broader markets, impacting business owners and consumers far from the land use changes. Such shifts can, in turn, potentially disrupt networks of capital and social connection critical to the long-term maintenance of the livelihoods in question. This framing of livelihoods opens up the analytic consideration of how networks of policy, economy, and power that might result in major livelihoods shifts in particular rural communities are intimately linked to, and productive of, the lives and livelihoods of those who we might commonly think of as "powerful." Thus, this reframing speaks to Scoones' concerns for better addressing agrarian change, globalization, and local power and politics, and at least opens the door to addressing several recent concerns for the impacts of local livelihoods decisions on processes operating beyond the local, and how that might impact how we assess the sustainability of livelihoods (Carr and McCusker, 2009; King, 2011; McCusker and Carr, 2006).

King's (2011) work on spatialized livelihoods further develops a political ecological understanding of livelihoods politics. His examination of livelihoods in Mzinti, a community in the KaNgwane bantustan in South Africa, weaves together highly spatialized resources and resource utilization impacts, complex sociopolitical networks, and translocal discourses of conservation in a manner that calls into question how we evaluate livelihoods decision-making and the environmental sustainability of livelihoods. For example, King shows that patterns of resource exploitation in Mzinti are the outcome of the interplay between resource availability and the local negotiation of resource management conflicts by state and traditional governance actors. The patterns of wood and other resource collection that result from this process of livelihoods decision-making place pressure on these resources, challenging some livelihoods more than others. For example, herders in Mzinti were once pressured by the expansion of agricultural livelihoods and related land use change. To address this challenge, they invoked discourses of cultural conservation that transcended the local, and mobilized conservation actors to preserve access to land and other livelihoods resources. The negotiation of this pressure and the sustainability of this livelihood were both shaped by the ability of herders to mobilize political and conservation discourses that transcended local social relations and social capital to ensure their continuing access to needed resources. The result of this analysis is a political ecological frame of explanation for observed decisions and outcomes.

Carr (2013, 2014) explicitly extends the social reframing of meaning and power in livelihoods advanced by Bebbington (1999) to consider how meaning in livelihoods is itself constructed through identities, discourses, and practices that transcend the community or local scale. Where Bebbington largely left the constitution of social categories and roles to the side of his discussion, Carr argues that these categories, and the ways in which their attendant roles and responsibilities shape livelihoods decision-making, are local mobilizations of broader identities that draw upon the meanings created through broader historical, ethnic, and spatial processes.

Carr (2013) argues that livelihoods are visible manifestations of intimate government, efforts by the community to manage their environment, economy, and one another to shifting but defined ends (Agrawal, 2005). This effort takes shape at the intersection of three spheres: tools of coercion (such as land tenure rules that, while limiting the decision-making latitude of all members of the community, have greater impacts on some more than others), discourses of livelihoods that contain the rationales people provide for selecting particular livelihoods activities and the actions they take related to those perceptions (for example, the need to balance market and subsistence production to maximize incomes while guarding against economic or environmental shocks), and the mobilization of identity to organize these activities (for example, by aligning particular agricultural roles, such as subsistence production, with wider understandings of appropriate roles and behaviors for those of a particular gender). These spheres are brought together through regimes of practices, in this case the routinized practices of making a living in which individuals, households, and communities participate every day (Dean, 1999: 18). These practices serve to naturalize the strategic mobilization of discourses of livelihoods, tools of coercion, and identity roles and responsibilities, placing them beyond question as what Gidwani (2001: 79) calls social facts. At the same time, Carr argues that the intimate government at the heart of livelihoods strategies is unstable, for such strategies merely mobilize aspects of larger processes that are both ever-changing, and outside the control of those who mobilize them to make a living. For example, identity categories are rarely completely determined by local social networks and factors, but generally reference much larger scales and longer histories that extend beyond the present local, and so can be mobilized but not fully controlled by actors to shape livelihoods decisions. In all cases, the explanation of particular livelihoods decisions and outcomes extends beyond the local scale. This is a deeply political ecological framing of power and decision-making within livelihoods that undermines efforts to explain particular decisions or strategies through narrowly economistic framings of behavior that reference local conditions.

Looking forward: future directions for livelihoods

While the study of livelihoods has long been a part of many different fields of inquiry, our understanding of livelihoods decision-making and outcomes remains relatively superficial (Carr,

2011). The bulk of the work conducted under the heading of livelihoods analysis and inquiry has used, and continues to use, frameworks of explanation limited in spatial scale, that produce limited and often flawed assessments of livelihoods decision-making and sustainability. Perhaps, then, the first and most obvious frontier for political ecologists engaged in livelihoods studies is to employ (or create) more political ecological framings of livelihoods in their analyses. Understanding how people make a living in particular places, and why they go about the activities they do in pursuit of this goal, remains a central consideration in everything from development studies to the ever-growing field of inquiry examining the human impacts of global change.

Second, despite Bebbington's (1999: 2022) suggestions that a wider livelihoods lens might address issues of regional or national sustainability, livelihoods studies rarely consider how local outcomes might filter back to larger, extralocal processes, thus altering the vulnerability context and potentially rendering the livelihoods in question unsustainable over a longer timeframe. This occurs, in part, because the cultural ecological tendencies of contemporary livelihoods approaches tend to look at communities as islands, operating independently of other communities around them. This is rarely true in practice. Communities interact with one another, and if they are engaged in similar livelihoods activities, often will respond in similar manners to the same shock or pressure. Thus, the market and environmental impacts of livelihoods outcomes in a given community might scale across many communities, multiplying their impact. To illustrate this point, I turn to a study of soil carbon sequestration via agriculture in the West African savannah in Senegal (Manlay et al., 2002). This study found that maize fields annually sequestered an average of 7.5 more tons of carbon per hectare than millet fields. If these measurements hold relatively constant across much of Sudanian Senegal, converting a mere 10 percent of Senegal's 121,235 hectares of maize to millet, whether due to environmental or market stresses, would release more than 900,000 tons of carbon into the atmosphere. While this is a small amount of carbon at the global scale, it is still significant at the regional scale. In neighboring Mali, where maize is more commonly grown, a similar agricultural shift would release roughly four million tons of carbon, or the equivalent of one year's emissions from a coal plant (Carr, 2012). This is the potential impact of a 10 percent shift in cropping, from one crop to one other crop, in some relatively small countries. What climate and climate-related environmental impacts of many shifts between many crops in many places are we likely to see, when will these impacts take place, and how will these shifts impinge upon the sustainability of the livelihoods that we are currently examining? These questions go to the heart of Scoones' (2009) concerns for greater engagement with the implications of climate change for livelihoods.

Third, while interesting work on the spatiality of livelihoods and on the multiscalar constitution of power and social relations have deepened our understanding of livelihoods strategies, decisions, and outcomes, these lines of inquiry exist on somewhat separate tracks. Both sets of work address the multiscalar, translocal character of livelihoods, but they tend to do so with different foci, with the spatial livelihoods literature more focused on the environmental outcomes of this complexity, and the literature on power, society, and livelihoods more concerned with the complex constitution of social drivers and outcomes. While there have been initial forays into the integration of these two emergent literatures (Carr and McCusker, 2009; McCusker and Carr, 2006), considerably more work should be done to connect the lessons of both into coherent livelihoods approaches in particular places.

Finally, there is the issue of implementing livelihoods approaches in the world of development and adaptation programming. As Scoones (2009) has ably demonstrated, livelihoods approaches came to prominence because a major donor (DfID) adopted them as a critical part of its mission, prompting attention to and use of these frameworks in both academia and the world of development implementation. The version of livelihoods analysis that became the standard in livelihoods studies, however, is greatly limited in its ability to explain observed decisions and outcomes. It is not enough to simply demonstrate these shortcomings to bring about changes in practice. Instead, the lesson of the history of livelihoods studies in development is that those interested in more political ecological livelihoods approaches must build constituencies for such approaches among the communities most likely to use them. How, then, can we build a constituency for efforts that would introduce complex, differentiated livelihoods approaches to development policy and implementation in a manner that can be taken up and used by those communities? The clearest opportunity for such work exists around a growing demand for serious monitoring and evaluation (M&E) tools among development donors. Where M&E might have been somewhat farcical in the context of previous development efforts, donors such as USAID (through its USAID Forward reforms) are currently attempting to shift M&E from simplistic reporting on outputs such as money spent, items purchased, and individuals trained to outcomes, actual changes in the quality of life for those whom development projects are ostensibly designed to benefit. These efforts are changing the incentives for program and project design, making project managers accountable for the actual workings of their projects. In this change there is an opening into which new and complex, but more effective, tools might emerge for understanding why people do what they do, and why particular efforts to make a living have the outcomes they do. Therefore, a productive frontier for political ecological livelihoods approaches might be the empirical demonstration of what these new approaches can render legible versus contemporary livelihoods approaches.

As the study of livelihoods fundamentally embraces the question of how people live in particular places, livelihoods will always be a part of conversations about development, climate change, and nature–society relations more generally. By shifting how we conduct livelihoods research to better engage with political ecological practice, we have the opportunity to better understand the world around us, and to shape ongoing efforts to address current and future challenges that the world will present in the Anthropocene.

References

Agrawal, A. (2005). Environmentality: Community, intimate government, and the making of environmental subjects in Kumaon, India. *Current Anthropology*, 46(2), 161–190.

Arce, A. (2003). Value contestations in development interventions: Community development and sustainable livelihoods approaches. *Community Development Journal*, 38(3), 199–212.

Bagchi, D.K., Blaikie, P., Cameron, J., Chattopadhyay, M., Gyawali, N., and Seddon, D. (1998). Conceptual and methodological challenges in the study of livelihood trajectories: Case-studies in Eastern India and Western Nepal. *Journal of International Development*, 10(4), 453–468.

Bassett, T.J. (2006). *The Peasant Cotton Revolution in West Africa: Cote d'Iviore, 1880–1995*. Cambridge: Cambridge University Press.

Batterbury, S. (2001). Landscapes of diversity: A local political ecology of livelihood diversification in south-western Niger. *Ecumene*, 8(4), 437–464.

Bebbington, A. (1999). Capitals and capabilities: A framework for analyzing peasant viability, rural livelihoods and poverty. *World Development*, 27(12), 2021–2044.

Bebbington, A.J., and Batterbury, S.P.J. (2001). Transnational livelihoods and landscapes: Political ecologies of globalization. *Ecumene*, 8(4), 369–380.

Brockington, D. (2002). *Fortress Conservation: The Preservation of the Mkomazi Game Reserve, Tanzania*. Bloomington: Indiana University Press.

Brookfield, H. (1964). Questions on the human frontiers of geography. *Economic Geography*, 40(4), 283–303.

Carney, D. (1998). *Sustainable Rural Livelihoods: What Contribution Can We Make?* London: Department for International Development.

Carney, J.A. (2004). Gender conflict in Gambian wetlands. In R. Peet and M. Watts (Eds.), *Liberation Ecologies: Environment, Development, Social Movements, Second Edition* (pp. 316–335). London: Routledge.

Carr, E.R. (2008). Between structure and agency: Livelihoods and adaptation in Ghana's Central Region. *Global Environmental Change*, 18(4), 689–699.

Carr, E.R. (2011). *Delivering Development: Globalization's Shoreline and the Road to a Sustainable Future*. New York: Palgrave Macmillan.

Carr, E.R. (2012). Small farmers, big impacts. In R. Shah and S. Radelet (Eds.), *Frontiers in Development* (pp. 92–97). Washington, DC: USAID.

Carr, E.R. (2013). Livelihoods as intimate government : Reframing the logic of livelihoods for development. *Third World Quarterly*, 34(1), 77–108.

Carr, E.R. (2014). From description to explanation: Using the Livelihoods as Intimate Government (LIG) approach. *Applied Geography*, 52, 110–122.

Carr, E.R., and McCusker, B. (2009). The co-production of land use and livelihoods change: Implications for development interventions. *Geoforum*, 40(4), 568–579.

Chambers, R., and Conway, G. (1992). *Sustainable Rural Livelihoods: Practical Concepts for the 21st Century*. Sussex: Institute of Development Studies.

De Haan, L.J., and Zoomers, A. (2005). Exploring the frontier of livelihoods research. *Development and Change*, 36(1), 27–47.

Dean, M. (1999). *Governmentality: Power and Rule in Modern Society*. London: Sage Publications.

Ellis, F. (2000). *Rural Livelihoods and Diversity in Developing Countries*. Oxford: Oxford University Press.

Farrington, J. (2001). Sustainable livelihoods, rights and the new architecture of aid. *Natural Resource Perspectives*, 69. Retrieved from www.odi.org.uk/sites/odi.org.uk/files/odi-assets/publications-opinion-files/2823.pdf.

Geoghegan, J., Villar, S.C., Klepeis, P., Mendoza, P.M., Ogneva-Himmelberger, Y., Chowdhury, R.R., ... Vance, C. (2001). Modeling tropical deforestation in the southern Yucatan peninsular region: Comparing survey and satellite data. *Agriculture Ecosystems & Environment*, 85(1–3), 25–46. Retrieved from <Go to ISI>://000169062400003.

Gidwani, V. (2001). The cultural logic of work: Explaining labour deployment and piece-rate contracts in Matar Taluka, Gujarat – parts I and II. *Journal of Development Studies*, 38(2), 57–108.

Goldman, M.J. (2011). Strangers in their own land: Maasai and wildlife conservation in Northern Tanzania. *Conservation and Society*, 9(1), 65.

Grossman, L. (1981). The cultural ecology of economic development. *Annals of the Assocation of American Geographers*, 71(2), 220–236.

Hanson, P.W. (2007). Governmentality, language ideology, and the production of needs in Malagasy conservation and development. *Cultural Anthropology*, 22(2), 244–284.

Horta, K. (2000). Rainforest: Biodiversity conservation and the political economy of international financial institutions. In P. Stott and S. Sullivan (Eds.), *Political Ecology: Science, Myth and Power* (pp. 179–202). London: Arnold.

Hussein, K. (2002). *Livelihoods Approaches Compared: A Multi-Agency Review of Current Practice*. London: Department for International Development.

Jakimow, T. (2012). Serious games in livelihood analysis : Reflections from the case of agricultural wage labourers in Andhra Pradesh. *Journal of Development Studies*, 48(9), 1274–1287.

Jakimow, T. (2013). Unlocking the black box of institutions in livelihoods analysis: Case study from Andhra Pradesh, India. *Oxford Development Studies*, 41(4), 493–516.

Jones, S. (2006). A political ecology of wildlife conservation in Africa. *Review of African Political Economy*, 33(109), 483–495.

Kaag, M., van Berkel, R., Brons, J., de Bruijn, M., van Dijk, H., de Haan, L., ... Zoomers, A. (2004). Ways forward in livelihoods research. *Globalization and Development*, 49–74.

Kea, P. (2013). "The complexity of an enduring relationship": Gender, generation, and the moral economy of the Gambian Mandinka household. *Journal of the Royal Anthropological Institute*, 19, 102–119.

King, B. (2010). Conservation geographies in Sub-Saharan Africa: The politics of national parks, community conservation and peace parks. *Geography Compass*, 4(1), 14–27.

King, B. (2011). Spatialising livelihoods: Resource access and livelihood spaces in South Africa. *Transactions of the Institute of British Geographers*, 36(2), 297–313.

Knutsson, P.E.R., and Ostwald, M. (2006). A process-oriented sustainable livelihoods approach – a tool for increased understanding of vulnerability, adaptation, and resilience. *Mitigation and Adaptation Strategies for Global Change*.

Little, P. (1994). Contract farming and the development question. In P.D. Little and M.J. Watts (Eds.), *Living Under Contract: Contract Farming and Agrarian Transformation in Sub-Saharan Africa* (pp. 216–247). Madison: University of Wisconsin Press.

Manlay, R.J., Masse, D., Chotte, J.-L., Feller, C., Kairé, M., Fardoux, J., and Pontanier, R. (2002). Carbon, nitrogen and phosphorus allocation in agro-ecosystems of a West African savanna: II. The soil component under semi-permanent cultivation. *Agriculture, Ecosystems & Environment*, 88(3), 233–248. Retrieved from www.sciencedirect.com/science/article/pii/S0167880901002195.

McCusker, B., and Carr, E.R. (2006). The co-production of livelihoods and land use change: Case studies from South Africa and Ghana. *Geoforum*, 37(5), 790–804.

McSweeney, K. (2004). The dugout canoe trade in Central America's Mosquitia: Approaching rural livelihoods through systems of exchange. *Annals of the Association of American Geographers*, 94(3), 638–661.

Murray, C. (2001). *Livelihoods Research: Some Conceptual and Methodological Issues. University of Manchester Background Paper 5* (pp. 1–16). Retrieved from www.chronicpoverty.org/uploads/publication_files/WP05_Murray.pdf.

Netting, R.M. (1993). *Smallholders, Householders: Farm Families and the Ecology of Intensive, Sustainable Agriculture.* Palo Alto: Stanford University Press.

Perreault, T. (2003). Making space: Community organization, agrarian change, and the politics of scale in the Ecuadorian Amazon. *Latin American Perspectives*, 30(1), 96–121.

Prowse, M. (2008). *Locating and Extending Livelihoods Research* (No. BWPI Working Paper 37). *BWPI Working Paper 37* (pp. 1–33). Manchester, UK.

Prowse, M. (2010). Integrating reflexivity into livelihoods research. *Progress in Development Studies*, 10(3), 211–231.

Ramutsindela, M. (2007). *Transfrontier Conservation in Africa: At the Confluence of Capital, Politics and Nature.* Boston: CABI.

Reenberg, A., Birch-Thomsen, T., Mertz, O., Fog, B., and Christiansen, S. (2008). Adaptation of human coping strategies in a small island society in the SW Pacific—50 years of change in the coupled human–environment system on Bellona, Solomon Islands. *Human Ecology*, 36(6), 807–819.

Schroeder, R.A. (1997). Re-claiming land in the Gambia: Gendered property rights and environmental intervention. *Annals of the Association of American Geographers*, 87(3), 487–508.

Schroeder, R.A. (1999). Geographies of environmental intervention in Africa. *Progress in Human Geography*, 23(3), 359–378.

Scoones, I. (1998). *Sustainable Rural Livelihoods: A Framework for Analysis* (No. 72). *IDS Working Paper 72.* Sussex, UK: IDS. Retrieved from http://200.17.236.243/pevs/Agroecologia/Sustainable Rural Livelihhods-Scoones.pdf.

Scoones, I. (2009). Livelihoods perspectives and rural development. *The Journal of Peasant Studies*, 36(1), 171–196.

Selfa, T., and Endter-Wada, J. (2008). The politics of community-based conservation in natural resource management: A focus for international comparative analysis. *Environment and Planning A*, 40(4), 948–965.

Small, L.-A.A. (2007). The sustainable rural livelihoods approach: A critical review. *Canadian Journal of Development Studies / Revue Canadienne D' Études Du Développement*, 28(1), 27–38.

Steward, J.H. (1977). *Evolution and Ecology: Essays on Social Transformation.* Urbana: University of Illinois Press.

Turner, M.D. (2004). Political ecology and the moral dimensions of "resource conflicts": The case of farmer-herder conflicts in the Sahel. *Political Geography*, 23(7), 863–889.

Valdés-Rodríguez, O.A., and Pérez-Vázquez, A. (2011). Sustainable livelihoods: An analysis of the methodology. *Tropical and Subtropical Agroecosystems*, 14, 91–99.

Vayda, A.P. (1969). Expansion and warfare among swidden agriculturalists. In A.P. Vayda (Ed.), *Environment and Cultural Behavior: Ecological Studies in Cultural Anthropology* (pp. 202–220). Garden City, NY: The Natural History Press.

Wilkie, D.S., Morelli, G.A., Demmer, J., Starkey, M., Telfer, P., and Steil, M. (2006). Parks and people: Assessing the human welfare effects of establishing protected areas for biodiversity conservation. *Conservation Biology*, 20(1), 247–249.

World Commission on Environment and Development. (1987). *Food 2000: Global Policies for Sustainable Agriculture, a Report of the Advisory Panel on Food Security, Agriculture, Forestry and the Environment to the World Commission on Environment and Development.* London: Zed Books.

26

POLITICAL ECOLOGIES OF DISEASE AND HEALTH

Brian King

Introduction

The HIV/AIDS epidemic in sub-Saharan Africa has taken on a new course in recent years that resonates with interests in political ecology and human health. Due to an aggressive push by international funding agencies, national governments, non-governmental organizations and activist groups, access to anti-retroviral (ARV) drugs has increased for those in need of these life-saving medications. The pace of this expanding access has been so dramatic that it prompted a recent cover story of *The Economist* to question whether Africa was entering a new era highlighted by "The end of AIDS" (*The Economist* 2011). South Africa is particularly notable in this regard. In the 2000s, the national government became the figure of international scorn for statements made by key officials, including former President Thabo Mbeki and his Health Minister Dr. Tshabalala-Msimang that questioned the scientific consensus on the link between HIV and AIDS and the efficacy of drug treatment protocols. Additionally, the government resisted a roll-out of a national ARV program, akin to what some other African countries had already accomplished. Invocations of traditional medicine as a course of treatment for the disease, coupled with assertions that HIV was a disease of poverty, complicated public health campaigns on the epidemic and the best ways to manage it. A lawsuit brought by the Treatment Action Campaign (TAC) through the South African Constitutional Court forced a change in the national position when it was decided in 2006 that preventing access to these drugs constituted a violation of the country's progressive Constitution. It is remarkable therefore that only six years later the country asserted that it had achieved universal access to ARVs (IRIN Humanitarian News and Analysis 2012) and is treating roughly two million patients each day (Beaubien 2013). Recent reports attest to the government's push for a new pill that can be taken once per day for just hundreds of dollars a year per patient. And by treating patients when their CD4 count (a critical measure of the health of the immune system) is 500 cells/mm3 or lower, in line with new recommendations from the World Health Organization (WHO), the government is reducing the chances of patients progressing from HIV disease to AIDS, thereby potentially extending a person's life for decades. A country whose government was once called a "lunatic fringe" by the United Nations ambassador to Africa for AIDS (Wines 2006) has now become a leader in an era in which HIV is becoming a more manageable disease and chronic condition.

I begin this chapter with this specific health transition to highlight the importance of socio-political factors in shaping human health and well-being. It is surprising, therefore that the field of political ecology has been limited in addressing the subject of human health. As existing research effectively demonstrates, political ecology's theoretical and methodological strengths are uniquely positioned to make contributions within geography and related fields. Yet even with existing work there are particular areas of emphasis and absence that warrant examination to identify future trajectories for political ecology research and practice. The intention of this chapter is to review past and current political ecology research on human health to identify future directions for the field. In the first section a review is provided of what I classify as the first phase of this work that emphasized a political ecology of disease approach. Primarily based within medical anthropology and medical geography, research during this phase tended to concentrate upon disease ecologies and the places of health. This helped contribute to a second phase of scholarship that coincided with an expansion of traditional domains of medical geography that included contributions from health geography. This has been reviewed more fully elsewhere (King 2010) but I draw upon this history to detail some of the emergent features for a political ecology of health. This work has been more insistent upon locating individual bodies within broader structures of power while attending to the disparities in vulnerabilities and exposure to conditions that produce poor health. Additionally, recent work has explored indoor environments and human bodies to unsettle assumptions about traditional concepts within political ecology (Biehler and Simon 2011; Sultana 2012; Guthman 2011). The chapter concludes by reviewing work within cognate fields, particularly medical anthropology and public health that decenters some assumptions within political ecology scholarship. I suggest that this represents an avenue of mutual exchange because political ecology has specific contributions to make in addressing the production of health within broader social and economic systems, discourses of disease and well-being, and the interactions between social and ecological systems.

Political ecologies of disease

> The division of knowledge into disciplines handicaps problem solving in public health. Social and cultural descriptions often ignore political and economic conditions, and vice versa.
>
> *(Turshen 1984: 17)*

From its origins, political ecology scholarship has highlighted the structural forces shaping political and environmental decision-making largely within the developing world, although this has shifted with scholars noting the benefits of the approach for addressing environmental decision-making and management in North America (cf. McCarthy 2002; Walker and Hurley 2011). Alternatively labeled an emblem (Blaikie 1999), an integrative approach (Zimmerer and Bassett 2003), or a text (Robbins 2012), political ecology has been notable for its divergent themes ranging from land degradation (Blaikie and Brookfield 1987) deforestation (Hecht 1985; Stonich 1993), soil erosion (Blaikie 1985; Zimmerer 1996), conservation (Neumann 1998), and critical analyses of development and social movements (Peet and Watts 2004; Bebbington 2000). The subject of human disease and health was less central to this first phase of political ecology scholarship with some notable exceptions coming from medical anthropologist Meredith Turshen's work that linked contemporary disease patterns in Tanzania with colonial and post-colonial political regimes (Turshen 1977, 1984). Providing a direct challenge to the disease ecology tradition, Turshen (1977: 48) argued that the field deemed economic and political processes to be irrelevant, and therefore suffered "from a failure to consider the relation of people to their environment in all its complexity." In this groundbreaking work, Turshen argued that the spread of infectious disease in the contemporary era could only be explained by addressing the colonial relationships and spatial patterns that were linked to political economic arrangements that advanced the power of particular stakeholders. In identifying some of the limitations of the biomedical model within medical anthropology and disease ecology at this time, she emphasized the importance of locating social actors within broader structural processes that generated disease for some and not for others.

Like others working within medical anthropology at the time (Doyal 1979; Morgan 1987; Packard 1989), Turshen employed a political economy framework that identified colonial and capitalist market relationships as central in producing disease vulnerabilities. This was familiar within medical anthropology studies, which were generally aligned with one of three theoretical perspectives: orthodox Marxist approaches, cultural critiques of medicine, and dependency theories (Morgan 1987). While dependency theory dominated the field, Morgan (1987: 132) argued that a political economy of health approach should include an "historical perspective, conflict or dialectical models of social change, and a theory of disease causation that is multifactorial and encompasses social etiology." Even within current research within the field of medical anthropology (Baer and Singer 2008), there is the recognition that capitalist economies produce disease in specific ways and that these systems warrant greater attention in addressing human health. These contributions, while not all explicitly political ecological, offered links to the emerging research field within geography and anthropology while establishing a foundation for future components of a political ecology of health that would typify the second phase of research discussed in the next section.

Central to this first phase of political ecology scholarship was the work of Jonathan Mayer, an epidemiologist situated in the medical geography and disease ecology traditions. In a much-cited paper, Mayer (1996: 449) asserted that a political ecology of disease approach would help demonstrate "how large-scale social, economic and political influences help to shape the structures and events of local areas." While noting the potential benefits from political ecology, Mayer emphasized the existing contributions from disease ecology, a field that he explained examines how "humanity, including culture, society and behavior; the physical world, including topography, vegetation and climate; and biology, including vector and pathogen ecology, interact together in an evolving and interactive system, to produce foci of disease" (1996: 441). In suggesting links between political ecology and disease ecology, Mayer argued that the role of culture, behavior, and other social and environmental factors would be taken into consideration to understand the spread of disease. In essence, Mayer advanced political ecology to remedy some of the limitations of disease ecology: overly localized and synchronic studies that do not always connect disease transmission to underlying structural processes that produce vulnerabilities and decision-making opportunities. Yet there is a particular reading of political ecology that seems more designed to reference the "rich work in disease ecology" (1996: 450). This remained the case in a later paper that argued that the political ecology of disease approach combines "the elements of traditional disease ecology with the concepts of political economy" (Mayer 2000: 948). As has been written elsewhere (King 2010), while influential in generating new research directions in geography, this approach elided the question as to whether structural analyses might challenge dominant assumptions within the field of disease ecology.

Mayer's promotion of a political ecology of disease approach deserves recognition for generating an expansion of work employing political ecology to address human disease. In examining the links between infectious disease and refugee camps, Kalipeni and Oppong (1998) showed how the transmission of diseases, such as tuberculosis and HIV, are shaped by social processes that increase vulnerability for certain populations. Oppong and Kalipeni (2005) detailed the historical use of landmines in Africa and the resulting impacts for social, economic, and health-care systems, noting that political ecology offers three central elements: context and scale, historical depth, and structural relationships. In a case study from Texas City, Cutchin (2007) applied several theoretical perspectives, including cultural ecology, political ecology, and territoriality, in order to advance a "new health geography" that demonstrated the ways that communities are exposed to toxic pollutants resulting from industrial production. He concluded that a combination of these approaches helps illustrate

> a way to collect information about place and landscape and then interpret how the processes that create and re-create them create the situation that an epidemiologist typically investigates. Although the concepts require an epistemological shift for epidemiologists, the processes they illustrate provide a *better explanation of underlying causes* for the patterns and associations discovered in epidemiological analyses.
>
> *(2007: 740, author's emphasis)*

As these examples evidence, the first phase of political ecology research on human health was theoretically broad, contextually rich, and multifaceted in the subjects with which it engaged. During this period human health tended to be theorized in terms of its relationships to particular disease patterns or toxic exposure, whether in terms of air pollution in south Texas or tuberculosis in refugee camps in Africa. Political ecology research on health was largely critical of the biomedical model that "takes individual physiology as the norm for pathology (as contrasted with broader social conditions) and locates sickness in the individual's body" (Turshen 1977: 46). Regardless, the first phase of political ecology research concentrated upon *disease patterns and health vulnerabilities* as opposed to more holistic and integrated perspectives on human health. The second phase of political ecology scholarship would begin to approach health in broader terms, addressing it less as the absence of disease and more in terms of long-term well-being and individual agency. While health is not always defined by these studies, and the definitions would likely diverge, they resonate with broader understandings such as the landmark Ottawa Charter for Health Promotion (1986) that identified the promotion of health as including the realization of aspirations, satisfaction of needs, and the ability to adjust to the environment. The Ottawa Charter continued by identifying the following as the fundamental conditions for human health: peace, shelter, education, food, income, ecosystem stability, sustainable resources, social justice, and equity. As the next section details, this second phase of political ecology research has continued to engage with the subject of human health in related ways while broadening central concepts, interrogating socio-ecological interactions, and examining bodies as sites of production.

Political ecologies of health

Scholarship from the first phase of political ecology work provided varied understandings on the relationships between disease and political economic systems. At this time, emerging insights from related fields have pushed scholarship on human health in new directions, some of which have been adopted by political ecology. This second phase of political ecology research has provided theoretical innovations and simultaneously broadened health across multiple spatial and temporal scales while centering these concerns at the site of the individual body. Some of this work has challenged existing categories within nature–society geography more broadly, such as the outdoor–indoor dichotomy or engaging with race and gender more explicitly.

Lastly, these studies have also provided concrete detail on the interactions between social and ecological systems in producing disease vulnerabilities and opportunities for human health and well-being. These new directions continue to highlight theoretical and methodological contributions from political ecology while incorporating insights from public health, ecology, medical anthropology, and social epidemiology.

Broadening the concepts

One defining feature of second phase political ecology of health research is the commitment to examining broadened conceptions of health and environment, and how they interlink with socio-economic class, race, gender, and ethnicity. As one example, Richmond *et al.* (2005) examine how aquaculture development in the 'Namgis First Nation in British Columbia, Canada has reduced access to environmental resources and cultural activities that contribute to the health and well-being of the community. Aquaculture development has been advanced by the government as a heavily industrialized and global form of production that utilizes licenses and quotas to allocate and regulate access to users, thereby challenging the traditional practices of indigenous populations. Situating this community population within broader political economic and development forces helps demonstrate the ways that health is connected to other elements, specifically political autonomy, use and enjoyment of environmental resources, and economic choice and opportunity. The relationships between socio-economic development and disease patterns are also a central theme in an historical assessment of malaria eradication in Argentina (Carter 2012). Carter details how the Northwest was cast as a malarial zone following the discovery of the disease in 1890 and how the resulting public health campaigns were promoted as necessary for the health and well-being of the nation. Identifying malaria in such broad and political terms enabled social elites to take advantage of development programs, such as agricultural intensification or urban sanitation that arose to eradicate the disease. Both of these studies contribute in demonstrating the power of historical assessments of the structural forces shaping health circumstances and how they collide with local needs and practices that are simultaneously economic, environmental, and cultural.

Other studies have worked to expand traditional themes in political ecology research, whether it is a focus upon external environments or emphasizing how humans and nature are co-produced. Noting the health threats from poor indoor air quality in the developing world, Biehler and Simon (2011) highlight the need for political ecology research on indoor environments. Asserting that indoor spaces have been treated as spatially fixed and static, they note that in fact they "teem with life, and are vital sites for production and reproduction of nature, scale, and environmental citizens. They also articulate with technologies of power while engaging in flows of matter, energy, capital, and knowledge" (2011: 174). Drawing upon political ecology's traditional strengths, they suggest that indoor environments and human health would be advanced by five themes: (1) social production of indoor environments, (2) indoor nature and the embodied subject, (3) animating indoor technologies and natures, (4) indoor environments as conduits of power and sites of governance, and (5) fluid indoor and outdoor boundaries. This work is notable for a number of reasons, such as the challenge to conventional domains of political ecology scholarship to consider indoor spaces but also what is contained within those spaces, such as human bodies, animals, microbes, and pollutants. Research in this vein resonates with a current desire in political ecology to integrate non-humans as agents shaping human behavior and health (Braun 2007; Kosek 2013).

The co-production of human and non-human species is highlighted in other studies that attend to shifting disease vectors of West Nile Virus (WNV). Scott *et al.* (2012) examine the mutual conditioning of humans and pathogens, which they assert is produced by continuously changing exposures (settlement and development patterns that modify pathogen and vector ecology) and institutional processes (legal, economic, and organizational contexts in which environments are modified and agents respond to risk). Similarly, Robbins and Miller (2013) examine competing state agency responses to the *Aedes aegypti* and *Culex quinquefasciatus* mosquitoes in the state of Arizona, noting that the counties of Yuma, Maricopa, and Pinal differ in their vector management strategies. Responses across these counties range from adulticide fogging, sophisticated GIS surveillance, and public education and source reduction. Interlinked with the specific approaches are the views of county managers that have to negotiate different population densities and growth patterns, in addition to differential operating budgets. The resulting public health decisions reveal much about the technocentric and bureaucratic approaches to disease management and also about the ways in which the mosquito contributes in producing the state through its understandings of vector ecology and idealizations of state responses.

Public health campaigns and state responses also collide with human health in the case of the arsenic crisis in Bangladesh. Sultana (2012) details how millions have been poisoned from naturally occurring arsenic that enters from tubewells used for domestic and irrigation purposes. The study details the "complex ways that the well-being of entire families is affected by having arsenicosis patients in the home, as well as from living with fear and uncertainty, dealing with rejection from society, and coping with the multifaceted lived experiences of ostracism and stigmatization" (2012: 1168). Sultana shows how the body becomes a site of the illness, marking its presence for others to see while causing hardship for those affected by arsenicosis. Additionally, the disease pattern is produced through structural political economies that are intended to generate socio-economic development. The public health campaigns are not only national but are also global in scope with development programs being initiated to promote awareness of the illness and advance mitigation strategies. Mitigation effectiveness is constrained by socio-economic poverty, in that millions are still vulnerable to consuming arsenic-laced water because they lack access to safe water sources or funding to invest in deeper tubewells. Sultana's study details how arsenic exposure and health are not only economic and cultural, but are also *socio-ecological*, which is a second theme of political ecology of health scholarship detailed in the next section.

Health as socio-ecological production

The first phase of political ecology research on human disease and health was notable for its positioning of social actors within broader structural relationships and systems of power. Turshen (1984) effectively demonstrated how colonial territorial arrangements continued to influence contemporary disease patterns within Tanzania while challenging the hegemony of medical and behavioralist positions for improved health. As she states, "[t]he question is how medicine can penetrate communities in order to manipulate and change behavior. Adherents of this position ignore the underlying causes of disease and death embedded in political and economic systems" (1984: 2). Yet Turshen's statement, while provocative and important, also reflected a bias of some studies in the first phase, specifically by focusing upon the socio-political dimensions of human health at the expense of addressing biophysical conditions. While capitalist meta-narratives are powerful in explaining how certain populations become vulnerable to disease transmission, in addition to the ways that institutions respond to the outbreak of infectious or non-infectious disease, they are limited in demonstrating the full spectrum of human health. Human disease spreads through ecological systems that structure disease vectors, and these biophysical conditions similarly constrain opportunities that contribute towards healthy decision-making. It is at the nexus between social and ecological processes that human health is shaped, and only in understanding the interactions between them can human health be properly understood (Crews and King 2013). Therefore, a second feature of current political ecology of health research has been to locate human health within social and ecological systems while attending to their multifaceted and multi-scalar interactions.

In a review of political ecology and health research, King (2010) argues that a political ecology of health is particularly well positioned to contribute to new insights on the political economy of disease, interrogate health discourses produced by actors and institutions, and show how health is shaped through the relationships between social and environmental systems. With regards to social and ecological dynamics, political ecology research that incorporates biophysical processes into the framework would help elucidate the ecological dimensions of infectious disease. The HIV/AIDS epidemic in sub-Saharan Africa is illustrative of this trend. While the disease is having tremendous impacts on demographic patterns, national economies, and gender dynamics (Drimie 2003; de Waal and Whiteside 2003; Love 2004; Masanjala 2007; UNAIDS 2008; Bolton and Talman 2010), recent research has shown that the spread of HIV/AIDS is also transforming ecological systems (Aldhous 2007; Hunter *et al.* 2008; Kaschula 2008; McGarry and Shackleton 2009). Even with these existing studies there remain significant gaps within the academic and policy literatures on the ecological impacts of HIV/AIDS (King 2013). In an exhaustive survey of existing research on the ecological effects of HIV/AIDS, Bolton and Talman (2010) conclude that among many possible impacts, HIV/AIDS sufferers exploit natural resources in the short-term through unsustainable practices, shift land use systems including leaving agricultural land fallow, and pressure currently protected natural resource areas due to increased need. It remains unclear whether increased reliance upon natural resources as a coping mechanism to HIV/AIDS presents a short-term, or a long-term, shift in ecosystem functioning. Additionally, the full gamut of natural resources extracted as a livelihood response remains unknown, as are the specific species, the resiliency of ecosystems in response to increased resource pressure, and potential land cover change. These dynamics emphasize the importance for political ecology research to attend to the ways that social and ecological processes, along with their coupled interactions, shape human health.

Healthy bodies

A third feature of current research is upon the individual body, which some scholars have argued has been "black-boxed" in health studies (Guthman and Mansfield 2012; see also Chapter 43, this volume). Attending to the processes associated with natural childbirth, Mansfield (2008a, 2008b) suggests that health is inherently a nature–society subject that is "biosocial." While not directly engaging with previous political ecology research, these studies help challenge biomedical perspectives that interpret "bodies and disease as nature" best managed through medical intervention; rather, Mansfield argues that natural childbirth should be interpreted as practicing a non-dualistic nature–society relation. In a similar vein, McSweeney and Pearson (2013) examine the ways that population narratives support state interventions into indigenous territories in lowland Latin America and the ways that indigenous groups resist through population expansion. In this case, fertility is employed as a political defense to secure access to land and natural resources by invoking the survival of indigenous peoples. Individual bodies that can be registered, vaccinated, or undergo family planning are therefore the sites of contestation between the state and its peoples as another form of territorial expansion. McSweeney and Pearson conclude that the "health of individuals can be deliberately modified in ways that are *at once* biological (the joining of gamete cells, the introduction of antigens) and social (part of a conscious policy of population expansion, the result of demands for state health support)" (2013: 152, authors' emphasis).

Guthman and Mansfield (2012) draw from epigenetics to detail the ways in which the social environment, in terms of nutrition and psychosocial stressors, triggers outcomes to which individuals are genetically predisposed. By analyzing the effects of external toxins or other environmental conditions in shaping bodily health and resulting responses, environmental epigenetics "highlights the activity of natural, material processes, through which the supposedly external environment actively enters, shapes, and becomes part of the body" (2012: 12). The authors state that this opens up the "black-box of the body" in showing how genetic, molecular, chemical, and physiological factors shape the lived experience of individuals while simultaneously shaping a body that is "always changeable" (2012: 12). Another intervention on the human body comes from Guthman's (2011) study of obesity in the United States. Challenging the obesity crisis and alternative food movement, she questions the hegemony of certain truth claims around obesity that make the condition into something "not all would agree is a problem" (2011: 9). Guthman's study advances a political ecology analysis of obesity that argues for examining the broader political economic and cultural context in which decisions that affect bodily ecologies are made. An example of this is her critical attention to the obesogenic environment thesis that couples obesity and the built environment. Guthman challenges the seemingly tight causal links between these variables and suggests instead that corporate behavior, state regulation, environmental toxins, and broader political economy are more significant in producing particular obesity discourses and bodily outcomes.

Future political ecologies of health: bodies, processes, and change

> From its origins the HIV/AIDS epidemic has been bound-up with far more than merely a battle of organisms and biology. Rather, the epidemic has also been associated with prior cultural understandings of what induces vulnerability to the disease, including a prescribed pathology of certain groups. HIV/AIDS therefore exacerbates existing stigmatisation and exclusion directed at less powerful groups and individuals.
>
> *(Jones 2005: 425)*

Political ecology research within geography, anthropology, and other fields has made important contributions towards understanding the structural determinants of human health with a particular focus upon historical spatial processes, power, social and ecological interactions, and embodiment. Political ecology's multi-scalar approach and combination of qualitative and quantitative methods has been well suited to analyzing disease patterns while situating them within broader political economic and socio-cultural systems. Much of this work has helped expand conceptualizations of health as being more than the absence of disease but also the achievement of dignity and well-being. I wish to conclude by returning to the HIV epidemic that I used to begin this chapter. Examinations of increasing access to ARVs in Southern Africa and elsewhere have underscored the political and pharmacological economy of providing life-saving drugs to those in need. While this is a remarkable turn of events it does not diminish the challenges for those living with the disease, and in fact it has the potential to complicate them in ways that are fundamentally political and ecological. In detailed ethnographic work from Central Mozambique, Ippolytos Andreas Kalofonos (2010) documented the experiences of those on anti-retroviral treatment (ART) and finds that food security is central to infected bodies given the caloric and nutritional needs for those on ARVs. The presence of the disease within the body has taken on a new course, with hunger becoming the embodied experience and central to the day-to-day life needs of infected individuals. Kalofonos details the mechanisms through which those infected access external sources of support, navigating through bureaucratic protocols in attempting to secure food. Individual bodies are regularly observed by others to interpret whether they are receiving medications and also food support. As he comments upon seeing a distribution of food aid by a local non-governmental organization (NGO), "[a]ll noted the bodies of those picking up food: who was fat, who was thin, who was dressed well, who was in rags, thus who deserved and did not deserve the food they received" (2010: 371). This dynamic has been noted elsewhere in Africa, whereby individual bodies are scrutinized to detect HIV's presence or the types of care they might be receiving. The body has become a window into the epidemic and the developmental infrastructure through which people negotiate political economic relationships with the state, international institutions, and each other. In a similar fashion to other work within medical anthropology (cf. Nguyen 2010 and Biehl 2011) this study shows how social relationship and practices are affected while simultaneously being mediated by national and global political and economic systems.

Managing HIV therefore is not simply about accessing desperately needed medications but embodying historical and spatial processes that determine which people have access to the many other things needed for survival. The access and use of these drugs occurs on a daily basis within multi-layered and multi-scalar political economies that are developmental, bureaucratic, and pharmacological. Yet these relationships are not simply political or economic but are also ecological. Access to natural resources, land allocation and tenure systems, and caloric and nutritional demands all intersect with biophysical processes to shape the trajectories of health and well-being. Differential vulnerabilities to HIV infection coupled with political and economic challenges in accessing medications and the full gamut of health needs help evidence the political ecologies of human health. Future political ecology work must continue to interrogate the multifaceted intersections between social and ecological systems in producing disease vulnerabilities and the possibilities for health. Lastly, political ecology's attention to historical processes and contemporary change in producing health outcomes remains critical. As one informant in Kalofonos' study states, many in Mozambique are "positivized" (2010: 373). This resonates with the attempts of the TAC in South Africa to reduce stigmas around the epidemic by promoting the expression "HIV Positive." This succinctly captures the need to speak openly about one's status, in order to recognize that it is possible to be positive about one's future. It also underscores a point made by some political ecology research that human health is never fixed or static; rather it is fluid over time and space. As a result of these dynamics, future political ecology research must approach human health in a holistic manner that continues to attend to its dynamic and embodied nature.

References

Aldhous, P. (2007) The hidden tragedy of Africa's HIV crisis. *New Scientist*, July 11, pp. 6–9.

Baer, H. and Singer, M. (2008) *Global Warming and the Political Ecology of Health: Emerging Crises and Systemic Solutions*. Walnut Creek: Left Coast Press, Inc.

Beaubien, J. (2013) After missteps in HIV Care, South Africa finds it way. *National Public Radio*, August 27.

Bebbington, A. (2000) Reencountering development: Livelihood transitions and place transformations in the Andes. *Annals of the Association of American Geographers* 90(3): 495–520.

Biehl, J. (2011) When people come first: Beyond technical and theoretical quick-fixes in global health. In: Peet, R., Robbins, P. and Watts, M.J. (eds.) *Global Political Ecology*. London: Routledge, 100–130.

Biehler, D.D. and Simon, G.L. (2011) The great indoors: Research frontiers on indoor environments as active political-ecological spaces. *Progress in Human Geography* 35(2): 172–192.

Blaikie, P. (1985) *The Political Economy of Soil Erosion in Developing Countries*. London: Longman.
Blaikie, P. (1999) A review of political ecology: Issues, epistemology and analytical narratives. *Zeitschrift fur Wirtschaftsgeographie* 43(3–4): 131–147.
Blaikie, P. and Brookfield, H. (eds.) (1987) *Land Degradation and Society*. London: Methuen.
Bolton, S. and Talman, A. (2010) Interactions between HIV/AIDS and the environment: A review of the evidence and recommendations for next steps. Nairobi: IUCN ESARO Office.
Braun, B. (2007) Biopolitics and the molecularization of life. *Cultural Geographies* 14: 6–28.
Carter, E.D. (2012) *Enemy in the Blood: Malaria, Environment, and Development in Argentina*. Tuscaloosa: University of Alabama Press.
Crews, K.A. and King, B. (2013) Human health at the nexus of ecologies and politics. In: King, B. and Crews, K.A. (eds.) *Ecologies and Politics of Health*. London: Routledge, 1–11.
Cutchin, M.P. (2007) The need for the "new health geography" in epidemiologic studies of environment and health. *Health and Place* 13(3): 725–742.
de Waal, A. and Whiteside, A. (2003) New variant famine: AIDS and food crisis in Southern Africa. *Lancet* 362(9391): 1234–1237.
Doyal, L. (1979) *The Political Economy of Health*. Boston: South End Press.
Drimie, S. (2003) HIV/Aids and land: Case studies from Kenya, Lesotho and South Africa. *Development Southern Africa* 20(5): 647–658.
The Economist (2011) The end of AIDS? June 2.
Guthman, J. (2011) *Weighing In: Obesity, Food Justice, and the Limits of Capitalism*. Berkeley: University of California Press.
Guthman, J. and Mansfield, B. (2012) The implications of environmental epigenetics: A new direction for geographic inquiry on health, space, and nature-society relations. *Progress in Human Geography* 37(4): 486–504.
Hecht, S. (1985) Environment, development and politics: Capital accumulation and the livestock sector in Eastern Amazonia. *World Development* 13(6): 663–684.
Hunter, L.M., De Souza, R.M. and Twine, W. (2008) The environmental dimensions of the HIV/AIDS pandemic: A call for scholarship and evidence-based intervention. *Population and Environment* 29(3–5): 103–107.
IRIN Humanitarian News and Analysis. (2012) South Africa: Revamped AIDS council makes its debut. Available at: www.irinnews.org/report/96492/south-africa-revamped-aids-council-makes-its-debut.
Jones, P.S. (2005) "A test of governance": Rights-based struggles and the politics of HIV/AIDS policy in South Africa. *Political Geography* 24(4): 419–447.
Kalipeni, E. and Oppong, J. (1998) The refugee crisis in Africa and implications for health and disease: A political ecology approach. *Social Science and Medicine* 46(12): 1637–1653.
Kalofonos, I.A. (2010) "All I eat is ARVs": The paradox of AIDS treatment interventions in central Mozambique. *Medical Anthropology Quarterly* 24(3): 363–380.
Kaschula, S.A. (2008) Wild foods and household food security responses to AIDS: Evidence from South Africa. *Population and Environment* 29(3–5): 162–185.
King, B. (2010) Political ecologies of health. *Progress in Human Geography* 34(1): 38–55.
King, B. (2013) Disease as shock, HIV/AIDS as experience: Coupling social and ecological responses in sub-Saharan Africa. In: King, B. and Crews, K.A. (eds.) *Ecologies and Politics of Health*. London: Routledge, 260–279.
Kosek, J. (2013) The natures of the beast: On the new uses of the honeybee. In: Peet, R., Robbins, P., and Watts, M.J. (eds.) *Global Political Ecology*. London: Routledge, 227–251.
Love, R. (2004) HIV/AIDS in Africa: Links, livelihoods, and legacies. *Review of African Political Economy* 31(102): 639–648.
McCarthy, J. (2002) First world political ecology: Lessons from the Wise Use movement. *Environment and Planning A* 34(7): 1281–1302.
McGarry, D.K. and Shackleton, C.M. (2009) Is HIV/AIDS jeopardizing biodiversity? *Environmental Conservation* 36(1): 5–7.
McSweeney, K. and Pearson, Z. (2013) Vaccines, fertility, and power: The political ecology of indigenous health and well-being in lowland Latin America. In: King, B. and Crews, K.A. (eds.) *Ecologies and Politics of Health*. London: Routledge, 139–158.
Mansfield, B. (2008a) The social nature of natural childbirth. *Social Science & Medicine* 66(5): 1084–1094.
Mansfield, B. (2008b) Health as a nature-society question. *Environment and Planning A* 40(5): 1015–1019.

Masanjala, W. (2007) The poverty-HIV/AIDS nexus in Africa: A livelihood approach. *Social Science & Medicine* 64(5): 1032–1041.
Mayer, J.D. (1996) The political ecology of disease as one new focus for medical geography. *Progress in Human Geography* 20: 441–456.
Mayer, J.D. (2000) Geography, ecology and emerging infectious diseases. *Social Science and Medicine* 50(7): 937–952.
Morgan, L.M. (1987) Dependency theory in the political economy of health: An anthropological critique. *Medical Anthropology Quarterly* 1(2): 131–154.
Neumann, R.P. (1998) *Imposing Wilderness: Struggles over Livelihood and Nature Preservation in Africa.* Berkeley: University of California Press.
Nguyen, V.K. (2010) *The Republic of Therapy: Triage and Sovereignty in West Africa's Time of AIDS.* Durham, NC: Duke University Press.
Oppong, J.R. and Kalipeni, E. (2005) The geography of landmines and implications for health and disease in Africa: A political ecology approach. *Africa Today* 52(1): 3–25.
Ottawa Charter for Health Promotion. (1986) Available at: www.who.int/healthpromotion/conferences/previous/ottawa/en/.
Packard, R.M. (1989) *White Plague, Black Labor: Tuberculosis and the Political Economy of Health and Disease in South Africa.* Berkeley: University of California Press.
Peet, R. and Watts, M. (eds.) (2004) *Liberation Ecologies: Environment, Development, Social Movements*, 2nd edn. London: Routledge.
Richmond, C., Elliott, S.J., Matthews, R. and Elliot, B. (2005) The political ecology of health: Perceptions of environment, economy, health and well-being among 'Namgis First Nation. *Health and Place* 11(4): 349–365.
Robbins, P. (2012) *Political Ecology: A Critical Introduction*, 2nd edn. Malden, MA: John Wiley & Sons Ltd.
Robbins, P. and Miller, J.C. (2013) The mosquito state: How technology, capital, and state practice mediate the ecologies of public health. In: King, B. and Crews, K.A. (eds.) *Ecologies and Politics of Health.* London: Routledge, 196–216.
Scott, C., Robbins, P. and Comrie, A. (2012) The mutual conditioning of humans and pathogens: Implications for integrative geographical scholarship. *Annals of the Association of American Geographers* 102(5): 1–9.
Stonich, S. (1993) *"I am Destroying the Land!" The Political Ecology of Poverty and Environmental Destruction in Honduras.* Boulder: Westview Press.
Sultana, F. (2012) Producing contaminated citizens: Toward a nature-society geography of health and well-being. *Annals of the Association of American Geographers* 102(5): 1165–1172.
Turshen, M. (1977) The political ecology of disease. *Review of Radical Political Economics* 9(1): 45–60.
Turshen, M. (1984) *The Political Ecology of Disease in Tanzania.* New Brunswick: Rutgers University Press.
UNAIDS (2008) *Report on the Global AIDS Epidemic.* Joint United Nations Programme on HIV/AIDS (UNAIDS).
Walker, P.A. and Hurley, P.T. (2011) *Planning Paradise: Politics and Visioning of Land Use in Oregon.* Tucson: University of Arizona Press.
Wines, M. (2006) AIDS cited in the climb in South Africa's death rate. *New York Times*, September 8.
Zimmerer, K.S. (1996) *Changing Fortunes: Biodiversity and Peasant Livelihood in the Peruvian Andes.* Berkeley: University of California Press.
Zimmerer, K.S. and Bassett, T.J. (eds.) (2003) *Political Ecology: An Integrative Approach to Geography and Environment-Development Studies.* New York: Guilford.

27

POLITICAL ECOLOGIES OF ENVIRONMENTAL DEGRADATION AND MARGINALIZATION

Tor A. Benjaminsen

Introduction

"Environmental degradation" and its link to "marginalization" has been a key concern and object of study within political ecology since the inception of this field. According to Robbins (2012), this is one of five recurrent themes (or theses) within the political ecology literature. The focus in this literature has usually been on the rural global South. More recently and increasingly, however, "Northern" environmental issues pertaining to "degradation and marginalization" have also been studied using a political ecology lens.

In this chapter I first discuss Malthusian and Marxist views on environmental degradation and marginalization and how Marxist critiques of Neo-Malthusian reports on "degradation" were prominent in the emergence of political ecology as a scholarly field in the 1980s. While early Marxist contributions to political ecology tended to accept mainstream narratives of environmental degradation associated with peasant production, political ecologists have since the 1990s challenged such narratives. This is the focus of the next section, which examines the "ecology" in political ecology through the presentation of a case study of cash crop cotton production and soil fertility in Mali. The case demonstrates the flaws of both Malthusian and Marxist interpretations of degradation and marginalization. It also shows the importance of an open-ended and empirically based approach to studying this issue where any "taken-for-grantedness" is questioned independently of whether it is inspired by Malthusianism or a critique of capitalism. Despite the flaws in early Marxist contributions to a political ecology of degradation and marginalization, there has recently been a revival of Marxist influence on this theme through other inroads. The last section in this chapter presents how degradation narratives serve to justify elite capture and the dispossession of smallholders as forms of "primitive accumulation" that Marx saw as a historical process of divorcing the producer from the means of production.

Malthus, Marx, degradation and marginalization

According to Malthus (1798), increasing population pressure on natural resources is the paramount cause of human misery. His "population law" postulated that the population growth of the poor inevitably will exceed the resource base represented by food production and cause widespread hunger and poverty. Marx later strongly rejected this idea, which he saw as an attempt to defend class relations under capitalism and blame poverty on the poor themselves (Ross 1998).

From the 1960s, the ideas of Malthus have seen a revival and are frequently used by scholars, policy-makers and the media to explain environmental problems. Neo-Malthusian ecologists such as Paul Ehrlich and Garrett Hardin were among the first leading scientists to use Malthus on the environment. Ehrlich's book *The Population Bomb* (Ehrlich 1968) and Hardin's article "The tragedy of the commons" (Hardin 1968) have in particular been influential. Early contributions to political ecologies of environmental degradation and marginalization consisted essentially of Marxist critiques of such views of population growth as the main factor behind environmental decline. These critiques came as a result of the political influence Neo-Malthusian views had on the international debate.

"Poor people make poor land" was a powerful slogan accepted by both Neo-Malthusian and Marxist inspired scholars alike. This slogan was also repeated by the Brundtland Report (WCED 1987), which used it as a political argument for the need to alleviate poverty in order to arrest environmental degradation. Hence, poverty was framed as the main cause of environmental decline and pollution.

However, while Marxist critics tended to accept the environmental impacts of human production described by the various Neo-Malthusian reports, they pointed at the inherent lack of social and political analysis in these studies and argued that Malthusian thinking would tend to lead to policies of "blaming the victims". In this way, as pointed out by Enzensberger (1974), ecologists and other natural scientists may pretend to be "objective" and "apolitical", but they become political actors when engaging in environmental debates, which are essentially about political choices with winners and losers. The presumed neutrality of ecology as a science is therefore illusory, and what Enzensberger referred to as "political ecology" corresponds to what political ecologists later have called "apolitical ecology" (Robbins 2012).

For instance in the case of the Sahelian famine of the 1970s and 1980s Neo-Malthusian presentations depicted this human disaster as a result of overpopulation and environmental mismanagement causing widespread desertification (e.g. Eckholm and Brown 1977; Eckholm et al. 1984; Timberlake 1985). Such descriptions of alleged serious environmental degradation and its causes have informed the views of Sahelian governments as well as of international aid donors, environmental organizations and the public at large.

In Mali, Moussa Traoré, who was president from 1968 to 1991, became a concerned "environmentalist" with the emerging success of the desertification discourse. To impress donors, the forest law was made even more severe in 1986 with massive fines for cutting branches for fuelwood and even for collecting dry wood (Benjaminsen 1997; Ribot 1999). The new forest law completely banned all bush fires and made wood saving stoves compulsory. This law was the product of a misguided policy that was threatening both rural livelihoods and savannah ecosystems. In fact, villagers in southern Mali practice a burning regime, which creates a mosaic of vegetation patches that prevent damaging late-season fires and increase biodiversity (Laris 2002).

The new forest law and a national plan to fight desertification were clearly influenced by Neo-Malthusian reports on the Sahel, and gave state forest agents a wide mandate to arbitrarily fine or imprison peasants and pastoralists in rural Mali (Benjaminsen 1993, 1997). This was justified by the necessity to fight an advancing desert, a process that was said to be caused by overgrazing, bush fires and cutting of wood for fuel.

Marxist scholars, on the other hand, generally accepted that desertification was taking place, but linked it to expansion of the market economy. Where a Malthusian narrative presents peasants and pastoralists as simultaneously "victims" and "villains", an alternative Marxist narrative sees local communities as victims and identifies as villains various agents of external intervention (e.g. investors and government agents) (Adger et al. 2001).

Franke and Chasin (1980) represented an expression of this view, stating that:

> the evidence from the Sahel famine shows that ecological deterioration and food shortages are not only linked with each other but also structurally related to a specific form of production – international capitalism – and the many secondary effects it produces in even the most marginal and faraway environments.
>
> *(Franke and Chasin 1980: 5)*

In a similar vein, Watts (1983), studying small-scale farming in northern Nigeria, found that commodification caused starvation and economic marginalization among peasants. Increasingly dependent on an unstable market, they became more vulnerable, and had to take up loans and generally take more risks. Previously self-sufficient, peasants gradually became underpaid farm workers. This in turn led to decreasing investments of labor on their own land, resulting in the degradation of soils on land where food crops were grown.

This literature reflects in some way Marx's idea that "all progress in capitalist agriculture is a progress in the art, not only of robbing the worker, but of robbing the soil" (Marx 1990: 638). Hence, market integration and the expansion of capitalism would automatically lead to environmental degradation, a link that was taken for granted and rarely investigated empirically in early Marxist contributions to political ecology. There was a general lack of engagement with ecology; a critique that has later been addressed to political ecology as a whole (Vayda and Walters 1999; Peterson 2000; Walker 2005). As we shall see, however, such critique would be unfair, if generalized to the whole field of political ecology.

Piers Blaikie's book *The Political Economy of Soil Erosion in Developing Countries* (Blaikie 1985) paved the way for an approach to political ecology that engages explicitly with natural science. The book was a critique of environmental conservation policies in the global South and presented three central arguments (Neumann 2008): First, there is often lack of sound scientific data on soil erosion and other environmental processes, which leads to a high level of uncertainty. Second, actors perceive environmental change differently depending on their "ideology". Blaikie argued "that *all* approaches to soil erosion and conservation are ideological – they are underpinned by a definite set of assumptions, both normative and empirical, about social change" (Blaikie 1985: 149). Third, environmental policies are about control over and rights to land and natural resources. A critical question that political ecology asks, therefore, is who wins and who loses from resource and conservation policies? This again leads us to study "where power lies and how it is used" (Blaikie 1985: 6).

Consequently Piers Blaikie as a social scientist proposed an approach to understanding environmental problems by, on the one hand, problematizing the quality and uncertainty of scientific data and, on the other hand, insisting that the production, interpretation and use of environmental data are inherently political. A process such as soil erosion could therefore only be fully understood with the help of the tools of political economy (Rigg 2006; Simon 2008).

These ideas were subsequently further discussed in Blaikie's next book, *Land Degradation and Society* (Blaikie and Brookfield 1987), where "land degradation" was presented as a perceptual term implying that environmental change will be perceived differently by the various actors involved. Hence, "degradation" is not a straightforward physical process that can be measured with natural science methods. Whether processes such as deforestation or soil erosion are perceived as "degradation" depends on the position of individual observers. Conservationists would tend to see loss of woody cover as "degradation", while the farmer who extends his field by clearing the bush will consider this to be an improvement of the land. Our understanding and interpretation of environmental change are thus guided by our norms, interests and values. However, while environmental data are constructed and subject to ideological interpretations, Blaikie and Brookfield still insisted on the necessity of improving techniques of measurement in order to obtain "those data which are beset with *least* uncertainty" (Blaikie and Brookfield 1987: 16).

Furthermore, Blaikie and Brookfield (1987) attempted to introduce a new understanding of "margin", "marginality" and "marginalization". Until then, a political economy-based approach to "marginalization" had been implicit in the few available studies in political ecology. This involves a focus on exclusion from access to land, natural resources, employment and decision-making. Blaikie and Brookfield combined, however, this perspective with two others; an economic concept of margin following Ricardo's classical theory of rent (the margin being a resource unit that produces its own costs and no more) and an ecological concept of margin ("where natural conditions will just permit the plant to survive" (Blaikie and Brookfield 1987: 20)). The effort to combine these three perspectives to create a new political ecological concept of marginality and marginalization has, however, been less successful than some of Blaikie and Brookfield's other contributions. The main reason seems to be that there is often an empirical mismatch between these three concepts. Or to put it in another way, Blaikie and Brookfield's idea that processes of degradation and marginalization overlap is often not supported by empirical evidence. As with the "poor people make poor land" slogan, this idea fits both a Neo-Malthusian and a Marxist agenda, although in different ways. As we shall see, it has later been challenged by empirical political ecology studies.

Blaikie and Brookfield (1987) also introduced a methodology that they called *chains of explanation*, which aimed to identify causes of environmental degradation. First, one would attempt to find out how the environment is changing. Is there deforestation, soil erosion, overgrazing, loss of soil nutrients or loss of biological diversity? Often there will be scientific studies available of ecological change that could be used as a starting point, so there could be cooperation with natural scientists. Or political ecologists might carry out their own investigations using aerial photos, satellite images, repeat landscape photography, data on agricultural and livestock production, or other ways to help measure or describe environmental change.

The next step is to investigate the rationality of the local land manager. This can be the individuals, households or institutions that directly use and manage natural resources. One would study how the use of land and natural resources takes place in practice. Who does what, and how are the resources used? What are the reasons for people using the resources in a particular way?

Thereafter the external influences on land management would be brought into the analysis. These influences can take the form of national laws, policies and bureaucratic practices as well as global forces (international conventions, discourses, funding agencies, investors).

More recently, some political ecologists have tried to move away from thinking in terms of causal chains running one-way from higher to lower geographical levels to focus on networks and webs as better frameworks for understanding complex interrelationships (e.g. Rocheleau 2008).

The "ecology" in political ecology

So, to what extent do poor people make poor land or marginalized peasants degrade their own environments? Blaikie and Brookfield (1987) insisted that explanatory models, also called "single-factor explanations", whether focusing on population pressure (Malthus), marginalization (Marx) or the Tragedy of the Commons (Hardin) have only partial validity when confronted with good empirical data. Therefore, it is important to carry out case studies in order to test these theories. From the late 1980s, a number of students and scholars who were inspired by the research agenda proposed by Blaikie (1985) and Blaikie and Brookfield (1987) have carried out empirical studies in the global South unpacking the "ecology" in the political ecology equation.

This approach to political ecology has meant extending the focus on peasant rationality and agency within peasant studies (e.g. Scott 1976, 1985) and cultural ecology (e.g. Netting 1993) to environmental dynamics. Many of these studies have focused on Africa and generated new knowledge and critiques of environmental orthodoxies in several fields.

The seminal contribution by Leach and Mearns (1996) was a collection of key critical contributions on various environmental issues in Africa (e.g. range ecology, desertification, deforestation, biodiversity conservation and soil erosion). A series of chapters challenged received wisdom on these issues and reflected a broader literature that had emerged during the late 1980s and early 1990s. Henceforth, a large number of case studies from different parts of the African continent and on various environmental issues have continued to question dominant (often Neo-Malthusian) narratives on environmental degradation through carefully collected environmental data.

On other continents, comparable studies on environmental degradation have been carried out. For instance, Forsyth (1996) studied deforestation in the Himalayas and questioned Neo-Malthusian explanations of this process; Robbins (2001) explored how state-sponsored modernization efforts produced various forms of unexpected land-cover change in Rajastan in India; and Zimmerer (1994) demonstrates the potential contribution of local soil knowledge and management to sustainable land management in highland Bolivia. These are just some examples of research in political ecology that actively have engaged with natural science and ecology. Most of this literature explicitly questions Neo-Malthusian positions. Some of it also implicitly qualifies or undermines a Marxist approach to degradation and marginalization. This is further discussed in the following case study on cash-crop cotton cultivation in Mali and its impact on soil fertility (Benjaminsen et al. 2010).

Cotton and soil fertility in Mali

Production of cotton as a cash crop in the Malian cotton zone has seen a rapid expansion during the last few decades. This has led cotton to be labeled "white gold" and the "motor of development", the development of cotton production as "Mali's white revolution", and the Malian cotton zone as a "success story".

The flip side of this relative economic success, however, is said to be environmental degradation including serious loss of soil fertility. But there has been little data available on soil fertility change in the region, despite its long involvement with international conservation and rural development projects.

Soils, land use and tenure vary along a landscape gradient. Close to village settlements there are manured and intensively cultivated gardens fenced with branches where different sorts of vegetables are cultivated. Outside these gardens and fields the areas with the big fields are located. Normally there is a continuous band of fields, which is almost permanently cultivated. Beyond these big fields, there is "the bush". It is often located at the upper parts of the slope. Extension of the cultivated area is taking place on this more marginal land, since the big fields are more intensively and permanently cultivated. Cotton is grown in rotation with cereals such as maize, sorghum and millet in alternate years. This land management model is common in farming communities in West Africa and involves the transfer of nutrients in the form of manure from the bush where livestock graze to the village fields where dung is used to fertilize soils.

In the cotton zone, large quantities of chemical fertilizers are also used. Until the beginning of the 1980s, these fertilizers were subsidized in Mali, which resulted in increasing harvests. As part of the Structural Adjustment Program, started in 1982, the government was, however, forced to phase out these subsidies. Since then production has continued to increase, while the yields (production per unit area) have stagnated. Instead, the farmers have compensated for the decline in the use of fertilizer per unit area by expanding the cultivated areas. Hence, this is a clear example of how the removal of a subsidy led to extensification and increased deforestation. But while yields leveled off, total production continued to increase. The total production of cotton increased from 150 tons in 1952 via 3,900 tons in 1958 to 68,000 tons in 1972 and 593,000 tons in 2003. Thereafter, production has varied, mainly due to fluctuating international cotton prices.

Food production in the cotton zone has also increased substantially in the same period. Paradoxically, the cotton zone exports food to other parts of the country and to neighboring countries. This is because the implementation of cotton production for sale gave the farmers capital to buy fertilizers. When they switch between cotton and millet or maize every other year, the food crops will also be able to enjoy the previous year's use of fertilizers.

The other side of the coin is, however, said to be large environmental problems. Cotton is generally a crop that requires large quantities of pesticides, and this may have health consequences for the farmers who do the spraying, and who rarely use protective equipment. Little is also known about the wider consequences of spraying on ecosystems.

In addition, the exhaustion of nutrients in the soil is an environmental problem often mentioned in connection with cotton production. Such nutrient budgets are a common method of studying soil fertility change. However, this method has certain weaknesses. Typically, data for such budgets are collected at the field, farm or village level and extrapolated to whole regions or countries. Results from such nutrient budgets tend to depend on the spatial scale used – the budgets being increasingly negative as the spatial scale increases from the farm to the country level. Aggregated nutrient budgets therefore tend not to be able to capture local dynamics in time and space.

Another criticism raised against studies of nutrient budgets in Mali is related to the quality of the output data of nutrients (erosion, leaching and volatilization). These estimates are not based on measurements in Mali, but rather on secondary data from other countries in West Africa. The uncertainty related to extrapolation of data to Malian conditions makes it difficult to give reliable estimates for nutrient budgets.

Due to the lack of a comparative baseline dataset, the study by Benjaminsen et al. (2010) had to settle for the spatial analogue method. It consists of comparing soil properties on different sites with varying histories of land use. The main independent variable used is length of fallow. The study did not focus on the effects of cotton production on soil fertility per se, but rather on the varying intensity of the cotton–cereal rotation and how this intensity impacts on soil fertility. Soil data from cultivated fields were compared with data from fallows of different lengths. In addition, samples were taken on woodland as well as in sacred groves (with permission from the village chief). The woodlands and sacred groves had never been cultivated, but the woodland sites had been grazed. It should also be noted that the sacred groves are usually centrally situated in the villages on some of the best soils. They therefore represent very valuable reference sites being as close to a natural system as one can get in terms of soil fertility. The strength of this method is its ability to account for spatial effects of cultivation. Soil fertility is typically higher in the vicinity of the homestead and decreases as the distance from the homestead increases. The soil samples taken from woodlands are distant from the homestead whereas samples taken from cultivated fields and fields under fallow represent soils that are more adjacent to the homestead.

All in all 273 soil samples were collected in 19 villages from north to south in the cotton zone. The villages were selected to cover diversities in agricultural development in the region. The 19 villages were grouped into five different "gradients", from 1 to 5, with 1 having the highest land use intensity and 5 the lowest.

In addition, soil samples were taken from land that had been continually cultivated for 5 years, land that had been fallow for 5 years, 8–10 years and 15–20 years, forest soil that had never been cultivated and soil from the sacred groves.

Not surprisingly, the results showed that the highest content of carbon, nitrogen, phosphorus, potassium and magnesium were found in the sacred groves. Beyond this, there was no clear tendency in the relationship between the intensity of cultivation and nutrients in the soil. Therefore, the study did not find support for the idea that smallholder cotton production in Mali results in exhaustion of nutrients in the soil. Soil studies in more subsistence-oriented production systems in other parts of Africa have come to similar conclusions (for example, Prudencio 1993; Ramisch 2005; Tittonell et al. 2007).

These findings support an open-ended and empirically based political ecology. In studies of environmental change, Blaikie and Brookfield's "chain of explanation" may still provide a useful point of departure with its step-wise process starting with an understanding of the actual environmental change taking place. Such an approach would also contribute to bringing ecology and natural sciences back into political ecology, if the ambition is to link social and natural sciences to address environmental change, and conflicts.

Without grounding studies of environmental change and associated conflicts in detailed natural science-based investigations leading to an understanding of the actual changes taking place, the end result will easily be structural determinism, whether inspired by Malthusianism or by a critique of capitalism. According to Forsyth and Walker (2008), conventional approaches to political ecology have *a priori* linked environmental degradation to the impacts of global capitalism. While economic growth may evidently degrade environments, such links should be investigated and not taken for granted by political ecologists. However

> the aims of questioning the role of capitalism are not to suggest that political ecologists should not be concerned about exploitation of people and resources, but to ask how the opposition to capitalism may have influenced the production of environmental explanations.
>
> *(Forsyth 2003: 7)*

This argument is also compatible with the conclusions from Tom Bassett's detailed historical account of cotton production in northern Ivory Coast. While not studying soil fertility, he focused on the agency of smallholder cotton producers – and his findings "stood in contrast to the passive image of peasants entrapped in exploitative production relations with foreign capital and African states" (Bassett 2001: 22).

Hence, a critical political ecology would imply a critical approach to all environmental narratives combined with a realist belief in science as a means to achieve a more accurate description and understanding of environmental realities.

Degradation narratives and marginalization

While Marx's idea about "the robbing of soils" in capitalist agriculture should be critically investigated like any "taken-for-grantedness", and while there is not an obvious connection between environmental degradation and political-economic marginalization, there has recently, through other inroads, been a revival of Marxist thinking in political ecology. Instead of marginalization necessarily leading to degradation, political ecology studies have documented how degradation narratives serve to justify elite capture and smallholders' dispossession of land and natural resources. The result may be seen as another example of "primitive accumulation", which Marx saw as a historical process of divorcing the producer from the means of production.

According to Harvey (2003: 149), "primitive accumulation as Marx described it ... entailed taking land, say, enclosing it, and expelling a resident population to create a landless proletariat, and then releasing the land into the privatized mainstream of capital accumulation". Since accumulation is an ongoing process, Harvey (2003) proposes the term "accumulation by dispossession" to describe current processes. The introduction of this term has sparked a renewed interest in the combination of dispossession and capital accumulation in development studies and in political ecology in particular (Glassman 2006; Sneddon 2007; Roberts 2008; Shivji 2009, Büscher 2009; Li 2009; Corson 2011; Kelly 2011; Benjaminsen and Bryceson 2012; Veuthey and Gerber 2012).

For instance, Benjaminsen and Bryceson (2012) use the lens of accumulation by dispossession to analyze enclosures in wildlife and coastal conservation in Tanzania. They show how recently established conservation initiatives steadily lead to local people's loss of access to land and natural resources. Dispossession has been gradual and piecemeal in some cases, while it involved violence in other cases, but does not primarily take the usual form of privatization of land. The spaces involved are still formally state or village land. It is rather the benefits from the land and natural resources that contribute to capital accumulation by more powerful actors (rent-seeking state officials, transnational conservation organizations, tourism companies and the state Treasury). In both wildlife and coastal management, restrictions on local resource use are justified by degradation narratives, while financial benefits from tourism are drained from local communities within a system lacking in transparent information sharing.

The violent evictions of pastoralists from the Usangu Plains in 2006 constitutes another example of how accusations of "degradation" play a key role to justify dispossession. Despite little research and poor empirical evidence supporting the idea of "overgrazing", the evictions in Usangu were justified on the grounds that livestock grazing in the catchment area of the Mtera Dam was the main cause of the water shortage and associated frequent power cuts. A full-scale military operation was launched on May 18, 2006 to evict pastoralists from Usangu using heavily armed police, anti-poaching units and game wardens. Hundreds of pastoralists with 300,000 cattle were cleared from the area (Benjaminsen et al. 2009).

Such processes of accumulation by dispossession in Tanzania have been made possible through the complex interaction of three different forces; a corrupt neopatrimonial state, international conservation interests working through big international NGOs that argue for environmental enclosures justified by degradation narratives, and neoliberal reforms implemented in Tanzania since the 1980s leading to increased commodification of nature (Benjaminsen et al. 2013).

Moving from the Equator to north of the Arctic Circle, Sámi reindeer pastoralism in Norway may serve as a parallel example from the Global North where a powerful degradation narrative contributes to the continued marginalization of a vulnerable group (Benjaminsen et al. 2015). Especially in the Finnmark region in the far north of the country, a narrative of overstocking and pasture degradation dominates public debates and policy implementation. The region has 150,000–200,000 reindeer and around 2800 reindeer owners among the Sámi ethnic minority. Pastoral production is based on seasonal migrations that take place between the winter pastures of the Finnmark plateau, which are dominated by lichen, and the green summer pastures at the coast.

The dominant narrative on reindeer herding says that the winter pastures in Finnmark are subject to an ecological crisis due to general overstocking and mismanagement. Reindeer numbers are said to exceed a carrying capacity set by the reindeer management authority. Overstocking is seen as the principal factor that causes a series of problems including ecological deterioration, economic inefficiency leading to declining carcass weights, and land-use conflicts. This narrative is shared by politicians, the Ministries of Food & Agriculture and of the Environment, environmental NGOs, most researchers and the media. It supports policies that have been in place for decades and which focus on reducing and stabilizing reindeer numbers as the main measure to avoid overstocking and achieve sustainability. While the policy has been resisted by reindeer herders through various "weapons of the weak" (Scott 1985), their own narrative resonates with non-equilibrium ecology – an approach to ecological thinking that was central to the development of political ecology (Turner 2013). The policy, on the other hand, seems to be based on the idea of establishing a system in equilibrium within a highly unstable Arctic environment.

While Norwegian biologists continue to reproduce the degradation narrative in public debates and in reports to the government – and continue to receive funding for more studies that confirm previous conclusions – they report in academic journals about "rapid recovery" of the winter pastures in Finnmark (Tømmervik et al. 2012). This recovery even took place in a period with rising numbers of reindeer. Hence, the question of "overgrazing" is not that clear-cut when confronted with empirical data.

The degradation narrative surrounding reindeer herding, which is based on a questionable empirical foundation, serves to reduce the intensity of herding as a prevailing land use. The narrative and the policy it results in are interpreted by herders as consequences of competition for land between herding and farming, tourism (ski resorts, holiday homes), mining, wind turbine development, and infrastructure (power lines, roads). These competing land-uses have more powerful advocates and spokespersons among members of the Norwegian society and parliament in particular. This has led to a steady encroachment on pastoral land for over a century.

Conclusions

Environmental degradation and the marginalization of social groups remain key themes in political ecology. The broad idea that "poor people make poor land" was shared by Neo-Malthusian and Marxist inspired accounts, while having different explanations as to why people were poor and marginalized. However, while Marxist critics of Neo-Malthusian studies tended to accept the environmental conclusions coming out of these reports, they pointed at the inherent lack of social and political analysis in these studies and argued that Malthusian thinking would lead to policies of "blaming the victims".

Since the 1990s, political ecologists have challenged the idea of a close link between degradation and marginalization in a series of empirical studies, mostly based on case studies from Africa and focusing on issues such as desertification, soil management and deforestation. These have shown that people who are economically and politically marginalized, even if they are integrated into the market, do not necessarily destroy nature. The link between degradation and marginalization remains an empirical question that should be investigated in critical political ecology studies, and not taken for granted.

More recently, a growing literature within political ecology has demonstrated how degradation narratives are used as tools in struggles over land and natural resources, which often lead to further marginalization of vulnerable groups. We see this in particular within attempts at arresting alleged degradation through the establishment of various types of environmental enclosures. Many of these environmental initiatives have recently been criticized in the political ecology literature for resulting in "accumulation by dispossession".

While "degradation and marginalization" is perhaps the first and "original" theme within political ecology that to a large extent initiated the emergence of this field, much remains to be done in terms of empirical research. Detailed case studies of environmental and land-cover change continue to be important to carry out, but political ecologists studying this theme could also to a larger extent pursue the ambition to link local environmental change with global political economic processes. This could for instance be done through studying environmental change as an integrated part of commodity chain analyses.

For instance, farmed salmon is a huge and rapidly expanding success in coastal Norway. About 10 percent of the feed consumed by Norwegian salmon comes from Brazilian soy, which may again have a direct or secondary effect on deforestation in the Amazon. Tracing the soy consumed in Norway back to its regional and local origin in Brazil and studying in detail the dynamics of the local soy/forest interface and possible impacts on local livelihoods would be a good example of political ecology coupled with commodity chain analysis. Such analyses would more explicitly investigate possible linkages between "Southern" environmental change and "Northern" political and economic interests. In addition, there is much unused potential in exploring such issues in a Northern context, both in rural and urban areas.

References

Adger, W.N., T.A. Benjaminsen, K. Brown and H. Svarstad. (2001) "Advancing a political ecology of global environmental discourses". *Development and Change* 32 (4): 681–715.

Bassett, T.J. (2001) *The Peasant Cotton Revolution in West Africa. Côte d'Ivoire, 1880–1995.* Cambridge: Cambridge University Press.

Benjaminsen, T.A. (1993) "Fuelwood and desertification: Sahel orthodoxies discussed on the basis of field data from the Gourma region in Mali". *Geoforum* 24 (4): 397–409.

Benjaminsen, T.A. (1997) "Natural resource management, paradigm shifts, and the decentralization reform in Mali". *Human Ecology* 25 (1): 121–143.

Benjaminsen, T.A. and I. Bryceson. (2012) "Conservation, green/blue grabbing and accumulation by dispossession in Tanzania". *Journal of Peasant Studies* 39 (2): 335–355.

Benjaminsen, T.A., J. Aune and D. Sidibé. (2010) "A critical political ecology of cotton and soil fertility in Mali". Geoforum 41: 647–656.

Benjaminsen, T.A., F.P. Maganga and J.M. Abdallah. (2009) "The Kilosa killings: political ecology of a farmer–herder conflict in Tanzania". *Development and Change* 40 (3): 423–445.

Benjaminsen, T.A., M. Goldman, M. Minwary and F. Maganga. (2013) "Wildlife management in Tanzania: state control, rent seeking and community resistance". *Development and Change* 44 (5): 1087–1109.

Benjaminsen, T.A., H. Reinert, E. Sjaastad and M.N. Sara. (2015) "Misreading the Arctic landscape: a political ecology of reindeer, carrying capacities and 'overstocking' in Finnmark, Norway". *Norwegian Journal of Geography*.

Blaikie, P. (1985) *The Political Economy of Soil Erosion in Developing Countries*. New York: Longman.

Blaikie, P. and H. Brookfield (eds.). (1987) *Land Degradation and Society*. London and New York: Methuen.

Büscher, B. (2009) "Letters of gold: enabling primitive accumulation through neoliberal conservation". *Human Geography* 2 (3): 91–94.

Corson, C. (2011) "Territorialization, enclosure and neoliberalism: non-state influence in struggles over Madagascar's forests". *Journal of Peasant Studies* 38 (4): 703–726.

Eckhom, E. and L. Brown. (1977) Spreading deserts – the hand of Man. Worldwatch Paper 13. Washington, DC.

Eckholm, E., G. Foley, G. Barnard and L. Timberlake. (1984) *Firewood: The Energy Crisis that won't go Away*. London: Earthscan.

Ehrlich, P. (1968) *The Population Bomb*. New York: Ballantine Books.

Enzensberger, H.M. (1974) "A critique of political ecology". *New Left Review* 8: 3–32.

Forsyth, T. (1996) "Science, myth and knowledge: testing Himalayan environmental degradation in Thailand". *Geoforum* 27 (3): 375–392.

Forsyth, T. (2003) *Critical Political Ecology. The Politics of Environmental Science*. London: Routledge.

Forsyth, T. and A. Walker. (2008) *Forest Guardians, Forest Destroyers. The politics of environmental knowledge in northern Thailand*. Seattle: University of Washington Press.

Franke, R.W. and B.H. Chasin. (1980) *Seeds of Famine. Ecological Destruction and the Development Dilemma in the West African Sahel*. Montclair: Allanheld, Osmun & Co.

Glassman, J. (2006) "Primitive accumulation, accumulation by dispossession, accumulation by 'extra-economic' means". *Progress in Human Geography* 30 (5): 608–625.

Hardin, G. (1968) "The tragedy of the commons". *Science* 162: 1243–1248.

Harvey, D. (2003) *The New Imperialism*. Oxford: Oxford University Press.

Kelly, A.B. (2011) "Conservation practice as primitive accumulation". *Journal of Peasant Studies* 38 (4): 683–701.

Laris, P. (2002) "Burning the seasonal mosaic: preventing burning strategies in the wooded savanna of southern Mali". *Human Ecology* 30 (2): 155–186.

Leach, M. and R. Mearns (eds.). (1996) *The Lie of the Land: Challenging Received Wisdom on the African Environment*. Oxford: James Currey.

Li, T. Murray. (2009) "To make live or die? Rural dispossession and the protection of surplus populations". *Antipode* 41 (1): 66–93.

Marx, K. (1990) *Capital: A Critique of Political Economy*, Volume 1. New York: Penguin.

Malthus, T.R. (1798) *An Essay on the Principle of Population*. London.

Netting, R.McC. (1993) *Smallholders, Householders: Farm Families and the Ecology of Intensive, Sustainable Agriculture*. Stanford: Stanford University Press.

Neumann, R.P. (2008) "Probing the (in)compatibilities of social theory and policy relevance in Piers Blaikie's political ecology". *Geoforum* 39 (2): 708–715.

Peterson, G. (2000) "Political ecology and ecological resilience: an integration of human and ecological dynamics". *Ecological Economics* 35: 323–336.

Prudencio, C.Y. (1993) "Ring managment of soil and crops in the West African semi-arid tropics. The case of the Mossi farming system in Burkina Faso". *Agriculture, Ecosystems and Environment* 47: 237–264.

Ramisch, J.J. (2005) "Inequality, agro-pastoral exchanges, and soil fertility gradients in southern Mali". *Agriculture, Ecosystems and Environment* 105: 353–372.

Ribot, J. (1999) "A history of fear: imagining deforestation in the West African Sahel". *Global Ecology and Biogeography* 8: 1–10.

Rigg, J. (2006) "Piers Blaikie (1942–)". In Simon, D. (ed.) *Fifty Key Thinkers on Development*. London: Routledge, pp. 35–39.

Robbins, P. (2001) "Tracking invasive land covers in India or why our landscapes have never been modern. *Annals of the Association of American Geographers* 91 (4): 637–654.

Robbins, P. (2012) *Political Ecology*. Oxford: Blackwell.

Roberts, A. (2008) "Privatizing social reproduction: the primitive accumulation of water in an era of neoliberalism". *Antipode* 40 (4): 535–560.

Rocheleau, D. E. (2008) "Political ecology in the key of policy: from chains of explanation to webs of relation". *Geoforum* 39 (2): 716–727.

Ross, E.B. (1998) *The Malthus Factor: Poverty, Politics and Population in Capitalist Development*. London: Zed Books.

Scott, J.C. (1976) *The Moral Economy of the Peasants: Rebellion and Subsistence in Southeast Asia*. New Haven: Yale University Press.

Scott, J.C. (1985) *The Weapons of the Weak: Everyday Forms of Peasant Resistance*. New Haven: Yale University Press.

Shivji, I. (2009) *Accumulation in an African Periphery. A theoretical framework*. Dar es Salaam: Mkuki na Nyota Publishers Ltd.

Simon, D. (2008) "Political ecology and development: intersections, explorations and challenges arising from the work of Piers Blaikie". *Geoforum* 39 (2): 698–707.

Sneddon, C. (2007) "Nature's materiality and the circuitous paths of accumulation: dispossession of freshwater fisheries in Cambodia". *Antipode* 39 (1): 167–193.

Timberlake, L. (1985) *Africa in Crisis: The Causes and Cures of Environmental Bankruptcy*. London: Earthscan.

Tittonell, P., B. Vanlauwe, N. de Ridder and K.E. Giller. (2007) "Heterogeneity of crop productivity and resource use efficiency within smallholder Kenyan farms: soil fertility gradients or management of intensity gradients." *Agricultural Systems* 94: 376–390.

Turner, M. (2013) "Political ecology I: an alliance with resilience?" *Progress in Human Geography* 38 (4): 616–623.

Tømmervik, H., J.W. Bjerke, E. Gaare, B. Johansen and D. Thannheiser. (2012) "Rapid recovery of recently overexploited winter grazing pastures for reindeer in northern Norway". *Fungal Ecology* 5: 3–15.

Vayda, A.P. and B. Walters. (1999) "Against political ecology". *Human Ecology* 27 (1): 167–179.

Veuthey, S. and J.-F. Gerber. (2012) "Accumulation by dispossession in coastal Ecuador: shrimp farming, local resistance and the gender structure of mobilizations". *Global Environmental Change* 22 (3): 611–622.

Walker, P.A. (2005) "Political ecology: where is the ecology?" *Progress in Human Geography* 29 (1): 73–82.

Watts, M. (1983) *Silent Violence: Food, Famine and Peasantry in Northern Nigeria*. Berkeley: University of California Press.

World Commission on Environment and Development (WCED). (1987) *Our Common Future*. Oxford: Oxford University Press.

Zimmerer, K. (1994) "Local soil knowledge: answering basic questions in highland Bolivia". *Journal of Soil and Water Conservation* 49 (1): 29–34.

28

INDUSTRIALIZATION AND ENVIRONMENTAL CHANGE

Stefania Barca and Gavin Bridge

Industrialization is one of the great markers for periodizing socio-ecological relations. It describes the second major ecological transition in the history of humankind, the first being the shift from hunting and gathering to agriculture that occurred in the Neolithic (Cipolla 1973; Steinberg 1986; Wrigley 1988). Industrialization is credited with driving the emergence of the 'Anthropocene', a term which likens the past 250 years to a geological era characterized by the ability of humans to alter geo-ecological processes on a global scale, and exemplified by the anthropogenic forcing of atmospheric CO_2 from 280 to 400 ppm (Steffen et al. 2011). Such readings of industrialization, however, highlight material transformations at the world-scale, and overlook equally significant processes of geographical and social differentiation. Since its inception in the second half of the eighteenth century in the English countryside, industrialization has been a continuous process of socio-ecological transformation and differentiation that has simultaneously pulverized and re-constituted social and geographical relations across all spatial scales. By accelerating the appropriation of raw materials and generation of wastes (Muradian et al. 2012), commodifying labour and land, and remaking human bodies through the consumption of industrial goods (Chapter 43, this volume) and exposure to radioactive, metal and persistent organic pollutants (Boudia and Jas 2014, Higgins 2010, Harremoes et al. 2002), industrialization may be considered a form of 'slow violence' (Nixon 2011) that both reproduces and transforms inequalities in economic and political power. Given industrialization's wrenching socio-ecological transformations – and the complex narratives, norms and subjectivities that variously sustain and query the social relations of which it is productive – it is somewhat surprising that political ecology has paid industry only limited attention. For the most part, political ecology has approached industrialization obliquely, and one needs to look to the field's fluid borderlands with environmental history, environmental sociology and ecological economics to find a more sustained engagement.

This chapter considers political ecology's limited engagement with industrialization. We suggest that although the field's foundational interest in livelihoods and modes of production has skewed strongly agricultural, some of political ecology's conceptual resources may be reworked to consider industry and industrialization. We share with recent writing in political ecology a wariness of the Anthropocene on account of its 'fetish of industrialization' (Moore 2014: 13) and attention to temporal differentiation rather than the production of socio-spatial difference. However, we make the case for a political ecology focused on the distinctiveness of industrialization's socio-ecological relations: the appropriation of ecological surplus in the form of inanimate energy, rapid growth in the technical composition of capital, and the geographical expansion and temporal acceleration of social metabolism that has sustained the growth of labour productivity over time. This framework, we suggest, can illuminate the political ecological significance of industrial labour regimes in wringing value from biophysical systems, and the environmental inequalities, ecological risks, and distribution conflicts consequent to an industrial mode of production (Martínez-Alier 2002). We conclude that a focus on industry and industrialization need not detract from understanding longer-term dynamics of the capitalist production of nature, and highlight the opportunities of a more thoroughly industrial political ecology.

The power and the machine: industrial capitalism

One of the core strands around which political ecology has evolved as a field is an interest in the social relations, technical practices and ecological conditions of conservation, agriculture and extractive industries. This interest has, in the main, led researchers to focus on struggles among different social groups for access to – or control of – the capacity of specific parcels of land to produce environmental goods and services. For the most part, this interest in understanding how forces, relations and conditions of production combine to appropriate surplus in ways that sustain economic and political power has not extended to the manufacturing sector (although see Walker et al. 1979 for a nascent political ecology's interest in the chemical industry). Yet processes of mechanization, the concentration of capital, and the socio-spatial distribution of economic surplus and pollution are no less significant in manufacturing than they are in primary production. Indeed, the amassing of capital and the fundamentally entropic character of this sector suggest it is arguably the primary driver of social and spatial differentiation across a range of scales (Bunker 1985; Hornborg 2006). This section outlines the significance of industry as a mode of production, focusing on the conjoined effects of mechanization, inanimate power and the elevation of labor productivity as a 'metric of wealth' (Moore 2014: 20).

At its core, the term 'industrialization' indicates a shift from manual to mechanized forms of production in all or most phases of the labor process. A consequence is a growing mass of machinery relative to labor time, what in Marxist terms is referred to as a rise in the technical composition of capital. However, this quite specific meaning has become indistinguishable from broader social meanings of the term. Since its early uses in the English language between the 1790s and 1830s, the word 'industrialism' has been associated with the idea of a new social order, a revolution 'based on organized mechanical production' (Williams 1983: 166–167). Here, two of the most pregnant meanings of the word power – in its social sense of command over people and in the physical sense of ability to carry out work – came to converge, leading to a new mode of production: industrial capitalism. Such convergence of meaning was the result of an underlying shift from manual to mechanized production, which first emerged in the industrial mills of England where a formerly autonomous and physically dispersed workforce came to be concentrated and disciplined in the factory system.

Contrary to what is often assumed, however, the core of the industrialization process is not the machine per se, but the energy source which moves it: what distinguishes industrial production from other manufacturing systems is that 'inanimate' power, not humans or animals, moves the machine (Wrigley 1988). Even if the use of inanimate energy pre-existed in a variety of activities (the most important being the grinding mill), a series of technical improvements allowed such automation to spread over the entire labor process rather than being isolated to a few phases, thus starting the era of mechanized mass production. In short, industrialization is synonymous with the use of non-living energy in the production of commodities.

Mechanization, however, does not imply that machines do the entire job that was previously done by human labor. Rather, the latter becomes the living component of a mechanized and automated process, made up of inanimate power sources, complex mechanical clusters, and organizational schemes aimed at regulating the input-output flow (energy and materials, labor, commercial products and waste). The industrial workplace can thus be seen as a peculiar kind of ecological system (McEvoy 1995), made up of biological processes (workers' bodies), thermodynamic properties (power sources and machines) and social regulation (engineering, labor relations, law). The history of the industrial workplace, however, shows that such particular ecologies have a tendency to become highly politicized and produce outcomes such as social unrest, reform or revolution (Barca 2014; Santiago 2006).

Industrialization was first set in motion by a peculiar form of inanimate energy: waterpower. It was water and not coal that moved the wheels of the textile mills of Yorkshire, Lancashire and Derbyshire in England in the late eighteenth century – and that first showed how much more money could be made out of mechanized production, due to its unprecedented ability to intensify the labor process (Malm 2014). Industrial capitalism was born and raised in the river valleys of Europe and North America, where the fundamental elements of a new system of ecological relations were first put into place. Those elements can be summarized as: (1) the appropriation of water as a form of 'natural capital' for the extraction of mechanical energy (waterpower) and the production of exchange value; (2) the mechanization of labor; and (3) the production of a new landscape – the early industrial riverscape – with its peculiar narrative and representation (Barca 2010).

With the later shift from waterpower to steam, however, industrial capitalism appropriated a far larger ecological surplus and set in motion a widespread experimentation with a third, crucial meaning of the word 'power': the thermodynamic property of concentrated energy which, while employed to move machines, is then dispersed and lost forever. Identified by French engineer Sadi Carnot in 1833 during his studies of the steam engine, the second law of thermodynamics – also known as the entropy law – is understood as a (meta)physical limit on the industrial economy (Daly 1991; Georgescu-Roegen 1976). Profoundly reconfiguring the organization of living and non-living matter in the biosphere, and the chemical composition of the atmosphere, industrialization has acted as a powerful accelerator of entropy for the last 200 years (McNeill 2000).

The political ecological significance of industrial assemblages for harnessing inanimate energy (like steam power) relates not only to the step change in the amount of energy available to societies. Such an expansion in energy availability also enabled a qualitative shift in the organizational logic of economic life, by allowing human labor to be substituted on a massive scale. In particular, industrialization changed the relationship to land in a significant way, from 'a direct relationship of surplus appropriation' to 'a condition for rising labor productivity within commodity production' (Moore 2014: 20). The significance of industrialization, then, lies in the coupling of machines and inanimate power, and their combined revolutionary impacts on labor productivity (Bridge and Bradshaw 2014).

By tapping inanimate sources of power, the industrial organization of work gains a seemingly 'automatic' character, in the sense that its pace and intensity do not depend anymore on those of living beings. This is an illusion, of course, a mystification of the productive forces and social relations of the industrialization process. Nevertheless, the technical ability to harness non-human forces to the machines, in ways which allowed the mechanization of the labor process from start to end, had vast and multiple consequences. The physical limits of labor could be overcome by simply replacing the workforce operating machines in prolonged or even continuous shifts (as in blast furnaces); inanimate sources of power do not eat and thus do not compete with humans for appropriation of biomass, so they constitute a net gain in the total amount of energy available in society. In short, the ecological limitations of the 'advanced organic economy' (Wrigley 1988) could be overcome. People now tended the machines as operators, not as movers. Coupled with the institution of private property and the historical process of capital accumulation, such energy shifts created a fundamental break with pre-industrial modes of production, for it set in motion a mechanism that was virtually unstoppable. As long as there was water running through the wheels, or coal burning in the steam engines, production could continue apace. All that was needed was a disciplined and subdued workforce to make sure that power did not go to waste, and a market demand large enough to absorb the sheer volume of goods coming out of the factory system. This is why the industrial revolution would be hardly understandable outside the historical context of the agrarian enclosures and dispossessions – which created a proletarian workforce – and of the geo-political context of European colonialism – which created a market for the industrial products of England and north-western Europe (Foster 1999; Hornborg 2001; Barca 2011).

Given this historical context, we suggest that a political ecology of industrialization focused on power and the machine may be read in two different ways. First, understood narrowly as an account of the environmental impacts of mechanization, a political ecology of industrialization highlights the accelerated throughput of energy and materials (including finished goods and waste) between a society and its biophysical environment. It focuses not on machines per se, but on the intensification of control over labor and revolutionary effects on labor productivity that lead to growing socio-ecological inequalities. So significant is the acceleration in the rate of industrial metabolism (Fischer-Kowalski and Haberl 2007) that it measurably alters the chemistry of the atmosphere; so extensive is the harnessing of biophysical forces that it becomes possible to speak of 'second nature' (Smith 1984: 4); and so profound are the inequalities generated via ecologically unequal exchange, resource exhaustion and widespread environmental degradation that the world becomes differentiated into industrial centers and their extractive peripheries (Bunker 1985). This first approach approximates quite well political ecology's relatively limited engagement with industrialization, which has focused largely on ecological distribution conflicts and the environmental risks associated with industrialization (as we illustrate below).

Second, when understood more broadly as an account of the economic rationales, social relations and subjectivities consequent to industrialization, a political ecology of industry extends beyond environmental impacts to include forms of economic and political life to which industrialization gives rise. This expanded approach enables one, for example, to interpret crises of over-accumulation and industrial strategies to restore profitability in political ecological terms (Desfor and Vesalon 2008); or, similarly, to understand struggles over the distribution of gains from improved labour productivity as a (Fordist) political ecological settlement extending beyond the workplace to permeate consumption and social reproduction (Huber 2009). This second approach remains a road less traveled within political ecology, although Huber (2013; see also Chapter 37, this volume) and Mitchell (2011) are important exceptions precisely because they seek to capture the systemic and ecological character of socio-political relations consequent to oil-fuelled industrialization. For Mitchell, for example, the enormous productive potential of oil as an inanimate energy source – specifically, the prospect of growth without limit – was an important condition of possibility for the emergence of the modern notion of economy. While for Huber, a political ecology of the 'golden years' of post-war industrialization in the United States acknowledges the role of oil in 'the alienated – seemingly autonomous –

power of capital over living labour' (2013: xiv), and in giving shape to forms of 'entrepreneurial life' and political identity readily conformable to neoliberalism through suburbanization and automobility. We return to these broader readings of the political ecology of industrialization in the conclusion. The next two sections address the social metabolism of industrialization and production of environmental risk.

The country and city: social metabolism and the treadmill of production

Industrial activity requires high-intensity inputs of energy, materials and human work. Structured by a productivist logic of continually expanding output, this anthropogenic flux of materials, energy and wastes simultaneously creates extractive frontiers and new markets for mass consumption. In this way the 'social metabolism' of industrialization drives processes of socio-spatial differentiation at all geographical scales, and underpins distinctive forms of ecological consciousness. For the most part, political ecology has approached these trans-local provisioning systems from the upstream end: research originates in fields, forests, mines, waters and other sites of raw material production and considers how the social relations of resource access and use are structured by wider processes. Urban political ecology, of course, takes a different tack and highlights the city as a site of commodity consumption and political power structuring and shaping socio-spatial flows of materials (Heynen et al. 2006). However, industrial processes – the dynamics of transforming and capturing value via the physical transformation and (re)assembly of raw materials into manufactured products – are often occluded in these accounts. In the case of agro-food, for example, political ecology accounts typically start with either agricultural practices or (to a lesser extent) the politics of consumption and food access, with the result that agricultural processing and food manufacturing appear primarily as contextual detail rather than an explanatory focus. A notable exception is Walker's (2004) account of 150 years of agribusiness in California, which highlights the central role of this sector in driving processes of innovation that, in turn, transformed both spaces of production and consumption. Similarly political ecologies of oil are weighted strongly towards processes and spaces of extraction: rarely has political ecology placed refining, plastics and petrochemicals at the center of its account, notwithstanding the tremendous scientific, technological and legal efforts of this sector towards the re-assembly and social proliferation of hydrocarbon products. Robbins (2007, see also Robbins and Sharp 2008) is noteworthy for its interest in the chemical production complex and the role of declining margins in driving the suburban lawn care economy. Huber's (2013: 61) explicit attention to petroleum refining and refinery workers as a 'central metabolic site' in the socio-ecological relations of twentieth-century capitalism is indicative of how political ecology might pay more attention to the dynamics of industrial production and to what effect.

Through growing labour productivity and product specialization, industrialization throws an unprecedented volume of commodities into circulation: such an enormous growth in manufacturing typically drives down unit exchange values, thereby opening up markets for mass consumption, while at the same time causing recurrent over-production crises. The logic of maximizing production is pervasive: it increases profits, state revenues (through taxation), and national, corporate and personal prestige and power. This implicit social critique of industrial capitalism is captured by the concept of 'the treadmill of production,' first theorized in the late 1970s based on observations of the Fordist system (Schnaiberg 1980). The treadmill argument describes a political-economic system based on manufacturing and driven by a fundamental belief that social welfare and wellbeing are advanced through economic growth: the constant expansion of production and consumption become the key instruments of social policy, around which there is a convergence of interests between capital, labor and the state. Mechanization to improve labor productivity expanded demand for energy and resources, while the fixed capital sunk in machines required high rates of throughput to be sustained (Gould et al. 2004). Since 'sustained ecosystem withdrawals and additions' are required to expand production and consumption 'the support of private capital, labor, and the state for economic growth (imply) conscious or unconscious support for ecological disruption and environmental degradation' (Gould et al 1996: 7). In the Fordist era this system was organized primarily at the national scale, although over time it has become increasingly transnational as a consequence of processes of economic globalization that, in turn, have accelerated the treadmill.

The political ecological significance of industrialization rests not only on the expansion of production and consumption. It also concerns the increasing differentiation between country and city – and between core and periphery – propelled by industrialization, a process captured in the notion of 'metabolic rift' (Foster 1999; Foster et al. 2010). In his discussion of 'Large-scale Industry and Agriculture' in Volume I of *Capital*, Marx noted how industrial capitalism collected population in large urban centers, thereby disturbing the return of nutrients derived from human, animal and organic waste to the soil: at the same time, capitalist agriculture undermined both the soil and the ability of workers to reproduce themselves by an unprecedented intensification of production. As a consequence, Marx and Engels were 'insistent about the need to transcend this form of alienation from nature upon which capitalism rested … the argument involved the abolition of the antagonistic relation between town and country through the integration of agriculture and industry, the dispersal of population, and what Marx referred to as 'the restoration' of the metabolic relation between human beings and the earth' (Foster 2000: 182–183). As a critical perspective, metabolic rift has captured the attention of those seeking to understand how spatial differentiation arises out of the political ecological relations of capitalism. For the most part this work has been taken up in environmental sociology and has focused on the restructuring of biogeochemical cycles consequent to industrialization: Clark and York (2005), for example, examine industrialization's disruption of the global carbon cycle and the flooding of atmospheric carbon sinks; while Clark and Foster (2009) develop a more richly geographical account of metabolic rift and unequal exchange in their work on the nineteenth-century trade in guano and nitrates from South America to restore European soil fertility (see also Chapter 29, this volume). Notwithstanding this important work, the 'eco-geographical logic' to which the concept of metabolic rift alludes – and which arguably constitutes 'one of critical political ecology's most powerful ideas' – has yet to be fully explored (Moore 2011: 39).

The eco-geographical differentiation of space represented by 'country and city' re-works socio-ecological relations at both material and symbolic levels, and is thus productive of new forms of environmental consciousness and politics. Indeed, the differentiation of country and city has had significant cultural and ideological repercussions. A number of scholars have pointed out how historically it gave rise to an elitist vision of the environment as a place for leisure and recreation, and for the conservation of an imagined wilderness devoid of human interaction and work (Cronon 1996; Merchant 1980). Originating as an expression of rural nostalgia on the part of English elites during the first industrial revolution (Guha 2000; Marx 1964; Smith 1984), such purifying urges towards 'wild nature' have frequently translated into authoritarian and racist conservation policies (Steinberg 2002; Kosek 2004). These largely elite versions of a white, Anglo-Saxon ecological consciousness do not automatically apply to other social classes and cultures, however: research on the ecological consciousness of the industrial working class, and of non-white communities, reveals quite different attitudes towards nature conservation and environmental pollution. Important discriminating elements are occupation and income: some environmental concerns have a much higher price for the working-class than for the middle or upper classes, as the former tend to be more directly dependent on dirty industrial jobs, and thus to be subjected to so-called 'job blackmail' (Obach 2004; Barca 2014).

The fertile notion of an 'environmentalism of the poor' expresses a distinctive environmental consciousness that arises from the socio-spatially unequal distribution of ecological goods and bads (Martínez-Alier 2002). Social movements calling for global environmental justice, including proposals for 'post-extractivism' (Gudynas 2013) and 'Buen Vivir' (Radcliffe 2012; Gudynas 2011) as alternative models of regional development, seek explicitly to politicize both the treadmill and eco-geographical character of industrial social metabolism. In this respect, they are part of a long line of social movements that have challenged the separation between country and city. These include, for example, anarchist and utopian organizations practicing the principle of 'back to the land' as a response not only to urban alienation, pollution and loss of economic autonomy (as in European and North-American neo-ruralism), but also to the proletarianization of the rural workforce and the environmental and public health threats posed by industrial monocultures (as in the agro-ecology movement of Latin America). Food security, food justice, and food sovereignty are common goals for a number of rural and urban-farming movements worldwide, which have acquired growing importance and self-awareness in the last decade.[1] The contemporary 'de-growth' movement – which shares its heterodoxy with political ecology and ecological economics, and to which European political ecology has been a significant contributor – is a strikingly normative project that seeks to re-politicize the socio-ecological relations of industrial growth in order to slow the treadmill of production and achieve an equitable and 'prosperous way down' (Kallis 2013). A self-consciously alternative proposition to sustainable development, the de-growth movement articulates a political ecological consciousness that 'affirms dissidence' with mainstream models of economy (Demaria et al. 2013: 192; D'Alisa et al. 2014).

Ecological modernization and the political ecologies of environmental risk

Industrialization and modernity are closely associated within social theory: the former is credited with the emancipation of pre-modern societies from the tyranny of nature to enable their full material and cultural development, and also with the growing domination of nature (including humans) expressed in the concerns of the Frankfurt school. For example, an increase in energy consumption per capita is a commonly accepted indicator of 'modern economic growth' as it testifies to the ability of sustaining production growth at a rate higher than the growth of population (Foster 1999; Peet et al. 2011). Historical increases in energy consumption per capita have, in practice, been inextricably linked to the harnessing of inanimate energy sources (mostly fossil, but also hydro and nuclear power). Mainstream economic discourse celebrates such trends in labour productivity and consumer welfare as a triumph of western modernity over pre-industrial resource supply crises, which brought recurrent famines as a consequence of societies' inability to sustain production. Ecological modernization theory replicates this progressive role for industrialization, but focuses on technological and managerial improvements to resource productivity and eco-efficiency that enable 'leaner' forms of industrial production: i.e. the production of more output with lower energy and material inputs (Mol 1995). Lauded as a process of 'de-materialization' observable at the level of individual products, this process ignores both the problem of aggregate resource consumption (which outstrips resource efficiency gains at the level of individual products) and specific environmental risks associated with the social metabolism of a putatively 'post-industrial' society.

Political ecology has a long record of critically challenging conventional accounts of modernity. Although relatively little work focuses on either the manufacturing sector or processes of industrialization, political ecology has developed an extensive critique of industrial forms of resource management and the field is characterized by a pervasive skepticism towards accounts that assign scarcity to nature (rather than society) and which present undifferentiated accounts of economic or environmental change. Political ecology has been slow to apply these critical lenses to the sites and spaces of manufacturing, but shows a growing interest in understanding the environmental and social consequences of ecological modernization through work on sectors like electronics manufacturing (Forsyth 2004; Little 2012); e-waste disposal (Pickren 2014a, 2014b); and renewable energy (Mulvaney 2013). Forsyth (2004) illustrates the potential of a 'brownfield' political ecology – as distinct from a greenfield focus on farms, forests and other forms of rural transformation – that examines the politics of environmental and social risk associated with rapid industrialization. In his work on Thailand, Forsyth highlights lead and solvent poisoning among electronics workers and the health effects arising from the combustion of lignite in power plants closely associated with the country's drive for industrial growth. His analysis centers on the politically-generative capacity of industrial pollution which, via the epistemology of class-based environmental movements in Thailand, successfully created an environmental consciousness around 'dirty development' as a way to challenge the state. He also points out how the cases of lead and lignite have become 'hegemonic environmental imaginaries in their own right' within the Thai environmental movement, subsequently structuring understandings of risk in restrictive ways. Ulrich Beck's concept of 'risk society' has been taken up by a number of researchers in political ecology to highlight a paradigmatic organizational shift arising from the increase in technological hazards associated with industrial production: a shift from the distribution of wealth to the allocation of risk (Beck 1992; Hannigan 2006). In a general sense the circulation of industrial toxins in the environment and their concentration in living organisms may be considered to be 'democratic' phenomena, as they affect society at large (Beck 1987). However, more careful observations reveal how pollutants tend to concentrate in specific spaces and thus affect the particular human groups that inhabit them (Faber 2008). In this way unequal exposure to the effects of industrialization gives rise to significant spatial and social differentiation. Even when toxic substances or carcinogens circulate more widely, as in the case of contaminated food or water, some human populations are more vulnerable to them than others because they do not have access to means of self-protection and 'inverted quarantine' – e.g. eating organic food – which are put in place by more affluent sectors of the population (Szasz 2007; Renfrew 2013). The political ecologies of industrial contamination have, therefore, frequently been interpreted through the lens of environmental justice (Little 2012; Pellow and Brulle 2005; Chapter 45, this volume).

Although there is now a rich body of work on environmental (in)justice associated with industrialization, there are surprisingly few political ecologies of the industrial workplace in what historians of occupational health and safety refer to as the 'dangerous trades' (Hamilton 1985). In modern industrial societies the tyranny of nature has been replaced with an 'industrial hazard regime' characterized by an unprecedented intensification of work hazards, leading to a contradiction between work and health – production and reproduction – common to both capitalist and centrally planned economies (Sellers and Melling 2012; Merchant 1980). Such a contradiction starts at the workplace: a crucial but often forgotten aspect of industrialization is the way it dramatically changes the work environment and the life conditions of the working classes, deeply altering disease patterns at the global level. This fact has been clearly perceived since the beginning of industrialization: the Marxian tradition of thought saw industrialization as a contradictory process bringing about an unprecedented advance in the forces of production, but with enormous social and environmental costs. In *The Condition of the Working Class in England* (1845), Engels exposed the enormous social cost of industrialization paid by the English working class through occupational hazards and the impairment of urban living conditions. A contemporary of Marx and Engels, William Morris, was alarmed by urban pollution and the toxic environment in which industrial workers were compelled to work and live. 'The proletariat thus became a universal class,' writes J.B. Foster, 'exposed to "universal pollution" and universal suffering, a class threatened with the total loss of humanity, and one that could emancipate itself only through the total emancipation of humanity' (Foster 2000: 119). Scholars in public health and historians of medicine have long identified the fundamental epidemiological shift which characterizes industrialized societies, that from a prevalence of infectious to degenerative diseases (Sellers 1998). Such an epidemiological shift is a major consequence of industrialization, carrying broad social, ecological and bio-political implications which, notwithstanding excellent work in environmental history and histories of public health (e.g. Nash 2006; Santiago 2006), deserves greater attention on the part of political ecologists. A political ecology of industrial hazards and environmental risk would start from the work environment, looking first at the effects of industrial work on human-nature (e.g. workers' bodies); it would then follow the flows of carcinogens and mutagenic particles from the shop-floor (or the farm-field) to the larger environment, in their meeting with the local landscape and living organisms; and finally their circulation through the atmosphere, the water cycle and the food chain. At the same time, however, the circulation of industrial hazards must be seen as a socio-technical process founded upon geo-political and economic inequalities, which is constantly renegotiated through the production and circulation of knowledge, and the possibilities for social subjects to become aware of and counteract those hazards.

Conclusion

As an empirical object of inquiry, political ecology has given industrial activity comparatively limited attention. Industrialization appears in political ecology primarily as a process that imperfectly penetrates the social relations of agriculture, as a motor of resource mobilization and social metabolism, and/or a source of new forms of technological and environmental risk. We have outlined in this chapter ways to build on these significant yet still limited engagements to develop a more thoroughly industrial political ecology. Centering an account on the political ecological relations that sustain labour productivity – and highlighting mechanization and the appropriation of ecological surplus in the form of inanimate energy, technological intensification, and the expansion and acceleration of social metabolism – can illuminate how industrial labor regimes extract value through the transformation of biophysical systems, and the environmental risks and distribution conflicts consequent to an industrial mode of production. Conceptual resources developed in other heterodox fields of inquiry – such as ecological economics, and sections of environmental history and environmental sociology – can be useful in this task: the concept of 'metabolic rift', for example, may be turned towards understanding how industrialization drives processes of spatial differentiation, and its implications for both socio-ecological outcomes and forms of political-ecological consciousness.

There is a tendency in environmentalist accounts of industry to fetishize machines and resources: what Moore (2014: 12) describes as a bourgeois distraction that 'it all began with coal'. We have argued, however, for a political ecology that acknowledges the distinctive character of industrial activity and which critically engages the environmental and social consequences of mechanization: here the focus is not on machines and inanimate energy in and of themselves, but on their revolutionary implications for labor productivity and socio-spatial differentiation. Importantly, however, a political ecology of industrialization should go further than an account of its environmental and social impacts, to consider the subjectivities, rationalities and habits of mind to which industrial activity gives rise.

Note

1 See, for example, Food Sovereignty: A Critical Dialogue. International Conference, Yale University, September 14–15, 2013 (http://yale.edu/agrarianstudies/foodsovereignty/papers.html).

References

Barca, S. (2010) *Enclosing Water. Nature and Political Economy in a Mediterranean Valley*. Cambridge: White Horse Press.

Barca, S. (2011) Energy, property and the industrial revolution narrative. *Ecological Economics* 70: 1309–1315.

Barca, S. (2014) Laboring the earth. Transnational reflections on the environmental history of work. *Environmental History* 19(1): 3–27.

Beck, U. (1987) The anthropological shock: Chernobyl and the contours of the risk society. *Berkeley Journal of Sociology*: 153–165.

Beck, U. (1992) *Risk Society. Towards a New Modernity*. London: Sage.

Boudia, S. and N. Jas (eds.) (2014) *Powerless Science? Science and Politics in a Toxic World*. New York: Berghahn.

Bridge, G. and M. Bradshaw (2014) Deepening globalisation: economies, trade and energy systems. In *Global Energy: Issues, Policy and Implications*. Ed. Paul Ekins, Michael Bradshaw and Jim Watson. Oxford: Oxford University Press, ch. 3.

Bunker, S. (1985) *Underdeveloping the Amazon: Extraction, Unequal Exchange and the Failure of the Modern State*. Chicago: University of Chicago Press.

Cipolla, C.M. (1973) *The Industrial Revolution*. Glasgow: Fontana/Collins.

Clark, B. and J.B. Foster (2009) Ecological imperialism and the global metabolic rift: unequal exchange and the guano/nitrates trade. *International Journal of Comparative Sociology* 50 3–4: 311–334.

Clark, B. and R. York (2005) Carbon metabolism: global capitalism, climate change and the biospheric rift. *Theory and Society* 34: 391–428.

Cronon, W. (ed.) (1996) *Uncommon Ground: Rethinking the Human Place in Nature*. New York: W.W. Norton.

D'Alisa, G., F. Demaria and G. Kallis (2014) *Degrowth: A Vocabulary for a New Era*. New York and London: Routledge.

Daly, H. (1991) *Steady-State Economics*, 2nd edition. Washington, DC: Island Press.

Demaria, F., F. Schneider, F. Sekulova and J. Martinez-Alier (2013) What is degrowth? From an activist slogan to a social movement. *Environmental Values* 22: 191–215.

Desfor, G. and L. Vesalon (2008) Urban expansion and industrial nature: a political ecology of Toronto's port industrial district. *International Journal of Urban and Regional Research* 32(3): 586–603.

Faber, D. (2008) *Capitalizing on Environmental Injustice: The Polluter-Industrial Complex in the Age of Globalization*. Lanham: Rowman & Littlefield.

Fischer-Kowalski, M. and H. Haberl (eds.) (2007) *Socioecological Transitions and Global Change: Trajectories of Social Metabolism and Land Use*. Cheltenham: Edward Elgar.

Forsyth, T. (2004) Industrial pollution and social movements in Thailand. In *Liberation Ecologies: Environment, Development, Social Movements*. Ed. R. Peet and M. Watts. London: Routledge, pp. 383–398.

Foster, J.B. (1999) *The Vulnerable Planet. A Short Economic History of the Environment*. New York: Monthly Review Press.

Foster, J.B. (2000) *Marx's Ecology. Materialism and Nature*. New York: Monthly Review Press.

Foster, J.B., B. Clark and R. York (2010) *The Ecological Rift: Capitalism's War on the Earth*. New York: Monthly Review Press.

Georgescu-Roegen, N. (1976) *Energy and Economic Myths: Institutional and Analytical Economic Essays*. New York: Pergamon Press.

Gottlieb, R. (2002) *Environmentalism Unbound: Exploring New Pathways for Change*. Cambridge, MA: MIT Press.

Gould, K., D. Pellow and A. Schnaiberg (2004) Interrogating the treadmill of production: everything you wanted to know about the treadmill but were afraid to ask. *Organization and Environment* 17(3): 296–316.

Gould, K.A., A. Schnaiberg and A. Weinberg (1996) *Local Environmental Struggles. Citizen Activism in the Treadmill of Production*. New York: Cambridge University Press.

Gudynas, E. (2011) Buen vivir: today's tomorrow. *Development* 54(4): 441–447.

Gudynas, E. (2013) Transitions to post-extractivism: directions, options, areas of action. In *Beyond Development: Alternative Visions from Latin America*. Ed. M. Lang and D. Mokrani. Amsterdam: Transnational Institute, pp. 165–188.

Guha, R. (2000) *Environmentalism. A Global History*. New York: Longman.

Hamilton, A. (1985) *Exploring the Dangerous Trades: The Autobiography of Alice Hamilton, M.D.* Boston: Northeastern University Press.

Hannigan, J. (2006) *Environmental Sociology*, 2nd edition. Abingdon: Routledge.

Harremoes, P., D. Gee, M. MacGarvin and A. Stirling (2002) *The Precautionary Principle in the 21st Century. Late Lessons from Early Warnings*. London: Routledge.

Heynen, N.C., M. Kaika and E. Swyngedouw (eds.) (2006) *In the Nature of Cities: Urban Political Ecology and the Politics of Urban Metabolism*. Abingdon and New York: Taylor & Francis.

Higgins, P. (2010) *Eradicating Ecocide: Laws and Governance to Prevent the Destruction of Our Planet*. London: Shepheard-Walwyn.

Hornborg, A. (2001) *The Power of the Machine: Global Inequalities of Economy, Technology, and Environment*. Walnut Creek: Rowman Altamira.

Hornborg, A. (2006) Footprints in the cotton fields: the Industrial Revolution as time–space appropriation and environmental load displacement. *Ecological Economics* 59: 74–81.

Huber, M. (2013) *Lifeblood Oil, Freedom, and the Forces of Capital*. Minneapolis: University of Minnesota Press.

Huber, M.T. (2009) The use of gasoline: Value, oil, and the 'American way of life'. *Antipode* 41(3): 465–486.

Kallis, G. (2013) Societal metabolism, working hours and degrowth: a comment on Sorman and Giampietro. *Journal of Cleaner Production* 38: 94–98.

Kosek, J. (2004) Purity and pollution: racial degradation and environmental anxieties. In *Liberation Ecologies: Environment, Development, Social Movements*. Ed. R. Peet and M. Watts. London: Routledge, pp. 115–152.

Little, P. (2012) Environmental justice discomfort and disconnect in IBM's tainted birthplace: a micro-political perspective. *Capitalism, Nature, Socialism* 23(3): 92–109.

Malm, A. (2014) Fleeing the flowing commons: Robert Thom, water reservoir schemes, and the shift to steam power in early nineteenth-century Britain. *Environmental History* 19(1): 55–77.

Martínez-Alier, J. (2002) *The Environmentalism of the Poor: A Study of Ecological Conflicts and Valuation*. Cheltenham: Edward Elgar.

Marx, L. (1964) *The Machine in the Garden: Technology and the Pastoral Ideal in America*. New York: Oxford University Press.

McEvoy, A. (1995) Working environments: an ecological approach to industrial health and safety. *Technology and Culture* 36: S145–73.

McNeill, J.R. (2000) *Something New Under the Sun: An Environmental History of the Twentieth-Century World*. New York: W.W. Norton.

Merchant, C. (1980) *The Death of Nature: Women, Ecology and the Scientific Revolution*. San Francisco: Harper & Row.

Mitchell, T. (2011) *Carbon Democracy Political Power in the Age of Oil*. New York and London: Verso Books.

Mol, A. (1995) *The Refinement of Production: Ecological Modernization Theory and the Chemical Industry*. The Hague: International Books.

Moore, J. (2011) Transcending the metabolic rift: a theory of crises in the capitalist world-ecology. *Journal of Peasant Studies* 38(1): 1–46.

Moore, J. (2014) The capitalocene: part 1, on the nature and origins of our ecological crisis. Available online at: www.jasonwmoore.com/uploads/The_Capitalocene__Part_I__June_2014.pdf.

Mulvaney, D. (2013) Opening the black box of solar energy technologies: exploring tensions between innovation and environmental justice. *Science as Culture* 22(2): 230–237.

Muradian, R., M. Walter and J. Martínez-Alier (2012) Hegemonic transitions and global shifts in social metabolism: implications for resource-rich countries. *Global Environmental Change* 23(3): 559–567.

Nash, L. (2006) *Inescapable Ecologies: A History of Environment, Disease and Knowledge*. Berkeley: University of California Press.

Nixon, R. (2011) *Slow Violence and the Environmentalism of the Poor*. Cambridge, MA: Harvard University Press.

Obach, B. (2004) *Labor and the Environmental Movement: The Quest for Common Ground*. Cambridge, MA: MIT Press.

Peet, R., P. Robbins and M. Watts (2011). Global nature. In *Global Political Ecology*. New York: Routledge, pp. 1–47.

Pellow, D.N. and R. Brulle (2005) *Power, Justice, and the Environment. A Critical Appraisal of the Environmental Justice Movement*. Cambridge, MA: MIT Press.

Pickren, G. (2014a) Political ecologies of electronic waste: uncertainty and legitimacy in the governance of e-waste geographies. *Environment and Planning A* 46(1): 26–45.

Pickren, G. (2014b) Geographies of e-waste: towards a political ecology approach to e-waste and digital technologies. *Geography Compass* 8(2): 111–124.

Radcliffe, S.A. (2012) Development for a postneoliberal era? *Sumak kawsay*, living well and the limits to decolonisation in Ecuador. *Geoforum* 43(2): 240–249.

Renfrew, D. (2013) 'We are not marginals:' the cultural politics of lead poisoning in Montevideo, Uruguay. *Latin American Perspectives* 40(2): 202–217.

Robbins, P. (2007) *Lawn People: How Grasses, Weeds, and Chemicals Make Us Who We Are*. Abingdon and New York: Temple University Press.

Robbins, P. and J. Sharp (2008) Turfgrass subjects: the political economy of urban monoculture. In *In the Nature of Cities: Urban Political Economy and the Politics of Urban Metabolism*. Ed. Nik Heynen, Maria Kaika and Erik Swyngedouw. Abingdon and New York: Routledge, pp. 110–128.

Santiago, M. (2006) *The Ecology of Oil: Environment, Labor, and the Mexican Revolution, 1900–1938*. Cambridge and New York: Cambridge University Press.

Schnaiberg, K. (1980) *The Environment: From Surplus to Scarcity*. Oxford: Oxford University Press.

Sellers, C. (1998) *Hazards of the Job: From Industrial Hygiene to Environmental Health Science*. Chapel Hill and London: University of North Carolina Press.

Sellers, C. and J. Melling (eds.) (2012) *Dangerous Trade. Histories of Industrial Hazards across a Globalizing World*. Philadelphia: Temple University Press.

Smith, N. (1984). *Uneven Development*. Oxford: Blackwell.

Steffen, W., J. Grinevald, P. Crutzen and J. McNeill (2011) The Anthropocene: conceptual and historical perspectives. *Philosophical Transactions of the Royal Society A* 369: 842–867.

Steinberg, T.L. (1986) 'An ecological perspective on the origins of industrialization'. *Environmental Review* 4: 261–276.

Steinberg, T. (2002) *Down to Earth. Nature's Role in American History*. New York: Oxford University Press.

Szasz, A. (2007) *Shopping Our Way to Safety: How We Changed from Protecting the Environment to Protecting Ourselves*. Minneapolis: University of Minnesota Press.

Walker, R. (2004) *The Conquest of Bread: 150 years of Agribusiness in California*. New York: New Press.

Walker, R., M. Storper and E. Gersh (1979) The limits of environmental control: the saga of dow in the delta. *Antipode* 11(1): 48–59.

Williams, R. (1983) 'Industry'. In *Keywords. A Vocabulary of Culture and Society*. New York: Oxford University Press.

Wrigley, E.A. (1988) *Continuity, Chance and Change. The Character of the Industrial Revolution in England*. Cambridge: Cambridge University Press.

29

CONCEPTUALIZING ECOLOGICALLY UNEQUAL EXCHANGE

Society and nature entwined

Alf Hornborg

Background: a brief review of theories of unequal exchange

Political ecology is concerned with environmental injustices and inequalities. This necessarily means combining perspectives and knowledge from the social and the natural sciences. A general question is how the use of natural resources both reflects and reinforces social relations of power. This question can be pursued in many different ways, but in this chapter I will focus on how ecology and economics interact in generating polarizing processes of accumulation and impoverishment in the world-system. Neither economic science nor ecology can by itself account for the global environmental inequalities exemplified by the uneven impacts of climate change, deforestation, biodiversity loss, soil degradation, overfishing, access to healthy food and water, exposure to toxins, pollution, and waste disposal. In order to understand such inequalities, there have been many attempts to integrate perspectives from economics and ecology. Although the history of such attempts goes back much further in time than the Industrial Revolution in early nineteenth-century Britain, this chapter concentrates on some of the problems of integrating economics and ecology since that time. In unraveling some of the conceptual pitfalls encountered by attempts to assimilate ecology into economics, we shall compare the approaches of conventional neoclassical economics, Marxist economics, and two varieties of ecological economics (Table 29.1). While mainstream economics is unable to conceptualize the material considerations that are prerequisite to any theory of ecologically unequal exchange, Marxist and much ecological economics instead risk reducing economics to physics. We shall argue for a non-reductionist ecological economics, building on the insights of Nicholas Georgescu-Roegen (1971), which keeps social and natural aspects analytically distinct precisely in order to grasp how they are intertwined.

A fundamental starting-point for thinking about unequal exchange is Karl Marx (1818–1883). He proposed that the accumulation of capital in industrialized Britain was made possible by the "dispossession" of the worker's labor-power, disguised as an equal exchange of labor-power for wages. As workers' wages were lower than the capitalists' income from sales, Marx (1967 [1867]) concluded that their labor-power contributed more to the value of the commodity than they were remunerated. This "surplus value" drawn from labor was the source of capitalist

Table 29.1 Some essential differences between four traditions of economic thought

Issues	*Neoclassical economics*	*Marxist economics*	*Neo-Physiocrat ecological economics (e.g., H.T. Odum, R. Costanza)*	*Non-reductionist ecological economics (e.g., N. Georgescu-Roegen, H. Daly)*
How is economic value defined?	By consumer preferences	By the quantity of embodied labor time	By the quantity of embodied energy or other natural values	By consumer preferences
Why are there environmental problems?	Environmental costs are insufficiently internalized in market prices	The capitalist mode of production generates environmental destruction	Natural values such as embodied energy are insufficiently internalized in market prices	Economic value creation generates entropy
What is unequal exchange?	A result of market power, such as monopoly	A result of the underpayment of labor	A result of the underpayment of natural values such as embodied energy	A result of the interaction of market valuation and physical laws

profits as well as investments in technology. A question that has generated much discussion is whether Marx's analysis reckoned only in money or conceived of labor-power also as a biophysical force measurable in energy. It is noteworthy that Marx at times considers its biophysical aspects relevant to the production of surplus economic value, but he does not reach a clear position on whether it is possible to specify the relation between money and the energy of labor. Although Sergei Podolinsky (2008 [1883]) perceived ways of phrasing the Marxian theory of surplus value in terms of energy, it seems that his ideas left little imprint on Marxist economic theory (Martinez-Alier and Naredo 1982). Marx was nevertheless well aware of the ecological foundations of labor-power and human life in general. Like the agrarian Physiocrats of the eighteenth century, he acknowledged that labor presupposed land, and he deplored the "metabolic rift" which separated cities from their rural hinterlands, draining soils of their nutrients while concentrating sewage in urban areas (Foster 2000). His criticism of modern society thus included some of its ecological shortcomings, but his vision of how to remedy the rifts and inequalities generated by capitalism was generally committed to the mainstream faith in technological progress. In this sense, he was constrained by the outlook of his time and social context (Benton 1989).

Much later, the Marxian understanding of unequal exchange as an underpayment of labor-power was applied to international trade. Arghiri Emmanuel (1972) calculated that the differences in labor costs between different parts of the world generated international transfers of surplus value, as a given quantity of commodities could be exchanged for another quantity of commodities with the same exchange value, but representing more embodied (i.e., invested) labor. In other words, although the two sets of commodities are equivalent in terms of monetary market value, the one embodying more past labor would contain more "real" value. This conclusion is based on the Marxian conviction that labor is the source of all economic value. As in Marx's own analyses, it is not clear whether the unequally traded hours of labor in Emmanuel's calculations have any specifiable relation to energy.

About the same time as Emmanuel attempted to account for global economic inequalities in terms of the unequal exchange of embodied labor, however, the systems ecologist Howard T. Odum (1971) presented a model for understanding social and economic processes in terms of flows of embodied energy. He coined the notion of "emergy" as shorthand for "embodied energy," but later redefined it as "energy memory" and explicitly claimed that it was a measure of economic value (e.g., Odum 1988). "Emergy" is a measure of the energy spent to generate a commodity, organism, or activity. Focusing on the causes of underdevelopment in Amazonia, the sociologist Stephen Bunker (1985) suggested that market prices concealed an unequal exchange of energy in the world-system, conceptualized as the underpayment of "natural values." In a paper co-authored with J.E. Arding, Odum similarly proposed that trade between the United States and Ecuador, when analyzed from an "emergy" perspective, represented an unequal exchange of embodied energy (Odum and Arding 1991). Odum's methodology for assessing such inequalities in international trade focused on the amount of "emergy" that, on average, was purchased per dollar exchanged between two countries (cf. Odum 1996). Like Podolinsky, Odum explicitly suggested that an energy perspective converged with the theory of surplus value offered by Karl Marx. Both Podolinsky and Odum have been categorized as belonging to the lineage of thought known as "ecological economics" (Martinez-Alier 1987).

Lonergan (1988) has found that the methods and models used to measure unequal exchange of embodied labor and embodied energy are "almost identical." In both frameworks, inequalities are conceptualized as deviations of market prices from "real" values, whether conceived in terms of embodied labor or embodied energy. Burkett (2005: 18) similarly observes that the embodied energy theory of value "closely and consciously parallels the Ricardian labor-embodied theory of value, with energy replacing labor as the primary factor of production." Theories of unequal exchange have ubiquitously been based on such notions of the existence of "real" (and underpaid) values.

In attempting to integrate perspectives from economic anthropology, ecological economics, and world-system analysis, I have approached the Industrial Revolution, i.e., the very phenomenon of industrial technology, as inextricably connected to global relations of unequal exchange (Hornborg 1992, 1998, 2001, 2006, 2011). Although initially persuaded that available energy was an underpaid "use value" with a pivotal role in the "thermodynamics of imperialism" (Hornborg 1992), I later realized that a coherent theory of ecologically unequal exchange requires a clear analytical distinction between physical, productive potential (such as labor-power or available energy) and economic value (Hornborg 1998). I focused on the logically *inverse* relation between productive potential and economic value, as revealed by Georgescu-Roegen's (1971) observation that production processes generating an increase in utility (economic value) simultaneously imply a dissipation of energy and matter, i.e., an increase in entropy. Rather than suggest a physical basis of economic value, in other words, we need to understand the physical *implications* of the production of value. I concluded that the market exchange of finished industrial products for fuels and raw materials will inexorably reward the dissipation of natural resources in core areas of the world-system with economic access to more resources to dissipate. In other words, economic growth means that the more resources the core has dissipated today, the more new resources it will be able to dissipate tomorrow. Moreover, the core uses the periphery not only as a source of resources, but also as a sink for its waste products.

This work was largely based on logical deduction. Given the widespread obsession with empirical quantification – as more persuasive than logic – and the methodological problems intrinsic to quantifying available energy, I turned to calculating unequal flows of embodied time and space. In a study of Britain's international trade in 1850, I showed that the British cotton textile industry at that time relied on an asymmetric transfer of embodied labor time and embodied ecological space appropriated from its colonies (Hornborg 2006). While recognizing that both labor time and natural space can be sources of energy and thus objects of what I called ecologically unequal exchange, I referred to such transfers as "time-space appropriation."

Sociologists Andrew Jorgenson and James Rice have focused their research efforts on developing methodologies for empirically identifying ecologically unequal exchange by tracing how wealthier nations appropriate environmental space from less wealthy ones (Jorgenson 2003; Jorgenson and Rice 2007; Rice 2007). A central parameter in these studies is the per capita ecological footprint of consumption (Wackernagel and Rees 1996). Jorgenson and Clark (2009) have edited a special issue of the *International Journal of Comparative Sociology* devoted to the theme "ecologically unequal exchange in comparative perspective." The eight articles in this special issue cover a wide range of global environmental inequalities generated by international trade, including differences in biodiversity loss, energy use, and climate change. Recent research has established empirically that major core areas of the world-system – USA, EU, and Japan – are all net importers of embodied resources from peripheral areas. This applies to materials and energy as well as to land and labor (Lenzen *et al.* 2012, 2013; Yu *et al.* 2013; Simas *et al.* 2014).

The conundrum of unequal exchange: underpaid material values?

The main analytical conundrum that has riddled the history of economic thought about unequal exchange is the relation between material and cultural aspects of production and trade. By "material" I mean the objective, metabolic properties evoked by the Marxian concept of "use value," and by "cultural" the subjective, contextual meanings which establish a commodity's exchange value (cf. Sahlins 1976). There is widespread agreement that mainstream, neoclassical economic science suffers from disciplinary myopia in largely ignoring insights from natural science. Although its focus on consumer value would suggest affinities with cultural analysis, neoclassical economics similarly ignores the extent to which human sciences such as anthropology have questioned many of its fundamental categories (cf. Carrier 2005).

Heterodox economists, particularly from the Marxist tradition and what is now known as ecological economics, over the past 150 years have attempted to transcend the myopia of mainstream economics by attributing crucial significance to biophysical aspects of production processes. The consideration of how biophysical and monetary factors are intertwined in economic processes can be traced from the discourse of eighteenth-century Physiocracy through classical and Marxian economics to contemporary deliberations within the trans-disciplinary field of ecological economics. Physiocracy, Marxism, and ecological economics all represent attempts to reconcile these two concerns. The ways in which Marxist and ecological economics have approached the notion of "value" are tightly linked with their ways of understanding "unequal exchange." In order to establish an analytically coherent theory of ecologically unequal exchange, we thus need to critically consider how these schools have conceptualized the foundations of economic value.

The Marxian distinction between exchange value and use value goes back to Aristotle's reflections on the monetarized economy of ancient Greece. In Marxist thought, use values are equated with "real wealth" and denote material quantities of resources such as (embodied) labor, energy, land, water, etc. that provide for human needs. As Sahlins (1976) has shown, however, human "needs" – beyond the bare metabolic requisites of keeping an isolated human organism alive – are impossible to extricate from their cultural context. (To illustrate this point, we might ask how much "use value" pork has for a Muslim?) If use values are culturally determined, it is difficult to see how they could be objectively identified with material quantities.

The Marxian economist Paul Burkett (2005) has attempted to reconcile Marxist theory and ecological economics. For instance, he agrees with Georgescu-Roegen (1971) that the laws of thermodynamics pose significant constraints on human economies. Burkett compares the perspectives of ecological economics, Physiocracy, and Marxism regarding in which sense nature can be considered a source of economic value. He finds that some ecological economists (such as Robert Costanza) treat nature as a direct source of value, while others (including Nicholas Georgescu-Roegen and Herman Daly) are content with observing that natural resources are consumed and dissipated in the production of valuable goods and services. In the words of Georgescu-Roegen (1971: 282), the definition of what is "valuable" is based on its contribution to the "enjoyment of life."

The difference between these two schools of thought is important. The former approach offers a physical (e.g., energy) theory of value, in effect equating economic value with quantifiable, past investments of some material resource (see Costanza 1980; Odum 1988). Because of the continuities linking it to eighteenth-century Physiocracy, I shall refer to it as "neo-Physiocrat" ecological economics. The latter approach actually accepts the mainstream perception among neoclassical economists of consumer preferences as fundamental to economic value, while adding the crucial observation that the production of economic value simultaneously increases entropy and environmental degradation. Rather than reducing economic value to embodied quantities of a physical force or flow, it makes it possible to show how these two phenomena are related to each other.

Although, as Burkett (2005) shows, eco-Marxists tend to reject narrowly defined energy theories of value, they hold that there are biophysical values in nature that are exploited in capitalism. In contrast to neoclassical economics, both eco-Marxist and ecological economics thus retain the concern of Physiocracy and classical economics with the physical, material aspects of economic activity. Prior to the Industrial Revolution, this concern focused on the productivity of agricultural land, which the Physiocrats recognized as the source of subsistence for all labor, including non-agricultural labor. After the turn to fossil fuels and steam engines – and the articulation of the laws of thermodynamics – what is currently known as ecological economics has increasingly focused on energy as defined by physics (cf. Martinez-Alier 1987). While most proponents of Marxism have maintained its emphasis on the generative capacity of labor, many of the "neo-Physiocrats" have specified their concern with land as a concern with energy. The latter shift seems a logical consequence of the transition from agrarian to industrial society, i.e., from a society deriving its energy resources from horizontal land surfaces to one drawing its energy from vertical shafts through the Earth's crust. Ultimately, the attention of Marxism, Physiocracy, and ecological economics to the physical aspects of economic processes all share a concern with energy, as both labor and land can be expressed as measures of available energy.

The continuities linking Physiocracy, classical economics, and Marxism are illuminating. Whereas the Physiocrats perceived land as the ultimate generator of economic value and growth, Adam Smith, David Ricardo, and Karl Marx shifted the focus to labor, but the structure of the argument is very similar. All sought to identify a factor of production with the special quality of being able to yield more value than is required for its maintenance. Burkett (2005: 25–37) demonstrates the extent to which Marx sympathized with the concerns of the Physiocrats, noting that Turgot in 1770 had referred to the ability of the agricultural laborer to "produce over and above the wages of his labor." Unlike modern economists, neither the Physiocrats nor Marx were content with identifying economic value with immaterial, subjective gratification, but struggled to relate it to the material basis and substance of human life. Marx thus praised the Physiocrats for conceptualizing value and surplus value in terms not of consumption, but of production, and for analyzing capitalist production in terms of "eternal natural laws of production." However, Marx simultaneously recognized that economic value in industrial capitalism could not simply be reduced to material parameters. In his view, the Physiocrats mistakenly equated value with material substance. Although surplus production as conceived by the Physiocrats seemed a material phenomenon, modeled after the physical processes of agricultural production, it was presented as fundamental to the business of making money (i.e., rent) from owning land. The attempt to account for monetary gain in terms of physical processes recurs in Marx's understanding of surplus production in industrial capitalism. He famously argued that the labor-power of workers had the ability to produce commodities containing more economic value than their wages. The interlacing of material flows of (labor) energy and flows of exchange values (money) pervades the Marxian labor theory of value. It builds on important intuitions about the connections between energy flows and economic processes, but ultimately does not clarify the nature of those connections.

Marx acknowledged that surplus production in an agricultural society is easier to conceptualize than in industrial society, primarily because it can be identified without the mediation of monetary measurement, but maintained the ambition to understand industrial capitalist profits using a materialist approach largely inspired by Physiocracy. At times, Marx's understanding of economic value formation strongly echoes that of the Physiocrats, as when he refers to the "naturally originating productivity of labor ... which of course rests on qualities of its inorganic nature – qualities of the soil, etc." (Marx 1967 [1867]: 511). Burkett (2005: 36) accounts for "Marx's endorsement of this kernel of truth in Physiocratic doctrine" as based on the point that "without an agricultural surplus, there can be no surplus labor in agriculture and no means of subsistence for nonagricultural workers, hence no surplus value in the economy as a whole." Marx's struggle to reconcile the material and monetary aspects of the economy resulted in inconsistencies such as the notion that the labor theory of value applies only to capitalist forms of production, not to non-capitalist forms. However, it is difficult to see why labor-power invested in export production in commercial, pre-capitalist civilizations should have been less significant for the creation of profit than in nineteenth-century capitalism.

Did Marx think of energy as implicated, in a specific, quantifiable way, in the creation of "surplus value"? Foster and Burkett (2008: 25) cite Rabinbach's (1990) observation that "Marx always emphasized the energetic basis of labor power and saw it connected to thermodynamics because labor involved mechanical work." According to these authors, there are phrasings of the Marxian labor theory of value which suggest that it is the excess of "productively expendable energy encapsulated in labor power" over the "caloric quantity of useful work needed to produce the worker's commodified means of subsistence" that "enables the capitalist to extract surplus value from the worker," implying that the worker's sale of his labor time is "an energy subsidy for the capitalist" (Foster and Burkett 2008: 26). The very concept of "labor power," it seems, "arose in part from the new thermodynamics" (2008: 29).

In order to ascertain the extent to which Marxian economic theory rests on an unclear connection between physics and economics inherited from Physiocracy, we can consider the famous formula $M\text{-}C\text{-}M_1$ and ask ourselves whether it is justified to posit a quantifiable relation between the material production of commodity C and the increase in economic value from M to M_1? For Marx, the surplus value is generated by embodied labor. To H.T. Odum (1988: 1136), recently endorsed by leading eco-Marxist scholar John Bellamy Foster (Foster and Holleman 2014), surplus value is generated by embodied energy. It is noteworthy that the thrust of Odum's argument is very similar to that of Podolinsky, whose attempt to persuade Marx and Engels about the energy basis of surplus value has been decisively dismissed in a series of articles by Foster and Burkett. The controversy about the so-called "Podolinsky business" (Martinez-Alier and Naredo 1982; Martinez-Alier 1987, 2011; Foster and Burkett 2004; Burkett and Foster 2006) has largely concerned the question of whether Marx and Engels were adequately versed in thermodynamics, but the crucial question is why Marxists should find it necessary, as the Marxian labor theory of value calculates in money, not energy. If Podolinsky was wrong about the derivation of economic value from energy, as Foster and Burkett have argued, it is difficult to see why they should need to attribute a cognate perspective to Karl Marx.

To the extent that labor-power is indeed a form of biophysical energy, it is reasonable to argue for an affinity between the ideas of Marx, Podolinsky, and Odum, but Foster's and Burkett's position on Podolinsky is contradictory. They dismiss him as an "energy reductionist" (Burkett and Foster 2006: 116) who "confused the physical with the economic" (2006: 137). Against this background, it is difficult to understand why it is so important for them to show that Marx had written that "labor-power itself is energy" (2006: 120) and that labor is an "energy subsidy for the capitalist" (Foster and Burkett 2008: 26), explicitly referring to Marx's "*energy income and expenditure approach to surplus value*" (Burkett and Foster 2006: 126; emphasis added). The question ultimately is whether they are arguing that Podolinsky was wrong, or whether his intervention was superfluous? Were there moments, in fact, when Marx himself "confused the physical with the economic"? A truly materialist account of surplus production cannot avoid implicating physics, as Podolinsky recognized, but precisely in not being able to assimilate this insight, the Marxian theory of surplus value as based on labor revealed itself to be entrenched not only in the *operation* of capitalism, but even in its fundamental analytical categories.

It is noteworthy that the net labor "value" that Arghiri Emmanuel (1972) and Samir Amin (1976) identify as having been transferred from less to more developed regions in the 1960s is measured in dollars (cf. Lonergan 1988: 135). If surplus labor value is indeed measurable in dollars, it should suffice to conclude simply that, for a successful capitalist, the price of labor (or energy) should be cheaper than the price of its products. There is no need, in other words, to ascribe to labor (or energy) a unique role in the creation of surplus value (cf. Martinez-Alier and Naredo 1982: 219). Consequently, it should be possible to acknowledge the exploitation of labor without subscribing to a labor theory of value, as well as to acknowledge the unequal exchange of embodied energy without subscribing to an energy theory of value. Moral and political indignation buttressed by theories of unequal exchange and exploitation, in other words, do not at all require the word "value."

The conventional understanding of unequal exchange in both Marxist and ecological economics, however, is in terms of "underpayment": flows of exchange values (money) are represented as not matching the flows of "real" resources conceptualized as use values or natural values. The implicit assumption is that use values and natural resources have a "real" monetary value that can be contrasted to actual market prices. However, it cannot be valid to quantify what Marxists refer to as "use values" (biophysical resources) in monetary terms. Hence, they cannot be "underpaid." It is noteworthy, given the thermodynamic definition of labor-power emerging from Foster and Burkett's (2008) reading of Marx, that this conclusion should be extended even to labor. The concept of "use value" is a misnomer, as it can neither be quantified in other than non-monetary, physical metrics nor extricated – as corresponding to a pure, metabolic need – from the cultural context. The existence of modern technology (the material form of capital accumulation) is certainly predicated on the discrepancy between flows of money and flows of matter-energy (Hornborg 2011), but it is analytically misleading to phrase this discrepancy in terms of an "underpayment" of "use values."

If unequal exchange is instead conceptualized as an asymmetric net transfer of material inputs in production (e.g., embodied labor, energy, land, water, etc.), rather than in terms of an "underpayment" of material inputs or an asymmetric transfer of "value," it will solve another conundrum that has plagued unequal exchange theory from the start, namely, how some extractive economies are able to thrive, rather than become impoverished. All processes of production and capital accumulation must build on net inputs and transfers of productive resources (e.g., from rural to urban areas), but whether the transfers imply impoverishment of a given population depends on circumstances of geography and history. Although it is undeniable that ecologically unequal exchange for centuries has implied exploitation of large segments of the world's population, and continues to do so today, the existence of historically privileged and sparsely populated nations richly endowed with natural resources (e.g., Canada, Australia, Scandinavia, Saudi Arabia) has enabled some extractive zones of the world-system to escape economic impoverishment. This in no way contradicts the definition of "unequal exchange" offered here.

Eco-Marxists Foster and Burkett have tirelessly sifted through Marx's writings in pursuit of every indication of ecological awareness. The quotations they have retrieved are an invaluable distillation of the extent of Marx's orientation in natural science, and their impressive exegetical efforts do not need to be duplicated. What these quotes and Foster's and Burkett's commentaries inadvertently reveal, however, is an inconsistency in the Marxian framework. A fundamental notion in Marxian economics, inherited from classical economics but abandoned in neoclassical economics, is that capitalist profits have a specifiable relation (i.e., are proportional to) inputs of one particular production cost, namely, labor. The labor theory of value is a survival, within Marxism, of nineteenth-century economics. It has been refuted by virtually all mainstream economists and even some Marxists (e.g., Keen 1993) but continues to be taken for granted by most Marxists, not as Marx's own conviction regarding the generative power of labor in general, but as his understanding of the specific mode of operation of the capitalist economy.

In relating the monetary cost of labor to the price of commodities, it is not clear why labor should be singled out among the various costs of production (including, e.g., fuels and raw materials) as the one factor which allows the capitalist to profit from the difference between costs of production and proceeds from sales. To reiterate Marx's well-known narrative about the worker who is only paid for a part of his work-day is no more convincing than to say that fuel costs only cover a part of their contribution to the production process. What labor and fuels have in common is that they are both forms of energy employed in production. To suggest that the use of labor energy has a specifiable relation to capitalist profits on the world market is analytically indistinguishable from the suggestion that energy in general has this ubiquitous connection to the augmentation of utility (and thus income from sales). It is thus not surprising that a leading theorist of Marxian economics, focusing particularly on the relation between Marxist theory and ecology, should now discover a fundamental agreement between labor and energy theories of value (Foster and Holleman 2014). As Lonergan (1988) long ago observed, Marxist and ecological approaches to unequal exchange (conceptualized as underpayment of labor and energy, respectively) are analytically identical.

Conclusions: money, energy, and unequal exchange

The confusion regarding the relation between biophysical factors of production (such as energy) and monetary, economic growth (i.e., capital accumulation) became particularly pronounced in the merchant capitalist states of early modern Europe. Whereas most societies until then had shared an intuitive acknowledgement of the sun's energy as the vital essence flowing through all living things, the experience of long-distance traders instead suggested that the essential flow was that of money. This certainly became a predominant world-view in the Portuguese, Dutch, and British trading empires, and to this day it no doubt remains a perplexing question for most people whether energy or money is ultimately the most important vital flow animating human society. A reasonable response today would be that the significance of money is precisely that it can provide access to energy, indicating that energy in the final instance is more indispensable than money. In the eighteenth and nineteenth centuries, however, the Physiocrats and Marxists had difficulties reconciling the physical and monetary aspects of economic growth. Both agricultural and industrial economies were based on material processes of production requiring physical inputs, yet the market valuation of their produce – and thus their income from sales – hinged on largely cultural processes determining people's willingness to pay. The concept of economic "value" belongs to the vocabulary of the market. The ambition to explain economic value in terms of physical inputs, whether of labor, land, or more generally energy, is to confuse two levels of reality that ought to be kept analytically distinct. The deliberations of Marxists and ecological economists reviewed here illuminate the persistence of this ancient source of confusion.

Rather than engage in further exegesis, it will suffice to outline the essential differences between the four main positions in these debates (Table 29.1). The differences reflect internally coherent frameworks of thought in each of the four schools and are reflected in their distinct approaches to unequal exchange. It will be noted that different foundational assumptions unite different traditions of economic thought. What I refer to as "neo-Physiocrat" ecological economics, which reduces economic value to physics, thus tends to share with neoclassical economics the understanding of environmental problems as the result of insufficiently internalized ecology, whether conceptualized as "environmental externalities," "ecosystem services," "natural values," or "embodied energy," and with Marxism a materialist approach to economic value and unequal exchange. Whereas neoclassical economics does not consider material constraints on economic processes, Marxism and both varieties of ecological economics aspire to unravel how their monetary and material aspects are related. However, only in the non-reductionist economics pioneered by Nicholas Georgescu-Roegen do we find a consistent analytical distinction between the semiotics of market valuation and its material consequences. His conclusion, that the products of economic processes simultaneously represent greater consumer value and greater entropy than the inputs in such processes, remains a formidable challenge to any advocate of economic growth.

Consistent with the different approaches to economic value and environmental problems embraced by the four schools in Table 29.1, each school offers a distinct perspective on unequal exchange. In neoclassical economics, unequal exchange is acknowledged only under conditions of market power, such as monopoly. In Marxist and "neo-Physiocrat" ecological economics, it is viewed as the result of underpayment of labor-power and natural values, respectively. Although not explicitly stated by non-reductionist ecological economists following Georgescu-Roegen, their theoretical framework should imply an approach to unequal exchange that views it as *a result of the interaction of market valuation and physical laws*. In articulating the cultural and political determination of commodity exchange values with inexorable processes of material resource degradation, it illustrates how social and natural realities must be kept analytically distinct if we are to grasp how they are intertwined.

Acknowledgement

Parts of this chapter overlap with my article in the journal *Ecological Economics*, 105: 11–18 (2014).

References

Amin, S. (1976) *Unequal development*. New York: Monthly Review Press.

Benton, T. (1989) "Marxism and natural limits: An ecological critique and reconstruction." *New Left Review* 178: 51–86.

Bunker, S.G. (1985) *Underdeveloping the Amazon: Extraction, unequal exchange, and the failure of the modern state*. Chicago: University of Chicago Press.

Burkett, P. (2005) *Marxism and ecological economics: Toward a red and green political economy*. Leiden: Brill.

Burkett, P. and J.B. Foster (2006) "Metabolism, energy, and entropy in Marx's critique of political economy: Beyond the Podolinsky myth." *Theory and Society* 35: 109–156.

Carrier, J.G. (ed.) (2005) *A handbook of economic anthropology*. Cheltenham: Edward Elgar.

Costanza, R. (1980) "Embodied energy and economic valuation." *Science* 210: 1219–1224.

Emmanuel, A. (1972) *Unequal exchange: A study of the imperialism of trade*. New York: Monthly Review Press.

Foster, J.B. (2000) *Marx's ecology: Materialism and nature*. New York: Monthly Review Press.

Foster, J.B. and P. Burkett (2004) "Ecological economics and classical Marxism: The 'Podolinsky business' reconsidered." *Organization & Environment* 17(1): 32–60.

Foster, J.B. and P. Burkett (2008) "Classical Marxism and the Second Law of Thermodynamics: Marx/Engels, the heat death of the universe hypothesis, and the origins of ecological economics." *Organization & Environment* 21(1): 3–37.

Foster, J.B. and H. Holleman (2014) "The theory of unequal ecological exchange: A Marx-Odum dialectic." *Journal of Peasant Studies* 41(2): 199–233.

Georgescu-Roegen, N. (1971) *The Entropy Law and the economic process*. Cambridge, MA: Harvard University Press.

Hornborg, A. (1992) "Machine fetishism, value, and the image of unlimited good: Towards a thermodynamics of imperialism." *Man* (N.S.) 27: 1–18.

Hornborg, A. (1998) "Towards an ecological theory of unequal exchange: Articulating world system theory and ecological economics." *Ecological Economics* 25(1): 127–136.

Hornborg, A. (2001) *The power of the machine: Global inequalities of economy, technology, and environment*. Walnut Creek: AltaMira Press.

Hornborg, A. (2006) "Footprints in the cotton fields: The Industrial Revolution as time-space appropriation and environmental load displacement." *Ecological Economics* 59(1): 74–81.

Hornborg, A. (2011) *Global ecology and unequal exchange: Fetishism in a zero-sum world*. London: Routledge.

Jorgenson, A. (2003) "Consumption and environmental degradation: A cross-national analysis of the ecological footprint." *Social Problems* 50: 374–394.

Jorgenson, A. and B. Clark (eds.) (2009) *Ecologically unequal exchange in comparative perspective*. Special issue of *International Journal of Comparative Sociology* 50(3–4).

Jorgenson, A. and J. Rice (2007) "Uneven ecological exchange and consumption-based environmental impacts: A cross-national investigation." In A. Hornborg, J.R. McNeill and J. Martinez-Alier (eds.), *Rethinking environmental history: World-system history and global environmental change*, pp. 273–288. Lanham: AltaMira Press.

Keen, S. (1993) "Use-value, exchange value, and the demise of Marx's labor theory of value." *Journal of the History of Economic Thought* 15(1): 107–121.

Lenzen, M. *et al.* (2012) Mapping the structure of the world economy. *Environmental Science & Technology* 46(15): 8374–8381.

Lenzen, M. *et al.* (2013) Building Eora: A global multi-regional input-output database at high country and sector resolution. *Economic Systems Research* 25(1): 20–49.

Lonergan, S.C. (1988) "Theory and measurement of unequal exchange: A comparison between a Marxist approach and an energy theory of value." *Ecological Modelling* 41: 127–145.

Martinez-Alier, J. (1987) *Ecological economics: Energy, environment and society*. Oxford: Blackwell.

Martinez-Alier, J. (2011) "The EROI of agriculture and its use by the Via Campesina." *The Journal of Peasant Studies* 38(1): 145–160.

Martinez-Alier, J. and J.M. Naredo (1982) "A Marxist precursor of energy economics: Podolinsky." *The Journal of Peasant Studies* 9(2): 207–224.

Marx, K. (1967 [1867]) *Capital*, vol. I. New York: International Publishers.

Odum, H.T. (1971) *Environment, power, and society*. New York: Wiley-Interscience.

Odum, H.T. (1988) "Self-organization, transformity, and information." *Science* 242: 1132–1139.

Odum, H.T. (1996) *Environmental accounting: Emergy and environmental decision making*. New York: John Wiley & Sons.

Odum, H.T. and J.E. Arding (1991) *Emergy analysis of shrimp mariculture in Ecuador*. Working paper prepared for Coastal Resources Center, University of Rhode Island. Narragansett, Rhode Island.

Podolinsky, S. (2008 [1883]) "Human labor and unity of force." *Historical Materialism* 16: 163–183.

Rabinbach, A. (1990) *The human motor: Energy, fatigue and the origins of modernity*. New York: Basic Books.

Rice, J. (2007) "Ecological unequal exchange: International trade and uneven utilization of environmental space in the world-system." *Social Forces* 85(3): 1369–1392.

Sahlins, M.D. (1976) *Culture and practical reason*. Chicago: University of Chicago Press.

Simas, M., R. Wood and E. Hertwich (2014) Labor embodied in trade: The role of labor and energy productivity and implications for greenhouse gas emissions. *Journal of Industrial Ecology*, 2014-09-22 [DOI: 10.1111/jiec.12187].

Wackernagel, M. and W.E. Rees (1996) *Our ecological footprint: Reducing human impact on the Earth*. Gabriola Island: New Society Publishers.

Yu, Y., K. Feng and K. Hubacek (2013) Tele-connecting local consumption to global land use. *Global Environmental Change* 23:1178–1186.

PART IV, SECTION C

Environmental governance

The ten chapters in this section reflect political ecology's broad and diverse concern with the *social arrangements and forms of rule through which people manage environments and resources*, and the *social, political, and ecological effects to which these arrangements give rise.* Overall, this section on environmental governance reflects the broad range of empirical contexts and conceptual frameworks through which political ecology has sought to understand how economic and political power are sustained through socio-ecological arrangements. The section opens with Rod Neumann's chapter on nature conservation – a topic occupying a significant place in the development of political ecology – and asks how efforts to maintain biodiversity are bound up with other cultural, political, and economic projects, such as the development of the nation state. This is followed by a pair of chapters that focus on the expansion of high-value export agriculture and aquaculture in the global South since the 1970s, and the rapid rise of environmental certification schemes. Derek Hall's chapter identifies how, by focusing on the social arrangements of agricultural production, political ecology raises distinctive questions about agro-industrial commodity systems, including how production systems are shaped by ecological characteristics of particular species, and how producers gain access to land, labor, and the other social relations necessary to sustain commodity production. Jon Otto and Tad Mutersbaugh examine the rapid emergence of environmental certification schemes – for coffee, shrimp, carbon sequestration and the like – as a form of governance, highlighting the "nature-workers" through whose labour environmental standards are applied, monitored, and enforced.

Scott Prudham's chapter picks up the theme of natural resource appropriation and the governance of socio-natural relations to consider the connections between commodification and property. He explores the legacy of Malthusianism and the influence of Marx in shaping political ecology's interest in property, the simultaneous commodification of land and labor, and the "extra-economic" mechanisms of social power through which rights to land and resources are reproduced. The following chapter by Karen Bakker focuses directly on the neoliberalization of nature. She highlights how political ecology raises important questions about the limits to nature's neoliberalization and its efficacy as a means of "sustainable development"; the distinctiveness of neoliberalism as a form of capitalism; and the implications of wide variation in how neoliberalism articulates with socio-natures for conceptualizing neoliberalism as a variegated yet global phenomenon. A central observation emerging from political ecology's engagement with the neoliberalization of nature concerns the changing role of the state in environmental governance, a core theme of the chapter by Morgan Robertson. By focusing on the ways in which political ecology has chosen to engage with theorizations of the state and state power, Robertson characterizes this relationship as an ambivalent one: while the environment regularly appears as an arena for state strategies and policies, rarely is research in political ecology explicit about how it conceptualizes the state.

Governments provide a central lever for environmental change, yet an enduring insight of political ecology's engagement with Foucauldian understandings of power is the way rule may be achieved not through the apparatus and authority of the state but through the rationalities and mentalities instilled in the population as "subjects." The chapter by Gabriela Valdivia explores political ecology's extensive engagement with Foucault's analysis of "governmentality." A key question here is how environmental objectives (preventing climate change, protecting nature, improving agricultural productivity) become a rationality of rule, such that the governance of populations is achieved by making environmental subjects. Matthew Huber's chapter on energy and social power takes seriously the economic and political subjectivities that are sustained by flows of energy. It critically reflects on an earlier tradition of "energetic" research within cultural ecology and asks how engaging energy more centrally within political ecology might broaden the range of empirical objects understood as "ecologies," beyond the fields, forests, mines, and conservation areas that have been the core-repertoire of political ecology. A similarly enlarged sense of the ecologically political is developed in the chapter by Celia Lowe, which considers what is at stake in the emergence of biosecurity as a rationality of rule. Lowe positions biosecurity as part of a broader expansion in practices of securing life through the registers of risk and preparedness. She explores how, in the context of food security and infectious disease, the rationality of biosecurity re-works social arrangements through practices of exclusion, preparation, regulation, and excitation. The final chapter, by Nathan Sayre, reflects on the centrality of scale to environmental governance and the relatively limited way in which political ecology has engaged with broader conceptual debates on scale within human geography. It examines the meaning and status of scale within political ecology, and how these diverse engagements with scale now provide political ecology with a series of epistemological and methodological guidelines for undertaking research.

30
NATURE CONSERVATION

Roderick P. Neumann

Introduction

Nature and conservation are words with complex etymologies. Conservation, in early modern English usage, referred broadly to the protection or preservation of existing conditions. In the realm of nineteenth-century natural history, it came to mean the careful husbanding of renewable resources such as water, soil, timber, and game. While still important, this broader usage was superseded in the late twentieth century by the arrival of the "biodiversity phase" of modern environmentalism (Zimmerer 2006: 5). The biodiversity phase of conservation is notable for territorializing practices intended to restrict human interaction with the non-human; specifically the establishment of state-controlled protected areas (PAs), such as national parks, wilderness areas, and marine reserves (Neumann 2005). In this phase, biodiversity is increasingly understood as a proxy for nature. Nature shares with natal, nation, and native a common root in the past participle of the Latin word, *nasci*, meaning to be born (Williams 1983). Such an etymology suggests a meaning-rich assemblage of interconnecting identities, histories, geographies, ecologies, and politics. These interconnecting meanings were baldly apparent in the earliest state efforts to establish national parks as symbols of national character and of nation-state origins rooted in nature (Nash 1982). Nature conservation, for the purposes of this chapter, thus refers to state-based initiatives to maintain existing biodiversity and its supporting ecological conditions through the establishment of PAs.

Nature conservation was a key focus in early political ecology inquiry and continues to be so today. As one group of scholars summarized, "interest in conservation started as a trickle ... [that] quickly became a stream" and more recently resulted in a "wave of articles" on a wide range of themes (Vaccaro et al. 2013: 255). One reason for the attention is undoubtedly the accelerated growth in the number, variety, and areal extent of PAs globally since World War II, a phenomenon political ecologists have labeled the "conservation boom" (Zimmerer 2000; Neumann 2002). Second, the number, size, and types of institutions involved in nature conservation and the amount of political and economic resources at their disposal have simultaneously increased. Third, the global expansion of PAs has generated social and political conflict among a range of actors, typically including agrarian communities, indigenous peoples, state agencies, non-governmental organizations (NGOs), corporations, and global and transnational governance institutions. PA establishment produces winners and losers, linking questions of social justice to conservation practices. Investigating and theorizing the material transfers, corporeal violence, displacements, and patterns of political cooperation and resistance associated with PAs have become the threads that weave together much of political ecologists' interest in conservation.

A fourth reason is suggested by the complexities in the meanings of the words, nature conservation, and the contradictions they generate. Their etymologies imply a rich and varied set of political-ecological inquiries. When we speak of nature conservation, what is it, exactly, that is feared will be lost through inaction? How does nature conservation relate to the emergence and development of the modern nation-state? How are state efforts to maintain biodiversity in specific places entangled with other cultural, political, and economic initiatives? How have conservation institutions helped construct and been constructed through contingent, relational meanings of the human and the non-human, nature and non-nature? What is won and lost and who wins and loses with the rise of modern governance institutions of nature conservation?

Though many new forms of conservation territories have emerged that feature "nature-society hybrids" (Zimmerer 2000: 356), much of the literature focuses on the more traditional forms of PAs such as national parks and wilderness areas. These represent the "fortress conservation" models that require the spatial segregation of the human and non-human and "reflect the priorities of national conservation agencies and international organizations" (Adams 2001: 272–273; Brockington 2002). Moreover, many of the newer nature–society hybrids, such as buffer zones, are located on PA perimeters and their *raison d'êtres* are to secure the core fortresses of wild nature. I thus focus in this chapter on the literature addressing traditional state PAs, which is large and theoretically diverse. In an effort to systematically survey this literature, I have ordered it under four headings; spatialities, institutions, capitalisms, and identities. I employ these categories for heuristic purposes, since they inevitably overlap and intersect in political-ecological analyses and in actual social life. For each I highlight the predominant lines of inquiry and main theoretical influences. The closing section critically evaluates the status and trajectory of political-ecology studies of nature conservation.

Spatialities

This section addresses the state's efforts to identify, order, and frame nature spatially through legal, technological, and scientific practices. Territorialization has been a key concept in political ecologists' analyses of these practices (Vandergeest and Peluso 1995; Igoe and Brockington 2007; Whitehead et al. 2007; Roth 2008; Guyot 2011; Peluso and Lund 2011; Holmes 2014). Territorialization refers to centralized, state-led processes of spatial demarcation for the purposes of controlling and regulating people and nature (Vandergeest and Peluso 1995; Whitehead et al. 2007). The power and authority of the state to designate boundaries, nationally and sub-nationally, is central to territorialization (Whitehead et al. 2007; Peluso and Lund 2011). Hence the state's nature conservation strategies are dependent on fixed boundaries that serve to rigidly separate land uses and rights, creating an unambiguous territorial binary of nature/non-nature in a process of "internal territorialization" (Vandergeest and Peluso 1995; Holmes 2014). Political ecologists have homed in on boundary-making processes and their interactions with ecologies, livelihood practices, property rights, and patterns of capital accumulation (Zimmerer 2000; Neumann 2001; Hazen and Harris 2007; Roth 2008; Laudati 2010; Sletto 2011).

State-imposed PA boundaries define nature in spatial terms while enclosing existing commons, eliminating competing property claims, and securing ownership for the nation-state. The internal space of state territory is produced through a process of dividing and containing, making both nature (resources) and people (citizen subjects) "legible" to centralized political authority (Scott 1998; Neumann 2001). The space of nature is thus constructed in PAs in "relation to the exercise of power and control" (Fairhead et al. 2012: 249). Some political ecology studies have emphasized how fixed PA boundaries may disrupt the more fluid, permeable, and overlapping boundaries of pre-existing socionatures (e.g., Zimmerer 2000). Stated in terms of political economy, PA boundary making initiates a process of property enclosure whereby existing property rights and access to land and resources are eliminated (Peluso 1993; Neumann 1998; Brockington 2002; Corson 2010; Kelly 2011; Benjaminsen and Bryceson 2012). Conservation enclosures may transfer the ownership of local commons to the state, which then controls the allocation of benefits from nature in all of its discursive-material-social manifestations (see Latour 1993). For this reason, many argue that boundary making, rather than reducing conflict through attempts to eliminate ambiguity and spatially segregate competing interests, actually has the effect of producing conflicts among diverse interests (Neumann 1998; Zimmerer 2000; Chatty and Colchester 2002; Roth 2008; Sletto 2011; Holmes 2014).

Social and political conflict is perhaps most evident when boundary making for PA establishment requires the spatial displacement of human populations, often by force and without adequate compensation. Hence political ecologists have often interpreted the socio-political effects of PAs through a vocabulary of human rights abuse, social justice, and social inequality (Peluso 1993; Neumann 1998; Kelly 2011). Most recently a new terminology has emerged in an attempt to generalize a widespread PA phenomenon. Authors have referred to "conservation displacement" (Brockington and Igoe 2006: 425), "conservation-related displacement" (Agrawal and Redford 2009: 56) "conservation-induced displacement" (Beazley 2011: 25), "park-induced displacements" (Hoole and Berkes 2010: 304), "PA-displaced people" (Cernea and Schmidt-Sotau 2006: 45), and, finally, "conservation refugees" (Dowie 2009: xxii). Dowie posits that conservation has produced a "new class of refugee," people removed from their homelands by the process of PA establishment (2009: xxi). Some political ecologists have conceptualized such spatial displacement as part of the state's larger efforts of surveillance and control of citizen subjects, thus understanding nature conservation through the Foucauldian (1979, 1991) concepts of disciplinarity and governmentality (Peluso and Vandergeest 2001; Neumann 2004; Moore 2005; Li 2007; Peluso and Lund 2011).

Another significant line of inquiry examines the technological and scientific practices that facilitate boundary making and the spatialization of nature. Whitehead et al. (2007), for example, focus on land use mapping as a power-laden exercise in extending state authority over territorialized nature. Following Scott (1998), they emphasize land use mapping as an abstraction and simplification of knowledge that makes space legible to centralized power. Cartographic visualization is particularly important to the processes of abstraction and simplification, especially when other practices (e.g., chemical tests of soils) are prohibitively expensive or unavailable. The latest advances in geospatial and biological technologies are reinforcing and expanding the spatial orientation of modern nature conservation (Zimmerer 2006; Hazen and Harris 2007; Campbell and Godfrey 2010). Advances in remote sensing and computerized Geographical Information Systems (GIS) have allowed the classification and mapping of nature on a global scale, while also facilitating a proliferation of competing classification schemes rather than a single standard (Redford et al. 2003; Brooks et al. 2006; Zimmerer 2006; Adams and Hutton 2007). The reliance on visualization practices in conservation mapping influences the identification and bounding of PAs and therefore both produces and reinforces particular visions of nature (Hazen and Harris 2007; Whitehead et al. 2007).

Other studies reveal how new technological developments in biology are used to remap and rescale nature conservation, with significant effects on patterns of resource access and control

(Campbell 2007; Campbell and Godfrey 2010). In particular, the increasing use of genetic testing to monitor wildlife populations has influenced the territorialization of nature in marine conservation. Campbell and Godfrey (2010) argue "that the science of genetics is implicated not only in delineating the physical space of conservation territories, but their governance structures" (2010: 905). Specifically, they demonstrate that managing nature at the molecular scale has implications for sovereignty claims, a topic that will be addressed in the next section. They show how new advances in the biological sciences—in their case genetics—may initiate a rescaling of conservation management with consequent effects on state sovereignty, governance, and ownership rights.

Institutions

Michel Foucault, reflecting on his career-long efforts to elaborate a theory of governmentality through the analyses of local, distinct institutions, such as prisons and hospitals, wrote that "by stepping outside" such institutions in order to conduct the analyses, one is immediately confronted "with the totalizing institution of the state" (2007: 119). This observation reminds us that the state is not just "one among other political ecology players," but rather "is in effect the embodiment of political hegemonies and the immanence of social relations inscribed" in environmental management (Ioris 2012: 126). A critical understanding of the state as a totalizing institution is essential to any political-ecological analysis of nature conservation and an appropriate starting point for this section (cf. Chapter 15, this volume). Moreover, nature conservation is both embedded in and produced by the state, an observation that is captured perfectly in the terms "state nature" and "political forests." State nature is "a form of territorially framed and administratively governed nature, which has been brought into existence as part of the processes that have resulted in the formation of modern nation-states" (Whitehead et al. 2007: 2). Whitehead et al. (2007) point out that the modern state has evolved a diversity of institutional subdivisions, such as agricultural ministries and forestry departments, to facilitate the definition, regulation, and management of and benefit distribution from nature. Peluso and Vandergeest's (2001, 2011) idea of political forests illustrates this process. Their term is meant to differentiate the administratively designated forest reserves from ecologically designated forests—some state forest reserve lands are not forested and not all forested lands are included in reserves—and to highlight the political nature of their creation. "The making of national political forests was intertwined with the violent making of nation-state territories and political subjects through common repertoires of violent state practices" (Peluso and Vandergeest 2011: 589). Political ecologists, then, have used PAs as a lens through which to investigate the co-constitution of nature and the state.

Territorial sovereignty is fundamental to the conceptualization of the modern nation-state and studies have illuminated the ways that nature conservation produces, extends, challenges, fragments, and weakens sovereignty claims. From a political ecology perspective, sovereignty claims are anchored in the mundane practices of nature conservation; land surveying, mapping, and the scientific cataloging and inventorying of life forms (Vandergeest and Peluso 1995; Bryant 1997; Peluso and Vandergeest 2001; Neumann 2004; Campbell and Godfrey 2010; Peluso and Lund 2011). A state's ability to conduct these practices across space both produces and confirms territorial sovereignty. Studies have demonstrated, however, that in the case of nature conservation state sovereignty claims are rarely absolute and complete, but rather are contingent, situational, and regularly renegotiated, reconfigured, and reasserted (Igoe and Brockington 2007; Campbell and Godfrey 2010; Büscher 2013; Lunstrum 2013) A series of neologisms and descriptors have accompanied efforts to understand the complexity of outcomes, including "selective," "articulated," and "privatized" sovereignty (Moore 2005; Lunstrum 2013; Igoe and Brockington 2007).

The worldwide proliferation of transboundary protected areas (TBPAs), also referred to as international peace parks or transfrontier conservation areas, has generated much discussion on state sovereignty outcomes (van Amerom 2002; Duffy 2006; Büscher 2013; Lunstrum 2013). Because TBPAs require two or more states to cooperate on land and resource management, these arrangements suggest a weakening of state sovereignty. Study findings, however, show that TBPAs remain enmeshed in networks of competing state and non-state actors to a degree that inhibits the transference of sovereignty (van Amerom 2002; Duffy 2006). Outcomes vary depending on whether internal (state–society relations) or external (state–state relations) sovereignty is being considered. Van Amerom (2002) argued that states viewed as relatively weak within the international community can strengthen external sovereignty claims by joining TBPA agreements while simultaneously ceding internal sovereignty to local communities. Similarly for Büscher (2013), the transfer of sovereignty to a supranational entity is often disconnected from facts on the ground, exposing the limits of states' internal sovereignty claims. In sum, opposing and contradictory state sovereignty outcomes in TBPA establishment often stand in tension. It is precisely this tension that Lunstrum (2013) chose to make the focus of her investigation of TBPAs. Through her concept of "articulated sovereignty" she posits sovereignty "as a set of powers articulated through interactions among various actors located within and beyond the state," rather than an undifferentiated authority grounded in the centralized, territorial state (2013: 3).

TBPAs logically extend from and are often composed of community-based conservation (CBC) initiatives, which emerged in the late 1980s (Neumann 1997, 1998). CBCs, as instituted in post-colonial states across the global South, are typically characterized by efforts to connect conservation to the range of neoliberal political-economic "reforms" that began with the International Monetary Fund's structural adjustment agreements of the early 1980s and which emphasized privatization and the devolution of political power from the state center (Watts 1994; McCarthy 2005). Building on Mbembe (2001) and Ferguson (2006), political ecologists have sought to use nature conservation as a means to understand how neoliberal reforms affect state sovereignty in Africa (Igoe and Brockington 2007; Igoe et al. 2010; Neves and Igoe 2012). They argue that while these reforms have reduced the capacity of African states to govern their territories, sovereignty nevertheless remains largely located in state centers, where the legitimacy and power it symbolizes can be leveraged to state actors' advantage. Non-state institutions such as NGOs and private companies benefit from the legitimacy that states bestow on their enterprises in the name of nature conservation. In these analyses, sovereignty is variously referred to as a "commodity" (Igoe and Brockington 2007: 440), a "currency" (Igoe et al. 2010: 494), and a form of "capital" (Neves and Igoe 2012: 173). Metaphors of fungibility and exchangeability thus structure this particular conceptualization of sovereignty in relation to nature conservation.

These latter investigations into sovereignty have highlighted the significance of non-state institutions, specifically environmental NGOs and transnational governance institutions. Early political ecology studies investigating the role of NGOs in the global South—particularly big international NGOs, or BINGOs—emphasized how these distantly located institutions invoked both moral and scientific authority in encouraging states to usurp local control and curtail local use of conservation territories (Neumann 1992; Peluso 1993; Bryant and Bailey 1997). Studies traced much of the social and political conflict around PAs to the colonial-era roots of BINGOs involved in conservation (Neumann 1996, 1998; Adams and Mulligan 2003). The end of formal colonialism heightened the importance of transnational governance institutions like the

United Nations Education, Scientific, and Cultural Organization (UNESCO), and the International Union for the Conservation of Nature (IUCN), which funneled money, expertise, and knowledge to support nature conservation in newly independent states (Adams 2001). UNESCO and IUCN subsequently supported the development of what eventually became some of the world's largest BINGOs, notably the World Wide Fund for Nature (WWF) (Neumann 2002). Reflecting their dominant-class roots, WWF and similar organizations benefited financially by cultivating personal associations with the world's business and political elite (Bonner 1993).

By the early twenty-first century, several BINGOs were able to distribute hundreds of millions of dollars annually to support nature conservation around the world as they internalized the logic of neoliberalization (Bryant 2009; Igoe et al. 2010; Büscher 2013). Neoliberal thought is manifested in BINGOs through partnerships with multinational corporations, through leadership positions filled from the ranks of corporate CEOs, and in their increasingly corporate organizational structures and cultures (Brosius 1999, Chapin 2004; Adams and Hutton 2007). As part of their effort to theorize the role of NGOs under neoliberalism, Brockington and Scholfield (2010) conducted a comprehensive survey of the conservation activities of 87 NGOs in sub-Saharan Africa. They found that both the number and size of conservation NGOs have increased dramatically in recent decades, though expenditures are greatly concentrated among a few of the largest. Moreover, the largest are powerful enough to produce or valorize certain forms of knowledge about nature and to influence nature conservation policies at the highest government levels (see also Bryant 2009; Corson 2010; Büscher 2013). Most significantly, Brockington and Scholfield argue that the biggest conservation NGOs are "facilitating economic growth, creating new commodities, promoting and legitimizing visions that require considerable alterations of nature and society" (2010: 570). In sum, NGOs, the state, and capital are intertwined and networked institutions in neoliberal conservation, a topic that will be examined thoroughly in the next section.

Capitalisms

Upon taking over the directorship of the Kenya Wildlife Service in 1994, the internationally renowned conservationist, David Western, soon reversed that country's decades-old conservation policy banning wildlife hunting. His premise was simple and rooted in neo-classical economics. African peasants and pastoralists belong to a universal species, *homo economicus*, and will adopt nature-conserving behaviors if they view conservation as more profitable than growing crops or grazing livestock. Money from sport-hunting fees, photo safaris, and wildlife products, he argued, have greater income potential than agriculture and if farmers and herders have a share of that income they will protect the goose that lays golden eggs. The policy became widely known as "use it or lose it" and initially generated heated debate within conservation circles worldwide (Baskin 1994). More than a decade earlier, however, leading conservation BINGOs, supported by the United Nations, had already outlined a similar logic for their *World Conservation Strategy* (WCS) (IUCN 1980). Not only did the WCS declare that conservation and economic development were compatible, it reasoned that they were symbiotically interlinked, a principle clearly evident in "use it or lose it."

In hindsight, we can recognize that use it or lose it and the WCS were among many early shudders worldwide that collectively announced a seismic shift toward market-based solutions in nature conservation. While the assumption of the self-interested rational actor remains central, recent market-based practices in conservation have moved well beyond the redistributive economics of use it or lose it to include speculative financing, derivatives markets, and new efforts to commodify nature (Garland 2008; Brockington and Scholfield 2010; Fairhead et al. 2012; Roth and Dressler 2012; Sullivan 2013). Political ecologists, many working in Africa, have labeled this shift "neoliberal conservation" and have suggested that it is the outcome of an underlying "shift in the conservation movement's own *conception* of [capitalist] practices" (Brockington and Duffy 2010: 470; Igoe and Brockington 2007; Büscher et al. 2012; Büscher 2013). By this they mean that the present conservation movement—or what is often referred to as "mainstream conservation" (Brockington et al. 2008: 9; Igoe et al. 2010; Neves and Igoe 2012)—is not positioned in opposition to capitalist development, as may be popularly assumed, but is materially and ideologically aligned with capitalism.[1] The unfolding twenty-first century has witnessed "a veritable explosion of scholarship examining the neoliberalization of ... conservation" (Fairhead et al. 2012) much of it building on Harvey's (2005, 2006) critique of neoliberalism and derivative of the literature on "neoliberalisation of nature" (Castree 2003, 2008a, 2008b; Heynen et al. 2007; McCarthy and Prudham 2004).

Political ecologists and others have long understood that PAs are embedded in and constituted by capitalist social relations, characterizing them as "neatly packaged cultural experiences of environment on which substantial profits are recorded" (Smith 1984: 57). Capitalist property relations, imposed initially through the enclosure of commons, are viewed as productive of an aestheticized and commodified nature that national parks epitomize (Neumann 1998). Germic traced the origins of the world's first national parks to crises within capitalism, arguing, "overvaluation and overproduction in economic and financial sectors, and railroads in particular, had everything to do with ... the *production* of Nature in America's first parks" (2001: 7, emphasis in the original). Moreover, as suggested by the WCS, conservation BINGOs have been explicitly pursuing closer relations with private capital for decades.

Hence the question arises, what is new about neoliberal conservation? The self-defined "epistemic community" (Brockington and Duffy 2010: 479) of neoliberal conservation scholars recognizes that conservation was never "a domain separate and set apart from capitalism" (Brockington and Scholfield 2010: 552) and they have endeavored to pinpoint the novelties of neoliberalism. They have suggested that neoliberal conservation is characterized by an "increase in the intensity and variety of forms of capitalist conservation" (Brockington and Duffy 2010: 470) with "many more players implicated, who are more deeply embedded in capitalist networks" (Fairhead et al. 2012). Parallel with Harvey's (1990) critique of postmodernism, the neoliberal conservation critique notes a marked shift toward "spectacularizing" nature "to open new conservation spaces for capitalist expansion" while simultaneously masking the resulting ecological and social contradictions (Büscher et al. 2012: 18). Existing features of capitalist nature conservation are "much more pervasive, precise and explicit" and efforts to derive exchange value from nature without consuming it "have become increasingly elaborate and detailed" (Neves and Igoe 2012: 175). Most importantly, the neoliberal literature has demonstrated that mainstream conservation has now internalized the logic of capitalism to an historically unprecedented extent (Igoe and Brockington 2007; Büscher et al. 2012; Büscher 2013; Sullivan 2013).

Many of the studies of neoliberal conservation have little to say about the effects on the daily lives of people resident near PAs or about the types of place-based politics generated by the neoliberal turn in conservation (cf. Büscher 2013). Some of these questions are answered by a recent spate of political ecology studies guided by Marx's (1967) concept of primitive accumulation and Harvey's (2003, 2005) reformulation, which he calls "accumulation by dispossession" (Kelly 2011; Corson 2011; Ojeda 2012; Neves and Igoe 2012; Benjaminsen and Bryceson 2012; Benjaminsen et al. 2013; Dressler et al. 2013). While significantly overlapping with neoliberal conservation critiques, this literature places greater emphasis on classical Marxist political economy in its analyses. Marx defined primitive accumulation as a process of separating producers from the means of production (land) through the enclosure of commons and privatization of property. However, though the creation of PAs also requires enclosure, it takes land out of production and does not involve privatization, raising the question of the relevance of primitive accumulation to PA analysis. Kelly (2011) suggests the answer to this paradox can be found in the way that acts of enclosure are often spatially and temporally distanced from sites of accumulation.

Several studies have confronted this paradox of taking land out of production for conservation as an accumulation strategy through empirical analyses of political-economic outcomes. PA enclosures with vastly different ecologies and spatial properties were found to funnel capital accumulation similarly to different sets of powerful actors distantly situated (Benjaminsen and Bryceson 2012). Dispossessed populations around Tanzanian PAs have resisted their establishment and attempted to cut off the flow of capital accumulation by disrupting ecotourist activities (Benjaminsen et al. 2013). Interestingly, and in contrast to the neoliberal literature emphasizing the state's diminished role, these studies suggest "a reconsolidation of wealth and rent-seeking power by the state" (Benjaminsen et al. 2013: 1089). Bringing concepts from agrarian political economy to bear on market-oriented conservation in a PA in Vietnam, researchers found that socioeconomic differentiation in the adjacent village accelerated and wealth disparities increased (Dressler et al. 2013). Some authors have adopted the phrase "green grabbing"—"the appropriation of land and resources for environmental ends"—to characterize such outcomes (Fairhead et al. 2012: 238; Ojeda 2012). Drawing from ethnographic work near a Colombian PA, Ojeda (2012) emphasizes that multiple forms of violence against local residents may accompany green grabbing. Lunstrum (2008), however, found that privatization schemes on the border of a Mozambican PA have strengthened local community land rights. Similarly, and in contrast to much of the political ecology literature on the neoliberal turn, Gardner (2015) found that Maasai pastoralists have responded to green grabs by aligning themselves with private capital as a defense against further land dispossessions by the state. According to Gardner, under the pressures of green grabbing, Maasai ethnic identity has been reoriented in relation to the state and private capital, a subject that is covered in depth in the next section.

Identities

Assessing the early twentieth-century conservation movement in the United States, Donna Haraway wrote, "The owners of the great machines of monopoly capital were, with excellent reason, at the forefront of nature work—because it was one of the means of production of race, gender, and class" (1989: 54). A central project in this nature work was the establishment of the great national parks of the western United States. These were meant to forever fix in space an idealized representation of American wilderness, which could serve as sites where American national identity could be renewed and revitalized in the decades to come (Nash 1982). As Haraway demonstrated, the identity in need of constant renewal was white, Anglo Saxon, bourgeois, and male; embodied in the political patron of American nature conservation, Theodore Roosevelt, and encapsulated in his ethos of following a "strenuous life" outdoors (see Neumann 2013). Thus PAs, especially national parks, have from the beginning functioned as important "meaning making machines" (Haraway 1989: 54) for producing identities of nationality, race, ethnicity, gender, and class. Sundberg's research question from her ethnographic work at a Guatemalan PA captures political ecology's agenda perfectly: "how is conservation-in-the-making constitutive of *identities-in-the-making?*" (2004: 43, emphasis in the original; see also Haraway 1997).

In formulating this question, Sundberg followed Butler's (1990, 1993) ideas of gender performativity. Sundberg was interested in how the practices of conservation might constitute race and gender identities and reinforce or challenge existing inequities among social groups. Informed by Foucault's (1979, 1991) studies of the constitution of modern subjectivities, Peluso and Vandergeest (2001) take a more discursive approach to demonstrate the role of PA establishment in constituting ethnic and racial identities. According to Peluso and Vandergeest, the "ways colonial governments resolved the question of native or 'minority' rights to land and forest resources contributed to the creation of racialized colonial categories for people (as 'natives,' 'Foreign Orientals,' 'primitives,' or 'minorities,') and frequently territorialized these identities as well as patterns of resource access" (2001: 800). Drawing inspiration from these insights, Ybarra (2012) links the discursive representation of nature as either "forest" or "jungle" to racializing processes that constitute social identities (e.g., indigenous, squatter) and determine who belongs and who must be removed or barred. Elmhirst advances Peluso and Vandergeest's work by employing feminist and queer theory "to explore the ways in which the 'political forest' might be interpreted as a gendered project," which can reveal much "about subjectivity, personhood and gendered citizenship in Indonesia" (2011: 173, 181; see also Chapter 40, this volume).

Such political ecology studies have elucidated, from different theoretical positionings, how "subject formation is at stake in conservation," whether by the subjects themselves or by others (Sundberg 2006: 242; see also Peluso and Lund 2011; Moore 2005; Kosek 2006). As suggested in this chapter's introduction, a key dimension of subject identity in PA establishment is nationality. Beginning with the establishment of the world's first in the western United States at Yosemite and Yellowstone, national parks have played a critical role in helping to define a national identity rooted in the landscapes of nature (Runte 1979; Nash 1982; Cronon 1995). Yellowstone and Yosemite linked ideas of landscape, nature, and national identity in a way that has been emulated in nearly every country of the globe (Olwig 2002; Schwartz 2006). In the United States, Finney (2014) has argued that national parks have been constituted as "white spaces," with African Americans rendered invisible in wild nature and therefore lacking full membership in American national identity.

The co-constitution of nature and indigeneity has been a particularly fruitful area of study that has revealed conservationist positions fraught with contradictions and produced complex and contrasting findings. The primary contradiction is found in conservationist discourses of indigenous peoples, which has essentialized them as either destructive or noble "savages" (Neumann 1995, 2004). Destructive savages must be removed and controlled, noble savages can stay as pre-modern caretakers of nature, an essentializing process Pulido refers to as an assessment of "ecological legitimacy" (1996: 37). Ecological legitimacy refers to whether or not a social group is considered able land stewards and resource managers and therefore is entitled to access and residence rights (see also Neumann 1997).

Marginalized ethnic groups have embraced the caretaker image of indigenous peoples as a means to gain support for their proprietary claims on land and resources. These efforts to gain political agency have been labeled "strategic essentialism" (Sylvain 2002: 1081) or "representational strategies" (Tsing 1999: 160) in recognition of people's agency in subject formation. Thus, indigenous peoples' identity politics help position them as "icons of resistance for environmentalists worldwide" (Brosius 1997: 48). Pulido (1996) has framed an ethnic group's appropriation of the caretaker identity as a form of resistance. In a study of a similar appropriation in Guatemala, Sundberg (2003, 2004) suggests that it is not a question of resistance but of the way competing segments of a society position themselves to take advantage of shifting power relations. Under the new PA power structure in Guatemala, certain long-term residents appropriate outside stereotypes as a means to position themselves as authentic caretakers. Furthermore, these discourses of authenticity are articulated most powerfully by educated men, two effects of which are the sharpening of internal divisions along class, gender, and ethnic lines and the conversion of large swaths of forest into men's space. In sum, as the etymologies of "nature" and "conservation" suggest, these investigations demonstrate how the work of nature conservation is entangled with the generation of race, ethnicity, gender, class, and national identities and their associated rights and privileges.

Conclusion

Already empirically and theoretically rich, the political ecology literature on nature conservation appears to be on a trajectory of continued expansion. As it has grown, this literature has contributed much to the maturation of political ecology in general. One of the most significant impacts of this literature has been to raise awareness of the centrality of nature conservation in the generation of new subjectivities and state formation, revealing a particular kind of environmental politics that had remained largely unexplored. Equally important are the interventions that focus on the spatial dimensions of nature conservation, offering contributions to the theorization of scale, territoriality, and the effects of boundary-making. Finally, this literature, in particular the critiques of neoliberal conservation, have added to our theoretical understanding of the value of nature under capitalism, a topic identified as a shortcoming in political ecology (Robertson and Wainwright 2013).

Yet, there is work ahead. In closing, I offer a few thoughts on potential future directions. Beginning with the project I labeled "Capitalisms," the work on neoliberal conservation has been extremely productive and necessary. It might be best viewed, however, as a starting point, rather than an end in itself. The literature too often conveys a sense of sudden discovery that nature conservation is thoroughly embedded in capitalist political economy. If the neoliberal turn of the last four decades has reoriented the logic of every conceivable cultural and political institution, including philanthropy, public education, medicine, and so on, it should come as little surprise that the same is also true of conservation. As key authors of the neoliberal critique have noted, "If anything, it would be more interesting to look for conservation strategies that are *un*touched by neoliberalism" (Brockington and Duffy 2010: 480). This may be one fruitful investigation to pursue. More generally, it might be productive to now look more closely at the *effects* of the neoliberal turn on subjectivity formation and the daily lives and life trajectories of people in place at these sites of neoliberal conservation (e.g., Gardner 2015).

Finally, when I initially outlined this chapter I had included a section labeled "Ecologies," which I subsequently abandoned for lack of material. The shortage of actual ecological work in political ecology is an old debate (Walker 2005). I was nevertheless struck by the paucity of it in a literature focused on a critique of mainstream conservation of biodiversity and habitats. Campbell's work showing how new genetic technologies shape the classification of species with direct implications for questions of scale, property rights, and sovereignty in marine conservation, while excellent in its own right, stands out precisely because there is not much else like it (Campbell 2007; Campbell and Godfrey 2010). This is a small example of the significant biotechnological and ecological developments unfolding that have direct relevance to nature conservation and could be productive areas for political ecological studies. For instance, the science of species reintroduction has matured and conservation biologists are currently debating the new field of resurrection biology—using DNA fragments of extinct species for re-assemblage or use in cloning. How might these manipulations of ecosystems relate to the questions raised in this chapter's introduction? The global-scale demographic shift from rural to urban is accelerating, leaving rural areas abandoned and altering existing socionatures in interesting ways, such as forest resurgence (Hecht 2014; Neumann 2014). Last, but not least, global climate change seems poised to deliver the most serious challenge to the territorialization of nature conservation. Species' biogeographies are shifting, hydro regimes are changing, new species migrations are appearing and existing ones fading, all of which collectively threaten to make current PA boundaries obsolete, making the politics they generated obsolete as well. What comes next in nature conservation is a question political ecologists might find worth pursuing through a closer engagement with both biological ecology and science and technology studies.

Note

1 While the foci of this chapter and of most of the political ecology literature are traditional state-based PAs, it must be noted that the neoliberal turn has also resulted in a greater emphasis on privatized conservation spaces, the Nature Conservancy being a prime example.

References

Adams, W. 2001. *Green Development: Environment and Sustainability in the Third World*. 2nd Edn. London: Routledge.

Adams, W. and M. Mulligan (eds.) 2003. *Decolonizing Nature: Strategies for Conservation in a Post-colonial Era*. London: Earthscan.

Adams, W. and J. Hutton. 2007. People, parks and poverty: political ecology and biodiversity conservation. *Conservation and Society* 5(2): 147–183.

Agrawal, Arun and Kent Redford. 2009. Place, conservation, and displacement. *Conservation and Society* 7(1): 56–58.

Baskin, Yvonne. 1994. There's a new wildlife policy in Kenya: use it or lose it. *Science* 265: 733.

Beazley, Kim. 2011. Spaces of opportunity: state-oustee relations in the context of conservation-induced displacement in Central India. *Pacific Affairs* 84(1): 25–46.

Benjaminsen, T.A. and I. Bryceson. 2012. Conservation, green/blue grabbing and accumulation by dispossession in Tanzania. *Journal of Peasant Studies* 39(2): 335–355.

Benjaminsen, Tor A., Mara J. Goldman, Maya Y. Minwary and Faustin P. Maganga. 2013. Wildlife management in Tanzania: state control, rent seeking and community resistance. *Development and Change* 44(5): 1087–1109.

Bonner, R. 1993. *At the Hand of Man. Peril and Hope for Africa's Wildlife*. London: Simon & Schuster.

Brockington, D. 2002. *Fortress Conservation. The Preservation of the Mkomazi Game Reserve, Tanzania*. Oxford: James Currey.

Brockington, D. and R. Duffy. 2010. Capitalism and conservation: the production and reproduction of biodiversity conservation. *Antipode* 42(3): 469–484.

Brockington, D. and J. Igoe. 2006. Eviction for conservation: a global overview. *Conservation & Society* 4: 424–470.

Brockington, Dan and Katherine Scholfield. 2010. The Conservationist Mode of Production and Conservation NGOs in sub-Saharan Africa. *Antipode* 42(3): 551–575.

Brockington, D., R. Duffy and J. Igoe. 2008. *Nature Unbound: Conservation, Capitalism and the Future of Protected Areas*. London: Earthscan.

Brooks, T.M., R.A. Mittermeier, G.A.B. da Fonseca, J. Gerlach, M. Hoffmann, J.F. Lamoreux, C.G. Mittermeier, J.D. Pilgrim, and A.S.L. Rodrigues. 2006. Global biodiversity conservation priorities. *Science* 313(5783): 58–61.

Brosius, P. 1997. Endangered forest, endangered people: environmentalist representations of indigenous knowledge. *Human Ecology* 25(1): 47–68.

Brosius, J.P. 1999. Green dots, pink hearts: displacing politics from the Malaysian rain forest. *American Anthropologist* 101: 36–57.

Bryant, R. 1997. *The Political Ecology of Forestry in Burma, 1824–1994*. Honolulu: University of Hawaii Press.

Bryant, R. 2009. Born to be wild? Nongovernmental organisations, politics and the environment. *Geography Compass* 3/4: 1540–1558.

Bryant, R.L. and Bailey, S. 1997. *Third World Political Ecology*. London: Routledge.
Büscher, Bram. 2013. *Transforming the Frontier: Peace Parks and the Politics of Neoliberal Conservation in Southern Africa*. Durham, NC and London: Duke University Press.
Büscher, Bram, Sian Sullivan, Katja Neves, Jim Igoe and Dan Brockington. 2012. Towards a synthesized critique of neoliberal biodiversity conservation. *Capitalism Nature Socialism* 23(2): 4–30.
Butler, Judith. 1990. *Gender Trouble. Feminism and the Subversion of Identity*. New York: Routledge.
Butler, Judith. 1993. *Bodies that Matter: On the Discursive Limits of "Sex"*. New York: Routledge.
Campbell, Lisa M. 2007. Local conservation practice and global discourses: a political ecology of sea turtle conservation. *Annals of the Association of America Geographers* 97(2): 313–334.
Campbell, Lisa M. and Matthew H. Godfrey. 2010. Geo-political genetics: claiming the commons through species mapping. *Geoforum* 41(6): 897–907.
Castree, N. 2003. Commodifying what nature? *Progress in Human Geography* 27(3): 273–297.
Castree, N. 2008a. Neoliberalising nature I: the logics of de- and re-regulation. *Environment and Planning A* 40(1): 131–152.
Castree, N. 2008b. Neo-liberalising nature II: processes, outcomes and effects. *Environment and Planning A* 40(1): 153–173.
Cernea, Michael and Kai Schmidt-Sotau. 2006. Poverty risks and national parks: policy issues in conservation and resettlement. *World Development* 34(10): 1808–1830.
Chapin, M. 2004. A challenge to conservationists. *World Watch Magazine* Nov/Dec: 17–31.
Chatty, D. and M. Colchester. (eds.). 2002. *Conservation and Mobile Indigenous Peoples: Displacement, Forced Settlement, and Sustainable Development*. New York and Oxford: Berghahn Books.
Corson, C. 2010. Shifting environmental governance in a neoliberal world: US aid for conservation. *Antipode* 42(3): 576–602.
Corson, C. 2011. Territorialization, enclosure and neoliberalism: non-state influence in struggles over Madagascar's forests. *Journal of Peasant Studies* 38(4): 703–726.
Cronon, William. 1995. The trouble with wilderness, or getting back to the wrong nature. In Cronon, W. (ed.), *Uncommon Ground: Toward Reinventing Nature*. New York: WW Norton & Co., pp. 69–90.
Dowie, M. 2009. *Conservation Refugees: The Hundred-Year Conflict Between Global Conservation and Native Peoples*. Cambridge, MA and London: The MIT Press.
Dressler, Wolfram H., Phuc Xuan To and Sango Mahanty. 2013. How biodiversity conservation policy accelerates agrarian differentiation: the account of an upland village in Vietnam. *Conservation and Society* 11(2): 130–143.
Duffy, R. 2006. The potential and pitfalls of global environmental governance: the politics of transfrontier conservation areas in Southern Africa. *Political Geography* 25: 89–112.
Elmhirst, Rebecca. 2011. Migrant pathways to resource access in Lampung's political forest: gender, citizenship and creative conjugality. *Geoforum* 42: 173–183.
Fairhead, James, Melissa Leach and Ian Scoones. 2012. Green grabbing: a new appropriation of nature? *The Journal of Peasant Studies* 39(2): 237–261.
Ferguson, J. 2006. *Global Shadows. Africa in the Neoliberal World Order*. Durham, NC and London: Duke University Press.
Finney, Carolyn. 2014. *Black Face, White Spaces: Reimagining the Relationship of African Americans to the Great Outdoors*. Chapel Hill: University of North Carolina Press.
Foucault, M. 1979. *Discipline and Punish: The Birth of the Prison*. New York: Vintage.
Foucault, M. 1991. Governmentality. In: Burchall, G., C. Gordon and P. Miller (eds.), *The Foucault Effect: Studies in Governmentality*. Chicago: University of Chicago Press, pp. 87–104.
Foucault, M. 2007. *Security, Territory, Population: Lectures at the College de France, 1977–1978*. New York: Picador.
Gardner, B. 2015. *Selling the Serengeti: The Cultural Politics of Safari Tourism in Tanzania*. Athens, GA: University of Georgia Press.
Garland, E. 2008. The elephant in the room: confronting the colonial character of wildlife conservation in Africa. *African Studies Review* 51(3): 51–74.
Germic, S. 2001. *American Green: Class, Crisis, and the Deployment of Nature in Central Park, Yosemite, and Yellowstone*. Lanham: Lexington Books.
Guyot, S. 2011. The eco-frontier paradigm: rethinking the links between space, nature and politics. *Geopolitics* 16(3): 675–706.
Haraway, Donna. 1989. *Primate Visions: Gender, Race, and Nature in the World of Modern Science*. New York: Routledge.

Haraway, Donna. 1997. *Modest_Witness@Second_Millenium.FemaleMan©_Meets_OncoMouseTM. Feminism and Technoscience.* New York: Routledge.
Harvey, D. 1990. *The Condition of Postmodernity: An Enquiry into the Origins of Cultural Change.* Oxford: Blackwell Publishers Inc.
Harvey, D. 2003. *The New Imperialism.* Oxford: Oxford University Press.
Harvey, D. 2005. *A Brief History of Neoliberalism.* Oxford: Oxford University Press.
Harvey, D. 2006. *Spaces of Global Capitalism: A Theory of Uneven Geographical Development.* London: Verso.
Hazen, Helen D. and Leila M. Harris. 2007. Limits of territorially-focused conservation: a critical assessment based on cartographic and geographic approaches. *Environmental Conservation* 34(4): 280–290.
Hecht, S. 2014. The social lives of transitions and successions: theories of forest resurgence. In: Hecht, S., K. Morrison and C. Padoch (eds.), *The Social Lives of Forests: The Past, Present, and Future of Woodland Resurgence.* Chicago: University of Chicago, pp. 97–113.
Heynen, N., J. McCarthy, S. Prudham and P. Robbins. 2007. *Neoliberal Environments: False Promises and Unnatural Consequences.* London and New York: Routledge.
Holmes, George. 2014. Defining the forest, defending the forest: political ecology, territoriality, and resistance to a protected area in the Dominican Republic. *Geoforum* 53: 1–10.
Hoole, Arthur and Fikret Berkes. 2010. Breaking down fences: recoupling social-ecological systems for biodiversity conservation in Namibia. *Geoforum* 41: 304–317.
Igoe, J. and D. Brockington. 2007. Neoliberal conservation: a brief introduction. *Conservation and Society* 5(4): 432–449.
Igoe, J., K. Neves and D. Brockington. 2010. A spectacular eco-tour around the historic bloc: theorising the convergence of biodiversity conservation and capitalist expansion. *Antipode* 42(3): 486–512.
Ioris, A.A.R. 2012. Applying the strategic-relational approach to urban political ecology: the water management problems of the baixada fluminense, Rio de Janeiro, Brazil. *Antipode* 44(1): 122–150.
IUCN. 1980. *World Conservation Strategy.* Geneva: International Union for the Conservation of Nature and Natural Resources.
Kelly, A. 2011. Conservation practice as primitive accumulation. *Journal of Peasant Studies* 38(4): 683–701.
Kosek, J. 2006. *Understories: The Political Life of Forests in Northern New Mexico.* Durham, NC and London: Duke University Press.
Latour, B. 1993. *We Have Never Been Modern.* Cambridge, MA: Harvard University Press.
Laudati, A. 2010. The encroaching forest: struggles over land and resources on the boundary of Bwindi Impenetrable National Park, Uganda. *Society & Natural Resources* 23(8): 776–789.
Li, T. 2007. *The Will to Improve: Governmentality, Development, and the Practice of Politics.* Durham, NC and London: Duke University Press.
Lunstrum, E. 2008. Mozambique, neoliberal land reform, and the Limpopo National Park. *Geographical Review* 98: 339–355.
Lunstrum, Elizabeth. 2013. Articulated sovereignty: extending Mozambican state power through the Great Limpopo Transfrontier Park. *Political Geography* 36(1): 1–11.
Marx, K. 1967. *Capital.* New York: International.
Mbembe, A. 2001. *On the Postcolony.* Berkeley: University of California Press.
McCarthy, J. 2005. Devolution in the woods: community-based forestry as hybrid neoliberalism. *Environment and Planning A* 37(6): 995–1014.
McCarthy, J. and S. Prudham. 2004. Neoliberal nature and the nature of neoliberalism. *Geoforum* 35: 275–283.
Moore, D. 2005. *Suffering for Territory: Race, Place, and Power in Zimbabwe.* Durham, NC and London: Duke University Press.
Nash, R. 1982. *Wilderness and the American Mind.* New Haven: Yale University Press.
Neumann, R. 1992. The political ecology of wildlife conservation in the Mount Meru Area, Northeast Tanzania. *Land Degradation and Rehabilitation* 3(2): 85–98.
Neumann, R. 1995. Ways of seeing Africa: colonial recasting of African society and landscape in Serengeti National Park. *Ecumene* 2(2): 149–169.
Neumann, R. 1996. Dukes, earls and ersatz Edens: aristocratic nature preservationists in colonial Africa. *Society and Space* 14: 79–98.
Neumann, R. 1997. Primitive ideas: protected area buffer zones and the politics of land in Africa. *Development and Change* 28(3): 559–582.
Neumann, R. 1998. *Imposing Wilderness: Struggles over Livelihood and Nature Preservation in Africa.* Berkeley: University of California Press.

Neumann, R. 2001. Africa's "last wilderness": reordering space for political and economic control in Colonial Africa. *Africa* 71(4): 641–665.

Neumann, R. 2002. The postwar conservation boom in British colonial Africa. *Environmental History* 7(1): 22–47.

Neumann, R. 2004. Moral and discursive geographies in the war for biodiversity in Africa. *Political Geography* 23(7): 813–837.

Neumann, R. 2005. *Making Political Ecology.* London and New York: Hodder Arnold.

Neumann, R. 2013. Churchill and Roosevelt in Africa: performing and writing landscapes of race, empire and nation. *Annals of the Association of American Geographers* 103(6): 1371–1388.

Neumann, R. 2014. Stories of nature's hybridity in Europe: implications for forest science and policy in the Global South. In Hecht, S., K. Morrison and C. Padoch (eds.), *The Social Lives of Forests: The Past, Present, and Future of Woodland Resurgence.* Chicago: University of Chicago Press, pp. 31–44.

Neves, K. and J. Igoe. 2012. Uneven development and accumulation by dispossession in nature conservation: comparing recent trends in the Azores and Tanzania. *Tijdschrift voor Economische en Sociale Geografie* 103(2): 164–179.

Ojeda, D. 2012. Green pretexts: ecotourism, neoliberal conservation and land grabbing in Tayrona National Natural Park, Columbia. *Journal of Peasant Studies* 39(2): 357–375.

Olwig, Kenneth. 2002. *Landscape, Nature and the Body Politic: From Britain's Renaissance to America's New World.* Madison: University of Wisconsin Press.

Peluso, N.L. 1993. Coercing conservation? The politics of state resource control. *Global Environmental Change* 3(2): 199–217.

Peluso, N.L. and C. Lund. 2011. New frontiers of land control: introduction. *Journal of Peasant Studies* 38(4): 667–681.

Peluso, N.L. and P. Vandergeest. 2001. Genealogies of the political forest and customary rights in Indonesia, Malaysia, and Thailand. *Journal of Asian Studies* 60(3): 761–812.

Peluso, N.L. and P. Vandergeest. 2011. Political ecologies of war and forests: counterinsurgencies and the making of national natures. *Annals of the Association of American Geographers* 101(3): 587–608.

Pulido, L. 1996. Ecological legitimacy and cultural essentialism: Hispano grazing in the Southwest. *Capitalism Nature Socialism* 7(4): 37–58.

Redford, K.H., P. Coppolillo, E.W. Sanderson, G.A.B. Da Fonseca, E. Dinerstein, C. Groves, G. Mace, S. Maginnis, R.A. Mittermeier, R. Noss, D. Olson, J.G. Robinson, A. Vedder and M. Wright. 2003. Mapping the conservation landscape. *Conservation Biology* 17(1): 116–131.

Robertson, M. and J. Wainwright. 2013. The value of nature to the state. *Annals of the Association of American Geographers* 103(4): 890–905.

Roth, R. 2008. "Fixing" the forest: the spatiality of conservation conflict in Thailand. *Annals of the Association of American Geographers* 98(2): 373–391.

Roth, R. and Dressler, W. 2012. Market-oriented conservation governance: the peculiarities of place. *Geoforum* 43(3): 363–366.

Runte, A. 1979. *National Parks: The American Experience.* Lincoln: University of Nebraska Press.

Schwartz, K. 2006. *Nature and National Identity after Communism: Globalizing the Ethnoscape.* Pittsburg: University of Pittsburg Press.

Scott, J. 1998. *Seeing Like a State: How Certain Schemes to Improve the Human Condition Have Failed.* New Haven: Yale University Press.

Sletto, Bjørn. 2011. Conservation planning, boundary-making and border terrains: the desire for forest and order in the Gran Sabana, Venezuela. *Geoforum* 43: 363–366.

Smith, N. 1984. *Uneven Development: Nature, Capital, and the Production of Space.* New York: Basil Blackwell.

Sullivan, Sian. 2013. Banking nature? The spectacular financialisation of environmental conservation. *Antipode* 45(1): 198–217

Sundberg, J. 2003. Strategies for authenticity and space in the Maya Biosphere Reserve, Petén, Guatemala. In Zimmerer, K. and T. Bassett (eds.), *Political Ecology: An Integrative Approach to Geography and Environment-Development Studies.* New York: Guilford, pp. 50–69.

Sundberg, J. 2004. Identities in the making: conservation, gender and race in the Maya Biosphere Reserve, Guatemala. *Gender, Place & Culture* 11(1): 43–66.

Sundberg, J. 2006. Conservation encounters: transculturation in the "contact zones" of empire. *Cultural Geographies* 13: 239–265.

Sylvain, R. 2002. "Land, water, and truth:" San identity and global indigenism. *American Anthropologist* 104(4): 1074–1085.

Tsing, A. 1999. Becoming a tribal elder and other green development fantasies. In Li, T. (ed.), *Transforming the Indonesian Uplands*. London: Harwood, pp. 159–202.
Vaccaro, I., O. Beltran and P.A. Paquet. 2013. Political ecology and conservation policies: some theoretical genealogies. *Journal of Political Ecology* 20: 255–272.
van Amerom, Marloes. 2002. National sovereignty and transboundary protected areas in Southern Africa. *GeoJournal* 58: 265–273.
Vandergeest, P. and N. Peluso. 1995. Territorialization and state power in Thailand. *Theory and Society* 24: 385–426.
Walker, P. 2005. Political ecology: where is the ecology? *Progress in Human Geography* 29(1): 73–82.
Watts, M. 1994. Development II: the privatization of everything? *Progress in Human Geography* 18(3): 371–384.
Whitehead, Mark, Rhys Jones, and Martin Jones. 2007. *The Nature of the State: Excavating the Political Ecologies of the Modern State*. Oxford: Oxford University Press.
Williams, R. 1983. *Keywords: A Vocabulary of Culture and Society*, Revised Edition. New York: Oxford University Press.
Ybarra, M. 2012. Taming the jungle, saving the Maya Forest: sedimented counterinsurgency practices in contemporary Guatemalan conservation. *Journal of Peasant Studies* 39(2): 479–502.
Zimmerer, K. 2000. The reworking of conservation geographies: nonequilibrium landscapes and nature-society hybrids. *Annals of the Association of American Geographers* 90(2): 356–369.
Zimmerer, K. 2006. Cultural ecology: at the interface with political ecology – the new geographies of environmental conservation and globalization. *Progress in Human Geography* 30(1): 63–78.

31

THE POLITICAL ECOLOGY OF INTERNATIONAL AGRI-FOOD SYSTEMS

Derek Hall

Introduction

Research on food and agriculture has played a vital role in the development of political ecology. Core studies in this interdisciplinary literature have explored the "agri-food systems" that connect agricultural production to the work of input provision, transportation, processing, and marketing that goes on in the food sector. Such systems link different parts of the world through flows of (among other things) cultivars, farming techniques, agrochemicals, capital, and food, fuel and fiber crops themselves. In studying international agri-food systems, political ecologists have focused especially on interactions between their environmental characteristics and dynamics of political economy, meaning, and conflict across scales. Their research has encompassed the history of agrarian structures, class and gender relations in agriculture, state and corporate power, the ecological conditions and consequences of different forms of agri-food organization, the ways in which food, agriculture, and the environment are understood and valued, and the profound inequalities that underpin the production, transformation, and consumption of food. It has shed light upon the complexities of agricultural restructuring, trade, and environmental degradation, and on the fundamental question of who gets to eat what (and how much of it).

Global connections in agriculture are not new. Environmental historians have explored the epochal agricultural transformations set in train by European imperialism that were central to the rise of capitalism (Pomeranz 2000; Moore 2010). These included the introduction of potatoes and tomatoes to Europe, wheat and cattle to the Americas, and sweet potatoes to China and Southeast Asia, and the rise of global trade crops like cod, coffee, cotton, pepper, sugar, and tobacco (Crosby 1986; Mann 2011). There has also been extensive critical research on post-World War II food exports and food aid from the North to the South and on the broader expansion and unsustainability of the industrial agricultural model developed in (especially) the United States (Friedmann and McMichael 1989; Clapp 2012). Such studies show how even farming for local or domestic consumption is usually embedded in international systems and exchanges. They also highlight the vast political-ecological consequences of five centuries of agricultural globalization, including deforestation and other types of landscape change, loss of biodiversity, and movements of peoples.

I focus specifically in this chapter on a subset of international agri-food systems research in political ecology: the analysis of the rapid expansion since the 1970s in Southern production of relatively high-value crops for export to (usually) the North. Long-established Southern export commodities like bananas, cocoa, coffee, palm oil, and soy have been joined by newer exports like fresh fruits and vegetables, cut flowers, and fresh or frozen meat and fish. Recent decades have seen spectacular booms (and some equally spectacular crashes) in the export-oriented production of these foodstuffs. The "biographies" of these commodities vividly illustrate the themes of a global political ecology, including the ways in which consumption in the North often drives environmental conflict and degradation in the South. Some of these stories are widely known: the consequences of shrimp farming for the coastal ecosystems of Southeast Asia and Central America (Stonich and Bailey 2000) and of salmon farming for those of Chile (Barton and Fløysand 2010); the spread of "the globalized soybean and burgeoning livestock sector" in the forests and grasslands of Brazil (Hecht 2011: 9; see also Walker *et al.* 2009); the deforestation, biodiversity loss, and dispossession attendant upon millions of hectares of new oil palm plantations in Indonesia and Malaysia (Hall *et al.* 2011: 90–98); and "the combined, interrelated livelihood and ecological crisis" in Meso-America that followed the collapse of coffee prices at the turn of the millennium (Goodman 2008: 3).

My goal is to assess how research in political ecology and closely related fields has analyzed and explained the expansion of this high-value agricultural production (for simplicity's sake, I refer to the products, plant or animal, generically as "crops"). I begin by surveying changes in the international political economy of agriculture since the 1970s that have been central to the process. I then discuss how political ecologists have studied the production of these crops, with a focus on environmental degradation and conflict in the places where they are farmed. The next section takes up processing, distribution, and consumption by asking how ecology and environmental politics have been integrated into the study of international commodity chains. The last main section raises the possibility that the basic parameters of international agri-food systems may be changing by introducing research on the "global land grab."

The studies I cite are just a portion of a vast, interrelated body of literatures that rely on diverse (and overlapping) frameworks and concepts. Research on the political ecology of agriculture and aquaculture is not always easily distinguishable from that on agrarian political economy more generally (see Chapter 42, this volume). It also shades into the broader study of the links between agricultural production and input provision, processing, distribution, and consumption. I consider research to be political ecology when it approaches agri-food systems by linking critical political economy to a concern for the specificity of biophysical processes and what might be called the agency of nature. I highlight that kind of work without worrying too much about what it calls itself, though I do return to the relationship between political ecology and agrarian political economy in discussing the "global land grab." One limitation of my analysis is that I refer only infrequently to capture fisheries. These fisheries play a prominent role in Southern food exports, but because their political ecology is quite distinct from that of agriculture and aquaculture, I mostly omit them from the discussion (though see Mansfield 2011; Campling *et al.* 2012).

Changes in the international political economy of agriculture since the 1970s

Recent changes in export crop production in the South, including rapid growth in relatively high-value crops like fresh or frozen meat, fish, fruits, vegetables, and flowers, have been driven by dramatic transformations in the international political economy of agriculture. These shifts helped to stimulate a resurgence in agrarian studies that began in the 1980s (for an influential overview by two political ecologists see Watts and Goodman 1997). Some critical scholarship has understood the changes in terms of the collapse of the US-centered postwar "second food regime" of the 1950s and 1960s (Friedmann and McMichael 1989; McMichael 2009). While it is not clear whether a coherent third food regime – perhaps a "corporate" one – has emerged (McMichael 2009: 148–154; Pritchard 2009), there is widespread agreement that three trends have contributed to the expansion of Southern high-value export-oriented agriculture. First, technological changes have been indispensable. Improvements in refrigeration and in transportation (particularly air freight) were necessary for many of the new relationships to exist at all – for green beans harvested in Burkina Faso to be sold fresh in French supermarkets (Freidberg 2004; Freidberg 2009), or for bluefin tuna caught off New England to be auctioned in Tokyo the next day (Bestor 2001: 78–80). Developments in communications technology, too, have allowed far more coordination and information exchange within commodity chains.

Second, there have been major shifts in national and international agricultural policies. In the South, Structural Adjustment Programs imposed by the International Monetary Fund and the World Bank in the 1980s and 1990s liberalized agricultural imports and eliminated state supports for farming like subsidies and marketing boards. Encouraging agricultural exports was an explicit goal of these policies (Clapp 2012: 60–63). Northern countries, however, have overall done little to reduce subsidies and protection for their main crops. While the launch of the World Trade Organization in 1995 brought the Agreement on Agriculture into effect, in practice the WTO has not opened up Northern markets for staple crops but *has* facilitated the continuing growth of highly-subsidized Northern staple exports to the South (Clapp 2012: 63–74). The combination of structural adjustment and the WTO's pathologies helped create what Philip McMichael calls "a politically constructed division of agricultural labour between Northern staple grains traded for Southern high-value products (meats, fruits and vegetables)" (McMichael 2009: 148). Within this context, many Southern governments (with donor encouragement and assistance) have promoted high-value agricultural exports as a source of foreign exchange revenue and "agro-industrialization." Promotion methods have ranged from support for technological transfer to encouraging foreign direct investment to looking the other way when expanding production violated environmental and land-use regulations. Chile and Thailand are among the biggest "success stories."

Third, transnational corporations (TNCs) have dramatically increased their power over agriculture (Clapp and Fuchs 2009). Corporate control of input production and provision, food processing, distribution, and retail has become far more concentrated, with a decreasing number of increasingly huge TNCs now determining what food is sold and at what price. At points in some agricultural commodity chains, three or four corporations account for 75 percent or more of the food traded; three companies, for instance, control more than 80 percent of US corn exports (Clapp 2012: 98–100; see also Patel 2007: 11–15). While TNCs have done less to extend their direct control over production, they have entered into wide-ranging contract arrangements with farmers (Little and Watts 1994), who thus continue to be directly exposed to the many risks (weather, disease, etc.) of agriculture. Another element of expanding corporate power has been the imposition onto farmers of intensifying safety, quality, social, and environmental standards (see below).

The political ecology of high-value export-oriented agriculture and aquaculture in the south

Political ecology research on agriculture is characterized by complex, fine-grained, historically embedded accounts of intertwined transformations in political economic, ecological, and social relations (for a terrific example see Schroeder 1993). In this section, I introduce some of the key themes political ecologists weave together in analyzing export-oriented crop production. Many accounts begin with the organization of farming, identifying three main forms: plantations, on which land, techniques, and inputs are controlled by non-farming owners or managers, while workers provide labor; contract farming, in which buyers promise in advance to purchase producers' crops, and usually provide some assistance with inputs and assert some control over techniques; and production by relatively independent smallholders, who farm land that they own or rent and do not have close, long-term connections with a buyer.

Much political ecology research explains the prevalence and distribution of these three forms of farming in part by reference to the biological and ecological particularities of different crops. How long crops take to come to maturity, how quickly they have to be processed after harvest, the extent to which they can be intercropped, how labor intensive they are and how they fit into seasonal labor cycles, how dependent they are on chemical inputs: these and other characteristics shape the organization of production (for two important statements see Mann and Dickinson 1978; Walker 2004). John McCarthy, for example, finds that "the commodity-specific characteristics of oil palm" make it "a rich farmer's crop that requires expensive inputs if it is to be farmed successfully" (McCarthy 2010: 826). Cocoa and coffee, meanwhile, tend to be smallholder crops. Most analyses also emphasize, however, that crop particularities do not determine production arrangements by themselves. Agrarian histories, the extent to which land is controlled by powerful actors, and state priorities all exert their influence (Grossman 1998: 4–8; Vandergeest *et al.* 1999: 585). Crop biology and ecology are not destiny; political economy is also key.

A second central question, then, is how producers access the resources, inputs, and relationships that they need to farm, including land, labor, capital, water, agrochemicals, distribution networks, and knowledge about techniques (Ribot and Peluso 2003). Access to labor is closely bound up with the three types of production arrangements discussed above. Access to land has been marked in part by producers taking up export-oriented crop production on land they were already using for other purposes: turning rice paddies into shrimp farms, for instance, or agroforests into cocoa plantations. Other techniques have included buying or leasing land, clearing "new" land not already under permanent cultivation, and seizing land from others (Hall 2011a). Contract farming has received substantial attention because it can allow small-scale farmers to access loans, technical advice, seeds, and chemical inputs. Studies by political ecologists have found, however, that it does so at the cost of the exclusion of many producers from export production and the increased subordination to capital of those who are included. More broadly, Lawrence Grossman's statement that "the study of contract farming involves a complex drama among peasants, capital and the state" (1998: 8) can be generalized to all three forms of agricultural production. Indeed, the question of access leads naturally to a third theme in the political ecology of agricultural production: the interventions of states and corporations to promote (or sometimes, in the case of states, to inhibit) export-oriented farming. State actors make land available in many ways, including by turning a blind eye to the clearance of land meant to be off-limits to agriculture; states and corporations often seek to impose quality standards on growers.

Research in political ecology seeks to unite a critical political economy approach to the preceding issues with the study of environmental change. How does export-oriented production affect local environments, and vice versa? Some work on this question inquires into the "positive" and "negative" implications of export-oriented agricultural intensification for the environment (on the "analytically mixed environmental effects" of agroindustrialization see Barrett *et al.* 2001: 428). To the extent that political ecologists frame things in this way, they largely emphasize the negative – though some accounts argue that negative consequences should not be assumed (Ariza-Montobbio *et al.* 2010: 878). The more characteristic focus, however, has been on the *winners* and *losers* from changes in the political ecology of agri-food systems.

Perhaps the most prominent environmental theme in the literature on export-oriented agriculture is landscape change, and especially deforestation: the millions of hectares of forests lost to cattle and soy in Brazil, for instance, or to shrimp in Southeast Asia and Central America, or to oil palm in Indonesia and Malaysia. A related theme is loss of biodiversity, both in the broader environment (as land is converted to agriculture) and in local agriculture itself (as diverse cropping practices make way for export-oriented monocropping). Much political ecology research also pays close attention to the overuse and pollution of water supplies. Water pollution may derive from the excessive agrochemical use that frequently accompanies export-oriented agricultural production (Barrett *et al.* 2001: 422; see also Grossman 1998). Export-oriented monocropping is not just a problem, finally, for the farms and ecologies around it; it is a problem for itself. Intensive production of (at least) bananas, cocoa, coffee, salmon, and shrimp has resulted in disease-driven collapses in many places, collapses that are central to shifts in world prices for these commodities (see for instance Barton and Fløysand 2010; Hall 2004).

Almost all political ecology research on export crop producing areas highlights the inequality and contention that can derive from the broad issues outlined above. Some conflicts occur within the production system itself, and relate especially to labor relations and to the contract arrangements that are meant to govern farming practices, quality standards, and what will be bought, when, and at what price (Grossman 1998; Vandergeest *et al.* 1999: 579). Others pit producers against people who live close to them. The introduction of new crops often results in some people rapidly becoming (much) wealthier, while others do not just miss out but also lose livelihood options. Loss of access to previously common resources like mangrove forests and locally-run water systems can have severely negative effects. Dispossession and loss of livelihood may be more catastrophic, as when a family has all of its land seized for an oil palm plantation or a cattle ranch. Conflicts over new export crops are not necessarily "local" ones: they can bring in "outside" actors like corporations, the state, and (as the next section shows) transnational activist groups.

Three final cross-cutting themes show up in much political ecology research on the above issues. The first is the question of meaning: the different values ascribed by different groups to different things, and how these valuations shape access to resources and the occurrence of contention. State actors have generally prioritized "modern," export-oriented, and foreign exchange-earning crops over other land uses, especially when they see the land in question as "waste" or "worthless." Producers, similarly, may see export crops as a potential ticket to prosperity, a route towards, in the words of a broccoli grower in Guatemala, *algo más* – something more, something better (Fischer and Benson 2006: 3). They can embrace new crops with such enthusiasm that states end up trying to slow down or stop the boom (Hall 2011b: 526). The second cross-cutting theme is gender. The questions of who grows which crops, who benefits from their growing, who suffers from environmental degradation and from lack of access to previously available resources, and who engages in contention are all highly gendered ones, and work on them has played a prominent role in political ecology (Schroeder 1993; Barndt 2002; Veuthey and Gerber 2012). It is also the case, however, that many studies devote little if any attention to gender dynamics.

The third cross-cutting theme is variation in practices and outcomes. Export crops can be produced and integrated into other agricultural and livelihood practices in many ways, and can have diverse environmental implications. McCarthy's work in four oil palm-growing villages in Sumatra demonstrates this variability "by analysing how similar processes in neighbouring locations led to contrasting outcomes" (McCarthy 2010: 823; see also Grossman 1998: 33; Hall

2004; Westphal 2008). Some political ecologists have also tried to explain variations in contention, which is not an inevitable result of the spread of new export crops (Stonich and Vandergeest 2001). Power inequalities in the producing area may make resistance very difficult, as may the enthusiasm for crop production discussed above. Overall, political ecologists generally emphasize the broad structural changes in the international political economy of agriculture discussed above, and the ways in which forces like technological development and market competition are increasing the sway of corporations over agriculture. Many also find, however, that those forces are not pushing towards the full homogenization of global agriculture or its final capture by capital. Their influence must, rather, be understood within the context of local agrarian histories, crop particularities, social organization, and ecological conditions. Political ecology thus often suggests that every situation is different – a stance that has been seen as both a strength and a weakness of the approach.

Political ecology from farm to fork

Research on Southern agri-food exports has also inquired into how markets for food, fuel, and fiber crops are created and how crops are processed, distributed, and sold. Work on these issues has gone on under different rubrics, including concepts like commodity chain, food chain, value chain, commodity system, and *filière* (see *inter alia* Busch and Bain 2004; Fold and Pritchard 2005; Gibbon and Ponte 2005). While the boundaries between theoretical approaches are even less clear in these areas than they were in the preceding section, I highlight here four key themes that resonate well with political ecology's concerns. First, many studies have approached agri-food commodity chains from the point of view of the work that has to be done, and the varied meanings that have to be created and negotiated, to get a food commodity through the many stages linking production to consumption. These studies often take an ethnographic approach to the constructed nature of desires for these commodities and the shifts in value and perception that occur as a commodity moves along the chain (Bestor 2001; Barndt 2002; Fischer and Benson 2006; Fougères 2008; Freidberg 2004; West 2012). Such work has addressed both the ecological conditions and consequences of processing and distribution, and the ways in which the environmental and other characteristics of specific commodities have been framed.

Second, some political ecology research seeks to understand the environmental implications of agri-food systems in their entirety. Some concepts employed for this purpose also feature in more positivist studies, and the distinctiveness of their use in political ecology again involves integrating their study with that of critical political economy (Weis 2013: 38–52). Studies that try to comprehend overall environmental impacts across scales have used terms like ecological shadow, footprint, or hoofprint (Dauvergne 2008; Weis 2013), metabolic rift (Foster 2000; McMichael 2009: 161), and distancing (Princen 2002). More specific concepts are "virtual water," which suggests that "the water needed to produce agricultural commodities" is, in a metaphorical but still important sense, traded when the commodities are (Allan 2003: 107, and see below); and "food miles," which highlights the air pollution, greenhouse gas emissions, and other problems associated with trading food across large distances (Iles 2005). I have argued in my own work (and used the shadow ecology concept to do so) that the intersections between political economy and the environmental characteristics and consequences of shrimp farming and fast-growing tree plantations have helped to shape agri-food regionalization in Eastern Asia (Hall 2003, 2006).

A third key theme has been the transnational politics of agri-food chains. This literature incorporates two often-intertwined strands. One focuses on activist campaigns against the negative social and environmental consequences of key Southern export crops. These campaigns have seen people directly affected by these consequences make common cause with activists from their own and other countries, and thus take the contentious politics discussed above to the national and transnational levels. Political ecologists have written about campaigns around farmed shrimp (Stonich and Bailey 2000; Veuthey and Gerber 2012), farmed salmon (Barton and Fløysand 2010), palm oil (Pye 2010), and tuna (Baird and Quastel 2011). This research focuses on the nature of the networks and alliances formed, the strategies used (including boycotts), and the impacts of mobilization. It also emphasizes the heterogeneity of and tensions within these alliances, especially those along North–South lines.

The other strand focuses on the responses to campaigns against Southern export crops by, especially, firms in the industry, but also state actors, international organizations, scientists, and large environmental NGOs. Such actors also create transnational networks and alliances to pursue their interests. Since about 2000, research in political ecology has been grappling with the most significant outcome of these alliances: the rapidly proliferating set of systems for certifying agri-food commodities as "sustainable," "responsible," "organic," "fair trade," and so on. Such systems seek to communicate social and environmental information across borders, and to reassure (mostly Northern) consumers that they can feel good about the commodities they are buying. Political ecology (and other) research on certification has thus sought to understand the implications of the formal incorporation of social and environmental standards into the governance of international agri-food chains.

Certification is discussed in Chapter 32 in this volume, so I only comment briefly on how political ecologists have approached its application in developing countries (see Bryant and Goodman 2004). While most political ecologists studying transnational activism around export crops implicitly or explicitly take the side of the activists, they have taken a much more critical stance towards certification schemes. One reason is that certification plays a double-edged role in shaping Southern producers' participation in international agri-food systems: it allows them to demonstrate compliance with the priorities of (usually Northern) export markets, while also working to exclude producers who for whatever reason are not certified. It can be difficult, for instance, for smaller producers to pay for certification without the technical and financial help of outside actors (including aid donors) (Bush *et al.* 2013). Political ecologists also point out that certification systems impose standards generated primarily in the North onto producers (and, effectively, regulators) in the South who may have quite different priorities (Vandergeest 2007; Baird and Quastel 2011; Vandergeest and Unno 2012). Finally, researchers have emphasized that the social and environmental implications of certification, and the extent to which the assumptions about social and ecological relationships embedded in certification schemes correspond to local conditions, are highly varied (Bacon *et al.* 2008; Galt 2010; Baird and Quastel 2011; Kusumawati *et al.* 2013).

Political ecology and the "global land grab"

Political ecology research on Southern agricultural exports has largely analyzed them within the contours of the post-1970s international political economy of agriculture described above. Since the food, fuel, and financial crises hit in 2007, however, it has been widely argued that those contours are rapidly changing. Many such arguments focus on what is (controversially) called the "global land grab," the rapid expansion since the mid-2000s in large-scale land acquisitions for, especially, food, fuel, and fiber crops like oil palm, soy, jatropha, wheat, and rice. Tens of millions of hectares may have been acquired for agriculture across the South by foreign and domestic investors. To the extent that the land grab is actually taking place (many announced projects fail to materialize), it points to at least two significant changes in global agriculture. First, while the high-value export crops focused on by political ecologists have generally been characterized by relatively small-scale production, land grab research suggests the intensification of a trend back towards plantation agriculture and large-scale contract farming that began in the 1990s (White *et al.* 2012: 624). Second, large-scale agricultural projects in the South are increasingly being undertaken by Southern corporations and states and/or targeting Southern export markets (Dauvergne and Neville 2010).

Research on the land grab has exploded since 2010 (for surveys see Cotula 2012; White *et al.* 2012; Wolford *et al.* 2013; Edelman *et al.* 2013; Hall 2013), and some of the approaches to the political ecology of agri-food systems discussed in this chapter have played prominent roles in efforts to theorize the phenomenon's broad features and causes. The idea of a global ecological crisis (or crises) has been invoked to help explain the push for large-scale land acquisitions. Some authors see this crisis as an objective force driving agricultural reorganization. McMichael understands land grabbing "as a reflex of changing conditions of accumulation" in the context of a "crisis of neoliberal accumulation [that] is deeply rooted in rising ecological contradictions," and analyzes it in part as a process in which "finance capital capitalizes offshore agro-food zones as (speculative) substitutes for ecologically exhausted Northern crop lands and as energy crop sites" (McMichael 2012: 681–682, 685; see also Moore 2010). A different (though complementary) argument emphasizes the *narratives* that frame land acquisitions as part of the solution to food, energy, and climate crises (White *et al.* 2012: 627–629). Promoters of large-scale agricultural projects assert that corporate investment can increase the productivity of Southern land variously referred to as "wasteland," "marginal" or "underutilized" (Ariza-Montobbio *et al.* 2010; White *et al.* 2012: 631–632; Wolford *et al.* 2013: 184–195). The painful ironies at work here are especially visible in the biofuels sector, where policies aimed at promoting shifts to "renewable" energy have driven dispossessory and environmentally destructive expansion in crops like oil palm (Pye 2010). The critical analysis of how environmental problems and their solutions are constructed and categories of land in need of "improvement" are created is a core political ecology element of land grab research. Third, some scholars have reconceptualized land grabbing in part as "water grabbing," and thus linked water-related dispossession and environmental degradation to far-away consumption. Mehta *et al.* use the virtual water concept to analyze this connection, writing that "global agricultural trade can also be seen as a massive transfer of water, in the form of commodities" (2012: 197).

If political ecology concerns thus feature in many attempts to conceptualize the broad features and causes of the "global land grab," what of their place in more locally focused research? As of early 2014, relatively few published studies have taken the particularities of local ecologies and their implications for how projects unfold as central topics of investigation, and most have been carried out more within an agrarian political economy than a political ecology framework. There are some exceptions. John McCarthy, Jacqueline Vel, and Suraya Afiff (2012) undertake what is essentially a political-ecological analysis (though they do not use the term) to explain why land acquisition projects in Indonesia often do not pan out as expected. Some studies use empirical research to question whether projects in fact take place on, or are even feasible on, marginal lands (Ariza-Montobbio *et al.* 2010). An important group of local studies of "water grabbing" emphasizes conflicts over water availability and quality. This work highlights processes of ecological change and their relationship to power and inequality, and Annabelle Houdret explicitly draws on political ecology in doing so (Houdret 2012; see also Arduino *et al.* 2012; Bues and Theesfeld 2012; Duvail *et al.* 2012; Williams *et al.* 2012). Overall, however, there is room for much more locally grounded work on the political ecology of individual land grab projects.

Conclusions

What, then, might be potentially fruitful directions for future research on the political ecology of international agri-food systems based on production in the South? The preceding section suggests that the study of the "global land grab" is one. Some of political ecology's strengths in this area are shared with other approaches, and here as throughout this chapter I do not want to make false distinctions between theoretical frameworks. It does seem to me, however, that political ecology's combination of attentiveness to global forces and processes with a concern for the specificities of the human metabolism with nature and for heterogeneity and complexity mean that the approach is particularly well suited for this study. It is already clear that "the" land grab is much more complex, multi-faceted, and locally variable than it is often understood to be (see Peluso and Lund 2011: 669; Visser *et al.* 2012; Fairbairn 2013; Wolford *et al.* 2013). Careful political ecology work on just what is and is not changing, and how changes vary across crops, locations, and commodity chains, can substantially add to this understanding. A critical issue here will be the extent to which emerging South–South chains are indeed changing the international political economy of agriculture rather than largely replicating the dynamics of North–South relations. The questions of whether we have entered a new era in the international political economy of agriculture, and of whether the concrete details of the emerging interconnections between ecology, political economy, and social relations match up with arguments in the land grab literature that have been pitched at the global scale, are important frontiers for new work in political ecology.

Acknowledgments

I am grateful to Gavin Bridge, Elizabeth Havice, Sarah Martin, and Tom Perreault for their comments on earlier drafts. All errors are my own.

References

Allan, J.A. (2003) "Virtual water – the water, food, trade nexus: useful concept or misleading metaphor?" *Water International*, vol. 28, pp. 106–113.

Arduino, S., Colombo, G., Ocampo, O.M., and Panzeri, L. (2012) "Contamination of community potable water from land grabbing: a case study from rural Tanzania." *Water Alternatives*, vol. 5, pp. 344–359.

Ariza-Montobbio, P., Lele, S., Kallis, G., and Martinez-Alier, J. (2010) "The political ecology of *Jatropha* plantations for biodiesel in Tamil Nadu, India." *Journal of Peasant Studies*, vol. 37, pp. 875–897.

Bacon, C.M., Méndez, V.E., Gliessman, S.R., Goodman, D., and Fox, J.A. (eds.) (2008) *Confronting the coffee crisis: Fair trade, sustainable livelihoods and ecosystems in Mexico and Latin America*. Cambridge and London: The MIT Press.

Baird, I.G., and Quastel, N. (2011) "Dolphin-safe tuna from California to Thailand: localisms in environmental certification of global commodity networks." *Annals of the Association of American Geographers*, vol. 101, pp. 337–355.

Barndt, D. (2002) *Tangled routes: Women, work, and globalization on the tomato trail*. Lanham: Rowman & Littlefield.

Barrett, C.B., Barbier, E.B., and Reardon, T. (2001) "Agroindustrialization, globalization, and international development: the environmental implications." *Environment and Development Economics*, vol. 6, pp. 419–433.

Barton, J.R., and Fløysand, A. (2010) "The political ecology of Chilean salmon aquaculture, 1982–2010: a trajectory from economic development to global sustainability." *Global Environmental Change*, vol. 20, pp. 739–752.

Bestor, T.C. (2001) "Supply-side sushi: commodity, market, and the global city." *American Anthropologist*, vol. 103, pp. 76–95.

Bryant, R.L., and Goodman, M.K. (2004) "Consuming narratives: the political ecology of 'alternative' consumption." *Transactions of the Institute of British Geographers*, vol. 29, pp. 344–366.

Bues, A., and Theesfeld, I. (2012) "Water grabbing and the role of power: shifting governance in the light of agricultural foreign direct investment." *Water Alternatives*, vol. 5, pp. 266–283.

Busch, L., and Bain, C. (2004) "New! Improved? The transformation of the global agrifood system." *Rural Sociology*, vol. 69, pp. 321–346.

Bush, S.R., Belton, B., Hall, D., Vandergeest, P., Murray, F.J., Ponte, S., Oosterveer, P., Islam, M.S., Mol, A.P.J., Hatanaka, M., Kruijssen, F., Ha., T.T.T., Little, D.C., and Kusumawati, R. (2013) "Certify sustainable aquaculture?" *Science*, vol. 341, pp. 1067–1068.

Campling, L., Havice, E., and Howard, P.M. (2012) "The political economy and ecology of capture fisheries: market dynamics, resource access and relations of exploitation and resistance." *Journal of Agrarian Change*, vol. 12, pp. 177–203.

Clapp, J. (2012) *Food*. Cambridge: Polity.

Clapp, J., and Fuchs, D. (eds.) (2009) *Corporate power in global agrifood governance*. Cambridge and London: The MIT Press.

Cotula, L. (2012) "The international political economy of the global land rush: a critical appraisal of trends, scale, geography and drivers." *Journal of Peasant Studies*, vol. 39, pp. 649–680.

Crosby, A. (1986) *Ecological imperialism: The biological expansion of Europe, 900–1900*. Cambridge: Cambridge University Press.

Dauvergne, P. (2008) *The shadows of consumption: Consequences for the global environment*. Cambridge and London: The MIT Press.

Dauvergne, P., and Neville, K.J. (2010) "Forests, food, and fuel in the tropics: the uneven social and ecological consequences of the emerging political economy of biofuels." *Journal of Peasant Studies*, vol. 37, pp. 631–660.

Duvail, S., Médard, C., Hamerlynck, O., and Nyingi, D.W. (2012) "Land and water grabbing in an East African coastal wetland: the case of the Tana Delta." *Water Alternatives*, vol. 5, pp. 322–343.

Edelman, M., Oya, C., and Borras, S.M. Jr. (2013) "Global land grabs: historical processes, theoretical and methodological implications and current trajectories." *Third World Quarterly*, vol. 34, pp. 1517–1531.

Fairbairn, M. (2013) "Indirect dispossession: domestic power imbalances and foreign access to land in Mozambique." *Development and Change*, vol. 44, pp. 335–356.

Fischer, E.F., and Benson, P. (2006) *Broccoli and desire: Global connections and Maya struggles in postwar Guatemala*. Stanford: Stanford University Press.

Fold, N., and Pritchard, B. (eds.) (2005) *Cross-continental food chains*. London and New York: Routledge.

Foster, J.B. (2000) *Marx's ecology: Materialism and nature*. New York: Monthly Review Press.

Fougères, D. (2008) "Old markets, new commodities: aquarian capitalism in Indonesia," in Nevins, J. and Peluso, N.L. (eds.), *Taking Southeast Asia to market: Commodities, nature, and people in the neoliberal age*. Ithaca: Cornell University Press, pp. 161–175.

Freidberg, S. (2004) *French beans and food scares: Culture and commerce in an anxious age*. Oxford and New York: Oxford University Press.

Freidberg, S. (2009) *Fresh: A perishable history*. Cambridge, MA and London: The Belknap Press of Harvard University Press.

Friedmann, H., and McMichael, P. (1989) "Agriculture and the state system: the rise and decline of national agricultures, 1870 to the present." *Sociologia Ruralis*, vol. 29, pp. 93–117.

Galt, R.E. (2010) "Scaling up political ecology: the case of illegal pesticides on fresh vegetables imported into the United States, 1996–2006." *Annals of the Association of American Geographers*, vol. 100, pp. 327–355.

Gibbon, P., and Ponte, S. (2005) *Trading down: Africa, value chains, and the global economy*. Philadelphia: Temple University Press.

Goodman, D. (2008) "The international coffee crisis: a review of the issues." In Bacon, C.M., Méndez, V.E., Gliessman, S.R., Goodman, D., and Fox, J.A. (eds.), *Confronting the coffee crisis: Fair trade, sustainable livelihoods and ecosystems in Mexico and Central America*. Cambridge and London: The MIT Press, pp. 3–25.

Grossman, L.S. (1998) *The political ecology of bananas: Contract farming, peasants, and agrarian change in the Eastern Caribbean*. Chapel Hill and London: University of North Carolina Press.

Hall, D. (2003) "The international political ecology of industrial shrimp aquaculture and industrial plantation forestry in Asia." *Journal of Southeast Asian Studies*, vol. 34, pp. 251–264.

Hall, D. (2004) "Explaining the diversity of Southeast Asian shrimp aquaculture." *Journal of Agrarian Change*, vol. 4, pp. 315–335.

Hall, D. (2006) "Regional shrimp, global trees, Chinese vegetables: the environment in Japan-Asia relations." In Katzenstein, P.J. and Shiraishi, T. (eds.), *Beyond Japan: East Asian regionalism.* Ithaca: Cornell University Press, pp. 188–210.

Hall, D. (2011a) "Land grabs, land control, and Southeast Asian crop booms." *Journal of Peasant Studies*, vol. 38, pp. 811–831.

Hall, D. (2011b) "Where the streets are paved with prawns: crop booms and migration in Southeast Asia." *Critical Asian Studies*, vol. 43, pp. 507–530.

Hall, D. (2013) "Primitive accumulation, accumulation by dispossession and the global land grab." *Third World Quarterly*, vol. 34, pp. 1582–1604.

Hall, D., Hirsch, P., and Li, T.M. (2011) *Powers of exclusion: Land dilemmas in Southeast Asia.* Singapore and Honolulu: National University of Singapore Press and University of Hawai'i Press.

Hecht, S.B. (2011) "From eco-catastrophe to zero deforestation? Interdisciplinarities, politics, environmentalisms and reduced clearing in Amazonia." *Environmental Conservation*, vol. 39, pp. 4–19.

Houdret, A. (2012) "The water connection: irrigation, water grabbing and politics in Southern Morocco." *Water Alternatives*, vol. 5, pp. 284–303.

Iles, A. (2005) "Learning in sustainable agriculture: food miles and missing objects." *Environmental Values*, vol. 14, pp. 163–183.

Kusumawati, R., Bush, S.R., and Visser, L.E. (2013) "Can patrons be bypassed? Frictions between local and global regulatory networks over shrimp aquaculture in East Kalimantan." *Society and Natural Resources*, vol. 26, pp. 898–911.

Little, P.D., and Watts, M.J. (eds.) (1994) *Living under contract: Contract farming and agrarian transformation in Sub-Saharan Africa.* Madison: University of Wisconsin Press.

Mann, C.C. (2011) *1493: Uncovering the new world Columbus created.* New York: Vintage Books.

Mann, S.A., and Dickinson, James M. (1978) "Obstacles to the development of a capitalist agriculture." *Journal of Peasant Studies*, vol. 5, pp. 466–481.

Mansfield, B. (2011) "'Modern' industrial fisheries and the crisis of overfishing." In Peet, R., Robbins, P., and Watts, M.J. (eds.), *Global political ecology.* London and New York: Routledge, pp. 84–99.

McCarthy, J.F. (2010) "Processes of inclusion and adverse incorporation: oil palm and agrarian change in Sumatra, Indonesia." *Journal of Peasant Studies*, vol. 37, pp. 821–850.

McCarthy, J.F., Vel, J.A.C., and Afiff, S. (2012) "Trajectories of land acquisition and enclosure: development schemes, virtual land grabs, and green acquisitions in Indonesia's outer islands." *Journal of Peasant Studies*, vol. 39, pp. 521–549.

McMichael, P. (2009) "A food regime genealogy." *Journal of Peasant Studies*, vol. 36, pp. 139–169.

McMichael, P. (2012) "The land grab and corporate food regime restructuring." *Journal of Peasant Studies*, vol. 39, pp. 681–701.

Mehta, L., Veldwisch, G.J., and Franco, J. (2012) "Water grabbing? Focus on the (re)appropriation of finite water resources." *Water Alternatives*, vol. 5, pp. 193–207.

Moore, J.W. (2010) "The end of the road? Agricultural revolutions in the capitalist world-ecology, 1450–2010." *Journal of Agrarian Change*, vol. 10, pp. 389–413.

Patel, R. (2007) *Stuffed and starved: From farm to fork, the hidden battle for the world food system.* London: Portobello Books.

Peluso, N.L. and Lund, C. (2011) "New frontiers of land control: introduction." *Journal of Peasant Studies*, vol. 39, pp. 667–681.

Pomeranz, K. (2000) *The great divergence: China, Europe, and the making of the modern world economy.* Princeton: Princeton University Press.

Princen, T. (2002) "Distancing: consumption and the severing of feedback." In Princen, T., Maniates, M., and Conca, K. (eds.), *Confronting consumption.* Cambridge, MA: The MIT Press, pp. 103–131.

Pritchard, B. (2009) "The long hangover from the second food regime: a world-historical interpretation of the collapse of the WTO Doha Round." *Agriculture and Human Values*, vol. 26, pp. 297–307.

Pye, O. (2010) "The biofuel connection: transnational activism and the palm oil boom." *Journal of Peasant Studies*, vol. 37, pp. 851–874.

Ribot, J.C., and Peluso, N.L. (2003) "A theory of access." *Rural Sociology*, vol. 68, pp. 153–181.

Schroeder, R.A. (1993) "Shady practice: gender and the political ecology of resource stabilization in Gambian garden/orchards." *Economic Geography*, vol. 69, pp. 349–365.

Stonich, S.C., and Bailey, C. (2000) "Resisting the blue revolution: contending coalitions surrounding industrial shrimp farming." *Human Organization*, vol. 59, pp. 23–36.

Stonich, S.C., and Vandergeest, P. (2001) "Violence, environment, and industrial shrimp farming." In Peluso, N.L. and Watts, M.J. (eds.), *Violent environments*. Ithaca: Cornell University Press, pp. 261–286.
Vandergeest, P. (2007) "Certification and communities: alternatives for regulating the environmental and social impacts of shrimp farming." *World Development*, vol. 35, pp. 1152–1171.
Vandergeest, P., Flaherty, M., and Miller, P. (1999) "A political ecology of shrimp aquaculture in Thailand." *Rural Sociology*, vol. 64, pp. 573–596.
Vandergeest, P., and Unno, A. (2012) "A new extraterritoriality? Aquaculture certification, sovereignty, and empire." *Political Geography*, vol. 31, pp. 358–367.
Veuthey, S., and Gerber, J.-F. (2012) "Accumulation by dispossession in coastal Ecuador: shrimp farming, local resistance and the gender structure of mobilizations." *Global Environmental Change*, vol. 22, pp. 611–622.
Visser, O., Mamonova, N., and Spoor, M. (2012) "Oligarchs, megafarms and land reserves: understanding land grabbing in Russia." *Journal of Peasant Studies*, vol. 39, pp. 899–931.
Walker, R., Browder, J., Arima, E., Simmons, C., Pereira, R., Caldas, M., Shirota, R., and de Zen, S. (2009) "Ranching and the new global range: Amazônia in the 21st century." *Geoforum*, vol. 40, pp. 732–745.
Walker, R.A. (2004) *The conquest of bread: 150 years of agribusiness in California*. New York and London: The New Press.
Watts, M.J., and Goodman, D. (1997) "Agrarian questions: global appetite, local metabolism: nature, culture, and industry in *fin-de-siècle* agro-food systems." In Goodman, D. and Watts, M.J. (eds.), *Globalising food*. London: Routledge, pp. 1–32.
Weis, T. (2013) *The ecological hoofprint: The global burden of industrial livestock*. London and New York: Zed Books.
West, P. (2012) *From modern production to imagined primitive: The social world of coffee from Papua New Guinea*. Durham and London: Duke University Press.
Westphal, S.M. (2008) "Coffee agroforestry in the aftermath of modernization: diversified production and livelihood strategies in post-reform Nicaragua." In Bacon, C.M., Méndez, V.E., Gliessman, S.R., Goodman, D., and Fox, J.A. (eds.), *Confronting the coffee crisis: Fair trade, sustainable livelihoods and ecosystems in Mexico and Central America*. Cambridge and London: The MIT Press, pp. 179–205.
White, B., Borras, S. M. Jr., Hall, R., Scoones, I., and Wolford, W. (2012) "The new enclosures: critical perspectives on corporate land deals." *Journal of Peasant Studies*, vol. 39, pp. 619–647.
Williams, T.O., Gyampoh, B., Kizito, F., and Namara, R. (2012) "Water implications of large-scale land acquisitions in Ghana." *Water Alternatives*, vol. 5, pp. 243–265.
Wolford, W., Borras, S.M. Jr., Hall, R., Scoones, I., and White, B. (2013) "Governing global land deals: the role of the state in the rush for land." *Development and Change*, vol. 44, pp. 189–210.

32

CERTIFIED POLITICAL ECOLOGY

Jonathan Otto and Tad Mutersbaugh

Introduction

Recent years have witnessed an explosion of certified environmental products and qualities ranging from agronomic and forestry activities such as 'bird-friendly®', shade-grown, sustainable, and organic – applied to diverse products such as coffee, shrimp, timber, carbon (sequestration), and hunting – to industrial processes such as waste management and mining, and to services such as ecotourism. Hundreds of thousands of 'nature workers' – farmers, foresters, fishers, and craftspeople – presently labor to apply 'environmental' standards, and they in turn find their compliance assessed by a burgeoning global certification service sector involving tens of thousands of personnel working in certification agencies, NGOs, national regulatory boards, and ISO working groups. As we detail in a subsequent section on certification types, certified qualities have drawn a good deal of scholarly attention as well, not least within political ecology.

The goal of our contribution is to examine the contours of an emerging political economy of nature work centered on inspections and audits, labor practices, and institutional frameworks, and examine its relevance to political ecology from three perspectives: First, from a policy standpoint, certification has become an ineluctable aspect of contemporary conservation initiatives. In both numbers of personnel and areal extent of certified conservation activities – linking price incentives to the performance of environmental labor – the participation in certified environmental activities is on par with other forms of conservation. Second, from a governmentality perspective, certification protocols are productive of nature. Certification may, in this sense, be viewed as a process through which environmental qualities are pegged to commodities, creating new arenas of environmental decision-making and valorizing particular environmental tasks and spaces while devaluing others. Third, and most importantly with respect to the perennial political ecology concern for environmental justice, the certification service economy organizes 'conservation work', providing jobs for millions of small producers, inspectors, and accreditors. However, the combination of low payments for this work and high certification costs often shifts the economic burden of conservation from (wealthier) consumers to (poorer) producers.

In the field, farmers have deemed environmental certification to be an 'ecological neocolonialism'. This pithy expression captures a commonly experienced sense of injustice:

much of the power to define which activities are environmentally appropriate is vested in the hands of standards boards whose environmental governance authority is backed by transnational economic organizations such as the ISO (International Organization for Standardization; e.g., Guide 65 – now ISO 17065) and the WTO (World Trade Organization; via the TBT agreement), and rests, in its north–south context, on deeper histories of neocolonial engagements (see Freidberg 2003). This use of regulatory standards to shape the work of nature is thus enabled by the use of certification to control market access: it is not possible to deal in internationally traded quality goods, such as organic food or carbon credits, without implementing certification protocols. Seen from a farmers' perspective, and within the context of these economic relations, the performance of activities necessary to comply with quality standards is not 'voluntary', but rather a question of economic necessity. In this chapter we provide an overview of certification and then explore the aspects noted above in three case studies. First, we examine the *policy implications* of the global expansion of certification as a form of environmental policy, then we examine *governmentality* in the context of certified organic coffee production, and finally we consider the *environmental justice* implications of carbon credits production in the context of certified carbon forestry.

What is certification?

Taking a schematic view (see Figure 32.1), certification triangulates between an (environmental) standards board, the site or field where the particular 'quality' (e.g., organic, shade-grown) is imparted into the product, and a site of 'consumption' (e.g., retail consumer or governmental agency). A 'chain of custody', using inspectors and governed by a certifying office, links sites of production with buyers.

Four elements of a quality certification structure include:

A *standards board* sets the norms that govern the production of environmental qualities. Given the importance of environmental policies, many environmental standards, like food standards, are governed at least in part by governmental agencies and linked to national and multilateral environmental policies. Such is the case for organic standards, which are also subject to governmental scrutiny as a food, and carbon offsets, which are a key component in global climate action. Other environmental standards, however, may be governed by NGOs, such as 'wild-caught' shrimp, dolphin-free tuna, and bird-friendly coffee among others. The content of standards may be set through participatory methods, as was the case with early organic food standards (Gonzalez and Nigh 2005), or through state action based upon scientific research and lobbying (Dietsch et al. 2004). Standards generally set a minimum 'bar', or lowest permissible standard, and then require producers to demonstrate compliance.

Commodity producers receive standards information from standards boards, such as the USDA NOSB (United States Department of Agriculture, National Organic Standards Board), and then perform the work required to have their commodities conform to environmental standards. Producers, particularly in the Global South, are typically organized into cooperatives or unions, for as detailed below, environmental qualities work is time-consuming, costly, and technically difficult, requiring strong organizations to facilitate norms compliance.

Certifying agencies inspect practices used to produce qualities in the field and along the supply chain, here described as a 'chain of custody'. It is important to note that environmental qualities are 'certified' because there is no way to tell whether the commodity contains an environmental 'value' except through inspections. A buyer cannot tell whether a coffee bean is organic, a shrimp 'wild-caught', or a tree has 'captured carbon' by looking at it: in this regard, a certification 'audit trail' provides the buyer with a guarantee that a purchased commodity (e.g., cup of

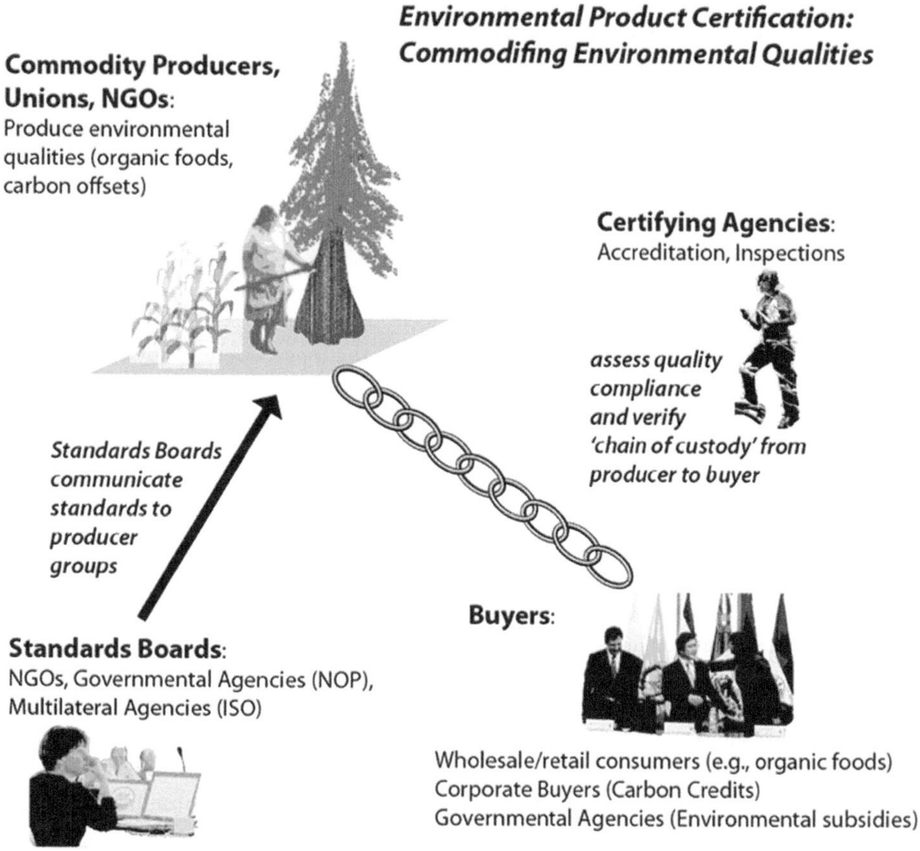

Figure 32.1 Components of a certification system for environmental qualities (source: by author).

coffee) was produced using practices that *also* produced environmental qualities. A field inspector undertakes certifications, checking to see whether producers are in compliance with the minimum standards set by the standards board.

Finally, *buyers* pay a price premium to producers to compensate them for their environmental work. Although we may typically imagine the buyer to be a retail consumer, perhaps sipping coffee in a roadside café, environmental qualities are more likely financed via purchases from wholesalers such as coffee roasters (e.g., Green Mountain, see Lyon 2010) or brokers tied to cap-and-trade schemes such as carbon offsets (Peters-Stanley and Yin 2013) and wetlands conservation (Lave 2012), and by subsidies from governmental agencies seeking to combine conservation work with social welfare and export promotion (e.g., McAfee and Shapiro 2010). In these instances, the premiums may be paid based on conservation work (e.g., environmental services), or as per-hectare crop payments to avoid WTO restrictions on commodity price supports (e.g., as in the case of Mexican organic acreages).

To summarize the relations set out in Figure 32.1, this particular configuration is known as 'third-party' certification, termed a 'Type 1' certification under ISO 17065 (ISO is the acronym for the Geneva-based International Organization for Standardization). This standard sets out 'norms' for certification that require a clear delineation of responsibilities for the production of qualities, a high degree of institutional separation between the various actors, and the absence of 'conflicts of interest'. However, as a practical matter things are rarely so simple. The variation in commodity types, in institutional settings, and in state and multilateral regulatory interest leads to a wide divergence in actually existing certification schemes that draw on additional forms of quality assessment including – to draw upon Gereffi et al.'s typology (2001, see also Dunn 2004) – *first-party quality control*, in which a company sets and verifies its own standards (e.g., McDonald's internal assessment of cheeseburger 'quality'); *second-party verification*, in which company-defined standards are *verified* by an external company (e.g., Starbucks' coffee 'practices'); and *fourth-party regulation*, in which a governmental agency sends auditors (e.g., meat packing inspections).

What are the policy implications of environmental product certification?

These messy, varying, real-world applications of certification have implications with respect to conservation policy. First, to give some sense of the wide diversity in environmental certifications, we analyzed the Ecolabel Index (www.ecolabelindex.com). Though far from comprehensive, and representing but a snapshot of a rapidly evolving certified goods sector, this list currently tracks over 400 ecolabels in 197 countries. Our tally (Figure 32.2) encountered a surprising diversity of items, with about 74 percent labeling specific products and the remaining 26 percent general sustainability and lifecycle certifications. Of specific products, food items are the most common environmental certification, while energy and building codes run a close second.

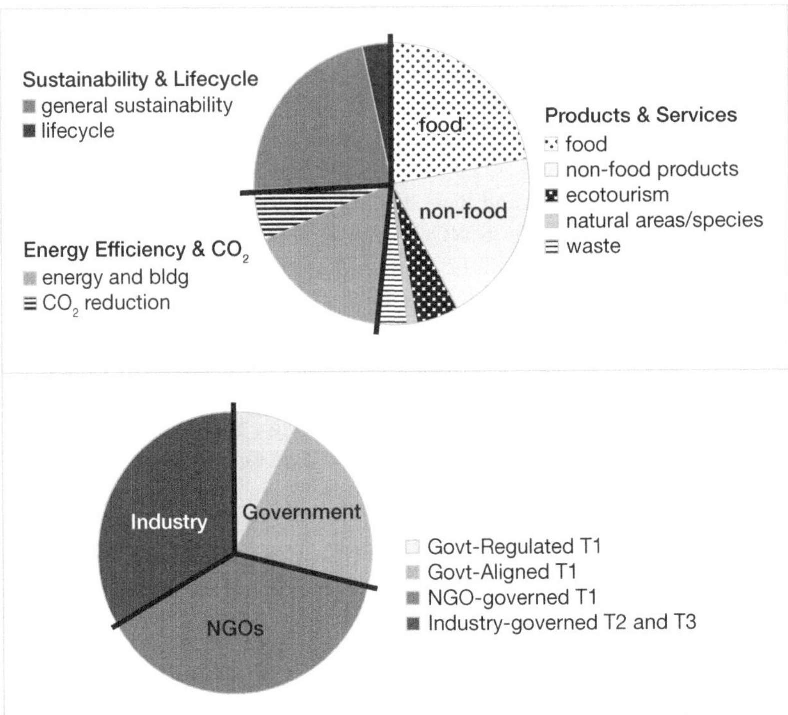

Figure 32.2 Environmental certifications by category (source: by author).

These are followed by diverse products including forest resources, flowers, cotton and wool fibers, and beauty products which comprise 20 percent, and environmental services such tourism, natural and green space protection, and waste management at 5 percent.

However, a categorical division such as that outlined above fails to account for the provenance and form of certifications, namely, *who sponsors* the certification (and whether they are subject to governmental oversight), and *how* the certification is performed. Even provided data limitations, the contemporary distribution is quite striking: a sector once dominated by NGO-based certifications such as Fairtrade, organic and biodiversity crop, and timber certifications (such as the FSC), and governmental initiatives such as Energy Star, is now increasingly populated by new certifications underwritten by governmental agencies (e.g., Estonian Tourism Ministry) or single companies (e.g., Starbucks Cafe Practices, Naturally Sephora). In addition, the second pie chart of Figure 32.2 identifies a second concern, namely, the difficulty in identifying the institutional arrangements that underlie ecolabels: although most of the labels ostensibly meet the highest, Type 1 certification level, only a third are NGO-backed. Another third are aligned with government agencies (primarily food, energy, and tourism) and about a third appear to be industry-backed second-party certifications. A key point of Figure 32.2, then, regards the lack of clarity in certification type.

The rise in certifications is matched by an increase in scholarship. Of these, many support earlier crtiques of 'voluntary' controls (Guthman 2007; McCarthy 2006), showing that certified products have at best an uneven track record with regard to conservation (Ponte and Cheyns 2013; Quaedvleig et al. 2014; Elder et al. 2013). However, although conservation goals go unmet, as our two case studies in the next section show, certification as a practice continues to shape environmental institutions, rural environmental interactions, and by extension biological and physical environments (Naylor 2014; Hatanaka 2014; Mancini 2013; Kimura 2012; Wilson and Curnow 2013; Arora and Hofman 2013; Brown 2013; Mansfield 2004). Does this proliferation lead, as Riisgaard (2012) argues, to a 'race to the bottom' in which lax standards are dominant? Whether laxity in standards does prevail may depend upon whether the environmental activist concerns that drove standards expansion in the 1990s – for instance, in certified foods, fibers, forest products, and fisheries (Wilson 2013; Raynolds 2012; Ponte 2012; Klooster 2010; Bacon 2010; Eden and Bear 2010; Dolan 2010; Lyon et al. 2010; Bassett 2010) – continues apace with new grassroots NGO Type 1 certifications, such as the recent cases of aquaculture (Vandergeest and Unno 2012), shrimp (Konefal and Hatanaka 2011), dolphin-free tuna (Baird and Quastel 2011), Japanese women's cooperative certification (Kimura 2012), and Tanzanite (Schroeder 2010).

However, as depicted in Figure 32.2, the bulk of new certifications in the Ecolabel Index are either government-associated (e.g., Slovak ecolabeling, Ukranian Green Crane, Estonian Eco-Tourism, Brazilian ABNT) or private initiatives (e.g., Sephora cosmetics, Nike shoes, FedEx *Earthsmart* label, Procter & Gamble *Future Friendly*). Government-aligned labels appear to be driven by two dynamics. In some cases, novel labels are tied into export-promotion programs (Rangnekar 2011) that vary from government-supported certifications (e.g., Japan (Hall 2010)), to fully government-organized 'third-party' certifications, for instance in China (Buckingham and Jepson 2013). In other cases, new certifications are prompted by government-sponsored health and conservation schemes, such as the cases of Chilean certified firewood (Conway 2013) and EU-wide, Kyoto Protocol prompted carbon-conserving building efficiency standards.

Private labels, on the other hand, are often initiated to enhance the 'green' credentials (greenwash) of particular products or to shield industries from scrutiny by substituting industry-controlled labels in the place of existing public labels. Pickren (2014) for instance makes a persuasive case that the electronics recycling industry worked to substitute a relatively lax 'Responsible Recycling' standard in the place of the more stringent NGO-backed e-Stewards standard, with the objective of reducing scrutiny and increasing industry profitability at the expense of environmental quality. This 'replace and market' strategy may explain the widespread duplication – and in some cases multiplication – of competing environmental certifications for cut flowers, tourism, cotton, industrial building materials, and mining among others. The next section analyzes these certification politics and practices in the context of case studies of organic coffee and carbon sequestration. Both represent important policy initiatives – reducing the environmental impacts of agriculture and combating global climate change, respectively – and yet each has also come to represent the extension of 'command and control' strategies that seek to prescribe 'appropriate' environmental practices and burden producers with new tasks.

Case studies: environmental certifications and working natures

In addition to a consideration of policy implications this chapter also examines the manner in which environmental certifications *as a labor practice* work to alter the geography of human–environment interactions. As Figure 32.1 shows, labor that shapes the particularities of environmental interactions occurs in an expanded number of locations, ranging from standards boards at which the specific environmental conservation tasks are determined (e.g., permissible fertilizer applications), to producer organizations that systematize required tasks and balance these against productivity-enhancement, to inspectors who assess the quality of conservation efforts, and finally to producers themselves. The following two case studies of organic coffee and carbon sequestration illustrate many of the tensions, ironies, and ultimately the political ecologies that converge to reconstruct human–environmental relations at the point of production.

Example 1: organic agriculture as a form of governmentality

Taken from a political ecology perspective, organic coffee farming, paradoxically, raises a number of social and ecological concerns. The processes of certification bind together networks of environmental workers – field inspectors, peasant technical workers, and farmers – into a social structure that performs organic farming activities, which in turn satisfy the policy dimensions of organic agriculture by putting a particular set of organic conservation norms into practice. This 'labor of nature' has a number of positive social and ecological effects. Socially, the participation in organic product networks provides new forms of paid, skilled labor, and for reasons noted below, is attractive to women farmers (Lyon et al. 2010). From a governmentality perspective, these new forms of labor are associated with changes in social identification: farmers appreciate their role in environmental production and the new class of professional eco-workers who coordinate tens of thousands of producers in a spatially-extensive conservation network that provides diverse environmental services (Mutersbaugh 2004).

However, if certified sustainable agriculture is to realize its full potential as a conservation strategy, it must be economically and ecologically sound and spread benefits to areas of high conservation value. In this regard, a number of points of tension are evident. From a *social* standpoint, the expectation that certified products will bolster conservation efforts has generally brought an expanded state role, for instance in subsidizing economic incentives to link organic agro-forestry schemes to carbon sequestration or biodiversity conservation (see also Buckingham and Jepson 2013). Nevertheless, despite, or perhaps in part *because* of this state involvement (and additional costs associated with bureaucratic liaisons) farmers still find costs high and returns low

(Bacon 2010; Mutersbaugh 2005; Jaffee 2007) – as signaled by high attrition rates for poorer producers (Mutersbaugh 2005) – and local administrators must perform unpaid work such as arranging inspections and managing documents. This combination of high farmer costs and high administrative overhead means that only relatively well-off farmers in wealthy villages – or those possessed of strong indigenous or cooperative governance mechanisms capable of providing low-cost administrative labor (see Mutersbaugh 2004; Wilson 2013) – are able to implement certified organic agriculture.

From a conservation standpoint, certification does not necessarily incorporate the best of ecological practices. To begin, cultivation practices built upon the norms created in the USDA National Organic Program, EU, or Japanese norms do not easily incorporate the biodiversity knowledge of indigenous producers who recognize the importance of cultivating biodiversity and have created local cultivation systems in which the horticultural complexity far outstrips any found in conventional agriculture. In an interesting illustration of governmentality, the institutional structure of certification, with its origins in state-sponsored regulatory frameworks, creates the conditions in which certifying agencies and inspectors operate outside of the rule of either standards boards or local agro-ecological knowledge, imposing tasks such as soil conservation measures, water quality control, and coffee cultivation activities not sanctioned by organic standards boards. These tasks may adversely impact existing biodiversity by disrupting existing indigenous systems of biodiversity management.

To conclude, the processes depicted in Figure 32.1 are open to interpretation and in practice become a question of negotiation in the fields, farmer organizations, and certifying agencies. This negotiation is productive of environmentalist subjectivities and our research has found that this engenders a strong sense of environmental citizenship (Otto 2014; Mutersbaugh 2004). Governmentality unfolds in a context of unequal power relations rooted in both to the north–south neocolonial dynamics noted above and in the power of large-scale organic producers (in coffee, these comprise organic coffee haciendas of more than 100 hectares) which advocate for simple, routinized, and uniform organic standards offering little protection to biological diversity. In this context, organic inspectors often lack sufficient information on coffee agro-forestry systems and local environmental variation, which leads them to suggest inappropriate conservation schemes or to favor increased production schemes at odds with those advocated by local farmers who, as the local 'makers' of political ecologies (Neumann 2005), do share a knowledge of biodiversity.

Example 2: environmental justice in certified carbon forestry: capturing carbon, losing respect

In our final case, we examine the environmental justice implications of 'carbon forestry' and related certification programs that originated from the need to certify projects designed to measure whether or not such projects are able to capture and limit CO2 emissions via afforestation and reforestation activities (see also Chapter 23, this volume). As we show below, however, forest carbon certification 'on the ground' has expanded from its original environmental purpose – certifying CO_2 reductions achieved by carbon projects – to verify 'pro-poor' elements that would ensure that global environmental benefits are matched to local community benefits. In this example, as in the previous section on organic agriculture, we are interested in a grounded political-ecological analysis that examines how community-level participation in carbon forestry projects can be both a benefit and a burden. Concluding with a brief example of a village-based community extension officer in southern Mexico, we make two points: first, we show how certification processes organize the 'nature work' of participants in a manner that not only reconstitutes their relationship to the natural environment – particularly the certified product – but also to their broader community; second, we portray a problem of environmental justice in which the marketization of the carbon captured in trees as carbon credits results in a transfer of risk from Global Northern consumers to Global Southern producers who can ill afford it (cf. Chapter 45, this volume).

The genesis of contemporary forest carbon certification may be found in the Kyoto Protocol's compliance-oriented Clean Development Mechanism (CDM), Joint Implementation (JI) and International Emissions Trading (ET) mechanisms, and parallel voluntary markets, the latter of which is of particular concern to us in our grounded political-ecological study. Voluntary carbon credits, referred to as Verified Emissions Reductions (VERs), are worth one ton of captured CO_2, and are, on the whole, significantly smaller than compliance markets in terms of the volume of carbon credits traded, having represented only 0.1 percent of total global carbon markets in 2010 (Hamilton et al. 2010). Within the voluntary markets context, REDD (Reducing Emissions from Deforestation and Forest Degradation), which is designed to pay local communities to engage in activities aiming to limit deforestation, to sustainably manage existing forests, and to promote reforestation, has emerged as an influential carbon forestry framework, capturing 9 percent of the voluntary market share for carbon credits transacted 'over the counter' (OTC) – i.e. not on a formal exchange – in 2012 (Peters-Stanley and Yin 2013).

The expansion of voluntary carbon markets has engendered a parallel unfolding of carbon certifications designed to ensure that forest carbon projects achieve their climate change mitigation imperative of sequestering CO_2. Within the landscape of voluntary carbon certification, however, so-called 'co-benefits' standards have emerged to meet the wishes of carbon credit buyers who desire to purchase credits that not only contribute to climate change mitigation, but also to social development (Peters-Stanley and Yin 2013). These standards are varied, and include names such as 'Climate, Community and Biodiversity Standards', 'The Gold Standard', 'Social Carbon', and 'Plan Vivo'. The certification work tied to these emergent co-benefits frameworks is complex and requires the formation of labor processes that organize the 'nature work' of participants, thereby reconstituting their relationship to the natural environment and to broader communities in which projects are implemented. In an effort to examine environmental justice issues tied to co-benefits carbon certification within voluntary markets we turn to the case of Scolel' Te in southern Mexico.

Scolel' Te (meaning 'Tree that Grows' in the local Tzeltal language) originated in 1994 and 1995, and currently enrolls over 1,100 farmers representing 77 communities and eight Mayan languages in Chiapas and Oaxaca. From 1998, the Scolel' Te project grew from 47 farmers in six communities located in Chiapas' Central Highlands region to, in 2014, over 1,100 farmers representing 77 communities and eight Mayan languages in the Central Highlands and Lacandón regions of Chiapas and including some in the neighboring state of Oaxaca (Otto 2014; Osborne 2010; Esquivel and Quechulpa 2010; Brown and Corbera 2003). Scolel' Te carbon forestry is managed by the NGO AMBIO and is organized under the Plan Vivo Standard, which – in a twist that has become characteristic of certification networks – is in turn governed by a Scottish charity, the Plan Vivo Foundation. While standards emphasize a variety of items across forest carbon projects (e.g., the protection of biodiversity and the promotion of clean water sources), the Plan Vivo Standard has been recognized for its design which promotes smallholder participation. It includes a variety of participatory exercises such as Plan Vivo mapping, for instance, in which farmers are asked to determine the parameters of their participation in the program. However, in the case of Scolel' Te, farmers are also recruited by AMBIO to work as extension officers, and the manner in which this work confronts the socio-political dynamics and power structures of participating villages is complex – a fact that is shown through the statement of one extension officer regarding his relationship to fellow community members who had become dissatisfied with the project:

> The [carbon forestry] project has not gone well. We have not received the money we are owed. When I leave the house, people ask me 'where is our money?' and 'what have you done with our money?' Some accuse me of having stolen it, and they won't stop asking about it. I prefer not to leave my house because I do not want to answer their questions anymore. It is easier to stay home.

This quote indicates the nature of the relationship of the community-based extension officer to disgruntled project participants within his community who had come to question the validity of the program. In the end, his relationship to fellow community members came to be mediated by his work in Scolel' Te in a manner that not only complicated his social interactions, but also his participation in the program (see Otto 2014). Ironically, from an environmental justice perspective, the same co-benefits standards that provided certainty to buyers also played a role in undermining the ideals of social development and environmental conservation that carbon credit purchasers desired.

So what *did* happen to the money? As the above quote indicates, farmers view their participation in the carbon capture project as environmental work, identifying Scolel' Te as an employment opportunity in the face of scarce alternatives and framing their relationship with AMBIO within an employer-employee dynamic. AMBIO, however, sells carbon credits in the voluntary market: although AMBIO's work is supported by Mexican government and international grants, the work of farmers and village-based inspectors is financed by the sale of carbon credits. If the credits are not sold, the producers are not paid. AMBIO strives to sell these units, but confronts a saturated market in which buyers have preferred to purchase credits from 'poorer' and more 'biodiverse' producers in other world regions. Thus while farmers may understand their relationship in wage terms, PES institutions constitute farmers as market subjects. Ironically, then, the acts that certify a carbon credit and bring it to market also result in a hidden transfer of risk to the most vulnerable actors in the carbon value chain – the carbon value may be guaranteed, but the return to labor is not (see also Beymer-Farris and Bassett 2012).

Conclusion

If one environmental justice critique leveled by organizations of the Global South were to be applied to this instance of carbon forestry, it would be the charge of 'ecological neocolonialism' (see Mutersbaugh 2005). As in the case with neocolonialism broadly writ, the 'ecological' variant operates to facilitate transfer of both financial and environmental risk from Global Northern consumers (who, needless to say, bear responsibility for the bulk of global environmental degradation) to peoples of the Global South. In this case, the marketized carbon credits produced by the community failed to sell, leaving community members with significant economic losses, and the extension officer noted in the quote above with a devastating loss of personal prestige: producers in the Global South incurred significant risk and losses in a market-based conservation project to the benefit Global Northern consumers.

What, then, does environmental certification bring to political ecology? First, and most simply, it calls attention to the need for empirical and theoretical engagement with the remarkable expansion of environmentally oriented qualities, institutionalities, certification agencies, and mechanisms, all of which are engaged in reimagining what constitutes environmental conservation. This rapid expansion of qualities parallels a sharp increase in political-ecologically informed studies of quality, yet, as we show in the introduction, much remains to be done. Political ecology would assert the necessity of environmental conservation practices based upon norms of social justice, without which conservation initiatives must inevitably fail, but as recent studies have shown, the rapid expansion in certified environmental qualities does not necessarily provide enhanced environmental protection: even highly regulated cases often shift the risks and costs of conservation from Northern consumers to Southern producers. As we show in this chapter, the marketization of environmental qualities transfers risk from Northern carbon credit consumers to Southern producers who, as a condition of market access, must perform environmental conservation tasks without a guarantee of compensation.

Second, certification studies would argue for the importance of governmentality to political ecology, that is to say, the need to examine the governmental institutions that set the conditions for those certification practices through which environmental qualities are constituted, produced, and assessed. For instance, political ecology has rightly championed the importance and theoretical sophistication of local environmental knowledge (see Chapter 18, this volume). Certification studies would argue that the notion of 'local knowledge' needs be expanded to include the inspectors, certifying agencies and standards boards that constitute quality, as well as sites such as ISO where certification protocols both constrain and enable the constitution of environmental quality. To carry this point forward, the knowledge politics of certification, a form of governmentality undertaken under ISO norms, makes no provision for democratic processes. Indeed, transnational certification norms forbid producers from engaging directly in the elaboration of standards on the premise that such exchanges would constitute a 'conflict-of-interest' in which farmers might attempt to alter norms and standards to their economic benefit (Mutersbaugh 2005).

Third and finally, certification studies favor the recent engagement in political ecology with theories of more-than-human geographies, assemblage, and environmental subjectivities, arguing for the inclusion of certification protocols as a form of governmentality that shapes the performance of labor and contributes to the formation of environmental subjectivities, albeit in often contradictory ways, as demonstrated in the conflicts over labor, environmental quality, and certification performance in our two case studies.

References

Arora, S. and Hofman, N.B. 2013, Cultivating compliance: governance of North Indian organic basmati smallholders in a global value chain. *Environment and Planning A* 45: 1912–1928.

Bacon, C.M. 2010, Who decides what is fair in fair trade? The agri-environmental governance of standards, access, and price. *Journal of Peasant Studies* 37(1): 111–147.

Baird, I.G. and Quastel, N. 2011, Dolphin-safe tuna from California to Thailand: localisms in environmental certification of global commodity networks. *Annals of the Association of American Geographers* 101(2): 337–355.

Bassett, T.J. 2010, Slim pickings: Fairtrade cotton in West Africa. *Geoforum* 41(1): 44–55.

Beymer-Farris, B.A. and Bassett, T.J. 2012, The REDD menace: resurgent protectionism in Tanzania's mangrove forests. *Global Environmental Change* 22: 332–341.

Brown, K. and Corbera, E. 2003, Exploring equity and sustainability in the new carbon economy. *Climate Policy* 3(S1): 41–56.

Brown, S. 2013, One hundred years of labor control: violence, militancy, and the Fairtrade banana commodity chain in Colombia. *Environment and Planning* A 45: 2572–2591.

Buckingham, K. and Jepson, P. 2013, Forest certification with Chinese characteristics: state engagement with non-state market-driven governance. *Eurasian Geography and Economics* 54(3): 280–299.

Conway, F.J. 2013, Firewood certification in Chile: equity in an innovative form of alternative trade. *Human Organization* 72(1): 55–64.

Dietsch, T.V., Philpott, S.M., Rice, R.A., Greenberg, R. and Bichier, P. 2004, Conservation policy in coffee landscapes. *Science* 303(5658).

Dolan, C. 2010, Virtual moralities: the mainstreaming of Fairtrade in Kenyan tea fields. *Geoforum* 41(1): 33–43.

Dunn, E.C. 2004, *Privatizing Poland: Baby Food, Big Business, and the Remaking of Labor.* Cornell University Press, Ithaca.

Eden, S. and Bear, C. 2010, Third-sector global environmental governance, space and science: comparing fishery and forestry certification. *Journal of Environmental Policy & Planning* 12(1): 83–106.

Elder, S.D., Zerriffi, H. and Le Billon, P. 2013, Is Fairtrade certification greening agricultural practices? An analysis of Fairtrade environmental standards in Rwanda. *Journal of Rural Studies* 32: 264–274.

Esquivel, E.B. and Quechulpa Montalvo, S. 2010, *Scolel' Te Programme Plan Vivo 2009 Annual Report.* Ambio: San Cristobal de las Casas, Mexico.

Freidberg, S.E. 2003, Culture, conventions and colonial constructs of rurality in south–north horticultural trades. *Journal of Rural Studies* 19: 97–109.

Gereffi, G., Garcia-Johnson, R. and Sasser, E. 2001, The NGO-industrial complex. *Foreign Policy* 125: 56–65.

Gonzalez, A.A. and Nigh, R. 2005, Smallholder participation and certification of organic farm products in Mexico. *Journal of Rural Studies* 21(4): 449–460.

Guthman, J. 2007, The Polanyian way? Voluntary food labels and neoliberal governance. *Antipode* 39: 456–478.

Hall, D. 2010, Food with a visible face: traceability and the public promotion of private governance in the Japanese food system. *Geoforum* 41: 826–835.

Hamilton, K., Sjardin, M., Peters-Stanley, M., and Marcello, T. 2010, Building bridges: state of the voluntary carbon markets 2010. Forest Trends' Ecosystem Marketplace and Bloomberg New Energy Finance.

Hatanaka, M. 2014, Standardized food governance? Reflections on the potential and limitations of chemical-free shrimp. *Food Policy* 45: 138–145.

Jaffee, D. 2007, *Brewing Justice: Fair Trade Coffee, Sustainability, and Survival.* 1st edn. University of California Press.

Kimura, A.H. 2012, Feminist heuristics: transforming the foundation of food quality and safety assurance systems. *Rural Sociology* 77(2): 203–224.

Klooster, D. 2010, Standardizing sustainable development? The Forest Stewardship Council's plantation policy review process as neoliberal environmental governance. *Geoforum* 41(1): 117–129.

Konefal, J. and Hatanaka, M. 2011, Enacting third-party certification: a case study of science and politics in organic shrimp certification. *Journal of Rural Studies* 27(2): 125–133.

Lave, R. 2012, *Fields and Streams: Stream Restoration, Neoliberalism, and the Future of Environmental Science.* University of Georgia Press, Athens.

Lyon, S. 2010, *Coffee and Community: Maya Farmers and Fair-Trade Markets.* University of Colorado Press, Boulder.

Lyon, S., Aranda Bezaury, J., and Mutersbaugh, T. 2010, Gender equity in fairtrade-organic coffee producer organizations: cases from Mesoamerica. *Geoforum* 41(1): 93–103.

McAfee, K. and Shapiro, E. 2010, Payments for ecosystem services in Mexico: nature, neoliberalism, social movements, and the state. *Annals of the Association of American Geographers* 100(3): 579–599.

McCarthy, J. 2006, Rural geography: alternative rural economies-the search for alterity in forests, fisheries, food, and fair trade. *Progress in Human Geography* 30(6): 803–811.

Mancini, M.C. 2013, Geographical indications in Latin America value chains: a 'branding from below' strategy or a mechanism excluding the poorest? *Journal of Rural Studies* 32: 295–306.

Mansfield, B. (2004) Organic views of nature: the debate over organic certification for aquatic animals. *Sociologia Ruralis* 44(2): 216.

Mutersbaugh, T. 2004, Serve and certify: paradoxes of service work in organic-coffee certification. *Environment and Planning D: Society and Space* 24: 533–552.

Mutersbaugh, T. 2005, Fighting standards with standards: harmonization, rents, and social accountability in certified agrofood networks. *Environment and Planning A* 37: 2033–2051.

Naylor, L. 2014, 'Some are more fair than others': fair trade certification, development, and North–South subjects. *Agric Hum Values* 31: 273–284.

Neumann, R.P. 2005, *Making Political Ecology*. Hodder Education, London.

Osborne, T.M. 2010, *Carbon Capital: The Political Ecology of Carbon Forestry and Development in Chiapas, Mexico*. PhD thesis, Energy and Resources, University of California, Berkeley.

Otto, J. 2014, Carbon development: pursuing climate change mitigation and poverty alleviation through market-based forest carbon schemes in Chiapas, Mexico. Dissertation, University of Kentucky.

Peters-Stanley, M. and Yin, D. 2013, Maneuvering the mosaic: state of the voluntary carbon markets 2013. Forest Trends' Ecosystem Marketplace and Bloomberg New Energy Finance.

Pickren, G. 2014, Political ecologies of electronic waste: uncertainty and legitimacy in the governance of e-waste geographies. *Environment and Planning A* 46(1): 26–45.

Ponte, S. 2012, The Marine Stewardship Council (MSC) and the making of a market for 'sustainable fish'. *Journal of Agrarian Change* 12(2–3): 300–315.

Ponte, S. and Cheyns, E. 2013, Voluntary standards, expert knowledge and the governance of sustainability networks. *Global Networks* 13(4): 459–477.

Quaedvlieg, J., García Roca, M. and Ros-Tonen, M.A.F. 2014, Is Amazon nut certification a solution for increased smallholder empowerment in Peruvian Amazonia? *Journal of Rural Studies* 33: 41–55.

Rangnekar, D. 2011, Remaking place: the social construction of a Geographical Indication for Feni. *Environment and Planning A* 43(9): 2043–2059.

Raynolds, L.T. 2012, Fair trade flowers: global certification, environmental sustainability, and labor standards. *Rural Sociology* 77(4): 493–519.

Riisgaard, L. 2012, Towards more stringent sustainability standards? Trends in the cut flower industry. *Review of African Political Economy* 38(129): 435–453.

Schroeder, R. 2010, Tanzanite as conflict gem: certifying a secure commodity chain in Tanzania. *Geoforum* 41(1): 56–65.

Vandergeest, P. and Unno, A. 2012, A new extraterritoriality? Aquaculture certification, sovereignty, and empire. *Political Geography* 31(6): 358–367.

Wilson, B.R. 2013, Breaking the chains: coffee, crisis, and farmworker struggle in Nicaragua. *Environment and Planning A* 45(11): 2592–2609.

Wilson, B.R. and Curnow, J. 2013, SolidarityTM: student activism, affective labor, and the fair trade campaign in the United States. *Antipode* 45(3): 565–583.

33

PROPERTY AND COMMODIFICATION

Scott Prudham

Introduction

Concern with the interconnections among property rights, commodification (specifically the commercial appropriation of natural resources), and conjoined dynamics of social and environmental change is one of the core features of the political ecology tradition. In this chapter, I discuss some fundamental connections between commodification and property. I start with the enduring legacy of Malthusianism, an approach whose basic premise of an inherent tension between human population growth and natural resource availability (initially food supply) deflects attention from questions of distribution that a focus on property rights brings to the fore. Second, I discuss Marx's analysis of primitive accumulation as a way of approaching the conjoined character of commodifying human "nature" (as wage labour) and non-human nature (primarily as non-human natural resources). From both Malthus and Marx emerges an important and enduring theme: private exclusive property rights link the commodification of means of subsistence with the commodification of labour. Thus, for both thinkers, we can see exclusive property regimes as part of what constitutes a distinct metabolism of conjoined socio-natural relations and transformations, and also as a key feature of a distinct liberal capitalist governmentality, one that continues to echo through natural resource and environmental policy in the neoliberal era. I close by connecting the themes established in the first two sections to contemporary political ecology scholarship as it concerns property, commodification, and governance of socio-natural relations broadly understood.[1]

Malthus

Thomas Malthus's (1798) influential and controversial essay on population has had a major impact on contemporary environmentalism and, indirectly, on political ecology scholarship. The primary legacy of the essay is a well-known (and highly problematic) thesis that there is an inherent tendency of human population growth to outpace increases in food production in the absence of "checks" on population growth. Prominent among these checks, according to Malthus, is famine. Malthus was also, however, quite overt in arguing that private property rights and markets should essentially determine who gets food during a shortage (via direct ownership over means of producing food and/or ability to pay for it). He was also explicit about enlisting the threat of food deprivation to induce the poor to work for wages (a principal motivation of his argument to eliminate food relief programmes). The overall thrust, then, is not merely an abstract argument about the inherent tensions between population increase and food supply (as is commonly understood), but a more complex portrait (and conservative defense) of a *metabolic totality* connecting food production and scarcity on the one hand with prevailing institutions of the emerging liberal order on the other, notably private property and the commodification of land and labour.

Malthus's (1798) basic formulation is (in)famous and immediately familiar. He postulated that human population growth had an inherent tendency to increase geometrically (e.g., 2, 4, 8, 16, 32, ... and so on). Food production, he argued, could only increase at best arithmetically (e.g., 1, 2, 3, 4, 5, ... etc.). Obviously, if this were true, then even an initially large supply of food in relation to a small population would develop into a situation of scarcity. Malthus then posited several "checks" on population, classifying these as "normative" (e.g., delayed marriages, religious or other cultural restrictions on reproduction, etc.) and "positive" (chiefly famine and disease outbreaks). "Checks" for Malthus had the effect of attenuating and even reversing population growth, offsetting the threat of scarcity.

Though there are nuances to the actual argumentation, Malthus concluded with the well-known laissez-faire prescription that famine and disease had to be accepted as unpleasant but necessary limits on population growth and that relief programs of the day merely made things worse. This argument was the basis of his well-known critique of utopian socialism in general, but also more specifically, of famine and poverty relief programs in eighteenth-century England. According to Malthus, providing food to the hungry meant that they would have the temerity not only to survive, but to breed. One cannot, sadly, make this stuff up.

Of course the original context for the essay has changed, and so too has its application. The basic Malthusian formula is a staple of what Robbins (2004) has called "apolitical ecologies": that is, portraits of environmental degradation and resource scarcity that obscure or ignore their particular social origins. The most widely known neo-Malthusian statement is the 1968 Ehrlich volume *The Population Bomb* (1968), but there are many variants. Neo-Malthusian formulations are often used to justify interventions of various kinds, from forest management to biodiversity conservation, and they remain potent and pervasive in both misreading the causes for scarcity and in providing a foundation (however flawed) for institutional and policy reforms (for analyses critical of the Malthusian and neo-Malthusian paradigm, see e.g. Benjaminsen et al., 2006; Fairhead and Leach, 1996; Harvey, 1974; Turner, 1993).

Usually, critics counter neo-Malthusian arguments with some reference to the basic institutions and power relations that shape who has access to resources. They do this in order to provide more sociologically (and ecologically) specific accounts of the origins of famine and of the institutional rather than purely demographic underpinnings of environmental degradation. These critiques of neo-Malthusianism may be more or less radical, but they certainly animated the environment and development debates of the 1970s and into the 1980s (Dryzek, 1997; Wisner et al., 1982). And these debates helped launch political ecology as a distinct area of scholarship (see e.g., Martinez-Alier, 2002; Watts, 1983).

And yet there is an irony in all this. Specifically, while critics of contemporary neo-Malthusianism often point to specific institutions, including property rights, in structuring resource access and in shaping the dynamics of environmental change, a core and explicit project of Malthus's was defense of the enclosure of means of subsistence in eighteenth-century England, and with it, the commodification of food. More specifically, a close reading of Malthus's entire essay reveals it to be a convoluted, indirect, and somewhat tortured defense of class privilege in early industrial England where private property rights structured who had direct access to land (and thus the means to produce food) and who did not. In his words:

> When an article is scarce, and cannot be distributed to all, he that can shew [sic] the most valid patent, that is, he that offers most money, becomes the possessor.
>
> *(Malthus, 1798, p. 24)*

> The rich might become poor, and some of the poor rich, but a part of the society must necessarily feel a difficulty of living, and this difficulty will naturally fall on the least fortunate members.
>
> *(Malthus, 1798, p. 25)*

> It seems highly probable, therefore, that an administration of property, not very different from that which prevails in civilized states at present, would be established, as the best, though inadequate, remedy for the evils which were pressing on the society.
>
> *(Malthus, 1798, p. 62)*

The argument is deeply conservative in the sense that Malthus essentially invokes the then prevailing order in England as being more or less inevitable, with private rights to land and uneven wealth the means of determining who should eat and who should starve. He does so without ever interrogating the ethical, political, and historical-geographical dimensions of whether indeed private ownership of the means of subsistence and the means of payment was the best way to settle things. Moreover, he ignored altogether whether or not enclosure might be one of the causes of hunger and poverty, the emergence of which coincided historically with alienation of people from direct access to land (Polanyi, [1944] 2001). Malthus plainly recognized that it was actually private property rights that made access to food a practical question in the England of his day. Moreover, he also understood that access to food for those without rights to agricultural land meant purchase (i.e., the commodification of means of subsistence). In fact, Malthus argued explicitly that the need to buy food might discipline the poor and force them to work for wages, no small problem in the emerging liberal capitalist order of the day.

Why is this still important? A careful reading of Malthus's original 1798 essay as an alternative to warmed over neo-Malthusian variants reveals it to be a trenchant defense of private exclusive property, commodification of the bases of social reproduction and of labour – in short, the emergent capitalist order of late eighteenth-century England. That alone is sobering in thinking about the ideological baggage that lies behind "scarcity talk" (Harvey, 1974). In this respect, Malthus's argument invokes (implicitly) an historically distinct *metabolism* of integrated socio-natural relations whereby private property rights over land as a means of subsistence are seen as an integral dimension of the conjoined commodification of the means of production/means of subsistence and the commodification of labour. All of the pieces move together, with the commodification of food and the means to produce it bound up in the production of the commodification of labour and the emergence of the modern working class. Malthus's politics are quite different from those of Marx, and yet he was no less aware of the far-reaching consequences of what Marx was later to call the process of primitive accumulation (see below).

Marx and primitive accumulation

In his critique of capitalism, Karl Marx also emphasized the historical significance of private claims to land and other resources. For Marx, the establishment of exclusive, private, transferable claims to land is one of the important ways in which capitalism, as the "production of commodities by commodities", took hold. In this respect, though their politics were certainly in contrast, and while Marx's analysis goes much further, both Malthus and Marx recognized the broader metabolism with which private property was linked, one as advocate, the other as critic.

Marx's theory of what he called the "so-called primitive accumulation" toward the end of volume I of *Capital* ([1867] 1977) constitutes a rejoinder, in part, to Adam Smith's (Smith and Cannan, 2003) notion of "original accumulation". Original accumulation, as the name would suggest, was for Smith an historical precondition for capitalism, one tied to the thrift of certain individuals (the first capitalists), allowing them to build up enough private wealth for the creation of capital to be invested in production for sale. Marx savaged this account and instead posited his own. In his words:

> In actual history it is notorious that conquest, enslavement, robbery, murder, briefly force, play the great part. In the tender annals of Political Economy, the idyllic reigns from time immemorial. Right and "labour" were from all time the sole means of enrichment, the present year of course always excepted. As a matter of fact, the methods of primitive accumulation are anything but idyllic.
>
> *(Marx, [1867] 1977, pp. 873–874)*

Reviewing the significance of Marx's insight (and associated critique) in detail is beyond the scope of this chapter, but three specific aspects are immediately relevant. First, Marx argued that the emergence of capitalism took place not only via putatively "economic" means alone (i.e., "normal" market forces, buying, producing, selling, working for wages, accumulating surpluses, reinvesting surplus in subsequent production, etc.) but rather also via so-called "extra-economic" means (Glassman, 2006). These extra-economic means include sometimes violent expropriations backed by the power of the state, but also and more generally, via the mobilization of elite authority to dissolve communal and other shared rights of access to land. For Marx, primitive accumulation was not a genteel, historical maturation of economic forms so much as a coercive revolution in land rights.

Marx's emphasis on the "extra-economic" character of primitive accumulation remains an important reminder to interrogate and examine the contours of political struggle and the exercise of power in "doing" both *political* economy and *political* ecology in general, and specifically in understanding how private, exclusive rights to socio-nature originate and are reproduced. This is true even if, as subsequent scholarship on historical enclosure (including work in political ecology) has shown, the actual extra-economic processes by which enclosure was undertaken may be: (1) quite protracted; (2) at least formally non-violent; and (3) achieved through cultural and political processes in addition to, or instead of, brute force. Nevertheless, the central point is that property rights are not endogenous creations of so-called "market" forces so much as they are exogenous political creations that act as extra-economic preconditions to accumulation.[2] It also follows that property rights are never politically or ethically neutral. Rather, any form of property, since it involves exclusions (the exception being true open access which is more accurately an absence of property), needs to be defended, by force real or implied, and by some means of discursive legitimation, explicit or implicit (Macpherson, 1978).

In general terms, then, property rights comprise aspects of the irreducibly social and political content of what are generally understood to be economic relations; they are made (and unmade).

A second reason for the enduring significance of Marx's account of primitive accumulation is that it opened a fascinating and highly germane debate about primitive accumulation in the ongoing history (and geography) of capitalism. The question in this debate has been whether or not primitive accumulation is only an historical precursor for capitalism per se, or is instead a "permanent" ontological condition and set of material-semiotic processes that must be re-enacted in order to reproduce conditions of accumulation. In the "classical" Marxist tradition, the likes of Rosa Luxembourg and Vladimir Lenin linked primitive accumulation as a capitalist imperative underpinning imperialism and colonialism fueled by a search for new markets, raw materials, and cheap labour outside the formal boundaries of the capitalist powers. Historically, this tendency may be seen as a foundation for commercial extraction of natural resources from imperial colonies (see, e.g., Mintz, 1985), including for instance via the introduction of colonial forest administration in South and Southeast Asia (Peluso, 1992; Vandergeest and Peluso, 2001). Primitive accumulation *qua* colonialism may also have led to the introduction of capitalist social relations more generally, as in the settler colonies. But the point here is that primitive accumulation is seen not so much as an isolated historical incident as an inherent historical-*geographical* tendency that propels uneven spatial development on a world scale (Harvey, 1982).

A resurgence of contemporary enclosures has helped to renew the salience of Marx's account of primitive accumulation in thinking about historical *and* contemporary relationships between private property forms and capitalism more generally (see De Angelis, 2001; Federici, 2004; Perelman, 1983, 2000). David Harvey (2003) has also helped re-energize and re-shape this debate via what he theorizes as "accumulation by dispossession" in neoliberal capitalism. These debates are highly relevant to understanding the dynamics of so-called "land" and "green" grabbing, and the enclosure of access rights to everything from fish to conservation parks, as I discuss in the third section of the chapter.

The third major reason (and most important for this discussion) why Marx's primitive accumulation framework remains germane to political ecology scholarship is because it provides, somewhat in parallel with Malthus's more abstract endorsement, a unified account of the relation between the enclosure of land (and other resources) on the one hand, and the production of a landless labouring class on the other. The primitive accumulation account links property rights to the commodification of the means (and conditions – see McCarthy, 2004) of production, the commodification of the means of subsistence and social reproduction, and the commodification of labour. As E.P. Thompson (1975, p. 207) wrote in reference to historical enclosures, their effects on "freeing up" land and labour, and the alienation of one from the other, in eighteenth-century England:

> In the seventeenth century labour had been only partly free, but the labourer still asserted large claims (sometimes as perquisites) to his [*sic*] own labour's products. As, in the eighteenth century, labour became more and more free, so labour's property came to be seen as something totally distinct, the property of landowner or employer.

Thompson goes on to note that via these processes, complex and overlapping traditional use rights to the land and its products that were constitutive features of the feudal social order dissolved into what we might now call a very different metabolic socio-natural regime, with on the one hand exchangeable rights to land increasingly seen objectively as "things" (rather than as social relations), and on the other hand, commodified wage labour cut loose from feudal integuments. Of these developments, Karl Polanyi was later to write:

> no people could forget that unless they owned their food and raw material sources themselves or were certain of military access to them, neither sound currency nor unassailable credit would rescue them from helplessness.
>
> *([1944] 2001, p. 199)*

While primitive accumulation remains a powerful lens for exploring the links between property and commodification (including via a resurgence of interest in political ecology scholarship – see below), there are questions and tensions (see Hall, 2012). It is apparent, for instance, that there are sometimes significant differences between the motivations underlying enclosures and their effects. Not all who seek to strengthen property rights do so in the service of capital. To say that the reproduction of the conditions of accumulation relies on historical and contemporary enclosures is not to say that capitalism per se can achieve those same enclosures. Indeed, for its complex portrait of the politics (and specific ecological materiality) of enclosure, the dissolution of communal use rights, and the resulting social implications (including the production of new subjectivities), E.P. Thompson's *Whigs and Hunters* (quoted above) remains something of a landmark.

Indeed, if the enclosers of land are not always the accumulators of capital, not all forms of enclosure are conducive to the commodification of that which has been enclosed, nor to the accumulation of capital at all. Primitive accumulation, as De Angelis (2004) points out, is the active creation of *capital* as a social relation, the ontological pre-condition for the exercise of capitalist power; enclosure per se is then but one moment in this. Moreover, enclosure can occur in ways that actually subvert the accumulation of capital (Rose, 1994). This includes, for instance, the establishment of collective (e.g., communal or state) forms of exclusive property, as well as individual forms of ownership rights that are not saleable.

In addition, those dispossessed and alienated from access to socio-natural means of subsistence or small holder production do not all end up working in industrial resource sectors where accumulation relies on more or less recently attenuated rights of access. Hundreds of millions dispossessed by the enclosure of agricultural and forest lands have fuelled a massive global wave of urbanization during the industrial era, but it is apparent, particularly in the global South, that many of the dispossessed are being rendered as "surplus", largely excluded from the formal economy and labour markets (Davis, 2007; Li, 2010). That disjuncture is a major contemporary qualifier to the general account of primitive accumulation offered by Marx: more importantly, it is also one of the massive "facts" of contemporary enclosure and commodification dynamics to be confronted.

Property, commodification, political ecology

There are two key themes relevant to political ecologies of commercial natural resource appropriation that emerge from this brief review of Malthus and Marx. First, commenting from very different normative positions, both recognized that private, exclusive rights to land as a means of both subsistence and production is one side of a coin whose other side is the commodification of labour. Using language more familiar in contemporary political ecology circles, we might say each points to what is in effect a metabolic integration of the commodification of specific socio-natures and the deployment of social labour broadly understood. Severing rights to land and other means of subsistence by means of enclosure "freed" up means of production, while having the dual effect of making means of subsistence available as an exchangeable commodity to an "unfree" labouring class whose lot, as Marx put it, was to sell themselves to live. While Malthus and Marx obviously held starkly different views, their holistic perspectives on the complex and integrated transformations caught up in the enclosure of land serve as important points of departure for thinking about the role of property rights in the commercial appropriation of natural resources today, situated within a broader historical metabolism by which conjoined social and ecological transformations take place.

Second, Malthus and Marx offer portraits of what Foucault (1991; Foucault et al., 2008) was later to theorize as the distinct governmentality[3] of liberal capitalism and of classical political economy, again, with Malthus as champion, and Marx as critic. Foucault was interested, among other things, in the development of ways of governing that would induce in people desired ways of being and acting in the world, particularly as forms of political authority shifted from autocratic modes of governance toward the modern liberal state (see Chapter 36, this volume). Foucault's interest in understanding the specifics of liberal and subsequently neoliberal modes of governmentality (Lemke, 2001) included tracing the emergence of an approach to governing that would best position the state to encourage disciplined, rational, self-interested utility maximizing subjects, effectively calling forth the *Homo economicus* first envisioned by classical political economists as the essence of human nature (Polanyi, [1944] 2001).

Though neoliberalism is a highly contested term used in reference to diverse and sometimes contradictory political economic and regulatory initiatives (Brenner and Theodore, 2002; Larner, 2003; Tickell and Peck, 1995), one of the core features of projects and processes to which this term has been applied is a renewed emphasis on the establishment of private exclusive and (usually) exchangeable property rights as a means to facilitate renewed or expanded capital accumulation and economic growth, but also as a means of securing governance reforms prioritizing individual decision-making, autonomy, and responsibility in relation to the self and/or to the administration of social affairs more generally (Brown, 2006; Harvey, 2005; Jessop, 2002; Peck, 2008; Peck and Tickell, 2002; see also Chapter 34, this volume). And though not always recognized as such, one of neoliberalism's primary and arguably constitutive fronts has been the extension of new or strengthened private claims to discrete socio-natures (Heynen and Robbins, 2005; Heynen et al., 2007; McCarthy and Prudham, 2004; Wolford, 2007), from land tenure reforms to genes and genetically modified organisms, and from water resources to rights to emit regulated substances into air and water, and so on. This has led some observers to suggest that, now more than ever, nature is being produced well and truly "all the way down" (Castree, 1995, 2003, 2008a, 2008b).

In this context, one reason Malthus's essay in particular remains germane is that it spoke on the one hand to property rights as a means to address scarcity questions, and on the other, to the role of property rights in encouraging wage labour with otherwise minimal state intervention; he provides something of a recipe for an efficient and efficacious mode of governmentality. Similar normative overtones run through mainstream prescriptions for the roll-out of new or strengthened exclusive property rights and market relations in contemporary environmental policy, as means by which to exploit resources, but also as approaches to securing conservation objectives and an ostensibly correct social ordering. H. Scott Gordon's (1954) seminal bio-economic analysis of overfishing, a much more intellectually rigorous and honest appraisal of the problems of open access than offered by Garett Hardin (1968), provides important intellectual grounding for this paradigm. Also foundational was the development of the so-called Hotelling rule prescribing efficient depletion of non-renewable resources (Hotelling, 1931). In environmental and ecological economics, establishing new and exclusive property rights over socio-nature also factor into discussions concerning efficient means of regulating emissions of pollutants (see, e.g., Tietenberg, 1980) as well as how to think about allocating rights to resources to ensure intergenerational equity (Howarth and Norgaard, 1990). These ideas continue to influence engagements in the climate policy debate (Bailey, 2007; Bumpus and Liverman, 2008; Ekins and Barker, 2001).

More recent forays into the conceptually messy world of payments for ecosystem services (Dempsey and Robertson, 2012; McAfee and Shapiro, 2010) rely on similar lines of reasoning when it comes to managing socio-nature as a form of stored wealth or what some call (problematically) "natural capital" (Costanza et al., 1997). Morgan Robertson's (2000, 2004, 2006) work on the important US wetland banking program, for instance, lays bare how vibrant (if problematic) the enclosure-commodification rubric remains as a form of environmental regulatory discourse. What is perhaps most remarkable is that this program's very foundational principles are those that lead to the greatest scepticism. That is, the wetland banking program, among the earliest institutional expressions of a payment-for-ecosystem services approach, and one that takes the commodification-as-conservation logic about as far as it can go, posits that an acceptable way to ensure no net loss of wetlands in the US is to allow investors to literally and intentionally produce new wetlands as exchange values (i.e., *for* sale) by means of securing credits for the new wetlands from the state regulator and then selling those credits to other developers looking to meet mitigation requirements. The basic premise is that one wetland can indeed be treated as both the functional and commercial equivalent of another; the commodification of wetlands demands no less. The idea is deceptively simple yet is haunted by disquieting ontological dilemmas. Among these is the precise articulation of science and capital in rendering equivalence between discrete wetlands, including for instance, when the two wetlands are not in the same watershed.

Other fronts likewise advance the privatization–conservation–commodification triad. For example, contemporary uptake of Individual Transferrable Quota (ITQ) systems in fisheries demonstrates that the basic parameters of H. Scott Gordon's aforementioned "no one's property is everyone's property" critique is a powerful ideological force helping to legitimate a globe spanning neoliberal re-regulation of wild fisheries (Mansfield, 2001, 2004a, 2004b; St. Martin, 2005). In all of this, Marx's account of the role of enclosure in capital formation (along with subsequent debates) constitutes an important historical and conceptual resource for those critics who are suspicious of contemporary enclosures, including in domains of interest frequented by political ecologists. Kloppenburg (2004), for instance, draws explicitly on the concept of primitive accumulation in tracing a long American century of converting seeds for agricultural production from a public to a private good through a combination of technical means (e.g., trade secrets and/or the use of technologies such as development of hybrid plant varietals) as well as regulatory and legislative ones (notably via roll-out of expanded plant patent rights).

In the US, these shifts have been marked by important legal and administrative decisions extending the boundaries of private property and the scope of the commodification of nature, including for instance, issuance of the Cohen-Boyer patent on rDNA technology in the 1970s, the 1980 Diamond v. Chakrabarty (447 U.S. 303 (1980)) US Supreme Court decision, and the 1987 US Patent Office decision regarding the Harvard oncomouse (Hughes, 2001; Kenney, 1998; Krimsky and Wrubel, 1996; Thackray, 1998). Internationalization of the US model of private rights to genes and GMOs has been a significant front in the uneven geography of neoliberal globalization (Featherstone, 2003; McAfee, 2003; Prudham, 2007; Prudham and Coleman, 2011; Schurman, 2004; Stone, 2007). Property rights (particularly patent rights) have been integral to shaping the contours of multinational firms in the biotechnology sector (Boyd, 2003). And while enclosure, transformation, and consolidation of access rights to land have been central, for instance, to the Green Revolution and the spread of capitalist agriculture during the latter half of the twentieth century (Bernstein, 1997; Bernstein and Byres, 2001; Ross, 2003), neoliberal enclosures in the food system represent an intensification of capitalist agriculture and an important expression of contemporary agrarian questions (Buck, 2007; Friedmann, 1993; Goodman and Watts, 1994, 1997; Kloppenburg, 2004; Mann, 1990; McMichael, 1997).

More generally, there is renewed interest in primitive accumulation per se in the context of so-called contemporary "land" or "green" grabbing (Bumpus and Liverman, 2008; Corson and MacDonald, 2012; Edelman et al., 2013; Fairhead et al., 2012; Leach et al., 2012; Snijders, 2012; Sullivan, 2013; White et al., 2012). Yet, it bears emphasizing that what is at stake, as discussed in McCarthy's important paper (2004), is not only enclosure and commodification of the *means*, but also the *conditions*, of production, including for instance, rights to use environmental milieu as dumping grounds for waste products from industrial processes. This is about more than accounting for externalities; it is about the relationship between primitive accumulation and the production of nature, and about examining how dynamics of primitive accumulation are constituted by specific socio-natural metabolisms.

And there are nuances. Clearly some enclosures, e.g., protected area creation, emphasize retention of socio-natural features for direct use or aesthetic appreciation (whether by elites or for more widespread enjoyment) and may be set up explicitly to impede certain kinds of commodification, including natural resource extraction. Yet even in these instances, complexities and contradictions may be involved as, for example, when protected areas have their amenity value come to underpin ostensibly non-extractive or non-consumptive forms of exchange-based production, e.g., international tourist travel and associated services. Moreover, some enclosures initially set aside for elite enjoyment, as in the case of colonial hunting preserves, are now the foundations of conservation areas increasingly commodified as "spectacular" natures or retained by states, including for the purposes of controlling access to and commodification of valuable natural resources (see, e.g., Brockington et al., 2008; Igoe et al., 2010; Neumann, 1998, 2001; Peluso, 1993).

While attending to the socio-ecological dynamics of contemporary enclosures, it bears remembering some of the core strengths and emphases of political ecology scholarship. One of these strengths and emphases is engagement with the specific institutions of property as they are being transformed (crudely, from what, to what, by whom, and why). Political ecologists have documented and attended to specificity in property regimes. They have shown repeatedly that, while there may well be explanatory power in broad classifications of property types (e.g., exclusive and transferable, communal, open access, state, public, etc.) as concrete abstractions, it is also the case that actual property rights are typically much more complex and hybrid in character, and that the more closely we look, sometimes the more complex the situation becomes. Communal rights and an un-extinguished messy "public" continues to haunt even modern, liberal property regimes emphasizing private, exclusive rights more generally (see, e.g., Rose, 1986). Political ecology scholarship tends to affirm in particular the insight that property in the form of "individual absolute dominion" (Gordon, 1996) is more market fundamentalist fantasy than fact, particularly (but not only) in those parts of the world where traditional and/or subsistence rights are prevalent in practice (see, e.g., Carney, 1993; Schroeder, 1993).

A second major theme in the political ecology literature is attention to both formal *and* informal claims. Commercial pressures and processes of commodification are often driven by formalization or hardening of exclusive property claims. And yet the specific trajectory of formalization may draw upon prior informal property regimes, including non-codified traditions and customs. Moreover, social struggle and conflict over contested rights of access may arise when attempts to develop more formal property rights eliminate or weaken traditional or customary claims (see, e.g., Peluso, 1995; Thompson, 1971, 1975). These dynamics continue to play out in the context of intensifying commercial pressures and, while they are certainly evident in parts of the global South where subsistence and/or peasant production regimes remain in place, contested "traditional" claims informed by distinct and stubborn "moral economies" (Scott, 1976) are also expressed, for example, in conflicts over access to public lands in the American west (McCarthy, 2002; Sheridan, 2001), in the context of intensifying claims on rural lands in Europe (see e.g. Benjaminsen and Svarstad, 2008; Lem, 1999), and in the transformation of post-socialist rural economies (Sikor, 2001; Sikor et al., 2009; Sturgeon and Sikor, 2004).

A third hallmark of the political ecological "take" on the property–commodification nexus is focus on the articulation between small-holder and sometimes subsistence and/or "traditional" production and resource management regimes on the one hand (often in some of the world's poorest countries), and broader political economic relationships of production, exchange, and social regulation on the other. The chain of explanation's critical ethos (Blaikie and Brookfield, 1987; Robbins, 2004) and methodological paradigm is motivated in part to capture these articulations via relentless contextualization. And while the integration of local production regimes into wider networks of social claims, including far-flung relations of commodification, may not be new or even unique to capitalism per se (Mintz, 1985; Peluso, 1992, 2012; Taussig, 1980), such dynamics are ongoing and shape life circumstances and the production of environmental change across the globe. Research continues to document how historically and geographically specific property regimes are displaced or subsumed, or conversely, how property rights and access regimes become sites of political struggle and resistance (see, e.g., Sikor and Lund, 2009).

This last point directs attention to a fourth key feature of the political ecology tradition; property is never just about rights to or ownership of "things" or "stuff". Many political ecologists have taken heed of the insight offered by Ribot and Peluso (2003) that questions of property sometimes emphasize rights at the expense of sufficient attention to *access*, as in who actually has the ability to derive benefits. And, influenced by the critical legal studies tradition and neo-Marxist perspectives (Blomley, 2003, 2005; Macpherson, 1978), political ecologists have recognized that property rights are social relations between people. This can be crucial in understanding how property and commodification are linked and speaks to a broader architecture of social power and control that can actually be obscured by a focus on rights per se.

In fact, situating property within a broader social field in understanding the dynamics of land-use change (including under the influence of commercial and non-commercial activities) reflects constitutive themes in political ecology scholarship. In an introduction to a collection of essays on socio-economic change in the European Alps, one of the first scholars to deploy the term "political ecology", Eric Wolf, stated it thus:

> The property connexion in complex societies is not merely an outcome of local or regional ecological processes but a battleground of contending forces which utilize jural patterns to maintain or restructure the economic, social and political relations of society.
>
> *(Wolf, 1972, pp. 201–202)*

And though he does not use the language of "metabolism", it is clear in the same essay that Wolf viewed claims to the land as fundamental to what we might now call the integrated socio-ecological order, namely the specific sets of relations and processes comprising and governing conjoined social and biophysical transformations (Swyngedouw, 1995, 1999). As Blaikie and Brookfield (1987) argued in their seminal volume, there is a local specificity to claims on and relations to land and other natural resources, including locally specific ways of understanding and representing "nature". Indeed, property rights governing access to and control over specific socio-natures are caught up in, and are constitutive of, the content of "society" more generally (see, e.g., Carney and Watts, 1990). These metabolic relationships and transformations help make "us" who we are but also help define important differences across space and time.

The particular meanings associated with institutions of access and control as well as the trajectories of commodification and socio-environmental change also lend – or should lend – the *materiality* of work in political ecology an inexhaustibly ethico-political dimension (Mann, 2009). And the long-standing emphasis on a knowledge–power nexus in political ecology scholarship points to the ways in which the power to *name* is bound up in the power to *claim* (and vice versa). Perhaps in no domain is an understanding of this dimension of the property–commodification nexus more pressing than in the aforementioned struggles over who owns life and its constituent parts in the ongoing commercialization of new biotechnologies. As Kaushik Sunder Rajan (2005, 2006) argues, material and semiotic transformations are so entwined in this arena that the commodification and redefinition of life go hand in hand; struggles over control are inevitably struggles over meaning, and vice versa.

Conclusion

These, then, are some of the main ways in which the nexus of property, commodification, and socio-environmental change have been and continue to be interrogated. The intensity of contemporary struggles over rights to land and other resources, whether those claims are animated by commercial claims or otherwise, and the diversity of settings and actors involved, means there is much left to do. As the chapters in this collection attest, political ecology is indeed a diverse, sprawling, and sometimes incomprehensible field defined sometimes by method, sometimes by themes, sometimes by questions and sometimes not at all! But a review of the evolution of the field over time shows that property rights have consistently provided a crucial lens through which to understand the complex socio-ecological, material and semiotic processes involved in propelling locally specific and conjoined environmental change and social differentiation, particularly in the context of articulation with broader commercial networks of commodity production and exchange and their social regulation. And political ecology scholarship reminds us that a focus on property rights is never merely about the allocation of "stuff", as important as that may be, but also provides a window through which to understand the specificity of relations between the human and the non-human world and how these comprise the conditions under which life is lived (or not lived) and what it means to live it.

Notes

1 Throughout this chapter I attempt, with admittedly uneven success, to move past a language reflecting commonly understood dualistic ontological notions of "nature" and "society". I do this by referring whenever possible to *socio-nature*, and by using the language of metabolism, to try to capture the conjoined, co-produced, and historically contingent character of integrated social and biophysical transformation. I draw inspiration for the terminology and the concepts from various sources including not only Smith's (2008) production of nature thesis, but also work by Swyngedouw (1995, 1999, 2004), Haraway (1991, 1997), Moore (2011), Foster (1999), Williams (1973), Latour (1993, 2004) and others.

2 For his discussion of the extra-economic as well as other aspects of primitive accumulation, I am grateful to Jim Glassman not only for his excellent article (Glassman, 2006) but also for conversations on the topic that helped me immensely.
3 As Lemke (2001) explains it, the term is best understood as a kind of mashing together of the terms "governing" and "mentality", pointing to how styles of governance and a rationality of governance are tied to particular kinds of subjectivities, ways of thinking and being.

References

Bailey, I. (2007). Neoliberalism, climate governance and the scalar politics of EU emissions trading. *Area*, 39(4), 431–442.

Benjaminsen, T.A., and Svarstad, H. (2008). Understanding traditionalist opposition to modernization: Narrative production in a Norwegian mountain conflict. *Geografiska Annaler, Series B: Human Geography*, 90(1), 49–62.

Benjaminsen, T.A., Rohde, R., Sjaastad, E., Wisborg, P., and Lebert, T. (2006). Land reform, range ecology, and carrying capacities in Namaqualand, South Africa. *Annals of the Association of American Geographers*, 96(3), 524–540.

Bernstein, H. (1997). Agrarian questions then and now. *Journal of Peasant Studies*, 24(1–2), 22–59.

Bernstein, H., and Byres, T.J. (2001). From peasant studies to agrarian change. *Journal of Agrarian Change*, 1(1), 1–56.

Blaikie, P.M., and Brookfield, H.C. (1987). *Land degradation and society*. London and New York: Methuen.

Blomley, N. (2003). Law, property, and the geography of violence: The frontier, the survey, and the grid. *Annals of the Association of American Geographers*, 93(1), 121–141.

Blomley, N. (2005). Remember property? *Progress in Human Geography*, 29(2), 125–127.

Boyd, W. (2003). Wonderful potencies: Deep structure and the problem of monopoly in agricultural biotechnology. In R. Schurman and D.D.T. Kelso (Eds.), *Engineering trouble: Biotechnology and its discontents* (pp. 24–62). Berkeley: University of California Press.

Brenner, N., and Theodore, N. (2002). Cities and the geographies of "actually existing neoliberalisms." *Antipode*, 34(3), 349–379.

Brockington, D., Duffy, R., and Igoe, J. (2008). *Nature unbound: Conservation, capitalism and the future of protected areas*. London and Sterling, VA: Earthscan.

Brown, W. (2006). American nightmare: Neoliberalism, neoconservatism, and de-democratization. *Political Theory*, 34, 690–714.

Buck, D. (2007). The subsumption of space and the spatiality of subsumption: Primitive accumulation and the transition to capitalism in Shanghai, China. *Antipode*, 39(4), 757–774.

Bumpus, A.G., and Liverman, D. (2008). Accumulation by decarbonization and the governance of carbon offsets. *Economic Geography*, 84(2), 127–155.

Carney, J. (1993). Converting the wetlands, engendering the environment: The intersection of gender with agrarian change in the Gambia. *Economic Geography*, 69(4), 329–348.

Carney, J., and Watts, M. (1990). Manufacturing dissent: Work, gender and the politics of meaning in a peasant society. *Africa: Journal of the International Africa Institute*, 60(2), 207–241.

Castree, N. (1995). The nature of produced nature: materiality and knowledge construction in Marxism. *Antipode*, 27(1), 12–48.

Castree, N. (2003). Commodifying what nature? *Progress in Human Geography*, 27(3), 273–297.

Castree, N. (2008a). Neoliberalising nature: Processes, effects, and evaluations. *Environment and Planning A*, 40(1), 153–173.

Castree, N. (2008b). Neoliberalising nature: The logics of deregulation and reregulation. *Environment and Planning A*, 40(1), 131–152.

Corson, C., and MacDonald, K.I. (2012). Enclosing the global commons: The convention on biological diversity and green grabbing. *Journal of Peasant Studies*, 39(2), 263–283.

Costanza, R., D'Arge, R., De Groot, R., Farber, S., Grasso, M., Hannon, B., ... Van Den Belt, M. (1997). The value of the world's ecosystem services and natural capital. *Nature*, 387(6630), 253–260.

Davis, M. (2007). *Planet of slums*. London and New York: Verso.

De Angelis, M. (2001). Marx and primitive accumulation: The continuous character of capitalist "enclosures." *The Commoner*, 2(September).

De Angelis, M. (2004). Separating the doing and the deed: Capital and the continuous character of enclosures. *Historical Materialism*, 12(2), 57–87.

Dempsey, J., and Robertson, M. (2012). Ecosystem services: Tensions, impurities, and points of engagement within neoliberalism. *Progress in Human Geography*, 36(6), 758–779.

Dryzek, J.S. (1997). *The politics of the earth: Environmental discourses.* Oxford and New York: Oxford University Press.

Edelman, M., Oya, C., and Borras Jr, S. M. (2013). Global land grabs: Historical processes, theoretical and methodological implications and current trajectories. *Third World Quarterly*, 34(9), 1517–1531.

Ehrlich, P.R. (1968). *The population bomb.* New York: Ballantine Publishers.

Ekins, P., and Barker, T. (2001). Carbon taxes and carbon emissions trading. *Journal of Economic Surveys*, 15(3), 325–376.

Fairhead, J., and Leach, M. (1996). *Misreading the African landscape: Society and ecology in a forest-savanna mosaic.* Cambridge: Cambridge University Press.

Fairhead, J., Leach, M., and Scoones, I. (2012). Green grabbing: A new appropriation of nature? *Journal of Peasant Studies*, 39(2), 237–261.

Featherstone, D. (2003). Spatialities of transnational resistance to globalization: The maps of grievance of the Inter-Continental Caravan. *Transactions of the Institute of British Geographers*, 28(4), 404–421.

Federici, S. (2004). *Caliban and the witch: Women, the body and primitive accumulation.* Brooklyn: Autonomedia.

Foster, J.B. (1999). Marx's theory of metabolic rift: Classical foundations for environmental sociology. *American Journal of Sociology*, 105(2), 366–405.

Foucault, M. (1991). Governmentality. In G. Burchall, C. Gordon and P. Miller (Eds.), *The Foucault effect: Studies in governmentality* (pp. 87–104). Chicago: University of Chicago Press.

Foucault, M., Senellart, M., and Collège de France. (2008). *The birth of biopolitics: Lectures at the Collège de France, 1978–79.* Basingstoke and New York: Palgrave Macmillan.

Friedmann, H. (1993). The political economy of food: A global crisis. *New Left Review*, I(197 January–February), 29–57.

Glassman, J. (2006). Primitive accumulation, accumulation by dispossession, accumulation by "extra-economic" means. *Progress in Human Geography*, 30(5), 608–625.

Goodman, D., and Watts, M. (1994). Reconfiguring the rural or fording the divide? Capitalist restructuring and the global agro-food system. *Journal of Peasant Studies*, 22(1), 1–49.

Goodman, D., and Watts, M. (Eds.). (1997). *Globalising food: Agrarian questions and global restructuring.* London and New York: Routledge.

Gordon, H.S. (1954). The economic theory of a common property resource: The fishery. *Journal of Political Economy*, 62, 124–142.

Gordon, R. (1996). Paradoxical property. In J. Brewer and S. Staves (Eds.), *Early modern conceptions of property* (pp. 95–109). London: Routledge.

Hall, D. (2012). Rethinking primitive accumulation: Theoretical tensions and rural Southeast Asian complexities. *Antipode*, 44(4), 1188–1208.

Haraway, D.J. (1991). *Simians, cyborgs and women: The reinvention of nature.* New York: Routledge.

Haraway, D.J. (1997). *Modest*Witness@Second*Millennium.FemaleMan *Meets*OncoMouse: Feminism and technoscience.* New York: Routledge.

Hardin, G. (1968). The tragedy of the commons. *Science*, 162, 1243–1248.

Harvey, D. (1974). Population, resources and the ideology of science. *Economic Geography*, 50, 256–277.

Harvey, D. (1982). *The limits to capital.* Oxford: B. Blackwell.

Harvey, D. (2003). *The new imperialism.* Oxford and New York: Oxford University Press.

Harvey, D. (2005). *A brief history of neoliberalism.* Oxford: Oxford University Press.

Heynen, N., and Robbins, P. (2005). The neoliberalization of nature: Governance, privatization, enclosure and valuation. *Capitalism, Nature, Socialism*, 16(1), 5–8.

Heynen, N. C., McCarthy, J., Prudham, S., and Robbins, P. (Eds.). (2007). *Neoliberal environments: False promises and unnatural consequences.* London and New York: Routledge.

Hotelling, H. (1931). The economics of exhaustible resources. *Journal of Political Economy*, 39(2), 137–175.

Howarth, R.B., and Norgaard, R.B. (1990). Intergenerational resource rights, efficiency, and social optimality. *Land Economics*, 66(1), 1–11.

Hughes, S.S. (2001). Making dollars out of DNA. The first major patent in biotechnology and the commercialization of molecular biology, 1974–1980. *Isis*, 92(3), 541–575.

Igoe, J., Neves, K., and Brockington, D. (2010). A spectacular eco-tour around the historic bloc: Theorising the convergence of biodiversity conservation and capitalist expansion. *Antipode*, 42(3), 486–512.

Jessop, B. (2002). Liberalism, neoliberalism, and urban governance: A state-theoretical perspective. *Antipode*, 34(3), 452–472.

Kenney, M. (1998). Biotechnology and the creation of a new economic space. In A. Thackray (Ed.), *Private science: Biotechnology and the rise of the molecular sciences* (pp. 131–143). Philadelphia: University of Pennsylvania Press.

Kloppenburg, J.R. (2004). *First the seed: The political economy of plant biotechnology* (2nd ed.). Madison: University of Wisconsin Press.

Krimsky, S., and Wrubel, R.P. (1996). *Agricultural biotechnology and the environment: Science, policy, and social issues*. Urbana: University of Illinois Press.

Larner, W. (2003). Neoliberalism? *Environment and Planning D: Society and Space*, 21, 509–512.

Latour, B. (1993). *We have never been modern*. Brighton: Harvester Wheatsheaf.

Latour, B. (2004). *Politics of nature: How to bring the sciences into democracy*. Cambridge, MA: Harvard University Press.

Leach, M., Fairhead, J., and Fraser, J. (2012). Green grabs and biochar: Revaluing African soils and farming in the new carbon economy. *Journal of Peasant Studies*, 39(2), 285–307.

Lem, W. (1999). *Cultivating dissent: Work, identity, and praxis in rural Languedoc*. Albany: State University of New York Press.

Lemke, T. (2001). "The birth of bio-politics": Michel Foucault's lecture at the College de France on neoliberal governmentality. *Economy and Society*, 30(2), 190–207.

Li, T.M. (2010). To make live or let die? Rural dispossession and the protection of surplus populations. *Antipode*, 41(s1), 66–93.

Macpherson, C.B. (1978). *Property, mainstream and critical positions*. Toronto: University of Toronto Press.

Malthus, T.R. (1798). *An Essay on the Principles of Population as it Affects the Future Improvement of Society with Remarks on the Speculations of Mr. Godwin, M. Condorcet, and other Writers*. London: J. Johnson.

Mann, G. (2009). Should political ecology be Marxist? A case for Gramsci's historical materialism. *Geoforum*, 40(3), 335–344.

Mann, S.A. (1990). *Agrarian capitalism in theory and practice*. Chapel Hill: University of North Carolina Press.

Mansfield, B. (2001). Property regime or development policy? Explaining growth in the US Pacific Groundfish Fishery. *The Professional Geographer*, 53(3), 384–397.

Mansfield, B. (2004a). Neoliberalism in the oceans: "Rationalization," property rights, and the commons question. *Geoforum*, 35, 313–326.

Mansfield, B. (2004b). Rules of privatization: Contradictions in neoliberal regulation of North Pacific Fisheries. *Annals of the Association of American Geographers*, 94(3), 565–584.

Martinez-Alier, J. (2002). *The environmentalism of the poor: A study of ecological conflicts and valuation*. Cheltenham: Edward Elgar Publishing.

Marx, K. ([1867] 1977). *Capital: A critique of political economy, Volume 1* (B. Fowkes, Trans.). New York: Vintage Books.

McAfee, K. (2003). Neoliberalism on the molecular scale. Economic and genetic reductionism in biotechnology battles. *Geoforum*, 34(2), 203–219.

McAfee, K., and Shapiro, E.N. (2010). Payments for ecosystem services in Mexico: Nature, neoliberalism, social movements, and the state. *Annals of the Association of American Geographers*, 100(3), 579–599.

McCarthy, J. (2002). First world political ecology: Lessons from the Wise Use movement. *Environment and Planning A*, 34, 1281–1302.

McCarthy, J. (2004). Privatizing conditions of production: Trade agreements as neoliberal environmental governance. *Geoforum*, 35(3), 327–341.

McCarthy, J., and Prudham, S. (2004). Neoliberal nature and the nature of neoliberalism. *Geoforum*, 35(3), 275–283.

McMichael, P. (1997). Rethinking globalization: The agrarian question revisited. *Review of International Political Economy*, 4(4), 630–662.

Mintz, S.W. (1985). *Sweetness and power: The place of sugar in modern history*. New York: Penguin.

Moore, J. (2011). Transcending the metabolic rift: A theory of crises in the capitalist world-ecology. *Journal of Peasant Studies*, 38(1), 1–46.

Neumann, R.P. (1998). *Imposing wilderness: Struggles over livelihood and nature preservation in Africa*. Berkeley: University of California Press.

Neumann, R.P. (2001). Africa's last "wilderness". Reordering space for political and economic control in colonial Tanzania. *Africa*, 71(4), 641–665.

Peck, J. (2008). Remaking laissez-faire. *Progress in Human Geography*, 32(1), 3–43.
Peck, J., and Tickell, A. (2002). Neoliberalizing space. *Antipode*, 34(3), 380–404.
Peluso, N.L. (1992). *Rich forests, poor people: Resource control and resistance in Java*. Berkeley: University of California Press.
Peluso, N.L. (1993). Coercing conservation? The politics of state resource control. *Global Environmental Change*, 3(2), 199–217.
Peluso, N.L. (1995). Whose woods are these? Counter-mapping forest territories in Kalimantan, Indonesia. *Antipode*, 27(4), 383–406.
Peluso, N.L. (2012). What's nature got to do with it? A situated historical perspective on socio-natural commodities. *Development and Change*, 43(1), 79–104.
Perelman, M. (1983). *Classical political economy: Primitive accumulation and the social division of labor*. Totowa: Rowman & Allanheld.
Perelman, M. (2000). *The invention of capitalism: Classical political economy and the secret history of primitive accumulation*. Durham, NC: Duke University Press.
Polanyi, K. ([1944] 2001). *The great transformation: The political and economic origins of our time*. Boston: Beacon Press.
Prudham, S. (2007). The fictions of autonomous invention: Accumulation by dispossession, commodification, and life patents. *Antipode*, 39(3), 406–429.
Prudham, S., and Coleman, W.D. (2011). Introduction: Property, autonomy, territory, and globalization. In W.D. Coleman (Ed.), *Property, territory, globalization: struggles over autonomy* (pp. 1–28). Vancouver, BC: UBC Press.
Ribot, J., and Peluso, N. L. (2003). A theory of access. *Rural Sociology*, 68(2), 153–181.
Robbins, P. (2004). *Political ecology: A critical introduction*. Malden, MA: Blackwell.
Robertson, M. (2000). No net loss: Wetland restoration and the incomplete capitalization of nature. *Antipode*, 32(4), 463–493.
Robertson, M. (2004). The neoliberalization of ecosystem services: Wetland mitigation banking and problems in environmental governance. *Geoforum*, 35, 361–373.
Robertson, M. (2006). The Nature that capital can see: Science, state, and market in the commodification of ecosystem services. *Environment and Planning D: Society and Space*, 24, 367–387.
Rose, C. (1986). The comedy of the commons: Custom, commerce, and inherently public property. *The University of Chicago Law Review*, 53(3), 711–781.
Rose, C.M. (1994). *Property and persuasion: Essays on the history, theory, and rhetoric of ownership*. Boulder: Westview Press.
Ross, E.B. (2003). Malthusianism, capitalist agriculture, and the fate of peasants in the making of the modern world food system. *Review of Radical Political Economics*, 35(4), 437–461.
Schroeder, R.A. (1993). Shady practice: Gender and the political ecology of resource stabilization in Gambian garden/orchards. *Economic Geography*, 69(4), 349–365.
Schurman, R. (2004). Fighting "Frankenfoods": Industry opportunity structures and the efficacy of the anti-biotech movement in Western Europe. *Social Problems*, 51(2), 243–268.
Scott, J.C. (1976). *The moral economy of the peasant: Subsistence and rebellion in Southeast Asia*. New Haven: Yale University Press.
Sheridan, T.E. (2001). Cows, condos, and the contested commons: The political ecology of ranching on the Arizona-Sonora borderlands. *Human Organization*, 60(2), 141–152.
Sikor, T. (2001). Agrarian differentiation in post-socialist societies: Evidence from three upland villages in north-western Vietnam. *Development and Change*, 32(5), 923–949.
Sikor, T., and Lund, C. (2009). Access and property: A question of power and authority. *Development and Change*, 40(1), 1–22.
Sikor, T., Stahl, J., and Dorondel, S. (2009). Negotiating post-socialist property and state: Struggles over forests in Albania and Romania. *Development and Change*, 40(1), 171–193.
Smith, A., and Cannan, E. (2003). *The wealth of nations/Adam Smith; introduction by Alan B. Krueger; edited, with notes and marginal summary, by Edwin Cannan* (Bantam classic ed.). New York: Bantam Classic.
Smith, N. (2008). *Uneven development: Nature, capital and the production of space* (3rd ed.). Athens: The University of Georgia Press.
Snijders, D. (2012). Wild property and its boundaries – on wildlife policy and rural consequences in South Africa. *Journal of Peasant Studies*, 39(2), 503–520.
St. Martin, K. (2005). Disrupting enclosure in New England Fisheries. *Capitalism, Nature, Socialism*, 16(1), 63–80.

Stone, G.D. (2007). Agricultural deskilling and the spread of genetically modified cotton in Warangal. *Current Anthropology*, 48(1), 67–103.
Sturgeon, J.C., and Sikor, T. (2004). Post-socialist property in Asia and Europe: Variations on fuzziness. *Conservation and Society*, 2(1), 1–17.
Sullivan, S. (2013). Banking nature? The spectacular financialization of environmental conservation. *Antipode*, 45(1), 198–217.
Sunder Rajan, K. (2005). Subjects of speculation: Emergent life sciences and market logics in the United States and India. *American Anthropologist*, 107(1), 19–30.
Sunder Rajan, K. (2006). *Biocapital: The constitution of postgenomic life*. Durham, NC: Duke University Press.
Swyngedouw, E. (1995). The city as a hybrid: On nature, society and cyborg urbanization. *Capitalism, Nature, Socialism*, 7(2), 65–80.
Swyngedouw, E. (1999). Modernity and hybridity: Nature, Regeneracionismo, and the production of the Spanish waterscape, 1890–1930. *Annals of the Association of American Geographers*, 89(3), 443–465.
Swyngedouw, E. (2004). *Social power and the urbanization of water: Flows of power*. Oxford: Oxford University Press.
Taussig, M.T. (1980). *The devil and commodity fetishism in South America*. Chapel Hill: University of North Carolina Press.
Thackray, A. (Ed.). (1998). *Private science: Biotechnology and the rise of the molecular sciences*. Philadelphia: University of Pennsylvania Press.
Thompson, E.P. (1971). The moral economy of the English crowd in the 18th Century. *Past and Present*, 50, 76–136.
Thompson, E.P. (1975). *Whigs and hunters: The origin of the Black act*. New York: Pantheon.
Tickell, A., and Peck, J. (1995). Social regulation after fordism: Regulation theory, neo-liberalism, and the global-local nexus. *Economy and Society*, 24(3), 357–386.
Tietenberg, T.H. (1980). Transferable discharge permits and the control of stationary source air pollution: A survey and synthesis. *Land Economics*, 56(4), 391–416.
Turner, M.D. (1993). Overstocking the range: A critical analysis of the environmental science of Sahelian pastoralism. *Economic Geography*, 69(4), 402–421.
Vandergeest, P., & Peluso, N.L. (2001). Genealogies of the political forest and customary rights in Indonesia, Malaysia, and Thailand. *The Journal of Asian Studies*, 60(3), 761–812.
Watts, M. (1983). *Silent violence: Food, famine, & peasantry in Northern Nigeria*. Berkeley: University of California Press.
White, B., Borras Jr, S.M., Hall, R., Scoones, I., and Wolford, W. (2012). The new enclosures: Critical perspectives on corporate land deals. *Journal of Peasant Studies*, 39(3–4), 619–647.
Williams, R. (1973). *The country and the city*. London: Verso.
Wisner, B., Weiner, D., and O'Keefe, P. (1982). Hunger: A polemical review*. *Antipode*, 14(3), 1–16.
Wolf, E. (1972). Ownership and political ecology. *Anthropological Quarterly*, 45(3), 201–205.
Wolford, W. (2007). Land reform in the time of neoliberalism: A many-splendored thing. *Antipode*, 39(3), 550–570.

34

NEOLIBERALIZATION OF NATURE

Karen Bakker

Introduction

The 'neoliberalization of nature' is one of the most controversial topics in contemporary environmental management. The past few decades have witnessed a rapid increase in the involvement of private corporations in resource ownership, biotechnological innovation, and the provision of ecosystem services. Simultaneously, markets (and market proxies) have been deployed as mechanisms of environmental governance at multiple scales. Advocates present these developments as a welcome 'greening' of capitalism that will resolve urgent environmental crises, and promise a virtuous fusion of goals of economic growth, efficiency, and environmental conservation. Opponents reject these developments as 'greenwashing' of the appropriation of resources and the environmental commons for private profit, which will deepen socio-environmental inequities.

The ensuing debate on 'neoliberal natures' has elicited sustained interest on the part of political ecologists (for edited collections, see Heynen et al., 2007; Mansfield, 2008; for a recent review, see Himley, 2008). Studies have been conducted of Bolivia's water wars, carbon emissions trading, the commodification of pets, bio-prospecting, wetland banking, international trade in human organs, and genetically modified organisms, to cite just a few examples (Bakker, 2004, 2005; Bridge, 2004; Higgins et al., 2008; Goldman, 2005; Guthman, 2004, 2007; Heynen et al., 2006; Heynen and Robbins, 2005; Liverman and Vilas, 2006; McAfee, 2003; McCarthy, 2005a, 2005b, 2005c, 2006a; McCarthy and Prudham, 2004; Mansfield, 2004, 2007a, 2007b; Perreault, 2005, 2006; Prudham, 2004, 2005, 2007; Robertson, 2004, 2007; St. Martin, 2005, 2007).

The neoliberalization of nature gives rise to a series of questions. Are markets and private ownership the solution to – or cause of – contemporary environmental crises? Does the increased involvement of private actors in environmental governance lead to improved environmental outcomes, or the opposite? Is it ethical for private companies to 'own' resources such as water, and is the sale of such resources a legitimate business transaction, or rather a matter of expropriation or 'accumulation by dispossession'? And how does the current phase of neoliberalization differ from previous phases of capitalist production of socio-natures? This chapter explores these questions, juxtaposing distinct (and at times divergent) conceptualizations of neoliberalism and nature, summarizing recent contributions by political ecologists to this debate, and mapping out a series of potential research agendas.

Defining neoliberalism

How might we define *neoliberalism*? And how to distinguish it from the cognate term *neoliberalization*? The two terms are related but distinct, and at times even incommensurate. To begin with a simple definition, neoliberalism may be understood as a doctrine: market exchange should serve as a guide for all human action (Harvey, 2005). This doctrine has intellectual roots stemming back to nineteenth-century liberalism. In the twentieth century, this doctrine first flourished in (and was actively promoted by) think tanks and specific academic disciplines (notably economics). Its uptake was expressed in policies that were adopted by 'radical' governments – such as Chile – before gradually disseminating (albeit unevenly, and subject to controversy and contestation) to countries around the world from the 1990s onwards (Peck, 2010). Neoliberalism is thus best understood as an ideal that expresses a utopian, ideational project of reorganizing capitalism at multiple scales – from the local to the global (Peck, 2010). Of course, this ideal is neither unitary nor uniform; the term 'neoliberalism' is perhaps best understood as a referent for a 'complex assemblage of ideological commitments, discursive representations, and institutional practices, all propagated by highly specific class alliances and organized at multiple geographical scales' (McCarthy and Prudham, 2004: 276).

Whereas neoliberalism is an *ideal*, neoliberalization is a *process* of reforms and ideological transformations that seeks (at least in theory) to implement the doctrine of neoliberalism. Although these neoliberalization processes are highly variegated (historically and geographically diverse and context-specific), a few generalizations can be made (Brenner et al., 2010a). First, at the heart of neoliberalization rests a set of political projects that seek to enhance conditions for capital accumulation and restore the power of economic elites at multiple scales (Harvey, 2005: 19; Peck, 2010). Second, establishing these conditions involves a (context-specific and historically and geographically variable) combination of strategies, including (although not necessarily limited to): privatization; marketization; deregulation and re-regulation ('re-tasking' the state and flexibilizing labour, for example); the implementation of market proxies in service provision; the strategic re-scaling of governance mechanisms; the exploitation of ecological and/or social fixes; and, the use of 'flanking mechanisms' (e.g. ideologically driven think tanks and campaigns) in civil society (Bakker, 2010; Castree, 2008a and 2008b; Peck et al., 2010). Third, neoliberalization is a process that discursively legitimates – while pursuing reforms in pursuit of the implementation of – neoliberalism; yet these reforms may at time conflict with the ideational norms of neoliberalism, while being subject to processes of 'counter-neoliberalization' as potential conceptual counterpoints (Brenner et al., 2010b).

An example is the set of reforms that took place in Latin America in the 1990s, which included: free trade agreements; the privatization of public utilities; the titling and marketization of property rights in land, water, forests, fisheries, and other previously publicly owned resources; relaxed environmental and labour regulations; and cuts in public expenditure (Liverman and Vilas, 2006). These reforms were influenced by other political and economic projects (such as popular nationalism in Argentina and indigenous neo-developmentalism in Bolivia), and were thus expressed differently in various countries. Across Latin America, neoliberalization generated controversy, and gave rise to counter-projects – whereby attempts have been made to '(re-)socialize' economies through processes such as welfare reform and participatory decision-making (Yates and Bakker, 2014). These counter-reforms have only been partly successful, leading to debate about whether they represent 'post-neoliberalism' or rather a kinder, gentler 'inclusive neoliberalism'. The debate in Latin America is thus reflective of debates around the world about neoliberalization, ranging from its underlying ethics and political legitimacy to its efficacy and impacts.

The neoliberalization of nature

In studying the 'neoliberalization of nature', scholars working at the nexus of political economy and political ecology take into account the definitions offered in the preceding section. But many also emphasize the fact that neoliberalization is inherently an environmental project. In part, this is because neoliberalization projects often entail environmental governance reform. More fundamentally, as explored below, there are important ideological interconnections between neoliberalization and (certain strands of) contemporary environmentalism (McCarthy and Prudham, 2004); simultaneously, environmentalist concerns often inform the most powerful sources of political opposition to neoliberalization. But the intersections between neoliberalization and the environment are more than merely discursive, ideological, or political: they are deeply material. Neoliberalization, from a political ecology perspective, is co-constituted by evolving relationships with biophysical natures. Of course, this is not a new insight: capitalism (of which neoliberalism is one variant) is predicated upon our metabolism of nature, as long recognized by Marxian political economists (e.g., Benton, 1996; Burkett, 1999; Foster, 2000; Smith, 2008). Indeed, there are obvious parallels between contemporary accounts of 'accumulation by dispossession' and historical enclosures under earlier phases of capitalism (see also Glassman, 2006; Harvey, 2003). Neoliberalism, just like capitalism, is inherently an environmental project; and neoliberalization is inherently metabolic (in the Marxian sense).

What (if anything) is distinctive about neoliberal natures? In other words, why talk about neoliberalism; why not just talk about capitalism? The regulation-theoretic accounts of neoliberalism as a political and economic project that rose to the fore of public policy in the 1970s as a (politically contested) response to the crisis of Fordism do not fully capture socio-technological innovations and political ecological factors that enabled neoliberalization of nature to emerge. To fully appreciate the uniqueness and unprecedented scope of the project of neoliberalizing nature, we also have to consider the history of the modern environmental movement – and to understand the rise of liberal environmentalism as a parallel ideational project.

A key turning point in the rise of neoliberal natures came about during the 1970s: a decade during which widespread awareness emerged that an instrumentalist approach to nature as a 'source' for resources and 'sink' for wastes was reaching (human-perceived) limits. As mass environmental movements emerged, oil prices peaked, and debates over the 'limits to growth' reached a zenith, governments around the world passed a series of environmental laws and regulations that constrained capital accumulation. At the same time, concerns over resource scarcity came to the forefront of public debate.

In the ensuing decades, environmental governance has been progressively captured by the doctrine of 'liberal environmentalism', which asserts the belief in the 'compatibility of environmental concern, economic growth, the basic tenets of a market economy, and a liberal international order' (Bernstein, 2000). This doctrine gradually achieved near-hegemonic status from the late 1980s onwards (albeit at different paces and in distinct ways across locales). Seminal moments included the 1984 Brundtland Report (*Our Common Future*), the Rio Environment and Development Summit in 1992, and the Johannesburg Summit in 2002 (Bernstein, 2000). Simply put, whereas market principles were often viewed in opposition to environmental protection and conservation in the mid-twentieth century, they had by the end of the century become reconciled with economic growth and entrenched in mainstream environmental policy as emblematized in the doctrine of sustainable development (Hartwick and Peet, 2003). As Margaret Thatcher famously stated: 'We are all greens now.'

The response of capital was diverse (and is ongoing), and included four noteworthy strategies. First, a new, intense phase of privatization and 'accumulation by dispossession' began. Its most recent, well-publicized iteration is a series of 'land grabs' and 'water grabs' by globalized corporations, but the phenomenon is actually much more widespread and persistent, covering a broad range of primary commodities, from forests to fisheries.

Second, corporations sought to reduce costs by globalizing 'ecological fixes' (also known as 'environmental externalities'); moving polluting industries, for example, to developing countries where regulations were weaker or unenforced. New commodity chains were extended worldwide, and flows of foreign direct investment followed, encouraged by the progressive deregulation of global trade. Environmental externalities accumulated in low-income countries (such as Mexico's *maquiladoras*), where labour costs were also cheaper.

Third, corporations sought to profit from these externalities, through developing new technologies and business models that (for example) turned waste into a resource. The corporate pursuit of sustainable development (to use Vandana Shiva's pithy phrase) began to frame environmental externalities as opportunities for profit. For example, the 'GreenTech' industry gave rise to a fascinating series of technological innovations (such as 'closed loop' water supply systems in which heat is extracted from sewage before it is turned into biofertilizer).

Fourth, corporations sought 'new frontiers', attempting to commodify new types of socio-natures, from the global climate to genes. Socio-technological innovation allowed for the development of new strategies of marketization. In other words, new practices of capital accumulation were brought to bear on new types of socio-natures (e.g., biocultural entities), such as human organs (Scheper-Hughes, 2005).

Moreover, and in support of the four preceding strategies, the neoliberalization of nature was supported by a global project of re-regulation, mediated by international financial organizations. Development assistance included projects that reformed national laws and policies, associated with new financing mechanisms such as the Global Environment Facility (created in 1991) (Goldman, 2005). Although locally differentiated, neoliberalization of the environment was (like neoliberalization more generally) underpinned and enabled by generic – and even formulaic macro-scale processes of re-regulation: for example, the so-called 'ten neoliberal commandments' of the Washington Consensus (Macdonald and Ruckert, 2009; Panizza, 2009; Ruckert, 2006).

The neoliberalization of nature may thus be defined as (1) an emergent regime of capital accumulation that redefines and co-constitutes socio-natures; and (2) a mode of regulation congruent (but not synonymous) with political economic neoliberalization. Its political traction arises, in part, because proponents purport to present a solution to the twinned economic-environmental crises which capitalism has played a role in creating. These crises arise from the social construction and political mediation of deteriorating (and interrelated) environmental, social, and economic conditions, raising questions about limits: on the one hand, the limits of the neoliberal natures approach to resolving environmental problems; and, on the other hand, the cognitive and material limits of our models of resource exploitation and our instrumentalist approach to nature, a point to which I will return in the conclusion.

The limits to the neoliberalization of nature

As explored above, political ecologists have been at the forefront of academic scholarship on the neoliberalization of nature. Their research has been characterized by detailed case studies that dispense with the notion of neoliberalism as an ideal-type, through careful specification of the specific processes at work in 'actually-existing neoliberalisms'. An important contribution of this body of research has been its elucidation of the 'limits' – political, technical, socio-economic – to the neoliberalization of nature.

These limits arise, in the first instance, because 'nature' is neither passive nor inert. Those studying the neoliberalization of nature frequently characterize resources and other socio-natures as 'uncooperative' and 'unruly', insofar as they impose constraints upon human action and/or barriers to profit – hence limiting capital accumulation. The most gripping case studies of the neoliberalization of nature are often those which present a nuanced analysis of the tentative, multi-faceted, and not always successful attempts to neoliberalize nature – a process often fraught with challenges because of the contradictions which arise through the need to articulate labour and accumulation strategies with ecological processes in specific biophysical settings. In other words, one might argue that scholars of neoliberal nature think 'relationally'; and, moreover, that they have a relational view of production which attempts to grapple with the mutual construction of the human and non-human (Robbins, 2004). This approach is evidenced, for example, in studies of 'adaptive' labour processes involving 'cultivated' resources – such as Robbins' account of the enrollment of grass seeds in the construction of the North American lawn (Robbins, 2007), or Kloppenberg's analysis of plant biotechnology and the progressive commodification of the seed (Kloppenberg, 2005). In other words, the limits to nature's neoliberalization arise not only because neoliberalism takes place within existing political economic formations with which it has an antagonistic relationship, but also because of the ecological/biophysical constraints within which political and economic processes of production must operate.

A second theme in this research explores the limits of the neoliberalization of nature as a project of sustainable development – through documenting the impacts of specific 'neoliberal nature' initiatives on both humans and ecosystems. This body of research offers a rebuttal to proponents of 'free market environmentalism' (or 'green neoliberalism', as opponents term it).[1] In adopting an interdisciplinary approach – which involves critically and selectively adopting methods and frameworks from both the biophysical and social sciences – political ecologists are able to provide a robust account of the ecological, economic, and political effects of nature's neoliberalization. For example, studies of the introduction of market regimes (such as tradeable quotas) in fisheries might incorporate the concerns of ecologists with depletion, extirpation, and decreasing mean trophic levels (the so-called 'fishing down the food chain' effect) with consumers' concerns about health and safety regarding fish consumption (Mansfield, 2004). Thus, rather than abstract macro-scale studies, political ecologists regularly engage in 'documentation, explanation and engagement in actually existing, complex and non-contiguous [socio-ecological-economic] formations' (Rocheleau and Roth, 2007). This has enabled political ecologists to build up an impressive body of scholarship that convincingly demonstrates that the neoliberalization of nature fails to achieve the objectives set forth by proponents. Political ecology research thus offers the insight that the neoliberalization of nature is a limited strategy for the pursuit of sustainable development goals.

A third insight from political ecological research into neoliberal natures stems from its conceptual contributions to debates over neoliberalization. A number of studies have causally linked the limits identified above to the process of re-regulation (or 'roll-out neoliberalization'), in which neoliberalization is accompanied by an intensification of regulatory intervention by the state – rather than deregulation. Studies of the neoliberalization of nature, in other words, offer a convincing explanation of why neoliberalization might be subject to 'limits': in order to sustain neoliberalization, it must be accompanied by an intensification of facilitative government activity, thereby countering widespread assumptions of the 'retreat of the state' under neoliberalism (e.g., Lockie and Higgins, 2007). In offering this insight, political scholarship on neoliberal natures contributes to refuting hegemonic, oversimplified understandings of neoliberalization writ large.

Future research directions

In this final section of the chapter, I explore three potential research directions for political ecological research on the neoliberalization of nature. The first topic arises as a response to a compelling critique: this literature, taken as a whole, is composed of disparate, unconnected case studies (Castree, 2005, 2008a, 2008b). Castree argues that the case study approach inhibits the ability of political ecologists working on neoliberal natures to 'connect the dots' between cases. In particular, he argues that little work has been done on identifying commonalities in drivers, patterns, and effects of neoliberalization processes – such as the privileging of private property rights, market-mimicking regulatory strategies, state-led market proxies, and commodification strategies. In his account, political ecologists are collectively unable to account for variegation; specifically, they are unable to generate convincing explanations of the neoliberalization of nature as a historically and geographically differentiated, yet global (or at least translocal) phenomenon.

This is troubling, because there is tremendous variation in the articulation of neoliberalism with different types of socio-natures (Bakker, 2009). Take, for example, the case of primary commodities: whereas neoliberalization has been widespread and far-reaching in some cases (such as coffee), it has been much more restricted in others (such as water) (Bakker, 2005; Daviron and Ponte, 2005; Talbot, 2004). Other cases, such as oil, are relatively mixed: nationalization has remained important (although often via 'hybrid' forms), but other dimensions of neoliberal economic strategies – a shift from state to private control of oil extraction, intensification of extraction rates, and price deregulation – have been widely implemented (Le Billon and Cervantes, 2009). In other words, the process of neoliberalization is variegated (Brenner et al., 2010a, 2010b; Peck and Theodore, 2007). Hence, we cannot adequately explain processes of neoliberalization if we have not accounted for the commonalities and differences in patterns of 'actually-existing neoliberalisms' across different types of resources in different places. Having identified these commonalities and differences, we need to theorize their emergence in the context of distinct neoliberal experiments. The problem with a case study approach (specifically, site- and resource-specific studies of neoliberalization) is that it enables differentiation to be empirically documented, but foregoes an analysis of the systematic production of geo-historical differentiation, insofar as it neglects to articulate local cases with translocal neoliberalization processes. As a result, political ecologists who study neoliberal natures limit their ability to explain when, where, and why neoliberal projects are implemented, succeed, and/or fail. An appropriate response to this critique, as I have argued elsewhere, is to develop conceptual frameworks in which the neoliberalization of nature is relationally defined – in articulation with the different biophysical characteristics of resources, as well as labour practices, consumption processes, and even affective relationships (Bakker, 2010). This implies cross-fertilization, for example, between political ecology, cultural geographies of commodities, and political economy.

A second, related future research direction entails closer engagement with a broader range of 'natures'. This suggestion arises from the critique that the 'neoliberal natures' literature is overly narrow, given that it focuses largely on primary commodities. A recent review of the literature (Himley, 2008) suggests that this pattern is relatively consistent: studies of neoliberal natures have tended to focus on the encroachment of capitalist economic relations on what we conventionally delimit as 'the environment' and 'resources'. Other types of 'nature' receive scant attention (although for an exception see Prudham, 2007). Moreover, studies of neoliberal natures implicitly adopt a humanist view of the subject, and an associated anthropocentric conception of political subjectivity: the neoliberalization of human bodies is barely addressed in this literature (in part, one suspects, because political ecologists implicitly define humans as separate from nature). The result, critics argue, is an overly narrow set of case studies, an overly constrained view of agency, and a failure to confront the political subjectivity of socio-natures (Braun, 2008). As Braun argues: the 'concerted attempt among political economists to understand the ways in which non-human nature resists its incorporation into particular political economic and spatial forms' is limited by its failure to incorporate the 'nonhuman ... as a constitutive element of social and economic life' (Braun, 2008: 668). His argument claims that it is no mere coincidence that the majority of scholars operating from a political economic perspective have generally shied away from studying biocultural entities or animals because they raise questions of agency of the co-constitution of humans and non-humans that are difficult to handle within the conceptual frameworks typically employed in this literature. The (inadvertent) consequence is a failure to address the full scope of environmental processes and socio-natural entities subsumed within processes of neoliberalization. Of course, notable exceptions to this critique include scholars who have sought to grapple with the ways in which the non-human (including plants, and by extension the ecological conditions of production) is constitutive of both economics and social/cultural forms (see, for example, Eaton, 2009; Henderson, 2003; Kloppenburg, 2005). Nonetheless, the majority of conventional framings of neoliberalism remain largely silent on questions of the materiality of neoliberalization; or, if alert to ecological issues, tend to focus rather narrowly on resources and primary commodities.

How might scholars extend the focus of neoliberal natures research beyond its focus on resources and primary commodities, and take up the post-humanist challenge (Bakker, 2010)? How might, in other words, we respond to Braun's challenge to scholars of neoliberal nature: to adopt a non-anthropocentric view of the agency of nature, and interrogate the status of non-humans as political subjects – while developing a more comprehensive account of nature's neoliberalization? We might consider turning for inspiration to work in animal geographies and cultural geographies of commodities. In these literatures, the neoliberalization of nature is framed more comprehensively: simultaneously economic and emotional, socio-cultural and material, and even libidinal and affective (see, for example, Adam, 2005; Guthman, 2009; Mansfield, 2003; Nast, 2006; Chapter 43, this volume; Chapter 9, this volume). This is, of course, only one suggestion: there are many potential strategies to consider. But it is likely, given this critique, that future scholarship will more centrally engage with the challenge to the humanist model of the subject upon which studies of neoliberal natures are conventionally predicated; and, in turn, the modern, Western notion of distributive justice (from which non-humans are generally excluded) with which it is associated. Of course, there is nothing novel in this insight; here, I simply note that this issue is one that scholars of neoliberal natures – particularly those of a political economic persuasion – have not yet addressed in a sustained manner.

Conclusion

My goal in this chapter has been two-fold: to summarize some of the most outstanding contributions of political ecologists to the analysis of the neoliberalization of nature; and to suggest fruitful future research directions. In the future, scholars will no doubt engage more comprehensively with the multiple entanglements between socio-natures and capital under neoliberal modes of governance, and grapple with ontologies and epistemologies for engaging with the co-presence of the human and non-human – both animate 'nature' and inanimate 'things'. It is to be hoped that this, in turn, will deepen our contributions to what Neil Smith termed a 'new political theory of nature', leading to more fruitful engagement with post-neoliberalism – not as an assertion of a putative political economic transition (of whose existence we should, in my opinion, be sceptical – cf. Peck et al., 2010), but rather as a thought-experiment fully attentive to emergent political, cultural, socio-economic, and socio-natural forms.

The urgency is undeniable in the face of multiplying environmental crises of rising intensity. There is nothing particularly new, of course, about fears of environmental threats. But the modern notion of environmental crisis derives its potency, in part, from the discursive mediation of popular perceptions of 'global' threats – albeit locally mediated and experienced – that have come to the fore in both policy and public discourse over the twentieth century (Guha, 2000). Framed in this way, the debate over neoliberal natures takes on a qualitatively new dimension in which psychological and political struggles over 'ecological fixes' play an important role (Bakker, 2004). In Beck's formulation, this is characterized as the negotiation of the risks posed by environmental 'bads' (rather than goods), spurred by the fear of their effects which comes to dominate collective politics and individual psyches in industrialized societies (Beck, 1992). The threat posed by capitalist modernity is not merely to an existing social order, but rather a deeply existential threat to the basis of life itself.

In making the connection between psyche and society, Beck's notion of the 'risk society' suggests a reinterpretation of Polanyi's concept of the 'double movement' – in which capitalism oscillates between dis-embedding and re-embedding economic activity from society (Polanyi, 2001). Simply put, Polanyi's account might usefully be supplemented by taking 'nature' into account. Capitalism threatens to alienate and dis-embed human economies and societies from the socio-natural relations through which they are constituted. If the neoliberalization of nature disembeds our economies from the socio-ecologies upon which we collectively depend, then the threat is existential – simultaneously material and psychological. The task of counter-neoliberalization – restraining technoscience, reinventing capitalism, and re-imagining our worldviews, scalar politics and scalar ontologies of socio-nature – thus takes on an added urgency. We might thus reframe the 'double movement' as a set of intertwined struggles over the material conditions of reproduction (in its broadest sense) as well as production.

Reframing the 'double movement' in this way is predicated upon what some might view as a heterodox conceptualization of neoliberalism, and an equally heterodox genealogical account of the emergence of 'green neoliberalism'. I would argue that this is necessary if we are to produce comprehensive accounts of the neoliberalization of socio-natures, broadly defined. In closing, let me emphasize that I am not suggesting that we should abandon the concepts of neoliberalism and nature, nor dilute them to the point where they are all-inclusive, and thus analytically unhelpful. Rather, my suggestions are aimed at stimulating dialogue on better strategies for confronting the conceptual polysemism of both 'neoliberalization' and 'nature', and on better equipping scholars to anticipate and explore alternatives to neoliberal imaginaries.

Note

1 While the majority of studies point to the negative impacts of the neoliberalization of nature, it is important to note that the literature is not one-sided. For example, some studies highlight the positive environmental outcomes of neoliberalization, in the context of tradeoffs between the environment, consumers, and labour (Bakker, 2004).

References

Adam, B. (2005) Constructing the neoliberal sexual actor: Responsibility and care of the self in the discourse of barebackers. *Culture, Health and Sexuality* 7(4): 333–346.

Bakker, K. (2004) *An Uncooperative Commodity: Privatizing Water in England and Wales.* Oxford: Oxford University Press.

Bakker, K. (2005) Neoliberalizing nature? Market environmentalism in water supply in England and Wales. *Annals of the Association of American Geographers* 95(3): 542–565.

Bakker, K. (2009) Neoliberal nature, ecological fixes, and the pitfalls of comparative research. *Environment and Planning A* 41: 1781–1787.

Bakker, K. (2010) Debating green neoliberalism: The limits of 'neoliberal natures'. *Progress in Human Geography* 34: 715–735.

Beck, U. (1992) *Risk Society: Towards a New Modernity.* Vol. 17. Sage.

Benton, T. (1996) *The Greening of Marxism.* New York: Guilford Press.

Bernstein, S. (2000) *The Compromise of Liberal Environmentalism.* New York: Columbia University Press.

Braun, B. (2008) Environmental issues: Inventive life. *Progress in Human Geography* 32: 667–679.

Brenner, N., Peck, J. and Theodore, N. (2010a) Variegated neoliberalization: Geographies, modalities, pathways. *Global Networks* 10(2): 1–41.

Brenner, N., Peck, J. and Theodore, N. (2010b) After neoliberalization? *Globalizations* 7: 327–345.

Bridge, G. (2004) Mapping the bonanza: Geographies of mining investment in an era of neoliberal reform. *The Professional Geographer* 56(3): 406–421.

Burkett, P. (1999) *Marx and Nature: A Red and Green Perspective.* New York: St. Martin's Press.

Castree, N. (2005) The epistemology of particulars: Human geography, case studies and 'context'. *Geoforum* 36(5): 541–544.

Castree, N. (2008a) Neoliberalising nature: The logics of deregulation and reregulation. *Environment and Planning A* 40: 131–152.

Castree, N. (2008b) Neoliberalising nature: Processes, effects, and evaluations. *Environment and Planning A* 40: 153–173.

Daviron, B. and Ponte, S. (2005) *The Coffee Paradox: Global Markets, Commodity Trade, and the Elusive Promise of Development.* London: Zed Books.

Eaton, E. (2009) Getting behind the grain: the politics of genetic modification on the Canadian Prairies. *Antipode* 41(2): 256–281.

Foster, J.B. (2000) *Marx's Ecology: Materialism and Nature.* New York: Monthly Review Press.

Glassman, J. (2006) Primitive accumulation, accumulation by dispossession, accumulation by 'extra-economic' means. *Progress in Human Geography* 30(5): 608–625.

Goldman, M. (2005) *Imperial Nature: The World Bank and Struggles for Social Justice in the Age of Globalization.* New Haven: Yale University Press.

Guha, R. (2000) *Environmentalism: A Global History.* New York: Longman.

Guthman, J. (2004) *Agrarian Dreams: The Paradox of Organic Farming in California.* Berkeley: University of California Press.

Guthman, J. (2007) The Polanyian way? Voluntary food labels as neoliberal governance. *Antipode* 39(3): 456–478.

Guthman, J. (2009) Teaching the politics of obesity: Insights into neoliberal embodiment and contemporary biopolitics. *Antipode* 41(5): 1110–1133.

Hartwick, E. and Peet, R. (2003) Neoliberalism and nature: The case of the WTO. *Annals of the AAPPS* 590: 188–211.

Harvey, D. (2003) *The New Imperialism.* Oxford University Press.

Harvey, D. (2005) *A Brief History of Neoliberalism.* Oxford: Oxford University Press.

Henderson, G. (2003) *California and the Fictions of Capital.* Philadelphia: Temple University Press.

Heynen, N. and Robbins, P. (2005) The neoliberalization of nature: Governance, privatization, enclosure and valuation. *Capitalism, Nature, Socialism* 16(1): 5–8.

Heynen, N., Perkins, H.A. and Roy, P. (2006) The political ecology of uneven urban green space: The impact of political economy on race and ethnicity in producing environmental inequality in Milwaukee. *Urban Affairs Review* 42(1): 3–25.

Heynen, N., McCarthy, J., Prudham, W.S. and Robbins, P. (eds) (2007) *Neoliberal Environments: False Promises and Unnatural Consequences.* New York: Routledge.

Higgins, V., Dibden, J. and Cocklin, C. (2008) Neoliberalism and natural resource management: Agri-environmental standards and the governing of farming practices, *Geoforum* 39(5): 1776–1785.

Himley, M. (2008) Geographies of environmental governance: The nexus of nature and neoliberalism. *Geography Compass* 2(2): 433–451.

Kloppenberg, J. (2005) *First the Seed: The Political Economy of Plant Biotechnology.* Madison: University of Wisconsin Press. 2nd ed.

Le Billon, P. and Cervantes, A. (2009) Oil prices, scarcity and geographies of war. *Annals of the Association of American Geographers* 99(5): 836–844.
Liverman, D. and Vilas, S. (2006) Neoliberalism and the environment in Latin America. *Annual Review of Environment and Resources* 31: 327–363.
Lockie, S. and Higgins, V. (2007) Roll-out neoliberalism and hybrid practices of regulation in Australia agrienvironmental governance. *Journal of Rural Studies* 23(1): 1–11.
Macdonald, L. and Ruckert, A. (2009) Post-neoliberalism in the Americas: An introduction. In: Macdonald, L. and Ruckert, A. (eds) *Post-Neoliberalism in the Americas*. Basingstoke: Palgrave Macmillan, 1–20.
Mansfield, B. (2003) From catfish to organic fish: Making distinctions about nature as cultural economic practice. *Geoforum* 34(3): 329–342.
Mansfield, B. (2004) Neoliberalism in the oceans: 'Rationalization', property rights, and the commons question. *Geoforum* 35: 313–326.
Mansfield, B. (2007a) Privatization: Property and the remaking of nature–society relations: Introduction to the special issue. *Antipode* 39(3): 393–405.
Mansfield, B. (2007b) Property, markets, and dispossession: The western Alaska community development quota as neoliberalism, social justice, both, and neither. *Antipode* 39(3): 479–499.
Mansfield, B. (ed.) (2008) *Privatization: Property and the Remaking of Nature-Society Relations*. Oxford: Blackwell.
McAfee, K. (2003) Neoliberalism on the molecular scale: Economic and genetic reductionism in biotechnology battles. *Geoforum* 34: 203–219.
McCarthy, J. (2005a) Commons as counterhegemonic project. *Capitalism, Nature, Socialism* 16(1): 9–24.
McCarthy, J. (2005b) Scale, sovereignty, and strategy in environmental governance. *Antipode* 37(4): 731–753.
McCarthy, J. (2005c) Devolution in the woods: Community forestry as hybrid neoliberalism. *Environment and Planning A* 37(6): 995–1014.
McCarthy, J. (2006a) Neoliberalism and the politics of alternatives: Community forestry in British Columbia and the United States. *Annals of the Association of American Geographers* 96(1): 84–104.
McCarthy, J. (2006b) Rural geography: Alternative rural economies – the search for alterity in forests, fisheries, food, and fair trade. *Progress in Human Geography*. 30(6): 803–811.
McCarthy, J. and Prudham, S. (2004) Neoliberal natures and the nature of neoliberalism. *Geoforum* 35: 275–283.
Nast, H. (2006) Loving ... whatever: Alienation, neoliberalism and pet-love in the twenty-first century. *Acme* 5(2): 300–327.
Panizza, F. (2009) *Contemporary Latin America: Development and Democracy beyond the Washington Consensus*. London: Zed Books.
Peck, J. (2010) *Constructions of Neoliberal Reason*. Oxford: Oxford University Press.
Peck, J. and Theodore, N. (2007) Variegated capitalism. *Progress in Human Geography* 31(6): 731–772.
Peck, J., Theodore, N. and Brenner, N. (2010) Postneoliberalism and its malcontents. *Antipode* 41(6): 1236–1258.
Perreault, T. (2005) State restructuring and the scale politics of rural water governance in Bolivia. *Environment and Planning A* 37: 263–284.
Perreault, T. (2006) From the Guerra del Agua to the Guerra del Gas: Resource governance, neoliberalism and popular protest in Bolivia. *Antipode* 38(1): 150–172.
Polanyi, K. (2001) *The Great Transformation: The Political And Economic Origins Of Our Time*. Beacon Press.
Prudham, S. (2004) Poisoning the well: Neoliberalism and the contamination of municipal water in Walkerton, Ontario. *Geoforum* 35: 343–359.
Prudham, S. (2005) *Knock On Wood: Nature as Commodity in Douglas-fir Country*. New York: Routledge.
Prudham, S. (2007) The fictions of autonomous invention: Accumulation by dispossession, commodification and life patents in Canada. *Antipode* 39(3): 406–428.
Robbins, P. (2004) *Political Ecology: A Critical Introduction*. Oxford: Wiley-Blackwell.
Robbins, P. (2007) *Lawn People: How Grasses, Weeds, and Chemicals Make Us Who We Are*. Philadelphia: Temple University Press.
Robertson, M. (2004) The neoliberalization of ecosystem services: Wetland mitigation banking and problems in environmental governance. *Geoforum* 35: 361–373.
Robertson, M. (2007) Discovering price in all the wrong places: The work of commodity definition and price under neoliberal environmental policy. *Antipode* 39(3): 500–526.

Rocheleau, D. and Roth, R. (2007) Rooted networks, relational webs and powers of connection: Rethinking human and political ecologies. *Geoforum* 38: 433–437.
Ruckert, A. (2006) Towards an inclusive-neoliberal regime of development: From the Washington to the post-Washington Consensus. *Labour, Capital, and Society* 39: 34–67.
Scheper-Hughes, N. (2005) The last commodity: Post-human ethics and the global trade in 'fresh' organs. In: Ong, A. and Collier, S. (eds) *Global Assemblages: Technology, Politics and Ethics as Anthropological Problems*. Oxford: Blackwell, 145–168.
Smith, N. (2008) *Uneven Development: Nature, Capital, and the Production of Space*. University of Georgia Press.
St. Martin, K. (2005) Disrupting enclosure in New England fisheries. *Capitalism, Nature, Socialism* 16(1): 63–80.
St. Martin, K. (2007) The difference that class makes: Neoliberalization and non-capitalism in the fishing industry of New England. *Antipode* 39(3): 527–549.
Talbot, J. (2004) *Grounds for Agreement: The Political Economy of the Coffee Commodity Chain*. Oxford: Rowman & Littlefield.
Yates, J. and Bakker, K. (2014) Neoliberalism and post-neoliberalism in Latin America. *Progress in Human Geography* 38(1): 62–91.

35

ENVIRONMENTAL GOVERNANCE

Political ecology and the state

Morgan Robertson

Introduction

Political ecologists have had an ambivalent relationship with the state and with what has been formalized as "state theory." They are generally happy to research and write about state employees, state policies, state economic and development strategies, global economic entanglements and migration between states – but have historically been reluctant to be explicit about what the state itself *is*. This is mostly due to a sense on the part of early political ecologists that the focus on states as such was precisely the problem that motivated the approach: to see the global political economy at the scale of governments and capital was to share the view of the bureaucrat in the airplane. This blinkered-ness was precisely what Blaikie and Brookfield (1987) wanted to remedy by providing an explanatory framework that began with the land manager and scaled out from there, as necessary and warranted by field-based ethnographic and environmental data.

This chapter traces the development of political ecology's approach to the state and governance, and how political ecologists understand both the formal apparatus of government and the distributed networks of governance in environmental issues and management. Political ecologists have engaged with a variety of state theories in attempting to account for environmental conflict: from neo-Weberian institutionalism to neo-Marxist "regulationist" approaches to ethnographic work on the people who inhabit and constitute state institutions. A common thread unites them: the task of constituting the environment as a field for state action is understood as a state project of the highest importance, and is likely to remain so under the growing threats of water conflicts, climate change, and the military concern for "environmental security." Nature is not simply the surface on which state strategy plays out, but is an active and internal part of these strategies. Political ecology's analytic tendency to challenge material and discursive dualisms will serve the analysis of governance very well as we seek to understand nature's involvement in states' political projects and the implications of governance for environmental outcomes.

Sauerian premonitions and the Mosquito Coast

> Land planning can certainly not be claimed as the geographer's discipline, nor as a discipline in any sense, since it must obviously be primarily projected from a specific theory of the state.
>
> *(Sauer 1941a, 3)*

Political ecology's Sauerian roots may go some way towards explaining its traditional ambivalence about the state and state theory. In a discipline looking back towards Sauer's anti-determinist materialism and dependence on anthropology for both its theoretical and fieldwork stance, the state and government were seen to be, at best, irrelevant units of analysis: "We are not concerned with universalized economic man, family, society, or economy, but with the comparison of localized patterns, or areal differentiations" (Sauer 1941a, 6). Abstract units of social organization were supposed to fall away to reveal relationships developed through material interactions with land and resources. The preferred spatial unit of analysis was of course the "culture area" and these were unlikely to be coextensive with state boundaries. But the concern with political economics was always there in Sauer, as he insisted that culture areas be defined not just with reference to ecology, but also with reference to economic flows, and nested ones at that – giving us a groundwork for economic globalization far *avant la lettre*: "A culture area of one order may be recognized by the dominance of a single economic complex. A culture area of a superior order may be determined by the interdependence of a group of areal economies" (1941a, 12). Of primary interest, he held, was the condition under which such economic zones achieved stability against one another. But despite his interest in fluid ecological-economic zones, the state was often the vessel into which he poured his analysis, as with the Aztec states and successor Mexican state in his analysis of "The personality of Mexico" (1941b). And in his famous essay on "The agency of man on the earth," the state's shepherding of capitalist transformation of landscapes stands in the background of the colonial and post-war changes he laments – his distaste for the changes in agriculture and industrial policy that flow from "the requisite growing power of government" (1956, 68) is palpable. One can also sense, by the essay's end, Sauer's relief that analysis of the state will be the next generation's problem and not his.

This tension between the site-focused analysis of material cultural ecology on the one hand, and acknowledgment of the site's connections to increasingly global scales of economic and political organization on the other, has kept political ecology circling state theory and political geography but rarely fully engaging with it. As political ecology developed from its roots in cultural ecology, the analytic site usually remained small, and its focus remained the land or resource manager. The analytic sites of political geography and political science, after all, were polities and abstractions whose inadequacy was the *raison d'être* of political ecology in the first place.

Sauer's unease with thinking beyond the traditional culture area, but the necessity of doing so, was the productive tension within two Central American studies, both published in 1979, and both enduringly important to political ecology. In Bernard Nietschmann's study of Miskito subsistence in coastal Nicaragua, one of his six principles for analyzing the disruption of traditional resource economies was that household livelihoods become "governed by distant economic and political decisions and conditions" (1979, 12), and he directly points at UN resolutions and the state interests of their signatories. But for Nietschmann, the fraying edge of Miskito identity trailed off into the global sea, and the pole of the Nicaraguan state was visible only as the lights of Managua over the horizon. For all that Nietschmann desired the integration of global economic perspective into political ecology, his state theory was, like Sauer's, one of absence and anxiety.

In William Durham's classic work subtitled "Ecological origins of the soccer war," however, the state and state interests are front and center. His convincing demolition of "overpopulation" narratives and characterization of ecological marginalization emerge within the nationalistic strutting and policymaking of Honduras and El Salvador as actor-like states. Village-level subsistence, for Durham, is shaped within the context of national land reform and the interplay between the Honduran government and the United Fruit Company. To understand the ecological degradation that caused mass migration and ultimately war, Durham moves between an institutionalist and interest-group analysis at the national scale and a textured village-scale resource-use study of Tenancingo, El Salvador. It was not enough to gesture at the heterogeneity of external forces pulling at Tenancingo – there had to be another pole organizing those forces, and for Durham it was the state: "one must ... consider the dynamics of resource availability in a given community against the larger national context" (1979, 100). Durham's theory of the state is that of something irreducibly separate from the village site, which is another way of constraining anxiety over the relationship between them. The state is analytically present in this work, albeit as a distant set of institutional actors, exogenous to the study site.

Government and first-generation political ecology

None of these early political ecologies show any attempt to wrestle with state theory in the way that they were clearly wrestling with Marxian economics or behavioral anthropology. When the moment came in 1985 to be explicit about what *kind* of state stood in relation to the land manager, it came with Piers Blaikie (1985, 83) endorsing the 1969 work of Marxist theorist Ralph Miliband as a "simple" and "serviceable" concept of the state. Miliband took a position in which the state is best understood anti-structurally as a collection of situated individuals, best approached at the level of personal ties and interests, rather than as an entity functioning coherently to reproduce capitalism. It is not clear whether Blaikie was actively siding with Miliband over the work of Nicos Poulantzas who (in what became known as the "Miliband–Poulantzas" debate on state theory; see Jessop 1990, 25–30) adopted a more structural but recognizably Gramscian concept of the state as strategic in securing hegemony, and having interests distinguishable both from capital and its own individual agents. Blaikie seems to alternate between the views: he describes the state as a thing that "speaks and acts," and yet he cites Miliband saying "the state is not a thing [and] does not, as such, exist" (1985, 84). Blaikie examines governance and the achievement of the consent of the governed, but his discussion of the state looks very much like the village-level sites of Nietschmann: the state, at all levels, is composed of individuals with contexts and interests, and it is the motivations of individuals and empirically documented collectives that must be documented. As an institution, the state is simply an abstraction invoked by those who attempt to secure hegemony – a distraction to the dedicated field-worker. Blaikie all but scoffs at liberal notions of disinterested bureaucrats and institution-building: the job of the political ecologist is to see past such fictions.

In 1987, Blaikie and Brookfield turn to the state in the second sentence after their famous definition of political ecology – "the concerns of ecology and a broadly-defined political economy" – and it becomes central to their focus on marginality, one of political ecology's major themes. The complexity of their socio-ecological concept of marginality, uniting as it does soil fertility with socioeconomic standing in the same analytic frame, can obscure the question of what the site or subject is marginal *to*. The answer, for them, is *state power*: immediately upon invoking the need to understand the state, they turn to introduce marginality by saying: "The allocation of state-controlled resources in rural development therefore usually disfavors the physical and social margin" (1987, 18). Marginality is created relative to the exercise of state power: this is a core–periphery continuum answer to the unanchored exogeneities of Sauer and Nietschmann and the bifurcation of Durham. Although Blaikie and Brookfield understood state power to be a key driver in the complicated concept of marginality, their attention was squarely on the spaces of marginality and therefore, by definition, not on the state.

Government to the fore

Two very different books were published in 1996, marking strikingly different paths forward in political ecology's second generation. The first was Raymond Bryant and Sinéad Bailey's book *Third World Political Ecology*, and it opened by gently scolding political ecologists for remaining so resolutely local in their focus. Bryant and Bailey urged attention to the scale of the state and the global, where political and economic forces were constituted and driven and about which political ecologists seemed to have little to say: "Although rarely stated, the lodestar of most political ecologists is a relatively decentralised political system that blends socialism and anarchism in the pursuit of social justice" (19). Bryant and Bailey are primarily concerned with reforming the naïve, or at least localist, notion that power is best understood at the level of the individual land manager. Their preference is to focus instead on institutions and the venues in which the setting of institutional priorities takes place. The physical environment and its management is a "manifestation of power relations" which sometimes can best be observed operating far from a specific site of interest – sometimes in a national capital, sometimes in a global city.

Bryant and Bailey's book is grounded on the idea that political ecology must leave the village and focus on state action, and to do so it must reach out to current work in political science. Quoting political scientist K.J. Walker (1989, 32) on the emerging idea of the fractured and polyvocal environmental state, they argue that:

> At the heart of any explanation of why states have been so destructive environmentally must be the recognition of a central paradox in the state's function. In effect, there is "an inherent, continuing potential for conflict between the state's roles as developer, and as protector and steward of the natural environment on which its existence ultimately depends."
>
> *(55)*

In other words, there is simply no point in trying to understand environmental change in an Angolan or Costa Rican or Italian town without understanding the institutional disposition in Luanda, San Jose, or Rome, or global economic strategies formulated in New York or Geneva. This is the strong form of an argument that political ecology is essentially still the political economics of the environment, and their insistence on the abandonment of land-manager-level ethnography was not (and could not yet be) paired with a notion of an ethnography of the state or global institutions.

The other book released that year was Richard Peet and Michael Watts' *Liberation Ecologies*, an edited volume that wound the threads of political ecology together with post-structural social theory. Peet and Watts sought to emphasize that, from the beginning, political ecology was "not inspired by the isolated rural communities studied by Rappaport but by peasant and agrarian societies in the throes of complex forms of capitalist transition" (5) – that is, concerned with global economic development from the outset. The problem was not that political ecology had failed to apprehend forces beyond "the local," but that these forces had been treated as exogenous to the land manager – replicating Durham's bifurcation between the institutions of governance and the people under study. As post-structuralist Marxists, Peet and Watts wanted to throw light on how local and global, individual and institutional, are socially produced and mutually constitutive: "In other words, one has to have a theory capable of explaining how the poverty of specific land mangers is reproduced through determinate structures and by specific relations of production" (7). They rather tartly criticize Blaikie and Brookfield's approach as being "atheoretical precisely when it most needs some kind of political theory" (8) concerning how the environmental choices made by the land manager are related to state and global power without resorting to flavors of voluntarism or determinism.

As the manifesto of post-structuralist political ecology, *Liberation Ecologies*' opening essay grapples with the state by focusing on its social constitution to the exclusion of examining the resultant power of the institutional form, however constructed. The state was still largely only an effect of other power relations, not an object of study in itself. Affirming the possibility of a dialectical state theory, and one that did not forget the need to iteratively perform the state's existence and powers, needed a merger with the Gramscian understanding of states articulated by Bob Jessop in *State Theory* (1990), and the aforementioned encounter with the sociology or ethnography of the state that came with Timothy Mitchell's *Colonising Egypt* (1999) and James Ferguson's *The Anti-Politics Machine* (1994).

In short, political ecologists were beginning to deal with govern*ment*, but not yet govern*ance*. These signal books convinced many political ecologists of the inadequacy of first-generation political ecology's approach to the state, but they did not clearly fill in the blank. Political ecologists could definitively describe what the state and government were *not*: the state was not exogenous, determinative, rational, the handmaiden of capital, and it was probably not even a thing at all (see Chapter 15, this volume). But affirmative concepts of states and state-ness, and the complex network of forces linking states and civil society, were missing or borrowed from mainstream political economy. Concern for the state permeates such landmark works as Peluso's *Rich Forests, Poor People* (1992) and Grossman's *Political Ecology of Bananas* (1998), but in different ways. Grossman recapitulates Durham in isolating a discussion of the state to a chapter setting the global context of banana trade – the Windwardian state as such disappears, its work done, when one descends to the village. In Peluso's work the identities and roles of the representatives of the state shimmer indeterminately with each scene: now they are amanuenses of the state, now fully local agents. And while this variability may be an accurate reflection of the lives of those who must perform state power, it was left unexplained. Peluso's work (unlike almost all before her) grappled explicitly with the role of state actors in Javanese forestry-dependent villages, and the conflicted role of local forest police who extend state power (see especially pp. 134–135), but "the state" itself remained in the wings, generically developmentalist and postcolonial. The apparatuses of village society and power are opposed to those of an exogenous state that arrives onstage fully-formed with policies and guns.

Political ecology and the first world state

One of the questions animating political ecology in the early 2000s was whether or not the approach was useful in understanding environmental management and change in the developed world, given its origins in the classic locations of ethnography. After all, Piers Blaikie was above all else a scholar of development in 1985, and his work was aimed at reforming development practice in the conventional locations of that activity. On the face of it, this question rested on a somewhat spurious distinction between the first and third worlds – is rural Kentucky "developed" while suburban Mexico City is "developing"? And one of the key first-generation works (Black 1990) took the concept of political ecology to rural Portugal, asking precisely this kind of question. However, we can reconceptualize this issue as two slightly more robust questions. The first is whether or not political ecology is best-suited to analyze the environmental transformations associated with the bleeding edge of capitalist expansion, the places where people and communities are being actively disarticulated from subsistence economies and moved into waged labor: in other words, development. This moment of disarticulation provides the driver for Nietschmann's work, for example, and it is possible to argue that the conceptual tools useful in that setting simply lack purchase in Omaha or Bristol. There is no environmental change wrought by the novel commodification of land and labor where regimes for transacting real estate have been stable for centuries. However, this is to be overly narrow about what constitutes a transformation of capitalism, and cannot survive contact with Harvey's concept of capital as restlessly redeveloping its own landscapes in successive spatial fixes – or indeed Schumpeter's older one of creative destruction. In this, Baltimore and Bahia are not qualitatively different.

The second question is more on point for us: even if capital is restless and dynamic, the first world state is markedly different in stability and form from the developmental state. One can see that where a land manager has access to (and responsibilities to) established property law, institutionalized environmental regulations, and a fairly static architecture of government, that resource conflicts will play out very differently from how they might where all of these things are constantly negotiated and in play. McCarthy probed exactly this distinction in 2002, but argued that the differences between first and third world state capacities were overdrawn. McCarthy found that the main difference lay in perceptions of marginality – and again, that means marginality to state power. Residents of first world states are, often mistakenly, assumed to be adnate to a powerful state that has

> the abilities to monitor resource use, enforce laws and policy, and run effective, professionalized resource-management bureaucracies, usually despite opposition by affected communities, corporations, and other interested parties.
>
> *(1287)*

But in the same way that capital's heartland should not be understood as free from development, McCarthy notes that state capacity in the first world should not be overestimated, as many of the same contradictions and negotiations over resource access and management occur: "the United States, often portrayed as the gold standard of sovereign state capacity, actually experiences many of the problems and limitations supposedly diagnostic of 'weak' states in controlling its own territory and population." It is precisely political ecology's power, he argues, to see these "gaps and contradictions in state control, and their consequences for resource management in the First World" (1287), where most policy analysts posit stability and rationality.

The development of first world political ecology throws into sharp relief the absence of a state theory underlying much political ecology. The state theory tacitly informing first and second generation political ecology was that of developmentalism, in which a weak central bureaucracy's task is to create the institutions of Western government in a national setting dominated by subsistence agriculture and primary sector economic production. Such a state typically lacks strongly developed relationships with civil society and hegemony, in the Gramscian sense, is poorly instilled in the population. The coming of political ecology to the first world requires, perhaps foremost, a different theory of the state to which land managers are (or are not) marginal. The household economy, local environmental knowledge, and struggles with distant forces over resource access can all be central to first world environmental problems as well (St. Martin 2001; Sayre 2002; Prudham 2004; McCarthy 2005; Robbins 2007): however, the state within and against which these struggles unfold is more accurately described in, for example, the *Régulation* Approach (RA) (see especially Huber 2013) than in the developmentalist tradition. The RA understands the late-capitalist state as a set of civil and governmental institutions managing (or "canalizing") the tendential crises of capitalism while maintaining consumption and accumulation – setting labor policy here, arranging for a spatial fix there. Whether or not this is a useful concept, it sets the first-world state a markedly different suite of tasks than it does the developmentalist state, while binding them together in responding to the dynamics of global capitalism.

Ethnography of the state

Robbins (2002) points out that one of the scholarly habits that students of first world environmental problems have is "looking up" – that is, understanding third-world environmental problems as local and contingent, while first-world problems are articulated "up" at the level of state or federal policy in which individual actors disappear. The answer, he says, is to look to the literature from "the sociology (or ethnography) of the state," which treats the institutions of the state itself as localities much the same as a Nicaraguan village or a Sahelian rangeland. In doing so, Robbins reinforces the strength of political ecology that Bryant and Bailey abandoned, which is to ground its claims in the exploration of global forces acting in local settings:

> Such an approach does not dispense with "local" political ecology in favor of the study solely of powerful organizations. Instead, it imagines those very organizations – forestry offices, World Bank evaluations departments, soil laboratories – as a part of local struggles, peering in to the mutual flows of formal state institutions and local social struggles.
>
> *(1511)*

The move toward an ethnographic understanding of state power has proven very productive for political ecologists who can understand their site as not just at the receiving end of national and global forces, but as also constituting the hegemony that allows national and global power to operate as such (Asher and Ojeda 2009). The relationships with resources and nature at the level of the land manager turn out to be crucial to the project of state-making, and this focuses our attention on the never-ending attempt to routinize and standardize (following Scott 1999) the idiosyncratic and contingent ways that environmental information is generated and taken up by states and their agents (Dunn 2003; Robertson 2006; Eden 2009; Whitehead et al. 2007). It is through the varied and daily activities of individual state agents – performing many of the same roles as the classic land manager of Blaikie and Brookfield – that nature achieves visibility and legibility to both state and capital. Rendering buried strata as a legible geological sequence (Braun 2000) or a field of native plants as a numeric "ecosystem service value" (Robertson 2004) cannot be achieved by actorless administrative fiat, and the failures and successes and shortcuts of state agents in the field give us vital information about the spaces of capital, the spaces of government, and the spaces in-between and beyond their reach. It also turns our attention to the tasks of creating hegemony through governmentality.

Governance and governmentality

Michel Foucault's famous definition of "governmentality" – "the conduct of conduct" – constituted a key shift in thinking about the state and governing that political ecologists began to absorb: a focus on the creation and sedimentation of conventions concerning how individuals relate to state power (see Chapter 36, this volume). The raw exercise of force is one kind of power, but a more pervasive and subtle one is the securing of hegemony by changing people's mentalities and attitudes. And as the Foucauldian concept of "biopower" suggests, this kind of hegemony can implicate our own biological processes and ecological relations (Wainwright and Mercer 2009; Patterson and Stripple 2010). When we discipline ourselves to manage our own carbon footprint or adopt particular attitudes toward GMO food, we are engaging in the long project of training and reproducing the capitalist consumer. The site of this disciplining is, appropriately for political ecology, often the household.

This line of inquiry about the role that "environmental governmentality" (or "environmentality" for Luke [1999]) plays in assuring (or failing to assure) the hegemony of capital interests over the control of resources again brings political ecology directly into contact with the Gramscian tradition in political economy, since Gramsci's concept of hegemony is the exercise of power through information, media, and culture. Mann (2009) goes so far as to say that an encounter with Gramsci is inevitable for political ecology, in that political ecology unifies the major problems of modern Marxism: social justice for marginalized communities, the crisis tendencies of capitalist development, and the need to grapple with often-misleading distinctions between natural and social forces. Gramsci is a necessary ingredient in using Foucauldian concepts to analyze environmental problems, says Mann, because "ideas like discourse, social construction, or desire are analytically useful, but it is difficult to imagine a way to discover their points of social purchase without the prior work that must be done to ground or historicize them" (340) – that is, work at which political ecologists are particularly adept. Ekers and Loftus (2008) argue that the study of how biopower and ecological relations are used to secure hegemony must not be left to those who assume hierarchically static relations between states and subjects, and rational behavior on all parts. A state theory is necessary: "We argue that state theory is a spectre (an elephant in the room?) that haunts much of the literature on water and social power. And yet it is never stated with the precision that we might expect" (708).

The pun of the word "governmentality" suggests that the success of strategies of government is in part the work of a mentality successfully instilled in subjects, such that they continue to conceive of themselves as subjects, often with a certain kind of nature as an external referent (e.g. Braun 2000; Agrawal 2005). This work helps us understand the complexities of subject-formation and agency on the part of the land-manager figure still at the center of political ecology's analysis (whether she is a peasant farmer or a staffer at an environmental agency). In Agrawal's work on village-level conservation policies in India, the distinction between the local official and the government itself is muted, and we nearly circle back to the localist bias in earlier political ecology, but this time around the state is an endogenous feature – integral to the dynamics of village life, rather than a shadow just over the horizon.

Conclusion

The state should trouble political ecologists. As an abstraction, a non-human actor, a force for environmental change, it is not amenable to study by many (or any) of the methods that emerge from the traditional sites of political ecology. And political ecology was established as an approach precisely to circumvent and improve on the kinds of explanations that seem to work only when you imagine states as coherent objects or even actors. Focusing on the land manager and ignoring the state (or at least showing how the state must wrestle with local social and environmental forces) was supposed to be the great virtue of political ecology.

But political ecologists should also trouble the state. As Whitehead et al. (2007: 14) explain:

> we are suspicious of discussions of states interacting with natures ... because such phrases tend to suggest two already established entities doing things to each other – the state affecting the natural environment, or changes in nature somehow influencing the policies and dictates of the state ... in order to analyze state-nature relations it is important to unpack both categories so as to understand how their constituent parts are intermixed and connected.

Unpacking the state is an essential part of understanding environmental change, and the first chapters of Whitehead's book contain a superb review and argument of this point for political ecologists. Whether the state is an effect of the search for hegemony by powerful elites, a venue for challenging capital through arguments over the valuation of nature, or an abstraction invoked by local actors seeking to jump scale, it is an essential platform for the exercise of changing the environment. Political ecologists cannot afford to bracket the state or governance; this was never an option from the moment Sauer began to mourn the demise of the culture area and Nietschmann watched Miskito villagers leaving on the road to the capital. Political ecology's strength will always be textured, site-level analysis, but this does not preclude the state as a subject of investigation – it only requires that we remember that the state is present and invested in nearly all settings, and is reproduced in part through the actions and subject-making of resource managers. With a rich array of choices concerning how to theorize and study state architectures, actions, and inhabitants, we can not only more completely understand the nature of socio-environmental change, environmental justice, and marginalization, but we can point effectively toward the governmental levers of change.

References

Agrawal, A. 2005. *Environmentality: Technologies of Government and the Making of Subjects*. Durham, NC: Duke University Press.

Asher, K. and D. Ojeda. 2009. Producing nature and making the state: Ordenamiento territorial in the Pacific lowlands of Colombia. *Geoforum* 40(3): 292–302.

Black, R. 1990. "Regional political ecology" in theory and practice: a case study from northern Portugal. *Transactions of the Institute of British Geographers, New Series* 15(1): 35–47.

Blaikie, P. 1985. *The Political Economy of Soil Erosion in Developing Countries*. New York: Longman.

Blaikie, P. and H. Brookfield. 1987. Defining and debating the problem. In P. Blaikie and H. Brookfield (eds.) *Land Degradation and Society*. London: Methuen, 1–26.

Braun, B. 2000. Producing vertical territory: geology and governmentality in late Victorian Canada. *Ecumene* 7(1): 7–46.

Bryant, R. and S. Bailey. 1996. *Third World Political Ecology*. New York: Routledge.

Dunn, E.C. 2003. Trojan pig: paradoxes of food safety regulation. *Environment and Planning A* 35: 1493–1511.

Durham, W.H. 1979. *Scarcity & Survival in Central America: Ecological Origins of the Soccer War*. Stanford: Stanford University Press.

Eden, S. 2009. The work of environmental governance networks: traceability, credibility and certification by the Forest Stewardship Council. *Geoforum* 40(3): 383–394.

Ekers, M. and A. Loftus. 2008. The power of water: developing dialogues between Foucault and Gramsci. *Environment and Planning D: Society and Space* 26(4): 698–718.

Ferguson, J. 1994. *The Anti-Politics Machine: "Development," Depoliticization, and Bureaucratic Power in Lesotho*. Minneapolis: University of Minnesota Press.

Grossman, L.S. 1998. *The Political Ecology of Bananas: Contract Farming, Peasants, and Agrarian Change in the Eastern Caribbean*. Chapel Hill: University of North Carolina Press.
Huber, M. 2013. Fueling capitalism: oil, the regulation approach, and the ecology of capital. *Economic Geography* 89(2): 171–194.
Jessop, B. 1990. *State Theory: Putting the Capitalist State in its Place*. University Park: The Pennsylvania State University Press.
Luke, T.W. 1999. *Capitalism, Democracy, and Ecology: Departing from Marx*. Chicago: University of Illinois Press.
Mann, G. 2009. Should political ecology be Marxist? A case for Gramsci's historical materialism. *Geoforum* 40(3): 335–344.
McCarthy, J. 2002. First World political ecology: lessons from the Wise Use movement. *Environment and Planning A* 34: 1281–1302.
McCarthy, J. 2005. Devolution in the woods: community forestry as hybrid neoliberalism. *Environment and Planning A* 37: 995–1014.
Mitchell, T. 1988. *Colonising Egypt*. Berkeley: University of California Press.
Nietschmann, B.Q. 1979. Ecological change, inflation and migration in the Far Western Caribbean. *The Geographical Review* 69(1): 1–24.
Patterson, M. and J. Stripple. 2010. My space: governing individuals' carbon emissions. *Environment and Planning D: Society and Space* 28(2): 341–362.
Peet, R. and M. Watts. 1996. Liberation ecology: development, sustainability, and environment in an age of market triumphalism. In R. Peet and M. Watts (eds.) *Liberation Ecologies: Environment, Development, Social Movements*. New York: Routledge, 1–45.
Peluso, N.L. 1992. *Rich Forests, Poor People: Resource Control and Resistance in Java*. Los Angeles: University of California Press.
Prudham, S. 2004. *Knock on Wood: Nature as Commodity in Douglas-Fir Country*. New York: Routledge.
Robbins, P. 2002. Obstacles to a First World political ecology? Looking *near* without looking *up*. *Environment and Planning A* 34: 1509–1513.
Robbins, P. 2007. *Lawn People: How Grasses, Weeds and Chemicals Make Us Who We Are*. Philadelphia: Temple University Press.
Robertson, M. 2004. The neoliberalization of ecosystem services: wetland mitigation banking and problems in environmental governance. *Geoforum* 35(3): 361–373.
Robertson, M. 2006. The nature that capital can see: science, state and market in the commodification of ecosystem services. *Environment and Planning D: Society and Space* 24(3): 367–387.
Sauer, C.O. 1941a. Foreward to historical geography. *Annals of the Association of American Geographers* 31(1): 1–24.
Sauer, C.O. 1941b. The personality of Mexico. *The Geographical Review* 31(3): 353–364.
Sauer, C.O. 1956. The agency of man on earth. In W.J. Thomas (ed.) *Man's Role in Changing the Face of the Earth*. Chicago: University of Chicago Press, 49–69.
Sayre, N.F. 2002. *Ranching, Endangered Species, and Urbanization in the Southwest: Species of Capital*. Tucson: University of Arizona Press.
Scott, J.C. 1999. *Seeing Like a State: How Certain Schemes to Improve the Human Condition Have Failed*. New Haven: Yale University Press.
St. Martin, K. 2001. Making space for community resource management in fisheries. *Annals of the Association of American Geographers* 91(1): 122–142.
Wainwright, J.D. and K. Mercer. 2009. The dilemma of decontamination: a Gramscian analysis of the Mexican transgenic maize dispute. *Geoforum* 40(3): 345–354.
Walker, K.J. 1989. The state in environmental management: the ecological dimension. *Political Studies* 35: 25–38.
Whitehead, M., R. Jones and M. Jones. 2007. *The Nature of the State: Excavating the Political Ecologies of the Modern State*. Oxford: Oxford University Press.

36
ECO-GOVERNMENTALITY

Gabriela Valdivia

Introduction

"Eco-governmentality" is a Foucaultian-inspired power analytic that political ecologists use to examine nature–society relationships. Since its early days, political ecologists have used Marxist-inspired critique to explain environmental domination and oppression (Watts 1983; Blaikie and Brookfield 1987; Hecht and Cockburn 2010; Peluso 1992). As Foucault's work became more broadly accessible and translated into English in the 1990s, it challenged some aspects of how political ecologists thought about history, change, and power (Bryant 1998). In some respects, Foucault's analytics parallel Marxist critiques of power, for example, like historical materialism, Foucault takes social practices as transitory and intellectual formations as connected with power and social relations. Things we consider universal, contends Foucault, are the result of very precise historical changes. In other respects, Foucault deviates from Marxist thought, moving away from "modes of production" as the site of social critique and towards "modes of information" (Poster 1984): how power works to produce structures of domination (and resistance) in modern society. His aim was to see power everywhere and in everything, not only in economic activity, and to investigate the microphysics of power rather than focusing on the macro-perspective of the state, or on class struggle as the venue for social change (Foucault 1980b). Doing so, Foucault argued, enabled recognizing the historical contingency of taken-for-granted concepts (e.g., madness, sickness, sexuality, class); the role of social practices in truth regimes; and how authorities and institutions that manage, rule, and control social life are socially produced.

Foucault's approach to power is critiqued for lacking an explicitly political position on how to change power relations. Foucault valorized certain forms of resistance and worked to undermine taken-for-granted arrangements of power, but he often skirted the question of how to act politically in relation to power. Instead, he suggested that some form of action is possible when individuals recognize the historical construction of power: "this work done at the limits of ourselves must ... open up a realm of historical inquiry and ... put itself to the test ... of contemporary reality, both to grasp the points where change is possible and desirable, and to determine the precise form this change should take" (Foucault 1984: 46).

Foucault's legacy in political ecology is unmistakable. While Foucault did not aim to theorize environmental relations, his insistence on tracing the hidden truths in power regimes has been taken up to ask about how ideas about nature and society direct what seems true and possible, and to address projects at the core of the field, for example, provincializing imperial environmental governance, decentering the location environmental power/knowledge, and exposing the contingent nature of power formations (Braun 2000; Escobar 1992; Neumann 1998; Peluso and Vandergeest 2001). Foucaultian-inspired political ecologists recognize that environmental truths are not ahistorical but produced in a variety of sites and through diverse actors, and that various social groups have the potential to act politically—not just the state, or the NGO, or the proletariat. Before delving into how political ecologists use and extend Foucault's work, the next section provides fundamental guidelines for understanding his contributions to eco-governmentality.

Foucault and modern power

> To begin the analysis with a "how" is to suggest that power as such does not exist … [I]t is to ask oneself what contents one has in mind when using this all-embracing and reifying term; it is to suspect that an extremely complex configuration of realities is allowed to escape when one treads endlessly in the double question: What is power? and Where does power come from? The little question, What happens?, although flat and empirical, once scrutinized is seen to avoid accusing a metaphysics or an ontology of power of being fraudulent; rather, it attempts a critical investigation into the thematics of power.
>
> *(Foucault 1982: 785–786)*

To move away from power as a "thing," Foucault worked between two targets: deconstructing power-as-domination and questioning the ethical construction of subjects (Balibar 1992; Lemke 2002). A few guidelines to Foucault's method for how to think about power can be discerned from his fragmented analyses. First, he challenged how power works. Power is not concentrated in specific social locations and bodies (e.g., it is not solely in people's heads, as representations, consciousness, acceptance, or interiorization) but in the material actions and re-actions of social exchanges within a particular society (Foucault 1980a: 10; Foucault 1990: 93; Foucault 1988b: 14). Power designates actions and relationships between partners—however uneven—that induce self and others to act in particular ways (Foucault 1982). Power, thus, is not necessarily repressive, prohibitive, or exclusionary (Foucault 1977a: 194) but produces social truths, reality, and individual subjects *and* structures.

Second, power is inseparable from knowledge. Following Nietzche, Foucault argued that knowledge "cuts" the social field into governable targets, and that disciplinary technologies naturalize differences as politically meaningful breaks within social blocks (Foucault 1977b). Knowledge about what constitutes recognizable difference forms through a historically produced system of communications, records, accumulation, and displacement—a language or system of exchange, or discourse—which itself is also a form of power. "There is not knowledge on one side and society on the other, or science and the state … only the fundamental forms of knowledge/power" (Sheridan and Foucault 1980: 283).

Third, Foucault developed an empirical method for tracing genealogies of power, investigations that chart the historical production of things that seem real and ahistorical. Foucault called the historical production of subject-structure realities "capacity blocks" (Foucault 1982), disciplinary assemblages, with defined space, position, methodology, and membership, and involving relations of communication and action that permit the exercise of power in particular ways. Blocks are diverse and adjust according to the circumstances in which interrelationships establish. Foucault initially investigated institutions such as the prison and the asylum as blocks with which to visibilize the tactics and procedures of power.

Fourth, in his later work, Foucault questioned how individuals see themselves and are seen as social beings (Foucault 1990). Within blocks exist subjugated knowledges that are buried and disguised, sometimes disqualified as inadequate or insufficient (e.g., the psychiatric patient, the delinquent, the ill person, the leper, etc.). These knowledges, embodied, practiced, and lived, form the basis for how a human being turns him- or herself into a subject (Foucault 1988b; Foucault 1980b: 83). Power is thus a question from 'below' as much as from 'above' (Foucault and Ewald 2003: 28). Subjugated knowledges circulate upwards and downwards in society, reflecting the struggles immanent to social blocks. Resistance comes from these knowledges, not in opposition to power but 'coextensive and contemporaneous with it ... as soon as there is a relationship of power, there's a possibility of resistance ... it's always possible to modify [power's] hold, in determined conditions and following a precise strategy' (Foucault 1980a: 13). A critique of society works through these unqualified and disqualified knowledges, which owe their force to the harshness with which they are opposed, but are not—nor need to be—unanimous (Foucault 1980b: 82).

Governmentality

"Governmentality" is one of Foucault's most highly engaged contributions to the analysis of modern power. It stands for an analytic which conceptualizes government as the contact point—or "field of action"—between power-as-domination and subject-formation. The most-often cited analyses come from Foucault's lecture series *Society Must be Defended* (1975–1976), *Security, Territory, and Population* (1977–1978) and *The Birth of Biopolitics* (1978–1979), which only became available to Anglophone scholars in the mid-2000s (though a translation of his famous lecture on governmentality was available in 1991). In his lectures, Foucault didn't stay with governmentality as a pivotal concept; rather, his ideas migrated towards concerns about power as historically situated formulations of how life is lived.

In the earlier lectures, Foucault suggests that ideas about the exercise of sovereign power began to change in the sixteenth century, fueled by changing conditions of existence in Europe, such as demographic expansion, the abundance of money, mercantilism, and the expansion of agricultural production (Foucault and Ewald 2003). Linked to these changing realities are several shifts in thought: debate about the reason of the state; ideas and knowledge about populations (statistics); the Reformation and the utility of pastoral power; and theories on self-government (Rutherford 2007). "[H]ow to govern oneself, how to be governed, how to govern others, by whom the people will accept being governed, how to become the best possible governor" (Foucault et al. 1991: 87) became defining questions of rule at the time.

The "governmentalization" of rule emerged from this crisis of sovereign truth (Foucault et al. 1991). Government does not transcend sovereign rule but decenters power, which makes the sovereign less vulnerable to the charge of oppressiveness and, hence, to resistance. Governmentalization breaks down the image of a sovereign state deciding on matters of life and death into a range of parties involved in the regulation of individual and collective life (Li 2007a). Further, it recognizes and encourages subjects to participate in their own rule (i.e., individuals acting upon themselves), according to pre-defined social areas of intervention, for example, education, health, religion, etc., that secure a better society. Governmental power works through "regimes of practice," where rules, plans of action, and rationalities (i.e., the "programs of conduct") meet what is actually done in practice (Foucault 1988a).

In governmental regimes, biological existence is reflected in political existence. While sovereign rule relies on juridical means to establish power over death (Foucault 1977a), governmental rule relies on the proliferation of techniques and knowledges to manage social life. Foucault called this the era of biopower (Foucault and Ewald 2003). From "above," the modern state administers life by regulating the institutions (family, army, schools, police, medicine, etc.) and habitual spaces of populations and individuals (and their deviances). Demography, state science (statistics), and insurance help place individuals and families within a larger population with measurable and manageable bio-social characteristics. Experts and state administrators voice truths about populations and aim to shape individual conduct to produce certain desired population effects and avert undesired ones, for example, through measures such as birth rates, infant mortality and longevity (Rose 1999: 52). Well-being and economic productivity also become calculable objects of rule (Legg 2005). From "below," individuals shape their own behavior, disciplining the actions and capacities of self and others toward proper—right—goals.

Governing is not force; it is an "equilibrium" between coercion and self-constitution that aims to optimize the life of populations and their relations to things such as wealth, health, happiness, prosperity, efficiency (Foucault 1993), so that the greatest possible quantity of wealth is produced, the people are provided with sufficient means of subsistence, the population is enabled to multiply, etc. (Foucault et al. 1991: 95). The rationality of government is the will to improve: the distance between what is discursively possible (a better society) and what is available for improvement (what actually exists). Improvement is about governing things according to their proper disposition, so that "an end which is 'convenient' for each of the things that are to be governed" (Foucault et al. 1991: 95) can be programmed:

> people … in their relations, their links, their imbrication with those other things which are wealth, resources, means of subsistence, the territory with its specific qualities, climate, irrigation, fertility, etc.; men [*sic*] in their relation to that other kind of things, customs, habits, ways of acting and thinking, etc.; lastly, men [*sic*] in their relation to that other kind of things, accidents and misfortunes such as famine, epidemics, death, etc.
>
> *(Foucault et al. 1991: 93)*

Foucault's power analytic raised transformative questions about the limits of subjugation and capacities to act. Foucaultian-inspired political ecologists examine the multi-scalar connections between global and local, to recognize the multiple locations of power (e.g., Collier and Ong 2005). Foucault's insistence on power/knowledge as a unit is not lost to political ecology either. In one of the most-often-cited texts about the field, Watts and Peet (1996) argue that discursive approaches to the analysis of environmental governance are fundamental to expanding the "frontiers" of political ecology, pointing to the sociology of science and knowledge, the history of policy and institutions, and the globalization of environmental knowledge as fruitful venues for analysis (e.g., Lowe 2006; Mathews 2011; Mitchell 2002). Political ecologists also paid attention to Foucault's method and adapted it to the analysis of global environmental institutions and state government offices, examining how these generate scientific and governmental knowledge that regulates the living spaces and functions of peoples around the world as subjects of government (e.g., Agrawal 2005a; Luke 1995; Goldman 2005; Peluso and Vandergeest 2011). And, to balance this "dark Foucaultian picture" (Robbins 2011: 227), scholars also extend the political possibilities of Foucaultian thought by blending it with Marxist-inspired frameworks that develop a "realist sense" (Ekers and Loftus 2008: 71) to the exercise of power, that is, linking power to everyday environmental relations and spaces of action (e.g., Braun 2007; Gandy 2004). More recently, scholars have linked Foucault's theorizations on neoliberal governance, a "type of relation" between governors and governed rather than a technique of governors on the governed (Foucault 2008: 218), to critiques of market-oriented practices, thought, and environmental action (Guthman 2008; Büscher et al. 2012; Fletcher 2010).

Eco-governmentality

Foucault's conception of power bolstered a monumental body of critical scholarship that dissects taken-for-granted ideas about nature–society relations, from biodiversity conservation, to sustainability, to resource extraction. This section, necessarily selective, examines works that engage with the two broad goals of Foucault's power analytic, deconstructing power/knowledge and questioning subject-making, to give a sense of the wide range of today's Foucaultian-inspired political ecology. I proceed with three cautions. First, we need to remain aware that prioritizing how power works can give the wrong impression that power is its own historical object and end (Spivak 2010: 274). Second, to expose the practices of environmental regimes, political ecologists blend Foucault's analytic with other frameworks. This blending is both supported because it expands the epistemological grounds of the field (Ekers and Loftus 2008) and critiqued for blurring theoretical and epistemological differences (Barnett 2005). Third, political ecologists have extensively used Foucault's power analytic to examine how environment, as a rationality of biopower, produces regimes of rule in the Global South. Taking Foucault's work beyond the geographic and historical boundaries of its production feeds the hopeful recognition and generation of alternative power/knowledges, but also runs the risk of misinterpretation of both theory and empirical observations.

Environmental power/knowledge

Political ecologists apply governmentality to examine how "the environment" has become a rationality of rule (Luke 1995). For instance, Michael Goldman examines juridical and legislative practices to demonstrate how the World Bank, through a series of new "green" practices, sponsored and mediated policies that produced new global regimes about nature, the right to use natural environments, and the formation of new, transnational state authorities (Goldman 2001). According to Goldman, the Bank re-organized what counts as "nature" and how to rule it, and redefined the state's sovereign relationship with the qualities of its territory (also see, Braun 2000; Sivaramakrishnan 1999). Goldman (2001: 501) argues that by bringing neocolonial conservationist ideas of enclosure and preservation together with neoliberal notions of market value and optimal resource allocation, the Bank "made particular natures and natural resource-dependent communities legible and accountable." New governing apparatuses produced new knowledge about natural and social bodies of proper conduct and rule. Watersheds, national biodiversity conservation areas, extractive reserves, and sustainable logging zones became objects to be managed by international experts, through action plans, surveys, and non-governmental organizations (NGOs) (Goldman 2004: 171). The new regulatory regime also created new population subjects in Laos (hill tribes, rice growers, technocrats, and entrepreneurs) who are global market-oriented, scientifically based, rational, and ecologically sustainable. The environmentalism that emerged from the Bank's "green" interventions re-calculated the potential for life in Laos, transforming its global identity into "the next Switzerland or Kuwait of South East Asia, a prospective engine for borderless commerce and energy-driven capital accumulation" (Goldman 2004: 170). Peluso and Vandergeest (2011) make a similar argument about how South East Asian states during the Cold War drew on imperial calculations of population resettlement, colonization, and surveillance to minimize forest-based insurgencies and secure natural resource extraction.

These works on the production of Global South natures and populations use Foucault's contributions to make explicit the relevance of juridical technologies of sovereign power. They show how populations and states are made and remade through legal apparatuses that settle environmental rationality, which speaks to the co-production of sovereign and governmental rule, and demonstrate how the production of environmental knowledge is internal to, and constitutive of, the exercise of power. The emphasis on how states and the Bank "see" and produce truths, however, reproduces the sort of domination-power that Foucault critiqued. It fixes the location and trajectory of power: new "green" discourses and practices originate and flow top-down, from the World Bank or the state, to subjects, thus reproducing power-as-domination. This reading reduces Foucault's governmental analytic to a disciplinary mode of eco-power over larger spatial scales under the auspices of the territorialized state.

Tania Li's (2007b) ethnography of landscape and livelihood improvement programs in Indonesia offers a different analysis of eco-power, examining how policy is *translated* into localized programs of action to produce local subjects of improvement. The focus is on the field of improvement: the space between what is thought/said and what actually happens in these regimes. Bringing together notions of enframing and expertise (Mitchell 2002), Li describes how certain actions and bodies are rendered deficient objects amenable to technical rectification through scientific practices involving a range of diagnoses, prescriptions, and techniques that build boundaries between those who are "trustees" (with capacity to diagnose deficiencies in others) and those who are subject to expert direction (Li 2007b: 7). These practices parallel the dividing practices that Foucault explored in his studies of madness, lepers, illness, and sexuality (also see Bäckstrand 2004), which objectify bodies and divide subjects into social groups with unequal relations of power.

Li's work illustrates how governmentality can be combined with Marxist critiques of power, such as Antonio Gramsci's work on hegemony and the practice of politics—another strong current within political ecology (Mann 2009). Gramscian-infused governmentality studies aim to avoid an overly diffused sense of power. They recognize not only the coercive state and the ability of elites to achieve the consent of non-elite groups, but also how subjugated knowledges intervene in eco-governmental rule (Ekers and Loftus 2008; Birkenholtz 2009). As Li observes, improvement is not merely a tactic to maintain the dominance of particular classes, or to assert control by the North over the South, but a vehicle that contains challenges to the status quo. Sometimes expert discourse is punctured by a challenge it cannot contain, such as when the targets of expert schemes reveal their own critical analysis of the problems that confront them, which makes visible the limits of discourses and reveals the incompleteness of environmental truths. Though Li pays attention to the struggles of villagers, highlanders, and indigenous peoples, a contradiction arises in her work. Li introduces a thought/action dichotomy, the divide between structure and agency that Foucault sought to deconstruct, into the analysis of eco-government. In the case of the Free Farmers Forum, for example, Li describes that Farmers found it difficult to "establish a positioning as legitimate political actors, not merely victims or dupes. The imperfections of their own conduct—their failure to obey the rules they had set for themselves—was part of the problem" (Li 2007b: 272). In other words, farmers resist but remain subjects working with logics and programs *external* to their desires and hopes and, thus, are rendered as subjects who inhabit structures, rather than co-produce them. Emily Yeh's (2005) analysis of state-led "ecological construction" in China similarly argues that articulation with global environmental regimes *deepens* state control over territory and that an oppressive, already-defined sovereign defines formulation of power in these accounts. She concludes with a question: what do subjects of the state think and do about changes in biopolitical life?

Environmental subjectivities

While some political ecologists focus on the disciplinary capacities of environmental regimes, others look at *who* is normed and disciplined. Everyday life, the mundane routines and spaces of social reproduction and interpersonal existence through which human multiplicity is transformed into a differentiated, classified society (De Certeau 1998), becomes the site of analysis in these cases. In his lectures on power and subjects, Foucault examined questions on how individuals learn to recognize themselves as subjects and how the subject "makes sense" of the distance between what they are and what they think they could be (Foucault 2005). This shift pivoted around the realization that theories of subjectivity can only be broken by developing an account of the active receptivity of people to being directed.

One highly engaged work on this topic is Arun Agrawal's (2005a) *Environmentality*. Agrawal takes up the question of *what do people actually think about* the environment in his study of shifts in forest governance in India, from centralized state control to evolving systems of local governance. Marrying Foucaultian subject-making with institutional frameworks, Agrawal proposes that environmental subjectivities are, partly, a learned response to the outcomes of environmental regulations, and vice versa. People may plan to act according to their preferences but plans are imperfect and actions may lead to unanticipated outcomes. Outcomes, in turn, incentivize individuals to reconsider their preferences *and* subjectivities. Agrawal asks: when and for what reason do individuals come to care for, act, and think of their actions in relation to the environment? His study goes beyond the deconstruction of power-as-domination and towards technologies of the self to propose that "environmental protection" is a governmental rationality that creates new environmental subjects. Technologies that regulate the practice of forest government at the community level (e.g., the division of space, the creation of local forest councils and constitutions, the introduction of self-elected groups that police and monitor forest uses, etc.) discipline environmental subjectivities. The more these practices are routinized, the more people participate in organized forest protection, the more they consciously organize their preferences, actions, and beliefs—a relationship of policy and subjectivity Agrawal terms "intimate government" (Agrawal 2005b). The disciplined bodies of individuals become ways of "grappling with the phenomena of population … to undertake the administration, control, and direction of the accumulation of men" (Foucault 1980b: 125).

In the quest for understanding how regulation, subjectivities, and autonomous intentionality intersect in environmental governance, however, Agrawal sacrifices the dispersed productivity of power. Instead of decentered power, we encounter rational, conscious subjects choosing future trajectories based on preferences and perceptions, which suggests that rationalities of power work primarily through governmental *mentalities*—i.e., power is in the mind that imagines power. Moreover, while providing evidence supporting the idea that some villagers "care about forests" through the practice of institution-building, Agrawal skirts matters of production and struggle that Gramscian-inspired political ecologists emphasize. As Leach and Fairhead (2000) argue, the relationships between subjugated knowledges produced in engagement with landscape and shaped by inhabitants' particular historical and social experiences are often "occluded" in the critique of scientific and social truth regimes.

Agrawal's *Environmentality* is not without critics. For example, drawing on Spinozean optics and thick ethnography, and responding directly to Agrawal's omission of everyday struggles over existence, Singh (2013) and Cepek (2011) propose that how individuals recognize themselves as subjects *existing* within truth regimes—e.g., as indigenous, black, Hispano, women, poor, etc.—is fundamental to environmental subject-making. Signaling another rapidly developing area in political ecology, the political ecology of affect, scholars argue that environmental subject-making is entangled with material geographies of belonging, exclusion, and subjugation (Ogden 2011; Raffles 2002). Individuals have "docile" capacities that make them available to government, their bodies can be put into practice, routinized and disciplined to shape subjectivities, but they have the potential to act in other ways, too. Michael Cepek's (2012) ethnographic exploration of how Amazonian peoples are not born indigenous but become indigenous by living with the land in particular ways, and within historically situated environmental regimes, illustrates this focus on struggle over everyday existence-as-knowledge. Cultural complexities, affective labor, and micro-political struggles are sites for subjugated knowledges to critically respond to external directives. In the Amazon, individuals deploy critical consciousness that produces possibilities for worlds to be otherwise for indigenous populations—one critical element of subject-making (Foucault 1977a) missing in Agrawal's account. In India, people's sense of self is shaped by their capacity to *respond* to other bodies, both human and non-human, as they labor, consume, sense, and move, that is, through their existence as subjects (Singh 2013). Similarly, in the United States, Chicano activists, environmentalists, nuclear scientists, and state workers in the US Southwest understand their lives through intimate and affective attachments to landscapes expertly marked as degraded and polluted (Kosek 2006). Bodies do not only represent and enact subjugated knowledges; as life runs through them they have capacities to act that exceed the rationalities of power that attempt to order and control life.

Also emphasizing the potential to act, Michael Watts (2004) takes eco-governmentality in complementary direction to highlight the *productive* incompleteness and plurality of environmental rule. Watts brings to bear categories that Agrawal (2005a: 166) dismisses as "static" (e.g., gender, ethnicity, generation, caste) to the analysis of oil governance in Nigeria. Drawing on Rose's (1999) spatialization of government, how social thought and practice territorialize, Watts examines how different forms of organizing civic power (chiefdoms, indigeneity, nationalism) form in Nigeria's petro-capitalist context, and how these multiple spatializations of local rule contest state efforts to normalize resource space. Recognizing the cultural politics of subject-making, he demonstrates that multiple spaces of rule operate simultaneously, sometimes overlapping, sometimes contesting each other in environmental regimes. Watts illustrates the "heteropia" (Foucault and Miskowiec 1986) of environmental rule: the simultaneity, juxtaposition, and dispersed expressions of eco-governmental rule. Governmental regimes might encourage a singular and authoritative reading of nature, resources, or territory, but different forms of rule and governable space operate simultaneously, each with their specific politics of scale and subjugated knowledges (Moore 2005).

The works outlined here raise questions about how existence is entangled with specific environmental truths and rationalities. Power travels in multiple ways through various institutions and differently racialized, classed, and geopolitical natures, enabling multiple forms of subjectivization. Technologies of power shape the conduct of individuals and submit them to certain societal ends (objectivizing the subject), as Agrawal suggests, yet technologies of the self also permit individuals to transform themselves and others—their own bodies and souls, thoughts, conduct, and way of being—in unexpected ways (Foucault 1988b). Taken together, these selected interventions remind us that the "self" is open to active engagement (Foucault 1988b), although there are limits "to the open-ended actualization of being" (Braun 2006: 210).

Where to now?

True to Foucaultian fashion, political ecologists borrow *and* build analyses of eco-power. Below, I discuss three epistemological and methodological provocations developed in the field: the ethical substance of eco-power; the primacy of human-centered relations; and the production of alternative environmental knowledges.

The ethical substance of eco-power

Development policies produce environmental objects and subjects of rule through discursive rationales and disciplinary technologies, in order to pursue a promise (sustainability, biodiversity conservation, protection, increased productivity, etc.). In political ecology accounts of eco-governance, the "ethical substance" of being (Foucault 1990)—the prime material of moral conduct and discipline—is often left unexamined. Yet not all objects and subjects are disciplined in the same way. Some bodies count more than others and some bodies are recognized as more raced, gendered, or ethnic than others. Recognizing how the biological form of socially constructed difference matters to power relations is fundamental to how we study and act. Foucault's discussion of state racism (Foucault and Ewald 2003) is a helpful place for further developing the connections between matter and environmental knowledge/power. In Foucault's formulation of modern power, the sovereign *and* the subject are already constituted—their relationship pre-exists the power trilogy of sovereignty–discipline–government. Giorgio Agamben (1998), responding to this enclosure of the subject, proposes to take a step back to look for the moment before the subject becomes an object of truth. What about the body becomes subject-truth (and what doesn't)? Through what economy of power is a divide established between the body that lives like *other* living things (*zoē*) and a body that lives properly (*bíos*)? For Agamben, ruling this spacing between *zoē* and *bíos* is the field of power.

Julie Guthman (2011, 2012) brings this insight on differentially lived bodies to bear in matters of environmental governance, by tracing how processes of racialization articulate in and through environmental formations and vice versa. Aiming her critique at the food justice movement, she asks why a critical movement that challenges structural obstacles to accessing nutritious foods nonetheless skirts another fundamental set of social justice issues: the universalization and normalization of certain bodies. Guthman deconstructs the abnormalization of fat bodies and the normalization of white ones, and emphasizes how the diversity of biological acts and environmental responses (e.g., biological adaptations, epigenetics, fat storage mechanisms) confound the ethical substance upon which truths about race and obesity are linked in food justice movements. Guthman's attention to the space between *bíos* and *zoē,* via her analysis of epigenetic genealogies and biological mechanisms, underscores the need to address occluded truths through which some populations are let die a premature death and others not.

Beyond human-centered relations

How governmental practices and rationalities enhance the rule of people in relation to environments and resources is now a hallmark of political ecology (Birkenholtz 2009; Valdivia 2008; Watts 2004). A human-centered perspective dominates these inquiries, however: people act upon resources, react to oppressive regimes, and deploy subjugated knowledges to effect change. Such anthropocentrism is but one possible formulation: one where "nature" and "society" belong to different ontological realms (Escobar 1996). What other socialities and politics are possible when the privileged, human location is questioned? How do "meetings" (Haraway 2008) with things as varied as climate, property, resources, and health produce new subjectivities, new histories, and new connections? Aligned with this reading of power, Moore et al. (2003) argue for conceptualizing nature as more than yet another docile body through which power operates. Juanita Sundberg (2013) explicitly takes us in this direction in her analysis of the government of bodies and environments at the US–Mexico borderlands (Sundberg 2011). She asks: how are nonhuman actors—including plants, animals, and biophysical processes—constitutive of boundary-making? What work do such "things" do in regimes of environmental truth? In her reading, geopolitical border practices are inflected by corporeal encounters with many species and not only humans. Thus, the capacity to affect, to act and be acted upon, which Foucault saw as the conditions of possibility of power, also applies to nonhumans. This is not an exercise in adding one more thing to the "witch's brew" of actual practices (Foucault 1977a) but a reformulation of how power/knowledge works, and an effort to take into account more-than-human actions and multi-sited power into the calculation of biopolitical life.

Alternative eco-knowledges

Foucault's (1988b) work on where to locate and animate the potential for change points to the continued need for situated objectivity. Feminist interventions in political ecology similarly urge us to remain attentive to the effects of how we construct and deploy knowledge. How is Foucaultian-inspired political ecology (re)producing specific regimes of truth and existence? Elizabeth Povinelli, for example, illustrates this reflexivity by using Foucault's work to critique how we narrate the connections between resistance, subjugated knowledges, and the potential for social change. Weaving her ethnography of endurance among Aboriginal Australians at the margins of "normal" society with a critique of liberalism, Povinelli (2011) tells stories of people who live in environments that slowly gnaw away at their lives—what she calls the heterotopias of modern power. Her goal is to incite ethical responses: to frustrate readings looking for hope, sacrifice, and adjudication in order to focus on the different manifestations and conditions of modern power. The point is not to provoke a feeling of injustice, vulnerability, or precarity, but to contribute to the cultivation of a politics of obligation that goes beyond amazement for sacrifice and suffering (Povinelli 2013). The epistemological challenge is to unlearn disciplinary habits that dismiss heterotopias and that assume knowledge about how the world works before the task of analysis starts.

Huntington and Watson (2012) offer a complementary provocation by decentering the sites of environmental authority and exploding notions of indigenous environmental "stewardship" and knowledge. Their goal is to encourage learning new habits that allow for practicing subject companionships and power/knowledge co-authorships. Like Povinelli, part of their intervention lies in *how* they produce environmental "truth-telling": they weave together multiple narrative locations, narrators, and narrating styles to emphasize plural standpoints and directions of power to demonstrate that environmental truths matter because they are *produced in* particular life worlds and from specific positions—and not because ahistorical power produces them. In terms of methodological challenges, these authors raise our attention to the obligations of writing about and with power.

Conclusions

> Everything is at stake—one should not change the tendencies of gravity and expect to remain the same. And if you wish to remain as an object affected by gravity, then what?
>
> *(Povinelli 2012: 470)*

One of Foucault's often-mentioned limitations is that he rarely discussed biology, environment, and resources. Political ecologists working in the Global South (though not exclusively) have often overlooked these limitations (or addressed them using other frameworks) and extended governmentality to examine how the government of populations is constituted through nature–society relations (Rutherford 2007). Foucault himself was aware of these limits. In the preface of *The Order of Things*, he reflects on a passage from a book by Borges, which he describes as a baffling list of classifications, a "Chinese Encyclopedia," that disrupted his thought (Foucault 1994: xv–xx). "What is it impossible to think" he asks, "and what kind of impossibility are we faced with here?" He concludes that the will to organize the world tames the "wild abundance of things" that exist *and* exposes some of the ways in which we seek to govern societies. What captured Foucault was not that there are "fabulous" things out there but that we put boundaries around our imagination of how the world works. His answer was to emphasize a dialectic of limits that generates tensions and provokes questions: between grand political considerations and mundane micropolitics, between structures and subjects, and between truth and change. Through its encounter with Foucault, political ecology also has pushed the theoretical limits of how we know, study, and reproduce environmental regimes. In turn, governmentality has been brought to bear into spaces and times not theorized by Foucault, helping political ecologists make "moralistic denunciations" (Ferguson 2011) of the power relations inherent to environmental regimes. Foucault also posed a methodological challenge: continuous empirical experimentation to trace the myriad ways that power is invested in the details of social action (Bouchard 1980). Foucault was not anti-power but against fixing the limits of political action to a few relations and actors. As political ecologists, we must be open to dispersed power without losing sight of the epistemological, ontological, and normative commitments through which we study power.

References

Agamben, G. (1998) *Homo sacer*, translated by M. Raiola. Stanford: Stanford University Press.

Agrawal, A. (2005a) *Environmentality: technologies of government and the making of subjects*. Durham, NC: Duke University Press

Agrawal, A. (2005b) Environmentality: community, intimate government, and the making of environmental subjects in Kumaon, India 1. *Current Anthropology* 46, pp. 161–190.

Bäckstrand, K. (2004) Scientisation vs. civic expertise in environmental governance: eco-feminist, eco-modern and post-modern responses. *Environmental Politics* 13, pp. 695–714.

Balibar, E. (1992) *Foucault and Marx: the question of nominalism*, New York: Routledge.

Barnett, C. (2005) The consolations of "neoliberalism." *Geoforum* 36, pp. 7–12.

Birkenholtz, T. (2009) Groundwater governmentality: hegemony and technologies of resistance in Rajasthan's (India) groundwater governance. *The Geographical Journal* 175, pp. 208–220.

Blaikie, P. and Brookfield, H. (1987) *Land degradation and society*. New York: Methuen.

Bouchard, D.F. (1980) For life and action: Foucault. S. *Oxford Literary Review* 4, pp. 20–28.

Braun, B. (2000) Producing vertical territory: geology and governmentality in late Victorian Canada. *Cultural Geographies* 7, pp. 7–46.

Braun, B. (2006) Towards a new earth and a new humanity: nature, ontology, politics. In N. Castree and D. Gregory (eds) *David Harvey: A Critical Reader.* Malden: Blackwell, pp. 191–222.

Braun, B. (2007) Biopolitics and the molecularization of life. *Cultural Geographies* 14, pp. 6–28.

Bryant, R.L. (1998) Power, knowledge and political ecology in the third world: a review. *Progress in Physical Geography* 22, pp. 79–94.

Büscher, B., Sullivan, S., Neves, K., Igoe, J. and Brockington, D. (2012) Towards a synthesized critique of neoliberal biodiversity conservation. *Capitalism Nature Socialism* 23, pp. 4–30.

Cepek, M.L. (2011) Foucault in the forest: questioning environmentality in Amazonia. *American Ethnologist* 38, pp. 501–515.

Cepek, M. (2012) *A future for Amazonia: Randy Borman and Cofán environmental politics.* Austin: University of Texas Press.

Collier, S.J. and Ong, A. (2005) Global assemblages, anthropological problems. In S.J. Collier and A. Ong (eds) *Global assemblages: technology, politics, and ethics as anthropological problems.* Malden: Blackwell, pp. 3–21.

De Certeau, M. (1998) *The practice of everyday life: living and cooking. Volume 2.* Minneapolis: University of Minnesota Press.

Ekers, M. and Loftus, A. (2008) The power of water: developing dialogues between Foucault and Gramsci. *Environment and Planning. D, Society and Space* 26, p. 698.

Escobar, A. (1992) Imagining a post-development era? Critical thought, development and social movements. *Social Text* 31–32, pp. 20–56.

Escobar, A. (1996) Construction nature: elements for a post-structuralist political ecology. *Futures* 28, pp. 325–343.

Ferguson, J. (2011) Toward a left art of government: from "Foucauldian critique" to Foucauldian politics. *History of the Human Sciences* 24, pp. 61–68.

Fletcher, R. (2010) Neoliberal environmentality: towards a poststructuralist political ecology of the conservation debate. *Conservation and Society* 8, pp. 171–181.

Foucault, M. (1977a) *Discipline & punish.* New York: Vintage Books.

Foucault, M. (1977b) Nietzsche, genealogy, history. *Language, counter-memory, practice: selected essays and interviews.* Ithaca: Cornell University Press, pp. 139–164.

Foucault, M. (1980a) The history of sexuality: interview. *Oxford Literary Review* 4, pp. 3–14.

Foucault, M. (1980b) *Power/knowledge: selected interviews and other writings, 1972–1977.* New York: Pantheon Books.

Foucault, M. (1982) The subject and power. *Critical Inquiry* 8, pp. 777–795.

Foucault, M. (1984) *The Foucault reader.* New York: Pantheon Books.

Foucault, M. (1988a) Power, moral values and the intellectual. *History of the Present* 4, pp. 1–2.

Foucault, M. (1988b) *Technologies of the self: a seminar with Michel Foucault.* Amherst: University of Massachusetts Press.

Foucault, M. (1990) *The history of sexuality: the use of pleasure.* Random House Digital, Inc.

Foucault, M. (1993) About the beginning of the hermeneutics of the self: two lectures at Dartmouth. *Political Theory* 21, pp. 198–227.

Foucault, M. (1994) *The order of things: an archaeology of the human sciences.* New York: Vintage Books.

Foucault, M. (2005) *The hermeneutics of the subject: lectures at the Collège de France 1981–1982.* New York: Macmillan.

Foucault, M. (2008) *The birth of biopolitics: lectures at the Collège de France, 1978–79.* New York: Palgrave Macmillan.

Foucault, M. and Ewald, F. (2003) *Society must be defended: lectures at the Collège de France, 1975–1976.* New York: Picador.

Foucault, M. and Miskowiec, J. (1986) Of other spaces. *Diacritics* 16, pp. 22–27.

Foucault, M., Burchell, G., Gordon, C., et al. (1991) *The Foucault effect: studies in governmentality.* University of Chicago Press.

Gandy, M. (2004) Water, modernity and emancipatory urbanism. *The emancipatory city*, London: Sage, pp. 178–191.

Goldman, M. (2001) Constructing an environmental state: eco-governmentality and other transnational practices of a "green" World Bank. *Social Problems* 48, pp. 499–523.

Goldman, M. (2004) Eco-governmentality and other transnational practices of a "green" World Bank. In R. Peet and M. Watts (eds) *Liberation ecologies: environment, development, social movements.* London: Routledge, pp. 166–192.

Goldman, M. (2005) *Imperial nature: the World Bank and struggles for social justice in the age of globalization.* New Haven: Yale University Press.

Guthman, J. (2008) Neoliberalism and the making of food politics in California. *Geoforum* 39, pp. 1171–1183.

Guthman, J. (2011) *Weighing in: obesity, food justice, and the limits of capitalism.* Berkeley: University of California Press.

Guthman, J. (2012) Doing justice to bodies? Reflections on food justice, race, and biology. *Antipode* 36, pp. 1153–1171.

Haraway, D.J. (2008) *When species meet.* Minneapolis: University of Minnesota Press

Hecht, S.B. and Cockburn, A. (2010) *The fate of the forest: developers, destroyers, and defenders of the Amazon.* Chicago: University of Chicago Press.

Huntington, O.H. and Watson, A. (2012) Interdisciplinarity, native resilience, and how the riddles can teach wildlife law in an era of rapid climate change. *Wicazo Sa Review* 27, pp. 49–73.

Kosek, J. (2006) *Understories: the political life of forests in northern New Mexico.* Durham, NC: Duke University Press.

Leach, M. and Fairhead, J. (2000) Fashioned forest pasts, occluded histories? International environmental analysis in West African locales. *Development and Change* 31, pp. 35–59.

Legg, S. (2005) Foucault's population geographies: classifications, biopolitics and governmental spaces. *Population, Space and Place* 11, pp. 137–156.

Lemke, T. (2002) Foucault, governmentality, and critique. *Rethinking Marxism* 14, pp. 49–64.

Li, T.M. (2007a) Governmentality. *Anthropologica* 49, pp. 275–281.

Li, T.M. (2007b) *The will to improve: governmentality, development, and the practice of politics.* Durham, NC: Duke University Press.

Lowe, C. (2006) *Wild profusion: biodiversity conservation in an Indonesian archipelago.* Princeton: Princeton University Press.

Luke, T.W. (1995) On environmentality: geo-power and eco-knowledge in the discourses of contemporary environmentalism. *Cultural Critique* 31, pp. 57–81.

Mann, G. (2009) Should political ecology be Marxist? A case for Gramsci's historical materialism. *Geoforum* 40, pp. 335–344.

Mathews, AS. (2011) *Instituting nature: authority, expertise, and power in Mexican forests.* Cambridge, MA: MIT Press.

Mitchell, T. (2002) *Rule of experts: Egypt, techno-politics, modernity.* Berkeley: University of California Press.

Moore, D.S. (2005) *Suffering for territory: race, place, and power in Zimbabwe.* Durham, NC: Duke University Press.

Moore, D.S., Kosek, J. and Pandian, A. (2003) *Race, nature, and the politics of difference.* Durham, NC: Duke University Press.

Neumann, R.P. (1998) *Imposing wilderness: struggles over livelihood and nature preservation in Africa.* Berkeley: University of California Press.

Ogden, L.A. (2011) *Swamplife: people, gators, and mangroves entangled in the Everglades.* Minneapolis: University of Minnesota Press.

Peluso, N.L. (1992) *Rich forests, poor people-resource control and resistance in Java.* Berkeley: University of California Press.

Peluso, N.L. and Vandergeest, P. (2001) Genealogies of the political forest and customary rights in Indonesia, Malaysia, and Thailand. *The Journal of Asian Studies* 60, pp. 761–812.

Peluso, N.L. and Vandergeest, P. (2011) Political ecologies of war and forests: counterinsurgencies and the making of national natures. *Annals of the Association of American Geographers* 101, pp. 587–608.

Poster, M. (1984) *Foucault, Marxism and history: mode of production versus mode of information.* Cambridge: Polity Press.

Povinelli, E.A. (2011) *Economies of abandonment: social belonging and endurance in late liberalism.* Durham, NC: Duke University Press.

Povinelli, E.A. (2012) The will to be otherwise/the effort of endurance. *South Atlantic Quarterly* 111, pp. 453–475.

Povinelli, E.A. (2013) The social projects of late liberalism. *Dialogues in Human Geography* 3, pp. 236–239.

Raffles, H. (2002) *In Amazonia: a natural history.* Princeton: Princeton University Press.

Robbins, P. (2011) *Political ecology: a critical introduction.* Malden: Blackwell.

Rose, N. (1999) *Powers of freedom: reframing political thought.* New York: Cambridge University Press.

Rutherford, S. (2007) Green governmentality: insights and opportunities in the study of nature's rule. *Progress in Human Geography* 31, pp. 291–307.

Sheridan, A. and Foucault, M. (1980) *The will to truth*. London: Tavistock Publications.
Singh, N.M. (2013) The affective labor of growing forests and the becoming of environmental subjects: rethinking environmentality in Odisha, India. *Geoforum* 47, pp. 189–198.
Sivaramakrishnan, K. (1999) *Modern forests*. Stanford: Stanford University Press.
Spivak, G.C. (2010) Can the subaltern speak? In *Can the subaltern speak? Reflections on the history of an idea*. New York: Columbia University Press, pp. 237–292.
Sundberg, J. (2011) Diabolic Caminos in the desert and cat fights on the Rio: a posthumanist political ecology of boundary enforcement in the United States–Mexico borderlands. *Annals of the Association of American Geographers* 101, pp. 318–336.
Sundberg, J. (2013) Decolonizing posthumanist geographies. *Cultural Geographies* 21, pp. 33–47.
Valdivia, G. (2008) Governing relations between people and things: citizenship, territory, and the political economy of petroleum in Ecuador. *Political Geography* 27, pp. 456–477.
Watts, M. (1983) On the poverty of theory: natural hazards research in context. In *Interpretation of calamity: from the viewpoint of human ecology*. Boston: Allen & Unwin, pp. 231–262.
Watts, M.J. (2004) Antinomies of community: some thoughts on geography, resources and empire. *Transactions of the Institute of British Geographers* 29, pp. 195–216.
Watts, M. and Peet, R. (1996) Towards a theory of liberation ecology. In *Liberation ecologies: environment, development and social movements*. London: Routledge, pp. 260–269.
Yeh, E.T. (2005) Green governmentality and pastoralism in western China: "converting pastures to grasslands." *Nomadic Peoples* 9, pp. 1–2.

37

ENERGY AND SOCIAL POWER

From political ecology to the ecology of politics

Matt Huber

Introduction

There is perhaps no better example of the inescapably political nature of nature–society relations than energy. The history of coal mining is one of working-class and union struggle over working conditions, wages, and benefits. The Organization of Petroleum Exporting Countries has the express purpose of shaping the price of oil for political aims. There is public anger toward the power of "Big Oil" corporations and their capacity to gouge consumers with high prices and despoil the environment. And electricity markets are deemed a "natural monopoly" in ways that often justify hybrid public–private regulated electric utilities. Thus, if political ecology is an attempt to correct the proliferation of "apolitical ecologies" within popular and policy discourse (Robbins 2011: 7), then energy presents an important case. Yet, there is considerable diversity in how "energy" has been understood and examined in the history (and pre-history) of political ecology. This chapter examines the role of energy in the development of political ecology and maps out some new directions.

In the first section I examine the roots of political ecology, through specific kinds of cultural ecological approaches (mainly in anthropology) that used systems ecology concepts of "energy" as a foundation for understanding human–environment relations. Second, I explain that while political ecology developed, in part, as a critique of the static and ahistorical concepts of equilibrium and adaptation at the heart of these "energetic" analyses of society, it subsequently lost its *foundational* focus on energy (and, in this respect, I will argue lost something important). In political ecology's "boom" period in the 1980s and 1990s, energy became "just another resource" prone to local conflict and struggle in the context of broader scalar forces of state policy and global capital. Third, as political ecology has broadened its geographical lens to include the First World and urban spaces, it has been forced to confront ecologies beyond extraction and examine the central role of energy in the operation of various kinds of critical infrastructures – pipelines, electricity grids, transportation networks – as other kinds of public "resources" that are objects of struggle over access and control. Finally, I argue energy has begun to be approached (mainly from scholars outside of "political ecology") as a material basis of politics more broadly – underpinning such powerful concepts as modernity, democracy, and freedom. A focus on energy (and other resources) and the "ecology of politics" perhaps opens up fresh new terrain for research in political ecology. Oddly enough, such a focus would bring us back to a focus on the *centrality* of energy flows that was a hallmark of early cultural ecology. I argue it is important to return to analyses that consider the foundational role of energy (when compared with "resources"), but such a focus comes with analytical dangers.

Energy in cultural ecology

With roots in the nineteenth-century ideas of Darwinian evolution (Stauffer 1957), the science of ecology envisions biological life as an intricate web of relationships and flows of matter and energy. By the 1970s, in a cultural context of increasing concern and knowledge of the importance of energy to modern industrial life, the work of Howard T. Odum (1971) placed "energy" at the core of an analysis that sought to measure the flows of energy through ecological systems. Odum's "systems ecology" approach was based on the simple premise that "the flow of radiant energy from the sun forms the foundation of most life in its familiar forms, so the capture of energy by the process of photosynthesis and the subsequent fate of the chemical energy thus 'fixed' is of considerable importance" (Simmons 1989: 10). Even more important for this perspective was the capacity to measure "energy" and plug it into complex quantitative models that were increasingly seen as a "scientific" pathway to examining and predicting diverse kinds of human and natural systems.

While the "systems ecology" school initially continued the ecological sciences' larger neglect of *human* interactions with ecological systems, by the early 1960s "cultural ecologists" began to apply similar insights to the study of cultural interactions with local environments. Although many scholars had recognized the centrality of energy to cultural history (White 1943; Cottrell 1955), the capacity to measure the flows of calories and nutrients through complex systems allowed for a "scientific" approach to measuring and classifying *different* kinds of local human-ecological systems (see Steward 1972). A classic example is Rappaport's (1968) study of war and pig sacrifice rituals amongst cultures in Papua New Guinea as functional responses to an increasing population and its draw upon local energy and nutrient resources.

As the energy crisis of the 1970s intensified, the energetic focus of systems ecology and cultural ecology grew in stature (Cook 1971; Rappaport 1971; Odum 1971; Odend'hal 1972). The wider "shock" of the crisis itself revealed that much of industrialized society (including the academy) had taken for granted the energetic basis of modern life. This led to a proliferation of scholarship that bordered on "energy reductionism" – including "energy theories of value" (Costanza 1980) and the reduction of political and social power in capitalist societies to energy flows (Adams 1975).

Perhaps overreaching, the rise of energetic approaches generated a critique in ways that directly led to the emergence of political ecology. Some critiqued the "calorific obsession" of cultural ecology (Brookfield 1972: 46), while others rejected the entire conceptual foundations of systems ecology based not only in energy flows, but also concepts of equilibrium, homeostasis, and adaptation (Hallpike 1973; Trimbur and Watts 1976; Bargatsky 1984). More importantly, an important critique emerged of the "system" or boundary-unit definition in ecological approaches. Ultimately such ecological approaches focused on the local/regional scale where energy and nutrient flows could be adequately measured, an approach that reflected the way initial studies addressed small-scale agricultural or even hunter-gatherer societies. Yet increasingly political ecologists argued that even peripheral, rural resource communities are profoundly entwined within broader forces of state policies and global markets (Wolf 1972; Watts 1983; Blaikie 1985; Blaikie and Brookfield 1987). Moreover, any local measures of energy or nutrient flows also could ignore the deep, and not always observable, *historical* processes of colonialism and dispossession that often constrained rural communities' livelihood strategies (e.g., Wolf

1982). All these critiques (including a quite separate critique of the risk/hazards school of geography – see Chapter 3, this volume), coalesced around the emergent field of political ecology. Yet, insofar as political ecologists examined energy at all, they approached it from quite a different perspective when compared with the cultural ecologists of their roots.

Just another resource? Energy, extraction, and power

In his monumental history of oil, Daniel Yergin (1991: 697–725) suggests that during the 1980s and 1990s oil became "just another commodity." He argues that the centralized forces that had shaped the price of oil during much of the twentieth century – from the Seven Sisters international oil companies to the Organization of Petroleum Exporting Countries (OPEC) – had given way to the power of markets – particularly financial oil futures traders on the New York Mercantile Exchange (among other sites). It could be argued that in political ecology, oil and other energy sources became "just another resource."

Political ecology's engagement with nature–society relations hinged upon the category of "resource." An often cited definition is provided by Peluso and Watts (2001: 24–25): "Political ecology provides tools for thinking about the conflicts and struggles engendered by the forms of access to and control over resources." Insofar as most political ecological work was about *place-based* investigations of resource *extraction*, several authors turned their focus to the geographies of energy extraction. To the extent that political ecology emerged to examine so-called "peripheral" resource geographies, it is important to point out that the energy geographies of such landscapes are not predominantly based on industrial fossil fuel combustion. Across the developing world, access to wood as fuel is an important object of struggle between local communities, the state, and various aspects of the timber commodity trade. Ribot (1990) examines how charcoal production in rural Senegal is shaped by extra-local political and economic forces in urban areas and beyond. Like Blaikie (1985) on soil erosion, a lot of early "classic" political ecological work on forests and fuelwood tended to critique more mainstream development narratives that blamed local peasants for deforestation (see, e.g., Bunker 1985; Hecht and Cockburn 1989; Peluso 1992).

While most people rely on "organic" sources of energy like wood and food, the developing world is also a central area for the extraction of fossil fuel energies. Michael Watts's (2004, 2008, 2012) extensive canon on the political ecology of oil focused attention on the troubling confluence of environmental devastation, human rights abuses, and corporate and state impunity in the Niger Delta. Such research served to make visible the devastating consequences of oil extraction on local communities. Moreover, Watts was also keen to trace specific linkages between flows of petroleum, money, and the constitution of state power (Watts 2004). Overall, he presents an uneven geography of environmental devastation and tremendous wealth generation that he calls "petrocapitalism" – a specific constellation of power between resource owning states and oil capital and, of course, local social movements and resistance.

As Bebbington (2012) has recently pointed out, with oil, gas, and other mining geographies becoming increasingly central to national and geopolitical projects, "underground political ecologies" are taking on increased significance. A decade of high energy prices, intensified debates over "peak oil" and "oil wars" have led to a resurgence of interest in uncovering the *specific* political ecologies of energy extraction. Such critical examination of energy extraction has often "scaled up" to look at wider geographies of geopolitical conflict (Le Billon and El Khatib 2004; Bradshaw 2010; Labban 2011; Johnson and Derrick 2012), imperialism (Labban 2008), or national identity (Bouzarovski and Bassin 2011). Yet, it is the *place-based* ethnographic detail that sets much political ecological work apart from large-scale "geopolitical" analysis (cf.

Klare 2012). Sawyer (2004) examines the political and legal tactics of indigenous groups in Ecuador in their struggles against Texaco (now Chevron) and its legacy of environmental destruction. Valdivia (2008) explains how struggles of settlers and oil workers in Amazonia conflict with a wider national discourse of oil as a central material ingredient in the body of the nation. Perreault (2013) shows how popular protests over natural gas extraction in Bolivia necessarily position imaginaries of national sovereignty against foreign involvement. Zalik (2011) examines the divergent forms of political protest over oil extraction in Nigeria and Alberta. Precisely at a moment where climate change is proving continued fossil fuel extraction untenable, we are also witnessing a massive global boom in unconventional fossil fuel extraction (e.g., shale gas/oil, Tar Sands, Deepwater and Arctic offshore oil). Such controversial and exceptionally destructive environmental geographies are already proving fertile ground for political ecologies of extraction (Kinne et al. 2014; Haluza-DeLay and Carter 2014; Willow and Wylie 2014).

Although much work is focused on the "holes" of fossil fuel extraction (Bridge 2010), perhaps even more political ecological work has been focused on sites of electricity generation – from nuclear power stations (Daultrey 1980) to hydropower (Bakker 1999; Desbiens 2004; Swyngedouw 2007; Sneddon and Fox 2008; Webber 2012). Such large-scale projects often *connect* to struggles over other resources like water and forests that must be destroyed to make way for electric power infrastructure. Furthermore, if political ecology is focused on *conflict* then one cannot ignore sizable struggles over *renewable* energy production. For example, there are striking resemblances between local/community resistance to wind energy and resistance to conventional fossil fuel extraction (cf. Kraus 2010; Pasqualetti 2011; Phadke 2011). Mulvaney (2013) examines environmental justice and occupational health concerns along the solar photovoltaic commodity chain. Over the last few years, agrofuel development has led to massive displacement of communities (Van der Horst and Evans 2010; Bailis and Baka 2011) and provided new regimes of accumulation for agribusiness (Gillon 2010).

In sum, all this work has been extremely important for uncovering the geographies of dispossession and conflict surrounding sites of energy extraction and generation. Yet, while Watts's (2004) work calls attention to the "mythic" qualities of oil's capacity for wealth generation, there was not much sense that "energy" should be analyzed any differently from any other resource (e.g., hard rock minerals, forests, land, water, etc.). Insofar as it literally fuels all life (all "ecology") from plants to animals – and eventually, machines – energy is foundational in a way that "resources" like aluminum are not.[1] Moreover, critiques of political ecology mounted to claim that political ecologists did not take "ecology" or the biophysical aspects of resources seriously enough (Vayda and Walters 1999; Walker 2005). It could be argued that the response to this critique was increased attention to the "materiality" of resources (Bakker and Bridge 2006) or the "difference that nature makes" (Boyd et al. 2001: 557). For example, Bridge (2003) provocatively argues that we cannot understand the political ecology of natural gas without examining the material constraints of the resource itself (e.g., the difficulty of ocean transport). Yet, there was not a return to the *ecological* notion – rooted in Odum et al. – that energy should be a foundational category of analysis in political ecological studies. There was, however, a different kind of shift in empirical directions to consider "First World" and "urban" geographies. This expanded the empirical possibilities for political ecologies of energy.

Energized infrastructures and urban political ecology

While "political ecology" grew up in the 1980s and 1990s focused on struggles over resource or land access and control in rural "Third World" contexts (e.g., Baily and Bryant 1997), by the 2000s, political ecology was increasingly applied to urban and First World contexts (e.g., McCarthy 2005; Heynen et al. 2006; Schroeder et al. 2006; Kennedy et al. 2011). Most significantly from an energy perspective, the development of urban political ecology (UPE) insisted the process of urbanization must be seen as simultaneously *ecological and social* (see Chapter 47, this volume). In their seminal paper, Swyngedouw and Heynen (2003: 899) assert:

> cities are dense networks of interwoven sociospatial processes that are simultaneously local and global, human and physical, cultural and organic. The myriad transformations and metabolisms that support and sustain urban life – such as … water, food, and computers – always combine physical *and* social processes as infinitely interconnected.

By way of example, Swyngedouw and Heynen (2003: 899) discuss a quintessential urban landscape: Piccadilly Circus in London and its lights, observing how "the neon lights are fed by energy coming from nuclear power plants and from coal or gas fired electricity generators." Beyond illustrating how even the most mundane aspects of urban life are inextricably tied to energy consumption, the UPE framework positions urbanization as an *energetic* process. At the heart of urban energy metabolism are particular social infrastructures for the acquisition, conversion, and distribution of energy – transportation networks, electricity grids, and gas distribution (Monstadt 2009). Nearly every aspect of urban life involves the metabolism of energy to sustain everyday practices that are simultaneously shaped by social and political forces – computing power in financial markets, diesel powered garbage trucks, and heating in public housing projects.

What is striking about the UPE literature, however, is that research has not often focused on energy *specifically*. For example, a major edited volume (Heynen et al. 2006) did not include a single case study on energy issues, but contained four on water and others on hunger, lawns, and wider political struggles over environmental justice. If "metabolism" is the key concept of UPE, it did not seem to matter whether the material being metabolized was energy or other forms of materials and the wastes they produce. Perhaps the problem is that energy systems often undergird *other* kinds of urban infrastructures in ways that get taken for granted. For example, what is a water distribution system without diesel powered pumping stations, or electricity powered monitoring systems (see Kaika 2005: 27–50)? What is a transportation network without petroleum-based fuels (gasoline and diesel) powering the cars, trucks, and buses that traverse it? Moreover, political struggles often focus on the material form of the infrastructure rather than the energy that makes it possible. For example, in Harlem in New York City there has been substantial political mobilization over the location of six of eight city (diesel-powered) bus terminals north of 96th St because of major local air quality problems (Hess 2007).

Although their work does not fly under the UPE flag, many have examined the politicized nature of urban (and rural) energy infrastructures. Solomon and Heiman (2004) show how policies of neoliberal "deregulation" that gripped virtually every sector of the economy – from banking to telecommunications – were also present in electric utility restructuring. Although deregulation of utilities was dressed up in the buzzwords of "green," "community," and "renewable," the result was less public and state involvement in a sector known for its "natural monopoly" status and centrality to the public interest. Further afield, Ahmed (2010) demonstrates how the Washington Consensus of liberalized markets took hold in India – only to allow the now notorious Enron to construct a shell electric grid of dubious viability and state-guaranteed profits. Harrison (2013) argues that the early twentieth-century development of electric utility regions in North Carolina is a prime example of the uneven development of capitalism and can still explain what some call "energy poverty" today. Fennell (2011) explains how residents experienced neoliberal reforms to public housing in Chicago through the "sensory politics" of heat. Henderson (2011) examines the contradictions between the Gulf of Mexico oil spill – and the wider societal desire to lessen oil dependence – and the auto-centric equations of "freedom" with driving in suburban southern geographies. Although lessening oil consumption makes sense in the abstract, in the US south – a region of widespread suburban sprawl – efforts to curtail driving were met with hostility.

Energy and nature are visibly constitutive of urban spaces, but urban environmental politics cannot be contained within cities. In an age of climate change and with over 50 percent of the human population living in cities (Davis 2010), it is increasingly important to *politicize* the energetic metabolism of cities and its *multiscalar effects* (cf. Bulkeley 2013). This means calling attention to the often invisible energy networks that make urban life possible – and spew carbon dioxide, soot, and other local pollutants in the process. For example, perhaps because energy extraction and capture frequently exist outside the metropolis, political ecologies of energy often do not discuss the *built environment itself*. Imagine the city without concrete and the cement industry (cf. Gandy 2002). Fry (2013) examines the political ecology of cement production in Mexico – a sector that is estimated to be responsible for 5–7 percent of global carbon emissions (coming not from fuel combustion, but a chemical reaction in the production process itself). Much overt politicization of the cement industry has focused on the point of production – specifically energy efficiency techniques and carbon sequestration as the basis of an overall cap and trade or carbon tax regime. However, Fry reveals that such measures would do nothing about rapidly rising demand for cement. The majority of this demand (over 50 percent) is based in housing construction. Of course, not all of this housing is constructed in cities, but it is safe to assume the majority will occur in urban and suburban metropolitan areas. Thus, by linking cement production to specific urban geographies, Fry raises wider political questions about urban political ecologies.

Efforts to make cities carbon neutral cannot only focus on the energy consumed in cities (Brown and Southworth 2008), but must also confront the wider geographies of energy consumption that make the built environment of the city possible. In the case of cement, it means focusing on alternative, less carbon-intensive building materials. Yet, as Ross (2011) points out in the case of Phoenix, "green" buildings are too often the self-satisfied domain of the rich while the poor are forced to live in areas susceptible to all kinds of environmental risks. As poor people's movements focus on "the right to the city" (Harvey 2012), this "right" can include wider access to a built environment that answers the challenges of climate change and sustainability. Indeed, new approaches to "energy justice" and "energy poverty" reveal stark inequalities when it comes to access to energy in both urban and rural environments (Walker and Day 2012; Hall et al. 2013).

As one explores the energetic metabolism of cities, it is quickly apparent that cities are *always already* energetic entities. In other words, from a dialectical perspective common to political ecology (cf. Harvey 1996), there is no thing or moment within the city that is not ultimately dependent upon relations to energy flows. Moreover, while one can find specific political struggles *over energy* in the city (over access to electricity or heat for example), struggles over things that appear disconnected from energy (urban gardens, highways) are *also* underpinned by energy relations. If energy underlies even our ideas of what constitutes "the urban," what does this mean for our conception of politics?

From political ecology to the ecology of politics

In traditional political ecology there is usually an empirical object that stands in for "nature" or "the environment" (e.g., mineral resources, forests, land, and water). Thus, some slice of empirical reality is quarantined as "nature" and becomes the *object* of political struggle and contestation. The task of political ecology then becomes explanation of the wider historical, social, and biophysical forces that shape such struggles. Because political ecology is tethered to those empirical objects we flag as "nature," research trajectories in political ecology have remained rooted within the Western constructions of "nature" that we are supposedly trying to escape. Why, for example, are there not political ecological studies of apparently denaturalized spaces such as a Wal-Mart? Has political ecology really taken seriously David Harvey's (1996: 186) infamous claim, "There is nothing unnatural about New York City"? I would argue that even UPE approaches often only focus upon naturalized aspects of the urban form (e.g., parks, gardens, hydrological systems).

Apart from empirical "spaces" we flag as nature or non-nature, a deeper question at the core of political ecology is implied in its nomenclature: what is the relationship between politics and ecology? What if we were to concede that our wider notions of politics are *always already ecological*? This corresponds to Jason Moore's (2011) recent writings on the relationship between capitalism and ecology. As opposed to thinking about capitalism as a fully social process that exploits or degrades some external realm called nature or the environment, Moore (2011: 34) contends, "Capitalism doesn't have an ecological regime; it is an ecological regime." This entails moving from studying the politics of nature-ecology-environment-resources toward understanding *the ecology of politics*. For example, not many would classify the US Tea Party as an "environmental social movement." Yet the popular forces of Tea Party politics come out of their own *lived ecologies* – rural and suburban landscapes, commodified access to food and energy and other material aspects of social reproduction, and, perhaps, an avid appreciation for outdoor (gun-based) recreation. Thus, an ecological analysis of Tea Party politics must examine the *socioecological relations* underpinning their specific political subjectivities – neoliberal ideologies of freedom, competition, and entrepreneurship and hatred of government and taxes.

It is not difficult to understand how this "ecology of politics" approach could be applied to energy. In the case of energy resources, it is clear that they are highly politicized objects that become objects of struggles over control (e.g., property relations) and meaning (e.g., nationhood). Yet, like cities, broader political subjectivities are also tied to the widespread consumption of energy resources. Even if not strictly classified as "political ecology," this movement has already begun in related fields. My own work examines oil and suburbanization as a material/ecological basis of neoliberal ideologies of "entrepreneurial life" (Huber 2013). John Urry (2013) has called for a "sociology of energy" in relation to automobility and climate change in particular (see also Paterson 2007). Timothy Mitchell (2011) suggests that we need to begin to consider current ideas of democracy (certainly an example of a wider view of politics) as inextricably tied to the industrial use of carbon-based fuels and the forms of sociality they made possible. In a call for anthropology to consider what he calls "energopolitics," Dominic Boyer (2011: 5) advocates studying not only conflicts over energy but also "energy as *the* undercurrent and integrating force for all other modes and institutions of modern power." For example, Boyer (2011: 5) suggests Foucault's (1977) classic analysis of penal institutions (prisons, schools, factories) and disciplinary power cannot be severed from energy relations: "[W]here would these exemplary modern institutions and their forms of expertise be without the harnessing and transformation of energy into their lighting and electricity, into their heat, even into their bricks and cement[?]" In environmental history – a parallel and complementary field to political ecology – many scholars have begun to understand history as the development of specific energy regimes each with their wider social, political, and environmental effects (McNeill 2000; Crosby 2006; Burke 2009).

For the most part these approaches see energy as "foundational" but *mediated* by history, culture, and politics. But, a "foundational" approach is fraught with analytical danger – many of the same dangers that afflicted early cultural ecology. A focus on energy as the basis of particular forms of politics not only runs the risk of what Boyer calls (2011: 5) a "naïve materialism," but also outright forms of energy reductionism. Environmental historians Russell et al. (2011) overreach by claiming, "Our thesis is that all power, social as well as physical, derives from energy." I would argue Mitchell (2011) overly relies on the energy transition between coal and oil to explain political declines in working-class power. Certainly the labor intensive nature of coal mining (when compared with oil extraction) mattered for the strategic capacity of workers to disrupt energy flows (and thus modern society), but the decline of unions and working-class solidarity is a much broader story than can be isolated in the mines or through energy flows (e.g., the rise of the neoliberal thought collective and the demonization of unions as monopolies).

While an analysis of energy and politics is much needed, we also cannot abandon "classic" examinations of the politics of energy. Political ecological researchers will still go to sites of energy extraction to understand the place-specific struggles and conflicts associated with the too often devastating environmental impacts and social dislocation of these activities. Yet, a focus on the ecology of politics not only suggests an important analytical shift in the focus of political ecology, but a methodological shift as well. Classic political ecology often relies on place-based field work where the "nature" to be examined is clear – a forest reserve, a plot of land, a mine. An "ecology of politics" approach may be tied to a particular place or region but, and this is the important point, it need not necessarily be so. To return to the example of the Tea Party, one could visit a local stranglehold of Tea Party politics – such as Petoskey, Michigan (This American Life 2010) – but the Tea Party is also an extra-local social movement that can be analyzed at a wider scale. The question is whether political ecology need necessarily be field work-based, local research focused on a specific "case study"? If political ecology is to answer critiques of fetishizing "the local" (Brown and Purcell 2005; Morris 2013) and failing to make generalizable analytical claims (Castree 2008), it is perhaps time to diversify the scale of research and analysis.

Conclusion

Ironically, the move from political ecology to the ecology of politics potentially brings us back full circle to early cultural ecological approaches to energy. Energy is once again being examined as a kind of foundation for nature–society relations. Again, the real question is what makes energy different from other resources and ecologies in tackling questions of political power? I believe we can retain early political ecology's focus on energy as a foundational resource underpinning food systems and more industrial geographies of electricity and transportation. Energy is different from and more all-encompassing than say copper or tin (i.e., "resources" in general). But, the dangers are threefold. First, any approach to understanding the centrality of energy to wider society and politics must avoid cultural ecology's penchant for "calorific obsession" – the reduction of particular cultural forms to expressions of energy flows. Such an obsession creates an ahistorical approach to energy as a fixed set of "thermodynamic laws," rather than a properly "political ecological" view that sees energy more as a "social relation" mediated by historically specific political struggles. Second, energy cannot be seen as determining any particular form of politics, but rather as laying the material conditions of possibility for manifold kinds of political forms and struggles: to put it crudely, energy creates historical possibilities, but people make history. Third, if the current energy boom in unconventional oil and gas production in the United States is as transformative as some suggest (Energy Information Administration 2013), there might be a return to stable and low energy prices. Thus, as happened in the 1980s and 1990s, there is a danger that the centrality of energy may recede from view and become "just another resource." Regardless of the price of energy, its importance to intractable environmental problems from local groundwater contamination to global climate change is not going away. As the necessity of an "energy transition" away from fossil fuels intensifies (Bridge et al. 2013), political ecology is well-equipped to examine the struggles that will shape our energy future.

Note

1 Of course, one must point out the energy as an abstract concept only emerged in the nineteenth century to explain similarities between things as varied as coal, wood, corn, and human muscles. Even so, this social construction is a powerful one that cuts to the heart of nature–society relations.

References

Adams, R.N. (1975) *Energy and Structure: A Theory of Social Power.* Austin: University of Texas Press.

Ahmed, W. (2010) Neoliberalism, corporations, and power: Enron in India. *Annals of the Association of American Geographers* 100(3), pp. 621–639.

Bailis, R. and Baka, J. (2011). Constructing sustainable biofuels: governance of the emerging biofuel economy. *Annals of the Association of American Geographers* 101(4), pp. 827–838.

Baily, S. and Bryant, R. (1997). *Third World Political Ecology: An Introduction.* London: Routledge.

Bakker, K. (1999) The politics of hydropower: developing the Mekong. *Political Geography* 18(2), pp. 209–232.

Bakker, K. and Bridge, G. (2006) Material worlds? Resource geographies and the "matter of nature." *Progress in Human Geography* 30(10), pp. 5–27.

Bargatsky, T (1984) Culture, environment, and the ills of adaptationism. *Current Anthropology* 25, pp. 399–415.

Bebbington, A. (2012) Underground political ecologies. *Geoforum* 43(6), pp. 1152–1162.

Blaikie, P. (1985) *The Political Economy of Soil Erosion in Developing Countries.* London: Longman.

Blaikie, P. and Brookfield, H. (1987) *Land Degradation and Society.* New York: Routledge.

Bouzarovski, S. and Bassin, M. (2011) Energy and identity: imagining Russia as a hydrocarbon superpower. *Annals of the Association of American Geographers* 101(4), pp. 783–794.

Boyd, W., Prudham, W.S. and Schurman, R. (2001) "Industrial dynamics and the problem of nature." *Society and Natural Resources* 14, pp. 555–570.

Boyer, D. (2011) Energopolitics and the anthropology of energy. *Anthropology News* 52(5), pp. 5–7.

Bradshaw, M.J. (2010). A new energy age in pacific Russia: lessons from the Sakhalin oil and gas projects. *Eurasian Geography and Economics* 51(3), pp. 330–359.

Bridge, G. (2003) Gas, and how to get it. *Geoforum* 35(4), pp. 395–397.

Bridge, G. (2010). The hole world: scales and spaces of energy extraction. *New Geographies 2: Landscapes of Energy*, pp. 43–51.

Bridge, G., Bouzarovski, S. Bradshaw, M. and Eyre, N. (2013) Geographies of energy transition: space, place and the low-carbon economy. *Energy Policy* 53, pp. 331–340.

Brookfield, H.C. (1972) Intensification and disintensification in Pacific agriculture: A theoretical approach. *Pacific Viewpoint* 13, pp. 30–48.

Brown, J. and Purcell, M. (2005) Against the local trap: scale and the study of environment and development. *Progress in Development Studies* 5(4), pp. 279–297.

Brown, M. and Southworth, F. (2008) Mitigating climate change through green buildings and smart growth. *Environment and Planning A* 40, pp. 653–675.

Bulkeley, H. (2013) *Cities and Climate Change.* New York: Routledge.

Bunker, S. (1985) *Underdeveloping the Amazon: Extraction, Unequal Exchange, and the Failure of the Modern State.* Chicago: University of Chicago Press.

Burke III, E. (2009) The big story: human history, energy regimes, and the environment. In *The Environment and World History*, ed. E. Burke III and K. Pomeranz. Berkeley: University of California Press, pp. 33–53.

Castree, N. (2008) Neoliberalising nature: processes, effects, and evaluations. *Environment and Planning A* 40(1), pp. 153–173.

Cook, E. (1971) The flow of energy in an industrial society. *Scientific American* 225(3), pp. 34–47.

Costanza, R. (1980) Embodied energy and economic valuation. *Science* 210, pp. 1219–1224.

Cottrell, F. (1955) *Energy and Society: The Relation Between Energy, Social Change, and Economic Development*. New York: McGraw Book Company.

Crosby, A. (2006) *Children of the Sun: A History of Humanity's Unappeasable Appetite for Energy*. New York: W.W. Norton & Co.

Daultrey, S. (1980) The political ecology of nuclear power in Ireland – or, how nuclear wastes. *Antipode* 12(1), pp. 108–111.

Davis, M. (2010) Who will build the ark? *New Left Review* 61(January–February), pp. 29–46.

Desbiens, C. (2004) Producing North and South: a political geography of hydro development in Québec. *Political Geography* 48(2), pp. 101–118.

Energy Information Administration (2013) Growing oil and natural gas production continues to reshape the U.S. energy economy. Available online at: www.eia.gov/pressroom/releases/press402.cfm.

Fennell, C. (2011) "Project heat" and sensory politics in redeveloping Chicago public housing. *Ethnography* 12 (1), pp. 40–64.

Foucault, M. (1977) *Discipline and Punish: The Birth of the Prison*. New York: Random House.

Fry, M. (2013) Cement, carbon dioxide, and the "necessity" narrative: a case study of Mexico. *Geoforum* 49, pp. 127–138.

Gandy, M. (2002) *Concrete and Clay: Reworking Nature in New York City*. Cambridge, MA: MIT Press.

Gillon, S. (2010) Fields of dreams: negotiating an ethanol agenda in the Midwest United States. *Journal of Peasant Studies* 37(4), pp. 723–748.

Hall, S.M., Hands, S. and Bulkeley, H. (2013). New approaches to energy: equity, justice and vulnerability. Introduction to the special issue. *Local Environment* 18(4), pp. 413–421.

Hallpike, C.R. (1973) Functionalist interpretations of primitive warfare. *Man* 8, pp. 451–470.

Haluza-DeLay, R. and Carter, A.V. (2014) Joining up and scaling up: Analyzing resistance to Canada's "Dirty Oil." *Cultural Studies of Science Education* 9, pp. 343–362.

Harrison, C. (2013) The historical–geographical construction of power: electricity in Eastern North Carolina. *Local Environment* 18(4), pp. 469–486.

Harvey, D. (1996) *Justice, Nature and the Geography of Difference*. Oxford: Blackwell.

Harvey, D. (2012) *Rebel Cities: From the Right to the City to the Urban Revolution*. London: Verso.

Hecht, S. and Cockburn, A. (1989) *The Fate of the Forest*. London: Verso.

Henderson, J. (2011) The politics of mobility in the south: a commentary on sprawl, automobility, and the Gulf oil spill. *The Southeastern Geographer* 51(4), pp. 641–649.

Hess, D. (2007) What is a clean bus? Object conflicts in the greening of urban transit. *Sustainability: Science, Practice, and Policy* 3(1). Available online at: http://sspp.proquest.com/archives/vol3iss1/0608-027.hess.html.

Heynen, N., Kaika, M. and Swyngedouw, E. (2006) Urban political ecology: politicizing the production of urban natures. In *In the Nature of Cities: Urban Political Ecology and the Politics of Metabolism*, ed. N. Heynen, M. Kaika and E. Swyngedouw. London: Routledge, pp. 1–20.

Huber, M.T. (2013) *Lifeblood: Oil, Freedom, and the Forces of Capital*. Minneapolis: University of Minnesota Press.

Johnson, C. and Derrick, M. (2012) A splintered heartland: Russia, Europe, and the geopolitics of networked energy infrastructure. *Geopolitics* 17(3), pp. 482–501.

Kaika, M. (2005) *City of Flows: Modernity, Nature and the City*. London: Routledge.

Kennedy, C., Pincetl, S. and Bunje, B. (2011) The study of urban metabolism and its applications to urban planning and design. *Environmental Pollution* 159(8–9), pp. 1965–1973.

Kinne, B., Finewood, M. and Yoxtheimer, D. (2014) Making critical connections through interdisciplinary analysis: exploring the impacts of Marcellus shale development. *Journal of Environmental Studies and Science* 4(1), pp. 1–6.

Klare, M. (2012) *The Race for What's Left: The Global Scramble for the World's Last Resources*. New York: Picadour.

Kraus, W. (2010) The "Dingpolitik" of wind energy in northern German landscapes: an ethnographic case study. *Landscape Research* 35(2), pp. 195–208.

Labban, M. (2008) *Space, Oil, and Capital*. London: Routledge.
Labban, M. (2011) The geopolitics of energy security and the war on terror: the case for market expansion and the militarization of global space. In *Global Political Ecology*, ed. R. Peet, P. Robbins and M. Watts. New York: Routledge, pp. 325–344.
Le Billon, P. and El Khatib, F. (2004) From free oil to "freedom oil": terrorism, war, and US geopolitics in the Persian Gulf. *Geopolitics* 9(1), pp. 109–137.
McCarthy, J. (2005) First world political ecology: directions and challenges. *Environment and Planning A* 37(6), pp. 953–958.
McNeill, J.R. (2000) *Something New Under the Sun: An Environmental History of the 20th Century*. New York: W.W. Norton.
Mitchell, T. (2011) *Carbon Democracy: Political Power in the Age of Oil*. London: Verso.
Monstadt, J. (2009) Conceptualizing the political ecology of urban infrastructures: insights from technology and urban studies. *Environment and Planning A* 41(8), pp. 1924–1942.
Moore, J. (2011) Transcending the metabolic rift: a theory of crisis in the capitalist world ecology. *Journal of Peasant Studies* 38(1), pp. 1–46.
Morris, J. (2013) The evolving localism (and neoliberalism) of urban renewable energy projects. *Culture, Agriculture, Food and Environment* 35(1), pp. 16–29.
Mulvaney, D. (2013) Opening the black box of solar energy technologies: exploring tensions between innovation and environmental justice. *Science as Culture* 22(3), pp. 214–221.
Odend'hal, S. (1972) Energetics of Indian cattle in their environment. *Human Ecology* 1, pp. 3–22.
Odum, H.T. (1971) *Environment, Power and Society*. New York: Wiley.
Pasqualetti, M.J. (2011) Opposing wind energy landscapes: a search for common cause. *Annals of the Association of American Geographers* 101(4), pp. 907–917.
Paterson, M. (2007) *Automobile Politics: Ecology and Cultural Political Economy*. New York: Cambridge University Press.
Peluso, N. (1992) *Rich Forests, Poor People: Resource Control and Resistance in Java*. Berkeley: University of California Press.
Peluso, N.L., and Watts, M. (2001) Violent environments. In *Violent Environments*, ed. N. Peluso and M. Watts. Ithaca, NY: Cornell University Press, pp. 3–38.
Perreault, T. (2013) Nature and nation: hydrocarbons, governance, and the territorial logics of "resource nationalism." In *Subterranean Struggles: New Dynamics of Mining, Oil, and Gas in Latin America*, ed. A. Bebbington and J. Bury. Austin: University of Texas, pp. 67–90.
Phadke, R. (2011) Resisting and reconciling Big Wind: middle landscape politics in the new American West. *Antipode* 43(3), pp. 754–776.
Rappaport, R.A. (1968) *Pigs for the Ancestors: Ritual in the Ecology of the New Guinea People*. New Haven: Yale University Press.
Rappaport, R.A. (1971) Nature, culture, and ecological anthropology. In *Man, Culture and Society*, ed. H.L. Shapiro. New York: Oxford University Press, pp. 237–267.
Ribot, J. (1990) *Markets, States and Environmental Policy: The Political Economy of Charcoal in Senegal*. Berkeley: University of California Press.
Robbins, P. (2011) *Political Ecology: A Critical Introduction*. 2nd Ed. Malden, MA: Wiley-Blackwell.
Ross, A. (2011) *Bird on Fire: Lessons from the World's Least Sustainable City*. New York: Oxford University Press.
Russell, E., Allison, J., Finger, T., Brown, J.K., Balogh, B. and Carlson, W.B. (2011) The nature of power: synthesizing the history of technology and environmental history. *Technology and Culture* 52, pp. 246–259.
Sawyer, S. (2004) *Crude Chronicles: Indigenous Politics, Multinational Oil, and Neoliberalism in Ecuador*. Durham, NC: Duke University Press.
Schroeder, R., St. Martin, K. and Albert, K.E. (2006) Political ecology in North America: discovering the Third World within? *Geoforum* 37(2), pp. 163–168.
Simmons, I.G. (1989) *Changing the Face of the Earth: Culture, Environment, History*. Oxford: Blackwell.
Sneddon, C. and Fox, C. (2008) Struggles over dams as struggles for justice: the world commission on dams (wcd) and anti-dam campaigns in Thailand and Mozambique. *Society and Natural Resources* 21(7), pp. 625–640.
Solomon, B. and Heiman, M. (2004) Power to the people: electric utility restructuring and the commitment to renewable energy. *Annals of the Association of American Geographers* 94(1), pp. 94–116.
Stauffer, R. (1957) Haeckel, Darwin and ecology. *The Quarterly Review of Biology* 32(2), pp. 138–144.

Steward, J. (1972) *Theory of Culture Change: The Methodology of Multilinear Evolution*. Champaign: University of Illinois Press.
Swyngedouw, E. (2007) Technonatural revolutions: the scalar politics of Franco's hydro-social dream for Spain, 1939–1975. *Transactions of the Institute of British Geographers* 32(1), pp. 9–28.
Swyngedouw, E. and Heynen, N. (2003) Urban political ecology, justice and the politics of scale. *Antipode* 35(5), pp. 898–918.
This American Life (2010) #417: this party sucks. Available online at: www.thisamericanlife.org/radio-archives/episode/417/this-party-sucks.
Trimbur, T.J. and Watts, M. (1976) Are cultural ecologists well adapted? A review of the concept of adaptation. *Proceedings of the Association of American Geographers* 8, pp. 179–183.
Urry, J. (2013) *Societies beyond Oil: Oil Dregs and Social Futures*. London: Zed.
Valdivia, G. (2008) Governing relations between people and things: citizenship, territory, and the political economy of petroleum in Ecuador. *Political Geography* 27(4), pp. 456–477.
Van der Horst, D. and Evans, J. (2010) Carbon claims and energy landscapes: exploring the political ecology of biomass. *Landscape Research* 35(2), pp. 173–193.
Vayda, A.P. and Walters, B.P. (1999) Against political ecology. *Human Ecology* 27(1), pp. 167–179.
Walker, G. and Day, R. (2012) Fuel poverty as injustice: integrating distribution, recognition and procedure in the struggle for affordable warmth. *Energy Policy* 49, pp. 69–75.
Walker, P. (2005) Political ecology: where's the ecology? *Progress in Human Geography* 29(1), pp. 73–82.
Watts, M. (1983) On the poverty of theory: natural hazards reviewed in context. In *Interpretations of Calamity: From the Viewpoint of Human Ecology*, ed. K. Hewitt. Boston: Allen Unwin, pp. 231–262.
Watts, M. (2004) Oil as money: the devil's excrement and the spectacle of black gold. In *Reading Economic Geography*, ed. T.J. Barnes, J. Peck, E. Sheppard and A. Tickell. Oxford: Blackwell, pp. 205–219.
Watts, M. (2008) *Curse of the Black Gold: 50 Years of Oil in the Niger Delta*. New York: Powerhouse.
Watts, M. (2012) A tale of two gulfs: life, death, and dispossession along two oil frontiers. *American Quarterly* 64(3), pp. 437–467.
Webber, M. (2012) The political economy of the Three Gorges Project. *Geographical Research* 50(2), pp. 154–165.
White, L. (1943) Energy and the evolution of culture. *American Anthropologist* 45(3), pp. 335–356.
Willow, A. and Wylie, S. (2014) Politics, ecology, and the new anthropology of energy: exploring the emerging frontiers of hydraulic fracking. *Journal of Political Ecology* 21(12), pp. 222–236.
Wolf, E. (1972) Ownership and political ecology. *Anthropological Quarterly* 45(3), pp. 201–205.
Wolf, E. (1982) *Europe and the People without History*. Berkeley: University of California Press.
Yergin, D. (1991) *The Prize: The Epic Quest for Oil, Money, and Power*. New York: Simon & Schuster.
Zalik, Anna (2011) Protest as violence in oilfields: the contested representation of profiteering in two extractive sites. In *Accumulating Insecurity*, ed. S. Feldman, C. Geisler and G. Menon. Athens, GA: University of Georgia Press, pp. 261–284.

38

FROM BIODIVERSITY TO BIOSECURITY

Celia Lowe

What is biosecurity?

Biosecurity is a contemporary rubric describing recent political and technical practices of securing life. Biosecurity positions biological materials in relation to risk and threat, and describes new sets of norms and programmatic responses to the hazards posed by and to life. Examples of recent concerns framed as problems of biosecurity include: (1) in Australia, the company Biosecurity Queensland fights off an invasive South American plant, *Miconia calvescens*, with drone helicopters; (2) in the UK, a survey finds Campylobacter on 59 percent of store-bought chicken meat; and (3) in the US, the government temporarily closes federally funded laboratories to investigate breaches of biosecurity after several laboratory safety incidents involving strains of smallpox, anthrax, and influenza. Biosecurity has been imagined and articulated most prominently in relation to issues of bioterror and biodefense, novel disease emergence, food safety, and invasive species.

The hyper-mobility of people, plants, animals, and microbes across borders is one empirical context for the emergence of biosecurity, as are specific events like the 2001 World Trade Center bombings, the release of Anthrax through the mail in that same year, and the emergence of Severe Acute Respiratory Syndrome (SARS) and H5N1 avian influenza. The genealogy of securitization has been traced to earlier moments in time, however, most specifically to the end of the Cold War (Peluso and Watts 2001) and the birth of the genome sciences (Collier et al. 2004).

The emergence of biosecurity reflects a securitization of the social where issues that, in other contexts, have been viewed as problems of public responsibility, care, or sovereignty are now filtered through the lens of security: natural resource management becomes "environmental security"; public health is reworked as "health security"; food sovereignty is set in opposition to "food security." The lens of security contrasts with other epistemic and ontological possibilities: Foucault's concept of pastoral power, in which care for life was modeled upon the care of a shepherd for her flock or a priest for his congregation, is but one example. Gilligan's (1982) feminist concept of an "ethic of care," which attends to specificities of context, human difference, and interdependence is another.

Anthropologist Carlo Caduff (2010) illustrates the shift from public health to health security in his examination of pandemic influenza vaccination campaigns. What was once a public health intervention intended to reduce morbidity and mortality changed in the 2000s to emphasize preserving essential services and ensuring that society would continue to function. With this change, the list of those who would first receive pandemic influenza vaccine shifted from the most medically vulnerable (health workers, the young and old, HIV+ patients, etc.) to those in "banking and finance, chemical, commercial facilities, communications, electricity, emergency services, food and agriculture, health care, information technology, nuclear, oil and natural gas, postal and shipping, transportation, water and waterwaste." Caduff's work illustrates the shifting terrain of efforts to secure life, and points to who wins and who loses in the biosecurity milieu.

Biosecurity also names the field of analysis where the securitization of life is studied. Collier et al. (2004), in a seminal essay on biosecurity, explore how security is being constituted as an object of thought, practice, and intervention in the emerging US biosecurity apparatus. In another early essay, Braun (2007) defines biosecurity as "a set of political technologies that seek to *govern biological disorder*, in the name of a *particular community*, through acts that are *extraterritorial*." Lakoff (2008) observes that as a new way of understanding and responding to threats to life, practices of biosecurity are emergent and unsettled. The human and animal bodies, narratives, and politics that surround the naturalcultural worlds known through models of biosecurity are cloudy and uncertain (Lowe 2010a). In the emerging literature on the topic, biosecurity has been read and theorized adjacent to Foucault's concepts of biopolitics and governmentality, Beck's concept of risk, and new work on multispecies interactions and on globalization. Following Foucault (2007), many have shown how the apparatus of security has the tendency to continually expand to encompass ever wider circuits of knowledge and practice; in other words, one can never be secure enough.

One way to understand biosecurity is to compare it with another "bio" form: biodiversity. Both biodiversity and biosecurity prefigure uncertain futures in which some outcomes are favored and some are guarded against, and a sense of urgency or even emergency pervades both terms. Each has singular characteristics, however. Life deemed biodiverse exists in "hotspots" that are relatively fixed in space, and those spaces are typically studied in the global South and remote rural areas. The imaginary of biodiversity is non-technological, requires distance from culture to be sustained, and has been conceptualized through the language of "natural resources." Biodiversity projects exist on the fringes of the state, and may satisfy the state's needs indirectly (for example the need for resource sovereignty or control over peripheral populations). Finally, nature is viewed as an innocent victim in discourses of biodiversity.

Biosecurity, on the other hand, takes up natures that exist at the heart of the technoscientific industrialized world. New biothreats are as likely to come from the educated garage biochemist as the forest-dwelling hunter, to be urban as rural, industrial as pastoral, Northern as Southern. The emergence of biosecurity responds to new scientific capacities in bioengineering and microbiology rather than the desire to set aside space for the archaic. Biosecurity is more directly aligned and integrated with state and national security than are projects of biodiversity. Further, biosecurity protects against undesirable natures—the invasive species or the deadly microbe. And finally, rather than victim, in biosecurity discourse nature has been named a "bioterrorist" for its capacity to evolve new diseases.

Biosecurity and political ecology

In the move from biodiversity to biosecurity, new terrains open up for the field of political ecology. Biosecurity is a theme appropriate to a form of political ecology that has appreciably broadened from original studies of biodiversity conservation, soil erosion, common property regimes, or the relationship between local natural resource husbandry and state policies that influence land use. Early studies were commonly set in parts of Asia and Africa where coercive state practices bearing a colonial legacy came up against indigenous and native property rights and land and species use. While some studies moved this work out of the global South and into the United States and Europe, many of the foundational studies in political ecology focused on settings and forms of life important to smallholder agricultural production, and the hunting and collecting practices of marginalized rural populations. They were also characterized by a standard set of state and local actors influenced in conventional ways by markets in a global political economy (Lowe 2006: viii–ix), and used anthropological and historical methods appropriate to agrarian economies and the "progressive contextualization" of local natural resource use.

The recent expansion of political ecology into fields like biosecurity has required an enlarging of the types of ecologies and forms of life at stake and the modes of expertise used to study them. Political ecology now supplements its Marxist origins with science and technology studies, anthropology of the contemporary, environmental humanities, and other post-human and new materialist ontologies. These modes of inquiry put life at the center of their analysis, demanding detailed practices of following laboratory life and scientists in action, tracking viruses across cloudy spaces of biosocial uncertainty, and following the networks of circulation that inform biological and physical processes, all the while cultivating ethnographic relationships with technical experts and other elites whose knowledge production centers on materiality.

These new forms of scholarship, most importantly, do not always center on human life; rather, some offer up to animals the compassion and curiosity that an older political ecology reserved for marginalized peoples (see, for example, Chapter 9, this volume). Biosecurity as a rubric has the ability to expand political ecology from its roots in studies of biodiversity and common property in the developing world to a framing of the ecologically political that encompasses life at divergent scales from the microbial to the charismatic species, and across divergent spatial geographies, from the human lung to the industrial hen house. This enlarged sense of the vital has allowed political ecologists to draw from and converse with others doing work on multispecies ethnography (Kirksey and Helmreich 2010), animal studies (Garcia 2010; Haraway 2008), and animal capital (Shukin 2009), among other fields.

Political ecology also has the capacity to contribute to and extend our analysis of biosecurity. Peluso's (1993) seminal article on "coercing conservation" illustrates how questions of securing life pre-date contemporary studies of biosecurity. Using examples from Kenya and Indonesia, Peluso showed how securing nature has included not only the efforts of conservationists to save wildlife, but also the state's own efforts to control populations and resources in ways that were ancillary to any direct conservation agenda. These arguments are extended in Peluso and Watts' (2001) critique of the neo-Malthusian "environmental security" paradigm which predicted a "coming anarchy." In place of an over-determined resource scarcity, they explain violence through historically situated studies of resource use, governmentality, and entitlements and inequality. Work in political ecology has demonstrated how efforts to secure nature are a facet of neoliberalism that pre-dates the terror events of the 2000s (see also Goldstein 2010); how conflicts occur in the name of securing nature often mask other societal troubles; and how violence and disaster are always situated and not determinable in advance by the form of nature in question.

Studies of biosecurity have entered the scholarly literature through a range of works that expand the sense of the vital at stake in late modernity. Although not many biosecurity studies self-consciously identify as works of political ecology, they are in fact both ecological and political. Biosecurity studies are ecological in that they draw in not only the individual species involved—from the anthrax spore to the invasive rat on the isolated island—but also imbricate the ecological relationships between and among species. They are political in that the ecological relationships at stake are encompassed by a global political economy that emerges from histories of interconnection. Moreover, the politics of biosecurity point to winners and losers as some are made safe and others are made to bear the costs of risk avoidance.

In looking at the ways that biosecurity expands the field of political ecology, and in examining what political ecology has to offer to studies of biosecurity, an agenda for studies in political ecology emerges. First, biosecurity can expand the forms of ecology and vital matter under examination in political ecology. Second, technoscientific and industrial natures, hybrid naturecultures, "more-than-human" natures, and frameworks that do not presume a separation between "human" and "natural resource" worlds are important to future studies in political ecology. And third, political ecologists can productively bring existing political ecological work on environmental security into dialog with recent work on biosecurity.

Biosecurity as practice

Contemporary practices of biosecurity are grounded by four empirical forms: *exclusion*, *preparation*, *regulation*, and *excitation*. First, practices of biosecurity are about allowing in, confining, or separating out biological materials that are perceived to pose risk. Thus, biosecurity is the business of taxonomizing, classifying, and then *excluding* hazardous from non-hazardous life. As described by Canguillhem (1991), separating the normal from the pathological is a political process inherently linked to the realm of value. Biological normativity is defined in relation to the abnormal and, likewise, the biosecure is defined in relation to the biohazardous.

Best characterized by its speculative and hypothetical nature, *preparation* is the second practice animating biosecurity. Once something enters the realm of calculation as a biological risk, preparedness comes into play (Caduff 2014; Lakoff 2008). "Pandemic preparedness" is an example of this. Whether or not something will actually become a global pandemic or not depends upon a reading of available data and information and then constituting it as a form that can be acted upon. Readings of risk rely upon expert knowledges and styles of reasoning, as well as political mandates to become prepared.

Moreover, in a point central to political ecology, some bear the responsibilities of preparedness more than others. Within the international community, countries, like China, Viet Nam, or Indonesia, that are identified as origination points for contagion must have preparedness practices in place, or face international sanction (Lowe 2010a and 2010b; Porter 2013). Women are positioned as protectors of unborn children who must regulate seafood intake in a gendered regime of biosecurity (Mansfield 2012).

Biosecurity also incites the practice of *regulation* and the integration of life with law; a wide variety of legal and bureaucratic control techniques regulate the practices of exclusion and preparedness. Old-fashioned plagues were contained through practices of quarantine and sanitation, and biosecurity still incorporates these measures. Additionally, however, overlapping juridical and preparedness functions of biosecurity respond to its speculative nature and are designed in anticipation of possible future events (Bingham and Lavau 2012; Redfield 2008).

Regulation can take place at a variety of scales, from the laboratory to the nation. While it is not a new thing for militaries to be concerned with disease, it was only during the Clinton presidency that an epidemiologist was first appointed to the US National Security Council. Different countries have different regulatory concerns in relation to biosecurity that interface with specific national anxieties. While in the United States security concerns are linked to terrorism and laboratory safety, in New Zealand and Australia biosecurity regulation primarily governs the entry of exotic species (Dobson et al. 2013).

And finally, biosecurity is greatly influenced by practices of *excitation*. Securing against rapidly emerging disease threats, and against bioterror, has relied upon enacting scenarios that prepare public health and defense workers to respond should a threat become actual. Lakoff (2008) writes of a seminal table top exercise, "Dark Winter," which took place at Andrews Airforce Base, and was a simulation of a smallpox attack on the United States. While the exercise demonstrated gaps in preparedness, it also excited a sense of vulnerability based upon a particular script that directed the fictive outcome. Imaginative reenactments are subjectifying events creating persons who give particular credence and attention to proposed threats.

Like technical experts who are enrolled into the world of biothreat through preparedness practices, the public is similarly enrolled into preparedness and an affect of anxiety. Recall, for example, the US Department of Homeland Security's advice after the anthrax attacks of 2001 to use duct tape to seal windows to keep biohazards from the home. The sense of threat and insecurity are further instilled through media portrayals of technoscientific and medical disaster. From scientific non-fiction, like Laurie Garrett's (1995) *The Coming Plague* or Alan Sipress' (2012) *The Fatal Strain*, to fictionalized accounts like Richard Preston's (1995) *The Hot Zone* and recent films like *World War Z* or *Contagion*, the public is interpellated into "outbreak narratives" (Wald 2008) that encourage readers and viewers to imagine a world imbued with particular kinds of biological danger, from bioterror to emerging infectious disease.

The overarching milieu for the processes of biosecurity is an epidemic of fear. Advanced neoliberalism has produced the demise of the middle class, a catastrophic rise in atmospheric carbon, and a predominance of diseases of affluence, as well as perpetual war. Actual scenarios of disaster, played out in real time, as with Hurricane Katrina or acts of terror, have demonstrated an inability to deal with events through extant security or preparedness measures. These experiences inform and mobilize a security-based normativity within a contemporary fear-based milieu. Ulrich Beck's (1992) concept of "risk society," or a society organized around responding to and containing risks, has been influential in studies of biosecurity. Beck sees the mode of risk-based organization as symptomatic of late modernity, with hazards stemming from the process of modernization itself.

The range of biosecurity practices described above, their politics, and their ecologies intersect prominently within two sets of interlocking issues: *securing food* and *emerging infectious disease*.

Securing food

Food is at the center of many attempts to secure life. What we call "foods" are those forms of life that bear a particular relationship to the human: food is life that is eaten. But what we want to eat as food can also come along with life that we want to avoid: the pathogen. Through practices of exclusion and preparedness, behaviors of individuals and animals and plants are spatially regulated into particular configurations whereby plants- and animals-as-food may circulate widely while limiting the travel of their companion pathogens. Recent research on the environmental politics of food indicates in outline what a focus on biosecurity offers to the field of political ecology, in terms of the forms of life at stake and the broader ecologies that are implied.

In its modern form, food is industrialized and bureaucratized. Industrialized food is typically corporate in scale, machine intensive rather than labor intensive, profit-driven, cheap, multinational, monocropped, regulated, and subsidized. These features introduce novel forms of risk into the living systems, ecologies, and species assemblages that are food. Through legal and illegal forms of mobility, the globalized and industrialized food supply is vulnerable in novel ways to contamination, infestation, and outbreak. These food assemblages create novel political ecologies within which an observer might explore and maneuver.

The application of the term biosecurity to practices to agriculture was relatively novel before 2000. In England, the term became prominent with the Bovine Spongeform Encephalitis (BSE, or Mad Cow disease) outbreak in 2001 (Dobson et al. 2013). In France the application of biosecurity to food became part of the conversation in debates over genetically modified organisms (Keck 2008). And in the United States, agricultural biosecurity came into prominence adjacent to concerns over avian influenza, and its implications for animal agriculture. These differences are illustrated in Freidberg's (2004) comparative political ecology of the ways anxieties around food safety have been differently mediated in the UK and France.

Two lines of biosecurity have emerged from the effort to secure the food supply: on-farm biosecurity and state biosecurity. On-farm biosecurity involves a set of technical measures to manage the risk of the spread of disease in agriculture. Most prominent in animal agriculture (though also present in grain and vegetable production), practices of on-farm biosecurity are designed to intervene between the animal (or crop) and the pathogen. They include tasks that isolate livestock into a sanitary bubble keeping workers and vermin from bringing pathogens onto the facility, and sanitizing litter, air, carcasses, etc. to prevent flows of pathogens between and off facilities. Preventing the mobility of pathogens between sheds, and in animal transport is also important. With enough control and management, the risks of pathogenic contaminations are deemed to be manageable.

The genetic homogeneity in industrial livestock production is designed to increase the ratio of meat to other parts of the animal. Homogeneity also creates disease risk, however. Industrial agriculture has responded to the risks of genetic homogeneity with more cleanliness within the bubble. While Confined Animal Feeding Operations (CAFOs) engage only what goes on within the bubble, and are concerned with what affects sales of meat, they are disconnected from the well-being of human and animal populations outside the bubble. The life that becomes meat is *zoe*, or "bare life" in Agamben's (1998) terms; the CAFO does not address itself to *bios*, or "political life."

Food is also made secure through processes that secure the nation-state. This happens at macro and micro scales. At the macro scale is the process known as "land grabbing," where national governments and private firms search for distant land parcels in order to secure and prepare for future food and other resource shortages. Political ecologists lead in studies of the land grab phenomena. In their introduction to a special issue of the *Journal of Peasant Studies*, Boras et al. (2011) call for increased systematic inquiry into the political ecology of contemporary land deals. An example of the work they call for is Li's (2011) study of how land deals create security for some while dispossessing others.

State practices of food security also occur at the micro scale in the form of food inspections. Bingham and Lavau (2012) write of British food safety practices by using the case of a singular restaurant inspection in which they follow "Allison," an inspector, through her work in a London restaurant kitchen. Using the principle of "safer food for the nation," in which the material object that is food is followed from "farm to fork," British food safety inspectors attempt to separate food from pathogen by maximizing the travel of food, while minimizing the travel of pathogens. Critically, the networks that food and pathogens travel are the same spatial networks, so this is not an easy task. Bingham and Lavau follow Allison's micropractices as she moves through the restaurant, spotting mouse droppings, finding uncovered meat next to an open mushroom container in the walk-in freezer, and looking for stamps that identify the chain of meat production. These techniques of securing the British food supply are precautionary:

they are not responding to particular events in the past or to known evidence but, instead, are responding to the possibility of future outbreak.

While plants and animals as food can become the vectors of pathogenic risk, the entire agricultural system as a whole is put at risk in the form of bee colony collapse disorder. One-third of global agriculture depends upon honeybee pollination. In 2006–2007, 40 percent of US honeybees were lost to the disorder, and observations of collapse are global. Jake Kosek (2010) discusses the irony of how US Homeland Security has put colony collapse disorder on its agenda while simultaneously bees and bee models are used in contemporary American warfare and national security research. Bees are used to detect radiation, explosive devices, and landmines, and are a model for drone warfare and swarming. Using a multispecies approach, Kosek documents how the bee has transformed over the past hundred years to be different from its historical ancestor. He records transformations in its exoskeleton, its nervous system, digestive tract, and its social organization (see also Tsing 1995). Like other multispecies ethnographers (Lowe [2010a] who owned chickens, Heather Paxson [2008] who knew *Lactobacilli* through making cheese), Kosek makes behavioral observations of the species he studies as part of his way of knowing them. Through the bee, Kosek shows how life is securitized when bees become both physically and metaphorically part of the military industrial complex, and when what it means to be human becomes attached in new ways to what it means to be bee.

As studies of industrial food ecologies enter into political ecology, the field is transformed in a myriad of ways. Inquiry into the securitization of life that is food offers political ecologists an agenda that expands the *bios* under discussion, and approaches materiality in a new way. Kosek, trained as a political ecologist, transforms the agenda of political ecology through the importance of what he calls "political entomology." His political entomology follows the shifting lines between the animal and the human to demonstrate material and symbolic transformations in the makeup of both. New animal and multispecies studies (Kirksey and Helmreich 2010) expand the interactions between economics, politics, technology, and ecology that are at the heart of the tradition of political ecology.

Emerging infectious disease

The scientific worldview of disease has transformed over the last half century. While, at least through the 1980s, there was the sense that we would conquer infectious disease through vaccination and sanitation, this optimism did not last long and quickly transformed into the comprehension that the evolutionary struggle between humans and microbes is not easily won. The concept of "disease emergence" was popularized in Laurie Garrett's 1994 book, *The Coming Plague*, where she connected it to globalization and environmental change. The HIV pandemic, likewise, was particularly influential in reinforcing a clinical and epidemiological awareness of the emergent nature of infectious diseases.

Biological warfare and bioterrorism have also been part of the emerging biosecurity approach to infectious disease and have played a central role in studies of biosecurity. The events of September 11, and the ensuing anthrax attacks in the United States were both contexts and pretexts for rethinking the risks of biological materials to life. Between 2001 and 2008 federal funding for biodefense was 40 billion dollars, and in 2004 President Bush issued a new directive, *Biodefense for the 21st Century*, that represented the United States and the "civilized world" as highly vulnerable to biological attack (Vogel 2008). Recall, as well, the discussion in 2003 on whether and who to vaccinate against smallpox, the only disease that has ever been successfully eradicated on a global scale (moreover, using conventional public health measures), and which now was re-envisioned as a threat to US national security (Rose 2008).

Most diseases are not engineered, however; they emerge through evolutionary processes. Viruses, in particular RNA viruses like influenzas or HIV, have high mutation rates. A combination of rapidly produced mutations and natural selection offer the opportunity for many new viral forms to emerge through genetic drift. Further, when two viruses infect the same cell, they can exchange genes, taking new forms in a process called genetic shift. A new relationship to microbial capacities has motivated new security practices, and uncertainty of the viral form mirrors the cloudiness of emergent social forms around viruses (Lowe 2010a).

Vaccination is an obvious line of defense against emerging disease. Extending back to practices of variolation,[1] vaccination has a history that pre-dates biosecurity. New issues are raised, however, in the contemporary moment. The pace at which viruses are able to transform and adapt creates problems for vaccine and anti-viral development as target microbes evolve away from technological fixes in as little as weeks or months. We now face the specter that lethal microbes will become resistant to the drugs we use to fight them, most notably antivirals and antibiotics. In part this is due to over- and unregulated use in human populations. More significant for antibiotic resistance is their blanket application in conventional animal agriculture where they are used to encourage rapid growth and to compensate for unsanitary conditions.

The view of diseases as emergent has occurred alongside a more recent shift in approaches to disease management. The emergence of SARS and H5N1 Avian Influenza were both addressed through the norms of biosecurity in addition to those of public health. The SARS outbreak in 2003 is exemplary of some of the features of this biosecurity. SARS affected people in 37 countries within weeks of its outbreak, exciting fear over the vulnerabilities of international travel and the speed with which disease can circle the globe. SARS also was enlivened by newly expanded disease surveillance systems that have global regulatory sovereignty. For example, SARS was first identified through a Canadian international health intelligence network (the Global Public Health Intelligence Network) which picked up an unusual flu outbreak using internet surveillance techniques, and then alerted the World Health Organization. Finally, the disease appeared to move from bats to civet cats to humans, highlighting the animal–human interface of zoonotic disease, and implicating many ecological actors in the disease's progress.

As SARS illustrates, emerging infectious disease brings together complex ecological narratives that have species moving beyond individual locales, and demonstrates the globalized nature of discourse and practice in biosecurity. Naming particular communities, not only as vulnerable to outbreak, but also as responsible for disease spread creates the geographies of risk and blame described by Sparke and Anguelov (2011) in relation to the H1N1 outbreak of 2009. In the case of H1N1, fear of the spread of the disease from Mexico into the United States overlapped with American racial discourse on Mexican border crossing. Similarly, D'Arcangelis (2008) has shown how the media portrayed Chinese animal hygiene in Orientalist terms aligning it with "tradition" and not "modernity" in both the SARS and H5N1 outbreaks.

Linking emergent disease with the discussion of food production outlined above, political ecologists who seek to track the causes of environmental change across globalized networks of causality will be interested in accounts that lay the source of recent influenza epidemics and other zoonoses at the door of the globalized filière of intensive farm animal production (Davis 2005; Greger 2006; Pew Commission 2008; Wallace 2009). Evolutionary biologist Rob Wallace and political ecologist Luke Bergmann (Wallace et al. 2010; Wallace 2009) describe the agroecological conditions in Southern China, and the links to global big agriculture, that are driving the emergence of many new highly pathogenic influenza varieties. Politicizing the normative explanations of influenza emergence that focus on China's liberalizing economy, the role of wild birds, or China's lack of agricultural modernity, Wallace and Bergmann document large population shifts, intensification of agricultural methods, pressure on wetlands and rural areas, and changes in the ownership structure of agriculture as the political ecological causes of the evolution to high pathogenicity of new influenza strains in the region.

Like Wallace and Bergmann, Allen and Lavau (2014) argue that practices of biosecurity also have the capacity to generate new ecologies. CAFOs present abundant characteristics that create opportunities for disease emergence: genetically homogeneous populations; stress from crowding and practices like de-beaking; metabolic energy used for growth and survival rather than health; the extensive use of antibiotics and antimicrobials in all parts of animal production; the sheer volume of production; inhalation of ammonia from urine that makes the respiratory tract vulnerable; and even vaccination of poultry against influenzas. These conditions have put disease emergence on an evolutionary fast track, and the global rise in zoonotic disease mirrors the livestock revolution of the past 30 years. These political disease ecologies are an example of the expanded sense of nature relevant to political ecology.

Ecology and the politics of biosecurity

Biosecurity, as an expansion in the practices of securing life, is part of a larger securitization of the social where the present is experienced through risk and preparedness, and the precariousness of modern existence appears to be best addressed through neoliberal, bureaucratic quasi- and fully militarized forms of sovereign intervention. In practices of biosecurity, life is integrated with law and regulation, and anxious subjectivities are an outcome.

A critique of efforts to secure life is emergent in the work on violence undertaken by political ecologists. While studies in biosecurity often seem to take the state of security for granted, as either successfully normalized or as something that can be achieved, violence is a counterintuitive outcome of some efforts at securitization. Studies by political ecologists of "forest wars" in Sierra Leone (Richards 2001), "ethnic" wars in Kalimantan (Peluso and Harwell 2001), or environmental enclosures in the American West (McCarthy 2001) all point to important ways that (1) security practices themselves can create violence, (2) states of violence rather than security can become normalized, and (3) security often relies upon older entrenched discourses of unruliness.

As recent research broadens the networks of materiality and sites of engagement that a political ecologist might invest in, ecology itself forms a node of possible critique in biosecurity studies. For example, viral evolution, and the emergence of disease despite efforts at on-farm biosecurity, point to a "vital materialism" in which things have forces and tendencies of their own. In political theorist Jane Bennett's (2010) "political ecology of things," she argues that rather than separate the human from the non-human, political theory needs to recognize the participation of non-human forces in events. Together with Bruno Latour's actor-network theory, Bennett and other new materialists indicate that efforts to secure life cannot be thought outside the agency of the life forms being secured or secured against. In relying on an object-oriented ontology or a multispecies approach that does not demand things to be passive, and through widening vitalities and widening ecological networks, political ecology can take up questions of life and security in new ways to ask what it might take to be secure, and not securitized.

Acknowledgements

The author would like to thank the editors of this volume, Luke Bergmann, the Simpson Center for the Humanities at the University of Washington, and the Rachel Carson Center at Ludwig-Maximilians Universitat, Munich for support in writing this entry of the *Political Ecology Handbook*.

Note

1 Variolation was the procedure of exposing healthy individuals to dried smallpox scabs in order to induce a mild case of the disease leading to life-long immunity. The practice originated in China and was introduced to Europe in the eighteenth century.

References

Allen, John and Stephanie Lavau (2014) "Just in Time Disease: Biosecurity, Poultry, and Power," *Journal of Cultural Economy*, April.

Agamben, Georgio (1998) *Homo Sacer: Sovereign Power and Bare Life*, Stanford: Stanford University Press.

Beck, Ulrich (1992) *Risk Society: Towards a New Modernity*, New York: Sage Publications.

Bennett, Jane (2010) *Vibrant Matter: A Political Ecology of Things*, Durham: Duke University Press.

Bingham, Nick, and Stephanie Lavau (2012) "The Object of Regulation: Tending the Tensions of Food Safety," *Environment and Planning A*, 44(1589–1606).

Boras, Saturnino M. Jr, Ruth Hall, Ian Scoones, Ben White, and Wendy Wolford (2011) "Towards a Better Understanding of Global Land Grabbing: An Editorial Introduction," *Journal of Peasant Studies* 38(2): 209–216.

Braun, Bruce (2007) "Biopolitics and the Molecularization of Life," *Cultural Geographies* 14(6): 6–28.

Caduff, Carlo (2010) "Public Prophylaxis: Pandemic Influenza, Pharmaceutical Prevention and Participatory Governance," *BioSocieties* 5(2): 199–218.

Caduff, Carlo (2014) "Pandemic Prophecy, or How to Have Faith in Reason," *Current Anthropology* 55(3): 296–315.

Canguillhem, Georges (1991 [1946]) *The Normal and the Pathological*, New York: Zone Books.

Collier, Stephen, Andrew Lakoff, and Paul Rabinow (2004) "Biosecurity: Towards an Anthropology of the Contemporary," *Anthropology Today* 20(6): 3–7.

D'Arcangelis, Gwen (2008) "Chinese Chickens, Ducks, Pigs, and Humans and the Technoscientific Discourses of US Empire," in *Tactical Biopolitics: Art, Activism, and Technoscience*, ed. Beatriz Da Costa and Kavita Phillip. Cambridge, MA: MIT Press, 429–442.

Davis, Mike (2005) *The Monster at our Door*, New York: Owl Books.

Dobson, Andrew, Kezia Barker, and Sarah Taylor, eds. (2013) *Biosecurity: The Socio-Politics of Invasive Species and Infectious Disease*, London: Routledge.

Foucault, Michel (2007) *Security, Territory, Population: Lectures at the Collège de France, 1977–78*, Basingstoke: Palgrave Macmillan.

Freidberg, Suzanne (2004) *French Beans and Food Scares: Food Security in an Anxious Age*, Oxford: Oxford University Press.

Garcia, Maria-Elena (2010) "Super Guinea Pigs?" *Anthropology Now* 2(2): 22–32.

Garrett, Laurie (1995) *The Coming Plague: Newly Emerging Diseases in a World out of Balance*, New York: Penguin Books.

Gilligan, Carole (1982) *In a Different Voice: Psychological Theory and Women's Development*, Cambridge: Cambridge University Press.

Goldstein, Daniel M. (2010) "Towards a Critical Anthropology of Security," *Current Anthropology* 51(4): 487–517.

Greger, Michael (2006) *Bird Flu: A Virus of Our Own Hatching*, New York: Lantern Books.

Haraway, Donna (2008) *When Species Meet*, Minneapolis: University of Minnesota Press.

Keck, Frédéric (2008) "From Mad Cow Disease to Bird Flu: Transformations of Food Safety in France," in *Biosecurity Interventions*, ed. A. Lakoff and S. Collier. New York: Columbia University Press, 195–226.

Kirksey, Eben, and Stephan Helmreich (2010) "The Emergence of Multispecies Ethnography," *Cultural Anthropology* 25(4): 545–576.

Kosek, Jake (2010) "Ecologies of Empire: On the New Uses of the Honeybee," *Cultural Anthropology* 25(4): 650–678.

Lakoff, Andrew (2008) "From Population to Vital System: National Security and the Changing Object of Public Health," in *Biosecurity Interventions*, ed. A. Lakoff and S. Collier. New York: Columbia University Press, 33–60.

Li, Tania (2011) "Centering Labor in the Land Grab Debate," *Journal of Peasant Studies* 38(2): 281–298.

Lowe, Celia (2006) *Wild Profusion: Biodiversity Conservation in an Indonesian Archipelago*, Princeton: Princeton University Press.

Lowe, Celia (2010a) "Viral Clouds: Becoming H5N1 in Indonesia," *Cultural Anthropology* 25(4): 625–649.

Lowe, Celia (2010b) "Preparing Indonesia: H5N1 Through the Lens of Global Health," *Indonesia*, 90(1): 147–170.

Mansfield, Becky (2012) "Environmental Health as Biosecurity: 'Seafood Choice', Risk, and the Pregnant Woman as Threshold," *Annals of the Association of American Geographers* 102(5): 969–976.

McCarthy, James (2001) "States of Nature and Environmental Enclosures in the American West," in *Violent Environments*. Ithaca: Cornell University Press, 117–145.

Paxson, Heather (2008) "Post-Pasteurian Cultures: The Microbiopolitics of Raw-Milk Cheese in the United States," *Cultural Anthropology* 21(1): 15–47.

Peluso, Nancy (1993) "Coercing Conservation: The Politics of State Resource Control," *Global Environmental Change* 3(2): 199–217.

Peluso, Nancy, and Emily Harwell (2001) "Territory, Custom and the Cultural Politics of Ethnic War in West Kalimantan, Indonesia," in *Violent Environments*, Ithaca: Cornell University Press, 83–116.

Peluso, Nancy, and Michael Watts (2001) *Violent Environments*, Ithaca: Cornell University Press.

Pew Commission (2008) *Putting Meat on the Table: Industrial Farm Animal Production in America*, Pew Charitable Trusts and Johns Hopkins Bloomberg School of Public Health.

Porter, Natalie (2013) "Bird Flu Biopower: Strategies for Multispecies Coexistence in Viet Nam," *American Ethnologist* 40(1): 132–148.

Preston, Richard (1995) *The Hot Zone: A Terrifying True Story*, New York: Anchor Books.

Redfield, Peter (2008) "Vital Mobility and the Humanitarian Kit," in *Biosecurity Interventions*, ed. A. Lakoff and S. Collier. New York: Columbia University Press, 147–171.

Richards, Paul (2001) "Are Forest Wars in Africa Resource Conflicts? The Case of Sierra Leone," in *Violent Environments*. Ithaca: Cornell University Press, 65–82.

Rose, Dale (2008) "How Did the Smallpox Vaccination Program Come About?" in *Biosecurity Interventions*, ed. A. Lakoff and S. Collier. New York: Columbia University Press, 89–120.

Shukin, Nicole (2009) *Animal Capital: Rendering Life in Biopolitical Times*, Minneapolis: University of Minnesota Press.

Sipress, Alan (2012) *The Fatal Strain: On the Trail of Avian Flu and the Coming Pandemic*, New York: Viking Press.

Sparke, Matthew, and Dimitar Anguelov (2011) "H1N1, Globalization, and the Epidemiology of Inequality," in *Health and Place* 18(4): 726–736.

Tsing, Anna (1995) "Empowering Nature, Or: Some Gleanings in Bee Culture," in *Naturalizing Power: Essays in Feminist Cultural Analysis*. London: Routledge, 113–143.

Vogel, Kathleen (2008) "Biodefense: Considering the Sociotechnical Dimension," in *Biosecurity Interventions*, ed. A. Lakoff and S. Collier. New York: Columbia University Press, 227–255.

Wald, Pricilla (2008) *Contagious: Cultures, Carriers, and the Outbreak Narrative*, Durham, NC: Duke University Press.

Wallace, Rob (2009) "Breeding Influenza: The Political Virology of Offshore Farming," *Antipode* 41(5): 916–951.

Wallace, Rob, Luke Bergmann, Lenny Hogerwerf, and Marius Gilbert (2010) "Are Influenzas in Southern China Byproducts of the Regions's Globalizing Historical Present?" in *Influenza and Public Health: Learning from Past Pandemics*, ed. Tamara Giles-Vernick and Susan Craddock. London: Earthscan, 101–144.

39

SCALES AND POLITIES

Nathan F. Sayre

Introduction

The issue of scale has long been central to political ecology. Efforts to explain local human–environment systems in relation to larger regional and historical factors characterized the field's roots in both anthropology and cultural ecology (see, e.g., Wolf 1972). In *Land Degradation and Society*, Blaikie and Brookfield (1987: 13) explicitly declared "the crucial considerations of geographical scale and the scale of social and economic organization" to be one of the three main characteristics of their approach to society–environment interactions. Their famous "chain of explanation" involved "links" at each of four scales: (1) individual land managers, (2) the local community ("their relations with each other, other land users, and groups in the wider society who affect them in any way"), (3) the state, and (4) the world economy (1987: 27). Only by attending to all four scales and their interactions could political ecologists account for the full range of factors affecting land management and degradation at any given site. A quarter-century later, Robbins (2012: 11) characterized political ecologists as employing "a mode of explanation that evaluates the influence of variables acting at a number of scales, each nested within another, with local decisions influenced by regional policies, which are in turn directed by global politics and economics."

Blaikie and Brookfield emphasized that this was a difficult task, because relations between scales were neither simple nor linear. Measurements of erosion made at a plot or field scale, for example, could not simply be multiplied by area to estimate erosion rates for a catchment or a country, because most soil lost from one plot would be deposited in another (1987: 53). Any measurement involved a scale, and how a process such as erosion worked was itself dependent on the scale at which one sought to measure and understand it. And *measuring* soil erosion, complex though it is, only began to indicate the challenges of *explaining* it, which involved myriad social, political, and economic processes that themselves operated in different ways at different scales. "There is seldom a neat one-to-one correspondence of geographical scale and 'level' of decision-making," Blaikie and Brookfield pointed out, and "the scale at which the analysis is pitched tends to affect the type of explanation given to land degradation" (1987: 64–65). Add to this the fact that scale is not only a spatial but also a temporal issue, with faster and slower processes—both social and biophysical—interacting with each other, and explanation becomes an historical as well as a geographical challenge. Blaikie and Brookfield summarized their approach this way, substituting regional for local:

> Clearly there is no "correct" scale for an investigation of land managers and their decisions, but there is an appropriate one for answering different questions. Frequently a comprehensive enquiry into land management will require an approach which employs a nested set of scales: local and site specific where individuals or small groups make the relevant decisions; the regional scale involving more generalized patterns of physiographic variation, types of land use, and property relations and settlement history; the national scale in which the particular form of class relations give the economic, political and administrative context for land-management decisions; and the international scale, which, in the most general manner, involves almost every element in the world economy, particularly through the commoditization of land, labour and agricultural production.
>
> *(1987: 68)*

In many ways, Blaikie and Brookfield's conceptualization of scale and their arguments regarding its importance for political ecology have stood the test of time remarkably well. As the field has grown and expanded its purview—from soil erosion to all manner of environmental problems, from farmers' decision-making to the national and international politics of land, water, wildlife, and conservation, and from the rural, developing world to urban and developed settings—political ecologists have continued to explore processes at diverse spatial and temporal scales. Scale is evidently an inherent feature of political ecology, at the very least because the political organization of today's world is fundamentally territorial—organized into discrete, bounded geographical spaces—whereas both ecological and economic processes routinely exceed or defy these boundaries. As we will see, modern polities depend on and produce scales in myriad ways.

In the intervening years, however, scale has emerged as a focus of inquiry in its own right, both within and beyond political ecology, provoking debates that have refined and in some respects challenged Blaikie and Brookfield's formulation. Five years after *Land Degradation and Society* came out, Neil Smith (1992: 72) complained that "The theory of geographical scale—more correctly the theory of the production of scale—is grossly underdeveloped." In the two decades since, the conventional categories of geographical research—such as households, cities, counties, provinces, nations, etc.—have been challenged as social constructs rather than natural or pre-given spatial orders. Thus one might question the precise boundaries between the land manager, the local/regional, the national and the international, and ask why one should begin with four scales at all (why not five, or three, or ten?). One might further ask how scales are produced, sustained, contested, and altered at particular places and times. The relative importance of different scales, and the relations among them, have also been subject to scrutiny: are scales necessarily "nested"—such that they "fit inside each other like a set of Chinese boxes" (Blaikie and Brookfield 1987: 69)—or is this assumption faulty? Must scales be hierarchically organized, with "higher" ones dictating or constraining action at "lower" ones, or can local agents provoke change in the other direction as well? Might one instead understand scales as emergent properties of non-hierarchical networks? What if scales aren't discrete, but fluid, overlapping, and mutually constitutive?

Questions such as these sparked intense debate in human geography in the 1990s and early 2000s, alongside growing interest in scale among political ecologists. The meaning of scale, its ontological and epistemological status, and even its existence and relevance to geography, were questioned and contested. Political ecologists engaged these debates to various degrees, giving rise to what Zimmerer and Bassett (2003) and Neumann (2009) termed "a political ecology of scale." In the first section of this chapter, I briefly describe scales' various meanings and dimensions in the hope of clarifying their relevance for political ecology and human geography more generally, beginning with Blaikie and Brookfield's third scale, the nation-state. In the second section, I apply these ideas in reviewing the debates about scale in human geography. I suggest that these debates proved irresolvable on the predominantly theoretical terms in which they took place, and that since about 2006 the prevailing approach to scale in human geography generally has been more inductive and empirical.

The final section shows how political ecologists had in many ways anticipated the shift and were already busy doing empirical research along the suggested lines, with or without express intent. Their methods and theories have been diverse, reflecting the varied conditions and questions at hand in each case, and no single paradigm, model, or methodological solution to the question of scale has yet emerged. Rather, several broad themes have developed, including the importance of multiple sites and scales of empirical inquiry; the problem of scale mismatches among political, economic, and ecological processes; temporal as well as spatial dimensions of scale; and the need for relational and process-based approaches to conceptualizing and studying social-environmental problems. I close with some thoughts on the strengths and weaknesses of this de facto strategy for the study of scales and polities in political ecology.

What is (the nature of) scale?

Blaikie and Brookfield's third scale—the national state—provides a useful starting point for exploring the meaning and status of scale. Three factors make the state an exemplar of geographical scale, while also highlighting the dangers of taking scales for granted as natural in geographical (and other) research.

First and most obviously, the state imposes explicit spatial and hierarchical patterns of social organization. It is fundamentally territorial, defining and relying on spatial boundaries that delimit its sovereignty. Moreover, the system of states has been comprehensive for more than a century, ensuring that Earth's entire terrestrial surface (excluding Antarctica) is incorporated into one or another nation-state's territory. Within their respective territories, states have typically imposed (or attempted to impose) sub-national scales of administration and law, such as the individual citizen or subject, the household, municipalities, counties, and provinces. The legal relationships between these scales have been codified, more or less precisely and effectively, usually in the form of exclusive hierarchies—meaning that larger (higher) scales dictate or impose what can or cannot be done at smaller (lower) ones. Notice that many of these nested, hierarchically organized scales can be (and often are) referred to as "levels" of government—although in a moment we will want to draw a distinction between scale and level.

Second, the state has material effects, independent of the observer, within and through this exclusive spatial hierarchy. As Blaikie and Brookfield noted, it was at the national scale that class relations determined "the economic, political, and administrative context" in which actors at the individual and local/regional scales made decisions regarding land management. The state's nested system of jurisdictions, statutes, and regulatory and fiscal agencies produces manifold observable effects on social reality. These effects are socially and historically produced, to be sure, but they are also ontologically real, and for this reason the state is an *operational scale*.

Third, the state also acts as an *observational scale*, collecting data according to its social and territorial divisions. With varying degrees of accuracy and detail, states measure land, property, people, and myriad economic activities and biophysical processes, and they typically do so in terms of the spatial categories that organize the state itself. These data carry with them the units of measurement (e.g., individuals, households, census tracts) in which they were collected, aggregated at various territorial scales (e.g., cities, counties, provinces, the nation); the former are the *grain* or resolution of the resulting data, while the latter are the *extent* over which the data are collected. A given grain and extent together constitute an observational scale—that is, a system of measurement that permits comparison of unlike things by abstracting away their dissimilar attributes. (Think, for example, of a meter stick, whose grain is a millimeter, extent is a meter, and which can be used to measure the length of all sorts of different things (Sayre and Di Vittorio 2009).) Data also carry temporal scales, with grains of, for example, days, months, or years, and extents of years or longer. Notice that here, scale is *not* synonymous with level, and no hierarchical relationship between scales is implied or presupposed.

It is important to distinguish between operational and observational scales: the former are real attributes of phenomena in the world, whereas the latter are epistemological tools, chosen and applied by the observer, to make sense of those phenomena (Sayre 2005). It is not surprising that so many social scientists have practiced "methodological nationalism" (Brenner 2004: 38), because the state is at once convenient and peculiar in the degree to which it performs both roles simultaneously: indeed, its authority to apply observational scales is a constitutive part of producing the effects that make it an operational scale (Bourdieu 1994). But failure to distinguish between them can mislead researchers into taking units of measurement (e.g., households or states) as given *a priori* because they *both* appear "natural" when viewed through the lens of state data *and* are made real through the social effects they help to produce. This social-epistemological puzzle lies at the heart of recent work in several fields, including political ecology, demonstrating that the operationalization of state scales is highly fraught and imperfect (e.g., Porter 1995; Robertson 2004; Prudham 2005; O'Neill 2006; Harris and Alatout 2010; Goldman et al. 2011; Mathews 2011).

The general lesson here is that operational and observational scale represent ontological and epistemological moments, respectively, in an ongoing dialectical relationship of material phenomena and human attempts to understand them. More specifically, researchers cannot avoid employing observational scales (this is true even of qualitative methods), and they ought to choose the grain and extent of their inquiries deliberately. Moreover, because material phenomena occur at operational scales, it is important to choose, insofar as possible, observational scales that match or fit the operational scales of the phenomena of interest.

The state provides a relatively clear instance of a geographical scale. But what about Blaikie and Brookfield's other scales: the individual, the local/regional, and the global? Are these operational or observational scales, or both? How are they produced? And what about phenomena that do not operate through or within the state's neatly defined territorialities? The state may try to dictate land management practices for farmers within its territory, for example, but what actually happens "on the ground" may deviate considerably from these prescriptions, and the relevant factors may not have such clearly demarcated spatial boundaries. The vague spatiality of Blaikie and Brookfield's local/regional scale is indicative of these considerations. The movements of people, goods, ideas, and capital, as well as all kinds of biophysical processes (e.g., the cycling of water, soils, and nutrients; the climate system; evolution) have operational scales, and they call for observational scales chosen to apprehend them appropriately. But the scale(s) of the state are often only contingently related, or entirely unrelated, to the scales of these phenomena (Figure 39.1). Moreover, whether there are "levels" that organize such phenomena is far less clear than it is in the case of the state.

Ecological science illustrates well the kinds of predicaments that scalar analysis of non- or extra-state phenomena must often confront. In an influential paper, Wiens (1989) analyzed empirical data from various ecological systems and demonstrated not only that patterns depended on the grain and extent of observation, but also that patterns discerned at different scales could contradict one another. This suggests that different processes are determinative at different scales, and that the *relations among scales* pose key challenges for scientific understanding. Two

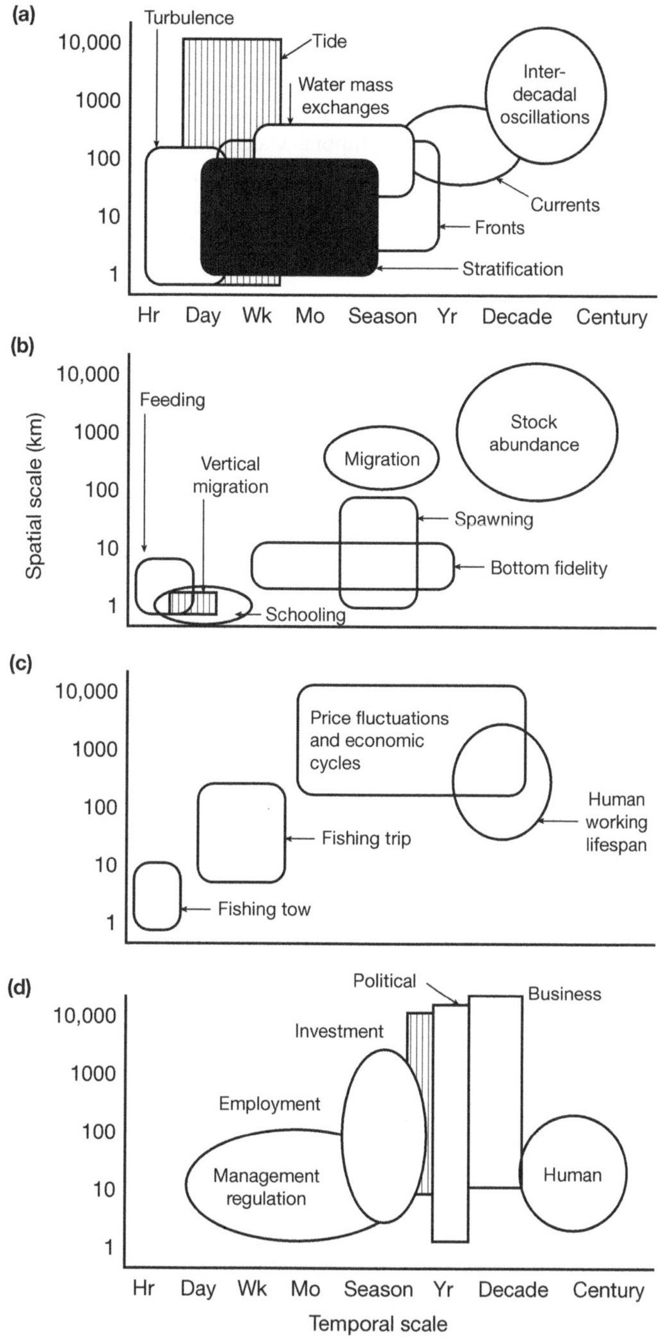

Space/timescale diagram of characteristic processes from the natural sciences: (a) physical; (b) biological; and from the social sciences: (c) fishing; (d) fishing communities.

Figure 39.1 The operational scales of biophysical and social processes affecting fisheries. From Perry and Ommer (2003).

points warrant emphasis here. First, ecologists frequently use hierarchy theory as a heuristic framework, with "levels" defined in a loosely functional sense (e.g., organism, population, community). But this is understood not as an exclusive hierarchy (in the state-bureaucratic, top-down sense discussed earlier) but as a constitutive hierarchy, in which phenomena at a "lower" or smaller scale may display different patterns when aggregated at a "higher" or larger scale—patterns that are irreducible to their smaller-scale components (so-called emergent properties, aka the whole is greater than the sum of its parts). Second, insights such as these have contributed to the broader shift away from equilibrium-based ecological theories, drawing attention to non-linear or threshold dynamics and helping to inform what Botkin (1990) termed "the new ecology" (cf. Zimmerer 1994). A key insight of the new ecology is that small differences in initial conditions can have large effects on subsequent dynamics, meaning that history becomes much more important than earlier ecological theories acknowledged (May 1977).

In summary, scale refers variously to size, level, and relation (Howitt 1998, 2003), and it is important to be clear about all three (Table 39.1). Scale as size is an observational-epistemological matter. Scale as level may be observational or operational (or both), as the nation-state example illustrates. Scale as relation is ontological, having to do with how processes operating at different scales interact in ways that are not simply aggregative or linear but instead produce qualitatively distinct material consequences (Sayre 2009; Sayre and Di Vittorio 2009).

Table 39.1 The multiple aspects of scale (from Sayre and Di Vittorio 2009: 22)

Scale as	*Also known as*	*Metaphysical status*	*Expressed*	*Consists of*	*Concerned to measure or understand*
Size	Observational scale; absolute scale	Epistemological	Quantitatively	Grain and extent	Weight, size, area, distance, duration, speed, etc.
Level	Conceptual scale. May be observational or operational; ongoing effort to reduce disparity between the two	Either epistemological or ontological	Qualitatively	Multiple scales-as-size arranged functionally and/or hierarchically	Different orders within one such metric
Relation	Operational scale; relative scale	Ontological	Both: where change in quantity becomes change in quality	Processes interacting across scales-as-levels; relations between scales (e.g., how to "scale up" or "scale down")	Scaling effects; thresholds or nonlinearities produced by cross-scale interactions; scale mismatches

"The scale question" in human geography

The distinctions identified above go a long way toward untangling the debates surrounding "the scale question" (Brenner 2001: 592, quoting Lefebvre 1976: 68) that erupted in human geography in the 1990s and early 2000s. The debate began from Peter Taylor's 1982 essay in *Transactions of the Institute of British Geographers* titled "A Materialist Framework for Political Geography." Calling for "a political economy of scale," Taylor identified the global as "the scale of reality," because it was at this scale that capital accumulation and circulation operated in the modern world-system. He argued that individuals did not experience this scale directly; rather, their lives unfolded at "the scale of experience," which Taylor equated with urban settings or systems. The contradictions between these two scales were managed (but not resolved) by the state, understood as "the scale of ideology," which separated and articulated reality and experience.

Taylor's intent was to question political geography's preoccupation with the nation-state while putting the field on a firm materialist basis, and he insisted that his three scales were not given *a priori* but were socially produced by the dynamic operations of the capitalist world-system. In particular, he stressed that "there is nothing 'natural' about the modern state" (Taylor 1982: 27). These points were acknowledged and broadly shared by early contributors to the subsequent debate, such as Smith (1984, 1992), Brenner (1997), Swyngedouw (1997), Delaney and Leitner (1997) and Marston (2000). But Taylor's choice of terms echoed the base-superstructure rubric of Marx's 1859 *Preface to a Critique of Political Economy*, and despite Taylor's explicit arguments to the contrary, subsequent scholars questioned the implication that local or urban experience was less "real" than the global, or that the state was merely ideological and not, again, real in its effects. World systems theory, others pointed out, "simply shifted the focus from one scale—the national—to another—the world system," whereas "what is needed is a multiscalar approach" (Mahon and Keil 2009: 10).

Taylor was proposing an operational scale and suggesting that political geographers ought to build their observational scale accordingly. Smith (1992: 74) recognized this distinction, albeit using different terms, and cautioned against equating "the local strictly with the concrete, the global with the general," and Swyngedouw (1997) and Brenner (1998) pointed out that the scales of capitalist (re-)production were dynamic; all three scholars identified the politics of scale as central to capitalist restructuring. Globalization, for example, did not render the nation-state less important but rather involved *rescaling* its relationships with both smaller and larger scales; Swyngedouw (1997) dubbed it "glocalisation." In a subsequent piece, Brenner (2001) further insisted that scale be distinguished from other core geographical concepts such as place, territory, and network. But others mistook Taylor (and Brenner) as imposing an epistemology that ignored or denigrated the experiences and agency of people in their everyday lives. Marston (2000), for example, challenged the priority assigned to capital accumulation and called for greater attention to processes of social reproduction rooted in non-wage labor (especially by women) at the scale of the household. Marston and Smith (2001; cf. Smith 1992) added the body as another scale worthy of inclusion.

As the debate escalated early in the 2000s, the distinction between scale and level was also overlooked, as almost everyone used the two terms interchangeably (Brenner 2001; Marston and Smith 2001; cf. Sayre 2005). No one disputed that scale was socially produced and therefore contestable and historically contingent, nor that conventional geographical scales such as the urban, the state, and the global should be questioned. But the debate foundered on whether scale is inherently hierarchical. Marston et al. (2005: 420) concluded that it was, and that scale and level could "be simply and effectively collapsed into" one another. Having done this, they reasoned that the concept was inextricable from domination and inequality in the world, and they therefore proposed to "expurgate scale from the geographic vocabulary" and replace it with a "flat ontology" (2005: 422). Curiously, they claimed that their flat ontology consisted of "self-organizing systems" (2005: 422) with emergent properties—precisely what ecologists associate with the constitutive hierarchies described above. Scale as relation went unmentioned in their argument.

Following numerous critical responses (e.g., Collinge 2006; Jonas 2006; Leitner and Miller 2007), the theoretical debate effectively ceased, apparently of exhaustion. It was replaced by empirical investigations in which the material reality of hierarchical relationships in the world—including but not limited to those of the state—was recognized as requiring methods and epistemologies attuned to the issue of scale. For example, in an edited volume of papers from the time, published several years later (Keil and Mahon 2009), scholars used scale to explore social movements such as anti-globalization, immigrants' and indigenous rights, the Black Panthers, and environmentalism, as well as child care provision and public health, so-called creative cities, and the global designer fashion industry. Many of the cases involved networks that operated across scales, connecting people and ideas in scalar but non-hierarchical ways. Several authors emphasized the need to delimit scale conceptually and to avoid privileging it over place, territory, networks, and mobility. Above all, they stressed attention to processes (which have scales or scaling effects) over scale per se. Some authors seemed to understand scale as hierarchical, while others did not—but this wasn't seen as problematic. The editors summarized the volume with the claim that "while the national state is no longer the pivotal scale, no other scale has succeeded in taking its place" (Mahon and Keil 2009: 12).

Political ecologies and scale

Nature and ecology were not prominent topics in the debates described above (Swyngedouw 2004). And as political ecology has grown over recent decades, scale has sometimes been invoked as a core problematic of the field (e.g., Zimmerer and Bassett 2003; Swyngedouw and Heynen 2003; Paulson and Gezon 2005) and sometimes remained in the background, relatively unremarked (e.g., Neumann 2005; Robbins 2012). The basic idea that social-environmental issues are simultaneously local and global—that there are both small- and large-scale factors at work—has remained a constant theme (e.g., Keil et al. 1998), even as the stability and coherence of "the local" and "the global" have been questioned. But explicit theorizations of scale itself have been relatively rare. Zimmerer and Bassett (2003: 3) challenged Blaikie's scales as "pregiven sociospatial containers" and suggested that scales are, instead, "social-environmentally produced." The chapters in their edited volume, they wrote, demonstrated "a variety of scalar configurations that display vertical (hierarchical, nested) and horizontal (networked) patterns," but they did not attempt to synthesize a theory of scale, instead merely highlighting "the central importance of ecological scale in shaping political-ecological dynamics" (2003: 4). Neumann (2009: 403) identified three themes from work in political ecology that together "suggest a richer theorization of scale: (1) the interactions of power, agency, and scale; (2) socioecological processes and scale; and (3) scaled networks." He saw these as incorporating "the key precepts of the politics of scale—scale as socially constructed, relational, contingent, and contested—into an existing framework that highlights power relations and a dialectical approach toward nature-society relations" (2009: 404).

Swyngedouw and Heynen (2003) present what is probably the most thoroughgoing theorization of scale in political ecology, building on Swyngedouw's (1997, 1999) earlier work on "glocalisation" and on the history of the Spanish waterscape. "The priority, both theoretically and politically," they write:

> never resides in a particular social or ecological geographical scale; instead, it resides in the socio-ecological process through which particular social and environmental scales become constituted and subsequently reconstituted. In other words, socioecological processes give rise to scalar forms of organisation—such as states, local governments, interstate arrangements and the like—and to a nested set of related and interacting socioecological spatial scales. In addition, these territorial scalar arrangements intersect—often in contradictory and conflicting ways—with the scalar networks of, for example, socioecological production and consumption systems.
>
> *(Swyngedouw and Heynen 2003: 912–913)*

It is perhaps not surprising that water and cities provided the empirical raw material for Swyngedouw and Heynen's reflections on scale. The geomorphological organization of watersheds is an excellent biophysical example of constitutive hierarchies, and the propensity of modern industrial societies to rearrange watersheds through massive engineering projects perfectly illustrates the idea of socioecological rescaling, as Swyngedouw (1999, 2004) shows to powerful effect in the cases of Spain and Ecuador. And the "metabolism" of cities—a metaphor that is itself an instance of rescaling, from the level of an organism to that of entire urban areas—likewise captures the complex intersection of social and biophysical processes in modern capitalism. Cities are at once sites of neoliberal state rescaling—in the devolution of regulatory authority from the national to the urban scale, for example—and embedded in networks of energy, water, raw materials, food, and waste that operate at all sorts of spatial and temporal scales. Urban political ecology sets itself the task of understanding how these socionatural processes are produced and how they interact with each other and with people, markets, built environments, and institutions.

Broadly speaking, Swyngedouw and Heynen see non-hierarchical scales in "the circulation of capital and its associated socioecological, metabolic transformation processes," and hierarchical scales in the state's territorializing processes "of regulation and governance in which these are embedded" (2003: 913). Capital in all its forms—money, means of production, raw materials, labor—flows through (and produces) networks of various scales, mobilizing and altering biophysical processes of all kinds. States attempt to guide or control these transformations through hierarchically organized scales of political organization, laws, and regulations. But the coherence and effectiveness of those efforts are tenuous and contested in the face of both networked and hierarchical maneuvering and resistance by firms, bureaucrats, activists, and NGOs. "[A] process-based approach to scale focuses attention on the mechanisms of scale transformation through social conflict and political struggle" (Swyngedouw and Heynen 2003: 913). In short, the scales of all these processes not only affect their operation and outcomes, but also become key stakes and strategies in their own right.

What this theory of scale offers, then, is a series of epistemological and methodological guidelines: Do not take the scales of one's analysis for granted; identify the key processes that produce a phenomenon, and induce their scales empirically; be alert to how processes are rescaled, and to the possibility of non-linear, qualitative change across scales; be reflexive and critical about how observational scales may affect the patterns one sees in the resulting data. Overall, these guidelines suggest an open-ended approach to scale, with the potential for a virtually limitless array of particular applications.

Although most work in political ecology has not emphasized scale conceptually, the analytical scope and content of the field is broadly compatible with these guidelines. Political ecologists routinely conduct research at multiple sites and scales of inquiry, following the processes empirically rather than positing their scales *a priori*. They often identify qualitative differences between sites connected by these processes, keyed to myriad cultural, ecological, political, and economic factors that affect and are affected by these connections. That the scales of political, economic, and ecological processes are often mismatched is unsurprising and frequently goes without saying. Historical legacies figure prominently in most political ecological accounts, bringing in temporal as well as spatial dimensions of scale. And in many cases, political ecologists rely on relational and process-based approaches to conceptualizing and studying environmental problems. The aggregate result is a wealth of empirically rich cases, which could lend themselves to closer comparative analysis and synthesis if examined more explicitly through the lens of scales and polities. How scientists conceive and study environmental problems—as local, regional, or global, for example—depends on both observational and operational scales, and this in turn shapes the political possibilities for addressing them (Taylor and Buttel 1992). Social movements and environmental activists can turn the state's exclusive hierarchies to their advantage by "jumping scales" to higher levels of government (McCarthy 2005; Urkidi 2010). Similarly, numerous indigenous groups have found that bypassing the nation-state altogether and asserting their rights at international levels can enable them to tilt the balance of power more in their favor (Perreault 2003).

The potential for such synthesis is further indicated by recent works that explore the intersection of markets, state agencies, and scientific practices in neoliberal environmental governance. If the state routinely imposes observational scales of measurement, it often does so specifically to render nature "legible" in terms suited to commodification and market-based regulation (Robertson 2006). Scientists may be enrolled in these efforts whether or not the underlying methods and metrics are coherent or robust, and the resulting contradictions and failures provide compelling examples of the difficulty (or impossibility) of harmonizing the logics of nature and capital. If academic scientists decline to produce the types of knowledge needed by regulators and investors, a kind of state-produced market may conjure extra-academic alternatives into prominence (Lave 2012). What emerges is a broad (if imperfect) homology between the state's use of observational scales to abstract, measure and regulate people and ecosystems, and capital's reliance on the value abstraction to convert labor and nature into commodities (Robertson and Wainwright 2013). As Mann (2007) shows, the value abstraction is itself at stake in many struggles over "the politics of measure," an insight that has abundant potential for application throughout a world in which—as Blaikie and Brookfield (1987: 68) put it—"the commoditization of land, labour and agricultural production" extends to "almost every element in the world economy."

References

Blaikie, P. and Brookfield, H. (1987) *Land degradation and society*. London and New York: Methuen.

Botkin, D.B. (1990) *Discordant harmonies: a new ecology for the twenty-first century*. New York: Oxford University Press.

Bourdieu, P. (1994) "Rethinking the state: genesis and structure of the bureaucratic field." Wacquant, L.J.D. and Farage, S. (trans.) *Sociological Theory*, vol. 12, pp. 1–18.

Brenner, N. (1997) "State territorial restructuring and the production of spatial scale: urban and regional planning in the Federal Republic of Germany, 1960–1990." *Political Geography*, vol. 16, pp. 273–306.

Brenner, N. (1998) "Between fixity and motion: accumulation, territorial organisation and the historical geography of spatial scales." *Environment and Planning D: Society and Space*, vol. 16, pp. 459–481.

Brenner, N. (2001) "The limits to scale? Methodological reflections on scalar structuration." *Progress in Human Geography*, vol. 25, pp. 591–614.

Brenner, N. (2004) *New state spaces: urban governance and the rescaling of statehood*. Oxford: Oxford University Press.

Collinge, C. (2006) "Flat ontology and the deconstruction of scale: a response to Marston, Jones and Woodward." *Transactions of the Institute of British Geographers*, vol. 31, pp. 244–251.

Delaney, D. and Leitner, H. (1997) "The political construction of scale." *Political Geography*, vol. 16, pp. 93–97.

Goldman, M.J., Nadasdy, P., and Turner, M.D. (2011) *Knowing nature: conversations at the intersection of political ecology and science studies*. Chicago and London: University of Chicago Press.

Harris, L.M. and Alatout, S. (2010) "Negotiating hydro-scales, forging states: comparison of the upper Tigris/Euphrates and Jordan River basins." *Political Geography*, vol. 29, pp. 148–156.

Howitt, R. (1998) "Scale as relation: musical metaphors of geographical scale." *Area*, vol. 30, pp. 49–58.

Howitt, R. (2003) "Scale." In Agnew, J., Mitchell, K. and Tuathail, G.O. (eds.), *A companion to political geography*. Oxford: Blackwell, pp. 138–157.

Jonas, A.E.G. (2006) "Pro scale: further reflections on the 'scale debate' in human geography." *Transactions of the Institute of British Geographers*, vol. 31, pp. 399–406.

Keil, R. and Mahon, R. (eds.) (2009) *Leviathan undone? Towards a political economy of scale*. Vancouver and Toronto: UBC Press.

Keil, R., Bell, D.V.J., Penz, P., and Fawcett, L. (1998) *Political ecology: global and local*. London and New York: Routledge.

Lave, R. (2012) *Fields and streams: stream restoration, neoliberalism, and the future of environmental science*. Athens and London: University of Georgia Press.

Lefebvre, H. (1976) *De l'Etat* (volume 2). Paris: Union Generale d'Editions.

Leitner, H. and Miller, B. (2007) "Scale and the limitations of ontological debate: a commentary on Marston, Jones and Woodward." *Transactions of the Institute of British Geographers*, vol. 32, pp. 116–125.

Mahon, R. and Keil, R. (2009) "Introduction." In Keil, R. and Mahon, R. (eds.), *Leviathan undone? Towards a political economy of scale*. Vancouver and Toronto: UBC Press, pp. 3–23.

Mann, G. (2007) *Our daily bread: wages, workers, and the political economy of the American West*. Chapel Hill: University of North Carolina Press.

Marston, S.A. (2000) "The social construction of scale." *Progress in Human Geography*, vol. 24, pp. 219–242.

Marston, S.A. and Smith, N. (2001) "States, scales and households: limits to scale thinking? A response to Brenner." *Progress in Human Geography*, vol. 25, pp. 615–619.

Marston, S.A., Jones III, J.P. and Woodward, K. (2005) "Human geography without scale." *Transactions of the Institute of British Geographers*, vol. 30, pp. 416–432.

Mathews, A.S. (2011) *Instituting nature: authority, expertise, and power in Mexican forests*. Cambridge, MA and London: MIT Press.

May, R.M. (1977) "Thresholds and breakpoints in ecosystems with a multiplicity of stable states." *Nature*, vol. 269, pp. 471–477.

McCarthy, J. (2005) "Scale, sovereignty, and strategy in environmental governance." *Antipode*, vol. 37, pp. 731–753.

Neumann, R.P. (2005) *Making political ecology*. London: Hodder Arnold.

Neumann, R.P. (2009) "Political ecology: theorizing scale." *Progress in Human Geography*, vol. 33, pp. 398–406.

O'Neill, K.M. (2006) *Rivers by design: state power and the origins of U.S. flood control*. Durham, NC and London: Duke University Press.

Paulson, S. and Gezon, L.L. (2005) *Political ecology across spaces, scales, and social groups*. New Brunswick: Rutgers University Press.

Perreault, T. (2003) "Making space: community organization, agrarian change, and the politics of scale in the Ecuadorian Amazon." *Latin American Perspectives*, vol. 30, pp. 96–121.

Perry, R.I. and Ommer, R.E. (2003) "Scale issues in marine ecosystems and human interactions." *Fisheries Oceanography*, vol. 12, pp. 513–522.

Porter, T.M. (1995) *Trust in numbers: the pursuit of objectivity in science and public life*. Princeton: Princeton University Press.

Prudham, W.S. (2005) *Knock on wood: nature as commodity in douglas-fir country*. New York: Routledge.

Robbins, P. (2012) *Political ecology: a critical introduction* (2nd edn). West Sussex: Wiley-Blackwell.

Robertson, M.M. (2004) "The neoliberalization of ecosystem services: wetland mitigation banking and problems in environmental governance." *Geoforum*, vol. 35, pp. 361–373.

Robertson, M.M. (2006) "The nature that capital can see: science, state, and market in the commodification of ecosystem services." *Environment and Planning D: Society and Space*, vol. 24, pp. 367–387.

Robertson, M.M. and Wainwright, J.D. (2013) "The value of nature to the state." *Annals of the Association of American Geographers*, vol. 103, pp. 890–905.

Sayre, N.F. (2005) "Ecological and geographical scale: parallels and potential for integration." *Progress in Human Geography*, vol. 29, pp. 276–290.
Sayre, N.F. (2009) "Scale." In Castree, N., Demerritt, D., Rhoads, B. and Liverman, D. (eds.), *A companion to environmental geography*. Oxford: Wiley-Blackwell, pp. 95–108.
Sayre, N.F. and Di Vittorio, A.V. (2009) "Scale." In Kitchen, R. and Thrift, N. (eds.), *International encyclopedia of human geography*, vol. 1. Oxford: Elsevier, pp. 19–28.
Smith, N. (1984) *Uneven development*. Cambridge, MA: Blackwell.
Smith, N. (1992) "Geography, difference and the politics of scale." In Doherty, J., Graham, E. and Malek, M. (eds.), *Postmodernism and the social sciences*. New York: St. Martin's Press, pp. 57–79.
Swyngedouw, E. (1997) "Neither global nor local: 'glocalization' and the politics of scale." In Cox, K.R. (ed.), *Spaces of globalization: reasserting the power of the local*. New York & London: Guilford Press, pp. 137–166.
Swyngedouw, E. (1999) "Modernity and hybridity: nature, regeneracionismo, and the production of the Spanish waterscape, 1890–1930." *Annals of the Association of American Geographers*, vol. 89, pp. 443–465.
Swyngedouw, E. (2004) "Scaled geographies: nature, place, and the politics of scale." In Sheppard, E. and McMaster, R.B. (eds), *Scale and geographic inquiry: nature, society, and method*. Oxford: Blackwell, pp. 129–153.
Swyngedouw, E. and Heynen, N.C. (2003) "Urban political ecology, justice and the politics of scale." *Antipode*, vol. 35, pp. 898–918.
Taylor, P.J. (1982) "A materialist framework for political geography." *Transactions of the Institute of British Geographers* (new series), vol. 7, pp. 15–34.
Taylor, P.J. and Buttel, F.H. (1992) "How do we know we have global environmental problems? Science and the globalization of environmental discourse." *Geoforum*, vol. 23, pp. 405–416.
Urkidi, L. (2010) "A glocal environmental movement against gold mining: Pascua-Lama in Chile." *Ecological Economics*, vol. 70, pp. 219–227.
Wiens, J.A. (1989) "Spatial scaling in ecology." *Functional Ecology*, vol. 3, pp. 385–397.
Wolf, E. (1972) "Ownership and political ecology." *Anthropological Quarterly*, vol. 45, pp. 201–205.
Zimmerer, K.S. (1994) "Human geography and the 'new ecology': the prospect and promise of integration." *Annals of the Association of American Geographers*, vol. 84, pp. 108–125.
Zimmerer, K.S. and Bassett, T.J. (eds.) (2003) *Political ecology: an integrative approach to geography and environment-development studies*. New York and London: Guilford Press.

PART IV, SECTION D

Environmental identities

The four chapters in this section examine *how social subjectivities are shaped through, and reflect, differential access to and control over nature.* Focusing on gender, race, class, indigeneity, and other forms of embodied difference, the chapters exemplify political ecology's long-standing concern with the ways identity and difference are socially constituted through struggles over natural resources. They also illustrate political ecology's more recent acknowledgment of the complex influences of nature's materiality on social difference. The section opens with a pair of chapters that outline the extensive contributions made by feminist political ecology to problematizing questions of identity. The opening chapter by Rebecca Elmhirst considers the growing internal differentiation of feminist political ecology, as it has evolved from an initial concern with gendered struggles over resource access to an interest in how gendered identities are produced as part of multiple and complex subjectivities. This chapter illustrates how feminist political ecology's openness to diverse theoretical influences has proven to be a significant strength in understanding the social constitution of difference.

In the following chapter, Emily Yeh and Joe Bryan explore contemporary political ecology's debt to postcolonial studies. The authors emphasize the political and intellectual work done by the concept of indigeneity as the term circulates through policy documents, social movements, and academic interventions. Focusing on the generative possibilities of indigeneity, the authors highlight the significant implications for political ecology of thinking with the ontologies and cosmologies mobilized by indigenous social movements. The chapter by Michael Ekers turns to examine class as a social identity produced through material and representational relationships with the non-human world. A key insight here is that class – like race, gender, and other forms of social difference – is not a category that pre-exists a relation to nature: instead, class is constituted through diverse forms of labor, including both waged and non-waged work. The final chapter in this section examines nature, difference, and the body. Julie Guthman and Becky Mansfield review political ecology's diverse engagements with the body before considering the implications of recent advances in epigenetics, which recognize the contingency of gene expression on environmental context. Medical research on obesity and autism, for example, highlights the possibility of "environmentally-induced difference" at the scale of the body. Advances in epigenetics, they argue, suggest political ecology needs not only address the biological and biochemical materiality of the body, but also the way biologically and chemically induced differences arise as a result of the ways that bodies interact with diverse environments.

40
FEMINIST POLITICAL ECOLOGY

Rebecca Elmhirst

Introduction

Feminist political ecology emerged as a subfield of Political Ecology in the 1990s, developing initially from gender and development studies, with which it shares a broad commitment to understanding the dynamics of gender in relation to the natural environment and in the context of natural resource-based livelihoods. As with Political Ecology more generally, Feminist Political Ecology (FPE hereafter) emphasizes politics and power at different scales, but goes further in highlighting gendered power relations, and in making an explicit commitment towards tackling gender disadvantage and inequality. FPE directs attention towards gendered processes within the politics of environmental degradation and conservation, the neoliberalization of nature and ongoing rounds of accumulation, enclosure and dispossession associated with each of these. Work within this field seeks to complicate arenas of assumed common interest, such as "community" and "household", and to explore the connections between nature, gendered subject formation and the body. Of central interest are the gender dimensions of struggles over nature and the environment, and how might these intersect and be informed by feminist objectives, strategies and practices. Whilst FPE embraces a diversity of approaches and subject matters, there is a shared (if often implicit) commitment to feminist epistemology, methods and values, where dominant, masculinist conceptions and practices of knowledge and authority are recognized and challenged, and where emphasis is given to research and practice that empowers and promotes social and ecological transformation for women and other marginalized groups.

One of the initial, and most widely cited, expositions of FPE is the collection edited by Dianne Rocheleau and colleagues, who set out a road map for FPE research and practice, by inviting political ecologists to extend their analysis of power to include gender relations, and to extend their consideration of politics to take in closer scales of analysis, spurred by an explicit and avowedly feminist concern to transform gender hierarchies and create more equitable outcomes for women (Rocheleau et al. 1996). Their volume, *Feminist Political Ecology: Global Issues and Local Experiences* provided a loosely configured framework that placed gender as "a critical variable in shaping resource access and control, interacting with class, caste, race, culture, and ethnicity to shape processes of ecological change, the struggles of men and women to sustain ecologically viable livelihoods, and the prospects of any community for "sustainable development" (Rocheleau et al. 1996: 4).

Since the publication of *Feminist Political Ecology*, this subfield has evolved in response to an almost seismic shift in theoretical approaches to gender in the social sciences more generally, and has gained also from an engagement with feminist science studies, which offers the conceptual language for both recognizing the situatedness of knowledge claims, and for tackling the kinds of epistemic privilege and authority within Political Ecology (and beyond) that render women's interests and gendered power relations invisible. Poststructuralist and performative approaches in feminist theory (e.g. Butler 2004) have challenged role-based and relational approaches to gender that rest predominantly on fixed notions of the autonomous subject, and that focus on men's and women's differentiated access to and control of environmental resources and socio-political processes. Empirically, FPE has advanced in new directions in response to the impacts of a changing (and increasingly neoliberal) policy climate. Economic reform programmes that favour market-led approaches to natural resource governance have deepened, whilst at the same time most rural populations are more mobile and more urbanized, as rural–urban and transnational linkages complicate and rework resource-based livelihood practices and institutions, often in gender-differentiated ways. The intensification of processes of environmental degradation (deforestation, desertification, climate change and urbanization) have brought new challenges which, in turn, have created new and often enlarged shocks and stresses to livelihoods, often in gendered ways. By reconfiguring patterns of natural resource use, environmental degradation has heralded new forms of development intervention and environmental governance that are themselves inflected with gendered discourses and assumptions that deepen differentiated and unjust life opportunities and exclusions.

It is in this context that there has been a renewed interest in the potential offered by FPE to further the wider transformative potential of feminism in gender and development studies, and to address the kinds of challenges posed by global environmental change and the neoliberal, marketized responses this engenders (Elmhirst and Resurrección 2008). This chapter provides an overview of recent work in this field, exploring the conceptual terrain underpinning FPE and opening up new questions and concerns for consideration. The chapter begins by outlining what counts as feminist political ecology before examining its development in four related strands: (1) analyses of gendered resource access and control which reflect feminist engagement with political ecology's Marxist heritage, (2) recent poststructuralist theorizations of gendered subjectivity and power, (3) emerging debates that draw on material feminist theories around the relationship between human and non-human nature, and (4) ideas around a feminist ethics of environmental care that suggest a renewed engagement with ecofeminism. Each of these areas is inspired by particular iterations of feminist theorizing and praxis, from feminist-inflected Marxist analyses of enclosure and resource access through to new material feminisms that are prompting a radical rethinking of the permeable boundaries between humans and nature, and in turn, raising new questions for political ecology more broadly. The chapter concludes by reflecting on the challenge for FPE of translating these critiques into emancipatory practice in order to address a widely held disquiet (e.g. Cornwall et al. 2007) that gender has lost its critical and politicized edge within mainstream natural resource management, having been institutionalized into a series of tools and techniques far removed from the wider goals associated with the label "feminist".

In what sense a "feminist" political ecology?

A review of the literature on "gender and environment" suggests relatively little work self-defines as "feminist political ecology". However, research bearing a "family resemblance" (Watts 2000: 271) to feminist political ecology may be found across a range of disciplines, in work on issues ranging from gendered resource access and property rights (water and land) to the dynamics of gender in policy discourses, collective action and social movements, much of which might be regarded as FPE but is not named as such (Elmhirst 2011a). There may be a reluctance in some instances to use the label "feminist" in gender and environment research, where it carries unhelpful resonances and unwanted political meanings (Wright 2008), particularly in some contexts in the Global South where both postcolonial critiques (Mohanty 1988) and new conservatisms (religious or otherwise) complicate feminism's meaning and intent (Cornwall et al. 2007). FPE may also have slipped from the agenda at the end of the 1990s, at which time gender was being "decentered" as a prime analytical category in poststructuralist theorizing (Radcliffe 2006), and when concerns emerged around the unintended consequences of feminist strategies based on notions of women's shared interests in the environment, as articulated around the time of the Rio Earth Summit. Many policy initiatives that arose at this time targeted women as an homogenous and undifferentiated grouping, charging them with "care" for degraded environments, resulting in greater burdens for women and exacerbating gender injustices in many documented instances (Leach 2007). More recently, various activist networks that focus on environments and social justice are adopting nuanced approaches to subjectivity and power, in which gender is seen neither as analytically central nor as the end point of critique and analysis. Scholars and activists are asking where and how gender might be considered a central concern, and how political ecology, articulated in such a way, might be regarded as "feminist". Awkward questions about the transformative possibilities of "feminism" within feminist political ecology remain.

Recently, a new wave of "material feminist theory" has grappled with such questions to suggest, in line with feminist epistemology more generally (Harding 2004), that there are always instances where there is an ethical imperative to "bring women into view" and to transform gender relations. In part, these questions form a useful counterpoint to new theoretical developments in environmental social theory where the influence of, for example, actor network theory goes even further in challenging notions of subjectivity, by unsettling understandings of the human (see Chapter 16, this volume). For some, this kind of analytical strategy runs the risk of reinstating an unmarked, disembodied but implicitly masculine subject (Sharp 2009). But rather than retreat to simplistic understandings of gender divisions and singular gendered power relations, the epistemological questions being raised by material feminist theory are providing inspiration for renewing a "feminist" perspective on environmental questions in FPE. As Barad puts it, "intra-active practices of engagement not only make the world intelligible in specific ways but also foreclose other patterns of mattering … therefore accountability and responsibility must be thought in terms of what matters and what is excluded from mattering" (Barad 2007: 394). The remainder of this chapter considers some of the avenues currently being pursued in FPE, and in which gender is shown to matter in important ways.

A feminist political ecology of resource access and control

To date, much FPE has centred on questions of resource access and control, drawing on political ecology's Marxist heritage and extending it to consider closer scales in which politics play out, i.e. within households and communities. Such studies seek to illuminate "the crucial role of family authority relations and property relations in structuring the gender division of labour and access to rural resources" such as land and labour (Carney 2004: 316). Work has detailed the gender-specific impacts of ecological change and/or environmental interventions, and how these are shaped by existing household divisions of labour and differing resource rights of men and women. More recently, this kind of work has been taken forward in studies of the gendered impacts of nature's neoliberalization (as showcased in the collection edited by Resurrección and Elmhirst 2008) and has been given added impetus by attention on the gendered impacts of dispossession and land/water grabbing in the Global South (Behrman et al. 2012; Wisborg 2014). A similar line of reasoning is opening up through studies of the gendered impacts of climate change, where FPE is usefully put to work to look not only at how climate change has gender-specific impacts, but also how knowledges, policies and practices that coalesce around climate change adaptation bring gendered effects (Arora-Jonsson 2011; Bee et al. 2013; Tschakert 2013). For example, market-based approaches to ameliorating climate change through "payment for ecosystem services" schemes, which reward resource users for avoiding deforestation, are based on formulations of property rights that may erase pre-existing informal modes of resource access on which women and other marginalized groups depend, thus deepening gender disadvantage in unanticipated ways. A common theme is that men and women hold gender-differentiated interests in the environment and natural resources through their distinctive roles, responsibilities and knowledge within household/family divisions of labour. Gender is thus understood as a critical variable in shaping processes of ecological change and the pursuit of viable livelihoods (Elmhirst and Resurrección 2008: 5).

This strand of FPE has also considered household and community *gender relations* as a critical and often overlooked site for politics, particularly where environmental interventions have brought about gender conflict within households (and across conjugal partnerships) generating in turn ecological effects. Both Schroeder (1999) and Carney's (2004) work in the Gambia documents the intra-household conflicts that arose following interventions to enhance the productivity of wetlands. Women's customary access to rice land, a key source of income, was undermined by donor-sponsored irrigation schemes and horticultural projects, which also brought new demands by men on female labour, both of which were widely contested within households and communities (Carney 2004). This kind of work has been extended to reveal the ways that gendered resource contestations are also historically and geographically situated struggles over meanings and identities, as gender categories and gender norms are themselves negotiated and socially produced in the course of environmental struggles (Resurrección 2006; Radel 2012).

The importance of gender in family authority structures and conjugal relations in shaping resource access and control is perhaps most clearly seen in settings where the ability to derive benefits from resources is contingent on social relationships that constrain or enable the realization of such benefits (Ribot and Peluso 2003). In much of South Asia, hierarchical social norms and practices associated with the conjugal partnership place women in a situation of dependence on male kin, who become a central conduit for access to resources (including land, labour and capital), and this creates gender-specific vulnerabilities for those experiencing marital breakdown or widowhood (Rao 2006; Agarwal 2003). FPE provides the conceptual tools necessary for revealing intra-household power dynamics of this kind. By problematizing the assumed division between public and private spheres, work has shown how gendered discourses and practices associated with national and international policies bleed into the reproductive realm. Early instances of this work linked FPE to a wider critique of the impacts of neoliberalism and marketization on gendered access to land or water, gendered labour and gendered livelihoods, particularly in Africa and parts of Southeast Asia such as Vietnam and Cambodia (Razavi 2003; Hue Le Thi Van 2008). Changes associated with liberalization and marketization are seen as playing a significant role in marginalizing women from natural resources when kin or community-based tenure systems are transferred into commoditized and individualized systems of resource tenure, which reduces women's access to land (and water) as community members (Tsikata and Golah 2011; Ahlers and Zwarteveen 2009).

Within these kinds of FPE analysis, conceptual weight is given to the ways in which capitalism transforms and produces nature. As these processes intersect with gender hierarchies at different scales, patterns of enclosure and marketization are seen to have important gender effects. As large scale, transnational investments in land – generally characterized as "land grabbing" – take hold across many contexts (White et al. 2012) and where land use is being transformed to mono-cultivated agro-fuels (e.g. palm oil), FPE is developing a critical feminist engagement with such processes and their implications for food security (Chu 2011; Behrman et al. 2012; Doss et al. 2014; Wisborg 2014). Recent work on the expansion of oil palm in Southeast Asia has shown how enclosure and commodification of resources impacts on resource access in highly gendered ways (Julia and White 2012; Elmhirst and Darmastuti 2014). In West Kalimantan, Indonesia, conversion of Hibun Dayak land into smallholder oil palm schemes has led to the loss of women's tenure rights (as customary practices give way to formal land registration in the name of the male household head) and as plantation expansion reduces the accessibility of forests for women, reducing their ability to derive income from non-timber forest products such as rattan and forest vegetables (Julia and White 2012).

Ecological politics: producing gendered subjectivities

Poststructuralist and performative theories of subjectivity and subject formation (e.g. Butler 2004) have taken debates in FPE in new directions. Work explores how gender is constituted in different contexts as a component of multiple and complex subjectivities, as the performance of masculinities and femininities construct and reconstruct the gendered subject through people's everyday practices. The emphasis on fluidity and "becoming" challenges essentialist and binary views of relations between men and women that may overemphasize difference and opposition, and that may also essentialize particular patterns of gendered disadvantage. Moreover, such an approach allows space to consider other kinds of gender relations that may be significant in peoples' lives beyond marital relationships, such as seniority and status.

From this starting point, attention is given to the ways in which ecologies and changing environmental conditions bring into existence categories of social difference, including gender. In other words, gender itself is re-inscribed in and through practices, policies and responses associated with changing environments and shifting modes of resource governance. FPE researchers working with such ideas have found them useful for explaining how and why the denial of women's access to development, land and other resources is maintained over space and time. By imagining gender as a process by which subjectivities are produced and shift over time, it is possible to analyse not only how gender structures differential constraints and opportunities for men and women in terms of access to knowledge, resources and political processes, but also why and how gender has come to matter at all (Nightingale 2006; Harris 2006).

Work in this strand of FPE shares an engagement with the concept of intersectionality – an approach to gender that studies the interconnections amongst various dimensions of social relationship and subject formation. As Lykke (2010) suggests, intersectionality works as a common platform for feminist theorizing in relation to identity formation and power relations. Subjectivities are produced through the ways axes of power (gender, race, ethnicity, class, sexual orientation, age, (dis)ability) intersect and emerge in relation to one another, rather than being based on stable or given understandings of social difference (Lykke 2010; Nightingale 2011). Within FPE, such ideas have been taken forward in specific contexts, where emphasis is given to different kinds of intersectional relations, such as caste (Nightingale 2011; Onta and Resurrección 2011), race (Sundberg 2004; Mollett and Faria 2013), sexuality (Elmhirst 2011b) and livelihood (Carr 2013). Whilst much of this work emphasizes the constitution of intersectional subjectivities in particular places, recent work is also exploring how identities are produced across space through migration and multi-local livelihoods (Elmhirst 2008; Carney 2014).

FPE, unlike many renderings of poststructuralist feminism which have been accused of signalling a "retreat from nature" (Alaimo and Hekman 2008), has extended intersectional gender theories to highlight the ecological contexts in which subjectivities emerge, and to illuminate the role of nature in producing particular bodies and identities. Nightingale, for example, describes how the work practices undertaken by community forest user groups in Nepal reveal how engagements with the physical environment (through leaf litter collection, and through timber work) play a fundamental role in the constitution of gender and caste (Nightingale 2011).

Elmhirst's (2011b) analysis of the experiences of landless migrants and their efforts to secure land in forest margin areas, develops an intersectional FPE approach to show how normative heterosexualities are in part produced through the materialities (labour requirements, positionings required to secure land) of access to particular kinds of nature at different historical moments. In other words, the materiality of nature and its production through cultural meanings (around access) and work (everyday labour practices) mean that people's engagement with forest ecologies also produce gendered subjectivities.

Theories of embodiment and intersectionality have also been used to develop more intimate feminist political ecologies, particularly in relation to water access and the governance of sanitation, thus extending work on urban socio-natures and the production of waterscapes (Sultana 2011; O'Reilly et al. 2009). As people engage with and reproduce the material "things" of the waterscape, they produce and reproduce power relations. Work in FPE thus examines how multiple social differences are reproduced through everyday and embodied water practices. Truelove (2011) describes how water shortage in Delhi leads girls and young women to experience a constriction of their spatial mobility in ways that shape life opportunities. She shows how a feminist political ecology view that centres on everyday experiences reveals the ways particular bodies bear the brunt of subsidizing and compensating for state water governance strategies, whilst at the same time, everyday practices around water re-produce multiple hierarchies of social difference. Each of these studies of gender, environment and the politics of subjectivity demonstrates how a FPE informed by theories of embodiment and intersectional subjectivity can avoid an unhelpful retreat to simplistic understandings of gender divisions and singular gendered power relations in relation to the environment.

Material feminism: nature's agency in feminist political ecology

Whilst recent FPE has provided a renewed focus on subjectivities in understanding gendered resource access and control, there is also a current of thinking which seeks to reconsider how "nature" is understood, not simply as an objectified backdrop against which social relations are played out, but as a subject in its own right, understood through a posthumanist relational ontology. Within new material feminism, as explored in the collection edited by Alaimo and Hekman (2008), the modern nature–society dualism is replaced by ontologies that reconceptualize nature in order to account for "intra-actions" (Barad 2007; see also Chapter 10, this volume) between material, discursive, human, more-than-human, corporeal and technological phenomena. In engaging with such ideas, FPE takes up a more radical cross-disciplinary stance (Lykke 2010: 22) to address human–nature power relations more explicitly, and to bring into question human dominance over non-human or more-than-human natures. As these ideas take hold, there is a sense in which some of the earlier tenets of ecofeminism are being reworked and reintroduced into FPE, after a long hiatus during which this body of ideas was rejected amidst charges of "essentialism". Instead, a distinction is drawn between essentialism and an "acknowledgment of embodied, material connections with the environment" (Gaard 2011: 42). The importance of these kinds of ideas in FPE can be seen in three emerging threads within this subfield.

First, the interaction of human and non-human nature through an intersectional theoretical approach is being explored by FPE researchers as they examine how the experiences of gender, race, ethnicity, class, age and so on often take shape through species-ist ideas of humanness vis-à-vis animality (Hovorka 2012; Sundberg 2011; Collard 2012; see also Chapter 9, this volume). Drawing inspiration from Donna Haraway's (2003) work on companion species and with links to animal geographies (Hobson 2007), the focus in this line of work is on the doing and becoming of social identities across species boundaries. As Hovorka puts it: "certain groups of humans become symbolically associated and materially related to certain other (non-human) species (and vice versa) – this process, together with hierarchical privileging and othering, reproduces the positionality and life chances of both humans and non-humans within society" (2012: 876). An example of this kind of work is Collard's study of human–cougar relations on Vancouver Island, which shows how gendered understandings of the relationship between humans and cougars, predator and prey, humans and animals, and culture and nature are informed by, and simultaneously produce, the historical figures of "Cougar Annie" (a woman settler famed for killing cougars), the contemporary trope of predatory "cougar women" and the animal cougars (Collard 2012). Similarly, such ideas are put to work in an applied development context in Hovorka's study of women chicken farmers in urban Botswana, in which the symbiotic relationship between women and chickens is linked to their mutually constituted positionality as marginalized beings compared with men and cattle respectively. In effect, feminist posthumanist thinking is used in an FPE context to consider the ways gender and species hierarchical arrangements work, materially, symbolically and through technologies of security, development and conservation, in a range of diverse settings.

Second, a posthumanist relational ontology that deals explicitly with the agency of non-human nature is also emerging in FPE, particularly in analyses of gender and climate change-related disasters. Tuana (2008) invokes the agency of non-human nature to analyse the impact of Hurricane Katrina in New Orleans, and considers the embodied agency of levees, hurricanes and swamps as well as the agency of the city's women and men. The significance of this for FPE is how a relational (or what Tuana calls an "interactionist") ontology challenges the division between nature and culture, things and people, and questions the privileging of the human subject and the centrality of human labour in understanding environmental change, instead of taking seriously the agency of the natural.

As climate change and exposure to large scale natural hazards place large numbers of people in the developing world "at risk", material feminist approaches in FPE provide a framework for rethinking the connections between nature and society, by moving away from a singular focus on human modification of the environment, or on the gender-specific impacts of climate change. This is seen as important for countering a tendency in some early work on gender and climate change to work with narrow man–woman binaries, and to depict women as vulnerable, marginalized victims, without accounting for the fluidity of subjectivities in relation to the material, discursive and technological dimensions of climate change and adaptation (Arora-Jonsson 2011; Tschakert 2013; Kaijser and Kronsell 2014). Arora-Jonsson (2011) suggests that gendered outcomes reflect existing forms of discrimination that arise from an intersection of different identity categories ranging from socioeconomic status to ethnicity, and as Hurricane Katrina showed, race, class and the materialities of non-human nature (Tuana 2008).

Finally, ideas around a posthumanist relational ontology are also being taken up in what might be described as feminist political ecologies of the body. Such work goes beyond the idea of bounded, interacting bodies to instead consider the flows between and through organisms, and between human and non-human natures. Those working in FPE have drawn on this kind of conceptualization to analyse the metabolic flows associated with food, making important links between the ecologies underpinning neoliberal globalized food systems, production and consumption practices, and the more traditional feminist terrain of gendered bodies (Heyman 2004; Guthman 2011; Hayes-Conroy and Hayes-Conroy 2013). The seepages of pollutants and carcinogens across and between human and non-human natures are seen as spatially uneven and associated with racialized and gendered processes of social and spatial marginalization (Guthman and Mansfield 2013; see also Chapter 43, this volume).

An example of this kind of work in FPE concerns analyses of activism and pedagogy around food security, principally amongst relatively marginalized communities in North America, where school garden and cooking programmes have been developed to encourage healthy eating habits in children through hands-on sensory experiences in gardens and kitchens. Hayes-Conroy and Hayes-Conroy (2013) propose a political ecology of the body (PEB), that involves an assessment of structural forces (the political economic forces that produce social inequalities in relation to food access), knowledge production and meaning-making (food and health discourses produced by people and institutions), and a relational ontology (showing the importance of the relationships between social and environmental systems and how these materialise in the affective, emotional dynamics of embodied everyday life). This extension of feminist political ecology into a relational ontology that takes seriously the emotional, affective dimensions of food and eating points to a promising avenue for gender and development work more generally: taking seriously everyday embodied practices and affective/emotive relationships "as processes in and through which broader political economic forces take shape and are constituted" (2013: 88; see also Carney 2014).

FPE that focuses on intimate scales has also emerged in recent analyses of the Fukushima nuclear accident in Japan, and the impact of radiation contamination on organic farmers (Kimura and Katano 2014). Kimura and Katano provide an intersectional analysis of divergent perceptions of risk amongst farmers and consumers, and the gendered power dynamics of expert and "lay" knowledge claims following the accident (Kimura and Katano 2014). What emerges is an account of gender difference in attitudes to radiation contamination, but also a deeper analysis of how identities and socially constructed notions of masculinity/femininity have mediated the mobilizations that followed the accident. This kind of work not only contributes to an embodied FPE centred on linking the intimate to wider scales of analysis, but also connects FPE with work on gender and environmental justice (Buckingham and Kulcur 2009; Tschakert and Machado 2012), thereby inviting promising synergies for future work.

A feminist ethics of care

In drawing on recent poststructuralist and posthumanist feminist theories, FPE is developing a suite of approaches for addressing critical environment and development challenges in the twenty-first century. The question remains, however: what animates a specifically "feminist" politics for those working in FPE? One pathway being pursued within FPE is to explore how a feminist ethics of care can be put to work to offer a post-capitalist alternative to neoliberal forms of natural resource-based development. Embracing a transnational perspective on environment and development, and in particular, the links between production and consumption worldwide, insights from FPE are being used to analyse practices of ethical consumption and cause-related marketing, and to problematize the ways in which such practices constitute the subjectivities of (women) consumers in the North, and human–environment relationships between and across the Global North and South. Hawkins (2012) links FPE to studies of ethical consumption to show how cause-related marketing tends to constitute everyday lives in the North as separate from natural environments except through consumption choices. This has the effect of suggesting the market is the only route for caring, and environmentally responsible actions, notions which are presented in highly gendered ways.

Whilst there is a sense in which some of these forms of ethical consumption draw on and reproduce forms of neoliberalism and do little to challenge material inequalities (including those associated with gender disadvantage), other possibilities for expressing an ethics of care are also being documented within FPE. Jarosz, for example, examines the motivations of women farmers involved in community-supported agriculture in the United States and concludes that these are expressive of an "ethics of care" that involves a sense of them nourishing themselves and others, nurturing people and the environment, as part of "an ethical positioning that challenges the processes of privatization, unfettered capital accumulation, competition and discourses of personal responsibility for inequality and poverty, which construct individuals as neoliberal subjects" (Jarosz 2011: 308). Jarosz is careful to avoid an essentialist connection between women and care for the environment and distant others. She suggests that, through their care work in community supported agriculture, the women in her study reveal motivations that are not primarily economic, but rather, are associated with social goals and desires to live a work-life that is satisfying and meaningful. This renewed interest in ethics brings current approaches back to the agenda first set out by Rocheleau et al. (1996), in which the transformative potential of feminist activisms is placed at the heart of a distinctively feminist political ecology.

Conclusion

Looking across all of these areas, what is clear is that FPE does not align closely with one, narrow analytical framework, and nor does it map onto a singular policy approach. Whilst FPE broadly comprises an eclectic mix of theoretical positions drawn from feminist theory more broadly, there is a common commitment to presenting a re-politicized recognition of gender as an optic for analysing the power effects of the social constitution of difference. From its beginnings in the early 1990s as a subfield of gender and development studies, and through its engagement with recent poststructuralist, posthumanist and postcapitalist feminist theory, feminist political ecology demonstrates the many ways that feminist theorizations and new understandings of gendered subjectivity can be taken forward within and through the permeable boundaries of an open-ended feminist political ecology. As with political ecology more generally, this theoretical ecumenism is a strength when it comes to addressing some of the more pressing environment and development challenges of our time. Also in alignment with this broader field, FPE offers sensitivity to complex assemblages of power that underlie environment and development issues, rather than a policy emphasis on managerial, problem-solving approaches.

FPE in all its iterations begins from the premise that environmental change is not a neutral process amenable to technical management but rather, arises through political processes and, as earlier sections of this chapter suggest, through nature's agency itself. FPE directs attention to various forms of political agency that arise from complex subjectivities (gender, race, class, sexuality), including those of academics, policy makers, practitioners and activists. By providing the tools for a nuanced and reflexive analysis of these political agencies, FPE offers a way past policy approaches commonly associated with gender and development in environmental contexts that foist ecological care upon those already burdened and disempowered women.

As a policy arena the environment is always subject to struggles around divergent and competing objectives, and FPE research can be used to legitimate courses of action far removed from the intentions of the researcher. Closing the gap between academics, policy makers and activists has been one approach to mitigate this risk: a hallmark of recent FPE work is its commitment to collaborations with other engaged people that span the worlds of academia, policy, practice and activism, where a feminist perspective requires self-reflexivity, an openness to multiple truths and more marginalized voices, and where feminist ethics guide everyday practices of research, engagement and "impact" (see Chapter 8, this volume). Many important avenues that align with what Braidotti (2009) has described as "affirmative politics" are currently being explored through new feminist political ecologies: the themes covered here are just part of a continued flowering of this revitalized and important realm of transformative debate, politics and praxis.

References

Agarwal, B. (2003) "Gender and land rights revisited: exploring new prospects via the state, family and market". *Journal of Agrarian Change*, vol. 3, no. 1–2: 184–224.

Ahlers, R. and Zwarteveen, M. (2009) "The water question in feminism: water control and gender inequities in a neo-liberal era". *Gender, Place and Culture*, vol. 16, no. 4: 409–426.

Alaimo, S. and Hekman, S. (eds.) (2008) *Material Feminisms*. Bloomington: Indiana University Press.

Arora-Jonsson, S. (2011) "Virtue and vulnerability: discourses on women, gender and climate change". *Global Environmental Change*, vol. 21, no. 2: 744–751.

Barad, K. (2007) *Meeting the Universe Halfway: Quantum Physics and the Entanglement of Matter and Meaning*. Durham, NC: Duke University Press.

Bee, B., Bermann, M., and Tschakert, P. (2013) "Gender, development and rights-based approaches: lessons for climate change adaptation and adaptive social protection". In Alston, M. and Whittenbury, K. (eds.), *Research, Action and Policy: Addressing the Gendered Impacts of Climate Change*. Dordrecht: Springer, pp. 95–108.

Behrman, J., Meinzen-Dick, R. and Quisumbing, A. (2012) "The gender implications of large-scale land deals". *Journal of Peasant Studies*, vol. 39, no. 1: 49–79.

Braidotti, R. (2009) "On putting the active back in to activism". *New Formations*, vol. 68, no. 3: 42–57.

Buckingham, S. and Kulcur, R. (2009) "Gendered geographies of environmental injustice". *Antipode*, vol. 41, no. 4: 659–683.

Butler, J. (2004) *Undoing Gender*. London: Routledge.

Carney, J. (2004) "Gender conflict in Gambian wetlands". In Peet, R. and Watts, M. (eds.), *Liberation Ecologies: Environment, Development, Social Movements*. Second edition. London: Routledge, pp. 316–336.

Carney, M. (2014) "The biopolitics of 'food insecurity': towards a critical political ecology of the body in studies of women's transnational migration". *Journal of Political Ecology*, vol. 21: 1–18.

Carr, E.R. (2013) "Livelihoods as intimate government: reframing the logic of livelihoods for development". *Third World Quarterly*, vol. 34, no. 1: 77–108.

Chu, J. (2011) "Gender and 'land grabbing' in sub-Saharan Africa: women's land rights and customary land tenure". *Development*, vol. 54, no. 1: 35–39.

Collard, R.C. (2012) "Cougar figures, gender, and the performances of predation". *Gender, Place and Culture*, vol. 19, no. 4: 518–540.

Cornwall, A., Harrison, E. and Whitehead, A. (2007) "Introduction: feminisms in development: contradictions, contestations and challenges". In Cornwall, A., Harrison, E. and Whitehead, A. (eds.), *Feminisms in Development: Contradictions, Contestations and Challenges*. London: Zed Books, pp. 1–20.

Doss, C., Summerfield, G. and Tsikata, D. (2014) "Land, gender, and food security". *Feminist Economics*, vol. 20, no. 1: 1–23.

Elmhirst, R. (2008) "Multi-local livelihoods, natural resource management and gender in upland Indonesia". In Resurrección, B.P. and Elmhirst, R. (eds.), *Gender and Natural Resource Management: Livelihoods, Mobility and Interventions*. London: Earthscan, Ottawa: IDRC, and Singapore: ISEAS Publications, 67–87.

Elmhirst, R. (2011a) "Introducing new feminist political ecologies". *Geoforum*, vol. 42: 129–132.

Elmhirst, R. (2011b) "Migrant pathways to resource access in Lampung's political forest: gender, citizenship and creative conjugality". *Geoforum*, vol. 42: 173–183.

Elmhirst, R. and Darmastuti, A. (2014 – in press) "Material feminism and multi-local political ecologies: rethinking gender and nature in Lampung, Indonesia". In Lund, R., Doneys, P. and Resurrección, B.P. (eds.), *Gendered Entanglements: Revisiting Gender in Rapidly Changing Asia*. Copenhagen: NIAS Press.

Elmhirst, R. and Resurrección, B.P. (2008) "Gender, environment and natural resource management: new dimensions, new debates". In Resurrección, B.P. and Elmhirst, R. (eds.), *Gender and Natural Resource Management: Livelihoods, Mobility and Interventions*. London: Earthscan, Ottawa: IDRC, and Singapore: ISEAS Publications, pp. 3–22.

Gaard, G. (2011) "Ecofeminism revisited: rejecting essentialism and re-placing species in a material feminist environmentalism". *Feminist Formations*, vol. 23, no. 2: 26–53.

Guthman, J. (2011) *Weighing In: Obesity, Food Justice, and the Limits of Capitalism*. Berkeley: University of California Press.

Guthman, J. and Mansfield, B. (2013) "The implications of environmental epigenetics: a new direction for geographic inquiry on health, space, and nature-society relations". *Progress in Human Geography*, vol. 37, no. 4: 486–504.

Haraway, D.J. (2003) *The Companion Species Manifesto: Dogs, People, and Significant Otherness*. Chicago: Prickly Paradigm Press.

Haraway, D.J. (2008) *When Species Meet*. Minneapolis: University of Minnesota Press.

Harding, S. (ed.) (2004) *The Feminist Standpoint Theory Reader: Intellectual and Political Controversies*. New York: Psychology Press.

Harris, L.M. (2006) "Irrigation, gender, and social geographies of the changing waterscapes of southeastern Anatolia". *Environment and Planning D: Society and Space*, vol. 24: 187-213.

Hawkins, R. (2012) "Shopping to save lives: gender and environment theories meet ethical consumption." *Geoforum*, vol. 43: 750–759.

Hayes-Conroy, J. and Hayes-Conroy, A. (2013) "Veggies and visceralities: a political ecology of food and feeling". *Emotion, Space and Society*, vol. 6: 81–90.

Heyman, J.M.C. (2004) "The political ecology of consumption: beyond greed and guilt". In Paulson, S. and Gezon, L. (eds.), *Political Ecology across Spaces, Scales and Social Groups*. Newark: Rutgers University Press, pp. 113–132.

Hobson, K. (2007) "Political animals? On animals as subjects in an enlarged political geography". *Political Geography*, vol. 26, no. 3: 250–287.

Hovorka, A. (2012) "Women/chickens vs men/cattle: insights on gender-species intersectionality". *Geoforum*, vol. 43: 875–884.

Hue Le Thi Van (2006) "Gender, Doi Moi and mangrove management in Northern Vietnam". *Gender, Technology and Development*, vol. 10, no. 1: 37–59.

Jarosz, L. (2011) "Nourishing women: toward a feminist political ecology of community supported agriculture in the United States". *Gender, Place and Culture*, vol. 18, no. 3: 307–326.

Julia and White, B. (2012) "Gendered experience of dispossession: oil palm expansion in a Dayak Hibun community in West Kalimantan". *The Journal of Peasant Studies*, vol. 39, no. 3–4: 995–1016.

Kaijser, A. and Kronsell, A. (2014) "Climate change through the lens of intersectionality". *Environmental Politics*, vol. 23, no. 3: 417–433.

Kimura, A.H. and Katano, Y. (2014) "Farming after the Fukushima accident: a feminist political ecology analysis of organic agriculture". *Journal of Rural Studies*, vol. 34: 108–116.

Leach, M. (2007) "Earth mother myths and other ecofeminist fables: how a strategic notion rose and fell". *Development and Change*, vol. 38, no. 1: 67–85.

Lykke, N. (2010) *Feminist Studies. A Guide to Intersectional Theory, Methodology and Writing*. New York: Routledge.

Mohanty, C.T. (1988) "Under Western eyes: feminist scholarship and colonial discourses". *Feminist Review*, vol. 30: 61–88.

Mollett, S. and Faria, C. (2013) "Messing with gender in feminist political ecology". *Geoforum*, vol. 45: 116–125.

Nightingale, A. (2006) "The nature of gender: work, gender and environment". *Environment and Planning D: Society and Space*, vol. 24: 165–185.

Nightingale, A. (2011) "Bounding difference: intersectionality and the material production of gender, caste, class and environment in Nepal". *Geoforum*, vol. 42: 153–162.

Onta, N. and Resurrección, B.P. (2011) "The role of gender and caste in climate adaptation strategies in Nepal". *Mountain Research and Development*, vol. 31, no. 4: 351–356.

O'Reilly, K., Halvorson, S., Sultana, F. and Laurie, N. (2009) "Introduction: global perspectives on gender-water geographies". *Gender, Place and Culture*, vol. 16, no. 4: 381–385.

Radcliffe, S.A. (2006) "Development and geography: gendered subjects in development processes and interventions". *Progress in Human Geography*, vol. 30, no. 4: 524–532.

Radel, C. (2012) "Gendered livelihoods and the politics of socio-environmental identity: women's participation in the conservation projects in Calakmul, Mexico". *Gender, Place and Culture*, vol. 19, no. 1: 61–82.

Rao, N. (2006) "Land rights, gender equality and household food security: exploring the conceptual links in the case of India". *Food Policy*, vol. 31: 180–193.

Razavi, S. (2003) "Introduction: agrarian change, gender and land rights". *Journal of Agrarian Change*, vol. 3, no. 1/2: 2–32.

Resurrección, B.P. (2006) "Gender, identity and agency in Philippine upland development". *Development and Change*, vol. 37, no. 2: 375–400.

Resurrección, B.P. and Elmhirst, R. (eds.) (2008) *Gender and Natural Resource Management: Livelihoods, Mobility and Interventions*. London: Earthscan, Ottawa: IDRC, and Singapore: ISEAS Publications.

Ribot, J. and Peluso, N.L. (2003) "A theory of access". *Rural Sociology*, vol. 68, no. 2: 153–181.

Rocheleau, D., Thomas-Slayter, B. and Wangari, E. (eds.) (1996) *Feminist Political Ecology: Global Issues and Local Experiences*. London: Routledge.

Schroeder, R.A. (1999) *Shady Practices: Agroforestry and Gender Politics in The Gambia*. Berkeley: University of California Press.

Sharp, J. (2009) "Geography and gender: what belongs to feminist geography?". *Progress in Human Geography*, vol. 33, no. 1: 74–80.

Sultana, F. (2011) "Suffering *for* water, suffering *from* water: emotional geographies of resource access, control and conflict". *Geoforum*, vol. 42, no. 2: 163–172.

Sundberg, J. (2004) "Identities-in-the-making: conservation, gender, and race in the Maya Biosphere Reserve, Guatemala". *Gender, Place and Culture: A Journal of Feminist Geography*, vol. 11, no. 1: 44–66.

Sundberg, J. (2011) "Diabolic caminos in the desert and cat fights on the Río: a posthumanist political ecology of boundary enforcement in the United States-Mexico Borderlands". *Annals of the Association of American Geographers*, vol. 101, no. 1: 1–19.

Truelove, Y. (2011) "(Re-)conceptualizing water Inequality in Delhi, India through a Feminist Political Ecology framework". *Geoforum*, vol. 42, no. 2: 143–152.

Tschakert, P. (2013) "From impacts to embodied experiences: tracing political ecology in climate change research". *Geografisk Tidsskrift-Danish Journal of Geography*, vol. 112, no. 2: 144–158.

Tschakert, P. and Machado, M. (2012) "Gender justice and rights in climate change adaptation: opportunities and pitfalls". *Ethics and Social Welfare*, vol. 6, no. 3: 275–289.

Tsikata, D. and Golah, P. (eds.) (2011) *Land Tenure, Gender and Globalization: Research and Analysis from Africa, Asia and Latin America*. New Delhi: Zubaan/Ottawa: IDRC/Pontypool: Merlin Press.

Tuana, N. (2008) "Viscous porosity: witnessing Katrina". In Alaimo, S. and Hekman, S. (eds.), *Material Feminisms*. Bloomington: Indiana University Press, pp. 188–212.

Watts, M. (2000) "Political ecology". In Barnes, T. and Sheppard, E. (eds.), *A Companion to Economic Geography*. Oxford: Blackwell, pp. 257–275.

White, B., Borras, S.M. Jr., Hall, R., Scoones, I. and Wolford, W. (2012) "The new enclosures: critical perspectives on corporate land deals". *The Journal of Peasant Studies*, vol. 39, no. 3–4: 619–647.

Wisborg, P. (2014) "Transnational land deals and gender equality: utilitarian and human rights approaches". *Feminist Economics*, vol. 20, no. 1: 24–51.

Wright, M.W. (2008) "Gender and geography: knowledge and activism across the intimately global". *Progress in Human Geography*, vol. 33, no. 3: 379–386.

41
INDIGENEITY

Emily T. Yeh and Joe Bryan

Indigeneity in political ecology

In a study of the cultural politics of nature conservation in Sardinia, anthropologist Tracey Heatherington meditates on the awkward productiveness of thinking about shepherds on the Italian island through the category of the indigenous. "If only the ambiguities of indigeneity can be recognized as wholesome possibilities," she writes, "the global dreamtimes of environmentalism can be healed with optimism" (2010: 53). Heatherington's unconventional use of indigeneity hints at the term's importance as a boundary concept in political ecology – malleable enough to be used to make sense of particular histories of inequality and struggle, yet capable of circulating across different social worlds (Star and Griesemer 1989; also Goldman 2009). That circulation conjoins three major theoretical strands of political ecology: cultural ecology's efforts to link the qualities of the physical environment with cultural identity, ecological anthropology's focus on the cultural politics of adaptation and response to environmental change, and Marxist-inspired agrarian studies' concern with the economic condition of peasants. Individually, each of these strands risks a reductive reading of indigeneity as cultural essence, as romantic "pulp fiction" (Ramos 2003), and as subterfuge for class identity. Taken together they draw attention to the open-ended qualities of indigeneity that lend room for self-determination and mobility, approximating its potential to enliven new futures.

While our emphasis is on the uses of indigeneity as a term, concept, and practice, we are aware that for many readers indigeneity refers first and foremost to indigenous peoples. Critiques of indigeneity are thus often misread as critiques of the very existence of indigenous peoples as distinct populations marked by differences of race, culture, geography, and history. We take a different course here, drawing attention to the political and intellectual work that the term does, underscoring its importance as a generative site for research and political struggle. Our effort is thus informed by the circulation of the term rather than any fixed definition, its indeterminacy opening a space for self-determination that is often foreclosed in academic and political debates where the term's meaning is assumed or left unexplored.

We begin by surveying the distinct valences that indigeneity takes in its circulation both within and beyond political ecology. The first of these concerns the concept's relevance to ongoing debates over the relationship between nature and culture. The second relates to the dynamics and shape of capitalism in frontiers and other regions regarded as economically peripheral. The third pertains to indigeneity's capacity to politicize "environmental" concerns, often through engaging a certain "coloniality of power" marked by questions of race and class. These three strands rarely exist independently of one another. Rather, they receive different analytical emphases that vary significantly with context, underscoring the political and intellectual work associated with the indeterminacy of the term. Following this, we shift our attention to the concrete, divergent contexts in which the concept of indigeneity is applied, surveying resulting debates about its risks and usefulness.

A political ecology of indigeneity

Nature/culture

Characterizations of indigeneity as defined by its relationship to nature and place trace cultural ecology's lasting influence on political ecology. Developed in the mid-twentieth century, cultural ecology sought to reject environmental determinist explanations of cultural traits as solely a function of the physical environment. It also rejected cultural determinist positions that made nature subservient to humans. Instead, some of the field's earliest practitioners advocated a more "ecological" approach to understanding how the physical environment shaped culture, and vice versa (e.g. Steward 1956). This argument built on the anthropological anti-racism of Franz Boas, which sought above all to explain cultural traits as responses to the physical qualities of the environment. Those adaptations, in turn, led to the production of culturally-specific forms of knowledge and modification of the physical environment (Denevan 2001; Sauer 1969; Stewart 2002). According to this view, limits to the improvement of peoples and cultures did not lie with natural limits such as biology or environment, but rather in social limits imposed by racism (Moore et al. 2003: 58).

Like Boas, many of his students turned to traditional or tribal societies to support their claims. Their efforts shaped the emergence of cultural ecology as a new academic field. They also bore a lasting impact on the understanding of these cultures as indigenous, their traits and practices expressly tied to particular environments and places. That relationship was further regarded as everywhere under threat of extinction by the spreading influences of modern society – settler states and capitalism by another name – that threatened to leave no people or place untouched. Destroy the place, or radically alter it, and the culture that had adapted to that place would disappear (Grinde and Johansen 1995). This approach was further modified by human ecologists and ecological anthropologists who documented processes of cultural change as adaptation to environmental change (Rappaport 1967; Vayda 1961). In spite of their divergent interpretations of the human–environment relationship, these approaches shared an ecological approach to understanding that relationship as shaped by complex systems of feedback and response (Bateson 1972).

This scholarly approach had far-reaching political effects. It helped consolidate, or at least rework, colonial stereotypes of indigenous peoples as living at one with nature. Saving the Indian thus became an integral part of saving nature, or so some claimed. During the 1990s in particular, this approach gained traction among conservationists and advocates of indigenous rights, many of whom portrayed native peoples as natural environmentalists (Conklin and Graham 1995; Hecht and Cockburn 2010; Tsing 1999). This importantly challenged decades of racist views of indigenous peoples as little more than animals, their presence an obstacle to development to be solved either through assimilation or extermination (e.g. Berger 1977; Price 1989; Turner 1995). It also contributed directly to understandings of indigenous peoples' occupancy and use of land and resources as a fundamental human right. Initially formulated in the early 1970s by social scientists studying indigenous populations, it had by the late 1980s become a key feature of international law. The International Labor Organization's (ILO) Convention 169 on the Rights of Indigenous and Tribal Peoples defined indigeneity in terms of a "territory" composed of "the total environment which the peoples concerned occupy or use." The Spanish version of the document translates "total environments" as "*hábitat*," further demonstrating the influence of cultural ecology's efforts to understand the relationship between indigenous peoples and their environment as systematic, their social expression constituting a "territory."

Peasants or indigenous peoples?

As much as the notion of territory found in Convention 169 speaks to cultural ecology, the institutional location of indigeneity within the ILO also connects it to Marxist-inspired agrarian studies. The ILO was founded at the same time as the League of Nations following World War I, due in part to efforts by the Socialist Second International. In keeping with the Second International's political agenda, the ILO's purpose was to provide a means for organizing labor outside of the state-dominated League of Nations. The ILO drafted Convention 107, the predecessor to 169, in 1957 as part of efforts to protect the rights of agricultural workers in former colonies in Africa, Asia, and Latin America. The Convention further distinguished peasants from "indigenous and tribal populations" either living outside of agrarian settings or practicing uniquely "indigenous" forms of subsistence agriculture (Rodríguez-Piñero 2005). This mirrored a contrast drawn in agrarian studies between "peasants" and their political economic place within capitalism and "indigenous peoples" living *outside* of that system or at least at its frontiers. This distinction was also made by national elites in Latin America, particularly in the Andes, separating *campesinos* subject to agrarian reforms from racially inferior *indios* treated as wards of the state – to the extent that they were considered "human" at all (Becker 2008; Healy 2001; Lucero 2008; Mallon 1995; Varese 2006). By the 1970s, that difference informed patterns of political mobilization. Where peasants fought for rights to land as a means of production, indigenous peoples struggled for territories.

Political ecologists drew on agrarian studies to critique the functionalism latent in cultural ecologists' conceptualization of indigenous or traditional societies as living close to nature. This analysis was further used to argue against idealist formulations of indigenous peoples as living outside of modern forms of economy and power – states and capital – in ways that ignored their embeddedness within colonialism, slavery, war, and capitalist exploitation. Drawing on agrarian studies, early political ecology recast these populations as peasants, analyzing the deep social inequalities and marginalization faced by rural "land managers" as a result of their integration in capitalist markets (Watts 2000). This emphasis on political economy led to a tendency to treat cultural politics with suspicion, characterizing claims to indigeneity as obscuring class-based inequalities (Brass 2002; Li 2001: 648; Hale 1994). Indeed, some political economy-grounded academic work gives "the impression that tribal stories are insincere fantasies while the peasant condition is real" (Tsing 2003: 126). That split was born out by groups in the Andes and in southern Africa who were divided into "tribe" and "peasant" under colonial rule (Mallon 1995; Sylvain 2002). Though the two categories may actually refer to the same people, their political and analytical separation creates difficulties for indigenous peoples to press both for recognition of cultural difference and also for full, fair incorporation into development processes as national citizens (Bebbington 2004: 402).

Indigeneity as political identity

A strict Marxist emphasis on class, however, was unable to account for the global proliferation of groups that took up indigeneity as a political identity. The origins of indigenism are generally traced to lobbying by "first peoples" in countries that experienced settler colonialism and are now dominated by populations of European descent: the United States, Canada, New Zealand, and Australia (Kingsbury 1998; Merlan 2009; Niezen 2003). By the 1990s, though, groups in every part of the world were using the term "indigenous peoples" to articulate claims to land, resources, and self-determination. Their efforts re-injected cultural politics into environmental debates, informing another thread of political ecology. While the focus of much of this work was social movements' politicization of the environment, these efforts often involved braiding cultural ecological notions of indigenous territory with agrarian idioms of struggle over land and resources. This use of indigeneity drew extensively from international human rights instruments such as ILO Convention 169. At the same time, it sought to actively use indigeneity as a means of further articulating historically and geographically situated experiences of dispossession and assimilation into modern societies. Rather than a purely racial identity, these efforts turned indigeneity into an explicitly political one, opening the concept up for mobilization by an ever-growing range of groups. In the Americas, Afro-descendent populations were among the first to use rights formerly associated with indigenous peoples to assert their control over lands and resources they had used and occupied collectively (Anderson 2007; Asher 2009; Gordon et al. 2003; Ng'weno 2007). Further afield groups as diverse as *adivasis* in India and *adat* communities in Indonesia availed themselves of the term, challenging state claims that all people living in both countries were indigenous in a formal sense of the term (Kingsbury 1998; Li 2000).

The circulation of indigeneity as a political identity suggests an infinitely more open interpretation of the term than strict adherents to cultural ecology or agrarian studies would allow. Through their appeals for self-determination in defining group membership and organization, their approaches use cultural difference as a resource for mobilizing political claims. Indigeneity thus becomes a relational category rather than an objective condition, one neither externally imposed nor created autonomously. Instead it is cast as a political identity that is at once historically based and emergent in relation to new political situations, its meaning drawn in relation to the non-indigenous. Self-identification is key. Though there is always a boundary politics of indigeneity, this view conceptualizes these precisely as politics to be analyzed in geographical and historical context, rather than a question to be adjudicated from the outside (de la Cadena and Starn 2007).

The focus of this line of analysis is on how and why indigeneity comes to be taken up, for what uses, and with what intended and unintended effects. Tania Li (2000) draws on Stuart Hall's (1996) use of the dual meaning of "articulation" to describe the process of defining political subjects (see also Clifford 2001). Articulation refers both to speaking and to a linkage that can be forged under some conditions, but is "not necessary, determined, absolute, and essential for all time" (Hall 1996: 141). A theory of articulation is thus a way to analyze "how ideological elements come, under certain conditions, to cohere together within a certain discourse, and a way of asking how they do or do not become articulated, at specific conjunctures, to certain political subjects" (Hall 1996: 141). This shifts attention away from questions of "invention" or authenticity, and instead sees indigeneity as a contingent, structured *positioning*, one that can be spoken and linked to other indigenous struggles only within particular geographically and historically specific conjunctures. Indigeneity is thus "without guarantees" (Hall 1983); there is no necessary correspondence between a subject and an identity, nor between indigeneity and its politics. Indigeneity's inherent tension between the singular and universal makes its effects uncertain. Its status as a general category allows those who claim it to link their struggles with other seemingly disconnected ones around the world. However, this can also lead to essentialism and romanticism, as well as an occlusion of the historical and geographical specificities at the heart of the politics at stake.

That indigeneity is without guarantees can be seen from the fact that it has been articulated with Hindu supremacists in a politics of hate and violence against Muslim minorities (Baviskar 2007) and with landowning and business elites in Bolivia who, like the otherwise very different *ayllu* movement, draw on essentialized understandings of indigeneity to legitimate claims to territorial rights (Perreault and Green 2013). In specific African contexts, indigeneity's claims of autochthony have been treated as fomenting "tribalism," sparking fears of economic and political instability and violence (Hodgson 2009; Nyamnjoh 2007).

As a category of representation, indigeneity can be effective only to the degree that an appropriate "field of attraction" is created (Tsing 1999). Only in spaces thus created can contingent collaborations be forged (Tsing 2005). This approach is fundamental to understanding how the concept of "indigeneity" travels as a "word in motion," a product of translation, understood as a "necessarily faithless appropriation," where the interaction of languages forges new meanings as texts are rewritten (Tsing 1997, 2009). Others have discussed indigeneity as a chameleon concept (Gray 1995), a term that invites particular peoples to be interpellated by it (Castree 2004: 153), and building on recent work by McFarlane (2009), as a "translocal assemblage" (Baird 2015). The success of this project is nowhere guaranteed. Claims to indigeneity remain subject to scrutiny by non-indigenous experts (Clifford 1988; Li 2000), who evaluate claims according to "hyper-real" definitions that equate lack of internal cohesion and political differences with illegitimacy and inauthenticity (Ramos 1998; Sparke 1998; Watts 2003). Such failures often occur despite abundant evidence of deprivation and dispossession, an outcome that informs emerging debates over the very coloniality of the term "indigeneity" (Cusicanqui 2012; Mamani Ramírez 2005).

Divergent indigeneities

Academic efforts to make sense of indigeneity are routinely driven, and even outpaced, by on-the-ground applications of the term. At its most basic, the use of indigeneity as a vehicle for making claims charts the shifting contours of colonialism, past and present, stretching it beyond "settler states" to include "post-colonial" states in Africa and Asia. Taken up by groups as divergent as pan-Mayan movements in Guatemala, *adat* organizations in Indonesia, *adivasis* in India, and communities in the oil-soaked Niger delta, these new applications of the term open it up to reinterpretations that help denaturalize understandings of indigeneity as an identity fixed by location, drawing attention to how the term is used to link governance and political economy with strategies of localization (Heatherington 2010; Li 2000; Sivaramakrishnan and Agrawal 2003). It also helps identify how indigeneity works as a thoroughly *modern* way of understanding, managing, and governing differences. These differences are no less generative of new possibilities for self-determination than for political ecological inquiry.

In its many manifestations, indigeneity is used to make a variety of claims, to ownership and control over territories, to self-determination, and to self-representation (Colchester 1995: 61). Two of the most common – and controversial – ways that indigenous peoples and non-indigenous activists have sought to reach those goals are through claims about environmental stewardship, and through land titling, particularly for collective land ownership. The implications of both have sparked considerable debate in political ecology.

The conjoining of indigeneity with the environment

The spread of the global indigenous people's movement paralleled the global emergence of concerns about sustainable, alternative, and environmentally friendly forms of development (Duhaylungsod 2011). At the same time, scholars drawing on traditions in cultural ecology demonstrated the environmental soundness of indigenous peoples' resource management strategies and argued for community-based resource management based on local control, knowledge, and tradition, in place of top-down coercive conservation (Berkes 1999; Brosius et al. 1998; Escobar 1995; Posey 1985). New political alliances were created between indigenous peoples and environmentalists who came together to oppose mining, hydroelectric dams, and other environmentally destructive activities (Conklin and Graham 1995; Kirsch 2006; Sawyer 2004). Out of this milieu emerged a complicated conjunction of ecology and indigeneity, in which it became assumed that safeguarding indigenous cultures and traditional practices would result in the protection of nature (Braun 2002).

The problems and risks of such representations are many, and have been the subject of much critique. Assuming that contemporary indigenous environmental practice always aligns perfectly with Western environmentalism has been dubbed the "Ecologically Noble Savage" (Krech III 1999) because it draws on a long-standing Western tradition of idealizing primitive native peoples as a foil to European social institutions (Conklin and Graham 1995). Such conceptions can conflate the preservation of cultural diversity with biodiversity, rendering indigenous peoples "part of [non-human] nature" as opposed to fully human. Hence it can become a form of eco-incarceration that denies the possibility of modernity to those who are deemed authentically indigenous.

The conjunction of indigeneity and environment also flattens and erases the rich complexity and diversity of practices, beliefs, and worldviews, rendering indigenous peoples generic and one-dimensional (Delcore 2004). Moreover, such representations generally anachronistically project modern conceptionalizations about the human world onto the past, presenting unattainable ideals (Huber and Pederson 1997). When mismatches are discovered between these standards of representation and actual indigenous practices, a harsh backlash can result, in which indigenous peoples are dismissed as having been contaminated by a "loss of culture" (Brosius 1997). Therefore, some argue that despite the usefulness of the idea of indigenous harmony with nature as a strategic essentialism, one that generates symbolic capital in the global environmental arena and new cross-scalar environmentalist-indigenous alliances for the advancement of political goals, the risks are simply too great: well-meaning struggles for indigenous rights only result in further marginalization (Shah 2010). At the same time, accusations of strategic essentialism are often grounded in their own essentializing moves that preserve nature as a constant and knowable by science, further defending the "West" as the ultimate source of authoritative knowledge (Mamani Ramírez 2011).

Other scholars accept these warnings about essentialism and romanticization but are more sanguine about the potential of an indigeneity without guarantees. Collaborations between indigenous and non-indigenous peoples around the environment can be more open-ended and flexible than ironic dismissals imply (Tsing 1999). To believe and act otherwise is to "offer a historical metanarrative of imperial modernization in which nothing can happen – good or bad – but more of the same" (Tsing 2005: 161). For example, the articulation of Tibetans with conservation that emerged in the late 1990s and early 2000s was romantic and essentializing, but for a brief time at least, it also opened a space for assertions of Tibetan cultural identity and practices otherwise designated as potentially threatening to state sovereignty and thus criminal (Yeh 2007, 2013, 2014a, 2014b). The coupling of indigeneity with discourses of environmentalism presents a double bind of essentialism, but also unpredictable political and cultural opportunities (Heatherington 2010: 53).

Propertization and collective land titling

Indigenous activism has increasingly turned toward the conceptualization of rights to land as property rights, particularly collective property rights based on customary occupancy and land use (Wainwright and Bryan 2009; see also Baird 2013; forthcoming). This has enabled some indigenous victories in court and in law, while simultaneously weakening indigenous demands and aligning with global capitalism.

The strategic articulation of indigenous claims as property rights to customary land is a product of the context in which indigenous groups operate, wherein "cultural objections to neo/colonialism are unheard within a political context that champions property as the harbinger of social good" (Blackburn 2009; in Coombes et al. 2012: 816). The dire political-economic situation for poor and marginalized indigenous groups, such as in Central America in the early twenty-first century, often propels activists to focus on concrete demands for immediate relief, with community land rights being viewed as an urgent and high priority (Hale 2006). Unfortunately, collective land titling efforts, most of which are being promoted by the World Bank, represent "a Faustian bargain: recognition of multicultural rights in return for endorsement, implicit or otherwise of the broader political project of neoliberalism" (Hale 2006: 108). Cultural differences are recognized while ongoing forms of racism that entangle formal processes of titling and demarcation and undermine the security of titles are elided (Hale 2011; Mollett 2013; Wainwright and Bryan 2009).

Though these problems are often regarded as emblematic of Hale's (2005) "neoliberal multiculturalism," they are no less prominent in "post-neoliberal" states across Latin America. Redoubling their efforts to achieve national economic growth through natural resource exports, states such as Bolivia and Ecuador have made titling communal properties a pivotal first-step toward neutralizing opposition to resource extraction (Bebbington and Bebbington 2011; Gudynas 2010). That outcome arguably illustrates the ease with which indigeneity as both concept and marker of cultural differences is absorbed by states and capitalism alike. That is, while indigeneity and its emphasis on collective attachment to place have been a successful defensive response to large-scale dispossession, it can simultaneously affirm structural forms of inequality and obscure everyday forms of dispossession under capitalism (Geschiere and Nyamnjoh 2000; Li 2010; Watts 2003).

The most commonly proposed solution – by both indigenous peoples and the World Bank – is to make communal land titles inalienable. While this answer is often defined in opposition to neoliberalism, its reliance on the state to safeguard property rights presents its own set of outcomes, intended and otherwise. In particular, it can lead to a reliance on "customary use" to allocate property claims, protecting community resources against privatization while institutionalizing hierarchies within and between communities that might otherwise be considered equally "indigenous" (Bakker 2008; Perreault 2008). After demonstrating that such strategies have a long history in colonial rule, Li (2010) argues that in contemporary contexts they are an expression of racialized paternalism, as they seek to "arborealize" or root indigenous peoples to particular places. Moreover, the assumption that collective landholding is the natural state of indigenous peoples perpetuates a colonial tendency to "overestimate the bonds of community and underestimate or misread the mechanisms through which dispossession occurs" (Li 2010: 388). As with the conjunction of indigeneity with environmentalism, then, Li's concern is that if indigenous peoples are found to desire individual title, to have dispossessed themselves of collective land, or to not be attached to place as a member of a group, their rights as indigenous peoples will not be visible or recognized (see also Merlan 2007).

While Li, building on research in Africa and Asia, focuses squarely on capitalism, work by Bryan (2009), Wainwright (2008; Wainwright and Bryan 2009), and Hale (2005, 2006, 2011) on different indigenous formations in Latin America focus on the relationship between property, capitalism, and the state. By transforming claims to territorial rights that directly challenge the state into rights to property based on customary use, this cartographic-legal strategy "extends liberal formulations of national identity founded on a social contract among property owners," deepening capitalist social relations as well as state power and the spatiality of the state (Wainwright and Bryan 2009: 154–155). Formal recognition of indigenous rights to property in North America and Latin America has also been a means to manage indigenous opposition to neoliberal economic reforms and the expansion of extractive industries like oil and gas (Bebbington and Bebbington 2011; Hale 2005; Perreault and Valdiva 2010; Dombrowski 2002).

Future directions: rethinking nature

In a provocative review, Coombes et al. (2012: 812) conclude that "notwithstanding … alignments between political ecology and Indigenous geographies, there is reason for hesitance about further engagement between them." From their perspective, political ecology fails to take *either* nature or culture seriously by reducing both to expressions of how external structures affect "local socioecological lives" (West 2005: 639). In particular they argue that any critique of indigeneity can slide into "universalizing disrespect for those Indigenes who organize themselves around the concept of community," reducing ethnic struggles to "mere resource conflicts" (Coombes et al. 2012: 813). While Coombes et al. rightly identify a peril that calls for constant vigilance, their characterization is also liable to paring political ecology down to a reductive reading of the Marxist strand of its approach. Indeed, there is a great deal of work in political ecology that attempts to do just the opposite, by working with indigenous peoples as interlocutors capable of creating new and generative forms of research (e.g. Bryan 2011; Coggins with Gesang 2014; de Leuw et al. 2012; Goldman 2007; Kirsch 2006; Nadasdy 2011; Povinelli 2011; Sletto 2009; Yeh 2014a). Still, as Coombes et al. suggest, considerable work remains to be done.

At stake is the extent to which political ecology is informed by insights from postcolonial studies. As it is, indigeneity has been one of the key ways in which postcolonial studies has been brought into political ecology, enriching it through the interrogation of how colonialism has worked over social, cultural, and political institutions and identities vis-à-vis nature (e.g. Braun 2002). Yet this attention has arguably not gone far enough, often remaining stuck in the critique of representation associated with Edward Said without moving onto the concepts and research practices that continually reinscribe inequalities (Wainwright 2008). A route forward takes claims made by indigenous movements at face value, thinking with the ideas and resources they mobilize to grasp the ontological differences that inform differing epistemologies. Specifically, these approaches have zeroed in on the natural construction of the social, directing attention to the more-than-human forces that shape lifeworlds (Blaser 2014; de la Cadena 2010). This approach would allow, for example, seriously engaging movements of people who protest because of their desire not to offend territorial deities – those who otherwise "cannot speak as 'theory' within the knowledge procedures of the university even when these knowledge procedures acknowledge and 'document' its existence" (Chakrabarty 2000: 41).

The question becomes, then, how to handle the agency of nature in the form of the presence of the divine or supernatural. This ontology must be accepted for the subaltern to be the subject of their own histories, but at the very same time, the interests and narratives of social justice often require "render[ing] this enchanted world into our disenchanted prose" (Chakrabarty 2000: 77). Indeed, many of the critical tools of political ecology depend on a disenchanted view of the world, one which fits into linear, empty time, but where those with "not yet" modern views risk being relegated to the imaginary waiting room of history (Yeh 2009). To address such dilemmas, proponents of the "ontological turn" invert this by suggesting that we learn from Amerindians' theory of translation: "[W]here our modern ... multiculturalist ontology is founded on the mutual implication of the unity of nature and the plurality of cultures, the Amerindian conception would suppose ... one 'culture,' multiple 'natures' ... a constant epistemology and variable ontologies" (Viveiros de Castro 2004: 4). There is much to be gained from engaging this line of thought. One of political ecology's strongest intellectual contributions has long been its ability to make nature itself into an object of inquiry. Rather than settling multiculturalism into the familiar epistemological grids of self and other, or nature and culture, a multiplicity of natures comes into view. This "multinaturalism," as Viveiros de Castro (1998) terms it, makes ontologies rather than epistemologies the emphasis of inquiry. This promises a shift that renders "indigeneity" and "nature" as colonial artifacts (Blaser 2014; de la Cadena 2010; Kohn 2013).

Can multinaturalism be compatible with political ecology? Can it productively engage with the indeterminacy of indigeneity as a concept and a practice, opening the "global dreamtime of environmentalism" to new natures? Or does it return to cultural ecology's limitations, treating culture as a constant and nature as a variable in ways that reproduce a generic indigeneity (Ramos 2012)? Certainly some ethnographies of indigenous ontologies seem to ignore or downplay situated histories and geographies of war, capitalist penetration, colonialism, state policies, development, and trade to define an abstracted indigenous ontology. Just as geographers and anthropologists of indigeneity (Blaser 2009; Coombes et al. 2012) criticize political ecology for missing the deeper significance of "resource conflicts" and focusing only on external forces, political ecologists might critique the former for not paying adequate attention to the coercive power of capital and the subject-making power of various forms of violence. Yet this impasse is not necessarily insurmountable. An attention to different worlds need not elide an analysis of state power or capitalist extraction (e.g. de la Cadena 2010; Yeh 2009). Nor can multinaturalism become the new generic measure of indigeneity, complete with its own terms of reference such as "pachamama" or "buen vivir" that run roughshod over other forms of knowledge (Gudynas 2011; Walsh 2010). Such a stance would only return us to new universals, missing the opportunity to reconfigure the very grounds of what can and cannot be included in the *political*. Rather than narrowing politics, this suggests a further widening that does not ignore the violence of accumulation by dispossession, but supplements it; it suggests that for liberatory projects, nature as a universal analytical category may be both an indispensable tool – a strategic essentialism – and also utterly inadequate. Without stretching political ecology too far or thin, such a dialogue will no doubt enrich the field through a sustained engagement with deeper postcolonial insights vis-à-vis "nature." For the foreseeable future, continuing engagement with indigeneity, in all its guises, will be indispensable to that effort.

References

Anderson, M. 2007, "When Afro becomes (like) indigenous: Garifuna and Afro-indigenous politics in Honduras," *The Journal of Latin American and Carribbean Anthropology*, vol. 12, pp. 384–413.

Asher, K. 2009, *Black and Green: Afro-Colombians, Development, and Nature in the Pacific Lowlands*, Durham, NC, Duke University Press.

Baird, I.G. 2013, "'Indigenous peoples' and land: Comparing communal titling and its implications in Cambodia and Laos," *Asia Pacific Viewpoint*, vol. 54, pp. 269–281.

Baird, I.G. 2015, "Considering the implications of the concept of indigeneity for land and natural resource mangement in Cambodia, Thailand and Laos," in Uddin, N., Gerharz, E. and Chakkarath, P. (eds.), *Futures of Indigeneity*, New York, Berghan Books.

Baird, I.G. 2015, "Translocal assemblages and the circulation of the concept of 'indigenous peoples' in Laos," *Political Geography*, vol. 46, pp. 54–64.

Bakker, K. 2008, "The ambiguity of community: Debating alternatives to private-sector provision of urban water supply," *Water Alternatives*, vol. 1, no. 2, pp. 236–252.

Bateson, G. 1972, *Steps to an Ecology of Mind: Collected Essays in Anthropology, Psychiatry, Evolution and Epistemology*, Chicago, University of Chicago Press.

Baviskar, A. 2007, "Indian indigeneities: Adivasi engagements with Hindu nationalism," in de la Cadena, M. and Starn, O. (eds.), *Indigenous Experience Today*, New York, Berg, 275–304.

Bebbington, A. 2004, "Movements, modernization, markets, and municipalities: Indigenous federations in rural Ecuador," in Peet, R. and Watts, M. (eds.), *Liberation Ecologies: Environment, Development, Social Movements*, second edition, New York, Routledge, 394–421.

Bebbington, A. and Bebbington, D.H. 2011, "An Andean avatar: Post-neoliberal and neoliberal strategies for securing the unobtainable," *New Political Economy*, vol. 16, pp. 131–145.

Becker, M. 2008. *Indians and Leftists in the Making of Ecuador's Modern Indigenous Movements*, Durham, NC, Duke University Press.

Berger, T.R. 1977, *Northern Frontier, Northern Homeland: The Report of the Mackenzie Valley Pipeline Inquiry*, Ottawa, Ministry of Supply and Services, Government of Canada.

Berkes, F. 1999, *Sacred Ecology: Traditional Ecological Knowledge and Resource Management*, Philadelphia, Taylor & Francis.

Blackburn, C. 2009, "Differentiating indigenous citizenship: Seeking multiplicity in rights, identity and sovereignty in Canada," *American Ethnologist*, vol. 36, pp. 66–78.

Blaser, M. 2009, "The threat of the Yrmo: The political ontology of a sustainable hunting program," *American Anthropologist*, vol. 111, pp. 10–20.

Blaser, M. 2014, "Ontology and indigeneity: on the political ontology of hetereogeneous assemblages," *Cultural Geographies*, vol. 21, pp. 49–58.

Brass, T. 2002, "On which side of what barricade? Subaltern resistance in Latin America and elsewhere," *Journal of Peasant Studies*, vol. 29, pp. 336–399.

Braun, B. 2002, *The Intemperate Rainforest: Nature, Culture, and Power on Canada's West Coast*, Minneapolis, University of Minnesota Press.

Brosius, J.P. 1997, "Endangered forest, endangered people: Environmentalist representations of indigenous knowledge," *Human Ecology*, vol. 24, pp. 47–69.

Brosius, J.P., Tsing, A.L. and Zerner, C. 1998, "Representing communities: Histories and politics of community-based resource management," *Society and Natural Resources*, vol. 11, pp. 157–168.

Bryan, J. 2009, "Where would be without them? Knowledge, space and power in indigenous politics," *Futures*, vol. 41, pp. 24–32.

Bryan, J. 2011, "Walking the line: Participatory mapping, indigenous rights, and neoliberalism," *Geoforum*, vol. 42, no. 1, pp. 40–50.

Castree, N. 2004, "Differential geographies: Place, indigenous rights and 'local' resources," *Political Geography*, vol. 23, pp. 133–167.

Chakrabarty, D. 2000, *Provincializing Europe: Postcolonial Thought and Historical Difference*, Princeton, Princeton University Press.

Clifford, J. 1988, *The Predicament of Culture: Twentieth Century Ethnography, Literature and Art*, Cambridge, MA, Harvard University Press.

Clifford, J. 2001, "Indigenous articulations," *Contemporary Pacific*, vol. 13, pp. 468–490.

Coggins, C. with Gesang Zeren. 2014, "Animate landscapes: Nature conservation and the production of agropastoral sacred space in Shangrila," in Yeh, E.T. and Coggins C. (eds.), *Mapping Shangrila: Contested Landscapes in the Sino-Tibetan Borderlands*, Seattle, University of Washington Press, 205–228.

Colchester, M. 1995, "Indigenous peoples' rights and sustainable resource use in South and Southeast Asia," in Barnes, R.H., Gray, A. and Kingsbury, B. (eds.), *Indigenous Peoples of Asia*, Ann Arbor, Association of Asian Studies, 35–58.

Conklin, B. and Graham, L.R. 1995, "The shifting middle ground: Amazonian Indians and eco-politics," *American Anthropologist*, vol. 97, pp. 695–710.

Coombes, B., Johnson, J.T. and Howitt, R. 2012, "Indigenous geographies I: Mere resource conflicts? The complexities in indigenous land and environmental claims," *Progress in Human Geography*, vol. 36, pp. 810–821.

Cusicanqui, S.R. 2012, "Ch'ixinakax utwxiwa: A reflection on the practices and discourses of decolonization," *South Atlantic Quarterly*, vol. 111, pp. 95–109.

de la Cadena, M. 2010, "Indigenous cosmopolitics in the Andes: Conceptual reflections beyond 'politics'," *Cultural Anthropology*, vol. 25, pp. 334–370.

de la Cadena, M. and Starn, O. (eds.) 2007, *Indigenous Experience Today*, New York, Berg.

Delcore, H. 2004, "Symbolic politics or generification? The ambivalent implications of tree ordinations in the Thai environmental movement," *Journal of Political Ecology*, vol. 11, pp. 1–30.

de Leeuw, S., Cameron, E. and Greenwood, M. 2012, "Participatory and community-based research, indigenous geographies, and the spaces of friendship: A critical engagement," *The Canadian Geographer/ Le Géographe Canadien*, vol. 6, no. 2, pp. 180–194.

Denevan, W.M. 2001, *Cultivated Landscapes of Native Amazonia and the Andes*, Oxford, Oxford University Press.

Dombrowski, K. 2002, "The praxis of indigenism and Alaska native timber politics," *American Anthropologist*, vol. 104, pp. 1062–1073.

Duhaylungsod, L. 2011, "Interpreting 'indigeneous peoples and resource use': The case of the T'Boil in the Southern Philippines," in Dove, M., Sajise, P.E. and Doolittle, A.A. (eds.), *Beyond the Sacred Forest: Complicating Conservation in Southeast Asia*, Durham, NC, Duke University Press, pp. 216–236.

Escobar, A. 1995, *Encountering Development: The Making and Unmaking of the Third World*, Princeton, Princeton University Press.

Geschiere, P. and Nyamnjoh, F.B. 2000, "Capitalism and autocthony: The see-saw of mobility and belonging," *Public Culture*, vol. 12, pp. 423–452.

Goldman, M. 2007, "Tracking wildebeest, locating knowledge: Maasai and conservation biology understandings of wildebeest behavior in Northern Tanzania," *Environment and Planning D: Society and Space*, vol. 25, pp. 307–331.

Goldman, M. 2009, "Constructing connectivity: Conservation corridors and conservation politics in East African rangelands," *Annals of the Association of American Geographers*, vol. 99, pp. 335–359.

Gordon, E.T., Gurdián, C.C. and Hale, C.R. 2003, "Rights, resources and the social memory of struggle: Reflections on a study of Indigenous and Black community land rights on Nicaragua's Atlantic Coast," *Human Organization*, vol. 62, pp. 369–381.

Gray, A. 1995, "The indigenous movement in Asia," in Barnes, H.R., Gray, A. and Kingsbury, B. (eds.), *Indigenous Peoples of Asia*, Ann Arbor, Association of Asian Studies, pp. 35–58.

Grinde, D.A. and Johansen, B.E. 1995, *Ecocide of Native America: Environmental Destruction of Indian Lands and Peoples*, Santa Fe, Clear Light.

Gudynas, E. 2010, "Si eres tan progresista ¿por qué destruyes la naturaleza? Neoextractivismo, izquierda y alternativas," *Ecuador Debate*, vol. 79, pp. 61–81.

Gudynas, E. 2011, "Buen Vivir: Today's tomorrow," *Development*, vol. 54, no. 4, pp. 441–447.

Hale, C. R. 1994, *Resistance and Contradiction: Miskitu Indians and the Nicarguan State, 1894–1987*, Stanford, Stanford University Press.

Hale, C. R. 2005, "Neoliberal multiculturalism: The remaking of cultural rights and racial dominance in Central America," *PoLAR: Political and Legal Anthropology Review*, vol. 28, pp. 10–19.

Hale, C.R. 2006, "Activist research v cultural critique: indigenous land rights and the contradictions of political engaged anthropology," *Cultural Anthropology*, vol. 21, pp. 96–120.

Hale, C.R. 2011, "Resistencia para que? Territory, autonomy and neoliberal entanglements in the 'empty spaces' of Central America," *Economy and Society*, vol. 40, pp. 184–210.

Hall, S. 1983, "The problem of ideology: Marxism without guarantees," in Matthews, B. (ed.), *Marx 100 Years On*, London, Lawrence and Wishart, 57–68.

Hall, S. 1996, "On postmodernism and articulation: An interview with Stuart Hall," in Morley, D. and Chen, K. (eds.), *Stuart Hall: Critical Dialogues in Cultural Studies*, London, Routledge, 131–150.

Healy, K. 2001, *Llamas, Weavings and Organic Chocolate: Multicultural Grassroots Development in the Andes and Amazon of Bolivia*, Notre Dame, IN, University of Notre Dame Press.

Heatherington, T. 2010, *Wild Sardinia: Indigeneity & the Global Dreamtimes of Environmentalism*, Seattle, University of Washington Press.

Hecht, S.B. and Cockburn, A. 2010, *The Fate of the Forest: Developers, Destroyers, and Defenders of the Amazon*, updated edition, Chicago, University of Chicago Press.

Hodgson, D.L. 2009, "Becoming indigenous in Africa," *African Studies Review*, vol. 52, pp. 1–32.
Huber, T. and Pederson, P. 1997, "Meterological knowledge and environmental ideas in traditional and modern societies: The case of Tibet," *Journal of the Royal Anthropological Institute*, vol. 3, pp. 577–598.
Kingsbury, B. 1998, "'Indigenous peoples' in international law: A constructivist approach to the Asian controversy," *American Journal of International Law*, vol. 92, pp. 414–457.
Kirsch, S. 2006, *Reverse Anthropology: Indigenous Analysis of Social and Environmental Relations in New Guinea*, Stanford, Stanford University Press.
Kohn, E. 2013, *How Forests Think: Toward an Anthropology Beyond the Human*, Berkeley, University of California Press.
Krech III, S. 1999, *The Ecological Indian: Myth and History*, New York, W.W. Norton & Co.
Li, T.M. 2000, "Articulating indigeneous identity in Indonesia: Resource politics and the tribal slot," *Comparative Studies in Society and History*, vol. 42, pp. 149–179.
Li, T.M. 2001, "Masyarakat adat, difference, and the limits of recognition in Indonesia's forest zone," *Modern Asian Studies*, vol. 35, pp. 645–676.
Li, T.M. 2010, "Indigeneity, capitalism and the management of dispossesion," *Current Anthropology*, vol. 51, pp. 385–414.
Lucero, J.A. 2008, *Struggles of Voice: The Politics of Indigenous Representation in the Andes*, Pittsburgh, University of Pittsburgh Press.
Mallon, F.E. 1995, *Peasant and Nation: The Making of Postcolonial Mexico and Peru*, Berkeley, University of California Press.
Mamani Ramírez, P. 2005, *Geopolíticas Indígenas*, El Alto, Boliva, CADES, Centro Andino de Estudios Estratégicos.
Mamani Ramírez, P. 2011, "Cartographies of indigenous power: Identity and territory in Bolivia," in Fabricant, N. and Gustafson, B. (eds.), *Remapping Bolivia: Resources, Territory and Indigeneity in a Pluri-National State*, Santa Fe, School for Advanced Research, 30–45.
McFarlane, C. 2009, "Translocal assemblages: Space, power and social movements," *Geoforum*, vol. 40, pp. 561–567.
Merlan, F. 2007, "Indigeneity as relational identity: The construction of Australian land rights," in de la Cadena, M. and Starn, O. (eds.), *Indigenous Experience Today*, New York, Berg, 125–150.
Merlan, F. 2009, "Indigeneity: Global and local," *Current Anthropology*, vol. 50, pp. 303–333.
Mollett, S. 2013, "Mapping deception: The politics of mapping Miskito and Garifuna space in Honduras," *Annals of the Association of American Geographers*, vol. 103, pp. 1227–1241.
Moore, D.S., Kosek, J. and Pandian, A. (eds.) 2003, *Race, Nature and the Politics of Difference*, Durham, NC, Duke University Press.
Nadasdy, P. 2011, "'We don't harvest animals; we kill them': Agricultural metaphors and the politics of wildlife management in the Yukon," in Goldman, M.J., Nadasdy, P. and Turner, M.D. (eds.), *Knowing Nature: Conversations at the Intersection of Political Ecology and Science Studies*, Chicago, University of Chicago Press, 135–151.
Ng'weno, B. 2007, *Turf Wars: Territory and Ciizenship in the Contemporary State*, Stanford, Stanford University Press.
Niezen, R. 2003, *The Origins of Indigenism: Human Rights and the Politics of Identity*, Berkeley, University of California Press.
Nyamnjoh, F.B. 2007, "'Ever-diminishing circles': The paradoxes of belonging in Botswana," in de la Cadena, M. and Starn, O. (eds.), *Indigenous Experience Today*, New York, Berg, 305–332.
Perreault, T. 2008, "Custom and contradiction: Rural water governance and the politics of Usos Y Costumbres in Bolivia's irrigators' movement," *Annals of the Association of American Geographers*, vol. 98, no. 4, pp. 834–854.
Perreault, T. and Green, B. 2013, "Reworking the spaces of indigeneity: The Bolivian ayllu and lowland autonomy movements compared," *Environment and Planning D: Society and Space*, vol. 31, pp. 43–60.
Perreault, T. and Valdiva, G. 2010, "Hydrocarbons, popular protest and national imaginaries: Ecuador and Bolivia in comparative context," *Geoforum*, vol. 41, pp. 689–699.
Posey, D. 1985, "Indigenous management of tropical forest ecosystems: The case of the Kayapo Indians of the Brazilian Amazon," *Agroforestry Systems*, vol. 3, pp. 139–158.
Povinelli, E.A. 2011, *Economies of Abandonment: Social Belonging and Endurance in Late Liberalism*, Durham, NC, Duke University Press.
Price, D. 1989, *Before the Bulldozer: The Nambiquara Indians and the World Bank*, Bethesda, Seven Locks Press.

Ramos, A. 1998, *Indigenism: Ethnic Politics in Brazil*, Madison, University of Wisconsin Press.
Ramos, A. 2003, "Pulp fictions of indigenism," in Moore, D.S., Kosek, J. and Pandian, A. (eds.), *Race, Nature and the Politics of Difference*, Berkeley, University of California Press, 356–379.
Ramos, A. 2012, "The politics of perspectivism," *Annual Review of Anthropology*, vol. 41, pp. 481–494.
Rappaport, R.A. 1967, *Pigs for the Ancestors: Ritual in the Ecology of a New Guinea People*, New Haven, Yale University Press.
Rodríguez-Piñero, L. 2005, *Indigenous Peoples, Postcolonialism and International Law: The ILO Regime, 1919–1989*, New York, Oxford University Press.
Sauer, C.O. 1969, "The morphology of landscape," in Leighly, J. (ed.), *Land and Life: A Section from the Writings of Carl Ortwin Sauer*, Berkeley, University of California Press, pp. 315–350.
Sawyer, S. 2004, *Crude Chronicles: Indigenous Politics, Multinational Oil, and Neoliberalism in Ecuador*, Durham, NC, Duke University Press.
Shah, A. 2010, *In the Shadows of the State: Indigenous Politics, Environmentalism and Insurgency in Jharkhand, India*, Durham, NC, Duke University Press.
Sivaramakrishnan, K. and Agrawal, A. 2003, "Regional modernities in stories and practices of development," in Sivaramakrishnan, K. and Agrawal, A. (eds.), *Regional Modernities: The Cultural Politics of Development in India*, New York, Oxford University Press, 1–62.
Sletto, B.I. 2009, "We drew what we imagined," *Current Anthropology*, vol. 50, no. 4, pp. 443–476.
Sparke, M. 1998, "A map that roared and an original atlas: Canada, cartography, and the narration of nation," *Annals of the Association of American Geographers*, vol. 88, pp. 463–495.
Star, S.L. and Griesemer, J.R. 1989, "Institutional ecology, 'translations' and boundary objects: Amateurs and professionals in Berkeley's Museum of Vertebrate Zoology, 1907–39," *Social Studies of Science*, vol. 19, pp. 387–420.
Steward, J.H. 1956, *The Concept and Method of Cultural Ecology*, Bobbs-Merill, College Division.
Stewart, O.C. 2002, *Forgotten Fires: Native Americans and the Transient Wilderness*, Norman: University of Oklahoma Press.
Sylvain, R. 2002, "'Land, water and truth': San identity and global indigenism," *American Anthropologist*, vol. 104, pp. 1074–1085.
Tsing, A.L. 1997, "Transition as translations," in Scott, J., Kaplan, C. and Keates, D. (eds.), *Transitions, Environments, Translations*, New York, Routledge, 253–272.
Tsing, A.L. 1999, "Becoming a tribal elder, and other green development fantasies," in Li, T.M. (ed.), *Transforming the Indonesian Uplands: Marginality, Power and Production*, Amsterdam, Harwood Academic Publishers, 157–200.
Tsing, A.L. 2003, "Agrarian allegory and global futures," in Greenough, P. and Tsing, A.L. (eds.), *Nature in the Global South: Environmental Projects in South and Southeast Asia*, Durham, NC, Duke University Press, 124–169.
Tsing, A.L. 2005, *Friction: An Ethnography of Global Connection*, Princeton, Princeton University Press.
Tsing, A.L. 2009, "Adat/indigenous: Indigeneity in motion," in Gluck, C. and Tsing, A.L. (eds.), *Words in Motion: Toward a Global Lexicon*, Durham, NC, Duke University Press, 40–66.
Turner, T. 1995, "An indigenous people's struggle for socially equitable and ecologically sustainable production: The Kayapo revolt against extractivism," *Journal of Latin American Anthropology*, vol. 1, pp. 98–121.
Varese, S. 2006, *Witness to Sovereignty: Essays on the Indian Movement in Latin America*, Copenhagen, International Work Group for Indigenous Affairs.
Vayda, A.P. 1961, "Expansion and warfare among swidden agriculturalists," *American Anthropologist*, vol. 63, pp. 346–358.
Viveiros de Castro, E. 1998, "Cosmological deixis and Amerindian perspectivism," *Journal of the Royal Anthropological Institute*, vol. 4, pp. 469–488.
Viveiros de Castro, E. 2004, "Perspectival anthropology and the method of controlled equivocation," *Tipití: Journal of the Society for the Anthropology of Lowland South America*, vol. 2, pp. 3–22.
Wainwright, J. 2008, *Decolonizing Development: Colonial Power and the Maya*, New York, Blackwell.
Wainwright, J. and Bryan, J. 2009, "Cartography, territory, property: postcolonial reflections on indigenous counter-mapping in Nicaragua and Belize," *Cultural Geographies*, vol. 16, pp. 153–178.
Walsh, C. 2010, "Development as buen vivir: Institutional arrangements and (de)colonial entanglements," *Development*, vol. 53, no. 1, pp. 15–21.
Watts, M. 2000, "Political ecology," in Sheppard, E. and Barnes, T. (eds.), *A Companion to Economic Geography*, Oxford, Blackwell, 257–274.

Watts, M. 2003, "Development and governmentality," *Singapore Journal of Tropical Geography*, vol. 24, pp. 6–34.

West, P. 2005, "Translation, value and space: Theorizing an ethnographic and engaged environmental anthropology," *American Anthropologist*, vol. 107, pp. 632–642.

Yeh, E.T. 2007, "Tibetan indigeneity: Translations, resemblances and uptake," in de la Cadena, M. and Starn, O. (eds.), *Indigenous Experience Today*, New York, Berg, 69–97.

Yeh, E.T. 2009, "From wasteland to wetland? Nature and nation in China's Tibet," *Environmental History*, vol. 14, pp. 103–137.

Yeh, E.T. 2013, "Tibet in China's environmental movement," in Brox, T. and Beller-Hann, I. (eds.), *On the Fringes of the Harmonious Society: Tibetans and Uyghurs in Socialist China*, Copenhagen, NIAS Press, 235–262.

Yeh, E.T. 2014a, "Reverse environmentalism: Contemporary articulations of Tibetan Buddhism, culture, and environmental protection," in van der Veer, P., Miller, J. and Smyer Yü, D. (eds.), *Religious Diversity and Ecological Sustainability in China*, New York, Routledge, 194–218.

Yeh, E. T. 2014b, "The rise and fall of the Green Tibetan: Contingent collaborations vs. the vicissitudes of harmony," in Yeh, E.T. and Coggins, C. (eds.), *Mapping Shangrila: Contested Landscapes in the Sino-Tibetan Borderlands*, Seattle, University of Washington Press, 255–278.

42

ON THE CONCRETENESS OF LABOR AND CLASS IN POLITICAL ECOLOGY

Michael Ekers

Introduction

This chapter examines how 'class', as an analytical and lived social identity, is forged and politicized through the material and representational relationships of social groups to various 'natures'. Class identities, it will be argued, are shaped as parts of the non-human world circulate through the conduits of political economic processes, and through cultural relationships to a range of landscapes. In making this argument, I track the relationship between class and nature in several key debates in the evolution of political ecology. As I assess the relevant literature, I focus on the concreteness of labor, that is the broad number of determinants and political possibilities articulated in people's working relationships to environmental landscapes, a point I examine in more detail below. The central claim is that political ecological studies of class should aspire to be more concrete, which entails more than a narrow call for more grounded studies of the realities of social and environmental life. Rather, following Marx and Gramsci I argue that a 'concrete' engagement involves a constant effort to understand the various constitutive forces that shape the relationship between class and nature and the political constraints and possibilities immanent in that relationship. The concrete then is not simply an empirical point of departure for political ecological studies but is achieved through processes of abstraction and theorization that enrich our understandings of socionatures.

I start by providing a brief account of where my argument fits into the quilt of political ecology and studies of class. Given the breadth of the field I highlight three central claims within political ecology that are germane to this chapter. First, binary understandings of nature and society are held to be untenable, leading scholars to insist on the imbrication of human and non-human processes (Braun, 2006; Castree, 1995; 2002; Castree and Braun, 1998; Smith, 2008[1984]). If we accept this claim, class and nature cannot be viewed as issues external to one another, but as necessarily connected and mutually constitutive. Second, is the claim that nature is constantly enrolled in various production and circulatory processes, giving rise to a politics of appropriating, contesting and representing natures (Prudham, 2005; Swyngedouw, 1999; Watts, 2000; Zalik, 2011). Building on this point, class identities might be constructed, represented and lived as parts of nature are metabolized through the labor process and as value circulates through nature. Third, various contributors highlight how social identities are constructed through reference to, and engagement with, human and non-human natures

(Anderson, 2001; Gururani, 2002; Haraway, 1991; Mortimer-Sandilands and Erickson, 2010; Nightingale, 2006; Williams, 1973). Following this claim, class cannot be seen as a stable identity that pre-exists a material and representational engagement with nature. To the contrary, class should be understood as being actively constructed and re-constituted as socionatural relations shift and change historically and geographically.

In the most narrow of terms, class has been defined structurally as one's position within the relations of production, those owning the means of production representing the bourgeoisie and those owning their labor power comprising the proletariat (for a discussion and critique of the 'two-class model' see Resnick and Wolff, 1981). It was E.P. Thompson's (1964: 9) project to provide a much more historical and humanistic account of class, remarking that he '[did] not see class as a "structure", nor even as a "category", but as something which in fact happens (and can be shown to have happened) in human relationships.' Here, class is lived and felt and made by those immersed in class relations and, importantly, economic relationships remain in the balance as a constitutive force in the making of such identities. It was the likes of Thompson (1964; 1978) and Raymond Williams (1973; 1980) that spawned a recognition that determinations related to the value-form of capital are but one 'relation of force' (Gramsci, 1971) and must be understood as constituting class in concert with a broader range of social, cultural and political relations. However, the project of developing an 'open' understanding of class remained, and remains, unfinished as feminist and anti-racist critiques have stressed that class as an analytic, political and lived social identity is always constructed by a broader number of relations of difference including 'race', gender and sexuality (Bannerji, 1995; McClintock, 1995; McDowell, 2006; Scott, 1988). These overlapping approaches to class require that we maintain an appreciation of how the value-form of capital shapes class identities in concert with a broad number of cultural and political forces and relations of difference, which, in sum, can provide a concrete understanding of class and nature.

Bringing together these preliminary remarks on political ecology and class, the goal of this chapter is to examine how questions of class come to be forged, materially and representationally, through the entanglement of nature and society and the enrollment of the environment into our political economic and social worlds. The field of political ecology is so expansive that my engagement with treatments of class is necessarily somewhat selective. However, I try to take a long view of political ecology starting with debates on agrarian capitalism, peasantry and class and then moving on to debates on the production of nature. I examine the concreteness of class and labor in these debates with an eye towards where future research should go.

Regarding the concrete

The concrete is often understood as the 'grounded,' the 'particular' or the 'empirical' manifestation of some broader process such as capitalism or colonialism. However, if we turn to Marx's methodological reflections in the *Grundrisse*, a more nuanced understanding of the concrete is possible which, arguably, is invaluable in thinking through the relationship between class and political ecology. Marx (1973[1939]: 100–101) writes:

> The concrete is concrete because it is the concentration of many determinations, hence unity of the diverse. It appears in the process of thinking, therefore, as a process of concentration, as a result, not as a point of departure, even though it is the point of departure in reality and hence also the point of departure for observation.

Marx is suggesting that the concrete is both a point of departure (the 'real concrete') and a site of arrival (the 'concrete in thought'). Crucially, because the concrete is 'the concentration of many determinations, hence unity of the diverse', Marx's methods require an appreciation of the different constitutive forces and lived relations that give form and content to something like 'class'. The concrete then allows us to appreciate the multiple determinants that shape class identities while also highlighting how relations of difference must be included within the set of determinants identified through a constant movement between the abstract and the concrete.

If Marx's method calls for an integral analysis of the determinants that shape something like class, Antonio Gramsci's (1971) work helps highlight both the humanistic and 'earthly' element of class (see also Thomas, 2009). Gramsci builds on Marx's method, calling for an 'absolute historicism' that insists on never settling on an 'overarching theory that endows the particular with a stable meaning', but rather, calls for a constant return to the fragments of historical and geographic life which 'induce new and more complex concepts' (Buttigieg, 1990: 66). Developing a concrete understanding of class in political ecology through the lens of Marx's and Gramsci's methods requires a move away from treating class as a structural and stable social position and a movement towards identifying the particular and immanent determinations that constitute class in relationship to the appropriation of resources and the making of natures. There is a restless historicism to such an approach that starts with what Gramsci might have referred to as the 'earthliness' of how class is lived and felt. Following Gramsci and Marx, one then identifies the generative relations that make class concrete: 'the unity of the diverse'. Setting class in motion through a ceaseless movement between the concrete and the abstract, as Geoff Mann (2007) does for the wage, provides a powerful method for grappling with the making of class in various environmental settings. In what follows, I examine the treatment of class within several key debates in political ecology and assess whether they meet the benchmark of being concrete and earthly, in Marx's and Gramsci's nuanced understandings of these terms.

The agrarian question

One of the foundations of political ecology has been studies of agrarian production, which highlight the difference that 'nature' makes in the capitalization of agriculture and the adoption of wage-labor in the sector – in short, 'the agrarian question' (see Bernstein, 1996; Henderson, 1999; Kautsky, 1988[1899]; Mann and Dickinson, 1978). Studies of agriculture in the 1970s and 1980s focused on understanding the presence and social character of the peasantry and the proletariat in agricultural settings. This literature represents the most sustained engagement with questions of class and nature in the general area of political ecology even if this literature pre-dated political ecology proper and grew out of a much longer debate on the peasantry and the proletariat initiated by Kautsky's (1988) *The Agrarian Question*, Lenin's (1964 [1899]) *The Development of Capitalism in Russia* and Chayanov's (1986[1925]) *The Theory of Peasant Economy*.

Kautsky emphasizes the effects of the capitalization of agriculture for the development of capitalism, the future of the peasantry and the significance of the former and latter for the socialist movement in Germany at the end of the nineteenth century. As Alavi and Shanin's (1988) introduction to the English edition of the *Agrarian Question* suggests, Kautsky initially stresses that the peasantry would be differentiated into capitalists and waged-workers as capitalist development progressed, only to argue later in the book that the particularities of agrarian production and access to land allow peasants to persist despite the advancement of agrarian capital. Kautsky tends to define peasants and the proletariat structurally with access to the means of production (including land) being the defining determinant. Peasants are seen as simple commodity producers that 'own their means of production' (1988: 63) and are 'self-sufficient' (15). Later, Kautsky (1988: 132) argues

> That the capitalist mode of production is responsible for bringing into existence a class of laborers for whom private property in the means of production has been swept away, but also by socializing the production process, and creating and accentuating the class antagonism between capitalists and wage-laborers.

Lenin's (1964: 174) position was not dissimilar although he was less certain that the peasantry would continue to exist as the countryside was capitalized:

> The old peasantry is not only 'differentiating,' it is being completely dissolved, it is ceasing to exist, it is being ousted by absolutely new types of rural inhabitants—types that are the basis of a society in which commodity economy and capitalist production prevail. These types are the rural bourgeoisie (chiefly petty bourgeoisie) and the rural proletariat—a class of commodity producers in agriculture and a class of agricultural wage-workers.

Both Kautsky and Lenin treat agricultural producers in a one-sided fashion in which their identity as peasants or proletariats is defined by their ownership, or lack thereof, of the means of production. It is not surprising that Kautsky and Lenin both relied on various quantitative statistics on capitalist enterprises, wage-labor and peasant production in advancing their positions, which arguably meshed well with their structural accounts of social class. The class character of the proletariat and the peasants, if we are to follow Marx's and Gramsci's method, is overly stable and understood through a theory of class differentiation tied to the development of capitalism. However, both Kautsky and Lenin acknowledge that multiple modes of production existed alongside of one another (see Lenin, 1964: 33 as cited by Bernstein, 2009; Kautsky, 1988: 63), but this insight is not stretched to recognize how the identities of agrarian producers are shaped by these modes of production.

Chayanov offers a different treatment of the peasantry from that of Kautsky and Lenin, although it could be argued that he picks up where Kautsky leaves off, specifically, regarding the persistence of the peasantry despite the expansion of capitalism throughout the countryside. Chayanov (1986) seeks to explain how the peasantry survived through a peasant economy focused on the use of family labor and other non-waged forms of work. The peasant economy, which Chayanov sees as distinct from capitalism, is structured by demographic cycles and peasant producers' self-exploitation, which allow such producers to out-compete capitalist enterprises. Whereas capitalist agriculturalists are concerned with covering their costs of production and making a profit, Chayanov suggests that peasants are guided by a labor-consumer balance in which the advantages of further production are balanced against further family consumption requirements. Chayanov views the peasantry as a particular class defined by the organizational structure of the family farm and labor. He conceptualizes 'the peasant farm as a family labor farm in which the family as a result of its year's labor receives a single labor income and weights its efforts against the material results obtained' (Chayanov, 1986: 41).

Chayanov has both vocal supporters (Banaji, 1976; 1977; Shanin, 1973; 1986; 2009; Thorner, 1986; van der Ploeg, 2013) and detractors (Cook and Binford, 1986; Deere, 1987; Friedmann, 1980; Patnaik, 1979), but I share Bernstein's (2009) critique that Chayanov tends to treat the peasant farm as overly autonomous, unchanging and free of any social relations that externally and internally constitute peasant production and the subjectivity of such producers.

Chayanov did not write from a Marxist perspective and so was unconcerned with the criticisms of his work, of which he was well aware: 'we are interested not in the system of the peasant farm and forms of organization in their historical development but, rather, in the mere mechanics of the organizational process. But this organizational analysis by its very nature ought to be static' Chayanov (1986: 44). The static quality of the peasantry in Chayanov's work is precisely the form of positivism and ahistoricism that Marx and Gramsci warned against, as there is little that is concrete in Chayanov's approach to the peasantry in agricultural settings: his work lacks both an earthly point of departure and an arrival point through which we can glimpse the 'unity of the diverse'.

The peasantry, the proletariat and agrarian production

Starting in earnest in the 1970s, scholars concerned with the capitalization of agriculture in the Global South focused on conceptualizing and studying the peasantry and processes of class differentiation. For many, Kautsky, Lenin and Chayanov were conceptual points of departure. I cannot possibly do justice here to the nuanced debates and numerous conceptual and methodological approaches for understanding 'class' and the 'peasantry' in agrarian production developed in the 1970s and 1980s. Instead, I focus on several key arguments rather than the entire debate on the agrarian question (for useful summaries of these debates see Akram-Lodhi and Kay, 2010a; 2010b; Deere, 1987; Goodman and Redclift, 1981).

Chayanov's intellectual legacy was carried on by a number of writers that attempted to conceptualize a distinctly peasant mode of production defined by an absence of waged-work and reliance on familial labor and control over the means and product of labor (see Harrison, 1977; Shanin, 1973; 1986; Thorner, 1965; 1986). The persistence of non-capitalist modes of production, such as subsistence production, is said to limit the development of capitalist production relations. Goodman and Redclift (1981) suggest that such an approach over-emphasizes the independence of different modes of production, while others such as Bernstein (2009) and Patnaik (1979) describe advocates of the peasant mode of production as 'agrarian populists'. The reliance on Althusser and Balibar (1970) in accounts of the peasant mode of production, noted by Goodman and Redclift (1981), gives this literature a structural focus ill-fitted for recognizing the humanistic elements of agrarian classes and their particular historical and geographical determinants.

More nuanced understandings of modes of production have subsequently gone well beyond what the 'agrarian populists' offered. Harold Wolpe's (1975; 1980) foundational work examines the articulation of capitalism and apartheid in South Africa through the lens of multiple modes of production. He develops the concept of 'extended modes of production', which includes the relations and forces of production together with an analysis of the state and broader questions of ideology and political-juridical 'superstructures'. Wolpe seeks to understand the articulation of modes of production within particular historical conjunctures, with an emphasis on the dominance of one particular mode existing alongside subordinated modes that continue to play a determining role in political economic processes and the perpetuation of 'internal colonialism' (for engagements with 'modes of production' see: Banaji, 1977; Bernstein, 1977; Goodman and Redclift, 1981; Hall, 1980). The overall contribution of studies of modes of production in agricultural settings is to suggest that capitalist development is determined by multiple forces beyond capital itself. Moreover, and for the purposes of this chapter, if multiple modes of production are imbricated with one another, adhering to a structuralist account of class and peasant identities becomes untenable as agrarian producers are often engaged in subsistence and simple commodity forms of production.

In reaction to the overly structuralist positions of Kautsky and Lenin regarding class differentiation and Chayanov's postulation of a static and autonomous peasantry, a number of approaches were developed for understanding social class within the broader agrarian question. The premise of much of the literature is that subsistence farming cannot be separated from participation in the market economy. As Gavin Smith (1985: 101; see also Smith, 1979; Watts, 1983) says of Bernstein's (1979) central insight: 'the reproductive cycle of... "peasants" inevitably passes through the market'. Similarly, Harriet Friedmann (1980: 160) advocates analyzing agrarian social relations through the concept of 'forms of production': 'The social formation provides the context for reproduction of units of production, and in combination with the internal structure of the unit, determines its conditions of reproduction, decomposition, or transformation.' Friedmann suggests that her approach makes it impossible to consider the peasantry in Bernstein's (1979: 437) words as 'singular' or 'essential' and 'that it is impossible to talk about peasants as a "class" in general terms'. As Bernstein (2009; 2010) suggests, the integration of subsistence and commodity production requires that we speak of 'classes of labor'.

The conceptualization of class and agrarian production developed by Friedmann and Bernstein is clearly distinct from the agrarian populists. There is an attempt to understand classes of labor as constructed out of multiple political economic relations both within and beyond sites of production. Moreover, in opposition to Chayanov's trans-historical account of the peasantry, analyses of simple commodity production suggest that the class character of producers must be discerned historically based on the layering of commodified and non-commodified relations. Importantly, the ontology of social class is held open. Here we begin to approach a concrete account of class that I advocated for at the outset. Several limitations still exist. The early work of Friedmann and Bernstein is staunchly materialist, but perhaps in an austere sense, insofar as economic processes are foregrounded at the expense of various cultural and ideological relations. Bernstein (1979) acknowledges this at the conclusion of his 'African Peasantries' article. In addition, the broad scale of analysis of many of the early studies of agrarian production is such that the lived dimensions of class are occluded. At the same time, the conceptual ambition and analytical scale of these early studies are among their strengths, as political ecology over time has become increasingly particularistic and localist in orientation.

By the mid-1980s historical materialist analyses of labor in agrarian settings began to emphasize the representational and ideological dimensions of social class in agricultural production processes. For instance, Gavin Smith (1985: 106) acknowledges that Friedmann's and Bernstein's accounts of simple commodity production tend towards being universalistic and suggests that more emphasis be placed on the 'complexity, variety and richness of local cultural identity as a feature of class consciousness'. Smith (1985: 106) goes on to argue that such class identities are the product of political struggle and are 'likely to take the form of territorially and ethnically particularistic expressions appropriate to [an ideology of a community]'. We should question whether territorial and 'ethnic' relations are simply particulars that shape the expression of class consciousness, as such an approach places considerations of 'race' and 'nationalism' in a secondary position to capital. With that caveat in mind, we can begin to see how class identities are shaped by a wide number of constitutive relations that exceed the economic and are cultural in orientation, but no less materialist.

The work of Carney and Watts (1990) is emblematic of attempts of understand how questions of culture and representational practices are expressed in women's and men's experiences of agricultural production. In an ethnographic study that traces the various chains of power from international donors, the colonial and post-colonial state, to household politics, Carney and Watts illustrate how labor demands associated with intensified capitalist rice production are worked through existing farming systems and labor arrangements and create a fabric of agrarian identities. Carney and Watts (1990) build on Burawoy's (1985: 7) argument that 'as men and women transform raw material into useful things, they also reproduce social relations as well as an experience of those relations'. In the hands of Carney and Watts, Burawoy's point yields important insights regarding how agricultural labor is lived, felt and organized through a range of social relations, with gender being a key determinant. More recently, Vinay Gidwani (2008: 161) has forwarded a similar argument, suggesting that more attention be placed on 'importance of the *meaning*fulness of lived conduct' within studies of agrarian labor. Like Carney and Watts, Gidwani draws on a lengthy ethnographic engagement with agrarian workers in Gujarat to argue that a series of cultural practices congeal to shape employment decisions and caste identities. In particular, he suggests that an individual's withdrawal from work does not reflect a rational choice or structural factors, but rather is a cultural practice that constitutes caste identities and social distinctions.

It is Carney and Watts' and Gidwani's ethnographic methods which allow them to identify both the external and internal social relations that shape agrarian production in contrast to the more universalizing debates outlined above. Moreover, we can see in these studies how men and women construct their own identities through ideologies, representations and material struggles and thus we get an appreciation of both the earthliness of social class and its humanistic elements. The concrete character of this research fills in the subjective dimensions of social class and agrarian production while also highlighting how concrete labor is, in Marx's words, fundamentally a 'concentration of many determinations'.

Class and the production of nature

In this final section I shift attention away from debates on agrarian capitalism and towards accounts of 'the production of nature'. Such an approach to political ecology is captured by Neil Smith's (2008: 50) famous remark: 'When the appearance of nature is placed in a historical context, the development of the material landscape presents itself as a process of the production of nature.' Whereas debates on agrarian production focused on the Global South and on how agricultural goods were produced through a shifting constellation of non-commodified and commodified labor relations, debates on social natures moved away from an explicit focus on labor and concentrated on overcoming dualistic understandings of nature and society through exploring the socionatural making of environmental landscapes. Nonetheless, labor remains in the picture, as the central claim of this literature is that labor represents a point of mediation between 'nature' and 'society'. The products made out of the labor process congeal human and non-human elements and can be said to be a form of produced nature (see Castree, 1995; 2000; Chapter 21, this volume; Eaton, 2011; Ekers and Loftus, 2013; Loftus, 2012; Prudham and Heynen, 2011; Sywngedouw, 2006).

Despite its foundational role, labor tends to be taken for granted in the more theoretically oriented literature on the production of nature (Ekers and Loftus, 2013). It is not uncommon for advocates of the production of nature approach to invoke the following passage written by Marx (1977: 283, 290; see also 1973: 87):

> The labor process is the universal condition for the metabolic interaction between man [*sic*] and nature, the ever-lasting nature imposed condition of human existence, and it is therefore independent of every form of that existence, or rather it is common to all forms of society in which human beings live.

The trans-historical understanding of labor that is employed in debates on the production of nature has meant that concrete understandings of labor and class, as a site of departure and arrival, are somewhat scarce (for exceptions see Heynen et al., 2007; Mitchell, 1996; 2003).

The social identity of the laborers that make nature receives more attention in Smith's work than the question of labor itself, but the engagement is not as intensive as in debates on agrarian capitalism. When discussing 'production in general', Smith (2008: 58) starts by suggesting that the production of a surplus creates the conditions of possibility for a society divided along class lines. He then goes on to discuss 'production for exchange' and argues that class becomes defined by a division between those who perform productive labor and those that appropriate the products of that labor. Smith (2008: 61) extends this point to the 'production of nature', arguing:

> With the development of social classes, access to nature is unequally distributed (both qualitatively and quantitatively) according to class. The ruling class, whether or not it directly controls the social means of production, certainly controls the surplus appropriated from nature through the human labor of others, while the laboring class works the means of production.

For Smith, social relations to nature are structured through a class relation, understood classically as a division between those who own the means of production and those who own only their labor power. However there is a more delicate understanding of class woven through this argument.

On several occasions *Uneven Development* considers an increasing social division of labor (Smith, 2008: 72), a technical division of labor (73) and a gendered division of labor (74–75). In this respect, the question of difference comes to bear on how various social groups engage in producing natures. For Smith, the social differentiation of society into two classes 'leads in the opposite direction, toward a differentiation of cultures along class lines and of course a further differentiation on the basis of gender and race' (Smith, 2008: 76). A relationship between class and broader relations of difference is highlighted as important, yet the claim lacks social content (reflecting the abstract character of *Uneven Development*, which Smith (2011) readily admits to in a retrospective piece). A gap thus exists between the socially textured and concrete approaches to class offered in the earlier debates on agrarian capitalism and Smith's conceptualization of labor within the production of nature.

A number of contributors have subsequently provided a more concrete account of class and labor and, tellingly, they engage with debates on agrarian capitalism. Don Mitchell (1996; 2003) builds on the work of Smith and works against the reification of landscapes as natural, stressing that agricultural landscapes are the product of living labor. He adds that 'labor in the abstract never made anything. Labor power is embodied. And in California those bodies have been marked by ideologies of race and gender, or more generally inferiority' (Mitchell, 1996: 10). Mitchell's (1996; 2003; 2012) detailed historical studies of agrarian labor in California focusing on the relationship of class, gender and race are an important complement to Smith's contributions. Similarly Heynen et al. (2007) detail how gendered and racialized divisions of labor in Milwaukee's Bureau of Forestry and the broader labor market lead to the uneven production of the city's forests in ways that advantage 'white' middle-class neighborhoods at the expense of alternative urban forests promoted by marginalized communities and workers.

In my own research, including an ethnography of reforestation work in British Columbia, Canada, I have drawn on Stuart Hall's (1996) work to suggest that we need an account of 'the production nature without guarantees': this requires letting go of the theoretical and empirical security that comes from having an unspecified 'quintessential worker' (Rose, 1997) doing the work of producing nature (Ekers, 2010; 2013; 2014). And although I have tried to develop such an approach through investigating how reforestation workers 'experience' the making of nature through the modalities of race, gender, sexuality and class, some of the theoretical and methodological clues as to how to do so may lie in older debates on agrarian production rather than the literature on the production of nature. Gidwani (2008: 180), discussed earlier, is a key ally in such a project arguing 'for a cultural political economy of workplace dynamics: the argument that abstract "labor" incarnates itself concretely as "work" within a lived cultural field that is regulative of individual conduct and rationality'.

Conclusion: the concrete as the possible

In the expansion of political ecology studies from the Global South to spaces in North America and Europe (see McCarthy, 2002) the ontology of labor has become secured as wage-labor and less emphasis has been dedicated to the relationship between commodified and non-commodified productive activity. In part, this may follow from a perception that waged-work is more generalized in settings such as North America and Europe compared with the Global South where 'wageless life' (Denning, 2010) is more normalized. However, un-waged work has always been important within full-fledged capitalist economies, as feminist scholars have continued to insist (Collins and Gimenez, 1990; Friedmann, 1990; McClintock, 1995; McDowell, 1999). Additionally, many indigenous communities labor (broadly understood as productive activity) outside of wage relations (Nadasdy, 2007) and forced labor in agricultural sectors is on the rise (Anderson and Rogaly, 2005; Strauss, 2013). More work is needed is to examine the range of class identities connected to the shifting relationship between waged- and non-waged work through which people come to labor in environmental landscapes. Attention to social differences beyond 'class' is key here, as what we could describe as non-waged environmental work has often been conducted by women on farms and racialized migrants and indigenous people living in colonial and post-colonial spaces. If labor is indeed how nature is produced, more work must examine the various ways in which labor is socially organized beyond the wage. Here a return to debates focused on the open ontology of labor, classes and peasantries in the 1970s and 1980s may provide important intellectual resources for such a project.

I end with one final point regarding the concreteness of labor. Diana Elson (1979) concludes her treatise on the value-theory of labor with a comment on the political possibilities inherent to Marx's analysis of value. She writes:

> Marx's analysis also recognizes the *limits* to the tendency to reduce individuals to bearers of value-forms. It does this by incorporating into the analysis the subjective, conscious, particular aspects of labor in the concepts of private and concrete labor; and the collective aspect of labor in the concept of social labor... Subjective, conscious and collective aspects of human activity are accorded recognition. The political problem is to bring together these private, concrete and social aspects of labor without the mediation of the value forms, so as to create particular, conscious collective activity directed against exploitation. Marx's theory of value has, built into it, this possibility.
>
> *(Elson, 1979: 174)*

If the concrete represents the unity of the diverse, part of that diversity includes immanent political possibilities. As a number of people have pointed out, Marx (1977) provides an analysis of the dual-nature of labor, which highlights living labor as both the source of appropriated surplus value and as creative activity that sustains life through the production of use values (Gidwani and Chari, 2004; Holloway 2010). More work needs to focus on the political possibilities within shifting class identities and forms of environmental labor, and doing so might help us realize Smith's provocative and unrealized call for the 're-enchantment of nature'. Smith (1998: 280) suggests that a re-enchanting of nature cannot be left to the right, the liberals, nor Disney, and writes that 'the raw materials for a revolutionary re-enchantment of nature are simultaneously scarce, yet all around'. Thinking through the concreteness of labor as Elson suggests might provide one means for prospecting the raw materials of a radical relationship to nature.

References

Akram-Lodhi, A. and Kay, C. (2010a) 'Surveying the agrarian question (part 1): Unearthing foundations, exploring diversity', *The Journal of Peasant Studies*, vol. 37, no. 1, pp. 177–202.

Akram-Lodhi, A. and Kay, C. (2010b) 'Surveying the agrarian question (part 2): Current debates and beyond', *The Journal of Peasant Studies*, vol. 37, no. 2, pp. 255–284.

Alavi, H. and Shanin, T. (1988) 'Introduction to the English edition: Peasantry and capitalism', *The Agrarian Question*, London and Winchester, MA: Zwan Publications, pp. xi–xxxix.

Althusser, L. and Balibar, E. (1970) *Reading Capital*, London: Nlb.

Anderson, B. and Rogaly, B. (2005) *Forced Labour and Migration to the UK*, London: COMPAS in collaboration with the Trades Union Congress.

Anderson, K. (2001) 'The nature of race', in Castree, N. and Braun, B. (eds.) *Social Nature: Theory, Practice, and Politics*, Oxford: Blackwell, pp. 64–83.

Banaji, J. (1976) 'Chayanov, Kautsky, Lenin: Considerations towards a synthesis', *Economic and Political Weekly*, vol. 11, no. 40, pp. 1594–1607.

Banaji, J. (1977) 'Modes of production in a materialist conception of history', *Capital & Class*, vol. 1, no. 3, pp. 1–44.

Bannerji, H. (1995) *Thinking Through: Essays on Feminism, Marxism and Anti-racism*, Toronto: Women's Press.

Bernstein, H. (1977) 'Notes on capital and peasantry', *Review of African Political Economy*, vol. 4, no. 10, pp. 60–73.

Bernstein, H. (1979) 'African peasantries: A theoretical framework', *The Journal of Peasant Studies*, vol. 6, no. 4, pp. 421–443.

Bernstein, H. (1996) 'Agrarian questions then and now', *The Journal of Peasant Studies*, vol. 24, no. 1–2, pp. 22–59.

Bernstein, H. (2009) 'V.I. Lenin and A.V. Chayanov: looking back, looking forward', *The Journal of Peasant Studies*, vol. 36, no. 1, pp. 55–81.

Bernstein, H. (2010) *Class Dynamics of Agrarian Change*, Halifax: Fernwood Publishing.

Braun, B. (2006) 'Towards a new Earth and a new humanity: Nature; ontology, politics', in Castree, N. and Gregory, D. (eds.) *David Harvey: A Critical Reader*, Oxford: Blackwell Publishing, pp. 191–222.

Burawoy, M. (1985) *The Politics of Production: Factory Regimes under Capitalism and Socialism*, London: Verso.

Buttigieg, J. (1990) 'Gramsci's method', *Boundary 2*, vol. 17, no. 2, pp. 60–81.

Carney, J. and Watts, M. (1990) 'Manufacturing dissent: Work, gender and the politics of meaning in a peasant society', *Africa*, vol. 60, no. 02, pp. 207–241.

Castree, N. (1995) 'The nature of produced nature: Materiality and knowledge construction in Marxism', *Antipode*, vol. 27, no. 1, pp. 12–48.

Castree, N. (2000) 'Marxism and the production of nature', *Capital & Class*, vol. 72, pp. 5–36.

Castree, N. (2002) 'False antithesis? Marxism, nature and actor networks', *Antipode*, vol. 34, no. 1, pp. 111–146.

Castree, N. and Braun, B. (1998) 'The construction of nature and the nature of construction: analytical and political tools for building survivable futures', in Braun, B. and Castree, N. (eds.) *Remaking Reality: Nature and the Millennium*, London: Routledge, pp. 3–42.

Chayanov, A.V. (1986[1925]) *A.V. Chayanov on the Theory of Peasant Economy*, Madison: The University of Wisconsin Press.

Collins, J.L. and Gimenez, M. (1990) *Work Without Wages: Comparative Studies of Domestic Labor and Self-Employment*, Albany: State University of New York Press.

Cook, S. and Binford, L. (1986) 'Petty commodity production, capital accumulation, and peasant differentiation: Lenin vs. Chayanov in rural Mexico', *Review of Radical Political Economics*, vol. 18, no. 4, pp. 1–31.

Deere, C. (1987) *The Peasantry in Political Economy: Trends of the 1980s*, International Area Studies Program, Amherst: University of Massachusetts.

Denning, M. (2010) 'Wageless life', *New Left Review*, vol. 66, Nov-Dec, pp. 79–97.

Eaton, E. (2011) 'On the farm and in the field: The production of nature meets the agrarian question', *New Political Economy*, vol. 16, no. 2, pp. 247–251.

Ekers, M. (2010) *Working the Landscape: Cultures of Labour in the British Columbia Tree Planting Sector*, University of Oxford.

Ekers, M. (2013) '"Pounding dirt all day": Labor, sexuality and gender in the British Columbia reforestation sector', *Gender, Place & Culture*, vol. 20, no. 7, pp. 876–895.

Ekers, M. (2014) 'Labouring against the grain of progress: Women's reforestation work in British Columbia, 1960–1975', *Journal of Rural Studies*, vol. 34, April, pp. 345–355.

Ekers, M. and Loftus, A. (2013) 'Revitalizing the production of nature thesis: A Gramscian turn?' *Progress in Human Geography*, vol. 37, no. 2, pp. 234–252.

Elson, D. (ed.) (1979) 'The value theory of labour', in *Value: The Representation of Labour in Capitalism*, London: CSC Books, pp. 115–180.

Friedmann, H. (1980) 'Household production and the national economy: Concepts for the analysis of agrarian formations', *The Journal of Peasant Studies*, vol. 7, no. 2, pp. 158–184.

Friedmann, H. (1990) 'A paradoxical relationship between unwaged and waged labor', in Collins, J.L. and Gimenez, M. (eds.) *Work Without Wages: Comparative Studies of Domestic Labor and Self-Employment*, New York: Suny Press, pp. 193–214.

Gidwani, V.K. (2008) *Capital Interrupted: Agrarian Development and the Politics of Work in India*, Minneapolis: University of Minnesota Press.

Gidwani, V.K. and Chari, S. (2004) 'Geographies of work', *Environment and Planning A*, vol. 22, no. 4, pp. 475–484.

Goodman, M. and Redclift, M. (1981) *From Peasant to Proletarian: Capitalist Development and Agrarian Transitions*, Oxford: Basil Blackwell.

Gramsci, A. (1971) *Selections from the Prison Notebooks of Antonio Gramsci*, London: Lawrence and Wishart.

Gururani, S. (2002) 'Forests of pleasure and pain: Gendered practices of livelihood in the forests of Kumaon Himalayas, India', *Gender, Place and Culture*, vol. 9, no. 3, pp. 229–243.

Hall, S. (1980) 'Race articulation and societies structured in dominance', in UNESCO (ed.) *Sociological Theories: Race and Colonialism*, Paris: UNESCO, pp. 305–345.

Hall, S. (1996) 'The problem with ideology: Marxism without guarantees', in Morley, D. (ed.) *Stuart Hall: Critical Dialogues in Cultural Studies*, London: Routledge, pp. 25–46.

Haraway, D. (1991) *Simians, Cyborgs, and Women: The Reinvention of Nature*, New York: Routledge.

Harrison, M. (1977) 'The peasant mode of production in the work of A. V. Chayanov', *The Journal of Peasant Studies*, vol. 4, no. 4, pp. 323–336.

Henderson, G.L. (1999) *California & the Fictions of Capital*, New York: Oxford University Press.

Heynen, N., Perkins, H. and Roy, P. (2007) 'Failing to grow 'their' own justice? The co-production of racial/gendered labor and Milwaukee's urban forest', *Urban Geography*, vol. 28, no. 8, pp. 732–754.

Holloway, J. (2010) *Crack Capitalism*, London: Pluto.

Kautsky, K. (1988[1899]) *The Agrarian Question*, London and Winchester, MA: Zwan Publications.

Lenin, V.I. (1964[1899]) *The Development of Capitalism in Russia*, 2nd rev. edition, Moscow: Progress Publishers.

Loftus, A. (2012) *Everyday Environmentalism*, Minneapolis: University of Minnesota Press.

Mann, G. (2007) *Our Daily Bread: Wages, Workers, and the Political Economy of the American West*, Chapel Hill: The University of North Carolina Press.

Mann, S.A. and Dickinson, J.M. (1978) 'Obstacles to the development of a capitalist agriculture', *The Journal of Peasant Studies*, vol. 5, no. 4, pp. 466–481.

Marx, K. (1973[1939]) *Grundrisse: Foundations of the Critique of Political Economy (Rough Draft)*, Harmondsworth, England; Baltimore, Maryland: Penguin Books.

Marx, K. (1977[1867]) *Capital volume 1*, New York: Vintage Press.

McCarthy, J. (2002) 'First world political ecology: Lessons from the wise use movement', *Environment and Planning A*, vol. 34, no. 7, pp. 1281–1302.

McClintock, A. (1995) *Imperial Leather: Race, Gender and Sexuality in the Colonial Contest*, New York; London: Routledge.

McDowell, L. (1999) *Gender, Identity & Place: Understanding Feminist Geographies*, Cambridge: Polity Press.

McDowell, L. (2006) 'Reconfigurations of gender and class relations: Class differences, class condescension and the changing place of class relations', *Antipode*, vol. 38, no. 4, pp. 825–850.

Mitchell, D. (1996) *The Lie of the Land: Migrant workers and the California landscape*, Minneapolis: University of Minnesota Press.

Mitchell, D. (2003) 'Dead labour and the political economy of landscape – California living, California dying', in Anderson, K. (ed.) *Handbook of Cultural Geography*, London: Sage Publications, pp. 233–248.

Mitchell, D. (2012) *They Saved the Crops: Labor, Landscape, and the Struggle Over Industrial Farming in Bracero-era California*, Athens: University of Georgia Press.

Mortimer-Sandilands, C. and Erickson, B. (eds.) (2010) 'Introduction: A genealogy of queer ecologies', in *Queer Ecologies: Sex, Nature, Politics, Desire*, Bloomington: Indiana University Press, pp. 1–39.

Nadasdy, P. (2007) 'The gift in the animal: The ontology of hunting and human-animal sociality', *American Ethnologist*, vol. 34, no. 1, pp. 25–43.

Nightingale, A. (2006) 'The nature of gender: Work, gender, and environment', *Environment and Planning D: Society and Space*, vol. 24, no. 2, pp. 165–185.

Patnaik, U. (1979) 'Neo-populism and Marxism: The Chayanovian view of the agrarian question and its fundamental fallacy', *The Journal of Peasant Studies*, vol. 6, no. 4, pp. 375–420.

Prudham, S. (2005) *Knock on Wood: Nature as Commodity in Douglas-fir Country*, New York: Routledge.

Prudham, S. and Heynen, N. (2011) 'Introduction: Uneven development 25 years on: Space, nature and the geographies of capitalism', *New Political Economy*, vol. 16, no. 2, pp. 223–232.

Resnick, S. and Wolff, R.D. (1981) 'Classes in Marxian theory', *Review of Radical Political Economics*, vol. 13, no. 4, pp. 1–18.

Rose, S. (1997) 'Class formation and the quintessential worker', in Hall, J. (ed.) *Reworking Class*, Ithaca: Cornell University Press, pp. 133–166.

Scott, J.W. (1988) *Gender and the Politics of History*, New York: Columbia University Press.

Shanin, T. (1973) 'The nature and logic of the peasant economy 1: A generalisation 1', *The Journal of Peasant Studies*, vol. 1, no. 1, pp. 63–80.

Shanin, T. (1986) 'Chayanov's message: Illuminations, miscomceptions, and contemporary "development theory"', in *A.V. Chayanov on the Theory of Peasant Economy*, Madison: The University of Wisconsin Press, pp. 1–24.

Shanin, T. (2009) 'Chayanov's treble death and tenuous resurrection: An essay about understanding, about roots of plausibility and about rural Russia', *The Journal of Peasant Studies*, vol. 36, no. 1, pp. 83–101.

Smith, G.A. (1979) 'Socio-economic differentiation and relations of production among rural-based petty producers in central Peru, 1880 to 1970', *The Journal of Peasant Studies*, vol. 6, no. 3, pp. 286–310.

Smith, G. (1985) 'Reflections on the social relations of simple commodity production', *The Journal of Peasant Studies*, vol. 13, no. 1, pp. 99–108.

Smith, N. (1998) 'Nature at the millennium: Production and reenchantment', in Braun, B. (ed.) *Remaking Reality*, New York and London: Routledge, pp. 271–285.

Smith, N. (2008[1984]) *Uneven Development: Nature, Capital, and the Production of Space*, 3rd edition, Athens: University of Georgia Press.

Smith, N. (2011) 'Uneven development redux', *New Political Economy*, vol. 16, no. 2, pp. 261–265.

Strauss, K. (2013) 'Unfree again: Social reproduction, flexible labour markets and the resurgence of gang labour in the UK', *Antipode*, vol. 45, no. 1, pp. 180–197.

Swyngedouw, E. (1999) 'Modernity and hibridity: Nature, regeneracionismo, and the production of the Spanish waterscape, 1890–1930', *Annals of the Association of American Geographers*, vol. 89, no. 3, pp. 443–465.

Swyngedouw, E. (2006) 'Circulation and metabolisms: (Hybrid) natures and (cyborg) cities', *Science as Culture*, vol. 15, no. 2, pp. 105–121.

Thomas, P.D. (2009) *The Gramscian Moment: Philosophy, Hegemony and Marxism*, Boston: Brill, Leiden.

Thompson, E.P. (1964) *The Making of the English Working Class*, 1st vintage edition, New York: Vintage Books.

Thompson, E.P. (1978) *The Poverty of Theory & Other Essays*, London: Merlin Press.

Thorner, D. (1965) 'A post-Marxian theory of peasant economy', *Economic and Political Weekly*, vol. XVII, no. 5–6–7, pp. 227–236.
Thorner, D. (1986) 'Chayanov's concept of the peasant economy', *A.V. Chayanov on the Theory of Peasant Economy*, Madison: The University of Wisconsin Press, pp. xi–xxiii.
van der Ploeg, J. (2013) *Peasants and the Art of Farming: A Chayanovian Manifesto*, Winnipeg: Fernwood Publishing.
Watts, M. (1983) *Silent Violence: Food, Famine, and Peasantry in Northern Nigeria*, Berkeley: The University of California Press.
Watts, M. (2000) 'Political ecology', in Sheppard, E. and Barnes, T. (eds.) *A Companion to Economic Geography*, Oxford: Blackwell, pp. 257–274.
Williams, R. (1973) *The Country and the City*, New York: Oxford University Press.
Williams, R. (1980) *Problems in Materialism and Culture: Selected Essays*, London: NLB.
Wolpe, H. (1975) 'The theory of internal colonialism: The South African case', in Oxhaal, I., Barnett, T. and Booth, D. (eds.) *Beyond the Sociology of Development*, Routledge & Kegan Paul, pp. 229–252.
Wolpe, H. (ed.) (1980) 'Introduction', in *The Articulation of Modes of Production: Essays from Economy and Society*, London, Boston and Henley: Routledge & Kegan Paul, pp. 1–43.
Zalik, A. (2011) 'Protest as violence in oilfields: The contested representation of profiteering in two extractive sites', in Feldman, S., Geisler, C. and Menon, G. (eds.) *Accumulating Insecurity*, Athens: University of Georgia Press, pp. 261–284.

43

NATURE, DIFFERENCE, AND THE BODY

Julie Guthman and Becky Mansfield

Introduction

Recent years have seen a raft of concern and even panic about changes in bodily morphology and functionality, such as an "obesity epidemic," an upsurge in autism, the emergence of previously undiagnosed ills such as gluten intolerance, or a gradual lowering of IQ. Many of these illnesses and conditions of concern share three remarkable commonalities. First, they are generally treated as biomedical or psychosocial problems, even though there is evidence for each of environmental etiologies, particularly related to environmental toxins. Second, these environmental etiologies often stem from sources that have a political economy. Not only are toxins, for example, under-regulated in the service of neoliberal capitalism; exposures are often differentiated along class, race, gender, and citizenship lines. Third, they are contested. They are contested not only because of competing claims about their etiologies, but also because not all would agree they are even problems. Instead, they may simply reflect human difference. Together, these three commonalities suggest the possibility of environmentally created human differences that are mediated by political economic and other power relations yet, perhaps, are subject to misplaced panic.

Currently, however, such conditions mainly are treated as what Paul Robbins (2004) has called "apolitical ecologies," referring to explanations of environmental degradation and/or resource depletion that that do not account for social power in either *producing* environmental changes or *defining* them as problems. Nor have they received much treatment within the field of political ecology. This is because as a field political ecology has for the most part retained a focus on landscapes, rather than bodies: landscapes of agriculture, natural resource industries (timber, mining, fisheries), conservation and recreation, and, more recently, waste. It has thus focused more on production than consumption (DuPuis, 1998), more on non-human ecology than human biology, and thus more on traditional environmental movements than environmental health movements.

To the extent that political ecology has taken up questions of health and illness, it is largely through examining the ecologies that produce disease and not the ecologies of disease per se. Political ecology has seemingly stopped at the presumed boundaries of the body. But what these cases also suggest is that bodies and environments are not all that separable. In cases of environmentally induced disease, for example, it is often the case that molecules from outside the body are coming into the body to re-work the body's form and function. That means that bodily ecologies cannot be treated as separate from environmental flows, a point that has been made by critics and scholars of biomedical paradigms (Alaimo, 2010; Braun, 2007; Braun, 2008; Landecker, 2011; Martin, 1998). In addition, political ecology accounts of health and disease have largely followed a declensionist narrative, assuming that changes are inherently bad (as much earlier work on the political ecology of environmental degradation once did). While this is understandable given the diseases on which political ecology has focused, the framework has given us little to think with for changes that might be adaptive, even healthful – or just different.

This chapter reflects on how political ecology and cognate fields have treated human health and illness, to suggest how political ecologists might better address environmentally induced human difference. Doing so necessarily entails engagement with human biology, an aspect that is surprisingly under-represented in work on the political ecologies of health and environmental justice. We thus concur with recent calls for "critical political ecology" (Forsyth, 2003) and "critical physical geography" (Lave et al., 2014) that emphasize the need to "combine critical attention to relations of social power [and] biophysical science or technology" (Lave et al., 2014: 2–3). To develop a critical political ecology of the body, we find much can be drawn from the emerging field of environmental epigenetics, a new paradigm in biology that is radically more open-ended and less reductionist than genomics and more ecological than traditional biomedicine. However, despite this open-endedness, it is being interpreted for quite normative ends. We therefore close by arguing that while epigenetics will be enormously useful for developing a political ecology of the body, we simultaneously need a critical political ecology of epigenetic knowledge.

On the political ecology of the body

Only a smattering of work explicitly references a "political ecology of the body," and much of it does so primarily programmatically. DuPuis (1998), for example, first used the term to suggest that commodity chain analysis (and political ecology) must take more seriously what happens at the "other end" of the chain, referring to the sphere of consumption. Sharp (2002), in an introduction to a special issue in *Medical Anthropology*, waved to a political ecology of the body in reference of the emergence of commodities that are either derived from human bodies or are designed to improve or alter their ailing states. For Sharp, political ecology is present in the analogy to processes of extraction, with language such as putting bodies "under siege" and making them "worthy of protection," presumably as humans do with non-human nature (p. 372). Lafferty's (2012) call for a political ecology of health and bodies turns on the insufficiency of biomedical explanation and the need to historicize and politicize the production of health disparities.

Hayes-Conroy and Hayes-Conroy (2013) have laid out the most comprehensive framework for a political ecology of the body. For them a political ecology of the body necessarily draws upon work on the political ecologies of health (discussed below), examining the structural forces that make or preclude health and the discursive practices by which health and disease come to be known and understood. But, they argue, a political ecology of the body must also engage with the material forces at work. For them, materiality refers to feeling and affect – or what they refer to as visceral knowledge – and relational ontologies and non-representational theory are their favored approaches to conceptualize matter. As they put it, "what makes a healthy body?" is also a question that turns on a person's "articulated bodily capacity to feel a certain level of comfort, excitement, affection, pride, and so on" (86).

We concur with many of these arguments, but wish to push them further. In the name of a critical political ecology as outlined above, we argue that engagements with the materiality of the body must go beyond the phenomenological to incorporate also the biological and biochemical. Contending with environmentally produced bodily difference in particular requires attending to changes both *within* the body itself as well as to the ecologies of bodies as they tangibly *interact with* environments (Guthman, 2005; Guthman, 2008). Nonrepresentational theory, including attention to affect, alone is inadequate to this task; many of these biological processes are not experienced directly by the individual and so must necessarily be represented in biological science – with utmost cognizance that translations of biological science are inherently socially mediated and, hence, representational, as Emily Martin (1991) has shown. We further argue that a political ecology of disparity requires querying how particular bodily differences become consequential, to whom, and with what effects. When we fail to attend to the biological dimensions of bodily materiality, we effectively leave it to biomedical experts (and their apolitical explanations) to determine what constitutes a problem.

Political ecologies of health and environmental justice

Within the broad tradition of political ecology and closely related cognate fields, such as environmental justice, a number of established approaches already undertake what might be considered a political ecology of the body. Work on the political ecology of health and disease, for example, has tended to emphasize the political economic contexts in which illnesses and diseases are made more virulent; while research on environmental justice has tended to assess the disproportional impacts of toxic environments on certain population groups. Here we review each in turn.

Scholarship on the political ecology of health and disease seeks to understand how socionatural ecologies and access to resources affect incidence, prevalence, and distribution of human health and disease. As with political ecology writ large, much of the work that comes under political ecology of health has focused on developing world environments and/or contagious or pathogenic disease (Collins, 2001; Craddock, 2000b; Hughes and Hunter, 1970; Kalipeni and Oppong, 1998; King, 2010; King and Crews, 2012; Mayer, 1996; Mkandawire et al., 2013; Scott et al., 2012; Singer and De Castro, 2001; Wagner et al., 2008) (cf. Craddock, 2000a) (see also, Chapter 26, this volume). That is in part because much of this scholarship has its roots in disease ecology, which treats people and disease as part of complex ecosystems, and treats diseases as contagious ones. As noted by Mayer (1996), disease ecology is an older field that was written without the benefit of the frameworks that political ecology inherited from cultural ecology, such as the attention to power and chains of explanation. In keeping with earlier political ecology, the focus tended to be tropical environments and diseases, and the narrative was characteristically one of disruption. More recently, political ecology of health has become more wide-ranging in its ambit, particularly taking up questions of the social etiologies of non-infectious illness – and even indoor environments (Day Biehler and Simon, 2011; Hanchette, 2008; Harper, 2004; Richmond et al., 2005; Sultana, 2012); there are also new calls to pay attention to the production of health knowledges and various non-human actors in and beyond the body (Jackson and Neely, 2014). Still, the basic framework is one in which human actions, and especially larger-scale political economic processes, change ecological processes in ways that create new health problems.

Much of this work is richly empirical, and it has certainly expanded understandings of health and disease as more than biomedical. However, in coming to terms with environmentally induced human difference it falls short in several ways. For one, the actual ecologies of concern are external to the body. Nature is addressed primarily at the landscape or, in the case of indoor environments, building scale, and there is little attention to the nature of the human body, which remains largely black-boxed (Guthman, 2012). Moreover, the ill health of the body is usually measured and assessed through traditional biomedical models – which, crucially, tend to use normative framings. Contending with difference requires understanding of the impact of ecological change on actual functionality – and thus awareness of how typical measures of this functionality (e.g., BMI, IQ, blood glucose levels) embed particular assumptions about health and well-being (Armstrong, 1996; Guthman, 2013; Mansfield, 2012c). In addition, the typical narrative follows in the footsteps of classic political ecology in its declensionist tale of human meddling in the natural environment causing poor health effects, and in that way implicitly assumes a perfect state away from which societies have veered. In political ecology writ large, such declensionist narratives have given way to new renderings in which human-changed nature is not necessarily damaged nature (Fairhead and Leach, 1995). Perhaps, then, a similar lens must be cast on bodily change. This is certainly the lesson of emerging scholarship on parasites and bacteria; even as exposure to these biological agents is shaped by political processes and can reflect socioeconomic marginalization, new research suggests that these exposures can *contribute to* health and well-being rather than simply reflect a form of neglect and damage (Clough, 2011; Hanski et al., 2012; Lozupone et al., 2012). This sort of work is generating altogether new conceptions of disease and health that also happen to reject clear boundaries between humans and their biotic environments.

Turning to environmental justice scholarship, it pays close attention to how illnesses affect certain groups disproportionately, especially as related to the spatiality of exposure; the field as we describe it here also tends to have a North American focus and the environments in question are largely urban and/or industrial (e.g., Bowen et al., 1995; Morello-Frosch et al., 2001; Pastor et al., 2002; Pulido, 2000; Reed and George, 2011). In its attention to the relationship between structural inequality and health outcomes, it shares some of the concerns of political ecology of health and disease – yet it also differs in important ways, even beyond geographical emphasis. Unlike political ecology, where the focus is on contagion, the focus in environmental justice is on exposure to environmental contaminants. Accordingly, the diseases in question are not communicable ones but rather chronic debilitating ones from asthma to breast cancer to various syndromes (Brown, 2007). As such, they tend to be what Brown calls "contested diseases": conditions and diseases that are either not recognized as illnesses or are medicalized in such a way that lifestyle factors (i.e., individual behavior) rather than environmental factors are regarded as the cause (think gluten intolerance or multiple chemical sensitivity). They are also contested because environmental causation of illness is harder to prove with epidemiological method, which Brown largely attributes to the weakness of apparatuses that monitor health status and investigate the environmental factors implicated in disease. In addition, environmental justice scholarship tends to employ different methodological approaches, understandably so. Where political ecology of disease has tended to use multi-sited ethnography to build chains of explanation, much environmental justice work has drawn on spatial epidemiology to draw connections between places and ill health. Indeed, the uncertainties associated with disease have become so important to environmental justice counter claims that people's epidemiology, such as the use of bucket brigades and drift catchers, has become a key activist strategy and area of research (Corburn, 2005; Harrison, 2011b; Sze, 2007).

This explicitly political work has been crucial for expanding our understanding of illness and health disparities. Particularly important for a critical political ecology of the body are the ways environmental justice attends to questions of space, inequality, exposure, and contested, chronic illness. Yet it, too, falls short for coming to terms with environmentally induced human difference. First, only sometimes has this sort of research been supplemented with deeper historical accounts of how some places have come to be sites of environmental injustices, as argued for by Pellow (2000), and exemplified in works like Sze's (2007) *Noxious New York*. Much environmental justice scholarship "has yet to become historical, conceptualizing the deeper socio-environmental histories that produce inequality in resource access and wealth in the construction of racial difference" (Lafferty, 2012). Second, like work in political ecology, there is very little attention to the ecologies of bodies as they contend with disease. Environmental justice scholarship rarely addresses the biochemical materiality of the body or the chemicals. To be clear, it is not that this work is disembodied: recent scholarship draws upon people's self-reports on the experience of being polluted or poisoned (Harrison, 2011a). This more phenomenological approach, however, still does not explicitly engage the biophysical processes of that poisoning or pollution, nor, for that matter, the mechanisms that give rise to the affective, embodied responses. Even the environment is treated rather statically, as a place/site where exposures occur. We get little sense of the dynamism with which the molecules of various chemicals, pollutants and such are interacting with bodily molecules to change bodily states – and therefore to produce difference, as suggested by Lafferty above.

Finally, even though this work is attentive to the politics of knowledge, given that these are often contested diseases, it does not always pays attention to "the ways in which biological disease processes associated with environmental change intersect with culturally mediated interpretations of health and disease" (Harper, 2004: 298). Environmental justice scholarship, along with the political ecology of health and disease, tends to accept biomedical constructions of disease. This is not to negate the severity of diseases such as HIV/AIDS, cholera, typhus, or even asthma, although, as Craddock (2000b) has argued, even being marked with such diseases can lead to denial of resources and other social goods. Our point is that many so-called conditions and diseases of today, from obesity to ADHD, are contested not due to a lack of support that they are real problems (cf. Brown, 2007). To the contrary, there are questions about whether they should be treated as diseases at all, much less subjects of tremendous moral panic. Drawing especially on disability scholarship (Crooks and Chouinard, 2006; Moss and Dyck, 1996), some scholars – and activists – of conditions like obesity, autism and so forth question the medicalization of difference as disease (Armstrong, 2010; Evans, 2006).

Environmental history and social studies of medicine

In light of these concerns, we argue that a critical political ecology of the body must attend to the politics of disease construction and interpretation, especially as it relates to difference. And to do that requires attentiveness to the ecology of health and disease itself. We thus turn to fields outside of political ecology, namely work in environmental history, including histories of occupational illness, and social studies of medicine, to add to this conversation.

Much of the work in environmental history as it relates to health and disease is virtually indistinguishable from political ecology of disease frameworks, with emphasis on how human-generated changes to the environment have generated illness, and, for that matter, how those illnesses have been viewed (e.g., Casper, 2003; Mitman, 2007; Mitman et al., 2004). A growing body of work on occupational health and disease is particularly nuanced in linking histories of industrial growth with particular occupational hazards, while giving voice to the embodied experience of those exposed (Murphy, 2006; Santiago, 2009; Sellers, 1999; Sellers and Melling, 2011). Two works, however, stand out for pushing the conversation forward in ways we find fruitful. One is Nash's *Inescapable ecologies* (2006). Specifically focusing on farmworker illness related to pesticide exposure, Nash writes about the difficulty of proof when farmworker bodies have already been made more vulnerable to illness owing to poor access to medical care, long term and cumulative exposures, and presence in certain environments. As she asks, how do you then statistically control for their "race" since these histories have helped produce race as a meaningful epidemiological category to begin with (p. 198)? The other is Langston's (2010) work on the legacy of DES, a drug that was given to women in the 1950s and 1960s to prevent miscarriages and promote lactation, and which ended up causing widespread infertility and reproductive cancers of their progeny. Along with laying out the social and political context in which DES was made available, Langston provides a compelling account of the biological pathways by which DES had these effects. At the same time, Langston is very careful not to pathologize change and difference per se, no easy feat when writing about the emblematic endocrine disrupting chemical! Rather than vilifying change itself, she notes that endocrine disrupters are harmful not because they make people different – human variation, she notes, is the norm – but because they transform the body's repair systems, making them less resilient.

Social studies of biomedicine provide additional tools for grappling with questions of difference. For example, critical medical anthropology, while "lack[ing] theoretical focus on nature-society relationships," "does provide a framework that highlights the role of socio-cultural histories in shaping wellness, and articulates how political economy patterns health outcomes" (Lafferty, 2012). Especially important for our purposes is current anthropological research on race and health. Resonating with the work of Nash, discussed above, this scholarship treats race and racialized difference in health as an effect (rather than a variable) of social and historical processes – including environmental pathways (Gravlee, 2009; Kuzawa and Sweet, 2009; Lock and Nguyen, 2010; Shim, 2005).

Also germane is related literature on "biosocieties." This scholarship has focused on new biomedical and broader life science understandings of life, health, and bodies. Attention has centered on advances at the frontiers of biomedicine, such as genomics, individualized medicine, and other biotechnologies – that is, the widely noted "molecularization" of life (Beck and Niewohner, 2006; Rabinow and Rose, 2006; Rose, 2007). This work makes many parallel arguments to political ecology, but applies them to bodies. First, scholarship of biosocieties draws from new knowledge in the life sciences to argue that nature – biological bodies – is not given, but is quite mutable and dynamic. Second, it understands that life forms are socionatures in which the line between the biological and social is erased. Third, it gives prominent attention to political economy (biocapital), albeit not necessarily as a causal force in the making of socionatures – but rather in the selling of them, for example as biomedicalization lends itself to new drugs (Cooper, 2008; Rajan, 2006). Finally, like at least some political ecology, this scholarship sees scientific knowledge as inherently political. Unlike political ecology (and environmental justice scholarship), however, this approach largely rejects declensionist thinking, so much so that it has been criticized for focusing too much on the ability to purposefully improve upon bodies through advances at the frontiers of biomedicine (Braun, 2007; Mansfield, 2012a; Roberts, 2009). And because it is highly attentive to ways that populations are constituted, divided, and managed (biopolitics), it is attentive to difference. Moreover, increasingly this literature is engaging with environmental pathways to bodily transformation in ways that political ecology of health has not (Braun, 2007; Landecker, 2011; Niewohner, 2011; Shostak, 2003; Shostak, 2013). Yet, there is much less attention to the environment in this literature than in political ecology; when attending to wider processes, these are largely socioeconomic and biopolitical, rather than biophysical and ecological.

In short, we contend that understanding environmentally induced bodily difference requires that we better integrate the material, biological body and knowledge about it into a political ecological analytical frame. Existing approaches in political ecology and allied fields provide tools to do this, but do not come together into the necessary framework. In making this call, we thus draw on Forsyth (2003), whose critical political ecology includes consideration of biophysical factors, including those that exist *outside of individual human experience* (p. 7). Since these are not known except through scientific discourse, the point is to challenge scientific (environmental) orthodoxies by "highlight[ing] as far as possible the implicit social and political models built into statements of supposedly neutral explanation" (p. 20). Our parallel task is to challenge biomedical orthodoxies, seeking not to falsify myths but to provide alternative explanations. We have found the emerging theory of "epigenetics" to provide fruitful ways of thinking for this project. Although a scientific discourse emerging from biomedicine, epigenetics provides for political ecology a non-deterministic yet still material model of environmentally induced bodily changes. Within epigenetics lies a potentially less normative way to understand such changes. Indeed, as with other "post-genomic" research (Fujimura, 2005) it is so much more open-ended that some have suggested it instantiates the socialization of the biological (Meloni, 2013).

Insights from epigenetics

Epigenetics is a relatively new science studying the mechanisms that affect how genes are expressed (rather than affecting the genes themselves). The term was first used in the 1940s to describe the "mechanisms necessary for the unfolding of the genetic programme for development" (Holliday, 2006: 76). It was not until the 1980s that epigenetics emerged as a field with concrete mechanisms and evidence of cellular processes that activate and deactivate genes, with a focus on cell stability and stem cell differentiation (Holliday, 2006; Jablonka and Lamb, 2002). The central discovery is that genes are not expressed in isolation, but rather in the context of their environments. For instance, DNA methylation involves combinations of carbon and hydrogen atoms attaching themselves along the DNA in ways that deactivate certain genes (Francis, 2011; Jirtle and Skinner, 2007; Kuzawa and Sweet, 2009). Methylation is one of several processes in which the cellular context shapes what genes will be expressed or suppressed and how that will affect protein synthesis and, hence, phenotype as well as emotional and intellectual development (Crews and McLachlan, 2006). Yet these environmental influences go beyond the immediate cellular environments to include many external agents – nutritional, psychosocial, and toxicological, all of which can enhance, alter, or silence genes in ways that override or even nullify the genetic code (Thayer and Kuzawa, 2011). As such, some in this field suggest that the environment has such a formidable influence on genes that "genomic activity is as much effect as cause" (Francis, 2011: 159).

A crucial aspect of this science is the discovery that these epigenetic markers can be heritable; changes in the cellular environment can not only persist throughout the lifetime of the organism but sometimes be passed from parent to child, and even to the child's offspring, thus affecting the phenotypical development of future generations (Crews and McLachlan, 2006; Jablonka and Lamb, 2002). It is this quality that has led some to equate epigenetics to a neo-Lamarkism, referring to Lamarck's theory of acquired characteristics countering the neo-Darwinism that holds the gene as the code for all life (Gorelick, 2004; Jablonka and Lamb, 2002; Weaver, 2007). Another important point is that epigenetic effects are stochastic and therefore do not determine an outcome (Faulk and Dolinoy, 2011; Heijmans et al., 2009).

A growing number of researchers are now arguing that the changes in bodies these days likely have epigenetic origins, with exposures occurring during gestation. The tragedies of DES, thalidomide, and methyl mercury in Minamata have all been attributed to epigenetic changes associated with endocrine disrupting chemicals (Crews and McLachlan, 2006), as have more contested diseases and subtle changes in body functionality. For instance, there is a good deal of evidence that the increase in obesity since 1980 is due to earlier emissions of environmental toxins acting epigenetically (Grun and Blumberg, 2009; Hatch et al., 2010; Newbold et al., 2008). Likewise, epigenetic pathways are now linked to autism (Persico and Bourgeron, 2006; Schanen, 2006), which has also been associated with methyl mercury exposure (Palmer et al., 2006). Twin studies have established that many autoimmune disorders, including celiac disease, have (unidentified) environmental origins that are evident in epigenetic markers (Ballestar, 2010). Importantly, it is not only toxins that are implicated in epigenetic change. There is a good deal of research on the role of nutrition and psychosocial cues in producing epigenetic changes (Landecker, 2011; Thayer and Kuzawa, 2011).

The implications of epigenetics for a critical political ecology of the body are manifold. First, it articulates a biological basis by which bodies are changing in relation to their biophysical, psychosocial, and nutritive environments. This theory necessarily incorporates the porosity of the human body (and especially the germ cells and placenta, since gestation is a central temporality of epigenetic processes), as chemical molecules, especially, flow into the body to alter its functionality. Second, it is providing new ways to think about human difference that do not reduce to pre-given "nature" (e.g., immutable genetics) or "culture." For, in this theory the *environment actually comes into the body* and shapes how genes express, rather than an organism adapting to an external environment. This means that the environment "is an inducer as well as selector of variation" and thus has an active role in making different bodies (Jablonka and Lamb, 2002: 94). More pointedly, because epigenetic changes often stem from the social environment, they are clearly at work in what Lock and Nguyen (2010) call biosocial differentiation, the making of all sorts of bodies and abilities. It is precisely this insight that is giving way to new conceptions of – and new debates about – race and health disparities. The idea here is that differential exposures to "social race" are potentially explanatory for broad population-based patterns in various health indicators such as hypertension, diabetes II, or breast cancer that are then inherited. Racial difference is then not reduced to a genetic code but rather a result of, as Happe (2013) puts it, "how oppression gets under the skin (p. 6), or, as Gravlee (2009) puts it, "race becomes biology" (p. 51) (also Kuzawa and Sweet, 2009, on this point).

Third, epigenetics *potentially* calls into question the idea that a changed body is a diseased one. Of course since much of the research on epigenetics comes from the biomedical field, it necessarily focuses on diseases and possible cures. Nevertheless, the existence of epigenetic processes (as well as undifferentiated stem cells) suggests that beings are always in a state of becoming. What this science suggests, in other words, is that the processes involved have been going on from the beginning of biological time, and thus there is no natural, perfect state from which humans have veered. This makes it untenable to view environmentally induced changes as necessarily pathological. To the contrary, epigenetics suggests that at least some developmental plasticity is functional, adaptive, and/or protective (Kuzawa and Sweet, 2009; Szyf, 2009). That means that the subtle ways in which bodies are being transformed may be worse or pathological – or they may be adaptive, healthful, or just different.

That said, we have no illusions that epigenetic knowledge is somehow beyond power. Although the knowledge embedded in epigenetics does not determine it politics, it is already evident that the post-genomic world of which epigenetics is a part is giving way to numerous issues with which political ecologists should be concerned. Certainly much of this knowledge is being garnered in the service of accumulation, as theorists of biocapital have discussed (Cooper, 2008; Rajan, 2006). But we are equally concerned about the discursive ways in which epigenetic knowledge is being mobilized. For example, as understandings of epigenetic processes have filtered into the popular news media, the messages that are being promulgated ask us, as individuals, to take more control. Premenopausal women and girls, especially, are asked to be even more vigilant in their self-practices to avoid various toxins and receive optimal nutrition, putatively to control what may be passed on to their offspring (Mansfield, 2012a; Mansfield, 2012b). Much of the research on epigenetics is being done in the name of curing diseases and conditions that are arguably not even problems. As epigenetic processes are implicated in conditions of difference, difference itself is being pathologized – even as epigenetics simultaneously suggests that variation is itself the norm (see also Langston, 2010: 144; Lock and Nguyen, 2010: 26).

Towards a more critical political ecology of the body

Political ecology's key advantage is its attention to the always power-laden socioecological environment, in which material processes matter even as they cannot be reduced to (apolitical) scientific discourse about them: knowledge of the biophysical world is always socially mediated. Yet the existing political ecological approaches we have discussed here have been less adept at applying this to understanding and challenging apolitical accounts of bodily difference. The problem as we see it is that current approaches to the political ecology of the body either neglect to engage the materiality of the body altogether or emphasize embodied experience at the expense of biological and chemical changes within the body and as they interact with diverse environments. We agree that embodied experience is important, but think we must also engage with the biological and medical sciences in order to assess the origin and character of those changes.

In this regard, environmental epigenetics is enormously useful for developing a critical ecology of the body because it is a biological science that is revolutionizing understandings about relationships between environments, bodies, and human health. As a science of biological becoming that is simultaneously environmental and molecular, it offers a non-determinist model of human health and bodily difference, allowing for other explanations for and interpretations of disease. With its emphasis on the uncontrollability, plasticity, and stochasticity of human development, it is suggesting the need to re-cast health as a quality related to dynamism, emergence, and even biodiversity.

At the same time, epigenetics provides an object lesson in why a critical political ecology of the body must be attentive to and critical of scientific discourse. As suggested above, epigenetics not only has ties to biomedical capital, but can also lend itself to projects of normalization, treating epigenetically influenced bodies not only as different but as degraded. Recent work in political ecology, however, indicates how environmental change is not always for the worse, and cautions against normative judgments about the quality of nature based on what it might have been in the past. Further, political ecology has long shown that knowledge about nature is always a political question. This is not just about how knowledge is put to use after the fact, but about how power relations underpin the production of that knowledge in the first place. In other words, a critical political ecology of epigenetics needs to query the very foundations of the knowledge on which it is built.

For political ecology writ large, this new research is another potent reminder that the nature that political ecology attempts to interrogate cannot coherently stop at the skin. What environmental epigenetics shows, in other words, is that bodies are environments and environments are made up of bodies, always and everywhere interacting.

References

Alaimo, S. (2010) *Bodily natures: Science, environment, and the material self.* Bloomington: Indiana University Press.

Armstrong, D. (1996) The rise of surveillance medicine. *Sociology of Health and Illness* 17(3): 393–404.

Armstrong, T. (2010) *Neurodiversity: Discovering the extraordinary gifts of autism, ADHD, dyslexia, and other brain differences.* Da Capo Lifelong Books.

Ballestar, E. (2010) Epigenetics lessons from twins: Prospects for autoimmune disease. *Clinical Reviews in Allergy & Immunology* 39(1): 30–41.

Beck, S. and Niewohner, J. (2006) Somatographic investigations across levels of complexity. *BioSocieties* 1(02): 219–227.

Bowen, W.M., Salling, M.J., Haynes, K.E. and Cyran, E.J. (1995) Toward environmental justice: Spatial equity in Ohio and Cleveland. *Annals of the Association of American Geographers* 85(4): 641–663.

Braun, B. (2007) Biopolitics and the molecularization of life. *Cultural Geographies* 14(1): 6–28.

Braun, B. (2008) Thinking the city through SARS: Bodies, topologies, politics. In: Ali, S.H. and Keil, R. (eds.) *Networked disease: Emerging infections in the global city.* Oxford: Wiley-Blackwell, 250–266.

Brown, P. (2007) *Toxic exposures: Contested illnesses and the environmental health movement.* New York: Columbia University Press.

Casper, M. (2003) *Synthetic planet: Chemical politics and the hazards of modern life.* New York: Routledge.

Clough, S. (2011) Gender and the hygiene hypothesis. *Social Science & Medicine* 72(4): 486–493.

Collins, A.E. (2001) Health ecology, land degradation and development. *Land Degradation & Development* 12(3): 237–250.

Cooper, M. (2008) *Life as surplus: Biotechnology and capitalism in the neoliberal era.* Seattle: University of Washington Press.

Corburn, J. (2005) *Street science: Community knowledge and environmental health justice.* Cambridge, MA: MIT Press.

Craddock, S. (2000a) *City of plagues: Disease, poverty and deviance in San Francisco.* Minneapolis: University of Minnesota.

Craddock, S. (2000b) Disease, social identity, and risk: Rethinking the geography of AIDS. *Transactions of the Institute of British Geographers* NS 25: 153–168.

Crews, D. and McLachlan, J.A. (2006) Epigenetics, evolution, endocrine disruption, health, and disease. *Endocrinology* 147(6): s4–10.

Crooks, V.A. and Chouinard, V. (2006) An embodied geography of disablement: Chronically ill women's struggles for enabling places in spaces of health care and daily life. *Health & Place* 12(3): 345–352.

Day Biehler, D. and Simon, G.L. (2011) The great indoors: Research frontiers on indoor environments as active political-ecological spaces. *Progress in Human Geography* 35(2): 172–192.

DuPuis, M. (1998) The body and the country: A political ecology of consumption. In: Gottdiener, M. (ed.) *New forms of consumption: Consumers, culture and commodification.* Lanham, MD: Rowman & Littlefield, 131–152.

Evans, B. (2006) "Gluttony or sloth": Critical geographies of bodies and morality in (anti)obesity policy. *Area* 38(3): 259–267.

Fairhead, J. and Leach, M. (1995) False forest history, complicit social analysis: Rethinking some West African environmental narratives. *World Development* 23(6): 1023–1035.

Faulk, C. and Dolinoy, D.C. (2011) Timing is everything: The when and how of environmentally induced changes in the epigenome of animals. *Epigenetics* 6(7): 791–797.

Forsyth, T. (2003) *Critical political ecology: The politics of environmental science.* London: Routledge.

Francis, R.C. (2011) *Epigenetics: The ultimate mystery of inheritance.* New York: W.W. Norton.

Fujimura, J.H. (2005) Postgenomic futures: Translations across the machine-nature border in systems biology. *New Genetics and Society* 24(2): 195–226.

Gorelick, R. (2004) Neo-lamarckian medicine. *Medical Hypotheses* 62(2): 299–303.

Gravlee, C.C. (2009) How race becomes biology: Embodiment of social inequality. *American Journal of Physical Anthropology* 139(1): 47–57.

Grun, F. and Blumberg, B. (2009) Minireview: The case for obesogens. *Molecular Endocrinology* 23(8): 1127–1134.

Guthman, J. (2005) Book review: Political ecology: A critical introduction. *Environment and Planning A* 37(4): 757–759.

Guthman, J. (2008) Book review: Inescapable ecologies: A history of environment, disease, and knowledge. *Annals of the Association of American Geographers* 98(2): 502–504.

Guthman, J. (2012) Opening up the black box of the body in geographical obesity research: Toward a critical political ecology of fat. *Annals of the Association of American Geographers* 102(5): 951–957.
Guthman, J. (2013) Fatuous measures: The artifactual construction of the obesity epidemic. *Critical Public Health* 23(3): 263–273.
Hanchette, C.L. (2008) The political ecology of lead poisoning in eastern North Carolina. *Health & Place* 14(2): 209–216.
Hanski, I., von Hertzen, L., Fyhrquist, N., Koskinen, K., Torppa, K., Laatikainen, T., Karisola, P., Auvinen, P., Paulin, L., Mäkelä, M.J., Vartiainen, E., Kosunen, T.U., Alenius, H. and Haahtela, T. (2012) Environmental biodiversity, human microbiota, and allergy are interrelated. *Proceedings of the National Academy of Sciences* 109(21): 8334–8339.
Happe, K.E. (2013) *The material gene: Gender, race, and heredity after the human genome project*. New York: New York University Press.
Harper, J. (2004) Breathless in Houston: A political ecology of health approach to understanding environmental health concerns. *Medical Anthropology: Cross-Cultural Studies in Health and Illness* 23(4): 295–326.
Harrison, J. (2011a) *Pesticide drift and the pursuit of environmental justice*. Cambridge, MA: MIT Press.
Harrison, J.L. (2011b) Parsing "participation" in action research: Navigating the challenges of lay involvement in technically complex participatory science projects. *Society & Natural Resources* 24(7): 702–716.
Hatch, E.E., Nelson, J.W., Stahlhut, R.W. and Webster, T.F. (2010) Association of endocrine disruptors and obesity: Perspectives from epidemiological studies. *International Journal of Andrology* 33(2): 324–332.
Hayes-Conroy, J. and Hayes-Conroy, A. (2013) Veggies and visceralities: A political ecology of food and feeling. *Emotion, Space and Society* 6(0): 81–90.
Heijmans, B.T., Tobi, E.W., Lumey, L.H. and Slagboom, P.E. (2009) The epigenome: Archive of the prenatal environment. *Epigenetics* 4(8): 526–531.
Holliday, R. (2006) Epigenetics: A historical overview. *Epigenetics* 1(2): 76–80.
Hughes, C.C. and Hunter, J.M. (1970) Disease and "development" in Africa. *Social Science & Medicine* 3(4): 443–493.
Jablonka, E. and Lamb, M.J. (2002) The changing concept of epigenetics. *Annals of the New York Academy of Sciences* 981(1): 82–96.
Jackson, P. and Neely, A.H. (2014) Triangulating health: Toward a practice of a political ecology of health. *Progress in Human Geography*.
Jirtle, R.L. and Skinner, M.K. (2007) Environmental epigenomics and disease susceptibility. *Nature Reviews Genetics* 8(4): 253–262.
Kalipeni, E. and Oppong, J. (1998) The refugee crisis in Africa and implications for health and disease. *Social Science & Medicine* 46(12): 1637–1653.
King, B. (2010) Political ecologies of health. *Progress in Human Geography* 34(1): 38–55.
King, B. and Crews, K.A. (2012) *Ecologies and politics of health*. London: Routledge.
Kuzawa, C.W. and Sweet, E. (2009) Epigenetics and the embodiment of race: Developmental origins of US Racial disparities in cardiovascular health. *American Journal of Human Biology* 21(1): 2–15.
Lafferty, J.L. (2012) Re-describing diabetes: Toward a political ecology of health and bodies. *Anthropologies*.
Landecker, H. (2011) Nutrition as exposure: Nutritional epigenetics and the new metabolism. *Biosocieties* 6(2): 167–194.
Langston, N. (2010) *Toxic bodies: Hormone disrupters and the legacy of DES*. New Haven: Yale University Press.
Lave, R., Wilson, M.W., Barron, E.S., Biermann, C., Carey, M.A., Duvall, C.S., Johnson, L., Lane, K.M., McClintock, N., Munroe, D., Pain, R., Proctor, J., Rhoads, B.L., Robertson, M.M., Rossi, J., Sayre, N.F., Simon, G., Tadaki, M. and Van Dyke, C. (2014) Intervention: Critical physical geography: Intervention en géographie physique critique. *The Canadian Geographer / Le Géographe Canadien* 58(1): 1–10.
Lock, M. and Nguyen, V.-K. (2010) *An anthropology of biomedicine*. West Sussex: Wiley-Blackwell.
Lozupone, C.A., Stombaugh, J.I., Gordon, J.I., Jansson, J.K. and Knight, R. (2012) Diversity, stability and resilience of the human gut microbiota. *Nature* 489(7415): 220–230.
Mansfield, B. (2012a) Environmental health as biosecurity: "Seafood choices," risk, and the pregnant woman as threshold. *Annals of the Association of American Geographers* 102(5): 969–976.
Mansfield, B. (2012b) Gendered biopolitics of public health: Regulation and discipline in seafood consumption advisories. *Environment and Planning D: Society and Space* 30(4): 588–602.

Mansfield, B. (2012c) Race and the new epigenetic biopolitics of environmental health. *Biosocieties* 7(4): 352–372.
Martin, E. (1991) The egg and the sperm: How science constructed a romance based on stereotypical male-female roles. *Signs: Journal of Women in Culture and Society* 16: 485–501.
Martin, E. (1998) Fluid bodies, managed nature. In: Braun B and Castree N (eds) *Remaking reality: Nature at the millenium*. Oxford, UK: Routledge, 64–93.
Mayer, J.D. (1996) The political ecology of disease as one new focus for medical geography. *Progress in Human Geography* 20(4): 441–456.
Meloni, M. (2013) Biology without biologism: Social theory in a postgenomic age. *Sociology* 48(4): 731–746.
Mitman, G. (2007) *Breathing space: How allergies shape our lives and landscapes*. New Haven: Yale University Press.
Mitman, G., Murphy, M. and Sellers, C. (2004) *Landscapes of exposure: Knowledge and illness in modern environments*. Washington, DC: Osiris.
Mkandawire, P., Richmond, C., Dixon, J., Luginaah, I.N. and Tobias, J. (2013) Hepatitis B in Ghana's upper west region: A hidden epidemic in need of national policy attention. *Health & Place* 23: 89–96.
Morello-Frosch, R., Pastor, M. and Sadd, J. (2001) Environmental justice and southern California's "riskscape." *Urban Affairs Review* 36(4): 551–578.
Moss, P. and Dyck, I. (1996) Inquiry into environment and body: Women, work, and chronic illness. *Environment and Planning D: Society and Space* 14(6): 737–753.
Murphy, M. (2006) *Sick building syndrome and the problem of uncertainty: Environmental politics, technoscience, and women workers*. Durham, NC: Duke University Press.
Nash, L. (2006) *Inescapable ecologies: A history of environment, disease, and knowledge*. Berkeley: University of California Press.
Newbold, R., Padilla-Banks, E., Jefferson, W. and Heindel, J. (2008) Effects of endocrine disruptors on obesity. *International Journal of Andrology* 31: 201–208.
Niewohner, J. (2011) Epigenetics: Embedded bodies and the molecularisation of biography and milieu. *Biosocieties* 6(3): 279–298.
Palmer, R.F., Blanchard, S., Stein, Z., Mandell, D. and Miller, C. (2006) Environmental mercury release, special education rates, and autism disorder: An ecological study of Texas. *Health & Place* 12(2): 203–209.
Pastor, J.M., Sadd, J.L. and Morello-Frosch, R. (2002) Who's minding the kids? Pollution, public schools, and environmental justice in Los Angeles. *Social Science Quarterly* 83(1): 263–280.
Pellow, D.N. (2000) Environmental inequality formation. *American Behavioral Scientist* 43(4): 581–601.
Persico, A.M. and Bourgeron, T. (2006) Searching for ways out of the autism maze: Genetic, epigenetic and environmental clues. *Trends in Neurosciences* 29(7): 349–358.
Pulido, L. (2000) Rethinking environmental racism: White privilege and urban development in southern California. *Annals of the Association of American Geographers* 90(1): 12–40.
Rabinow, P. and Rose, N. (2006) Biopower today. *BioSocieties* 1(2): 195–217.
Rajan, K.S. (2006) *Biocapital: The constitution of postgenomic life*. Durham, NC: Duke University Press.
Reed, M.G. and George, C. (2011) Where in the world is environmental justice? *Progress in Human Geography* 35(6): 835–842.
Richmond, C., Elliott, S.J., Matthews, R. and Elliott, B. (2005) The political ecology of health: Perceptions of environment, economy, health and well-being among 'Namgis First Nation. *Health & Place* 11(4): 349–365.
Robbins, P. (2004) *Political ecology*. Oxford: Blackwell.
Roberts, D. (2009) Race, gender, and genetic technologies: A new reproductive dystopia. *Signs: Journal of Women in Culture and Society* 34(4): 783–804.
Rose, N. (2007) Molecular biopolitics, somatic ethics, and the spirit of biocapital. *Social Theory and Health* 5: 3–29.
Santiago, M.I. (2009) *The ecology of oil: Environment, labor, and the Mexican revolution, 1900–1938*. New York: Cambridge University Press.
Schanen, N.C. (2006) Epigenetics of autism spectrum disorders. *Human Molecular Genetics* 15(suppl 2): R138–R150.
Scott, C.A., Robbins, P.F. and Comrie, A.C. (2012) The mutual conditioning of humans and pathogens: Implications for integrative geographical scholarship. *Annals of the Association of American Geographers* 102(5): 977–985.

Sellers, C. (1999) *Hazards of the job: From industrial disease to environmental health science*. Chapel Hill: University of North Carolina Press.

Sellers, C. and Melling, J. (2011) *Dangerous trade: Histories of industrial hazard across a globalizing world*. Philadelphia: Temple University Press.

Sharp, L.A. (2002) Bodies, boundaries, and territorial disputes: Investigating the murky realm of scientific authority. *Medical Anthropology* 21(3–4): 369–379.

Shim, J.K. (2005) Constructing "race" across the science-lay divide: Racial formation in the epidemiology and experience of cardiovascular disease. *Social Studies of Science* 35(3): 405–436.

Shostak, S. (2003) Locating gene-environment interaction: At the intersections of genetics and public health. *Social Science & Medicine* 56(11): 2327–2342.

Shostak, S. (2013) *Exposed science: Genes, the environment, and the politics of population health*. Berkeley: University of California Press.

Singer, B.H. and De Castro, M.C. (2001) Agricultural colonization and malaria on the Amazon frontier. *Annals of the New York Academy of Sciences* 954(1): 184–222.

Sultana, F. (2012) Producing contaminated citizens: Toward a nature–society geography of health and well-being. *Annals of the Association of American Geographers* 102(5): 1165–1172.

Sze, J. (2007) *Noxious New York: The racial politics of urban health and environmental politics*. Cambridge, MA: MIT Press.

Szyf, M. (2009) Implications of a life-long dynamic epigenome *Epigenomics* 1: 9–12.

Thayer, Z. and Kuzawa, C. (2011) Biological memories of past environments: Epigenetic pathways to health disparities. *Epigenetics* 6(7): 798–803.

Wagner, T., Benbow, M.E., Burns, M., Johnson, R.C., Merritt, R., Qi, J. and Small, P.C. (2008) A landscape-based model for predicting mycobacterium ulcerans infection (buruli ulcer disease) presence in Benin, West Africa. *EcoHealth* 5(1): 69–79.

Weaver, I.C.G. (2007) Epigenetic programming by maternal behavior and pharmacological intervention. *Epigenetics* 2(1): 22–28.

PART IV, SECTION E

Environmental politics

This section of the Handbook considers the question, *In what ways and for what reasons do people mobilize politically around nature?* Such questions of environmental politics – struggles over who has access to, use of, control over, or benefit from the environment; over how the environment is understood and valued; over how environmental issues are politicized or depoliticized – are central to political ecology. Indeed, many define the field in opposition to purportedly "apolitical" analyses of human–environment relationships: analysis of ways in which social relationships with the environment are always political, always shaped by power relationships, and typically structured by and reproductive of social difference and inequality, have been major and distinguishing themes in political ecology.

The chapters in this section explore four key themes central to political ecology and its constitutive engagement with understandings of environmental politics. The first, by Wendy Wolford and Sara Keene, focuses on social movements. It argues that while social movements seem logically central to the concerns of political ecology, including its professed desire for progress toward more equitable and sustainable human–environment relationships, the field has in fact paid far more attention to informal and less organized politics than to environmental social movements as such. The authors suggest that this omission may be explained by political ecology's roots in agrarian studies and a focus on peasant rebellions and protests. The next chapter, by Ryan Holifield, examines the parallel genealogies of political ecology and environmental justice scholarship, considering the extent to which the objects of concern, theoretical and political frameworks, and methodologies of the two fields align or overlap with one another. By clarifying what is shared and what is distinct between two fields that can seem continuous or interchangeable to newcomers, it lays productive ground for conversation and interchange. The third chapter, by Philippe Le Billon, examines environmental conflicts, which are absolutely central to political ecology. Yet the chapter makes the key point that while conflicts over access to, use of, and the distribution of benefits and hazards associated with environmental resources are rightly central objects of study within political ecology, another of the field's key contributions has been to expose and analyze the many ways in which environmental conflicts are often "naturalized" – inaccurately interpreted as direct or unavoidable social outcomes of given natural conditions or processes. As the author points out, such explanations are themselves profoundly political maneuvers. This theme is elaborated in this section's final chapter, by Erik Swyngedouw, which carefully explicates some of the ways in which the overly narrow or constrained social framing of environmental issues – e.g., by treating them as the proper remit of scientists, of official policy makers, of technocrats or other experts of various sorts – can serve to occlude the inherently political nature of decision-making about them, and to foreclose the possibility of broader participation in such decisions.

44
SOCIAL MOVEMENTS

Wendy Wolford and Sara Keene

Introduction

Perhaps more than any other field, political ecology grew out of a desire to understand marginalization and contestation from the perspective of those who were impoverished, excluded or exploited; the field today retains this commitment. Political ecologists have broken new ground analyzing the ways in which the social relations of production (or, the mechanisms, associations and norms of surplus creation and extraction) shape, and are themselves shaped by, unequal location *within*, access *to* and control *over* the environment. Given the "preference for the poor" that marks the field, it is not surprising that emphasis is placed on politics – on understanding the political (and thoroughly un-natural) nature of both exclusion and incorporation. What is more surprising is the marked preference among political ecologists for studying informal, unorganized politics rather than studying organized groups such as the state or social movements. There are of course excellent studies of social movements in political ecology, but fewer than one might imagine, especially given the argument that social movements will be the primary vehicle for long-term progressive change (Escobar 1995, 2008; O'Connor 1988). Almost all those who identify as political ecologists have worked on marginalization, contestation and even resistance – all of which are crucial for understanding the formation, organization and work of social movements – but the work has generally been done at the level of individuals, households, communities, groups broadly defined (ethnic, regional, national or other) or organizations (NGOs, associations, etc.). Social movements themselves are invoked but not regularly studied.

Part of the explanation for this surprising paucity of social movement studies in an otherwise expansive field may lie in the difficulty of identification, of knowing exactly who is a political ecologist or what studies can be defined as such. Wendy Wolford (one of the authors of this chapter) has worked with social movements and incorporated the tools of political ecology (2004, 2010) but she is rarely identified as a political ecologist. On the other hand, James McCarthy is identified very closely with the field of political ecology but few people think of him as having worked on or with social movements, even though his early work was an incisive examination of the Wise Use Movement in the rural US West (2002).

Beyond the issue of identification, there are perhaps four other factors that explain why political ecology has not engaged more with movements. We briefly outline them here before turning to the substantive purpose of the chapter. First, although political ecology began as a study of land managers in the so-called global south, as a discipline it is dominated by academics in the global north, particularly by scholars in the United States and Great Britain. For better or for worse, in these two countries the study of social movements (and even of mobilization more broadly) is dominated by sociology, and there is very little interaction between sociology and political ecology. American sociology is very U.S.-centric and discussions of 'the environment' tend to be situated in either environmental sociology or environmental justice (but see the ambitious attempt to bridge political ecology and environmental justice organized by David Carruthers 2008, also see Holifield et al. 2009, and Chapter 45, this volume).

A second reason why there might not be as much political ecology work on social movements is because although social movements are highly visible, active and organized, they are in the minority. As important as it is to know when and why people organize to protest or celebrate and defend their conditions, it is potentially even more important to know when and why they do not. One of political ecology's real contributions has been to show how seemingly isolated, reactionary acts of violence are in fact often deliberate responses to historically situated inequalities. Christian Kull's (2004) work on forest fires in Madagascar and Nancy Peluso's (1993) work on community resistance to state control over forest resources in Indonesia are excellent examples of work that helps to redefine the meaning of the political, and to bring resistance to the fore, though neither author focuses on social movements (but see Peluso et al. 2008).

A third reason that may help to explain why political ecologists haven't engaged with movements as much as one might expect is the ethical dilemma of focusing on movements as objects of study. Work in political ecology tends to focus on the everyday and to employ the very intrusive tools of ethnography and in-depth participant observation. If such intimate translation and interpretation is always a violent act, it is even more so in analyses of social movements, organizations that have very specific messages they wish to transmit. Ironically, as Peet and Watts (2004) suggest, work on social movements in political ecology has often "[exposed] the limits of a naïve invocation of the local community as a theatre of governance" (18; see especially Rangan 2000). It turns out that movements are people too! And there are many different ways to be traditional (Bebbington 2004), indigenous (Li 2000, 2004; Valdivia 2005) and landless (Peluso et al. 2008; Wolford 2003) such that it isn't always clear for – or to whom – the research ought to speak (Wolford 2010). As a result, many movements (particularly new or small ones) choose not to participate in academic research, and researchers may choose to work *with* social movements rather than on them.

Finally, the fourth reason that may explain what we see as limited engagement with social movements is that our review of the literature is necessarily partial; it is partial in the sense of being incomplete and partial in the sense of privileging a particular reading of the field. We have no doubt missed many good studies or perhaps been too narrow in our definition of both social movements and political ecology.

In the rest of this chapter, we first outline the roots of social movement work in political ecology, arguing that the field's perspective on mobilization grew out of critical agrarian studies and a focus on peasant rebellions and protests. We then outline what we see as the main contributions of political ecology for analyzing social movements and, more broadly, resistance. We focus on four contributions: first, political ecologists understand all struggles to be a struggle over objective and subjective conditions – in other words, political ecologists bring together materiality and meaning, focusing on both the grounded conditions of production and social reproduction as well as the ways in which people make sense of – or bring meaning to – their situations. Second, political ecology treats movements as produced in and through particular environments; the research emphasizes the importance of place (the spatial context) and conjuncture (the temporal context) for analyzing the formation and maintenance of protest. Third, political ecology shows particular strength in analyzing the discourses or narratives that frame contestation over natural resources, and in situating these within broader structural relationships, such as global trade flows, state power, globalizing movements and regional or global processes of land use or environmental change. Fourth, although, as noted above, political ecology's focus has tended to speak more to informal political processes than to institutional actors such as national states, important work has been done that sheds light on the relationship between the interests of the state (and state actors) and political activists on the ground. Most political ecology work on social movements incorporates some combination of all of these contributions, but in this chapter, we outline the four separately and illustrate each with key examples.

The roots of political ecology as a study of contestation

Early work in political ecology took up Blaikie and Brookfield's (1987) injunction to study the "land managers," situating them within communities, regions and states (and, eventually, within households). For those working on the topic of social movements, however, the inspiration came more directly from the field of agrarian studies. As amorphous as political ecology, classical agrarian studies dates back to the early twentieth century when socialist theorists analyzed the (expected) penetration and diffusion of capitalism into agriculture and, therefore, into rural production and society. The key question at that time was the fate of the then-numerically predominant but historically marginalized and under-appreciated peasantry (Chayanov 1925; Gramsci et al. 1971; Kautsky 1899; Lenin 1956 [1925]; Shanin 1981).

While this original agrarian question was not directly concerned with the formation of social movements, scholars took up the theoretical questions and tools again in the 1970s with the return of research on rural transformation, solidarity and revolution (Moore 1966). Although the peasantry had been declared obsolete on numerous occasions throughout the twentieth century (cf. Hobsbawm 1994), the 1970s marked a period of intense political activity among rural classes. Much more than the last gasp of a dying class, revolutions from Bolivia to Nicaragua and Peru to Kenya, Algeria and Vietnam (Paige 1975; Wolf 1969) all seemed to be fought over agrarian issues (whether this was in fact an accurate representation was debated, as in the cases of the socialist revolution in Vietnam, cf. Wolf 1969; the communist advance in China, cf. Potter and Potter 1990; and the Cuban revolution, cf. Mintz 1974). Scholars within the tradition of agrarian studies incorporated an eclectic set of tools from classical political economy (Smith 2010 [1776]; Marx et al. 1990 [1867]; Lewis 1954) and contemporary social sciences to examine the historical roots of these radical grievances, actors and movements.

As movements and revolutions for independence in Africa and Asia gained force in the second half of the twentieth century, research in critical development studies merged with agrarian studies to investigate the role of ideology, brokers and leaders, class relations, and the state in political activity from foot dragging to revolutions to nation building (Scott 1985; Starn and Fox 1997). Theories of moral economy (Thompson 1971; Scott 1976) were important in suggesting that norms and popularly held ideas of fairness were shaped by convention, custom and structures that, when violated, led to resistance. Therefore, the violation of these perceived moral economies was as, if not more, important to understand in analyzing the rise of food riots or social mobilization than unfairness or grievances per se (Shanin 1972). At the same time, social injustice was situated firmly within broader structural conditions, such as those created by the world market system, from dependency to inequality (Wallerstein 1974; Frank 1969;

Cardoso and Faletto 1979; Amin 1994). Jim Ferguson's (1990) classic study of development projects in Lesotho helped to illuminate the material effects of universalizing assumptions about the objects of development (cf. Escobar 1995).

The rise of subaltern studies in the 1980s also figured prominently in a new theoretical framing of resistance and social movements (Guha 1997; Spivak 1988a). As a project to rethink history (especially national histories) from the perspective of the subaltern, subaltern studies was a reaction to both Marxist and liberal interpretations of history as linear and neat, written from the perspective of elites. Subaltern studies scholars insisted that multiple histories lay hidden in the silences and cracks of official narratives, and that a proactive agenda was required to ferret out the meanings of (and from) the margins (Spivak 1988b, 1993, 2004).

All of these concerns influenced the study of social mobilization within political ecology, a field that itself came into being in tandem with the proliferation of new social movements around the world. Unlike many other fields of study, political ecology retains its focus on the struggles of agrarian and marginalized populations, highlighting the complex ways in which power relations condition mobilization and resistance, and with what effects. We explore these specific contributions in the overview that follows, beginning with the centrality of meaning – norms, values, customs and ideologies – in ecological conflicts and contestations.

Struggles over resources: materiality and meaning

Drawing inspiration from E.P. Thompson's seminal essay, "The moral economy of the English crowd in the eighteenth century" (1971), work on social mobilization in political ecology emphasizes the importance of norms and customs in shaping struggles over access to resources and the environment and thus the co-constitution of cultural meanings and "material forces" (Gramsci et al. 1971: 165; Moore 1993). Thompson's claim that food riots in eighteenth-century England were not simply "rebellions of the belly" (1971: 77), but instead premised on a shared sense of outrage over prices perceived to be unfairly high and new forms of exchange that prejudiced the poor, compelled scholars to take seriously not only the motivations and objectives of protestors, but also – and perhaps even more importantly – the historical conditions and relationships that conditioned collective protest. James Scott's (1976) work situated the moral economy more firmly in the peasant economy with his analysis of a 'subsistence ethic' among the rural poor that he argued created "standards of justice and equity" that applied to all peasants (Scott 1976: 157). Like Thompson, Scott argued that rebellion was directly linked to normative conceptions of obligation, right and reciprocity.

Thompson's influence is evident in a range of social movement studies within political ecology that emphasize the importance of moral economies produced through the social relations of production (or, property relations) in shaping contestation. Michael Watts' (1983) classic study of Hausa peasants in Nigeria emphasized the transformation of a pre-capitalist moral economy by a predatory colonial state and local elites in ways that not only generated recurrent famines but also naturalized the condition of food insecurity as a function of ecological and cultural incompetence. Judith Carney and Michael Watts (1990, see also Carney 2004) built on this perspective to illustrate how pre-colonial moral economies guiding access to land shaped the repeated "failure" of intensive rice production schemes in the Gambia. Women, in particular, resisted attempts to intensify production on land to which they would no longer have a claim, given its incorporation into "community resources." Matt Turner's (2004) work among pastoralists in the African Sahel also sheds light on the moral overtones of the Malthusian narratives that dominate understandings of farmer-herder conflict. As he argues, "It is only through a full and critical engagement with both the materiality which underlies all social life and the moral claims that implicate natural resource use that the etiology of resource-related conflict can be better understood. Struggles over resources are often only superficially so—they in fact reflect not only broader tensions (with ethical dimensions) between social groups but also tensions within these groups" (866). Anthony Bebbington's work in Latin America has brought together a livelihoods framework with political ecology to focus on how people "[make] a living and [make] living meaningful" (2000: 498). Bebbington has worked with communities, non-governmental organizations and social movements and argues that there are many different ways "to be an Indian" in the highlands of Ecuador (1991, 1993, 2004) and, as such, research should focus not on romanticized notions of tradition but should analyze and situate the moral economies of *modernization* (Valdivia 2005).

In Wendy Wolford's analysis of the Rural Landless Workers Movement (the MST) in Brazil (2003, 2005), she highlights the production of moral economies of access to land and argues that claims to land by the agrarian elite in Brazil were generated through a "narrative that attributed their traditional rights to land to hard work, personal responsibility, and reliance on the market rather than on 'politics'" (2005: 251). Likewise, the moral economy of the MST, expressed as "land for those who work it," was generated through historical experiences of displacement and marginalization. In short, these competing moral economies were born out of the inherently relational practice of struggle in a moment of capitalist entrenchment. Another emblematic case of the political ecology of social movements from the vantage point of customs, values and norms is James McCarthy's (1998, 2002) study of the Wise Use Movement. Wise Use was a group of organizations and rural commodity producers in the U.S. American West who fought against state control of public land and mobilized for the right to commodity production on federally-owned lands from the late-1980s to mid-1990s. McCarthy illustrated the ways in which Wise Use activists and campaigns drew upon similar sorts of populist claims as the movements studied in the global south, with appeals to self-determination, local knowledge and local rights. While this moral economy was not framed as anti-capitalist, it did constitute an "ongoing struggle over nature" and "resistance to the perennial dynamics of capitalism" through their articulation of an alternative set of economic relations that maintained the conditions, livelihoods and culture of rural Western communities (2002: 1291).

Though diverse in their geographic, historical and cultural locations, these studies illustrate the centrality of meaning in contestations over resources and environments in the context of capitalist change. These meanings of course are not created in a vacuum; they are constituted in and by particular people, places and times. Political ecology's attentiveness to conjuncture and place is thus another of its important contributions to our understanding of social struggles.

Grounding mobilization: the importance of place

In part because of its close connection to the disciplines of geography and anthropology and in part because of its focus on land managers and material practice, political ecology has always emphasized the importance of place in shaping the conditions of exploitation and of protest or mobilization; if nature and society are co-constituted then by definition location matters. Political ecologists such as Christian Kull working on social mobilization situate movements and protest in historically rich descriptions of local environments and ecologies (2004). Arturo Escobar has been particularly important for urging social scientists to study the political construction of place (2001), which he has recently re-conceptualized as territory in his magisterial study of the Pacific Coast Black Communities (PCN) in Colombia (2008). The book brings together Escobar's focus on modernity/coloniality (in which the project of modernity is understood as predicated on coloniality) with an interest in alternative knowledges.

He argues that the peoples of the Pacific coast region have been shaped by an articulation of processes that have simultaneously produced the region, including historical processes of geological and biological formation, the daily practices of local black, indigenous and mestizo groups, capital accumulation, incorporation into the state, and cultural political practices of social movements (2008: 31). Redefining the liberal conceptions of rights, the PCN has called for collective rights to land; cultural, political and economic autonomy (on the basis of their group identity); and the right to a 'shared vision of the future' based on their cultural autonomy. Through this case study, Escobar demonstrates how identity and place are 'dialogic and relational' (2008: 203), being constantly (re)created through material and political practices that mutually condition one another.

This focus on the importance of place is visible in studies of indigenous peoples and indigeneity more generally. Tom Perreault's (2008) work in particular examines the material and symbolic importance of traditional rights and norms for governing and shaping livelihoods and political claims in Bolivia and Ecuador. In Cocahabamba, Bolivia, where the water wars erupted in 2003, most observers focused on urban movements in organizing the protests but Perreault argues that peasant movements were actually far more important and organized in utilizing the power of traditional discourses around *usos y costumbres* (customs and habits) to manage water and shape new forms of governance. With Gabriela Valdivia, Perreault has also done important work on the role of place in shaping mobilization around new resource imaginaries (2010; see also Wolford 2005). Valdivia and Perreault compare mobilizations against the privatization of natural gas in Bolivia and oil in Ecuador to demonstrate the importance of historically situated, place-based notions of citizenship and nation. The importance of place is also stressed in Donald Moore's (1998, 2005) examination of histories of settlement, freedom and resistance in Zimbabwe. Moore argues that colonial and post-colonial forms of governance called upon fixed lines and spatial concentrations – government spaces and settlement areas – that violated the fluid spaces of house and field in traditional societies and in the newly created squatter areas. Moore's work highlights what he calls the "sedimenting" of multiple spaces in any given place, all shaped by contending and racialized practices of inhabiting, laboring and suffering.

These studies help us to ground movements and mobilization in particular historical and geographic locations, without neglecting the broader global processes within which they are constituted. But what of the discourses and narratives that give life to movements? How do these inform social struggles and the environments in and for which they are waged? While movement discourses are inextricably connected to the cultural values and norms that give them meaning, they do additional work of shaping the contours of resistance – defining what (and who) is to be included and excluded, and the terms of their inclusion. Narratives and discourses are deeply political, as demonstrated below, and can have unintended effects as they travel back and forth through time and space.

Traveling narratives: of myths, discourses and representations

Movement narratives represent an amalgam of origin stories, principles, goals and visions that animate the movement and provide direction for political action. As such, movement narratives articulate specific notions of justice, modes of political participation and engagement, and collective identities. Such narratives – particularly when backed by a broad spectrum of participants – provide the ideological space to reconsider power relations, economic organization, land tenure, social relationships and rights (to name a few). In other words, movement narratives provide the impetus (and possibility) for meaningful social change. However, there are theoretical and practical limitations posed by "movement narratives." On the one hand, having the appearance of a "united front" is necessary to movement mobilization, legitimacy and efficacy. Movements often engage in what Spivak has called the "strategic use of positivist essentialism" (1996: 214). The notion of "strategic essentialism" is fundamentally different from "a substantive or real essentialism" (Spivak 1993: 3). While the latter refers to a flattening of difference, whereby master labels such as "woman" subsume and conceal difference, strategic essentialism involves the political mobilization of such master labels alongside theoretical critique. The notion of "strategic essentialism" thus offers an intervention that recognizes difference without relinquishing the political salience of normative discourses. This approach is particularly useful in understanding how actors emphasize and mobilize categories of identification, and simultaneously re-shape those categories to define the parameters of their engagement. Yet, this sort of "strategic essentialism" can obscure differences and make it difficult to maintain the kind of ethical negotiation that is critical to democratic inclusion (and the valorization and promotion of difference).

Peter Brosius has provided insight into how environmental discourses in Malaysia – and the campaigns through which they are mobilized – have served to simplify and translate "local" or "indigenous" knowledges, thereby defining and structuring the contours of social and political debates around environmental actions (1997, 1999; Brosius et al. 1998). Brosius argues that the institutionalization of environmental politics and their accompanying discourses may in fact "obstruct meaningful change" through the naturalization, simplification and depoliticization of discourses used to advance particular political objectives and projects (1999). Haripriya Rangan documented a similar process in her study of the Indian Chipko movement and activists' attempts to halt private and national deforestation schemes (1993, 2000, 2004). While an elite group of activists was able to achieve popular and state support through discourses of ecological degradation, the narrative of the movement changed as it traveled more broadly and became "detached from its specific demands regarding access to forest resources and local economic development" (Rangan 2004: 383). Increasingly, the Chipko movement was framed in environmentalist terms emphasizing the problems of deforestation and ecological degradation. The state responded with a suite of policies that addressed environmental degradation but impeded the communities' ability to achieve forest-based livelihoods and exacerbated labor conflicts between local residents and migrant Nepalese workers.

At the same time, discourses are re-invented in different ways across time and space and Anna Tsing (2005) shows how the Chipko movement helped to serve as the basis for social change amongst environmental activists in Indonesia. There, activists brought together the story of Chipko with the story of Chico Mendes at a moment of political restructuring (i.e. the fall of Indonesia's New Order government), inspiring the production of national and transnational alliances to mobilize for ecological and social justice. In this way, allegories – or universal narratives – served to advance movement interests through the production of new forms of political subjectivities and possibilities for social action. Likewise, Tania Li (1996, 2000) analyzes the ways in which Lauje swidden farmers in Central Sulawesi strategically appropriated representations of "community" to secure land tenure rights and local control of natural resources. Although this strategy failed to address internal inequalities (i.e. gender and class) and resulted in outcomes that were not as equitable as might otherwise have been the case (1996: 521), Li's work highlights the ways in which familiar cultural categories can be discursively re-configured to serve counter-hegemonic aims.

Due to the contingent and unanticipated effects discourses can have, movement leaders carefully manage the interplay between movement narratives and objectives and the diverse ideologies and experiences of movement members. This has been particularly evident within the MST in Brazil, where movement leaders have had to negotiate competing notions of land, property and agrarian reform in order to sustain collective mobilization. As Wolford demonstrates, members of the MST reflect diverse "work economies, family practices and community traditions," all of which condition how they understand, evaluate and participate in the movement (Wolford 2010: 17). In this case, MST leaders have carefully crafted a coherent movement narrative that emphasizes peasant unity and communalism, agricultural sustainability, horizontal governance and, importantly, opposition to the Brazilian state. Such representations are reproduced in the popular media, as well as through internal forms of knowledge production, such as oral histories and movement publications (Wolford 2010). Importantly, such efforts are never finished – narratives are ongoing and dynamic processes of movement-building.

These illustrations of translating, appropriating and reconstituting "universal" discourses illuminate the ways in which movements and social mobilization disrupt singular notions of knowledge by transforming them and imbuing them with new meanings. What political ecology contributes to the understanding of social movements is thus not only a critique of the challenges and limitations of appropriating and or imposing Western discourses and knowledge in Southern or subaltern social movements (Forsyth 2001, 2003, 2004; Brosius 1997), but also how actors strategically engage, transform, construct and deploy discourses and representations to advance specific political goals – from claims to land and resources to the "right" to livelihoods and cultural practices (Nygren 2004; Valdivia 2005). Movement narratives are powerful – in that they instigate material effects – yet like the agents of movements themselves, narratives are enmeshed in complex webs of power that continually shape and reshape their meaning and content. In attending to movement discourses, political ecology highlights the simultaneously fluid and fragmented character of social struggles, and the historical and political processes within which they are constituted.

Mobilization and the state

Political ecology's concern with power, marginalization and contestations over resources has generated a perspective on state–society relations that focuses less on "the state" per se, and more on "how power works" (Li 2005: 383) through practices of governance, discourses, state agencies and officials, and territories or communities of rule and resistance (Watts 2004), to name only a few. Indeed, one could argue that the state becomes harder (and less productive) to trace in the context of increasing transnational connections between movements (Edelman 2005; Tsing 2005) and new and fluid relations between (and within) states, capital, multilateral and non-governmental organizations (Peluso 1993; Brosius 1997). And so political ecology's focus on the everyday, messy practices of rule and resistance over formal, organized politics and institutions helps to de-center the state (Mallon 1995; Gupta 2006; Wolford forthcoming), focusing on the ways in which political power is constituted, experienced and (re)produced at multiple levels.

Judith Carney's (2004) work on gender conflicts over resources in the Gambian wetlands serves as an illustrative example. While she traces the historical development of colonial and national rice cultivation policies, she does not produce an analysis of "the state," but rather of how a confluence of state policies and global economic restructuring re-shaped gendered relations of production and household dynamics, ultimately leading to women's economic marginalization and intra-household conflicts. In Carney's work, a complex interplay between Gambian governmental officials, international donors and financial institutions, institutional legacies of colonialism and changes in the global political economy contributed to a reconfiguration of not only land use and labor relations, but also resource access within households – engendering new forms of resistance and protests against state policies. In a similar fashion, Tom Bassett (1988) shows how conflicts seemingly between peasants and herders in Cote d'Ivoire are actually shaped by the state's privileging of herder claims over indigenous land tenure regimes.

Such approaches to mobilization and resistance in political ecology have also led to a re-working of 'the state' not as a monolithic entity but as a terrain of struggle in which multiple and shifting interests collide, converge or are transformed. In the wake of the Zapatista uprising and movement formation in southern Mexico, Aaron Bobrow-Strain (2007) provides a subtle analysis of power – including but not simply state power – with detailed work on landowners in the state of Chiapas. He argues that the landowning elite simultaneously worked to advance the agenda of the state, "subordinating themselves and their indigenous workers to the rationales of a liberal government" and negotiated their own spaces of authority that allowed them to resist later incursions by the same state (2007: 79). Tom Perreault's work on indigenous mobilization in Bolivia demonstrates how a convergence of processes – neoliberal restructuring, a shift toward export-oriented natural gas production, increasing inequality and new discourses of indigeneity – coalesced to generate social mobilization and protest in Bolivia, and ultimately resulted in the election of Evo Morales and the integration of indigenous and *campesino* movements into formal political processes (2008). As Perreault's work illustrates, states are not only fluid and dynamic, they are composed of locatable actors, agencies and institutions that are part of broader social, political and economic structures.

Concluding thoughts

This brief chapter has attempted to evaluate and summarize the contributions of political ecology to the study of social movements. Although only a handful of movements were mentioned in the piece, it is clear that social mobilization is a key subject in political ecology. The de-centering, re-working and situating of social movements reflects political ecology's broader concern with the *workings* of power – the specific and grounded ways in which struggles over resources are enacted. This approach to power treats everyday practices, cultural meanings and discourses and representations as constitutive of movements and indeed processes of state formation. Of crucial importance in all of these analyses is the contingent nature of social mobilization; outcomes are never certain, and the effects of environmental discourses and actions can have unanticipated consequences. Struggles – over meanings, representations, environments and political outcomes – are always situated within pre-existing relations of power; the social and historical contexts in which collective struggles are waged both constitute and are in turn transformed by collective actions.

References

Amin, S. (1994) "The origin and development of underdevelopment." In R.K. Kanth (ed.), *Paradigms in Economic Development: Classic perspectives, critiques, and reflections*. Armonk: M.E. Sharpe, pp. 161–164.

Bassett, T.J. (1988) "The political ecology of peasant-herder conflicts in the northern Ivory Coast." *Annals of the Association of American Geographers*, vol. 78, pp. 453–472.

Bebbington, A. (1991) "Indigenous agricultural knowledge systems, human interests, and critical analysis: reflections on farmer organization in Ecuador." *Agriculture and Human Values*, vol. 8, pp. 14–24.

Bebbington, A. (1993) "Modernization from below: an alternative indigenous development?" *Economic Geography*, vol. 69, pp. 274–292.

Bebbington, A. (2000) "Reencountering development: livelihood transitions and place transformations in the Andes." *Annals of the Association of American Geographers*, vol. 90, pp. 495–520.

Bebbington, A. (2004) "Movements and modernizations, markets and municipalities: indigenous federations in rural Ecuador." In R. Peet and M. Watts (eds.), *Liberation Ecologies: Environment, development, social movements*. New York: Routledge, pp. 358–382.

Blaikie, P.M. and Brookfield, H.C. (1987) *Land Degradation and Society*. New York: Methuen.

Bobrow-Strain, A. (2007) *Intimate Enemies: Landowners, power, and violence in Chiapas*. Durham, NC: Duke University Press.

Brosius, J.P. (1997) "Endangered forest, endangered people: environmentalist representations of indigenous knowledge." *Human Ecology*, vol. 25, pp. 47–69.

Brosius, J.P. (1999) "Green dots, pink hearts: displacing politics from the Malaysian rain forest." *American Anthropologist*, vol. 101, pp. 36–57.

Brosius, J.P., Tsing, A.L. and Zerner, C. (1998) "Representing communities: histories and politics of community-based natural resource management." *Society and Natural Resources*, vol. 11, pp. 157–168.

Cardoso, F.H. and Faletto, E. (1979) *Dependency and Development in Latin America*. Berkeley: University of California Press.

Carney, J. (2004) "Gender conflict in Gambian wetlands." In R. Peet and M. Watts (eds.), *Liberation Ecologies: Environment, development, social movements*. New York: Routledge, pp. 316–336.

Carney, J. and Watts, M. (1990) "Manufacturing dissent: work, gender and the politics of meaning in a peasant society." *Africa*, vol. 60, pp. 207–241.

Carruthers, D. (2008) *Environmental Justice in Latin America: Problems, promise, and practice*. Cambridge, MA: MIT Press.

Chayanov, A.V. (1966 [1925]) *The Theory of Peasant Economy*. Madison: University of Wisconsin Press.

Edelman, M. (2005) "Bringing the moral economy back in ... to the study of 21st century transnational peasant movements." *American Anthropologist*, vol. 107, pp. 331–345.

Escobar, A. (1995) *Encountering Development: The making and unmaking of the third world*. Princeton: Princeton University Press.

Escobar, A. (2001) "Culture sits in places: reflections on globalism and subaltern strategies of localization." *Political Geography*, vol. 20, pp. 139–174.

Escobar, A. (2008) *Territories of Difference: Place, movements, life, redes*. Durham, NC: Duke University Press.

Ferguson, J. (1990) *The Anti-Politics Machine: "Development," depoliticization, and bureaucratic power in Lesotho*. New York: Cambridge University Press.

Forsyth, T. (2001) "Environmental social movements in Thailand: how important is class?" *Asian Journal of Social Science*, vol. 29, pp. 35–51.

Forsyth, T. (2003) *Critical Political Ecology: The politics of environmental science*. New York: Routledge.

Forsyth, T. (2004) "Industrial pollution and social movements in Thailand." In R. Peet and M. Watts (eds.), *Liberation Ecologies: Environment, development, social movements*. New York: Routledge, pp. 383–398.

Frank, A.G. (1969) *Capitalism and Underdevelopment in Latin America*. New York: Monthly Review Press.

Gramsci, A., Hoare, Q. and Nowell-Smith, G. (eds.). (1971) *Selections from the Prison Notebooks of Antonio Gramsci*. New York: International Publishers.

Guha, R. (1997) *A Subaltern Studies Reader, 1986–1995*. Minneapolis: University of Minnesota Press.

Gupta, A. (2006) "Blurred boundaries: the discourse of corruption, the culture of politics, and the imagined state," in A. Sharma and A. Gupta (eds.), *The Anthropology of the State: A reader*. Malden: Blackwell Publishing, pp. 211–242.

Hobsbawm, E.J. (1994) *The Age of Extremes: A history of the world, 1914–1991*. New York: Pantheon Books.

Holifield, R., Porter, M. and Walker, G. (2009) "Spaces of environmental justice: frameworks for critical engagement." *Antipode*, vol. 41, no. 4, pp. 591–612.

Kautsky, K. (1988 [1899]) *The Agrarian Question: In two volumes*. London: Zwan Publications.

Kull, C.A. (2004) *Isle of Fire: The political ecology of landscape burning in Madagascar*. Chicago: University of Chicago Press.

Lenin, V.I. (1956 [1925]) *The Development of Capitalism in Russia: The process of the formation of a home market for large-scale industry*. Moscow: Foreign Languages Publishing House.

Lewis, W.A. (1954) "Economic development with unlimited supplies of labor." *Manchester School*, vol. 22, pp. 139–191.

Li, T.M. (1996) "Images of community: discourse and strategy in property relations." *Development and Change*, vol. 27, pp. 501–527.

Li, T.M. (2000) "Articulating indigenous identity in indonesia: resource politics and the tribal slot." *Comparative Studies in Society and History*, vol. 42, pp. 149–179.

Li, T.M. (2004) "Environment, indigeneity and transnationalism." In R. Peet and M. Watts (eds.), *Liberation Ecologies: Environment, development, social movements*. London: Routledge, pp. 309–337.
Li, T.M. (2005) "Beyond 'the state' and failed schemes." *American Anthropologist*, vol. 107, pp. 383–394.
Mallon, F. (1995) *Peasant and Nation: The making of postcolonial Mexico and Peru*. Berkeley: University of California Press.
Marx, K., Engels, F., Mandel, E., Fowkes, B. and Fernbach, D. (1990 [1867]) *Capital: A critique of political economy*. London: Penguin Books.
McCarthy, J. (1998) "Environmentalism, Wise Use and the nature of accumulation in the rural West." In B. Braun and N. Castree (eds.), *Remaking Reality: Nature at the millennium*. New York: Routledge, pp. 125–148.
McCarthy, J. (2002) "First World political ecology: lessons from the Wise Use movement. *Environment and Planning A*, vol. 34, pp. 1281–1302.
Mintz, S.W. (1974) "The rural proletariat and the problem of rural proletarian consciousness." *The Journal of Peasant Studies*, vol. 1, pp. 291–325.
Moore, B. (1966) *Social Origins of Dictatorship and Democracy: Lord and peasant in the making of the modern world*. Boston: Beacon Press.
Moore, D.S. (1993) "Contesting terrain in Zimbabwe's eastern highlands: political ecology, ethnography, and peasant resource struggles." *Economic Geography*, vol. 69, pp. 380–401.
Moore, D.S. (1998) "Subaltern struggles and the politics of place: remapping resistance in Zimbabwe's eastern highlands." *Cultural Anthropology*, vol. 13, pp. 344–381.
Moore, D.S. (2005) *Suffering for Territory: Race, place, and power in Zimbabwe*. Durham, NC: Duke University Press.
Nygren, A. (2004) "Contested lands and incompatible images: the political ecology of struggles over resources in Nicaragua's Indio-Maiz Reserve." *Society & Natural Resources*, vol. 17, pp. 189–205.
O'Connor, J. (1988) "Capitalism, nature, socialism: a theoretical introduction." *Capitalism, Nature, Socialism*, vol. 1, pp. 11–38.
Paige, J.M. (1975) *Agrarian Revolution*. New York: Free Press.
Peet, R. and Watts, M. (2004) *Liberation Ecologies: Environment, development, social movements*. London: Routledge.
Peluso, N.L. (1993) "Coercing conservation: the politics of state resource control." *Global Environmental Change*, vol. 3, pp. 199–217.
Peluso, N.L., Afiff, S. and Rachman, N.F. (2008) "Claiming the grounds for reform: agrarian and environmental movements in Indonesia." *Journal of Agrarian Change*, vol. 8, pp. 377–407.
Perreault, T. (2008). "Custom and contradiction: rural water governance and the politics of *usos y costumbres* in Bolivia's irrigators' movement." *Annals of the Association of American Geographers*, vol. 98, pp. 834–854.
Perreault, T. and Valdivia, G. (2010) "Hydrocarbons, popular protest, and national imaginaries: Ecuador and Bolivia in comparative context." *Geoforum*, vol. 41, pp. 689–699.
Potter, S.H. and Potter, J.M. (1990) *China's Peasants: The anthropology of a revolution*. New York: Cambridge University Press.
Rangan, H. (1993) "Romancing the environment: popular environmental action in the Garhwal Himalayas." In J. Friedmann and H. Rangan (eds.), *In Defense of Livelihood: Comparative studies on environmental action*. West Hartford: Kumarian Press, pp. 155–181.
Rangan, H. (2000) *Of Myths and Movements: Rewriting Chipko into Himalayan history*. London: Verso.
Rangan, H. (2004) "From Chipko to Uttaranchal: the environment of protest and development in the Indian Himalaya." In R. Peet and M. Watts (eds.), *Liberation Ecologies: Environment, development, social movements*. New York: Routledge, pp. 371–393.
Scott, J.C. (1976) *The Moral Economy of the Peasant: Rebellion and subsistence in Southeast Asia*. New Haven: Yale University Press.
Scott, J.C. (1985) *Weapons of the Weak: Everyday forms of peasant resistance*. New Haven: Yale University Press.
Shanin, T. (1972) *The Awkward Class: Political sociology of peasantry in a developing society Russia 1910–1925*. Oxford: Clarendon Press.
Shanin, T. (1981) "Marx, marxism and the agrarian question: I Marx and the peasant commune." *History Workshop*, vol. 12, pp. 108–128.
Smith, A. (2010 [1776]) *Inquiry into the Nature and Causes of the Wealth of Nations*. [S.L.]: Simon & Brown.
Spivak, Gayatri Chakravorty (1988a) "Can the subaltern speak?" In C. Nelson and L. Grossberg (eds.), *Marxism and the Interpretation of Culture*. Chicago: University of Illinois Press, pp. 271–313.

Spivak, G.C. (1988b) *In Other Worlds: Essays in cultural politics*. New York: Routledge.
Spivak, G.C. (1993) *Outside in the Teaching Machine*. New York: Routledge.
Spivak, G.C. (1996 [1985]) "Subaltern studies: deconstructing historiography." In D. Landry and G. MacLean (eds.), *The Spivak Reader*. London: Routledge, pp. 203–236.
Spivak, G.C. (2004) "Righting wrongs." *The South Atlantic Quarterly*, vol. 103, pp. 523–581.
Starn, O. and Fox, R.G. (1997) *Between Resistance and Revolution: Cultural politics and social protest*. New Brunswick: Rutgers University Press.
Thompson, E.P. (1971) "The moral economy of the English crowd in the eighteenth century." *Past and Present*, vol. 50, pp. 76–136.
Tsing, A.L. (2005). *Friction: An ethnography of global connection*. Princeton: Princeton University Press.
Turner, M.D. (2004) "Political ecology and the moral dimensions of resource conflicts: the case of farmerherder conflicts in the Sahel." *Political Geography*, vol. 23, pp. 863–889.
Valdivia, G. (2005) "On indigeneity, change and representation in the northeast Ecuadorian Amazon." *Environment and Planning A*, vol. 37, pp. 285–303.
Wallerstein, I. (1974) *The Modern World-System*. New York: Academic Press.
Watts, M. (1983) *Silent Violence: Food, famine and peasantry in northern Nigeria*. Berkeley: University of California Press.
Watts, M.J. (2004) "Antinomies of community: some thoughts on geography, resources and empire." *Transactions of the Institute of British Geographers*, vol. 29, pp. 195–216.
Wolf, E.R. (1969) *Peasant Wars of the Twentieth Century*. New York: Harper & Row.
Wolford, W. (2003) "Producing community: the MST and land reform settlements in Brazil." *Journal of Agrarian Change*, vol. 3, pp. 500–520.
Wolford, W. (2004) "Of land and labor: agrarian reform on the sugarcane plantations of northeast Brazil." *Latin American Perspectives*, vol. 31, pp. 147–170.
Wolford, W. (2005) "Agrarian moral economies and neoliberalism in Brazil: competing worldviews and the state in the struggle for land." *Environment and Planning A*, vol. 37, pp. 241–261.
Wolford, W. (2010) *This Land is Ours Now: Social mobilization and the meanings of land in Brazil*. Durham, NC: Duke University Press.
Wolford, W. (forthcoming) "'*Sitting by the Pool, Having Drinks with the Fazendeiros*': state-society dynamics in contemporary Brazilian land reform." *Latin American Perspectives*, accepted November 2013.

45

ENVIRONMENTAL JUSTICE AND POLITICAL ECOLOGY

Ryan Holifield

Introduction

Environmental marginalization and inequality are at the heart of both the research program of political ecology and the concept of environmental justice. And yet until relatively recently, political ecology and environmental justice appeared to be traveling down quite different paths. What took these two—seemingly such a perfect match—such a long time to meet? What finally brought them together, and what has come out of this encounter so far? And what might be some possible paths for their shared future?

This chapter will venture some answers to these questions. Since the story of political ecology runs throughout the handbook, the chapter emphasizes the trajectories of environmental justice, along with related concepts like environmental equity and environmental racism, which are then brought into explicit conversation with political ecology. Environmental justice is sometimes presented as a distinct "approach" or "framework" within human–environment research, but I contend that it is better considered as a concept, topic, or phenomenon, for which there are numerous possible approaches to analysis. The chapter begins by tracing the early history of environmental justice, which emerged in a very different geographic setting from that of political ecology, and with very different purposes: specifically, to mobilize activism and shape policy and law within the United States (see Chapter 14, this volume, on such policy orientations and goals). Only in the late 1990s and early 2000s did the two traditions begin explicitly to cross-fertilize, as political ecologists began looking more closely at the "Global North" and the concept of environmental justice began circulating in the "Global South."

But in bringing the concept and subfield together, just as important as this geographical shift was the emergence of alternative trajectories within environmental justice scholarship. These trajectories emphasized qualitative methods, radical political economy, critical social theory, and normative political theory, departing from the quantitative approach dominant among the field's pioneering studies. The closing sections of the chapter assess selected outcomes of this more explicit engagement between political ecology and environmental justice, suggesting ways in which each has enriched the other, and introduces emerging frameworks, agendas, and approaches that might take the partnership in new directions in the future.

Different origins, different paths

During the 1970s and early 1980s, while geographers and anthropologists in predominantly rural Third World settings were establishing the field of political ecology, US grassroots activists and their academic allies in sociology and related disciplines were commissioning and conducting the first analyses of phenomena variously identified as environmental inequity, environmental injustice, or environmental racism—often but not always associated with cities and urbanization. In many cases, the primary purpose of these analyses was to provide empirical support for the claims of civil rights and anti-toxics activists that environmental hazards were disproportionately located in areas with predominantly minority populations, and thereby to influence legislators, policymakers, and sometimes specific legal cases. For example, sociologist Robert Bullard's (1983) classic study of the relationship between race and the distribution of solid waste sites in Houston initially served to support plaintiffs in a lawsuit alleging racial discrimination in municipal landfill siting (*Bean* v. *Southwestern Waste Management, Inc.*). Other important early environmental justice studies were commissioned by civil rights leaders, including a regional analysis of hazardous waste sites by the US General Accounting Office (1983) and a nationwide study of toxic sites by the United Church of Christ Commission for Racial Justice (1987). These studies eventually played key roles in convincing the US Environmental Protection Agency to place environmental equity on its policy agenda in the early 1990s. In 1994, President Bill Clinton signed an executive order directing all federal agencies to identify and address environmental inequities based on low income or minority status.

As environmental justice studies subsequently proliferated, their overriding concern was a practical question of little interest to most political ecology research: that is, whether environmental inequity in the US was indeed a problem serious enough to require federal resources, policies, and regulations. Some analysts directly challenged the early studies' findings of racial inequities in the spatial distribution of environmental hazards (e.g., Anderton et al. 1994; Lambert and Boerner 1997), and others found that results varied with the scale and resolution of analysis (e.g., Bowen et al. 1995; Cutter et al. 1996). Other studies accepted the presence of inequitable patterns, but they drew on longitudinal analyses of demographic change to challenge the claim that these patterns were results of discrimination in siting (e.g., Been 1993). Unsurprisingly, subsequent research challenged the skeptics, and debates increasingly focused on fine points of methodology (e.g., McMaster et al. 1997). Since the 1990s, quantitative environmental justice analysis has become steadily more sophisticated and complex, but its empirical, methodological, and practical preoccupations remain quite different from those of most political ecology, which has focused instead on expanding and refining its theoretical and political repertoire.

Despite considerable progress, quantitative environmental justice research has struggled to move beyond inferences of human health impact based on simple residential proximity to hazards, and its capacity to establish causal relationships between environmental conditions and observed health outcomes remains limited (Chakraborty et al. 2011). In part because this goal remains immensely challenging, many environmental justice activists and advocates have embraced community-based participatory research, which replaces the concern with verifying patterns of inequity with efforts to empower citizens and find practical solutions to community health problems (Shepard et al. 2002). In either case, the emphasis on human health in mainstream environmental justice research marks another important difference from most political ecology (although see Chapters 26 and 43, this volume). While political ecology has traditionally emphasized the dynamics of local-scale biophysical conditions, environmental justice research (with some important exceptions described below) has historically focused on the impacts of environmental conditions on human health.

Political ecologists have sometimes criticized the absence of theory in mainstream environmental justice research, noting the latter's predominantly empirical and methodological orientation (e.g., Swyngedouw and Heynen 2003). Critical and radical social and political economic theory has indeed had little impact on this dominant trajectory, and overt theoretical considerations are absent from most of this research. In the exceptional studies that do apply theory to environmental justice analysis, we can frequently see parallels to what Paul Robbins (2012) calls "apolitical ecology." Early efforts to theorize aspects of environmental inequality drew on classic sociological models of neighborhood change (e.g., Liu 1997) or rational choice models from economics (e.g., Viscusi and Hamilton 1999). More recently, interdisciplinary urban ecology projects have theorized environmental justice using concepts from systems ecology and landscape ecology, such as disturbance and patch dynamics (Clark et al. 2007). So although mainstream environmental justice research has not been entirely atheoretical, its gravitation towards rational choice and systems models has been an important reason why this trajectory remains largely divorced from political ecology.

The emergence of alternative trajectories in environmental justice research

The mainstream approach to environmental justice analysis described above remains dominant within a number of disciplines, at least in the US. However, in the late 1980s and early 1990s, several intellectual developments laid the groundwork for alternative trajectories in environmental justice research, which in turn helped pave the way for deeper, more explicit engagements between the concept of environmental justice and the field of political ecology at the turn of the twenty-first century. Although a complete account of these developments is beyond the scope of this chapter, in this section I highlight the influence of three important touchstones: the resurgence and rethinking of normative theories of social justice; the influence of this rethinking on political-economic analyses of environmental change and conflict; and the extension of constructivist social movement theory and conceptions of discourse to environmental justice activism and Third World "environmentalism of the poor."

Rethinking normative theories of social justice

One important development that set the stage during the 1990s for a closer encounter between political ecology and the concept of environmental justice was the resurgence of interest within Marxian political economy and geography in normative theories of social justice. In the wake of his classic *Social Justice and the City*, Marxist geographer David Harvey (1973) departed from an explicit focus on social justice, and the concept remained in the background within critical and radical geography during the formative years for both political ecology and environmental justice. Of course, justice and injustice were central themes animating political ecological research from the beginning (Forsyth 2008; Robbins 2012). As Robbins (2012: 87) puts it, "Political ecology stories are stories of justice and injustice," in particular for the marginalized populations at the heart of much classical political ecology. However, justice remained an implicit theme in much of this research, and many radical scholars of the time followed Harvey in setting justice aside as a bourgeois concept, by this time associated closely with the political liberalism of John Rawls' (1971) *A Theory of Justice*.

Although multiple sources resurrected interest in normative theories of social justice in the early 1990s, the most influential was probably Iris Marion Young's (1990) *Justice and the Politics of Difference*. Young critiqued the dominant *distributive* paradigm in liberal theories of justice, arguing that theories of justice should attend to deeper institutional and structural conditions, such as dynamics of domination and oppression, which generated and sustained unequal distributions. She emphasized *procedural* justice, and especially inclusion and participation in decision-making processes. But she also argued that by focusing on the abstract individual, other theories had overlooked the significance of embodiment, including race, gender, and sexuality, and its inseparability from group identity and membership. Along similar lines, she contended that political philosophy had paid inadequate attention to the discourses and experiences of marginalized groups and social movements themselves, urging normative theories of justice to engage with them directly. Inspired by Young's feminist critiques of liberal theories of justice, David Harvey and his students, along with David M. Smith, played leading roles in restoring justice to the agenda of critical and radical geography (Harvey 1992, 1996; Smith 1994; Merrifield and Swyngedouw 1996).

Justice and the political economy of environmental conflict and change

A parallel source of inspiration that had begun to crystallize during the 1980s was new Marxist theorizing of environmental problems and conflicts, including Allan Schnaiberg's (1980) concept of the "treadmill of production," Neil Smith's (1984) theorization of the production of nature, and the inauguration of the journal *Capitalism, Nature, Socialism* with James O'Connor's (1988) elaboration of the second contradiction of capitalism. Harvey brought the threads of justice, political economy, and environment together in *Justice, Nature and the Geography of Difference* (1996). In one influential chapter, he argued that the grassroots environmental justice movement's conception of justice—although in many ways problematic and parochial—derived distinctive power and value from its grounding in the embodied, place-specific positionality of the groups most marginalized with respect to global processes of capital accumulation.

Meanwhile, environmental justice began to appear as a topic in *Capitalism, Nature, Socialism* and other radical journals like *Antipode* (a special issue in 1996). Some articles extended Young's critique of the distributive paradigm to the analysis of environmental injustice. For instance, although the question of procedural justice came up even in some of the earliest research on environmental racism—including Bullard's (1983) study of solid waste sites in Houston—Robert Lake (1996) argued that most environmental justice scholarship continued to overlook this dimension and to overemphasize distributive patterns. Renewed interest in normative theories of justice also influenced political ecology, which in turn helped broaden the scope of the environmental justice concept beyond hazards to human health. Frequently this work continued to prioritize distributive notions of justice (e.g., Gleeson and Low 1998; see Schlosberg 2007). But Young's exhortation to look beyond the distributive paradigm left a clearer mark on other political ecologists. For example, the case studies and essays in *People, Plants, and Justice* (Zerner 2000) showed in various ways how practices and norms of nature conservation and biodiversity protection involved dimensions of justice other than the distributive (e.g., Schroeder 2000).

Another prominent critic of the distributive paradigm implicit in mainstream environmental justice analysis was geographer Laura Pulido (1996, 2000), whose influential historical-geographical research conceptualized environmental justice and racism in ways that resonated clearly with political ecology. She critiqued the limited conception of racism implicit in longitudinal environmental justice analyses—those designed to determine whether minorities or hazards arrived first to a particular area—arguing that their narrow focus on siting and intentionality obscured structural and hegemonic forms of racism, including white privilege. In addition, she criticized quantitative analyses for their limited conception of the *spatiality* of environmental injustice and racism, contending that their preoccupation with the scale and resolution of localized spatial distributions led them to overlook the complex, multi-scalar processes that generated environmental inequalities. Although Pulido's research in Los Angeles focused on toxic pollution instead of land degradation, the latter critique resonated with political ecology's long-standing emphasis on tracing local environmental changes to political, economic, and social dynamics at broader scales. Alongside the important work of urban environmental historian Andrew Hurley (1995), Pulido's research helped usher in a historical-geographical tradition within environmental justice studies (e.g., Pulido et al. 1996; Boone and Modarres 1999; Colten 2002; Gandy 2003).

Social movements, interpretive frames, and discourses

A third influence that helped nudge the concept of environmental justice closer to the field of political ecology was social movement theory, and in particular the constructivist approach of *frame analysis* initially developed by Erving Goffman (1974). Sociologists Robert Benford and David Snow and their colleagues argued that prevalent approaches in social movement theory, including the resource mobilization approach, had overlooked interpretive schemes as catalysts for mobilizing social movement activists against particular grievances (Snow et al. 1986). The US environmental justice movement was rising to prominence at the time, and it was not long before the first analyses of a distinctive "environmental justice frame" emerged (Čapek 1993). The research that grew out of this constructivist approach shifted attention from patterns of environmental inequality themselves to the distinctive ways that communities and activist networks translated these patterns into grievances, attributed blame, and advocated remedies (e.g., Taylor 2000). The goal of this research, then, is not to build normative theories of environmental justice, but to understand the strategies, dynamics, and actions of environmental justice movement and policy actors.

The rise of interest in the dynamics of environmental justice as a social movement and interpretive frame provided still another source of resonance with political ecology research, in which various peasant and indigenous movements have long been important topics of study (Watts and Peet 1996; Neumann 2005). In *Varieties of Environmentalism*, to take one of the most prominent examples, Ramachandra Guha and Joan Martínez-Alier (1997) used a political ecology framework to investigate "vocabularies of protest" and ideological underpinnings distinguishing environmental movements in the Global South from those in the Global North. Building on their earlier work (Guha 1990; Martínez-Alier 1991), and leading up to a subsequent volume, *The Environmentalism of the Poor* (Martínez-Alier 2003), this research was among the first to explore continuities between the US environmental justice movement and the Third World environmental mobilizations that had long interested political ecologists. Although political ecology, much of which gravitated towards poststructuralist approaches to discourse analysis, drew from a wider theoretical repertoire than sociological studies of social movement frames, there is significant overlap in that both traditions emphasize struggles over meaning as central dimensions of environmental movements throughout the world (e.g., Escobar 1998).

Environmental justice and political ecology: a deepening encounter

By the turn of the millennium, although mainstream quantitative environmental justice analysis continued on a path separate from political ecology, critical and radical scholars had set the stage for deeper engagements between political ecology and alternative approaches to environmental justice. First, the traditional geographic division that confined environmental justice to the US and political ecology to poorer countries had largely broken down. On the one hand, the concept of environmental justice began to travel from its largely urban US origins to circulate in some of the Third World settings that had long been of primary interest for political ecology (Schroeder et al. 2008; Holifield et al. 2009). In the first decade of the twenty-first century, articles on environmental justice activism in other countries began to proliferate, and scholars published anthologies on environmental justice in Latin America (Carruthers 2008) and South Africa (McDonald 2002), to name two examples. Meanwhile, a growing body of writing in the US attended to struggles over justice in rural land and natural resource management—the traditional purview of political ecology—especially within American Indian reservations (e.g., LaDuke 1999; Mutz et al. 2002). On the other hand, political ecologists began devoting more attention to First World countries, including the urban environments previously neglected in political ecology but long emphasized in environmental justice activism and scholarship (McCarthy 2002; Schroeder 2005; Schroeder et al. 2006).

Second, alternative approaches to environmental justice have continued to build on the theoretical and empirical developments that initially brought the concept explicitly within political ecology's orbit. Normative political theorists have elaborated on additional dimensions of environmental justice, and social scientists have investigated these dimensions in practice. As political ecology has ventured into cities and urbanization processes, the flourishing subfield of urban political ecology has introduced new models for the conceptualization of environmental inequalities (see Chapter 47, this volume). At the same time, environmental justice has traveled into new topical domains, such as climate justice and food justice. One thread that now connects much research in environmental justice and political ecology is an emphasis on science and the politics of knowledge production, which has led to shared interest in such approaches as actor-network theory—and debates over the merits of such approaches (see Chapter 16, this volume). Finally, scholarship on environmental justice movements, interpretive frames, and discourses has continued to proliferate and develop.

Dimensions of environmental justice as normative concept

Normative political theorists of justice have continued to complicate the distributive paradigm by introducing additional dimensions. Nancy Fraser (1997) has been among the most influential theorists of the dimension of *recognition*, conceived as the dismantling or overcoming of institutionalized subordination preventing particular groups from full participation in social life. Meanwhile, Amartya Sen (2009) and Barbara Nussbaum (2003) have elaborated arguments about justice grounded in Sen's long-standing concept of *capabilities*, referring to people's liberty and capacity to achieve states of well-being. Following Young's (1990) recommendation to listen closely to the discourses of activists themselves, David Schlosberg (1999, 2007) has sought to articulate a conception of environmental justice that incorporates distributive, procedural, recognition, and capabilities dimensions. He also joins such theorists as Peter Wenz (1988) and Andrew Dobson (1998) in arguing for the extension of principles of justice to the nonhuman world.

Empirical research on environmental justice and political ecology has begun to show the influence of Schlosberg's work, investigating ways in which these other dimensions are expressed in specific conflicts and conditions. For example, Petra Tschakert's (2009) scholarship on artisanal gold mining in Ghana emphasizes the misrecognition of small-scale, usually unlicensed miners as a form of environmental injustice. Government policy in Ghana casts these miners as trespassers and criminals, who harm themselves and others by using toxic mercury to extract gold. Tschakert argues that this devaluation and ostracism—deployed to justify the exclusion of the miners from both gold-rich land and state programs for health and welfare—constitutes injustice in the form of what Nancy Fraser called "status injury." As a potential corrective, Tschakert implemented a model of collaborative, participatory research aimed at fostering the miners' capabilities. Indeed, the capabilities approach to justice also implicitly underlies community-based participatory research, as described above (see also Schlosberg and Carruthers 2010), and it has been an important influence on political ecology (Forsyth 2008).

Urban political ecology, environmental justice, and scientific knowledge

The subfield of urban political ecology has rapidly become a fertile intellectual terrain for the analysis of environmental injustice. Although urban political ecology takes multiple forms, the dominant thread has been grounded in traditions of Marxist political economy (Keil 2003; Heynen 2014). As Nik Heynen (2014) notes, urban political ecology has in a short time and in a wide variety of geographic settings generated empirical research on processes underlying numerous patterns of environmental inequality, from pollution to tree canopy cover. However, since urban political ecology is considered in another chapter (Chapter 47, this volume), the discussion here will be brief and will focus on two questions, one straightforward and the other more complex.

The first of these questions is whether urban political ecology is an approach to be distinguished from environmental justice, or instead is one of several possible approaches to analyzing environmental justice. Some essays distinguish urban political ecology from environmental justice on the grounds that the former emphasizes processes and the latter emphasizes patterns. Ian Cook and Erik Swyngedouw (2012: 7), for instance, classify environmental justice as a "school of thought" or an "approach" separate from urban political ecology:

> Whereas the EJ [environmental justice] literature is primarily focused on the *patterns* of socio-spatial environmental inequality and the political *procedures* through which they are mediated, the urban political ecology (hereafter UPE) literature is primarily concerned with the political-economic processes involved in the reworking of human–nonhuman assemblages and the *production* of socio-environmental inequalities.

In contrast, Roger Keil (2003) identified environmental justice as one of four "clusters" or "exemplary strongholds" *within* what was then a still-emerging paradigm of urban political ecology. My own position, which I have sought to support through the structure of this chapter, comes closer to that of Keil: that is, that urban political ecology is not an approach distinct *from* environmental justice, but a distinctive approach *to* environmental justice. Although mainstream quantitative environmental justice research has indeed concentrated primarily on analyzing patterns of inequality, the alternative trajectories that have brought environmental justice to its deeper encounter with political ecology have long been concerned with political-economic processes and the production of inequalities.

The second, and more complex, question concerns the relationship between Marxist political economy and actor-network theory, an influential ontology and methodology that emerged initially within science studies. Marxist political economists frequently cast actor-network theory as inattentive to power relations and structural inequalities, while actor-network theorists often criticize Marxists for appealing to such power relations and inequalities as given, pre-assembled explanatory contexts, hidden to the actors but not the analysts (for more on this debate, see Chapter 16, this volume). But others have sought to synthesize elements of the two approaches, and urban political ecology has proven to be a particularly popular testing ground (see, e.g., Swyngedouw 1996; Castree 2002; Robbins 2007; Perkins 2007).

Although I have argued against such a synthesis elsewhere (Holifield 2009), my aim here is not to repeat or revisit that argument, but instead to highlight the significance of this relationship for research on environmental justice and for political ecology more broadly. Specifically, actor-network theory (ANT) presents an approach to analyzing the production, circulation, and application of environmental *knowledge*, which has long been recognized as a central axis of controversy and struggle in environmental conflict. It has become increasingly clear in environmental justice and political ecology research that attending to the production of environmental inequalities requires engaging with knowledge controversies, but it has proven challenging to theorize these two phenomena together.

ANT is by no means the only contemporary approach to the study of science and technology. Other recent US-based environmental justice scholarship focused on the production of scientific knowledge has turned, for example, to Sandra Harding's (1992) influential conception of "strong objectivity" (Allen 2003), Donna Haraway's (1991) notion of the cyborg (Sze 2006), or Nikolas Rose's (2001) analyses of molecular biopolitics (Shostak 2004). Political ecologists studying other parts of the world engage with a similarly wide range of approaches from science and technology studies (see Goldman et al. 2011). Nonetheless, at least some research on environmental justice has begun to experiment with actor-network theory as an approach to tracing the negotiations and translations that resolve controversies over environmental knowledge (e.g., Holifield 2012).

Social movement frames and discourses

Since the turn of the millennium, research on interpretive frames and discourses of environmental justice has continued to develop, further enriching the encounter between environmental justice and political ecology. In the US, new monographs based on ethnographic and participant observation research, such as Julie Sze's (2003) *Noxious New York* and Melissa Checker's (2005) *Polluted Promises*, have provided empirically detailed accounts of the ways local residents and activists in particular places come to interpret grievances in the terms of environmental justice and racism. Much recent research on environmental justice politics builds on the work of Hilda Kurtz (2003), who developed the concept of *scale frames* to account for how actors fighting for or against the siting of a PVC facility in Louisiana deployed the scalar ambiguity inherent to quantitative environmental justice analysis. Scholars have found competing framings of scale to be crucial elements of conflicts over environmental management throughout the world, from unequal access to Brazilian agricultural land (Wolford 2008), to struggles over how to define and manage a freshwater ecosystem in California (Sze et al. 2009), to disagreements over whether a city in the UK Rust Belt is an appropriate site for disassembling and recycling toxic "ghost ships" (Bickerstaff and Agyeman 2009). Although not all of this research explicitly adopts a political ecology framework, it resonates with political ecology nonetheless, by situating the drivers of and constraints on these localized conflicts with respect to political-economic processes that extend far beyond the local.

Other recent research has examined the circulation of the "environmental justice frame" beyond the US grassroots environmental justice movement. Some studies have focused on the ways that environmental justice comes to be framed within US public policy discourse, highlighting discontinuities with the interpretations of activists and advocates (Sandweiss 1998; Holifield 2004). Others have traced the trajectories of environmental justice frames as they have traveled to different parts of the world (e.g., Debbané and Keil 2004; Walker and Bulkeley

2006; Walker 2009b). On the one hand, environmental justice has not caught on as a grassroots discourse in as many different places as one might expect, perhaps due to its origins in the distinctive racialized politics of the US civil rights movement (see, e.g., Reed and George 2011). In some countries, the early adopters of the language have been elites and policymakers, rather than activists (Walker and Bulkeley 2006). On the other hand, as it has traveled, the concept of environmental justice has taken on local inflections that in many cases have consolidated the connection with the traditional concerns of political ecology.

Conclusion

So has environmental justice been a part of political ecology all along? And has political ecology always addressed environmental justice? If we consider environmental justice in its broadest sense—injustice with respect to environmental conditions, both desired and undesired—and if we understand "political ecology" to refer to political struggles and conflicts over these environmental conditions, then the answer to both questions is yes. But as I have sought to show in this chapter, the answers are less straightforward when we consider the more specific meanings that the two terms have taken on in academic debate. Political ecology stories may be stories of justice and injustice, but it is only relatively recently that the academic field of political ecology has engaged explicitly with the concept and language of environmental justice, in its specific guise as activist and policy rhetoric that emerged within the United States. And although much critical and radical environmental justice scholarship undoubtedly qualifies as political ecology research—even if it does not explicitly identify with the approach—the methods and assumptions of the dominant form of environmental justice analysis have kept this mainstream trajectory at a distance from political ecology.

I will close with a few suggestions for continuing to deepen and enrich the shared path of environmental justice and political ecology. First, Gordon Walker's (2009a) call to examine the distinctive *spatialities* of procedural, recognition, and capabilities dimensions of justice has inspired several recent empirical studies (e.g., Urkidi and Walter 2011; Holifield 2012; Gibson-Wood and Wakefield 2013). However, this line of inquiry remains in its infancy, and political ecology provides appropriate conceptual frameworks for carrying it forward. For instance, political ecologists have long engaged with Sen's concept of capabilities (Forsyth 2008), and they are thus well positioned to analyze its spatial dimensions in struggles for environmental justice.

Second, there is much more research to be done to examine the relationship between environmental justice and the complex politics of scientific knowledge, especially with the rise of concerns about climate justice. I support the pluralism that characterizes current engagements between political ecology and science and technology studies, so long as the dialogues and debates among competing perspectives continue. For scholars interested in actor-network theory but skeptical of its approach to the "political" in political ecology, Bruno Latour's (2013) "modes of existence" project may provide fruitful new conceptualizations of the ways that conditions of environmental inequality circulate within and among the worlds of science, politics, law, and even religion (which remains sorely neglected, despite its historical significance to environmental justice activism).

Finally, with respect to the environmental justice frame, an interesting question that has emerged recently is why this language does *not* resonate in all places or in all struggles over environmental inequities. On the one hand, some suggest that the term *environmental justice*, by virtue of its breadth, longer history, and origins within influential activist networks, has influenced a wider range of struggles worldwide than related concepts like climate justice

(Martínez-Alier et al. 2014). However, recent scholarship has shown how racial politics and positionality can trigger local resistance to the language of environmental justice (Little 2012), or how such a framing may fail to take hold within national political cultures less receptive than others to the linking of social justice and environmental concerns (Davies 2006; see also Benford 2005; Pellow and Brulle 2005). Political ecology could thus be at the forefront not simply of analyzing environmental justice as value, process, and discourse, but also of identifying strengths and limitations of environmental justice as mobilizing concepts.

References

Allen, B.L. (2003). *Uneasy alchemy: Citizens and experts in Louisiana's chemical corridor disputes*, Cambridge, MA: MIT Press.

Anderton, D.L., A.B. Anderson, J.M. Oakes, and M.R. Fraser. (1994). "Environmental equity: the demographics of dumping." *Demography*, vol. 31, pp. 229–248.

Been, V. (1993). "Locally undesirable land uses in minority neighborhoods: disproportionate siting or market dynamics." *Yale Law Journal*, vol. 103, pp. 1383–1422.

Benford, R. (2005). "The half-life of the environmental justice frame: innovation, diffusion, and stagnation." In Pellow, D. and R. Brulle (eds.), *Power, justice, and the environment: A critical appraisal of the environmental justice movement*, Cambridge, MA: MIT Press, pp. 37–53.

Bickerstaff, K., and J. Agyeman. (2009). "Assembling justice spaces: the scalar politics of environmental justice in north-east England." *Antipode*, vol. 41, pp. 781–806.

Boone, C.G., and A. Modarres. (1999). "Creating a toxic neighborhood in Los Angeles County: a historical examination of environmental inequity." *Urban Affairs Review*, vol. 35, pp. 163–187.

Bowen, W.M., M.J. Salling, K.E. Haynes, and E.J. Cyran. (1995). "Toward environmental justice: spatial equity in Ohio and Cleveland." *Annals of the Association of American Geographers*, vol. 85, pp. 641–663.

Bullard, R.D. (1983) "Solid waste sites and the Black Houston community." *Sociological Inquiry*, vol. 53, pp. 273–288.

Čapek, S.M. (1993). "The 'environmental justice' frame: a conceptual discussion and an application." *Social Problems*, vol. 40, pp. 5–24.

Carruthers, D. (ed.). (2008). *Environmental justice in Latin America: Problems, promise, and practice*, Cambridge, MA: MIT Press.

Castree, N. (2002). "False antitheses? Marxism, nature and actor-networks." *Antipode*, vol. 34, pp. 111–146.

Chakraborty, J., J.A. Maantay, and J.D. Brender. (2011). "Disproportionate proximity to environmental health hazards: methods, models, and measurement." *American Journal of Public Health*, vol. 101, pp. S27–S36.

Checker, M. (2005). *Polluted promises: Environmental racism and the search for justice in a southern town*, New York: New York University Press.

Clark, W.C., R.W. Kates, J.F. Richards, J.T. Mathews, W.B. Meyer, B.L. Turner II, S.T.A. Pickett, C. Boone, and M.L. Cadenasso. (2007). "Relationships of environmental justice to ecological theory." *Bulletin of the Ecological Society of America*, vol. 88, pp. 166–170.

Colten, C.E. (2002). "Basin Street blues: drainage and environmental equity in New Orleans, 1890–1930." *Journal of Historical Geography*, vol. 28, pp. 237–257.

Cook, I.R., and E. Swyngedouw. (2012). "Cities, social cohesion and the environment: towards a future research agenda." *Urban Studies*, vol. 49, pp. 1959–1979.

Cutter, S.L., D. Holm, and L. Clark. (1996). "The role of geographic scale in monitoring environmental justice." *Risk Analysis*, vol. 16, pp. 517–526.

Davies, A.R. (2006). "Environmental justice as subtext or omission: examining discourses of anti-incineration campaigning in Ireland." *Geoforum*, vol. 37, pp. 708–724.

Debbané, A.M., and R. Keil. (2004). "Multiple disconnections: environmental justice and urban water in Canada and South Africa." *Space and Polity*, vol. 8, pp. 209–225.

Dobson, A. (1998). *Justice and the environment: Conceptions of environmental sustainability and dimensions of social justice*, Oxford: Oxford University Press.

Escobar, A. (1998). "Whose knowledge, whose nature? biodiversity, conservation, and the political ecology of social movements." *Journal of Political Ecology*, vol. 5, pp. 53–82.

Forsyth, T. (2008). "Political ecology and the epistemology of social justice." *Geoforum*, vol. 39, pp. 756–764.

Fraser, N. (1997). *Justice interruptus: Critical reflections on the "postsocialist" condition*, New York: Routledge.

Gandy, M. (2003). *Concrete and clay: Reworking nature in New York City*, Cambridge, MA: MIT Press.

Gibson-Wood, H. and S. Wakefield. (2013). "'Participation,' white privilege and environmental justice: understanding environmentalism among Hispanics in Toronto." *Antipode*, vol. 45, pp. 641–662.

Gleeson, B., and N. Low. (1998). *Justice, society and nature: An exploration of political ecology*, New York: Routledge.

Goffman, E. (1974). *Frame analysis: An essay on the organization of experience*, Cambridge, MA: Harvard University Press.

Goldman, M.J., P. Nadasdy, and M.D. Turner. (eds.). (2011). *Knowing nature: Conversations at the intersection of political ecology and science studies*, Chicago: University of Chicago Press.

Guha, R. (1990). *The unquiet woods: Ecological change and peasant resistance in the Himalaya*, Berkeley: University of California Press.

Guha, R. and J. Martínez-Alier. (1997). *Varieties of environmentalism: Essays North and South*, New York: Earthscan.

Haraway, D. (1991). *Simians, cyborgs, and women: The reinvention of nature*, New York: Routledge.

Harding, S. (1992). "After the neutrality ideal: science, politics, and 'strong objectivity.'" *Social Research*, vol. 59, pp. 567–587.

Harvey, D. (1973). *Social justice and the city*, Baltimore: Johns Hopkins University Press.

Harvey, D. (1992). "Social justice, postmodernism and the city." *International Journal of Urban and Regional Research*, vol. 16, pp. 588–601.

Harvey, D. (1996). *Justice, nature and the geography of difference*, Oxford: Wiley-Blackwell.

Heynen, N. (2014). "Urban political ecology I: The urban century." *Progress in Human Geography*, vol. 38, pp. 598–604.

Holifield, R. (2004). "Neoliberalism and environmental justice in the United States environmental protection agency: translating policy into managerial practice in hazardous waste remediation." *Geoforum*, vol. 35, pp. 285–297.

Holifield, R. (2009). "Actor-network theory as a critical approach to environmental justice: a case against synthesis with urban political ecology." *Antipode*, vol. 41, pp. 637–658.

Holifield, R. (2012). "Environmental justice as recognition and participation in risk assessment: negotiating and translating health risk at a Superfund site in Indian Country." *Annals of the Association of American Geographers*, vol. 102, pp. 591–613.

Holifield, R., M. Porter, and G. Walker (2009). "Introduction: spaces of environmental justice: frameworks for critical engagement." *Antipode*, vol. 41, pp. 591–612.

Hurley, A. (1995). *Environmental inequalities: Class, race, and industrial pollution in Gary, Indiana, 1945–1980*, Chapel Hill: University of North Carolina Press.

Keil, R. (2003). "Urban political ecology 1." *Urban Geography*, vol. 24, pp. 723–738.

Kurtz, H.E. (2003). "Scale frames and counter-scale frames: constructing the problem of environmental injustice." *Political Geography*, vol. 22, pp. 887–916.

LaDuke, W. (1999). *All our relations: Native struggles for land and life*, Cambridge, MA: South End Press.

Lake, R.W. (1996). "Volunteers, NIMBYs, and environmental justice: dilemmas of democratic practice." *Antipode*, vol. 28, pp. 160–174.

Lambert, T. and C. Boerner. (1997). "Environmental inequity: economic causes, economic solutions." *Yale Journal on Regulation*, vol. 14, pp. 195–228.

Latour, B. (2013). *An inquiry into modes of existence*. Porter, C. (trans.), Cambridge, MA: Harvard University Press.

Little, P. (2012). "Environmental justice discomfort and disconnect in IBM's tainted birthplace: a micropolitical ecology perspective." *Capitalism Nature Socialism*, vol. 23, pp. 92–109.

Liu, F. (1997). "Dynamics and causation of environmental equity, locally unwanted land uses, and neighborhood changes." *Environmental Management*, vol. 21, pp. 643–656.

Martínez-Alier, J. (1991). "Ecology and the poor: a neglected dimension of Latin American history." *Journal of Latin American Studies*, vol. 23, pp. 621–639.

Martínez-Alier, J. (2003). *The environmentalism of the poor: A study of ecological conflicts and valuation*, Cheltenham: Edward Elgar Publishing.

Martínez-Alier J., I. Anguelovski, P. Bond, D. Del Bene, F. Demaria, J.-F. Gerber, L. Greyl, W. Haas, H. Healy, V. Marín-Burgos, G. Ojo, M. Porto, L. Rijnhout, B. Rodríguez-Labajos, J. Spangenberg, L.

Temper, R. Warlenius, and I. Yánez. (2014). "Between activism and science: grassroots concepts for sustainability coined by Environmental Justice Organizations." *Journal of Political Ecology*, vol. 21, pp. 19–60.

McCarthy, J. (2002). "First World political ecology: lessons from the Wise Use movement." *Environment and Planning A*, vol. 34, pp. 1281–1302.

McDonald, D.A. (ed.). (2002). *Environmental justice in South Africa*, Athens, OH: Ohio University Press.

McMaster, R.B., H. Leitner, and E. Sheppard. (1997). "GIS-based environmental equity and risk assessment: methodological problems and prospects." *Cartography and Geographic Information Systems*, vol. 24, pp. 172–189.

Merrifield, A. and E. Swyngedouw (eds.). (1996). *The urbanization of injustice*. London: Lawrence and Wishart.

Mutz, K., G. Bryner, and D. Kenney (eds.). (2002). *Justice and natural resources: Concepts, strategies, and applications*, Washington, DC: Island Press.

Neumann, R.P. (2005). *Making political ecology*, Oxford: Oxford University Press.

Nussbaum, M. (2003). "Capabilities as fundamental entitlements: Sen and social justice." *Feminist Economics*, vol. 9, pp. 33–59.

O'Connor, J. (1988). "Capitalism, nature, socialism: a theoretical introduction." *Capitalism Nature Socialism*, vol. 1, pp. 11–38.

Pellow, D., and R.J. Brulle. (2005). "Power, justice, and the environment: toward critical environmental justice studies." In Pellow, D. and R.J. Brulle (eds.), *Power, justice and the environment: A critical appraisal of the environmental justice movement*, Cambridge, MA: MIT Press, pp. 1–19.

Perkins, H.A. (2007). "Ecologies of actor-networks and (non) social labor within the urban political economies of nature." *Geoforum*, vol. 38, pp. 1152–1162.

Pulido, L. (1996). "A critical review of the methodology of environmental racism research." *Antipode*, vol. 28, pp. 142–159.

Pulido, L. (2000). "Rethinking environmental racism: white privilege and urban development in Southern California." *Annals of the Association of American Geographers*, vol. 90, pp. 12–40.

Pulido, L., S. Sidawi, and R.O. Vos. (1996). "An archaeology of environmental racism in Los Angeles." *Urban Geography*, vol. 17, pp. 419–439.

Rawls, J. (1971). *A theory of justice*, Cambridge, MA: Belknap Press.

Reed, M.G., and C. George. (2011). "Where in the world is environmental justice?" *Progress in Human Geography*, vol. 35, pp. 835–842.

Robbins, P. (2007). *Lawn people: How grasses, weeds and chemicals make us who we are*, Philadelphia: Temple University Press.

Robbins, P. (2012). *Political ecology: A critical introduction*. 2nd edition. Oxford and Malden, MA: Wiley-Blackwell.

Rose, N. (2001). "The politics of life itself." *Theory, Culture & Society*, vol. 18, pp. 1–30.

Sandweiss, S. (1998). "The social construction of environmental justice." In Camacho, D. (ed.), *Environmental injustices, political struggles: Race, class, and the environment*, Durham, NC: Duke University Press, pp. 31–58.

Schlosberg, D. (1999). *Environmental justice and the new pluralism: The challenge of difference for environmentalism*, Oxford: Oxford University Press.

Schlosberg, D. (2007). *Defining environmental justice: Theories, movements, and nature*, Oxford: Oxford University Press.

Schlosberg, D., and D. Carruthers. (2010). "Indigenous struggles, environmental justice, and community capabilities." *Global Environmental Politics*, vol. 10, pp. 12–35.

Schnaiberg, A. (1980). *Environment: From surplus to scarcity*, Oxford: Oxford University Press.

Schroeder, R.A. (2000). "Beyond distributive justice: resource extraction and environmental justice in the tropics." In Zerner, C. (ed.), *People, plants and justice: The politics of nature conservation*, New York: Columbia University Press, pp. 52–66.

Schroeder, R.A. (2005). "Debating the place of political ecology in the First World." *Environment and Planning A*, vol. 27, pp. 1045–1048.

Schroeder, R.A., K. St. Martin, and K.E. Albert. (2006). "Political ecology in North America: discovering the Third World within?" *Geoforum*, vol. 37, pp. 163–168.

Schroeder, R.A., K. St. Martin, B. Wilson, and D. Sen. (2008). "Third world environmental justice." *Society and Natural Resources*, vol. 21, pp. 547–555.

Sen, A.K. (2009). *The idea of justice*, Cambridge, MA: Harvard University Press.

Shepard, P.M., M. Northridge, S. Prakash, and G. Stover. (2002). "Preface: advancing environmental justice through community-based participatory research." *Environmental Health Perspectives*, vol. 110 (Suppl 2), pp. 139–140.
Shostak, S. (2004). "Environmental justice and genomics: acting on the futures of environmental health." *Science as Culture*, vol. 13, pp. 539–562.
Smith, D.M. (1994). *Geography and social justice*, Oxford: Blackwell.
Smith, N. (1984). *Uneven development: Nature, capital and the production of space*, Oxford: Blackwell.
Snow, D.A., E.B. Rochford, S.K. Worden, and R.D. Benford. (1986). "Frame alignment processes, micromobilization, and movement participation." *American Sociological Review*, vol. 51, pp. 464–481.
Swyngedouw, E. (1996). "The city as a hybrid: on nature, society and cyborg urbanization." *Capitalism Nature Socialism*, vol. 7, pp. 65–80.
Swyngedouw, E. and N.C. Heynen. (2003) "Urban political ecology, justice and the politics of scale." *Antipode*, vol. 35, pp. 898–918.
Sze, J. (2003). *Noxious New York: The racial politics of urban health and environmental justice*, Cambridge, MA: MIT Press.
Sze, J. (2006). "Boundaries and border wars: DES, technology, and environmental justice." *American Quarterly*, vol. 58, pp. 791–814.
Sze, J., J. London, F. Shilling, G. Gambirazzio, T. Filan, and M. Cadenasso. (2009). "Defining and contesting environmental justice: Socio-natures and the politics of scale in the Delta." *Antipode*, vol. 41, pp. 807–843.
Taylor, D.E. (2000). "The rise of the environmental justice paradigm: injustice framing and the social construction of environmental discourses." *American Behavioral Scientist*, vol. 43, pp. 508–580.
Tschakert, P. (2009). "Digging deep for justice: A radical re-imagination of the artisanal gold mining sector in Ghana." *Antipode*, vol. 41, pp. 706–740.
United Church of Christ, Commission for Racial Justice. (1987). *Toxic wastes and race in the United States: A national report on the racial and socio-economic characteristics of communities with hazardous waste sites*, New York: Public Data Access.
United States General Accounting Office. (1983). Siting of hazardous waste landfills and their correlation with racial and economic status of surrounding communities. Washington, DC: US General Accounting Office. RCED-83-168.
Urkidi, L. and M. Walter. (2011). "Dimensions of environmental justice in anti-gold mining movements in Latin America." *Geoforum*, vol. 42, pp. 683–695.
Viscusi, W.K. and J.T. Hamilton. (1999). "Are risk regulators rational? evidence from hazardous waste cleanup decisions." *The American Economic Review*, vol. 89, pp. 1010–1027.
Walker, G. (2009a). "Beyond distribution and proximity: exploring the multiple spatialities of environmental justice." *Antipode*, vol. 41, pp. 614–636.
Walker, G. (2009b). "Globalizing environmental justice: the geography and politics of frame contextualization and evolution." *Global Social Policy*, vol. 9, pp. 355–382.
Walker, G. and H. Bulkeley. (2006). "Geographies of environmental justice." *Geoforum*, vol. 37, pp. 655–659.
Watts, M. and R. Peet. (1996). "Liberating political ecology." In Peet, R. and M. Watts (eds.) *Liberation ecologies: Environment, development, social movements*, New York: Routledge, pp. 3–47.
Wenz, P.S. (1988). *Environmental justice*, Albany, NY: SUNY Press.
Wolford, W. (2008). "Environmental justice and the construction of scale in Brazilian agriculture." *Society and Natural Resources*, vol. 21, pp. 641–655.
Young, I.M. (1990). *Justice and the politics of difference*, Princeton: Princeton University Press.
Zerner, C. (ed.). (2000). *People, plants, and justice: The politics of nature conservation*, New York: Columbia University Press.

46
ENVIRONMENTAL CONFLICT

Philippe Le Billon

Introduction

A notion of conflict is arguably at the core of political ecology. Originally coined in the 1970s by anthropologists and cultural ecologists, the term political ecology first appeared within a leading geography journal through Bassett's (1988) demonstration that peasant-herder conflicts in northern Ivory Coast resulted from the productivist policies of the state rather than resource scarcity and ancient hatred. Since then, political ecologists have continued to work on environment-related conflicts, broadening the scope of their enquiries through the range of actors, contexts, motives and "objects" as well as types of conflicts considered. Deploying a range of methods and conceptual frameworks, political ecologists are united by their commitment to offering critical perspectives on often taken-for-granted processes. Using multi-scalar, historically informed and culturally sensitive entitlement analyses, political ecology aims to complicate – if not overturn – simplistic narratives of environmental conflicts driven by "scarcity" or "greed". By 2013, about two-thirds of political ecology studies had used the concept of conflict in their analyses, the second most frequent conceptual term after "power" according to a Google Scholar search.

That conflict is at the core of the discipline is first reinforced by a number of definitions. Martinez-Alier (2003: 71), among the most notable examples, classically defines political ecology as "the study of ecological distribution conflicts". More broadly, political ecology is about politics, and about recognizing the political character of environmental and resource issues. Conventional definitions of politics include collective decision-making processes contesting a pre-existing status quo or consensus; as such, politics can be broadly understood as defining and resolving such contestations and disagreements, which can take the form of *conflicts* when they are strong, entrenched and perceived to be irreconcilable. Political ecology is thus in large part about the "conflictual" character of political processes around ecological issues.

Second, political ecologists have also been keen to explore the *politicization* of the environment via conflicts, rather than naturalizing conflicts through environmental analysis. This perspective represents a crucial departure from neo-Malthusian concepts of "environmental conflict" supporting depoliticized concepts of environmental scarcity (or abundance, see below) "naturally" triggering conflicts – generally of the "violent ethnic" kind. Political ecologists seek to understand conflicts *around* or *through* the environment, and not simplistically explain conflicts as resulting *from* the environment. This means notably that political ecologists are at least as interested in studying the political factors and conflict dimensions of what comes to shape environmental conditions as they are in studying the conflict dimensions of the effects of these environmental conditions. Many political ecologists examine both sides, and their dialectical relations, giving particular attention to the ways in which such environmental conditions become politicized, and how resources and the environment come to participate in the reification of conflicts. A classic example here is that of farmer-herder conflicts in the Sahel, a "repeated game" in which "conflicts that are waged over the long term with the conflict's history being invoked and reworked to make moral claims in the present" (Turner 2004: 878). It is in part through such conflicts that identities and (exclusionary) social ties become consolidated (Rikoon 2006).

Third, many political ecologists take as given a stratified notion of society structured by uneven power relations. From this starting point, many political ecologists understand that conflicts are either inevitable, or at least ought to occur to bring about environmental justice (see Chapter 45, this volume). Such stratification may not systematically result in conflicts, understood in the form of open struggles. Durable stratification may reflect the absence of "effective" conflicts, or at least their failure to deliver more egalitarian outcomes. It is thus important to recognize different expressions of conflicts, and forms of struggle – as demonstrated by Scott's (1985) "weapons of the weak". As such, by understanding conflicts in a broad sense – and not waiting to see violent manifestations to recognize conflicts – political ecologists express a sensitivity that better captures the unfair or tense character of social relations, and associated processes of legitimation and resistance.

Fourth, and following from the previous point, political ecologists do not systematically treat conflicts as nefarious processes with only negative outcomes but acknowledge or even promote their emancipatory role in challenging structural and cultural forms of violence (Galtung 1990), on both people and the non-human. Seeing conflict in a positive light sharply contrasts with mainstream representations depicting conflicts as simply negative, and using for example terms such as "riots" instead of "demonstrations", in an attempt to criminalize aggrieved victims of inequalities as "troublemakers" and delegitimize their struggles (Zalik 2011). Even in cases where inequalities are apparently legitimated by a dominant social order, political ecologists seek to denounce such structural forms of violence. As discussed below, a major concern of some political ecologists is thus to avoid a depoliticization of environmental issues.

Finally, in his impressive introduction to the field, Robbins (2004: 14) identifies "environmental conflict" as one of the five central theses of political ecology, through which he sees an effort to demonstrate that the actors and causes of conflicts over environmental access "are part of larger gendered, classed and raced struggles and vice versa" – the other four being: degradation and marginalization; conservation and control; environmental subjects and identity, and political objects and actors. These conflicts not only take place over the environment, but within the context of economic, ecological and cultural differences (Escobar 2006: 8). As such it is often through the recognition and respect of differences, but also through the reduction of inequalities, that political ecologists see a resolution of conflicts.

Defining "environmental conflicts": multiple views

Environmental conflict can be broadly understood as a social conflict relating to the environment. This relation can take several forms and directions. It can be a conflict *over* the environment, most notably in terms of access to and control over environmental resources (Ribot and Peluso 2003). These so-called "resource conflicts" are defined by Turner (2004: 863) as consisting of "social conflict (violent or nonviolent) associated with both struggles to gain access to natural resources and struggles resulting from the use of natural resources". From a neo-Malthusian perspective, environmental conflicts also consist of conflicts resulting *from* environmental processes – especially resource scarcity supposedly putting strains on social relations – even if the conflict per se is not over those "scarce" environmental resources (Dalby 2002). A variant to this scarcity-driven argument is the so-called "resource curse", according to which the exploitation of abundant resources in undiversified economic contexts results in high levels of large revenues and resource dependence that would increase vulnerability to conflicts by undermining the quality of institutions, exposing societies to economic shocks, and exacerbating tensions over the distribution of resource rents and more generally the costs and benefits of dominant resource sectors (Le Billon 2012).

Closer to the core interests of political ecologists, Robbins (2004: 173) identifies two major facets of the environmental conflict thesis, according to which "increasing scarcities produced through resource enclosure or appropriation by state authorities, private firms, or social elites accelerate conflict between groups (gender, class, or ethnicity)". The first one consists in the "*politicization*" of environmental problems "when local groups … secure control of collective resources at the expense of others by leveraging management interventions by development authorities, state agents, or private firms". The second consists in the "*ecologization*" of pre-existing conflicts as a result of "changes in conservation or resource development policy". This argumentation, according to Robbins (2004), is based on three lessons drawn by political ecologists from feminist theory pointing at the effect of labor and power divisions distributing unevenly "access and responsibility for natural goods"; from property research understanding "property systems as complex bundles of rights that are politically partial and historically contingent"; and from critical development studies showing that development activities are "rooted in specific assumptions about the class, race, and gender of participants in the development process, often resulting in poorly formed policy and uneven results". In this respect, Turner (2004: 866) points out that,

> moral and material motivations are often strongly intertwined in "resource conflicts" … It is only through a full and critical engagement with both the materiality which underlies all social life and the moral claims that implicate natural resource use that the etiology of resource-related conflict can be better understood.

If political ecologists acknowledge the significance of conflict in the politics of socio-environmental relations, conflicts matter differently among them. I highlight here three main motivations: the pursuit of justice, the politicization of socio-environmental interactions and the fight against the "naturalization" of environmental conflicts. Interestingly, relatively few political ecologists are actually motivated by the resolution of environmental conflicts, possibly because they see conflicts as emancipatory for marginalized people. As a prominent political ecologist mentioned in this regard to the author, "we are here to document conflicts, not to solve them". Seeking to bring about compromise, trade-offs and compensations can indeed be understood as being complicit in processes that are often perceived to be at the advantage of the most powerful groups (either between the opposing parties, or within the aggrieved group). Monetary compensations for environmental damage or loss of access to resources, for example, extend a colonial logic of commodification and monetization. Not only do such "compensations" assert commensurability between money and a vast range of socio-environmental relations, but they also often result in further distributional conflicts among (un)compensated communities and households.

What justice for slugs?

Closely tied to discussions of ethics, the search for justice is a prominent motivation of many political ecologists. Environmental conflicts, from that perspective, are struggles for environmental justice. In turn, environmental justice encompasses two aspects: the justice of ecological distribution among people, and the justice of relationships between humans and the non-human world (see Chapter 45, this volume). As noted by Low and Gleeson (1998: 1), who put justice at the core of their conceptualization of political ecology, it is in the justice found through and towards the environment that we "define who and what we are and who and what 'the other' is". By far the most attention has been paid to environment-related issues of *social justice* as redress against unfair ecological distribution processes; most notably whereby less powerful groups come to bear ecological costs as a result of racial prejudices. Ecological distribution conflicts mostly consist of resistance against the imposition of "externalities" (i.e. the "cost-shifting" of environmental exploitation). As discussed by Martinez-Alier (2001: 161), environmental justice issues in the Third World have been mostly about the "defense of common property resources against the state or the market"; whereas in the United States it was mostly a struggle against the "disproportionate allocation of toxic waste to Latino or African-American communities". The concept of environmental racism, based on discriminatory practices undermining human dignity according to racial or ethnic criteria, has thus been central to some environmental justice movements, with environmental conflicts being added and interpreted through broader conflicts articulated around the politics of race and rights.

Issues of environmental justice towards the non-human are also gaining greater attention (Chapter 9, this volume). Among early studies figure Ted Benton's reflections of the extension of social justice to the non-humans and the parallels between political ecology and animal rights movements on the "moral significance of non-human beings" (Benton 1993: 23). Here the politics of recognition are key, whether around the recognition of *non-human rights* as part of a broad fight against *environmental speciesism* – a set of values privileging human entitlement on the environment over that of other species. The non-human can include "close others", such as furry little monkeys captured from tropical forests to be traded as pets, or returned to the wild after going through processes of (de)commodification and alienation from humans (Collard 2013). As well as "distant others", such as slugs, with Ginn (2013) asking how many of the 20 billion slugs in British gardens are slaughtered every year by humorous garden-lovers following a still discriminating more-than-human ethics of gardening.

Conflicts matter to highlight (in)justice, whereby conflict becomes the symptom and revelatory crisis of underlying unfairness. Seeing environmental conflicts through the lens of ethics and justice means questioning selective recognition of rights and pursuing a politics of difference that remains inclusive. It is also about due process and the possibility of a fairer future. Environmental conflict, in this view, is often a step in the right direction, the conflict opening new avenues for justice struggles and the hope of more universal fairness. In this respect Turner (2004: 886) cautions that,

> [Political ecologists] are well placed to understand the fuller politics of not only resource-related conflicts but of their own active and passive roles as researchers in the international debates about conservation and development. It is important that [they] develop the language and analytical tools to present the fuller complexity of resource-related conflict … to counter the overly simplified depictions that greatly reduce the social, political and moral lives of rural peoples in the pursuit of policy prescription.

In other words, political ecologists need to do justice to the people and issues they are engaging with, working in solidarity without falling into the trap of seductive but counterproductive policy.

Socio-environmental relations, politics and conflicts

For many political ecologists, the prevalence of conflicts around environmental issues demonstrates that all human–environment interactions are unavoidably political (see above). Not only are "ecological issues … politicized through local and regional conflict, [but] political questions are increasingly cast in ecological terms" (Robbins 2004: 173). In his account of conflicts around forest in northern New Mexico, Kosek (2006: x) demonstrates how "forest management, protection, exploitation, degradation, and restoration are inseparably tied to the social conflicts and cultural politics of class, race, and nation … Polluted soils are related to degraded souls; national forests to be protected from foreign bodies; board-foot quotas become the site of intense class politics".

Conflicts, from this perspective, are understood as a prime form and expression of politics. Building on the idea of "post-politics" (Zizek 1999), several political ecologists have pushed this perspective further, arguing that politics without conflicts would not be politics. This is not to say that human–environment interactions have been "apolitical", but rather that they have actively become "depoliticized". Such depoliticization rest on two main processes characterizing changes in modes of government, both broadly shaped by neoliberalism and a general shift from government to "governance" (Swyngedouw 2007). First is the *managerial* approach to particular demands – for example mining project or pipeline construction – through a combination of expert knowledge and public consultation (though with strict inclusion criteria and limited participation in actual decision-making), rather than through emancipatory forms of conflict that would offer possibilities of a "metaphorical universalization of particular demands" and result in systematic changes, including in the ways politics work (Swyngedouw 2007: 24). Second is the *populism* of environmental views – for example on sustainability and climate change – that constitute an exclusionary form of consensus that avoids critical debates by characterizing alternative viewpoints as "radicalism". Conflicts, from this perspective, are intrinsically constitutive of politics, and thus politics without conflict (i.e. politics through consensus) is "post-political".

The end of adversarial politics, from this perspective, would thus represent the end of politics in its possibilities of radical outcomes and the pursuit of utopias (for a critique, see McCarthy 2013). A perspective rejoins the critique of neoliberalism as "TINA" (There Is No Alternative) – an ideology seeking to achieve hegemony through the denial of possibilities. Examining questions of urban environmental justice, Swyngedouw (2009) argues that the consolidation of an urban post-political condition runs "parallel to the formation of a postdemocratic arrangement that has replaced debate, disagreement and dissensions with a series of technologies of governing that fuse around consensus, agreement, accountancy metrics and technocratic environmental management". In other words, these less political and democratic forms of decision-making have displaced conflicts. This "age of 'post-politics'" is, for Zizek (2005: 115), a time "when politics proper is progressively replaced by expert social administration, [and] the sole remaining legitimate sources of conflict are cultural (religious) or natural (ethnic) tensions". Such "postpolitical consensual police order", as Swyngedouw (2009: 6045) defines it, not only depoliticizes the environment and threatens democracy, but by doing so "must, of necessity, lead to an ultra-politics of violent disavowal, radical closure and, ultimately, to the tyrannies of violence and of foreclosure of any real spaces of engagement".

Post-political spaces can thus be characterized as the "house of reasonable politics" (Blaser 2013a), within which only "minor" differences amenable to compromises are allowed, with the threat of expulsion should differences become "unreasonable". Outside of the house, reigns spaces of criminalization and forceful policing – another form of depoliticization whereby adversaries become delegitimized "rebels", "bandits" and "criminals". Conflicts, in this view, become both exacerbated (e.g. with an escalation into war-like rhetoric and use of force) and restricted (e.g. in terms of opportunities for adversarial debates).

Not all political ecologists share this view, some suggesting that adversarial politics in fact lacks radicalism and hollows a "progressive middle-way". Examining conflicts over rural landscapes in the American West, Sheridan (2007: 121) not only presents the "ideological clashes and political manoeuvring among interest groups who claim access to those lands", but also the "struggle to move beyond polemics and dualities and mobilize, in the words of [a local group, the Quivira Coalition] a 'radical center' committed to 'foster ecological, economic, and social health on western landscapes'". So while some political ecologists have denounced the dangers of depoliticizations and warned of the dangerous backlash that would see re-emerging conflicts escalate into the "tyrannies of violence", others examine the search for consensual forms of politics upon which solutions can be found. In both cases, however, political ecologists have maintained that there is no such thing as "apolitical" socio-environmental relations, but that conflicts come to play distinct roles within those politics.

Denaturalizing conflicts

In addition to seeking to expose injustices and demonstrate the political character of socio-environmental relations, political ecologists have also sought to challenge the ontological status of and deterministic arguments regarding environmental conflicts, and to insist on the social rather than "natural" character of conflicts. Political ecology rejects the simplistic association, widespread in the popular literature, that conflicts are most frequently associated with absolute resource scarcity. In other words, it refutes the notion that the likelihood of conflict increases as resources become scarcer (whether through depletion, increased degradation, more uneven capture or allocation, or rising demand). According to the "conflict-resource scarcity" argument, widening the scope of the (international) security agenda to include environmental breakdown and livelihood resource access could help address widespread, chronic, low-intensity and intra-state conflicts, and provide a basis for more peaceful relations (Conca and Dabelko 2002).

Scholarly studies of so-called "green-wars" have generally distanced themselves from a simple and direct causal relation model between resource scarcity and conflicts. Rather, they have identified indirect linkages with increased poverty, social segmentation, migrations and institutional disruptions (Baechler and Spillman 1996; Homer-Dixon 1999). Much of this work has received potent critiques for its methodological approach, with Gleditsch (1998) pointing notably at definitional and case selection issues, as well as reverse causality or speculation on future outcomes used as evidence. Critiques coming from political ecology have stressed the neo-Malthusian assumptions, reductionist and essentializing character of these studies (Hartmann 2001), as well as the naturalizing of an environment–insecurity nexus in the South exonerating (Northern-led) modernity and development (Dalby 2002). As such many studies echo Harvey's (1974: 256) warning about the "profound political implications" of supposedly ethically neutral scientific discussions of the population–resources relationship, especially a projection of neo-Malthusian views that invited "repression at home and neo-colonial policies abroad". Ironically, these essentialist views also mean that false expectations become the foundation of wrongly headed programs, the failures and unintended consequences of which result in frustrations, grievances and (further) conflicts (Robbins 2012).

While "scarcity-induced" conflict arguments have received the most attention, the new paradigm of the resource curse has also come under some attention. Arguing that abundance rather than scarcity breeds conflicts (de Soysa 2002), the resource curse paradigm often ends up pathologizing resource producing regions (as being under the supposedly inescapable negative influence of resource sectors), the social conduct in relation to resource control (people being "naturally" driven to fight over resources rather than find cooperative solutions), and the conduct of belligerents (resources shaping their motivations and behaviours). Among the consequences of such pathologization are political de-legitimization of protest and popular (armed) resistance (Zalik 2011); the criminalization of small-scale mineral exploitation by local communities and regional migrants, which undermines livelihoods and coping mechanisms (Le Billon 2008); and the prioritization of a certain types of economic activity (such as large-scale mining or logging) over local livelihoods, as well as environmental and cultural practices. For Kuntala Lahiri-Dutt (2006: 15), resource wars theories, especially the resource curse argument, (re)produces:

> a picture of complete lack of control and disorder in the Third World, whose inhabitants – by some irrational logic of nature – have found themselves endowed with resources that they cannot or do not know how to deal with in an orderly manner. They envisage a paranoid fear about the unruly Third World, a landscape of apprehension, risk and insecurity where conflicts could only be resolved for one and all if either state-owned or multinational corporations take over the control and ownership of mineral resources, and manage them in a systematic manner – in the process putting their profits first and taking over the control of what should rightfully belong to the communities.

Such a picture is deeply anchored into neo-colonial mindsets, while being instrumental in processes of "accumulation by dispossession" (Harvey 2003). Yet it is increasingly challenged by affected communities, which in part explain the prominence and frequency of environmental conflict – as seen for example in the case of resistance to large-scale mining in the Andean region (Bebbington et al. 2009).

Studying environmental conflicts

Rather than following a deductive approach based on linear models linking environmental scarcity to social effects such as forced migration and social segmentation or associating resource wealth with institutional breakdown and greedy rebellions, political ecology opens up research to a wider array of historically and geographically contingent actors and processes – something that Watts (2004) terms the "resource complex". Such opening up not only broadens the number of "variables" while avoiding the pitfall of reductionist "hypotheses"; it also acknowledges the hybrid "socio-natural" character of resources themselves, the importance of situated perspectives, and the historicity and contingency of conflicts.

Generally following an inductive and multi-scalar approach, political ecology understand conflicts and more specifically the various forms of violence associated with them "as a site specific phenomenon rooted in local histories and social relations yet connected to larger processes of material transformation and power relations" (Peluso and Watts 2001 : 5). Through a focus on uneven power relations around the environment and the ecological dimensions of resource-based political economies, political ecologists have emphasized the "many violent ecologies of global inequalities" (Robbins 2012: 1), with violence being broadly understood through its physical, structural and symbolic forms – to use the typology of Galtung (1990). This involves studying variations in property rights and documenting "movements of resistance to resource capitalism and the legitimacy of the state in matter of resource access and control" (Allen 2012: 158). It also involves studying the different processes and forms of exclusion in access and control over resources at the micro-scale, including "intimate exclusion" at the community and family level (Hall et al. 2012). In this regard, ethnographic approaches within political ecology allow identifying divides along gender lines (e.g. the vulnerability of matrilineal inheritance to resource capitalism), between generations (e.g. the selling-out and squandering of birth rights entitlements to land), and modes of production (e.g. advocates of large-scale exploitation versus traditionalist resource users) (Allen 2012).

Anthropologist and political ecologist Arturo Escobar (2006) has rightly pointed to the importance of accounting for cultural differences in explaining environmental conflicts, an attention that is frequently absent from environmentally deterministic mainstream accounts. Conflicts relating to the environment often start with distinctive ontologies about the environment and what come to constitute "resources". For Escobar (2006: 9):

> many communities in the world signify their natural environment, and then use it, in ways that markedly contrast with the more commonly accepted way of seeing nature as a resource external to humans and which humans can appropriate in any way they see fit.

Such "worlding" extends to the register of expressions involved in conflicts. Persuasively arguing for a blending of political ecology and ecological economics that acknowledges values incommensurability, Juan Martinez-Alier (2003: viii) has emphasized in this regard that "ecological conflicts are fought out in many languages".

Beyond questions of how certain cultures see, value and fight over "nature" and "resources" differently, political ecologists also consider how transformations bring about "new worlds" and, to use a crude binary divide, how transformed natures affect cultures through new socio-natural worlds. Environmental and resource conflicts are thus inescapably cultural conflicts through worldviews and representations but also through their material implications. Blaser (2013b) makes several suggestions to address these dimensions, including: taking time to understand what the conflict is about (it may or may not be about the environment, while "the environment" itself may be understood very differently); recognizing the possibility of ontological conflict, while not assuming that because cultural differences exist, ontologies must differ; focusing on performance rather than group ascription; seeking to maintain a "pluriverse" and openness of outcomes rather than accurate accounts that risk providing "just another cultural perspective".

Conclusion

Conflicts are at the core of many political ecology studies. The conflicts studied are generally over the defense of the environment as a source of livelihood for indigenous and marginalized communities, constituting what Martinez-Alier (2003) defined as the "environmentalism of the poor". This focus reflects a tradition based in large part on anthropology, cultural ecology and agrarian studies, which when combined with a Marxian political economy yields a concern for the emancipation of historically oppressed groups from the forces of capitalism and colonialism

– whether these destroy, conserve or technically render more "productive" the environments under study. These studies are driven by a normative concern for the exacerbation of ethically and morally undesirable distribution of burdens, rights and responsibilities resulting from ecological change. Political ecology thus departs from environmental sciences through this concern for social justice, but also conceptually and empirically for its attention to the political dimensions of socio-environmental change. By denaturalizing socio-environmental change, political ecologists strive to understand (and sometimes reduce) the political marginality of groups who are ignored until they become visible through the unexpected ecological changes and frequent conflicts that erupt as a result of lack of inclusion (Robbins 2012).

Environmental conflicts are contextualized by and played out through cultural differences, discursive representations and material practices. Cultural differences have sharp material, or physical consequences, with the (re)construction of socionatures through modern capitalists interests often implying a major reshaping of landscapes and dispossession of resources for "traditional" livelihoods. For many political ecologists, "what is at stake is a redefinition of production and the economy in line with both the ecological and cultural dimensions of the environment" (Escobar 2006: 10). In this respect "development" is a major topic, with critical reflections on essentialized portrayals of "beneficiaries", and classed, gendered and raced assumptions that guide development interventions, and lack of sensitivity to the many differences among "target" households and divergences of interests within communities and stakeholders. By changing the conditions in which people make a living, by misrepresenting the subjects of development, or by ignoring divergences of interests and perspectives, development interventions are likely to ignite, become enmeshed and exacerbate local struggles.

New directions for political ecology research on environmental conflicts remain largely open, but some topics and approaches are demonstrating growing prominence, greater urgency or promises of new theoretical insights. Within the confines of traditional topics, conflicts associated with the "global land grab" and more generally with the "green grab" are receiving increasing attention (Peluso and Lund 2011), notably with respect to the dispossession of "smallholders" by large-scale agro-industrial investments and food production regimes. More recent topics likely to receive growing attention include conflicts associated with urban political ecology, broadly defined biopolitics in the "Anthropocene", the political ecology of "de-growth", and dispossession and environmental degradation in "*emerged* economies". Debates around "speciesism" and the ethics of the "non-human", as well as conflicts around the environmental dimensions of "new technologies" such as nanotechnologies also offer avenues for further research. Methodologically, quantitative tools including GIS and statistic analysis seem to be making some headway to help further analyze mostly ethnographic material, while many bridges remain to be built with the approaches and findings of "natural sciences". Theory-wise, political ecology will likely build upon and further contribute to feminism, post-colonialism, as well as Science and Technology Studies (STS). At a scholarly level, political ecology studies of environmental conflicts could perhaps most significantly contribute by further politicization within mainstream ecology studies and sustainability science.

Political ecology studies of environmental conflicts demonstrate the value of historical enquiries into the interplay of environmental and political economy changes. Rather than seeking to draw universal laws or make prophesies of future "resource wars", political ecology strives to expose political marginality, processes of dispossession and the truth regimes that "naturalize" them. By better understanding historically contingent assemblages of matter, actors and discourses – what Watts (2004) terms "resource complexes and their systems of rule" – political ecology can better resist both neoliberal assumptions that drive contemporary forms of environmental transformation, as well as securitization processes that see to reconfiguration of issues through a narrow and often violent and historical oppressive security lens. A progressive move, here, is towards "worlding" – an approach through which the recognition of distinct ontologies, values and desires is privileged over the mobilization of security narratives and the institutional imperative of self-righteous intervention. This worlding not only broadens horizons to diverse cosmologies and "ways of being", but also produces new understandings and solidarities.

References

Allen, M. (2012). "Melanesia's violent environments: Towards a political ecology of conflict in the western Pacific". *Geoforum* 44: 152–161.

Baechler, G. and K.R. Spillman (1996). *Environmental Degradation as a Cause of War*. Zurich, Ruegger.

Bassett, T.J. (1988). "The political ecology of peasant-herder conflicts in the Northern Ivory Coast". *Annals of the Association of American Geographers* 78(3): 453–472.

Bebbington, A., L. Hinojosa, D. Humphreys Bebbington, M. L. Burneo and X. Warnaars (2009). "Contention and ambiguity: Mining and the possibilities of development". *Development and Change* 39(6): 887–914.

Benton, T. (1993). *Natural Relations: Ecology, Animal Rights and Social Justice*, London, Verso.

Blaser, M. (2013a). "Caring for caribou/killing atîku: The common world and radical alterity". *Social Justice Noted Scholars Lecture*. University of British Columbia.

Blaser, M. (2013b). "Notes towards a political ontology of 'environmental' conflicts". *Contested Ecologies: Nature and Knowledge*. L. Green, ed. Cape Town: HSRC Press, pp. 13–27.

Collard, R.-C. (2013). "Putting animals back together, taking commodities apart". *Annals of the Association of American Geographers* 104(1): 151–165.

Conca, K. and G.D. Dabelko (2002). *Environmental Peacemaking*. Baltimore, Johns Hopkins University Press.

Dalby, S. (2002). *Environmental Security*. Minneapolis, University of Minnesota Press.

de Soysa, I. (2002). "Ecoviolence: Shrinking pie, or honey pot?" *Global Environmental Politics* 2(4): 1–27.

Escobar, A. (2006). "Difference and conflict in the struggle over natural resources: A political ecology framework". *Development* 49(3): 6–13.

Galtung, J. (1990). "Cultural violence". *Journal of Peace Research* 27(3): 291–305.

Ginn, F. (2013). "Sticky lives: slugs, detachment and more-than-human ethics in the garden". *Transactions of the Institute of British Geographers* 39(4): 532–544.

Gleditsch, N.P. (1998). "Armed conflict and the environment: A critique of the literature". *Journal of Peace Research* 35(3): 381–401.

Hall, D., P. Hirsch, and T. M. Li (2012). "Powers of exclusion: Land dilemmas in Southeast Asia". *Asian Theatre Journal* 29(1).

Hartmann, B. (2001). "Will the circle be unbroken? A critique of the project on environment, population, and security". *Violent Environments*. N.L. Peluso and M. Watts, eds. Ithaca, Cornell University Press, pp. 39–62.

Harvey, D. (1974). "Population, resources, and the ideology of science". *Economic Geography* 50(3): 256–277.

Harvey, D. (2003). *The New Imperialism*. Oxford, Oxford University Press.

Homer-Dixon, T.F. (1999). *Environment, Scarcity and Violence*. Princeton, Princeton University Press.

Kosek, J. (2006). *Understories: The Political Life of Forests in Northern New Mexico*. Durham, NC, Duke University Press.

Le Billon, P. (2008). "Diamond wars? Conflict diamonds and geographies of resource wars". *Annals of the Association of American Geographers* 98(2): 345–372.

Le Billon, P. (2012). *Wars of Plunders: Conflicts, Profits and the Politics of Resources*. New York, Columbia University Press.

Low, N. and B. Gleeson (1998). *Justice, Society, and Nature: An Exploration of Political Ecology*. London, Routledge.

Lahiri-Dutt, K. (2006). "'May God give us chaos, so that we can plunder': A critique of 'resource curse' and conflict theories". *Development* 49(3): 14–21.

Martinez-Alier, J. (2001). "Mining conflicts, environmental justice, and valuation". *Journal of Hazardous Materials* 86(1): 153–170.

Martinez-Alier, J. (2003). *The Environmentalism of the Poor: A Study of Ecological Conflicts and Valuation.* Cheltenham, Edward Elgar Publishing.
McCarthy, J. (2013). "We have never been 'post-political'". *Capitalism Nature Socialism* 24(1): 19–25.
Peluso, N.L. and M. Watts (2001). *Violent Environments.* Ithaca, NJ, Cornell University Press.
Peluso, N.L. and C. Lund (2011). "New frontiers of land control: introduction". *Journal of Peasant Studies* 38(4): 667–681.
Ribot, J.C. and N.L. Peluso (2003). "A theory of access". *Rural Sociology* 68(2): 153–181.
Rikoon, J.S. (2006). "Wild horses and the political ecology of nature restoration in the Missouri Ozarks". *Geoforum* 37(2): 200–211.
Robbins, P. (2004). *Political Ecology: A Critical Introduction.* Oxford, Blackwell.
Robbins, P. (2012). *Political Ecology: A Critical Introduction*, 2nd edn. Oxford, John Wiley & Sons.
Scott, J.C. (1985). *The Weapons of the Weak: Everyday Forms of Peasant Resistance.* New Haven, CT, Yale University Press.
Sheridan, T.E. (2007). "Embattled ranchers, endangered species, and urban sprawl: the political ecology of the new American West". *Annu. Rev. Anthropol.* 36: 121–138.
Swyngedouw, E. (2007). "Impossible sustainability and the post-political condition". *The Sustainable Development Paradox: Urban Political Economy in the United States and Europe.* R. Krueger and D. Gibbs, eds. New York, Guilford Press, pp. 13–40.
Swyngedouw, E. (2009). "The antinomies of the postpolitical city: in search of a democratic politics of environmental production". *International Journal of Urban and Regional Research* 33(3): 601–620.
Turner, M.D. (2004). "Political ecology and the moral dimensions of 'resource conflicts': the case of farmer–herder conflicts in the Sahel". *Political Geography* 23(7): 863–889.
Watts, M. (2004). "Resource curse? governmentality, oil and power in the Niger Delta, Nigeria". *Geopolitics* 9(1): 50–80.
Zalik, A. (2011). "Protest-as-violence in oil fields. The contested representation of protests in two extractive sites". *Accumulating Insecurity: Violence and Dispossession in the Making of Everyday Life.* S. Feldman, C. Geisler and G.A. Menon, eds. Athens, GA, University of Georgia Press, pp. 261–284.
Zizek, S. (1999). "Carl Schmitt in the age of post-politics". *The Challenge of Carl Schmitt.* C. Mouffe, ed. New York, Verso, pp. 18–37.
Zizek, S. (2005). "Against human rights". *New Left Review* 34: 115.

47

URBANIZATION AND ENVIRONMENTAL FUTURES

Politicizing urban political ecologies

Erik Swyngedouw

> Urbanism is the mode of appropriation of the natural and human environment by capitalism.
>
> *(Debord 1994: 121)*

Introduction

Some time in 2013, the earth passed the symbolic threshold of 400 ppm of CO_2 in the atmosphere. The 5th report of the IPCC concluded that 'most aspects of climate change will persist for many centuries even if emissions of CO_2 are stopped' (IPCC 2013: 27). Despite the migrating circuses of the UN's Climate Summits and their dismal record of failures, precious little has been achieved in lowering total greenhouse gas emissions. In the meantime, cities in both the global North and South are choking as the concentration of small particles and other forms of pollution reach dangerously high levels.

We have now truly entered what Paul Crutzen in 2000 tentatively named the Anthropocene, the successor geological period to the Holocene (Crutzen and Stoermer 2000), and planetary urbanization is not only its geographical form: more importantly, it is also the socio-spatial process that shapes the intimate and accelerating fusion of social and physical transformations and metabolisms that gave the Anthropocene its name (Swyngedouw 2014a). Planetary urbanization refers to the fact that every nook and cranny of the earth is now directly or indirectly enrolled in assuring the expanding reproduction of the urbanization process. Indeed, the sustenance of actually existing urban life is responsible for 80 percent of the world's greenhouse gas emissions (Bulkeley and Betsill 2005), for the accelerating mobilization of all manner of natures, and for producing most of the world's waste.

From this perspective, we are here not primarily concerned with the city as a dense and heterogeneous assemblage of accumulated socio-natural things and gathered bodies in a concentrated space, but rather with the particular forms of capitalist urbanization as a socio-spatial process whose functioning is predicated upon ever longer, often globally structured, socio-ecological metabolic flows. These flows not only weld together things, natures and peoples, but do so in socially, ecologically, and geographically articulated, but uneven, manners (Swyngedouw 1996; Cook and Swyngedouw 2012; Angelo and Wachsmuth 2014). The key question is, therefore, not about what kinds of natures are present *in* the city, but rather about the capitalist form of urbanization *of* natures: the process through which all manner of non-human 'stuff' is socially mobilized, discursively scripted, imagined, economically enrolled (commodified), and physically metabolized/transformed to produce socio-ecological assemblages that support the urbanization process (Heynen et al. 2005). Consider, for example, how dependent are the purportedly de-materialized affective economies that animate much of contemporary urban social and cultural life (IT-networks, social media, smart networks, eco-architecture, informatics, and the like) upon mobilizing a range of minerals (like Coltan (columbite–tantalite)); upon feverish resource grabbing, often through tactics of dispossession, in socio-ecologically vulnerable places; upon production chains that are shaped by deeply uneven and often dehumanizing socio-ecological metabolisms (material and immaterial production processes) to render it useful in ITC hardware; and upon a 're-cycling' process that returns much of the e-waste to the socio-ecologically dystopian geographies of Mumbai's or Dhaka's suburban informal wastelands. Indeed, the excesses of urbanization – from (e-)waste to CO_2 – are customarily decanted onto the socio-ecological dumping grounds of the periphery of cities.

The capitalist form of planetary urbanization and the socio-ecological and political-economic processes that animate its combined and uneven socio-ecological development on a world scale are now generally recognized as key drivers of anthropogenic climate change and other socio-environmental transformations such as biodiversity loss, soil erosion, large eco-infrastructures like dams, deforestation, resource extraction and deep-geological mining, pollution, and the galloping commodification of all manner of natures. Our urban fate and natures' transformations are irrevocably bound up in an intimate and intensifying metabolic – but highly contentious – symbiosis, one characterized by extraordinarily uneven socio-ecological patterning choreographed by the power relations that animate the reproduction of neoliberal capitalism. The configuration of this urban metabolic relationship has now been elevated to the dignity of global public concern, and a feverish search for all manner of eco-prophylactic remedies has entered the standard vocabulary of both governmental and private actors.

Indeed, a global urban intellectual and professional technocracy has spurred a frantic search for a 'smart' socio-ecological urbanity and seeks out the socio-ecological qualities of eco-development, retrofitting, sustainable architecture, resilient urban governance, the commodification of environmental 'services', and innovative – but fundamentally market-conforming – eco-design (Mostafavi and Doherty 2010). Nonetheless, 'sustainable' eco-technological urban developments are often predicated upon mobilizing precarious labor and dispossessing local people from their resources and livelihoods (Caprotti 2014), while still further expanding the mobilization of the earth's resource base. This techno-managerial dispositive has now been consensually established as the frontier of architectural, planning, and urban design, theory and practice, presumably capable of saving both city and planet, while assuring that civilization as we know it can continue for a little longer. Under the banner of radical techno-managerial restructuring, the focus is now squarely on how to sustain capitalist urbanity so that nothing really has to change!

Nature as the externally conditioning frame for urban life has indeed come to an end. The Anthropocenic inauguration of a socio-physical historical and thoroughly globally urbanized nature forces a profound reconsideration and re-scripting of both nature and urbanization in political terms. The question is not any longer about bringing environmental issues into the domain of urban politics, but rather about how to bring the political into the urban environment.

In what follows, we shall explore first the ways in which urban thought and research have begun to incorporate political matters within urban theory and practice. Particular attention will be paid to urban environmental justice perspectives on the one hand and urban political ecology on the other. While we fully endorse the extraordinary progress that has been made in recent years, we shall insist, in a second part, that urban political-ecology needs to take the question of 'the political' in 'political ecology' much more seriously. A number of pointers for a politicized urban political ecology will conclude the chapter.

UrbanNatural

The urbanization process as a constituent part of the world's socio-ecological predicament was foregrounded in the 1970s as part of the broader concern with the accentuation of deteriorating environmental conditions. While the voices of eco-urban visionaries like Murray Bookchin went largely unnoticed (see White 2008), the Malthusian clarion call of pending resource depletion foreshadowed by the Club of Rome's *Limits to Growth*, raising the spectre of immanent scarcity in nature, really got the global elites worried about the allegedly feeble prospects for sustaining capitalist accumulation for much longer, and pointed to urbanization as the main culprit of the world's accelerating resource depletion (Meadows et al. 1972). In addition, the environmental movement, particularly active around contesting nuclear energy use in the Global North, and hyper-urbanization in the Global South, propelled environmental matters to the top of the urban policy agenda.

Urban thought and practice followed suit. Urban scholars and activists began to dissect the urbanization of nature as a process of continuous de- and re-territorialization of socio-ecological metabolic circulatory flows, organized through predominantly capitalist social relations sustained by privately or publicly managed socio-physical conduits and networks (Swyngedouw 2006), and nurtured by particular imaginaries of what nature is or should be. Under capitalism, so the argument went, the mobilization and transformation of non-human 'stuff' into a commodified form under the impetus of capital circulation and accumulation shape these socio-ecological processes and turn the city into a metabolic socio-environmental process that stretches from the immediate environment to the remotest corners of the globe (Heynen et al. 2005).

Through this conceptual lens, urbanization is viewed as a process of geographically arranged socio-environmental metabolisms that fuse the social with the physical. In so doing, a 'cyborg' urbanity is produced that mixes distinct physical forms with geographically highly uneven socio-ecological consequences (Swyngedouw 1996; Gandy 2005). A proliferating body of scholarly work began to explore, both empirically and theoretically, how urbanization and its human and non-human inhabitants across the globe are linked through networks and flows of technology, and social relations of power for the extraction, circulation, and disposal of matter such as water (Swyngedouw 2004), energy (Bouzarovski 2014; Verdeil 2014), fat (Marvin and Medd 2006), chemicals and e-waste (Pellow 2007), household waste (Njeru 2006), infrastructures (Monstadt 2009; Graham and Marvin 2001), or redundant ships (Buerk 2006; Hillier 2009). Burrowing into the metabolic process of less visible, yet powerfully important socio-natural actants, Ali and Keil mapped how the SARS epidemic challenged global networks of urban governance (Ali and Keil 2011). Bulkeley searched for the urban roots of CO_2 (Bulkeley and Betsill 2005), and Robbins reconstructed the global networks of production, pollution, and toxic waste which sustain the insatiable drive to nurture the 'green' lawns that feed the suburban middle-class dream (Robbins 2007).

A series of exciting urban monographs explored the political-ecological dynamics that undergirded the historical-geographical production of particular cities. For example, William Cronon's seminal monograph rewrites Chicago's urbanization process through an examination of how wheat and hog production shaped the city's metabolic and spatially expanding transformation process (Cronon 1991). Brechin narrates how San Francisco's elites rummaged through nature in search of earthly gain and power (Brechin 2001) while Matthew Gandy's *Concrete and Clay* undertakes the archeology of New York's urbanization process as a political-ecological construct (Gandy 2003). In some studies, water became an emblematic entry into the excavation of socio-ecological flows. Maria Kaika's *City of Flows* considers the cultural, socio-economic, and political relations through which the urbanization of water is cast and recast during modernity (Kaika 2005). Swyngedouw's *Social Power and the Urbanization of Nature* excavates the relationship between planetary urbanization and nature's transformation through the lens of Guayaquil's water (Swyngedouw 2004), while Karen Bakker follows the flow of water through the privatization politics of England and Wales (Bakker 2003), and Saurí et al. explore the political-ecological dynamics, conflicts and struggles around Barcelona's urban water supply (Masjuan et al. 2008; March and Saurí 2013). It is also worth pointing to how Mike Davis's dystopian *Dead Cities and other Tales* excavates the peculiar ecologies of cities that should not be where they are (Davis 2002), and to Freidberg's majestic study of how green beans link African cities to Paris and London, thereby exploring how urbanization is indeed sustained by planetary socio-ecological networks and relations driven by particular dynamics of urbanization and city life (Freidberg 2004).

The above narratives in urban political ecology and cognate research demonstrated, in a variety of ways and from a range of theoretical perspectives, how the matter of matter becomes an active moment in the political-ecological transformations that shape planetary urbanization. These authors have argued convincingly that the urban process has to be theorized, understood, and managed as a socio-natural process that goes beyond the technical-managerial mediation of urban socio-ecological relations. By doing so, they contributed to delegitimizing dominant twentieth-century perspectives on the city that ignored nature, without falling into the deadlock of nature fetishism or ecological determinism. Moreover, by transcending the binary division between nature and society the urban metabolism perspective has shown that socio-ecological processes are intensely political, and confirmed that urban theory without nature cannot be but incomplete.

However, this body of thought has paid relatively little attention to the political opportunities such re-natured understandings of urbanization could bring, or to imagining radically different future urban socio-ecological assemblages. Thus, although we may now be able to trace, chart, follow, and narrate the multiple socio-ecological lines that shape the globalizing urban process, preciously little has been said about how to produce alternative, more equitable and enabling, urban socio-ecological constellations. In what follows we shall briefly explore two perspectives that have galvanized more politicized thinking and practice around the urban environmental question.

1. Urban environmental justice (UEJ): the distribution of environmental bads

The urban environmental justice perspective opens up a politicizing view in which the unequal distributional characteristics of urban metabolism take centre stage (see Chapter 45, this volume). UEJ is sensitive to the conflicting and power-laden processes of urbanizing nature through elite-led techno-managerial fixes (Walker 2012). Originating in the United States, urban environmental justice emerged both as a normative concept and a social movement, sustained by new insights into the highly uneven distribution of environmental 'goods' and 'bads' in the city. Early work in the 1980s had already begun to recognize that poor, often predominantly African American, neighborhoods were overwhelmingly located in areas characterized by environmentally hazardous conditions (Bullard 1990). Significant positive correlations were found between the presence of toxic dumps, waste processing facilities, ground pollution, hazardous chemicals, and absence of green zones on the one hand, and concentrations of low-income households on the other. In other words, the spatial distribution of environmental goods and bads mirrored the socio-spatial distribution of political power, wealth, and income (Schlosberg 2007).

Urban environmental justice became defined and understood as a question of Rawlsian distributional (in)justice, choreographed and structured by the highly uneven political and economic power relations through which decisions over environmental distributional conditions are made and implemented. Emphasis is put on the socially and economically uneven positions, recognition and capabilities of different urban dwellers in the urban political and economic decision-making machinery that allocates the distribution of environmental goods and bads throughout the city, showing that the partitioning of environmental 'goods' mainly benefit urban elites, whilst environmental 'bads' are decamped to areas where the powerless and disenfranchised live. It became clear that sustainable urban lives are primarily the privilege of the rich, and that environmental havens are sustained on the back of deteriorating socio-ecological conditions elsewhere.

More recent explorations of urban environmental injustices have extended the earlier focus on race to other social categories such as gender, class, age, ability, and geographical scale (Walker 2009). Nonetheless, the emphasis of UEJ remains clearly on foregrounding liberal notions of procedural and distributional justice as fairness, and expressing a distinct form of NIMBY-ism (Not in My Back Yard).

UEJ is primarily concerned with the procedures through which smart and other environmental technologies, infrastructures, and amenities are partitioned throughout the city and highlights the socially highly uneven patterning of ecological qualities and hazards. However, this perspective succeeded in socializing nature and ecology by excavating the intricate mechanisms through which nature, ecological processes, and socio-environmental conditions in the city are highly interwoven in deeply unjust manners. The latter become etched in the urban landscape through a combination of highly elitist decision-making procedures on the one hand and their cementation into the architecture of eco-technological infrastructures and technologies on the other. Nonetheless, UEJ tends to be symptomatically silent about the particular ways in which political forms of power interweave with the concrete modalities through which nature becomes enrolled in processes of capital circulation and accumulation. Its place-based focus too needs to be augmented by considering the global metabolic assemblages and flows that produce combined and uneven socio-ecological change.

2. Urban political ecology (UPE): re-asserting the capitalist production of planetary urbanization

While UEJ focuses primarily on patterns of socio-ecological injustice within the city, urban political ecology shifts the interpretative gaze to the socio-ecological inequalities embodied in and shaped by the production and reproduction of capitalist urbanization itself (Keil 2003, 2005). Under capitalism, natures become increasingly enrolled in the circuits of capital accumulation through which they are both transformed and de-/re-territorialized. This is a socio-metabolic process whereby 'physical matter such as water or cows is transformed into useable, ownable and tradable commodities' (Coe et al. 2007: 161). From this perspective, 'Nature' as the homogenized collective name for all manner of non-human things, organisms and processes does not exist, but rather there is a highly diverse and continuously changing collection of all sorts of very different non-human imbroglios that become historically and geographically produced in specific and decidedly urbanized manners (Swyngedouw 2010). It is such a conceptualization that led David Harvey, for example, to argue that 'there is nothing unnatural about New York City' (Harvey 1996).

UPE is decidedly anti-Malthusian. In contrast to the doom-laden spectre of Malthusian limits to the earth's resource base and the menace of pending absolute scarcity, urban political ecology considers scarcity as socio-ecologically produced through the twin imperative of 'accumulation for accumulation's sake' on the one hand and 'market' forces as naturalized and privileged instruments for the social allocation and distribution of (transformed) natures on the other. Furthermore, UPE rejects the apocalyptic imaginary that customarily accompanies attempts at politically foregrounding a public concern with nature as inherently depoliticizing and reactionary (Swyngedouw 2013a, 2013b). For UPE, the socio-ecological catastrophe is already present and reflects the combined and uneven socio-ecological patterns produced by the specifically capitalist form of globalization of urban metabolisms.

Indeed, a political-economic configuration – usually called capitalism – whose 'sustainability' is predicated upon growth for growth's sake necessarily hits the physical and social limits of its own pre-conditions of existence, thereby ushering in continuous and highly uneven dynamics of continuous socio-ecological transformation. More importantly, such produced urban socio-physical environments embody and reflect the unequal power and associated asymmetrical socio-ecological living conditions inscribed in socio-ecological metabolisms. 'Scarcity' or 'socio-ecological disintegration' resides, therefore, not in Nature but in the socially constructed and utterly contingent modalities of its spatially and socio-ecologically variegated enrolling within urbanizing circuits of capital circulation and accumulation.

The production of urban environments, and the 'metabolic vehicles' (such as infrastructures of all kinds, the technical conditions that permit the flow and metabolization of energy, food, information, bodies, and things) that secure its functioning are of course mediated by institutional arrangements that are often nominally democratic, but are nonetheless necessarily deeply committed to assuring the uninterrupted expansion of the capital circulation process (Virilio 1986). 'Metabolic vehicles' are the hard and soft infrastructures through which non-human matter becomes transformed, and express in their techno-political functioning multiple relations of power in which social actors strive to create and defend socio-physical environments that serve their interests and satisfy their desires. It is precisely this articulation between state, class, and environmental translation that renders urban socio-ecological processes, including the question of 'sustainability', highly conflictive and subject to intense political and social struggle. Consider, for example, how the urban rebellion that engulfed Turkey with rarely seen intensity in the summer of 2013 emblematically sparked off with a conflict over a park and a few trees on Istanbul's Taksim Square. Or how climate summits meet with increasingly intense street protests (Swyngedouw 2013b).

The urbanization of nature is decidedly multi-scaled and spatially networked in an extended manner. Multi-scalar governance arrangements, from Agenda 21 to the Kyoto Protocol, suggest how the global span of socio-ecological transformation processes are articulated with multi-scaled governance ensembles, each of which expresses particular power relations whereby struggles for control, access, and transformation of nature and the distribution of ecological goods and bads are carefully negotiated and intensely contested. From this political-ecological perspective, urban ecological conditions and the configurations of their governance are never just local, but are attached to processes that operate in diverse ecologies across the world. Such urban political-ecological approaches foreground the political core of environmental change and transformation and insist on the fundamentally political nature of the modes of socio-technically organizing the metabolic transformation of nature.

Therefore, urban political ecology is concerned with the democratic and emancipatory political process through which such politically embedded ecological transformation takes place. Rather than invoking a normative notion of environmental justice or of an idealized (balanced) nature, UPE insists on focusing on the realities of the presumed democratic political equality in the decision-making processes that organize socio-ecological transformation and choreograph the management of the commons.

Politicizing the political ecology of planetary urbanization

Despite the extraordinary leap forward in critical understanding of the urban environmental condition and the consensual attention to 'sustainable' and 'smart' eco-technologies, global ecological conditions continue to deteriorate at an alarming rate as planetary urbanization intensifies. This is a veritable paradoxical situation that can only be rendered legible in strictly ideological terms. As Slavoj Žižek put it: 'Despite the fact we know very well (the ecological predicament that we are in), we continue to act as if we do not know' (Žižek 2008). While the techno-managerial elites desperately attempt to micro-engineer socio-ecological conditions in ways that permit both sustaining economic growth indefinitely into the future and turning environmental technologies into a 'green' accumulation strategy, the depth and extent of environmental degradation gallops further in what Williams calls 'a combined and uneven apocalypse' (Williams 2011).

It is also becoming abundantly clear that the early ecologists' clarion call, borrowed from the twentieth-century Italian communist, Amadeo Bordiga, that 'when the ship goes down, the first class passengers drown too' is manifestly untrue. The earth's first class urban passengers are busily building trans-planetary rescue vessels while ecological and political refugees drown in the Mediterranean, and many others continue to live in the proliferating socio-ecological wastelands of their degrading socio-ecological environments. Planetary urbanization, unfolding through the universalization of the commodification and accumulation of natures within a neo-liberalizing political configuration, accelerates the process of combined and uneven ecological apocalypse, one increasingly sustained by the mythical promise of technologically mediated sustainability and post-democratic forms of consensual governance that do not tolerate radical dissent or the pursuit of real political-ecological alternatives. The de-politicizing techno-managerial endeavours that characterize dominant modes of environmental governing suture the ideological landscape, and foreclose more politically grounded modes of producing a more egalitarian socio-ecological mode of governing and transforming the urban commons (Swyngedouw 2009). Transgressing this 'deadlock' between the real and present dangers of combined and uneven socio-ecological planetary urbanization on the one hand, and the impotent acting out of post-democratic 'sustainable' management of resources and people on the other, demands a serious intellectual and political engagement with some of the most intractable conditions our cities are in.

As argued elsewhere (see Swyngedouw 2007, 2010, 2011), consensually established concerns such as 'sustainability' nurture a politically reactionary stance. They are an expression of the current process of post-politicization and post-democratization, one that is arranged around distinct bio-political gestures. Post-politicization refers to a politics in which techno-managerial planning and intervention, expert management, and bio-political administration displace ideological or dissensual contestation and struggles. Such post-politicizing arrangements signal a profound narrowing of democratic agonistic struggles over the content and direction of socio-ecological life, and institute a public space where the terrain of politics has been reduced to policy-making in which expert knowledge, interest intermediation, and administration through governance have begun to replace dissensual debate and agonistic encounter (Marquand 2004). Debate and contentious argument are restricted to questions of techno-managerial management whereby the neoliberal frame of market-led and growth-centered development cannot be legitimately questioned. This depoliticized consensual arrangement is organized through post-democratic institutions of managerial governance that are increasingly replacing the political institutions of government (see Crouch 2004) and are embedded in a geographically variegated, but broadly naturalized, neo-liberalizing political–economic order.

These arguments point to the vital importance of grappling with this process of post-politicization and for moving from an urban politics of the environment to urbanizing environmental politics. If the aim of politics – including urban politics and policies, city design, planning, and architecture – is intervention that can change the given socio-environmental ordering and partitioning in a certain direction, then such intervention often constitutes a violent act, in the sense that it erases what is there (at least in part) in order to erect something new and different. The central point is to recognize that political acts are singular interventions that produce particular socio-ecological arrangements and milieus and, in doing so, foreclose the possibility of others emerging. An intervention enables the formation of certain socio-ecological assemblages while closing down others. The 'violence' inscribed in such choice has to be fully acknowledged. For example, one cannot have simultaneously a truly carbon-neutral city and permit unlimited car-based mobility. They are mutually exclusive. Even less can an egalitarian, democratic, solidarity-based, and ecologically sensible urban future be produced without marginalizing or excluding those who insist on the private appropriation of the commons of the earth and its mobilization for accumulation, personal enrichment, and hereditary transfer of accumulated resources.

Such violent encounters always constitute a political act, one that can be legitimized only in political terms, and not through an externalized legitimation that resides in a fantasy of Nature or Sustainability. Any political act is one that reorders socio-ecological coordinates and patterns, reconfigures uneven socio-ecological relationships (while foreclosing others), often with unforeseen or unforeseeable consequences. Such interventions that express a choice and take sides are invariably somehow exclusive. Any instituted order, including a liberal-democratic one, produces its own inequalities. This is precisely what critical urban theory has successfully explored and analyzed over the past few decades. This suggest that there is an irreducible gap or abyss between the democratic as a political given predicated upon the presumption of the equality of each and all, on the one hand, and the instituted/institutionalized forms of policy-making that invariably suspend this axiomatic equality. In other words, any form of policy-making is to certain degree oligarchic. This gap between 'politics' and 'the political' needs to be endorsed fully.

Most importantly, it pits those who are bent on maintaining the current trajectory that produces a combined and uneven socio-ecological apocalypse radically against those who prefigure an inclusive and egalitarian production of socio-ecological urban commons. Rather than invoking a normative notion of environmental justice or of an idealized (balanced) nature, our perspective insists on focusing on the inegalitarian realities of the presumed democratic political equality in the decision-making processes that organize socio-ecological transformation and choreograph the management of the commons. In doing so, the attention shifts from a techno-managerial, physico-ecological, or ethical perspective to a resolutely political vantage point – articulated around the notion of equality – that considers the ecological conundrum to be inexorably associated with democratic political acting and focuses on the fundamentally politicized conditions through which natures become produced (Swyngedouw 2014b).

While a pluralist democratic politics insists on difference, disagreement, radical openness, and exploring multiple possible futures, concrete spatial–ecological intervention is necessarily about relative closure (for some), definitive choice, singular intervention and, thus, certain exclusion and occasionally even outright silencing. For example, tar sand exploitation and fracking cannot coincide with a climate policy worthy of its name. Climate justice requires a process that keeps the coal in the hole, the oil in the soil and tar sand in the land. This would strictly outlaw dominant practices and produce extraordinary distributional effects that can only be attended to politically.

While 'traditional' democratic policies are based on majoritarian principles, the democratic–egalitarian perspective insists on foregrounding equality and socio-ecological solidarity as the foundational gesture for a green urban future. Politicizing environments democratically, then, becomes an issue of enhancing the urban democratic political content of socio-environmental construction by means of identifying the strategies through which a more equitable distribution of social power and a more egalitarian mode of producing urban natures can be achieved. This requires the nurturing of processes that enable the production of spaces of democratization (i.e. spaces for the enunciation of agonistic dispute) as a foundation and condition for more egalitarian urban socio-ecological arrangements, and the naming of positively embodied ega-libertarian socio-ecological futures that are immediately realizable. Agonism refers here to the process by which oppositional positions between enemies become articulated and organized as oppositional encounter between adversaries (Mouffe 2013).

In other words, egalitarian urban ecologies are about demanding the impossible and realizing the improbable, often in the face of radical and powerful opposition, and this is exactly the challenge the Anthropocene poses. In sum, the politicization of the environment is predicated upon the recognition of the indeterminacy of nature, the constitutive split of the people, the unconditional democratic demand of political equality, and the real possibility of the inauguration of public and collective urban socio-ecological futures that express the democratic presumptions of freedom and equality.

Ultimately, the intellectual challenge posed by the socio-environmental conditions shaped by planetary urbanization must be to extend the intellectual imaginary and the powers of thought and practice to overcome the contemporary cultural impasse identified by Fredric Jameson that it is easier to imagine the end of the world than changes in the eco-capitalist order and its inequities (Jameson 2003: 76). This is the courage of the intellect that is now required more than ever, a courage that takes us beyond the impotent confines of a sustainability discourse and leaves the existing combined and uneven, but decidedly urbanized, socio-ecological dynamics fundamentally intact. It is a courage that charts new politicized avenues for producing a new common urbanity. There is an urgent task ahead, therefore, to delve into the complex linkages between politicizing discourse and practice, post-political eco-management, and the reproduction of environmental socio-ecological inequalities. It is necessary to ask questions about what visions of 'Nature' and what socio-environmental relations are being promoted; what quilting points are being used and how they are being stitched together; and who are promoting these visions and why. In this respect, there is an urgent need to consider the eco-politicizing movements and discourses such as those of the environmental political movements or the various '*indignados*' and other insurgent political mobilizations that over the past few years have been demanding a new constituent democratic process. The articulation between urban political ecological thought and democratizing urban practices with a view towards thinking whether an ecologically sensible, equal, free and solidarity based form of planetary urbanization can still be imagined for the twenty-first century is, I believe, the greatest intellectual challenge for an urban political ecology that desires to be politically performative.

Acknowledgements

I gratefully acknowledge the support of the People Program (Marie Curie Actions) of the European Union's Seventh Framework Program; under REAS agreement No 289374 – 'ENTITLE'. I would also like to thank Maria Kaika for her comments and suggestions.

References

Ali, S.H. and R. Keil (2011) *Networked Disease: Emerging Infections in the Global City*. Oxford: Wiley-Blackwell.

Angelo, H. and D. Wachsmuth (2014) Urbanizing Urban Political Ecology: A Critique of Methodological Cityism. *International Journal of Urban and Regional Research* (forthcoming) DOI:10.111/1468-2427.1205.

Bakker, K. (2003) *An Uncooperative Commodity – Privatizing Water in England and Wales*. Oxford: Oxford University Press.

Bouzarovski, S. (2014) Energy Poverty in the European Union: Landscapes of Vulnerability. *WIREs Energy and Environment*, 3, 276–289.

Brechin, G. (2001) *Imperial San Francisco: Urban Power, Earthly Ruin*. Berkeley: University of California Press.

Buerk, R. (2006) *Breaking Ships: How Supertankers and Cargo Ships are Dismantled on the Beaches of Bangladesh*. New York: Chamberlain Books.

Bulkeley, H. and M. Betsill (2005) *Cities and Climate Change: Urban Sustainability and Global Environmental Governance*. London: Routledge.

Bullard, R.D. (1990) *Dumping in Dixie: Race, Class, and Environmental Quality*. Boulder: Westview.

Caprotti, F. (2014) Eco-urbanism and the Eco-City, or Denying the Right to the City? *Antipode*, 46, 1285–1303.

Coe, N., Kelly, P.F. and Yeung, H. (2007) *Economic Geography: A Contemporary Introduction*. Oxford: Blackwell.

Cook, I. and E. Swyngedouw (2012) Cities, Social Cohesion and the Environment: Towards a Future Research Agenda. *Urban Studies*, 49(9), 1938–1958.

Cronon, W. (1991) *Nature's Metropolis: Chicago and the Great West*. New York: Norton.

Crouch, C. (2004) *Post-Democracy*. Cambridge: Polity Press.

Crutzen, P.J. and E.F. Stoermer (2000) The Anthropocene. *Global Change Newsletter*, 17–18.

Davis, M. (2002) *Dead Cities, and Other Tales*. New York: New Press.

Debord, G. (1994) *The Society of the Spectacle*. New York: Zone.

Freidberg, S. (2004) *French Beans and Food Scares: Culture and Commerce in an Anxious Age*. New York: Oxford University Press.

Gandy, M. (2003) *Concrete and Clay: Reworking Nature in New York City*. Cambridge, MA: MIT Press.

Gandy, M. (2005) Cyborg Urbanization: Complexity and Monstrosity in the Contemporary City. *International Journal of Urban and Regional Research*, 29, 26–49.

Graham, S. and S. Marvin (2001) *Splintering Urbanism: Networked Infrastructures, Technological Mobilities and the Urban Condition*. London: Routledge.

Harvey, D. (1996) *Justice, Nature, and the Geography of Difference*. Oxford: Blackwell.

Heynen, N., M. Kaika and E. Swyngedouw (eds.) (2005) *In the Nature of Cities – The Politics of Urban Metabolism*. London: Routledge.

Hillier, J. (2009) Assemblages of Justice: The "Ghost Ships" of Graythorp. *International Journal of Urban and Regional Research*, 33, 640–661.

IPCC (2013) Summary for Policymakers. In: Climate Change 2013: The Physical Science Basis. Contribution of Working Group I to the Fifth Assessment Report of the Intergovernmental Panel on Climate Change [Stocker, T.F., D. Qin, G.-K. Plattner, M. Tignor, S.K. Allen, J. Boschung, A. Nauels, Y. Xia, V. Bex and P.M. Midgley (eds.)]. Cambridge: Cambridge University Press.

Jameson, F. (2003) Future City. *New Left Review*, 65–79.

Kaika, M. (2005) *City of Flows*. London: Routledge.

Keil, R. (2003) Urban Political Ecology. *Urban Geography*, 24, 723–738.

Keil, R. (2005) Progress Report – Urban Political Ecology. *Urban Geography*, 26, 640–651.

March, H. and D. Saurí (2013) The Unintended Consequences of Ecological Modernization: Debt-induced Reconfiguration of the Water Cycle in Barcelona. *Environment and Planning A*, 45, 2064–2083.

Marquand, D. (2004) *Decline of the Public: The Hollowing Out of Citizenship*. Cambridge: Polity Press.
Marvin, S and W. Medd (2006) Metabolisms of Obecity: Flows of Fat through Bodies, Cities, and Sewers. *Environment and Planning A*, 38(2), 313–324.
Masjuan, E., H. March, E. Domene and D. Saurí (2008) Conflicts and Struggles over Urban Water Cycles: The Case of Barcelona 1880–2004. *Tijdschrift Voor Economische En Sociale Geografie*, 99, 426–439.
Meadows, D.H., G. Meadows, J. Randers and W.W.I. Behrens (1972) *The Limits to Growth*. New York: Universe Books.
Monstadt, J. (2009) Conceptualizing the Political Ecology of Urban Infrastructures: Insights from Technology and Urban Studies. *Environment and Planning A*, 41(8), 1924–1942.
Mostafavi, M. and G. Doherty (eds.) (2010) *Ecological Urbanism*. Zurich: Lars Müller.
Mouffe, C. (2013) *Agonistics – Thinking the World Politically*. London: Verso.
Njeru, J. (2006) The Urban Political Ecology of Plastic Bag Waste Problem in Nairobi, Kenya. *Geoforum*, 37, 1046–1058.
Pellow, D. (2007) *Resisting Global Toxics: Transnational Movements for Environmental Justice*. Cambridge, MA: MIT Press.
Robbins, P. (2007) *Lawn People: How Grasses, Weeds, and Chemicals Make Us Who We Are*. Philadelphia: Temple University Press.
Schlosberg, D. (2007) *Defining Environmental Justice – Theories, Movements and Nature*. Oxford: Oxford University Press.
Swyngedouw, E. (1996) The City as a Hybrid – On Nature, Society and Cyborg Urbanisation. *Capitalism, Nature, Socialism*, 7(1), 65–80.
Swyngedouw, E. (2004) *Social Power and the Urbanization of Water: Flows of Power*. Oxford: University Press.
Swyngedouw, E. (2006) Circulations and Metabolisms: (Hybrid) Natures and (Cyborg) Cities. *Science as Culture*, 15, 105–121.
Swyngedouw, E. (2007) Impossible/Undesirable Sustainability and the Post-Political Condition. In *The Sustainable Development Paradox*, eds. J.R. Krueger and D. Gibbs, 13–40. New York: Guilford.
Swyngedouw, E. (2009) The Antinomies of the Post-Political City. In Search of a Democratic Politics of Environmental Production. *International Journal of Urban and Regional Research*, 33, 601–620.
Swyngedouw, E. (2010) Trouble with Nature – Ecology as the New Opium for the People. In *Conceptual Challenges for Planning Theory*, eds. J. Hillier and P. Healey, 299–320. Farnham: Ashgate.
Swyngedouw, E. (2011) Interrogating Post-Democracy: Reclaiming Egalitarian Political Spaces. *Political Geography*, 30, 370–380.
Swyngedouw, E. (2013a) Apocalypse Now! Fear and Doomsday Pleasures. *Capitalism, Nature, Socialism*, 24, 9–18.
Swyngedouw, E. (2013b) The Non-Political Politics of Climate Change. *ACME – An International E-Journal for Critical Geographies*, 12, 1–8.
Swyngedouw, E. (2014a) Anthropocenic Politicization: From the Politics of the Environment to Politicizing Environments. In *Green Utopianism: Politics, Practices and Perspectives*, eds. J. Hedrén and K. Bradley, 23–37. London: Routledge.
Swyngedouw, E. (2014b) The Violence of Sustainable Urbanity. *Harvard Design Magazine*, 37.
Verdeil, E. (2014) The Energy of Revolts in Arab Cities: The Case of Jordan and Tunisia. *Built Environment*, 40(1), 128–139.
Virilio, P. (1986) *Speed and Politics: An Essay on Dromology*. Semiotext(e). Cambridge, MA: The MIT Press.
Walker, G. (2009) Beyond Distribution and Proximity: Exploring the Multiple Spatialities of Environmental Justice. *Antipode*, 41, 614–636.
Walker, G. (2012) *Environmental Justice: Concepts, Evidence and Politics*. New York: Routledge.
White, D. (2008) *Bookchin – A Critical Appraisal*. London: Pluto Press.
Williams, C.E. (2011) *Combined and Uneven Apocalypse*. Washington, DC: Zero Books.
Žižek, S. (2008) Nature and its Discontents. *SubStance*, 37, 37–72.

48
EDITORS' CONCLUSION

James McCarthy, Tom Perreault, and Gavin Bridge

Working on this volume has been exciting and instructive. We are struck by the exceptional vibrancy and dynamism of contemporary political ecology: alongside continuing excellent work on long-established and still critically important themes and topics, current research extends the field in new empirical and theoretical directions. In so doing, it continues political ecology's history of challenging and expanding both dominant understandings of how humans interact with their environments, and the methods and frameworks we use for constructing alternative explanations of such relations. With such ferment, the meaning of "political ecology" within geography and cognate disciplines continues to expand rapidly, moving as it has from a term for a relatively narrowly focused and arguably counterhegemonic stream of research within certain key Anglophone institutions, to an institutionally sanctioned umbrella term for critical analyses of "environment"-related research in multiple disciplines, regions, and languages.

Such rapid growth almost inevitably raises theoretical, empirical, and normative questions: What is political ecology now? What currently are the major developments and directions in the field? And what should political ecology do more of, or do better? In considering these questions, we have no interest in easy definitions, border policing, or prescriptive agendas: such exercises are neither interesting nor productive. And we are well aware that much ink has been spilled on the question of "what is political ecology," without the various inductive, deductive, and prescriptive responses proffered settling the question. Nor, indeed, do we wish to provide such closure: we are delighted by the fact that political ecology has become such a diverse and productive field for critical inquiry into and analysis of what we still call – in a self-consciously crude and ontologically inadequate shorthand – nature–society relations. Nevertheless, we do believe that it is interesting and productive, indeed vital, to be reflexive, explicit, and rigorous about where the field stands and where it is going. So, at the end of the process of working on this volume into which the many contributors put so much work, we believe it is important to consider these three legitimate and important questions. Without doubt, others will have different responses to them, or choose to consider other questions altogether. Nevertheless, we offer here our reflections on the roots and characteristics of political ecology, and possible directions for future research.

What is political ecology?

The wide range of empirical topics, theoretical frameworks, and methodological approaches evident not just in this volume, but in other publications, conferences, and initiatives labeled as "political ecology," understandably raises the question of what, if anything, holds this tremendously diverse body of work together conceptually and methodologically. Political ecology is certainly not a discipline or sub-discipline in a conventional academic sense: its objects of study and analytical frameworks and approaches are too disparate, and indeed its explicitly interdisciplinary character is one of its defining hallmarks and strengths. We could take a purely inductive approach and say simply that political ecology is whatever anyone is doing and calling political ecology. And we recognize that in practice this is how the field is largely identified at any given moment, as well as how its parameters are changed over time. Yet this response seems to us to beg an important question: we might still reasonably and usefully ask whether the self-defined participants in that field have important things in common – whether their having chosen to self-identify and participate under the sign of "political ecology" indicates anything significant about the substantive content of their work and their approaches to it.

We believe that it does, and that there are important commonalties across this diverse body of work. We see political ecology as a theoretical and political lens through which to understand, challenge, and structure further inquiry into nature–society relationships in the contemporary world, with certain methodological preferences following from its dominant theoretical perspectives and normative commitments. While this lens was applied first and most famously to topics such as agrarian dynamics in the context of postcolonial articulations with the global economy, centralized conservation, and resource conflicts and governance (as illustrated in Chapters 2, 3, and 30 by Watts, Wisner, and Neumann respectively, and others in this volume), we believe that it has also proved to have broad and enduring utility, offering analytical insights into and purchase upon a wide range of human–environment relationships. The chapters we have assembled in this volume support this interpretation. Moreover, we contend that some inquiries not necessarily labeled as "political ecology" have used effectively the same lens.

What, then, are the major shared elements of political ecology, to be found in almost all work in this wide field? While different research surely displays the elements below in various proportions and forms, and realizes their goals more or less fully, we believe that these elements deeply inform the great majority of work described as political ecology. More significantly, perhaps, we find it difficult to imagine research that explicitly rejected or argued against them fitting with our sense of the field. Below, we point to five such shared elements that characterize political ecology as a field.

First, political ecology is deeply shaped by the encounter between Marxism and contemporary environmental questions. The field takes as given that capital accumulation and the defining social relations of capitalism, such as private property, commodification, and class structures, produce and drive much environmental transformation, degradation, and conflict in the modern world. It was precisely such a focus on and structural understanding of political-economic connections, relationships, and processes that first and most decisively distinguished political ecology from cultural ecology and other ways of thinking about human–environment relationships. In this context, it is critical to acknowledge that political ecology drew from the beginning on vibrant debates in economic geography and political economy that were emerging from geography's engagement with Marxism at the time (an argument developed more fully in McCarthy 2012). An important contribution was the distinctive engagement on the part of some Marxist geographers with the environment: on the one hand, works such as Harvey's (1974) critique of dominant neo-Malthusian environmentalism demonstrated how a Marxist perspective offered a sharply different, and dialectical, way of thinking about "nature-society" questions, while on the other hand, works such as Walker's (1973, 1974) analyses of wetland valuation practices wrestled with the ways in which environmental questions presented challenges to elements of Marxist theory. At the same time, geographers sought to develop a specifically geographical Marxism, inventing and refining concepts and approaches such as dependency theory, uneven development, uneven exchange, spatial fixes, and commodity chain analysis, all of which became important components of political ecology's intellectual framework. Key thinkers in the development of political ecology were all deeply versed in and indeed directly engaged in this conversation between economic geography and Marxism (see, for example, the reflections by Ben Wisner in Chapter 3 in this volume, as well as Chapter 21 by Castree, this volume). Thus, while one common explanation for the emergence of political ecology lies in its immanent critique of cultural and human ecology, we must recognize that this critique was not *sui generis*. Rather, it emerged from a very specific theoretical and political position formed in conversation with Marxist political economy. This is significant in that it means that works by Harvey (1974), Walker (1973, 1974), and other Marxist scholars examining the socionatural metabolism of capitalism in industrialized countries and at global scales are as much a part of the wellsprings of political ecology as the canonical works by Watts (1983a, 1983b), Blaikie (1985), and Blaikie and Brookfield (1987), more commonly cited in this context. While much current work in political ecology may be tacitly post- or neo-Marxist rather than explicitly Marxist, the debt and shared assumptions are clear: can we imagine a work of political ecology framed in terms of neoclassical economics?

Second, political ecology is a form of critique, which is to say it is explicitly normative, and in that context is committed specifically to siding with the marginalized and less powerful in the situation in question. This may seem so obvious as to hardly be worth stating, but it is in fact one of the great divides between political ecology and more mainstream approaches that remain committed to the ideal of objective, value-neutral science and analysis. This commitment also marks one of the ruptures with cultural ecology, which, however strongly its practitioners might have felt personally, retained a professional commitment to objective analysis of human–environment systems as that field understood them. Another critical facet of this commitment is that political ecologists therefore typically make normative judgments about the actors and systems they are studying: taking the side of one group in an inherently agonistic situation means opposing others. Thus, the aim of political ecology is to make an argument that is thoroughly political, as much as one that is theoretical or empirical.

Third, feminist theory and politics have become part of the core of political ecology, shaping central intellectual and political assumptions and commitments of the field. Through a critical engagement with the field's early Marxism, feminist political ecology by Rocheleau et al. (1996), Carney (1996), Schroeder (1999), and others emphasized the ways in which power relations always operate through multiple, intersecting axes and categories of social difference in any social setting; that people experience differential outcomes depending upon their relationships to those categories; and that those categories are social constructs (see Chapter 40 by Elmhirst, this volume). Disaggregating and denaturalizing analytical units such as "the household," research in this vein explores the consequences of those dynamics with respect to resource access and control, differential experiences of environmental costs and benefits, attitudes regarding legitimate environmental stewardship, and more. While this line of work focused initially on gender, it led directly to a broader engagement with questions of how identities were socially produced and with what consequences for environmental politics and outcomes, with the categories and consequences of indigeneity, ethnicity, and race becoming prominent topics (see, e.g., Moore et al. 2003; Perreault 2003; Mollet and Faria 2013). Feminist theory and anthropological approaches to identity alike have informed this line of work, which has increasingly focused on ideas of intersectionality: the effort to understand all of the above categories as fluid and political creations that always exist and operate in combination. At this point, then, even in political ecological work not explicitly identified as feminist or focused on gender relations, it is commonly accepted that identity categories are socially constructed, dynamic, always shaped by diverse relations of power, and consequential for all environmental interactions and issues. Such insights are now as fundamental a part of the fabric of the field as those originating from Marxist theory, as evident in many of the chapters in this volume.

Fourth, political ecology remains committed to and characterized by largely qualitative and interpretive methods and methodologies (although, to be sure, much work has demonstrated that these can often be fruitfully combined with other research methods). Such approaches dominate in part because, studying marginalized people, political ecologists realize that official records tell only a partial story: what is often at stake is precisely the facts and motivations of resistance, of extra-legal activity, of the grievances and collective desires that give rise to social movements, and so on. It is very difficult to find out what marginalized people think and how they are affected by transformed socionatural relationships without actually talking with them, or attempting to recover or reconstruct such perspectives and effects through painstaking historical work and interpretation (e.g., the pathbreaking work of E.P. Thompson [1975]). In short, as befits a field with deep roots in cultural geography and anthropology, questions of meaning are as important as questions of fact. Such meanings are typically not captured in surveys sent out by central governments or in images taken from orbit. Indeed, for political ecologists, the representations formed on the basis of these latter methods are as much objects of study as they are sources of data (see Chapter 19 by Bryan, this volume). This points to the related fact that political ecology appears committed to a distinctly post-positivist view of science, accepting some of its products as inputs, yet also believing that knowledge production is always inextricably bound up with social relations and operations of power (see Chapters 6 by Robbins and 11 by Zimmerer, this volume).

Fifth, political ecology is likewise attentive to historical and social context. As Diana Davis (Chapter 20, this volume) argues, while relatively focused and intensive case studies may be the norm in the field, a necessary complement to an intensive focus is that those cases must be understood within their broader social and historical contexts. The depth of that temporal lens varies, and has been the subject of some debate: political ecology's specific focus on the transformation of nature–society relations in the context of capitalist modernity has meant that historical attention often focuses on the ways in which nature–society relations have been reshaped through and by specifically modern colonial and post-colonial dynamics.

As we acknowledge above, this constellation of commitments is arguably not unique to work that self-identifies as "political ecology": much work in, for instance, anthropology, environmental sociology, environmental history, science and technology studies, and other cognate fields shares many or perhaps even all of these attributes, which helps to explain the frequent and fertile interchanges among them and geography – sometimes under the sign of political ecology, and sometimes under other labels. Yet, we still find it useful to sketch the broad commonalties within political ecology as a way to register what we have learned, before moving on to consider where the field is going and what it might consider adding to its agenda.

Current developments and directions

The rapid ferment and expansion so evident in political ecology makes cataloguing or characterizing major new directions in the field challenging. Nonetheless, a handful of themes and topics stand out as representing clearly new, distinctive, and significant developments in the field's evolution.

The first is that political ecology is an increasingly international and polylinguistic field. As several of the chapters in this volume both represent and discuss (see in particular Chapters 4, 5, and 24 by Leff, Gautier and Kull, and Ulloa respectively), "political ecology" is now a recognized term and organizing principle for research, criticism, and activism in non-Anglophone countries and research traditions, with new journals, conferences, and research networks emerging under the explicit heading of political ecology. These conversations have their own intellectual and political trajectories, rooted in particular literatures, politics, and problems, and it would be a mistake to understand them as merely regionalized expressions of the Anglophone tradition. But all these varieties of political ecology broadly share an intellectual and political commitment to social change and to critiquing dominant structures of political and economic power. Political ecology's growing popularity as a term and as a political and intellectual position is surely rooted in some significant commonalties and cross-fertilization, and the opportunity to develop a more diverse, representative, and multi-faceted political ecology is critical to the field's future. It is also very much in keeping with its intellectual and normative commitments.

Attention to the significance of categories and axes of social difference and their consequences for environmental politics and outcomes, broadly understood, continues to expand and deepen. The recent turn towards what we can term, following Sundberg (2011), "posthumanist political ecology" represents an important new direction. The increasing interest in animals and other non-human entities as not merely objects of study or functional elements of ecosystems, but actors in (and perhaps with) their own rights, in investigations of how humans act in and interact with a heterogeneous world, represents an important break with earlier, unabashedly anthropocentric political ecology (see, for example, Chapter 9, this volume; Collard 2014; Shaw et al. 2010; Kosek 2010). There are especially challenging but also rich exchanges on this front between political ecology and political theory – a conversation also developing with respect to topics such as the "post-political" (Chapter 47 by Swyngedouw, this volume) and conceptualizations of climate politics (see, for example, Baldwin 2013).

The past decade or so has also seen an intense research focus on extractive industries and regions, and particularly mining – what Tony Bebbington (2012) has termed "underground political ecologies." While much in this research is quite familiar to political ecology – the focus on primary commodity producing regions and communities in the global south, investigation of how livelihoods and access shift with increasing production for global markets, an emphasis on tensions between the goals of national governments and the impacts on local, often indigenous, communities and territories with respect to policies around natural resources, investment, and exports – the growing focus on subterranean, mineral resources is arguably a significant departure from the studies of agrarian and forest dynamics and conflict so central to the first few decades of political ecology. Among other considerations, mining and fossil fuel extraction are nearly always directly and strongly connected to global political economic relations of exchange and consumption of raw materials and energy. Whereas farming and herding systems, or even conservation efforts may be relatively localized, and only indirectly influenced by broad-scale political economic processes, minerals and hydrocarbons enter directly into global capital flows. So while these processes are inextricably rooted in particular sites of extraction, they also immediately and explicitly connect to national and global scales. Resource extraction has been a research topic in political ecology for decades, but it has become far more central to the field in recent years, at least in part because of mounting global concerns regarding continued extraction of fossil fuels. This interest in the accelerating "torrent of raw materials" drawn into industrial economies links directly to a small but rapidly growing interest within political ecology in shifting geographies of energy production and consumption – a point of potential rich interchange with economic and resource geographies, industrial ecology (Chapter 28 by Barca and Bridge and 37 by Huber, this volume) and, of course, to climate change.

Climate change is now a central topic in political ecology, and one likely to permeate inquiry in the years ahead (Chapter 23 by Liverman, this volume). The relationship between climate change and political ecology as a field is complex. On the one hand, climate change appears to be just the sort of topic political ecology is tailor-made to study: a host of wrenching, profoundly unjust transformations of nature–society relations, driven by centuries of capitalist dynamics with global consequences, in which the poor and otherwise marginalized will suffer most, but in ways deeply shaped and complicated by local circumstances and specificities. Indeed, political ecology's rapid rise in popularity, as measured by things like specialty group memberships and job descriptions, surely owes much to the fact that it offers a rich, theoretically rigorous, and explicitly political framework through which to investigate and understand such dynamics. There are explicit parallels with and echoes of political ecology's genesis in the 1970s, when it began in large part as an effort to formulate a critical and more nuanced alternative to the sweeping global neo-Malthusian diagnoses of and solutions to environmental crisis. It seems, at first glance, that perhaps this time around more people are listening to its hard-won lessons regarding structurally produced differential vulnerability, the social origins of alleged environmental "drivers," and the like.

However, there are also reasons to be skeptical of this quick embrace. As Watts (Chapter 2, this volume) argues forcefully, many recent efforts to "mainstream" some elements of political ecology fail to grasp either its original critiques of adaptive and systems thinking (see also Bassett and Fogelman 2012), or the careful and comprehensive social theories out of which they were born. It is impossible, for instance, to reconcile a structural critique of capitalism with an embrace of the "green economy" and the yet more extensive and intensive incorporation of nature into circuits of capital accumulation as a "solution" to climate change. And it is equally impossible to adequately theorize complex social dynamics within the functionalist, putatively universal framework of the now-ubiquitous "resilience cycle" diagrams (Resilience Alliance 2014). Yet, this is precisely what is on offer by a still-hegemonic neoliberalism, the analysis of which with respect to environmental governance has become another major research theme in contemporary political ecology. Thus, Watts' argument, with which we are in full agreement, is that the current popularity of resilience and systems thinking with respect to climate change (and indeed financial markets and a host of other referents) makes some of political ecology's original analytical insights and critiques newly and urgently relevant. It is in part for this reason that we think it worthwhile to articulate and emphasize some of the field's central commitments and points of consensus.

It is relatively easy to list what is new in terms of topics and approaches in political ecology (if hard to do them all justice). What is more difficult, but arguably more critical, is to characterize the broader context in which these trends are occurring, as well as the essential elements of a political ecological critique of that moment. In other words: if a critical and specifically Marxist critique of the particular conjuncture of Cold War geopolitics, early postwar and postcolonial development interventions, and neo-Malthusian environmentalism was integral to the genesis of political ecology as we now understand it, what are the analogous contextual contours to which we respond today, and the key theoretical and political elements of our critical responses? As in the 1970s, we see a global capitalist economy struggling with dramatic shifts in its geographies of production and consumption, with the energetic basis of its metabolism and ongoing expansion, and with potentially dramatic reconfigurations in the location and techniques of hegemonic power. If anything, those struggles are more pronounced and severe, with the challenges of climate change amplifying and cutting across the standard neo-Malthusian refrain of looming environmental scarcities and conflicts. What is notably different, though, is how thoroughly decades of neoliberalism and neoliberalization have transformed the terms of contestation and debate, even within, sadly, environmental politics themselves (McCarthy and Prudham 2004). One aspect of neoliberalism, we argue, has been an effort to reinvigorate capital accumulation not so much by an internationalization of production and the development of new markets – although those processes, so prominent in the 1970s and following decades, surely continue apace – but by drawing life itself and its creative capacities more directly into circuits of capital, in ways ranging from the emergence of biotechnology as a central sector in contemporary regimes of accumulation (Sunder Rajan 2006) to the marketization and indeed fabrication of "ecosystem services" (Robertson 2004, 2012; Lave 2012). We see the rapidly growing prospect of industrial-scale efforts to geoengineer the biosphere as the next step along this trajectory. In short, rather than accept that living organisms or ecosystems may present limits to capital accumulation, contemporary capitalism seeks instead to rework those organisms and ecosystems in ways conducive to the continued expansion of capital. Finally, it is notable that contemporary responses to the chronic social and environmental insecurity and vulnerability brought about by the expansion of this crisis-ridden system eschew the explicit geopolitical ambitions, centralized planning, and explicit coercion so prominent in the political imaginaries of the *Limits to Growth* era responses to crisis. Rather, current responses, shaped by decades of neoliberal political imaginaries, turn instead on the securitization of privileged lives through permanent states of undeclared war, while encouraging the rest of humanity to utilize adaptation and the cultivation of resilience to survive in a global market economy whose volatility, inequality, and unpredictability are naturalized via references to ecological theories of complex systems (see Chapter 2 by Watts, this volume). In short, while many of the specific contours of the present moment have changed, political ecology's critiques and contributions remain as relevant as ever: an emphasis on the political economic roots of environmental problems; a rejection of facile and apolitical understandings of human–environment relationships; an insistence on the complexity, historicity, and malleability of social structures and processes; and a commitment to siding with the marginalized continue to be not only relevant but also vitally necessary. In the depths of the neoliberal era, one of the main tasks and contributions of political ecology (as of many related critical agendas) has been to insist, and to demonstrate at times, that alternative and non-capitalist human–environment relationships are possible.

What should political ecology do more of, or do better?

We wish to close by suggesting several areas where we think political ecology as a field could make important contributions, and perhaps stretch itself in new directions. These grow directly out of the chapters in this volume and our conversations about the field, as well as out of conversations over the years with other colleagues.

First, we think it is critical to continue, and to actively foster, conversations among political ecologists working in different national, regional, and linguistic traditions. This will not be easy and involves far more than just overcoming linguistic barriers (though this is, of course, a crucial first step). The relative lack of interchange between the various traditions in political ecology is rooted in part in the uneven geographies of knowledge production (shorthanded in the inadequate and increasingly dated spatial imaginary of global north and south). This unevenness, in turn, has everything to do with colonial histories: for example, many government-funded regional studies programs in the USA emerged directly out of Cold War politics and the global aspirations of northern elites. At the same time, critical traditions of scholarship on the mobilization of environment and resources have in many instances been actively suppressed as part of colonial projects to consolidate the national state, as in Chile. One consequence of this and other legacies of colonialism and uneven development is that intellectuals in the global south often remain dependent in important ways upon forms and relations of knowledge production strongly centered in the global north, whether it be the dominance of journals and books published mainly or solely in English, expensive electronic access to restricted library systems, or the locations and structures of conferences and professional networks. Such conditions have tremendous influence on the ways that academic literature – including political ecology – is produced, disseminated, read and taught. Yet simple dependence is far from a complete or accurate characterization of the legacies of the histories above: another consequence is that, by necessity or by choice, some research trajectories and forms of social engagement undertaken by researchers based within the global south on the urban and rural livelihoods of marginalized groups have evolved relatively independently of formal research programs and scholarly trajectories originating in the global north. Often political ecology in all but name by the substantive criteria we lay out above, such work nonetheless has its own dynamics, institutional contexts, and political rationales, which cannot and should not be defined by its relationship to the academic institutions and infrastructures of the global north. The post-colonial encounter between these contextually evolved practices of research and political ecology programs emanating from the global south and the more commonly recognized "political ecology" written and circulated primarily by academics in the global north is a complex one, characterized by relations of dominance, independence, and hybridization. Finding ways to overcome these power relations and engage in these cross-language conversations will be difficult, and will involve much more than just reading works in translation. It also requires an acknowledgement on the part of Anglophone researchers of their privileged position in relation to the means of knowledge production. We are emphatically *not* suggesting such scholars turn away from an engagement with the global south, or what others have characterized as the majority world. Far from it. Rather, we call for those whose academic practices are both a consequence of, and constitutive of, the dominance of the global north to acknowledge the structural inequities within with they work, and for political ecologists from the whole range of geographic and institutional locations to find meaningful ways to transcend these obstacles.

Second, we would encourage political ecologists to explore more direct engagements with policy and political practice alike – from work with state agencies or NGOs, to work with social movements and direct action activists. While the specific forms, decisions, and commitments of this co-production of environmental knowledge cannot be discussed in general terms, we are deeply sympathetic to the arguments made by Loftus (Chapter 13, this volume) regarding the need for "political ecologies of practice" and Bebbington (Chapter 15, this volume) on the instability of the distinction often made between political critique and the actual construction of policy. Following from these points, political ecologists may find they can further their objectives by being more willing to speculate about the future as part of their work (in addition to their more traditional role in critiquing), and to suggest, endorse, and contribute to the development of specific visions and plans (see Chapter 7, this volume). The field's ability to travel beyond the academy has been limited, in part, by a tendency to eschew specific contributions to discussions, beyond the articulation of general principles, regarding how the future might or should look.

An example of a domain in which political ecologists could make just such contributions is in the consideration of future energy geographies, another topic ripe for attention. As Bridge et al. (2013) and others have argued, the coming years are likely to see dramatic reconfigurations of energy complexes at every scale and throughout the globe. Such reconfigurations will be inevitably political, and they will be about questions at the heart of political ecology: who will make decisions about how to use the environment; who will benefit and who will bear the costs of such uses; how will historical patterns of land uses, claims, and rights shape what happens in the present and future; how will such reconfigurations work through and either reinforce or alter existing categories or axes of difference within social formations; and more. Political ecologists are extremely well suited to contribute to the investigation of such questions, and to develop convincing arguments for more equitable and sustainable versions of new energy complexes.

Related to this point, we believe political ecology has to engage substantively with the shifting configurations of the global economy, including the rise of the so-called 'BRICSAM' (Brazil, Russia, India, China, South Africa, ASEAN states, and Mexico) countries and the implications for global resource flows, environmental conflicts and social movements, and more. To name one prominent but hardly isolated example, Chinese investments rival (and in some instances have surpassed) those of the US and EU in some Latin American, African, and Asian countries. This global reconfiguration of capital flows demands a deep re-thinking of the north–south conceptual scaffolding upon which much of political ecology has been constructed. We have used and reproduced the terms and imaginary of "global north and south" here, in part because other alternatives such as "minority and majority worlds" are equally and problematically binary, while also glossing over important historical and geographical configurations. And, like all authors in this domain, we must at times use shorthand terms to refer to vastly complicated social realities. Yet we find such dichotomous frameworks increasingly inadequate for contemporary political ecology, and we especially suggest that a geographical imaginary organized around northern, industrialized domination and exploitation of a predominantly agrarian global south, while still capturing much, is no longer adequate to the world in which we live, or to the range of topics that contemporary political ecology ought to investigate.

Closing thoughts

While much has changed about the world in the 40-plus years since the emergence of political ecology as a field of research and praxis, the core commitments of the field have never been more relevant. Political ecology's theoretical commitments to Marxist, feminist, and postcolonial analysis; its commitment to intensive, post-positivist research methods; and its strong normative commitment to social justice, are all the more urgent in the face of contemporary capitalism and diverse state projects of territorialization and securitization in the name of environment, energy, and resources. The specific empirical foci of political ecology have changed considerably in the past 40 years, as scholars have trained their attention on emerging ecologies and scales of social and ecological relations. Given the restless nature of global capitalism and socio-environmental relations, together with the changing (and increasing) demands of the academy, we have no doubt that political ecology will continue to evolve in numerous ways. Indeed, we welcome these changes and are excited by the prospect of what is to come. Whatever directions the field may take in the future, our hope and expectation is for a political ecology that is at once more global in its orientation – embracing the field's diverse linguistic and regional traditions, scales of analysis, empirical foci, and epistemological approaches – and more thoroughly engaged in practice, policy, and activism. In short, the more political ecology changes, the more we believe its core commitments remain the same. To paraphrase Marx (1975; see also Chapter 13 by Loftus, this volume), we call for a political ecology that strives not only to interpret the world as it is, but which continues to work actively to change it.

References

Baldwin, A. 2013. Vital ecosystem security: emergence, circulation, and the biopolitical environmental citizen. *Geoforum* 45: 52–61.

Bassett, T. and C. Fogelman 2012. Deja vue or something new? The adaptation concept in the climate change literature. *Geoforum* 48: 42–53.

Bebbington, A. 2012. Underground political ecologies: the second annual lecture of the Cultural and Political Ecology Specialty Group of the Association of American Geographers. *Geoforum* 43(6): 1152–1162.

Blaikie, P. 1985. *The Political Economy of Soil Erosion in Developing Countries*. London: Routledge.

Blaikie, P. and H. Brookfield 1987. *Land Degradation and Society*. London: Methuen.

Bridge, G., S. Bouzarovski, M. Bradshaw and N. Eyre 2013. Geographies of energy transition: space, place and the low-carbon economy. *Energy Policy* 53: 331–340.

Carney, J.A. 1996. Converting the wetlands, engendering the environment: the intersection of gender with agrarian change in Gambia. In: Peet, R. and Watts, M. (Eds.), *Liberation Ecologies: Environment, Development, and Social Movements*. London: Routledge, pp. 165–187.

Collard, R.C. 2014. Putting animals back together, taking commodities apart. *Annals of the Association of American Geographers* 104(1): 151–165.

Harvey, D. 1974. Population, resources and the ideology of science. *Economic Geography* 50(3): 256–277.

Kosek, J. 2010. Ecologies of empire: on the new uses of the honeybee. *Cultural Anthropology* 25(4): 650–678.

Lave, R. 2012. *Fields and Streams: Stream Restoration, Neoliberalism, and the Future of Environmental Science*. Athens: University of Georgia Press.

Marx, K. 1975. *Early Writings*. London: Pelican.

McCarthy, J. 2012. The political economies of political ecology. In: J. Peck, T. Barnes and E. Sheppard (Eds.), *The New Companion to Economic Geography*. Oxford: Wiley-Blackwell, pp. 612–625.

McCarthy, J. and S. Prudham 2004. Neoliberal nature and the nature of neoliberalism. *Geoforum* 35(3): 275–283.

Mollett, S. and C. Faria 2013. Messing with gender in feminist political ecology. *Geoforum* 45: 116–125.

Moore, D., J. Kosek and A. Pandian, eds. 2003. *Race, Nature, and the Politics of Difference*. Durham, NC: Duke University Press.

Perreault, T. 2003. "A people with our own identity": toward a cultural politics of development in Ecuadorian Amazonia. *Environment and Planning D: Society and Space* 21(5): 583–606.

Resilience Alliance 2014. Adaptive cycle. www.resalliance.org/index.php/adaptive_cycle. Accessed October 2, 2014.

Robertson, M.M. 2004. The neoliberalization of ecosystem services: wetland mitigation banking and problems in environmental governance. *Geoforum* 35(3): 361–373.

Robertson, M. 2012. Measurement and alienation: bringing ecosystems to market. *Transactions of the Institute of British Geographers* 37(3): 386–401.

Rocheleau D., B. Thomas-Slayter and E. Wangari, eds. 1996. *Feminist Political Ecology: Global Issues and Local Experiences*. London and New York: Routledge.

Schroeder, R. 1999. *Shady Practices: Agroforestry and Gender Politics in The Gambia*. Berkeley: University of California Press.

Shaw, I., P. Robbins and J.P. Jones. 2010. A bug's life and the spatial ontologies of mosquito management. *Annals of the Association of American Geographers* 100(2): 373–392.

Sundberg, J. 2011. Diabolic caminos in the desert and cat fights on the Rio: a posthumanist political ecology of boundary enforcement in the United States-Mexico borderlands. *Annals of the Association of American Geographers* 101(2): 318–336.

Sunder Rajan, K. 2006. *Biocapital: The Constitution of Post-Genomic Life*. Durham, NC: Duke University Press.

Thompson, E.P. 1975. *Whigs and Hunters: The Origin of the Black Act*. London: Allen Lane.

Walker, R.A. 1973. Wetland preservation and management on Chesapeake Bay: the role of science in natural resource policy. *Coastal Zone Management Journal* 1(1): 75–101.

Walker, R.A. 1974. Wetland preservation and management: a rejoinder – economics, science and beyond. *Coastal Zone Management Journal* 1(2): 227–233.

Watts, M.J. 1983a. On the poverty of theory: natural hazards research in context. In: K. Hewitt (Ed.), *Interpretations of Calamity: From the Viewpoint of Human Ecology*. Boston: Allen & Unwin, pp. 231–262.

Watts, M.J. 1983b. *Silent Violence: Food, Famine, and Peasantry in Northern Nigeria*. Berkeley: University of California Press.

INDEX